Smad Signal Transduction

PROTEINS AND CELL REGULATION

Volume 5

Series Editors: Professor Anne Ridley
Ludwig Institute for Cancer Research
and Department of Biochemistry and Molecular Biology
University College London
London
United Kingdom

Professor Jon Frampton
Professor of Stem Cell Biology
Institute for Biomedical Research,
Birmingham University Medical School,
Division of Immunity and Infection
Birmingham
United Kingdom

Aims and Scope

Our knowledge of the ways in which a cell communicates with its environment and how it responds to information received has reached a level of almost bewildering complexity. The large diagrams of cells to be found on the walls of many a biologist's office are usually adorned with parallel and interconnecting pathways linking the multitude of components and suggest a clear logic and understanding of the role played by each protein. Of course this two-dimensional, albeit often colourful representation takes no account of the three-dimensional structure of a cell, the nature of the external and internal milieu, the dynamics of changes in protein levels and interactions, or the variations between cells in different tissues.

Each book in this series, entitled "*Proteins and Cell Regulation*", will seek to explore specific protein families or categories of proteins from the viewpoint of the general and specific functions they provide and their involvement in the dynamic behaviour of a cell. Content will range from basic protein structure and function to consideration of cell type-specific features and the consequences of disease-associated changes and potential therapeutic intervention. So that the books represent the most up-to-date understanding, contributors will be prominent researchers in each particular area. Although aimed at graduate, postgraduate and principle investigators, the books will also be of use to science and medical undergraduates and to those wishing to understand basic cellular processes and develop novel therapeutic interventions for specific diseases.

Smad Signal Transduction

Smads in Proliferation, Differentiation and Disease

Edited by

Peter ten Dijke
Leiden University Medical Center, Leiden, The Netherlands

and

Carl-Henrik Heldin
Ludwig Institute for Cancer Research, Uppsala University, Sweden

Springer

A C.I.P. Catalogue record for this book is available from the Library of Congress.

ISBN-10 1-4020-4542-5 (HB)
ISBN-13 978-1-4020-4542-4 (HB)
ISBN-10 1-4020-4709-6 (e-book)
ISBN-13 978-1-4020-4709-1 (e-book)

Published by Springer,
P.O. Box 17, 3300 AA Dordrecht, The Netherlands.

www.springer.com

Printed on acid-free paper

Cover Legend: Phylogenetic tree of the SMAD protein family: members from human, Drosophila melanogaster (fruit fly) and Caenorhabditis elegans (nematode) are shown. The topology of the tree is the same as the tree shown in Chapter 1 by Newfeld and Wisotzkey (Figure 1); see the figure legend for details. Here each of the seven subfamilies is shown in a different color. Note that human and fly proteins are clustered together in four subfamilies while three subfamilies contain only nematode sequences.

Dedicated to the memory of Dr. Anita Roberts who sadly passed away on May 25, 2006, at the age of 64. Dr. Roberts was a leader in the field of TGF-β research, and through her remarkable personality she was an inspiration to us all.

CONTENTS

CONTRIBUTORS

Dr. Tamara Alliston Department of Cell and Tissue Biology, University of California at San Francisco, 513 Parnassus Avenue, Room HSW-658, SAN FRANCISCO, CA 94143-0512, USA. Phone: +1-415-476-0320, Fax: +1-415-502-7338, E-mail: tamara.alliston@ucsf.edu

Dr. Liliana Attisano Department of Biochemistry, Donnelly CCBR Building, Room 1008, 160 College Street, University of Toronto, TORONTO, ONTARIO M5S 3E1, Canada. Phone: +1-416-946-3129, Fax: +1-416-978-8548, E-mail: liliana.attisano@utoronto.ca

Dr. Brandi Berry Oncology Division, Lilly Research Laboratories, Lilly Corporate Center, INDIANAPOLIS, INDIANA 46285, USA. Phone: +1-317-433-3008, Fax: +1-317-276-9666, E-mail: bsberry@lilly.com

Dr. Rohit Bose Dr. Jeff Wrana's lab, Room 1070, Samuel Lunenfeld Research Institute, Mount Sinai Hospital, 600 University Avenue, TORONTO, ON M5G 1X5, Canada. Phone: +1-416-586-4800 ext. 2363, Fax: +1-416-586-8869, E-mail: Christine@mshri.on.ca

Prof. Erwin P. Böttinger Department of Medicine, Mount Sinai Medical Center, One Gustave L. Levy Place, Box 1118, NEW YORK, NEW YORK 10029, USA. Phone: +1-212-241-0800, Fax: +1-212-849-2643, E-mail: Erwin.bottinger@mssm.edu

Dr. Lisa Choy Department of Cell and Tissue Biology, University of California at San Francisco, 513 Parnassus Avenue, Room HSW-658, SAN FRANCISCO, CA 94143-0512, USA. Phone: +1-415-476-0320, Fax: +1-415-502-7338, E-mail: Lisa.Tomlinson@ucsf.edu

Dr. Andrew R. Conery University of California, Berkeley, Department of Molecular and Cell Biology, 16 Barker Hall, mail code 3204, BERKELEY, CA 94720-3206, USA. Phone: Fax: E-mail:

Dr. Chu-Xia Deng Mammalian Genetics Section, GDDB, NIDDK, National Institutes of Health, 10/9N105, 10 Center Drive, BETHESDA, MD 20892, USA. Phone: +1-301-402-7225, Fax: +1-301-480-1135, E-mail: chuxiad@mail.nih.gov

Prof. Rik Derynck Institute for Stem Cell and Tissue Biology, Departments of Cell and Tissue Biology, and Anatomy, University of California at San Francisco, 513 Parnassus Avenue, Room HSW-613, SAN FRANCISCO, CA

94143-0512, USA. Phone: +1-415-476-7322, Fax: +1-415-502-7338, E-mail: rik.derynck@ucsf.edu

Prof. Xin-Hua Feng Department of Molecular & Cellular Biology and Michael E. DeBakey Department of Surgery, Baylor College of Medicine, One Baylor Plaza, Rm. 137D, HOUSTON, TX 77030, USA. Phone: +1-713-798-4756, Fax: +1-713-798-4093, E-mail: xfeng@bcm.tmc.edu

Dr. Fang Liu Center for Advanced Biotechnology and Medicine, Rutgers University, 679 Hoes Ln., PISCATAWAY, NJ 08854, USA. Phone: +1-732-235-5372, Fax: +1-732-235-4850, E-mail: fangliu@cabm.rutgers.edu

Prof. Carl-Henrik Heldin Ludwig Institute for Cancer Research, Uppsala University, Box 595, SE-751 24 UPPSALA, Sweden. Phone: +46-18-160401, Fax: +46-18-160420, E-mail: C-H.Heldin@LICR.uu.se

Dr. Caroline S. Hill Laboratory of Developmental Signalling, Cancer Research UK, London Research Institute, 44 Lincoln's Inn Fields, LONDON WC2A 3PX, U.K. Phone: +44-20-7269-2941, Fax: +44-20-7269-3093, E-mail: Caroline.Hill@cancer.org.uk

Dr. Stéphane Huet GlaxoSmithKline R&D, Biology Department, 25 avenue du Québec, 91951 LES ULIS, France. Phone: +33-1-6929-6079, Fax: +33-1-6929-6000, E-mail: stephane.r.huet@gsk.com

Dr. Takeshi Imamura Department of Biochemistry, The Cancer Institute of the Japanese Foundation for Cancer Research (JFCR), 3-10-6 Ariake, Koto-ku, TOKYO 135-8550, Japan. Phone: +81-3-3570-0459, Fax: +81-3-3570-0459, E-mail: timamura-ind@umin.ac.jp

Dr. Delphine Javelaud INSERM U697 "Molecular Bases of Skin Homeostasis", Hôpital Saint-Louis, Pavillon Bazin, 1, Avenue Claude Vellefaux, PARIS 75010, France. Phone: +33-1-5372-2078, Fax: +33-1-5372-2051, E-mail: delphine.javelaud@stlouis.inserm.fr

Dr. Wenjun Ju Department of Medicine, The Mount Sinai Medical Center, One Gustave L. Levy Place, Box 1118, NEW YORK, NEW YORK 10029, USA. Phone: +1-212 241-0800, Fax: +1-212 849-2643, E-mail: Wenjun.ju@mssm.edu

Dr. Jing Cao Cutaneous Biology Research Center, Massachusetts General Hospital East 3rd Floor, Building 149 13th Street, CHARLESTOWN, MA 02129, USA. Phone: +1-617-724-8290, Fax: +1-617-726-4354, E-mail: jing.cao@cbrc2.mgh.harvard.edu

Dr. Seong-Jin Kim The Laboratory of Cell Regulation and Carcinogenesis, The Center for Cancer Research, The National Cancer Institute, The National Institutes of Health, Building 41, Room B1106, 9000 Rockville Pike, BETHESDA,

MARYLAND 20892-5055, USA. Phone: +1-301-496-8350, E-mail: kims@mail.nih.gov

Dr. Susanne Kloeker Oncology Division, Lilly Research Laboratories, Lilly Corporate Center, INDIANAPOLIS, INDIANA 46285, USA. Phone: +1-317-277-7541, Fax: +1-317-276-9666, E-mail: rhoadess@lilly.com

Dr. Svetlana Korochkina CIMIT-Russia, Massachusetts General Hospital East 3rd Floor, 50 Staniford St., Suite 801, BOSTON, MA 02114, USA. Phone: +1-617-722-3000, ext. 4459, Fax: +1-617-726-2901, E-mail: skorochkina@partners.org

Dr. Marcin Kowanetz Genentech, Inc., 1 DNA Way, SOUTH SAN FRANCISCO, CA 94080, USA. Phone: +1-650-467-5955, Fax: +1-650-225-6327, E-mail: kowanetz.marcin@gene.com

Dr. Etienne Labbé Department of Biochemistry, Donnelly CCBR Building, Room 1050, 160 College Street, University of Toronto, TORONTO, ONTARIO M5S 3E1, Canada. Phone: +1-416-978-1359, Fax: +1-416-978-8548, E-mail: etienne.labbe@utoronto.ca

Dr. Michael Lahn Oncology Division, Lilly Research Laboratories, Lilly Corporate Center, INDIANAPOLIS, INDIANA 46285, USA. Phone: +1-317-433-9786, Fax: +1-317-276-9666, E-mail: mlahn@lilly.com

Dr. Nicholas J. Laping GlaxoSmithKline Pharmaceuticals, UW2521, 709 Swedeland Road, PO Box 1539, KING OF PRUSSIA, PA 19406-0939, USA. Phone: +1-610-270-5310, Fax: +1-610-270-5681, E-mail: Nicholas.J.Laping@gsk.com

Dr. John J. Letterio The Laboratory of Cell Regulation and Carcinogenesis, The Center for Cancer Research, The National Cancer Institute, The National Institutes of Health, Bldg 41, Rm B702, 41 Library Drive, BETHESDA, MD 20892-5055, USA. Phone: +1-301-496-8348, Fax: +1-301-496-8395, E-mail: letterij@mail.nih.gov

Dr. Xia Lin Michael E. DeBakey Department of Surgery, Baylor College of Medicine, One Baylor Plaza, HOUSTON, TX 77030, USA. Phone: +1-713-798-4899, Fax: +1-713-798-4093, E-mail: xialin@bcm.edu

Prof. Kunxin Luo University of California, Berkeley, Department of Molecular and Cell Biology, 16 Barker Hall, mail code 3204, BERKELEY, CA 94720-3206, USA. Phone: +1-510-643-3183, Fax: +1-510-643-6334, E-mail: kluo@berkeley.edu

Dr. Shingo Maeda Department of Biochemistry, The Cancer Institute of the Japanese Foundation for Cancer Research (JFCR), 3-10-6 Ariake, Koto-ku, TOKYO 135-8550, Japan. Phone: +81-3-3570-0459, Fax: +81-3-3570-0459, E-mail: shingo.maeda@jfcr.or.jp

Dr. Alain Mauviel INSERM U697 "Molecular Bases of Skin Homeostasis", Hôpital Saint-Louis, Pavillon Bazin, 1, Avenue Claude Vellefaux, PARIS 75010, France. Phone: +33-1-5372-2069, Fax: +33-1-5372-2051, E-mail: alain.mauviel@stlouis.inserm.fr

Dr. Kohei Miyazono Department of Molecular Pathology, Graduate School of Medicine, University of Tokyo, Hongo 7-3-1, Bunkyo-ku, TOKYO 113-0033, Japan. Phone: +81-3-5841-3345, Fax: +81-3-5841-3354, E-mail: miyazono-ind@umin.ac.jp

Dr. Aristidis Moustakas Ludwig Institute for Cancer Research, Box 595, SE-751 24 UPPSALA, Sweden. Phone: +46-18-160-414, Fax: +46-18-160-420, E-mail: Aris.Moustakas@LICR.uu.se

Dr. Atsuhito Nakao Dept. of Immunology, Faculty of Medicine, University of Yamanashi, 1110, Shimokato, Chuo City, YAMANASHI 409-3898, Japan. Phone: +81-55-273-6752, Fax: +81-55-273-9542, E-mail: anakao@yamanashi.ac.jp

Prof. Stuart J. Newfeld School of Life Sciences, Life Sciences C-wing Rm. 274, Arizona State University, TEMPE, AZ 85287-4501, USA. Phone: +1-480-965-6042, Fax: +1-480-965-6899, E-mail: newfeld@asu.edu

Prof. Richard W. Padgett Waksman Institute, Room 133, 190 Frelinghuysen Road, Rutgers University, PISCATAWAY, NJ 08854-8020, USA. Phone: +1-732-445-0251, Fax: +1-732-445-5735, E-mail: padgett@waksman.rutgers.edu

Dr. Garth I. Patterson Department of Molecular Biology and Biochemistry, A231 Nelson Laboratories, 604 Allison Road, Rutgers University, PISCATAWAY, NJ 08854, USA. Phone: +1-732-445-7181, Fax: +1-732-445-4213, E-mail: patterson@biology.rutgers.edu

Prof. Laurel A. Raftery Cutaneous Biology Research Center, Massachusetts General Hospital East 3rd Floor, Building 149 13th Street, CHARLESTOWN, MA 02129, USA. Phone: +1-617-726-1825, Fax: +1-617-726-4453, E-mail: lraftery1@partners.org

Dr. Arvind Rao University of Michigan, Dept. of Electrical Engineering and Computer Science (EECS) and Bioinformatics, 4230 EECS Building, 1301 Beal Avenue, ANN ARBOR MI 48109-2122, USA. Phone: None, Fax: +1-734-763-8041, E-mail: ukarvind@umich.edu

Dr. Christine Le Roy Dr. Jeff Wrana's lab, Room 1070, Samuel Lunenfeld Research Institute, Mount Sinai Hospital, 600 University Avenue, TORONTO, ON M5G 1X5, Canada. Phone: +1-416-586-4800 ext. 2363, E-mail: Christine@mshri.on.ca

Dr. Bernhard Schmierer Laboratory Of Developmental Signalling, Cancer Research UK, London Research Institute, 44 Lincoln's Inn Fields, LONDON WC2A 3PX, U.K. Phone: +44-20-7269-2940, Fax: +44-20-7269-3093, E-mail: bernhard.schmierer@cancer.org.uk

Dr. Serhiy Souchelnytskyi Ludwig Institute for Cancer Research, Box 595, SE-751 24 UPPSALA, Sweden. Phone: +46-18-160-411, Fax: +46-18-160-420, E-mail: Serhiy.Souchelnytskyi@LICR.uu.se

Prof. Yigong Shi Department of Molecular Biology, Princeton University, Lewis Thomas Laboratory, Washington Road, PRINCETON, NEW JERSEY 08544, USA. Phone: +1-609-258-6071, Fax: +1-609-258-6730, E-mail: ygshi@Princeton.EDU

Dr. Peter ten Dijke Molecular Cell Biology, Building 2, Room R-02-022, Leiden University Medical Center, Postzone S-1-P, Postbus 9600, (Einthovenweg 20), 2300 RC LEIDEN, The Netherlands. Phone: +31-71-526-9271, Fax: +31-71-5268270, E-mail: P.ten_Dijke@lumc.nl

Dr. Muneesh Tewari Fred Hutchinson Cancer Research Center, 1100 Fairview Avenue North, Mailstop D4-100, P.O. Box 19024, SEATTLE, WA 98109-1024, USA. Phone: +1-206-667-4748, Fax: +1-206-667-4023, E-mail: mtewari@fhcrc.org

Dr. Sylvie Thuault Ludwig Institute for Cancer Research, Box 595, SE-751 24 UPPSALA, Sweden. Phone: +46-18-160-415, Fax: +46-18-160-420, E-mail: Sylvie.Thuault@licr.uu.se

Dr. Michael Weinstein Department of Molecular Genetics, Room 216 Biological Sciences Building, 484 West Twelfth Avenue, Ohio State University, COLUMBUS, OH 43210-1292, USA. Phone: +1-614-688-0161, Fax: +1-614-292-4466, E-mail: weinstein.41@osu.edu

Dr. Robert G. Wisotzkey Ingenuity Systems, Redwood City, CA, USA. Phone: +1-510-484-3699, Fax: E-mail: wisotz@pacbell.net

Dr. Jeffrey L. Wrana Samuel Lunenfeld Research Institute, Mount Sinai Hospital, 600 University Avenue, Room 1070, TORONTO, ONTARIO M5G 1X5, Canada. Phone: +1-416-586-4800, ext. 2791, Fax: +1-416-586-8869, E-mail: wrana@mshri.on.ca

Dr. Ihor Yakymovych Ludwig Institute for Cancer Research, Box 595, SE-751 24 UPPSALA, Sweden. Phone: +46-18-160-412, Fax: +46-18-160-420, E-mail: Ihor.Yakymovych@licr.uu.se

Dr. Jonathan M. Yingling Angiogenesis & Tumor Microenvironment Biology, Eli Lilly and Company, DC0546, Room H4320C, Lilly Research Labs - Oncology Division, INDIANAPOLIS, IN 46285, USA. Phone: +1-317-433-6087, Fax: +1-317-277-3652, E-mail: yingling_jonathan@lilly.com

PREFACE

THE SMAD FAMILY

PETER TEN DIJKE[1] AND CARL-HENRIK HELDIN[2]
[1] *Department of Molecular and Cell Biology, Leiden University Medical Center, Leiden, The Netherlands*
[2] *Ludwig Institute for Cancer Research, Uppsala University, Sweden*

1. INTRODUCTION

About 10 years ago, our understanding of how signals from transforming growth factor-β (TGF-β) family members and their specific serine/threonine kinase receptors are transduced from the plasma membrane to the nucleus, was a black box. Yeast two-hybrid screening approaches with the intracellular domain of TGF-β superfamily receptors as baits, were initiated by several laboratories, but failed to identify critical intracellular downstream effectors. A breakthrough came through genetic studies in *Drosophila*; in screens for dominant enhancers of weak *dpp* alleles (dpp is the TGF-β homolog in *Drosophila*) *Mothers against dpp (Mad)* and *Medea* were discovered (Raftery et al., 1995; Sekelsky et al., 1995). Homozygous *Mad* and *Medea* mutants are phenotypically similar to dpp mutants. In *C. elegans, daf-4* encodes a serine/threonine kinase receptor and *daf-4* mutants are dauer-constitutive and smaller than wild-type. Screening for mutants with the same small *daf-4* phenotype revealed three genes, *sma-2, sma-3* and *sma-4* (Savage et al., 1996). Mad, Medea and Sma proteins were found to be essential components downstream of TGF-β receptor signaling pathways in these lower invertebrates (Newfeld et al., 1996; Wiersdorff et al., 1996) (see Chapters 2 and 3). Shortly thereafter, homologous *Mad* and *sma*-related genes were identified in *Xenopus*, mouse and man, and shown to function as principal effectors downstream of serine/threonine kinase receptors in vertebrates (Eppert et al., 1996; Graff et al., 1996; Hoodless et al., 1996; Liu et al., 1996; Thompson et al., 1996) (see Chapter 1). The designation Smad was then suggested for the vertebrate homologues of Sma and Mad (Derynck et al., 1996).

After the discovery of Smads, several laboratories independently, and at about the same time, identified additional members of the Smad gene family through their

P. ten Dijke and C.-H. Heldin (eds.), Smad Signal Transduction, 1–13.

homology with Sma and Mad genes by PCR cloning and/or mining of expressed sequence tags (EST) databases. Since then intensive work has been devoted to elucidate which Smads are activated by specific type I serine/threonine kinase receptors, the subcellular localization of Smads before and after ligand stimulation, the role of Smads as transcriptional factors, modulation of Smad function by interacting proteins, and the *in vivo* function and role in disease of Smads. These efforts led to a remarkably quick progress in our understanding of the mechanism of action of Smads; their role as principle intracellular downstream effectors for TGF-β family members is now firmly established (Attisano and Wrana, 2002; Derynck and Zhang, 2003; Shi and Massagué, 2003; ten Dijke and Hill, 2004) (see Chapters 2-8). Smad activation, subcellular distribution and stability have been shown to be intricately regulated (see Chapters 9-16), and Smads have been found to function as signal integrators within an extensive intracellular network (see Chapters 14-18).

This volume provides an in-depth review of the rapidly developing field of Smad research, in which structures are integrated with *in vivo* functions (see Chapter 11). Moreover, the impact of functional genomics and systems biology approaches on Smad signaling (see Chapters 17 and 18), links between alterations in Smad signaling and disease (see Chapters 19 and 20) and how this knowledge may come to be applied in the clinic (see Chapters 21 and 22), will be discussed. In this preface, we will start by reviewing the TGF-β family and their specific type I and type II serine/threonine kinase receptors, and will subsequently introduce the Smad family.

2. TGF-β FAMILY MEMBERS AND THEIR SIGNALING RECEPTORS

2.1 TGF-β Family Members are Multifunctional Cytokines

TGF-β family members, which include TGF-βs, Activins, and bone morphogenetic proteins (BMPs)/growth and differentiation factors (GDFs), are structurally related secreted dimeric cytokines (Roberts and Sporn, 1990). They are produced by cells as larger precursor proteins that are processed within the Golgi apparatus by endoproteases of the convertase family (*e.g.* furin) (Dubois et al., 1995) (Fig. 1A). Upon cleavage, the amino-terminal remnant, also termed latency-associated peptide (LAP), remains non-covalently associated to the carboxy-terminal part that contains the mature protein. LAP prevents binding of mature ligand to the receptor and thus keeps the ligand inactive (Annes et al., 2004). The mature TGF-β can be released from the inactive complexes by several mechanisms, including cleavage of LAP by proteases, such as plasmin (Lyons et al., 1988), and through action of LAP binding proteins, such as thrombospondin (Crawford et al., 1998). This mechanism of latency imposed by LAP has been mainly investigated and demonstrated for TGF-βs; whether it also occurs for the many other TGF-β family members remains to be investigated.

The TGF-β family members share most similarity in their mature domains that have a characteristic cystine knot motif. At least 34 family members have been

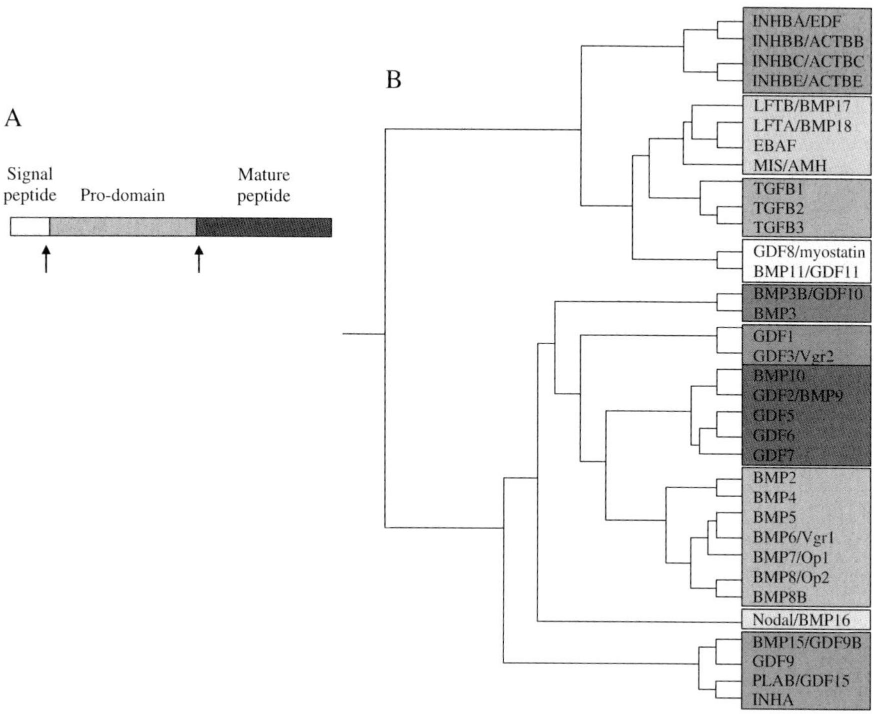

Figure 1. TGF-β family members: multiple cytokines with pleiotropic functions. (A) Schematic structure of TGF-β family members. The precursor of TGF-β family members is composed of a signal peptide, an amino-terminal propeptide (also termed latency-associated peptide) and a carboxy-terminal mature ligand. (B) Phylogenetic analysis of the TGF-β family. Reproduced with permission and copyright © of the Britisch Editorial Soceity of Bone and Joint Surgery (ten Dijke et al., J. Bone Joint Surg 2003, 85-B:34-8) (A color version of this figure is freely accessible via the website of the book: http://www.springer.com/1-4020-4542-5)

Abbreviations: ACT, Activin; BMP, bone morphogenetic protein; GDF, growth and differentiation factor; INH, inhibin; MIS/AMH, müllerian inhibiting substance/anti-müllerian hormone; OP, osteogenic protein; TGF-β, transforming growth factor β

identified in the human genome (Fig. 1B). TGF-β, the founding member of this family, was discovered in the late 70's/early 80's as a factor produced by virus-transformed cells with the ability to induce the growth of normal rat kidney cells in soft agar (DeLarco and Todaro, 1976; Roberts et al., 1981). Subsequent studies demonstrated that TGF-β has a potent growth inhibitory activity (Tucker et al., 1984), and, in fact, is a pleiotropic molecule that regulates cell proliferation, differentiation, apoptosis, migration, adhesion of many different cell types (Moses and Serra, 1996; Roberts and Sporn, 1990). Activin was originally identified as a factor that stimulates the secretion of follicle stimulating hormone from the pituitary gland (Mason et al., 1985), and as a stimulator of erythroid differentiation (Murata et al., 1988). BMP was first known for its ability to induce cartilage and bone (Wozney et al., 1988). Subsequent studies on Activins and BMPs revealed that these, like TGF-β, also are multifunctional proteins (Massagué, 1990). Certain members of the TGF-β family, such as müllerian inhibiting substance (MIS)/anti-müllerian hormone (AMH), nodal, myostatin, and GDF-9, appear to have a restricted expression pattern and have been implicated in specific biological responses. However, it is likely that future studies will ascribe additional functions to these factors.

Often TGF-β family members act as homodimers, but heterodimers between different isoforms can also occur. In the case of Activins, four β chains (βA through βE) have been identified, of which βA and βB can form homo- as well as heterodimers. Moreover, inhibins that antagonize the activity of Activins, are heterodimers of inhibin α chains and Activin β chains (Mathews, 1994). In addition, a BMP2/7 heterodimer has been isolated from bone (Sampath et al., 1990) and shown to be more potent in bone induction than their respective homodimers (Israel et al., 1996).

TGF-β family members, and their downstream signaling components, can be found in species as diverse as nematodes, fruit flies, frogs, fish and mammals (see Chapters 1-3). Gene targeting approaches of TGF-β family ligands have revealed their pivotal roles in embryogenesis and in maintaining tissue homeostasis (Chang et al., 2002). Disruption of TGF-β signaling has been linked to various developmental disorders and numerous human diseases, including cancer, fibrosis and autoimmune diseases (see Chapters 19-22) (Blobe et al., 2000; Siegel and Massagué, 2003) (Fig. 2).

The multifunctional characteristics of TGF-β family members imply the need for tight control of their activities. Such control is exerted at different levels. TGF-β is synthesized as an inactive precursor form and is activated in a controlled manner (see above). In addition, the activity of TGF-β family members is kept in check by

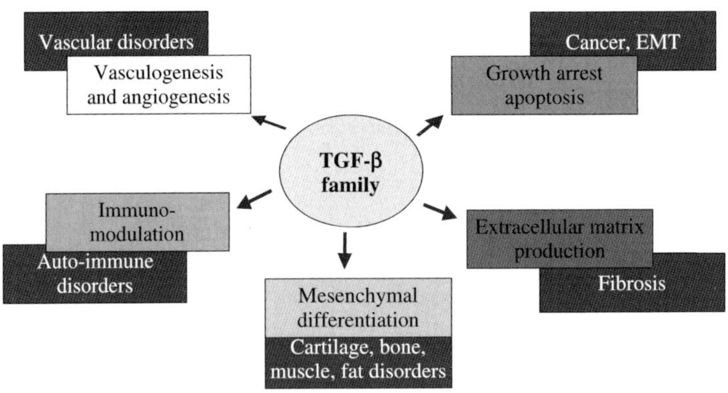

Figure 2. TGF-β family members are multifunctional proteins with crucial roles in embryonic development and in maintaining tissue homeostasis. For example, TGF-β inhibits proliferation of epithelial, endothelial and immune cells, stimulates mesenchymal cell proliferation and extracellular matrix production, regulates the migration and differentiation of many different cell types. Deregulation of their signaling has been implicated in several developmental disorders and in various human diseases including cancer, fibrosis, connective tissue diseases, auto-immune diseases and vascular diseases (A color version of this figure is freely accessible via the website of the book: http://www.springer.com/1-4020-4542-5)

interaction of specific extracellular inhibitors that prevent ligands from binding to signaling receptors (Balemans and Van Hul, 2002). For example, noggin strongly interacts with BMPs with affinities that resemble BMP binding to BMP receptors (Groppe et al., 2002), and follistatin sequesters Activins and prevents them from binding to Activin receptors (de Winter et al., 1996). Moreover, the TGF-β family members are rapidly cleared through the action of scavenger proteins, such as α2-macroglobulin (O'Connor-McCourt and Wakefield, 1987), which results in a short biological half-life of TGF-β family members (Wakefield et al., 1990).

2.2 TGF-β Serine/threonine Kinase Receptors

TGF-βs transduce their signals across the plasma membrane into the cell by inducing heteromeric complexes of type I and type II receptors with intrinsic serine/threonine kinase activity (Attisano and Wrana, 2002; Derynck and Zhang, 2003; Shi and Massagué, 2003; ten Dijke and Hill, 2004). The receptor types are structurally similar with short cysteine-rich extracellular domains, single transmembrane spanning regions, and intracellular parts with serine/threonine kinase domains (Fig. 3A). At least two type II receptors and two type I receptors are needed for signaling (Luo and Lodish, 1996; Weis-Garcia and Massagué, 1996), and probably form a heterotetrameric receptor complex (Yamashita et al., 1994). Five type II receptors and seven type I receptors, also termed Activin receptor-like kinases (ALKs), are present in the human genome (Fig. 3B). The TGF-β type II receptor, MIS type II receptor and BMP type II receptors only bind TGF-β, MIS/AMH and BMPs/GDFs, respectively, but Activin type IIA and type IIB receptors bind Activins, nodal as well as BMPs. Different members of the BMP family thusbind to different type II receptors. The type II receptor has constitutive kinase activity and upon ligand-induced heteromeric complex formation, the type II receptor kinase phosphorylates the type I receptor on particular serine and threonine residues in the juxtamembrane region (also termed GS-domain) (Wrana et al., 1994). Thus, type I receptors act downstream of type II receptors; consistent with this notion, type I receptors have been shown to determine the specificity of the heteromeric receptor complex (Cárcamo et al., 1995).

In most cells, ALK4 and ALK5 are type I receptors for Activin and TGF-β, respectively. Recently, GDF9 and myostatin have been shown to bind to ALK5 in cooperation with BMP type II receptor and Activin type II receptor, respectively (Mazerbourg et al., 2004; Rebbapragada et al., 2003). ALK4 and ALK7 are nodal type I receptors. BMPs (and possibly also MIS/AMH) generally signal via ALK2, ALK3 and ALK6 (Miyazono et al., 2005), but surprisingly, BMP3 has been shown to signal via Activin type II and ALK4 receptors (Daluiski et al., 2001). Thus, homodimeric and heterodimeric forms of individual BMPs are capable of recruiting different type I and type II receptors in the signaling receptor complex; moreover, individual receptors can bind several different ligands. Furthermore, ALK1, in addition to ALK5, is a signaling type I receptor for TGF-β in endothelial cells and neurons (Goumans et al., 2002; Konig et al., 2005). Interestingly, TGF-β signaling

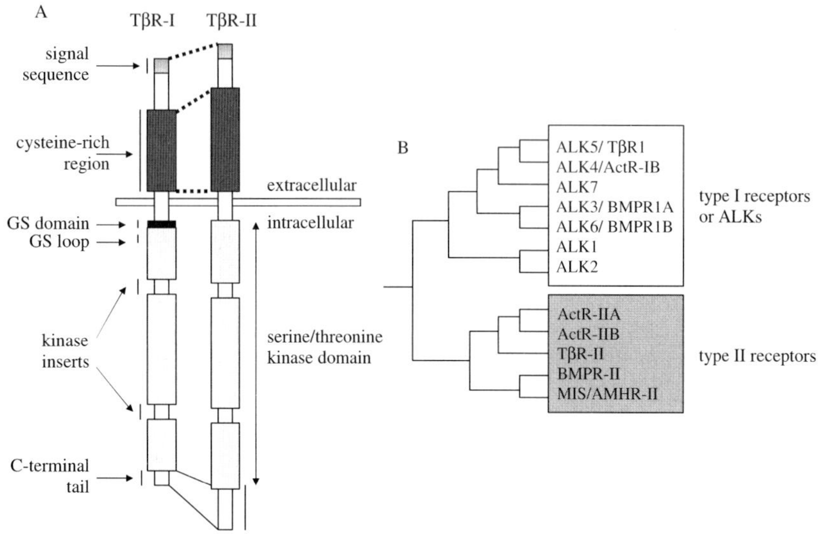

Figure 3. TGF-β family type I and type II receptors. (A) Schematic representations of human TGF-β type I and type II serine/threonine kinase receptors. TβR-I and TβR-II are single transmembrane proteins with short cysteine-rich extracellular domains and intracellular serine/threonine kinase domains with two short kinase inserts. The carboxy-tail of TβR-II is longer than that of TβR-I, and TβR-I contains a domain rich in glycine and serine amino acid residues (termed GS domain) in which particular serine and threonine residues are phosphorylated by TβR II kinase. The L45 loop in TβR I is an exposed nine-amino acid residue region within the kinase domain that is an important determinant for R-Smad interaction. Modified with permission from Figure 1A from Cardiovascular Research, 65(3):599-608, Lebrin, et al. © 2005 European Society to Cardiology. (B) Phylogenetic analysis of human type I and type II receptors of two distinct subfamilies. Five type II receptors and seven type I receptors (also termed ALKs) have been identified in humans (A color version of this figure is freely accessible via the website of the book: http://www.springer.com/1-4020-4542-5)

Abbreviations: ActR, Activin receptor; ALK, Activin receptor-like kinase; MIS/AMH, MIS/AMH receptor; BMPR, BMP receptor; TβR, TGF-β receptor

via ALK1 was shown to be dependent on the kinase activity of ALK5, thereby providing a lateral mode of signaling (Goumans et al., 2003). Whether similar lateral signaling occur for other ALKs or other ligands, needs to be investigated. The availability of more and more ligands as recombinant proteins, will allow a detailed determination of their preferences for type I and type II receptor partners.

3. THE SMAD FAMILY OF SIGNAL TRANSDUCERS

3.1 Nomenclature and Structure of Smads

The Smad family can be divided into three distinct subfamilies: receptor-regulated (R)-Smads, common partner (Co)-Smads and inhibitory (I)-Smads. Activated type I receptor kinases transiently interact with and phosphorylate particular R-Smads at their extreme C-terminal serine residues (see Chapter 12) (Fig. 4A). Whereas Smad2 and Smad3 act downstream of ALK4, ALK5 and ALK7, Smad1, Smad5 and Smad8 are phosphorylated by ALK1, ALK2, ALK3 and ALK6 (Fig. 4C). The L45 loop in the type I receptor kinase domain determines the specificity of Smad isoform activation (see Chapter 11). Phosphorylated Smads form heteromeric complexes with Co-Smads that are shared components in signal transduction by TGF-β family members. Whereas one Co-Smad, *i.e.* Smad4, has been identified in mammals, two C-Smads, *i.e.* Smad4α and Smad4β (also termed Smad10), have been identified in *Xenopus*. Smad complexes accumulate in the nucleus (see Chapter 10), where they can bind to DNA directly or indirectly through other DNA binding proteins (see Chapter 15), and thus control the expression of target genes in a cell type-specific manner through interaction with co-activators and co-repressors (see Chapter 14).

R-Smads and Co-Smads have two highly similar regions at their amino terminal and carboxy terminal regions, termed Mad homology 1 (MH1) domain and MH2 domain, respectively (Fig. 4B). The two MH domains are separated by a less conserved linker region of variable length that is rich in proline residues. The MH1 domain of R-Smads, except Smad2, can bind through a protruding 11-residue β-hairpin directly to specific DNA sequences (Fig. 4B). The MH1 regions in R- and Co-Smads contain a nuclear localization signal-like (NLS-like) sequence (Fig. 4B), which in Smad3 and Smad4 has been shown to interact with importin β and α, respectively. Mutation of these NLS sequences prevent Smad nuclear accumulation

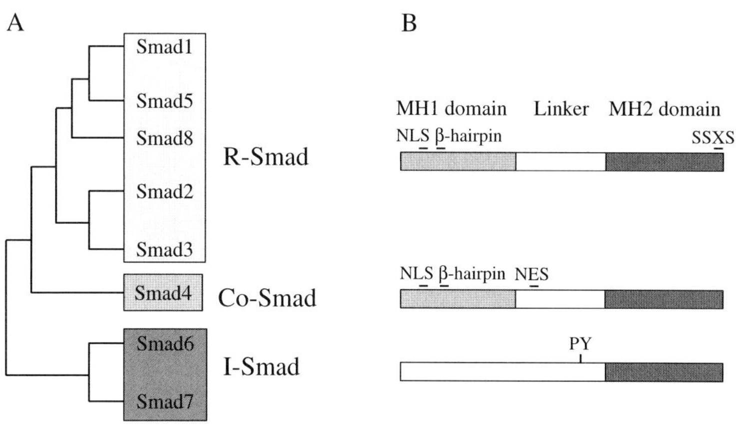

Figure 4. The Smad family. Phylogenetic analysis of human Smads and schematic representations of human Smad structures. The Smad family can be divided into three distinct subfamilies: Receptor-regulated (R)-Smads (*i.e.* Smad1, Smad2, Smad3, Smad5 and Smad8), Common-partner (Co)-Smad (*i.e.* Smad4) and Inhibitory (I)-Smads (*i.e.* Smad6 and Smad7). Conserved Mad Homology (MH)1 and MH2 domains are indicated. The β hairpin and the PPxY motif (PY) that mediates binding to DNA and Smad ubiquitin regulatory ligases (Smurfs), respectively, are indicated. Nuclear localization signal (NLS) and nuclear export signal (NES) important for nuclear-cytoplasmic translocations are also shown. The serines in the C-terminal SXS motif of R-Smads can be phosphorylated by type I receptor kinases (A color version of this figure is freely accessible via the website of the book: http://www.springer.com/1-4020-4542-5)

in response to TGF-β. Smad4 contains a nuclear export signal (NES) in the linker region (Fig. 4B), which interacts with the nuclear exporter CRM1. Formation of homomeric and heteromeric complexes among R- and Co-Smads are mediated via their MH2 domains (see Chapter 11). The MH2 domains of R- and Co-Smads can also recruit transcriptional co-activators and co-repressors (see Chapter 14).

I-Smads have a conserved MH2 domain, but their amino-terminal regions show only weak similarity to the MH1 domains of R- and Co-Smads (see Chapter 19) (Fig. 4B). I-Smads interact via a PY-motif with WW-domain containing HECT-domain ubiquitin ligases (Smurfs) (Fig. 4B). Upon recruitment of an I-Smad-Smurf complex to the activated receptor, Smurf induces receptor degradation via proteosomal and lysosomal pathways (Kavsak et al., 2000). Additional mechanisms by which I-Smads antagonize signaling have been described and are discussed in Chapter 19.

3.2 Activation and Regulation of Smad Function

The recruitment of Smads to activated TGF-β receptor complexes is carefully controlled. Several proteins with scaffolding, anchoring and/or chaperone activity have been identified. Smad anchor for receptor activation (SARA) is localized in early endosomes and, by interacting with non-activated Smads and receptor complexes, presents Smad2 or 3 for the type I receptor and promotes their phosphorylation and activation (see Chapter 9). In their non-activated state, the MH1 and MH2 domains interact and inhibit each others functions, *i.e.* the MH1 domain represses MH2-domain-mediated recruitment of transcriptional co-activators and the MH2 domain inhibits MH1-domain-mediated DNA binding. Upon C-terminal phosphorylation, R-Smads form homo- and hetero-oligomeric complexes of different stoichiometry with each other and with Smad4. Upon Smad complex formation, nuclear import sequences may become exposed and nuclear export sequences shielded, thereby inducing the nuclear accumulation of these complexes (see Chapter 10) (Fig. 5). The affinity of R- and Co-Smads for binding to DNA is relatively low and Smads therefore require other DNA sequence-specific binding factors to bind efficiently to promoters of target genes. Some of these Smad-interacting transcription factors are expressed in a cell-type specific manner and their activation state is subject to specific stimuli, thereby providing integration with other signaling pathways (see Chapter 14-18). In addition, signal integration is achieved through various post-translation modifications of Smads (in addition to the C-terminal phosphorylation by the activated type I receptor), including phosphorylation, poly- and mono-ubiquitination, sumoylation and acetylation. These modifications were found to change interaction of Smads with partner proteins or DNA, stability and/or subcellular localization (see Chapter 12 and 13). The transactivation or repression properties of Smads are mediated through interaction with co-activators and co-repressors that recruit, or contain intrinsic, histone acetyltransferase (HAT) or histone deacetylase (HDAC) activities, respectively, and thereby regulate chromosome condensation and accessibility of Smads with the basal transcription machinery (see Chapter 14).

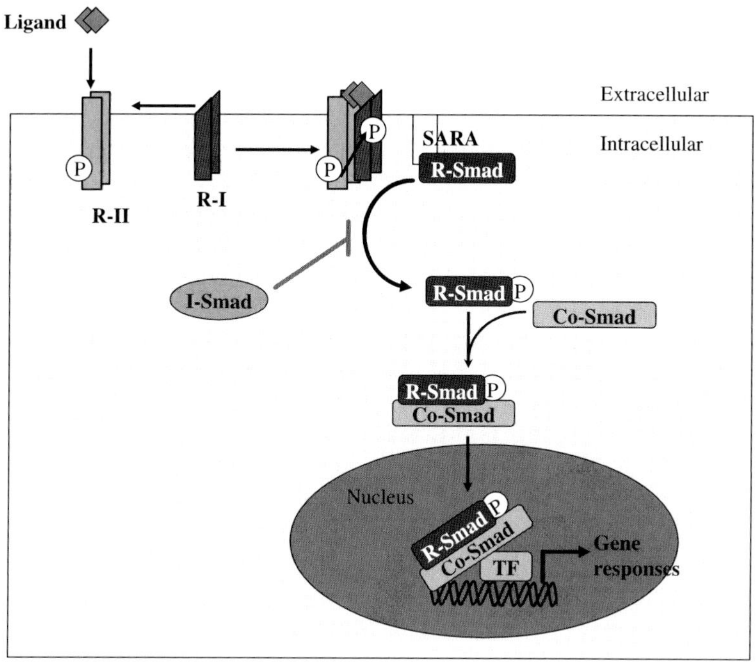

Figure 5. The canonical TGF-β/Smad signaling pathway. TGF-β binds to and stabilizes heteromeric complexes of type I and type II serine/threonine kinase receptors. The type II receptor is endowed with constitutively active kinase activity and phosphorylates the type I receptor on specific serine and threonine residues in the juxtamembrane region (also termed GS domain). Upon this activation, the type I receptor propagates the signal inside the cell through phosphorylation of R-Smads at two C-terminal serine residues found in an SXS motif. R-Smads can be recruited to the activated type I receptor through auxiliary proteins, such as Smad anchor for receptor activation (SARA). Activated R-Smads form heteromeric complexes with Smad4 that in combination with transcription factors can bind to promoters of target genes. These complexes regulate, together with co-activators and co-repressors, specific transcriptional responses (A color version of this figure is freely accessible via the website of the book: http://www.springer.com/1-4020-4542-5)

3.3 The Future of Smad Research

The field of Smad research has diversified enormously as is exemplified by the many aspects of Smad research covered in this book. The multifunctional character of TGF-β family members is reflected in the many positive and negative modes of regulation of Smads. The investigation of cross-talk with other signaling pathways will be a recurring theme in future studies. While the TGF-β/Smad pathway has been implicated in many responses, an important issue that largely remains to be explored is the requirement of particular Smad isoforms in these responses, *e.g.* whether responses require specific R-Smads and/or Smad4. In addition, the recent results which have demonstrated transcription-independent functions of Smads, such as recruitment, sequestration and enzyme activation, need to be further investigated (ten Dijke and Hill, 2004).

Studies in genetically accessible model organisms, such as *C. elegans*, *Drosophila* and the vertebrate Zebrafish, will continue to be important in elucidating the mechanisms that underlie TGF-β/Smad signaling. The combination of these efforts with unbiased large scale genetic, biochemical and/or proteomic interaction screens (Vidal, 2005), functional genomic approaches using siRNAs (Paddison et al., 2004) or morpholino's (Ekker, 2000), and transcriptional profiling using micro arrays (Ideker, 2004) (see Chapters 2, 3, 17 and 18), will be particularly powerful. It will also be important to validate the patho-physiological significance of the identified biochemical and genetic interactions between TGF-β signaling components using transgenic mouse models.

An important challenge for the future will be to translate our current knowledge into clinical applications. Specific TGF-β receptor kinase inhibitors have recently been generated, and shown to block ligand-induced Smad-dependent responses (see Chapters 21 and 22). However, like TGF-β, Smads are multifunctional proteins; they have been implicated in the anti-proliferative response of TGF-β (see Chapter 4), but also in TGF-β-induced invasion and metastasis of tumor cells (see Chapter 7 and 20) and in TGF-β-induced extracellular matrix formation leading to fibrosis (see Chapter 22). Further dissection of Smad-driven responses, and identification of specificity determinants for these various responses, may allow for specific intervention of diseases with perturbed TGF-β/Smad signaling.

ACKNOWLEDGEMENTS

We would like to thank Ingegärd Schiller for her expert secretarial assistance. Research on TGF-β/Smad signaling in our laboratories is supported by grants from the Dutch Cancer Society (NKI 2000-2217), the European Community (BRECOSM and EpiPlastCarcinoma) and the Ludwig Institute for Cancer Research.

REFERENCES

Annes, J., Vassallo, M., Munger, J.S., and Rifkin, D.B., 2004, A genetic screen to identify latent transforming growth factor β activators. *Anal Biochem* 327: 45-54.

Attisano, L., and Wrana, J.L., 2002, Signal transduction by the TGF-β superfamily. *Science* 296: 1646-1647.

Balemans, W., and Van Hul, W., 2002, Extracellular regulation of BMP signaling in vertebrates: a cocktail of modulators. *Dev Biol* 250: 231-250.

Blobe, G.C., Schiemann, W.P., and Lodish, H.F., 2000, Role of transforming growth factor β in human disease. *N Engl J Med* 342: 1350-1358.

Cárcamo, J., Zentella, A., and Massagu, J., 1995, Disruption of transforming growth factor β signaling by a mutation that prevents transphosphorylation within the receptor complex. *Mol Cell Biol* 15: 1573-1581.

Chang, H., Brown, C.W., and Matzuk, M.M., 2002, Genetic analysis of the mammalian transforming growth factor-beta superfamily. *Endocr Rev* 23: 787-823.

Crawford, S.E., Stellmach, V., Murphy-Ullrich, J.E., Ribeiro, S.M.F., Lawler, J., Hynes, R.O., Boivin, G.P., and Bouck, N., 1998, Thrombospondin-1 is a major activator of TGF-β1 in vivo. *Cell* 93: 1159-1170.

Daluiski, A., Engstrand, T., Bahamonde, M.E., Gamer, L.W., Agius, E., Stevenson, S.L., Cox, K., Rosen, V., and Lyons, K.M., 2001, Bone morphogenetic protein-3 is a negative regulator of bone density. *Nat Genet* 27: 84-88.

de Winter, J.P., ten Dijke, P., de Vries, C.J.M., van Achterberg, T.A.E., Sugino, H., de Waele, P., Huylebroeck, D., Verschueren, K., and van den Eijnden-van Raaij, A.J.M., 1996, Follistatins neutralize activin bioactivity by inhibition of activin binding to its type II receptors. *Mol Cell Endocrinol* 116: 105-114.

DeLarco, J., and Todaro, G.J., 1976, Membrane receptors for murine leukemia viruses: characterization using the purified viral envelope glycoprotein, gp71. *Cell* 8: 365-371.

Derynck, R., Gelbart, W.M., Harland, R.M., Heldin, C.-H., Kern, S.E., Massagué, J., Melton, D.A., Mlodzik, M., Padgett, R.W., Roberts, A.B., Smith, J., Thomsen, G.H., Vogelstein, B., and Wang, X.-F., 1996, Nomenclature: Vertebrate mediators of TGFβ family signals. *Cell* 87: 173.

Derynck, R., and Zhang, Y.E., 2003, Smad-dependent and Smad-independent pathways in TGF-β family signalling. *Nature* 425: 577-584.

Dubois, C.M., Laprise, M.H., Blanchette, F., Gentry, L.E., and Leduc, R., 1995, Processing of transforming growth factor β1 precursor by human furin convertase. *J Biol Chem* 270: 10618-10624.

Ekker, S.C., 2000, Morphants: a new systematic vertebrate functional genomics approach. *Yeast* 17: 302-306.

Eppert, K., Scherer, S.W., Ozcelik, H., Pirone, R., Hoodless, P., Kim, H., Tsui, L.-C., Bapat, B., Gallinger, S., Andrulis, I.L., Thomsen, G.H., Wrana, J.L., and Attisano, L., 1996, MADR2 maps to 18q21 and encodes a TGFβ-regulated MAD-related protein that is functionally mutated in colorectal carcinoma. *Cell* 86: 543-552.

Goumans, M.-J., Valdimarsdottir, G., Itoh, S., Lebrin, F., Larsson, J., Mummery, C., Karlsson, S., and ten Dijke, P., 2003, Activin receptor-like kinase (ALK)1 is an antagonistic mediator of lateral TGFβ/ALK5 signaling. *Mol Cell* 12: 817-828.

Goumans, M.-J., Valdimarsdottir, G., Itoh, S., Rosendahl, A., Sideras, P., and ten Dijke, P., 2002, Balancing the activation state of the endothelium via two distinct TGF-β type I receptors. *EMBO J* 21: 1743-1753.

Graff, J.M., Bansal, A., and Melton, D.A., 1996, Xenopus Mad proteins transduce distinct subsets of signals for the TGFβ superfamily. *Cell* 85: 479-487.

Groppe, J., Greenwald, J., Wiater, E., Rodriguez-Leon, J., Economides, A.N., Kwiatkowski, W., Affolter, M., Vale, W.W., Belmonte, J.C., and Choe, S., 2002, Structural basis of BMP signalling inhibition by the cystine knot protein Noggin. *Nature* 420: 636-642.

Hoodless, P.A., Haerry, T., Abdollah, S., Stapleton, M., O'Connor, M.B., Attisano, L., and Wrana, J.L., 1996, MADR1, a MAD-related protein that functions in BMP2 signaling pathways. *Cell* 85: 489-500.

Ideker, T., 2004, Systems biology 101 – what you need to know. *Nat Biotechnol* 22: 473-475.

Israel, D.I., Nove, J., Kerns, K.M., Kaufman, R.J., Rosen, V., Cox, K.A., and Wozney, J.M., 1996, Heterodimeric bone morphogenetic proteins show enhanced activity *in vitro* and *in vivo*. *Growth Factors* 13: 291-300.

Kavsak, P., Rasmussen, R.K., Causing, C.G., Bonni, S., Zhu, H., Thomsen, G.H., and Wrana, J.L., 2000, Smad7 binds to Smurf2 to form an E3 ubiquitin ligase that targets the TGFβ receptor for degradation. *Mol Cell* 6: 1365-1375.

Konig, H.G., Kogel, D., Rami, A., and Prehn, J.H., 2005, TGF-β1 activates two distinct type I receptors in neurons: implications for neuronal NF-κB signaling. *J Cell Biol* 168: 1077-1086.

Liu, F., Hata, A., Baker, J.C., Doody, J., Cárcamo, J., Harland, R.M., and Massagué, J., 1996, A human Mad protein acting as a BMP-regulated transcriptional activator. *Nature* 381: 620-623.

Luo, K.X., and Lodish, H.F., 1996, Signaling by chimeric erythropoietin-TGF-β receptors: Homodimerization of the cytoplasmic domain of the type I TGF-β receptor and heterodimerization with the type II receptor are both required for intracellular signal transduction. *EMBO J* 15: 4485-4496.

Lyons, R.M., Keski-Oja, J., and Moses, H.L., 1988, Proteolytic activation of latent transforming growth factor-β from fibroblast-conditioned medium. *J Cell Biol* 106: 1659-1665.

Mason, A.J., Hayflick, J.S., Ling, N., Esch, F., Ueno, N., Ying, S.-Y., Guillemin, R., Niall, H., and Seeburg, P.H., 1985, Complementary DNA sequences of ovarian follicular fluid inhibin show precursor structure and homology with transforming growth factor-β. *Nature* 318: 659-663.

Massagué, J., 1990, The transforming growth factor-β family. *Annu Rev Cell Biol* 6: 597-641.

Mathews, L.S., 1994, Activin receptors and cellular signaling by the receptor serine kinase family. *Endocr Rev* 15: 310-325.

Mazerbourg, S., Klein, C., Roh, J., Kaivo-Oja, N., Mottershead, D.G., Korchynskyi, O., Ritvos, O., and Hsueh, A.J., 2004, Growth differentiation factor-9 signaling is mediated by the type I receptor, activin receptor-like kinase 5. *Mol Endocrinol* 18: 653-665.

Miyazono, K., Maeda, S., and Imamura, T., 2005, BMP receptor signaling: transcriptional targets, regulation of signals, and signaling cross-talk. *Cytokine Growth Factor Rev* 16: 251-263.

Moses, H.L., and Serra, R., 1996, Regulation of differentiation by TGF-β. *Curr Opin Genet Dev* 6: 581-586.

Murata, M., Eto, Y., Shibai, H., Sakai, M., and Muramatsu, M., 1988, Erythroid differentiation factor is encoded by the same mRNA as that of the inhibin β_A chain. *Proc Natl Acad Sci U S A* 85: 2434-2438.

Newfeld, S.J., Chartoff, E.H., Graff, J.M., Melton, D.A., and Gelbart, W.M., 1996, *Mothers against dpp* encodes a conserved cytoplasmic protein required in DPP/TGF-β responsive cells. *Development* 122: 2099-2108.

O'Connor-McCourt, M.D., and Wakefield, L.M., 1987, Latent transforming growth factor-β in serum: A specific complex with a_2-macroglobulin. *J Biol Chem* 262: 14090-14099.

Paddison, P.J., Silva, J.M., Conklin, D.S., Schlabach, M., Li, M., Aruleba, S., Balija, V., O'Shaughnessy, A., Gnoj, L., Scobie, K., Chang, K., Westbrook, T., Cleary, M., Sachidanandam, R., McCombie, W.R., Elledge, S.J., and Hannon, G.J., 2004, A resource for large-scale RNA-interference-based screens in mammals. *Nature* 428: 427-431.

Raftery, L.A., Twombly, V., Wharton, K., and Gelbart, W.M., 1995, Genetic screens to identify elements of the *decapentaplegic* signaling pathway in Drosophila. *Genetics* 139: 241-254.

Rebbapragada, A., Benchabane, H., Wrana, J.L., Celeste, A.J., and Attisano, L., 2003, Myostatin signals through a transforming growth factor β-like signaling pathway to block adipogenesis. *Mol Cell Biol* 23: 7230-7242.

Roberts, A.B., Anzano, M.A., Lamb, L.C., Smith, J.M., and Sporn, M.B., 1981, New class of transforming growth factors potentiated by epidermal growth factor: isolation from non-neoplastic tissues. *Proc Natl Acad Sci U S A* 78: 5339-5343.

Roberts, A.B., and Sporn, M.B., 1990, The transforming growth factor-βs, in *Peptide Growth Factors and Their Receptors, Part I*, Sporn, M. B. and Roberts, A. B., eds. Springer-Verlag, Berlin, pp. 419-472.

Sampath, T.K., Coughlin, J.E., Whetstone, R.M., Banach, D., Corbett, C., Ridge, R.J., Özkaynak, E., Oppermann, H., and Rueger, D.C., 1990, Bovine osteogenic protein is composed of dimers of OP-1 and BMP-2A, two members of the transforming growth factor-β superfamily. *J Biol Chem* 265: 13198-13205.

Savage, C., Das, P., Finelli, A.L., Townsend, S.R., Sun, C.-Y., Baird, S.E., and Padgett, R.W., 1996, *Caenorhabditis elegans* genes *sma-2*, *sma-3*, and *sma-4* define a conserved family of transforming growth factor β pathway components. *Proc Natl Acad Sci U S A* 93: 790-794.

Sekelsky, J.J., Newfeld, S.J., Raftery, L.A., Chartoff, E.H., and Gelbart, W.M., 1995, Genetic characterization and cloning of *Mothers against dpp*, a gene required for *decapentaplegic* function in *Drosophila melanogaster*. *Genetics* 139: 1347-1358.

Shi, Y., and Massagué, J., 2003, Mechanisms of TGF-β signaling from cell membrane to the nucleus. *Cell* 113: 685-700.

Siegel, P.M., and Massagué, J., 2003, Cytostatic and apoptotic actions of TGF-β in homeostasis and cancer. *Nat Rev Cancer* 3: 807-821.

ten Dijke, P., and Hill, C.S., 2004, New insights into TGF-β-Smad signalling. *Trends Biochem Sci* 29: 265-273.

Thompson, D.A., Spies, A., Stephan, R.N., Brooks, S.P., Grande, C.C., and Tomasi, T.B., 1996, Prolongation of survival of rat kidney allografts by transforming growth factor-β2. *Transplant Proc* 28: 1948-1951.

Tucker, M.R., Guilford, W.B., and Howard, C.W., 1984, Coronoid process hyperplasia causing restricted opening and facial asymmetry. *Oral Surg Oral Med Oral Pathol* 58: 130-132.

Wakefield, L.M., Winokur, T.S., Hollands, R.S., Christopherson, K., Levinson, A.D., and Sporn, M.B., 1990, Recombinant latent transforming growth factor β1 has a longer plasma half-life in rats than active transforming growth factor β1, and a different tissue distribution. *J Clin Invest* 86: 1976-1984.

Weis-Garcia, F., and Massagué, J., 1996, Complementation between kinase-defective and activation-defective TGF-β receptors reveals a novel form of receptor cooperativity essential for signaling. *EMBO J* 15: 276-289.

Vidal, M., 2005, Interactome modeling. *FEBS Lett* 579: 1834-1838.

Wiersdorff, V., Lecuit, T., Cohen, S.M., and Mlodzik, M., 1996, *Mad* acts downstream of Dpp receptors, revealing a differential requirement for *dpp* signaling in initiation and propagation of morphogenesis in the *Drosophila* eye. *Development* 122: 2153-2162.

Wozney, J.M., Rosen, V., Celeste, A.J., Mitsock, L.M., Whitters, M.J., Kriz, R.W., Hewick, R.M., and Wang, E.A., 1988, Novel regulators of bone formation: molecular clones and activities. *Science* 242: 1528-1534.

Wrana, J.L., Attisano, L., Wieser, R., Ventura, F., and Massagué, J., 1994, Mechanism of activation of the TGF-β receptor. *Nature* 370: 341-347.

Yamashita, H., ten Dijke, P., Franzén, P., Miyazono, K., and Heldin, C.-H., 1994, Formation of hetero-oligomeric complexes of type I and type II receptors for transforming growth factor-β. *J Biol Chem* 269: 20172-20178.

CHAPTER 1

MOLECULAR EVOLUTION OF SMAD PROTEINS

STUART J. NEWFELD[1] AND ROBERT G. WISOTZKEY[2]

[1] *School of Life Sciences, Arizona State University, Tempe, AZ 85287-4501, USA*

[2] *Ingenuity Systems, Redwood City, CA, USA*

Abstract: To date, Smad family members have been found only in eumetazoan animals. To understand the evolutionary relationship between family members we conducted a phylogenetic analysis. To simplify the analysis but retain its explanatory power, we focused on Smad proteins from organisms in three distinct phyla: human, fly, and nematode. Overall, we found that human and fly proteins always cluster together in four subfamilies while three subfamilies contain only nematode proteins. Sequence alignments of distinct regions of were also analyzed. Data from the alignments confirmed that the MH1 (DNA-binding) and MH2 (protein-protein interaction) domains are highly conserved family-wide. The linker region between these domains is also highly conserved but only within subfamilies. Conservation in the C-terminal receptor phosphorylation region provides new insight into a unique subfamily containing three interacting nematode proteins that signal for DAF-7. From a larger perspective, our analysis strongly supports the traditional view that flies are more closely related to humans than to nematodes

Keywords: multigene family; SMAD proteins; phylogeny; amino acid alignments; evolutionary conservation; developmental-evolution; signal transduction

1. INTRODUCTION

The evolutionary relationships between members of a multigene family are ascertained through a phylogenetic analysis involving three steps. First, one must calculate the amount of amino acid similarity between each family member by aligning the protein sequences (Thompson et al., 1997). Second, one applies an amino acid similarity matrix and the extent of similarity between each protein and all of the others are prioritized with the most similar proteins clustered together. These clusters are depicted as the familiar phylogenetic tree (Kumar et al., 2001). Third, the relationships between pairs of proteins are tested for robustness using statistical methods such as bootstrap analysis (Felsenstein, 1985).

P. ten Dijke and C.-H. Heldin (eds.), Smad Signal Transduction, 15–35.

Here we describe a new phylogenetic analysis of the Smad protein family. In order to simplify the analysis but retain its explanatory power, we focus on Smad sequences from organisms in three distinct phyla: human (deuterostome), fly (protostome) and nematode (pseudocoelomate; see Raff 1996, for example, for a taxonomic description of these phyla). We include other species as necessary to add confidence to individual results. Our studies of the MH1 and MH2 domains support the long-standing view that they are highly conserved. Our analysis of the linker region between the MH1 and MH2 domains, previously dismissed as a highly divergent and potentially non-functional part of the protein, reveals surprising levels of sequence conservation within Smad subfamilies. This suggests the hypothesis that distinct functions associated with each subfamily involve linker sequences. An analysis of the receptor phosphorylation domain provides new insights into a unique subfamily containing only nematode proteins that signal for the TGF-β/Activin subfamily member DAF-7.

Recently it has become possible to test phylogenetically derived hypotheses using an approach known as functional genomics. In this technique, interspecies experiments are conducted that evaluate the ability of a family member from one species to mimic the activity of another family member either by rescuing mutant phenotypes (*e.g.* Padgett et al., 1993) or in parallel over-expression experiments (*e.g.* Marquez et al., 2001). We have conducted a number of such tests and review those results here.

2. SMAD FAMILY MEMBERS

To date, Smad family members have been found only in animals. Within the animal kingdom they have been identified in eumetazoans (multicellular organisms with many types of cells) but not yet in metazoans such as sponges (multicellular organisms with very few cell types). However, several transmembrane receptors with similarity to both type I and type II TGF-β receptors have been identified in a freshwater sponge (Suga et al., 1999). A phylogenetic analysis showed that the sponge receptors are very similar to the unusual *C. elegans* receptors DAF-1 and SMA-6 that also fall between receptor types (Herpin et al., 2004). The similarity between sponge and nematode receptors suggests that Smad-like proteins will eventually be found in sponges. Thus, ancestral TGF-β family members and their signaling pathways predate the metazoan/eumetazoan divergence roughly 1.5 billion years ago (Hedges and Kumar, 2003).

The simplest eumetazoans with definitive Smad family members are cnidarians (animals with two germ layers – diploblasts). A sequence similar to Smad1/Mad in the BMP signaling subfamily has been identified in coral (Samuel et al., 2001) and in hydra. The simplest eumetazoan with Smad proteins similar to both Smad1/Mad and Smad2/3 is the blood fluke *Schistosoma mansoni* – an acoelomate with three germ layers but no digestive cavity (Beall et al., 2000). From this it is reasonable to conclude that BMP signaling Smads, and by extension their cognate ligands and receptors, represent the oldest of the TGF-β pathways found in higher animals such as flies and mammals.

Nevertheless, one word of caution: gene discovery in simple organisms is not always simple. Insuring that DNA samples are free from contamination from higher organisms is difficult. For example, parasites like *Schistosoma* may be contaminated with human white blood cells or cnidarians may contain shrimp larvae from their last meal. Reproducibility is essential to insuring confidence in these studies.

In order to achieve easily interpretable results but to maintain maximum confidence in our phylogenetic analysis, we focused on three species with fully sequenced genomes. These species belong to three distinct phyla allowing us maximum discriminatory power in the analysis. Humans (deuterostome) and the fruit fly *D. melanogaster* (protostome) belong to sister taxa at the top of the animal kingdom. They are coelomates – animals with three germ layers and a digestive tract with two openings. Our third species is the nematode *C. elegans* (a pseudocoelomate – animals with three germ layers and a digestive cavity with only one opening). *C. elegans* is the simplest organism with a full set of Smad proteins (R-Smad, Co-Smad and I-Smad subfamilies; Newfeld et al., 1999). Molecular evolution studies indicate that the split between deuterostomes and protostomes occurred 990 million years ago and the split between coelomates and pseudocoelomates occurred 1.2 billion years ago (Hedges and Kumar, 2003). Any amino acids conserved over this enormous span of time are clearly subject to strong positive selection that is most likely due to an essential role in either protein structure or function.

Table 1 describes the 19 Smad sequences we examined. These sequences were utilized for the phylogeny (Fig. 1) and for the MH1, MH2 and receptor phosphorylation domain alignments (Figs. 2, 3 and 5). There are eight Smad proteins in humans (hSmads). hSmad1, hSmad5 and hSmad8 (also known as Smad9 in the Entrez Gene database) transduce DPP/BMP subfamily signals. hSmad2 and hSmad3 transduce TGF-β/Activin subfamily signals. hSmad4 participates with the other Smads to transduce signals of both subfamilies. hSmad6 and hSmad7 antagonize signals of both subfamilies (reviewed in Massagué et al., 2000).

There are four Smad proteins in *Drosophila melanogaster* (DmSmads). Mothers Against Dpp (Mad) transduces Dpp signals. DmSmad2 (also known as SMOX in the Entrez Gene database) transduces DmActivin signals. MEDEA (MED) participates with the other Smads to transduce signals of both subfamilies. Daughters Against Dpp (DAD) antagonizes DPP and possibly DmActivin signals (reviewed in Raftery and Sutherland, 1999).

There are seven Smads in the *Caenorhabditis elegans* (CeSmads). SMA-2, SMA-3 and SMA-4 transduce DBL-1 (a BMP subfamily member) signals. DAF-14 and DAF-8 (also known as Ce1J160 in the Entrez Gene database) transduce DAF-7 (a TGF-β/Activin subfamily member) signals. DAF-3 antagonizes DAF-7 signals. Ce1L81 is a predicted open reading frame that has not yet been assigned to a gene (reviewed in Inoue and Thomas, 2000).

Table 1. Representative Smad Family Members[a]

Entrez Gene		Reference Sequence		
Symbol	GeneID	DNA	Protein	Synonyms[b]
Human				
SMAD1	4086	NM_001003688	NP_001003688.1	JV4-1, MADH1
SMAD2	4087	NM_001003652	NP_001003652.1	JV18-1, MADH2
SMAD3	4088	NM_005902	NP_005893.1	MADH3, JV15-2
SMAD4	4089	NM_005359	NP_005350.1	DPC4, MADH4
SMAD5	4090	NM_001001419	NP_001001419.1	JV5-1, MADH5
SMAD6	4091	NM_005585	NP_005576.2	MADH6, MADH7
SMAD7	4092	NM_005904	NP_005895.1	MADH7, MADH8
SMAD9	4093	NM_005905	NP_005896.1	**SMAD8**, MADH9
Drosophila melanogaster				
DAD	42059	NM_057912	NP_477260	CG5201, EP3196
MAD	33529	NM_057669	NP_477017	CG12399, K00237
MEDEA[c]	43725	NM_079871	NP_524610.1	CG1775
		NM_170559	NP_733438	l(3)SG70
SMOX	31738	NM_078524	NP_511079.1	**SMAD2**, CG2262
Caenorhabditis elegans				
1J160	187612	NM_059920	NP_492321.1	**DAF-8**, R05D11.1
1L81	172930	NM_060345	NP_492746.1	F37D6.6
DAF-14	177908B	NM_069479	NP_501880.1	F01G10.8
DAF-3	180431	NM_001029434	NP_508161.2	F25E2.5
SMA-2	176229	NM_066530	NP_498931	ZK370.2, Cem-1
SMA-3	175955	NM_066092	NP_498493	R13F6.9, Cem-2
SMA-4	175815	NM_065855	NP_498256	R12B2.1, Cem-3
Danio rerio				
SMAD1	30628	NM_131356	NP_571431	madh1, fb39h09
SMAD2	30639	NM_131366	BC044338	madh2, fj43c06
SMAD3a	50892	NM_131571	NP_571646	madh3, smad3
SMAD3b	326283	NM_175083	NP_778258	madh3b, gc92234
SMAD4[d]	30640	XM_694498	XP_699590	madh4
SMAD5	30641	NM_131368	NP_571443	madh5, fb67b04
SMAD7	326282	NM_175082	NP_778257	madh7
Xenopus tropicalis				
SMAD1	493211	NM_001007480	NP_001007481	MGC89254
Xenopus laevis				
SMAD3	378633	AJ311059	CAC38118	madh3-A
SMAD7	394331	AF026125	AAC17489	madh7, smad10
Anopheles gambiae				
ENSANGG000 00003874	127-5428	XM_314661	XP_314661	**AgMAD**
SMAD2	–	AY578801	AAT07306	
ENSANGG000 00016187[c]	127-3058	XM_312001	XP_312001	**AgMEDEA**
		XM_311999	XP_311999	

(Continued)

Table 1. (Continued)

Entrez Gene		Reference Sequence		
Symbol	GeneID	DNA	Protein	Synonyms[b]
Apis mellifera				
LOC409301	409301	XM_392819	XP_392819	**AmMAD**
LOC412601	412601	XM_396056	XP_396056	**AmSMAD2**
LOC409321	409321	XM_392838	XP_392838	**AmMEDEA**
LOC413371	413371	XM_396816	XP_396816	**AmDAD**

[a] Entrez Gene = www.ncbi.nih.gov/entrez/query.fcgi?db = gene.
[b] Synonyms in bold are utilized in this chapter.
[c] *D. melanogaster* and *A. gambiae* MEDEA have two splice variants.
[d] *D. rerio* SMAD4 reference sequences are unofficial. LOC560317 also has regions of identity with human SMAD4.

As part of this study we conducted the first detailed analysis of sequences in the linker region of Smad family members. Perhaps as an artifact of our inability to align this region of CEM-1, CEM-2 and CEM-3 (*C. elegans* Mad-like genes) with Mad, this domain appeared to us as a highly divergent stretch that was easily dismissed as non-functional (Sekelsky et al., 1995). We now know that CEM-1 (SMA-2), CEM-2 (SMA-3), CEM-3 (SMA-4) and Mad all belong to distinct subfamilies of the Smad family (discussed below). We were able to align the linker region from sequences belonging to the same subfamily and found surprisingly high levels of conservation (Fig. 4). However, no subfamily contains more than four sequences. Therefore, to add confidence to our linker region alignments we added sequences from three vertebrate species (two frogs and zebrafish) and two insect species (mosquito and honey bee) as described in Table 1.

3. SMAD FAMILY TREE

Figure 1 shows a phylogenetic tree consistent with previous reports (*e.g.* Newfeld et al., 1999) that that there are four distinct subfamilies of Smads. Clusters of sequences corresponding to two subfamilies of R-Smads, a subfamily of Co-Smads and a subfamily of I-Smads are observed. However, seven subfamilies are present overall. Human and fly genes always cluster together in the four known subfamilies while three subfamilies contain only nematode sequences. Further, human and fly proteins belonging to the same subfamily have been shown to function similarly in transgenic experiments (Marquez et al., 2001).

The Smad1/Mad subfamily contains signal transducing R-Smads dedicated to DPP/BMP subfamily ligands. The Smad2/3 subfamily contains signal transducing R-Smads dedicated to TGF-β/Activin subfamily ligands. The hSmad4/MED subfamily contains signal transducing Co-Smads that form complexes with R-Smads of both subfamilies. One nematode protein (SMA-4) also belongs to this subfamily.

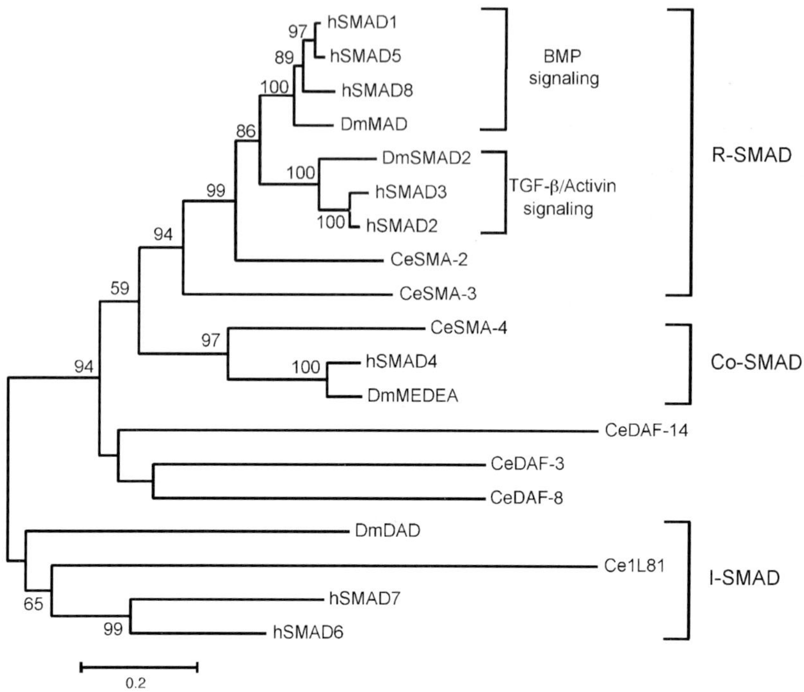

Figure 1. Phylogenetic analysis of the Smad family. Note that human and fly genes cluster together into four major subfamilies: 1) Receptor-associated Smads involved in signaling by DPP/BMP proteins, 2) Receptor-associated Smads involved in signaling by TGF-β/Activin proteins, 3) Co-Smads involved in signaling by DPP/BMP and TGF-β/Activin proteins, 4) Inhibitory Smads. Three subfamilies contain only nematode sequences. Smad sequences were aligned using Clustal-X (Thompson et al., 1997). The neighbor-joining method was utilized (Saitou and Nei, 1987) in the program Mega2 (Kumar et al., 2001) to generate an unrooted phylogeny from the alignment. The length of the alignment (including all unique insertions) is 994 amino acid residues. Branch lengths are drawn to scale. The scale bar shows the number of amino acid substitutions per site between two sequences. Bootstrap values (the percent of trees containing the indicated branch during 1000 trials) above 50 are shown

The hSmad6/7/DAD subfamily contains I-Smads that antagonize signal transduction by R-Smads of both subfamilies. One nematode sequence (Ce1L81) belongs to the I-Smad subfamily.

Of the three subfamilies that include only nematode proteins two contain a single sequence and the third contains three proteins. Even though they signal for a DPP/BMP subfamily member and they are clearly R-Smads, SMA-2 and SMA-3 are different enough from other R-Smads (and each other) that they each constitute a distinct subfamily. Interestingly, the three-member nematode subfamily contains proteins that cooperate in the same pathway but have distinct functions. Each of these proteins influences dauer formation, an alternative third-stage larva specialized for survival and dispersal activated by environmental stress (Cassada and Russell, 1975). In addition, they all function downstream of the TGF-β/Activin subfamily member DAF-7. The constitutively active DAF-3 antagonizes TGF-β signal transduction by binding to DNA and repressing gene expression (Thatcher et al., 1999), a mechanism not used by other I-Smads. Alternatively, the TGF-β-inducible proteins DAF-8 and DAF-14 stimulate the expression of DAF-7 target genes by inhibiting DAF-3 function (Inoue and Thomas, 2000). This is the only subfamily containing proteins that function as both agonists and antagonists in the same pathway.

The tree generates two overall impressions. First, for R-Smads and Co-Smads confidence in the clusters is very high – particularly between human and fly sequences (bootstrap values over 70% are considered statistically significant; Hillis and Bull, 1993). This impression is supported by a study utilizing transgenes expressing human Smad genes in flies. That study showed that human and fly Smads that cluster together in the tree generate the same phenotype (Marquez et al., 2001). Taken together the amino acid similarity and functional conservation studies indicate that human and fly proteins in the same subfamily are encoded by homologous genes. Further, they indicate that one or more gene duplications have occurred in the vertebrate lineage after the split with arthropods leading to multiple human Smad proteins in each R-Smad subfamily.

The second impression is that Smad signaling clearly works the same in flies and humans but is different in many ways in nematodes. For example, there is a nematode specific subfamily composed of agonists and antagonists for the same ligand where the antagonist binds DNA (DAF-3) and the signal transducers (DAF-8 and DAF-14) do not (Thatcher et al., 1999; Inoue and Thomas, 2000). This mechanism is the opposite of that utilized by human and fly Smads (signal transducers bind DNA and inhibitors do not).

Overall these two impressions (homology of fly and human Smads and distinctions between Smad signaling mechanisms utilized in humans and flies versus nematodes) strongly argue against the existence of an "Ecdysozoan" phylum containing nematodes and flies (*e.g.* Aguinaldo et al., 1997). All functional genomics and phylogenetics studies of the Smad family support the traditional view (*e.g.* Hedges and Kumar, 2003) that flies are more closely related to humans than they are to nematodes.

4. SMAD FAMILY DOMAINS

Previous studies have shown that Smad family members that transduce signals (R-Smads and Co-Smads) contain well conserved MH1 domains near their N-terminus and MH2 domains near their C-terminus. Inhibitory Smads have highly divergent MH1 domains but have conserved MH2 domains (*e.g.* Newfeld et al., 1999). This data fits well with experiments showing that the MH1 domain is required for DNA-binding and transcriptional activity while the MH2 domain is involved in a variety of protein-protein interactions including forming multi-Smad complexes (*e.g.* Lagna et al., 1996). Between the MH1 and MH2 domains is a proline-rich linker region not previously characterized in detail. At the C-terminus of R-Smads there is a receptor phosphorylation domain containing serine residues (SSXS) targeted for phosphorylation by TGF-β type I receptor kinases. This domain has typically been included in MH2 domain analyses but in our view it deserves scrutiny as an independent domain. In the analysis that follows, for easy reference, an amino acid residue number described in the text refers to that residue's location in an alignment rather than its location in a given Smad protein.

4.1 MH1 Domain

Figure 2 shows an alignment of the Smad family MH1 domain. Although sequence variation is evident, subsets of the N-terminally located MH1 domain are recognizable in every sequence except for DAF-14. For DAF-14, its N-terminal region is so short that although the MH1 alignment begins at amino acid residue 7, the last 59 amino acid residues of the alignment actually belong to its MH2 domain. We conclude that DAF-14 simply has no MH1 domain. DAD has an extensive amino terminus that only very weakly resembles an MH1 domain. There is just one readily recognizable region in DAF-8: a seven amino acid residue stretch containing the most highly conserved amino acids in the alignment (between amino acid residues 120 and 130). As reported previously (Newfeld et al., 1999), the human I-Smads (hSmad6 and hSmad7) align reasonably well.

The MH1 domain is divided into subregions by unique amino acid insertions in a number of Smads. If there is a biological function for these insertions it is unknown. Two Smad2 proteins, one with and one without the insert, are present in mice. However, mice engineered to express only the short form of Smad2 (without the insert) appear completely normal suggesting that the insert is non-functional (Dunn et al., 2005).

Given its documented role in DNA binding and transcriptional activation (Liu et al., 1997) it is somewhat surprising that there are no absolutely invariant amino acid residues in the MH1 domain. Here we examine the extent of conservation for a number of amino acid residues with known functions. A crystal structure of hSmad3 bound to DNA showed that an 11 amino acid residue region forms a β-hairpin that fits into the major groove of DNA (Shi et al., 1998). The DNA-contacting hairpin is contained within a conserved 20 amino acid residue region beginning with Arg54 and ending with Pro74. The three residues that contact DNA are Arg61, Gln63 and Lys71. Arg59 is present in all R-Smads and Co-Smads (except DmSmad2, an R-Smad that inexplicably contains a unique stretch of nine amino acid residues in this region) and in DAF-3, hSmad6 and hSmad7. Gln63 and Lys71 are present in all R-Smads and Co-Smads (except DmSmad2) and in DAF-3. All three DNA-contacting residues are absent from DAD, DAF-8 and DAF-14.

A more detailed crystal structure of hSmad3 bound to DNA (Chai et al., 2003) found a bound zinc atom. The zinc-contacting residues are Cys44 (present in all but DAD and DAF-14), Cys105, Cys122 and His127 (these three are present in all but DAF-14). Surprisingly, the four zinc-binding amino acids are more highly conserved than the DNA-contacting residues suggesting that zinc-binding is essential to all Smad functions.

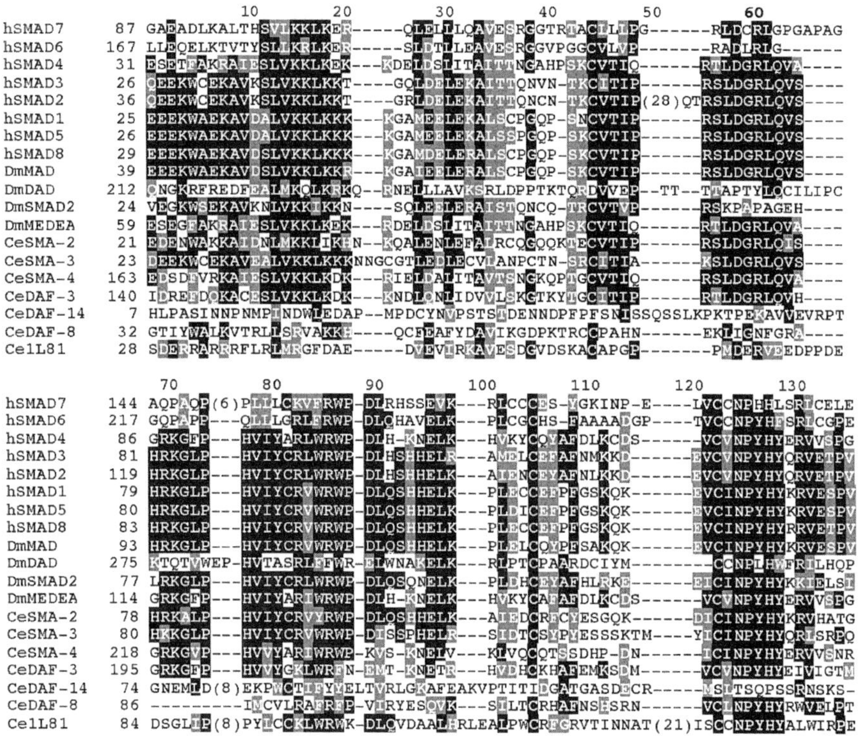

Figure 2. Smad family MH1 domain. This domain is located near the N-terminus of Smad proteins. This domain is highly conserved in R-Smads and Co-Smads. The domain was defined by Pfam (www.sanger.ac.uk/Software/Pfam/index.shtml) based on the crystal structure. Here we show the evolutionarily conserved portion beginning at Glu39 in DmMad and ending at Val144 in DmMad. The length of the alignment is 194 amino acid residues. Regions were removed when insertions were present in three or fewer sequences and the number of residues shown instead. Residues were shaded if 40% of them were identical (black) or similar (grey) by Boxshade3.21 (www.ch.embnet.org/software/BOX_form.html). Numbers above the alignment begin with the first amino acid and run consecutively. Residue number 60 in bold indicates the location of the DNA-binding region

An alignment of the DNA-binding domain of the NFI/CTF family and the MH1 domain of the Smad family identified 22 highly conserved residues, including the four zinc-binding residues (Stefancsik and Sarkar, 2003; Sadreyev and Grishin, 2003). The nuclear factor I (NFI) and CCAAT box-binding transcription factor (CTF) family is composed of vertebrate nuclear proteins that bind a palindromic DNA sequence. The 22 conserved amino acid residues are present in all NFI/CTF family members and all DNA-binding Smads (except DmSmad2). All 22 residues are present in hSmad6 and hSmad7 but not in fly or nematode I-Smads. The majority of conserved amino acids are located in two regions. Four are located between residues 10 and 20 and eleven are located between residues 80 and 100. In contrast, none of the three Smad DNA-binding residues are conserved in the NFI/CTF family. The authors' data is consistent with their hypothesis that the NFI/CTF family diverged from the Smad family after the split between flies and mammals.

A number of conserved residues in the MH1 domain have had their functional importance demonstrated via mutation. For example, we recently conducted a transgenic analysis of MH1 point mutations in Mad (DNA-binding residue), Med (Zinc-binding residue) and hSmad4 (two residues conserved in the NFI/CTF family). We showed that they elicit a variety of mutant phenotypes (Takaesu et al., 2005).

To explicitly examine the relationship between Smad MH1 domains we generated a phylogenetic tree (not shown) from the MH1 domain alignment in Fig. 2. The MH1 domain tree places most *C. elegans* sequences in subfamilies distinct from their placement in the full-length tree shown in Fig. 1. First, SMA-2 and SMA-3 were unique R-Smad subfamilies in the full-length tree but in the MH1 tree they now cluster with the Smad1/Mad subfamily. This fits with the fact that their ligand (DBL-1) belongs to the DPP/BMP subfamily. Second, the subfamily containing the three DAF-7 signaling pathway components (DAF-3, DAF-8 and DAF-14) in the full-length tree breaks up. In the MH1 domain tree, DAF-3 now clusters with SMA-4 in the Co-Smad subfamily. This fits with the fact that both proteins can bind DNA. In the MH1 domain tree, DAF-8 and DAF-14 each form unique subfamilies due to their highly divergent or absent MH1 domains respectively. Another difference is that DAD moves out of the I-Smad subfamily to become a unique subfamily in the MH1 tree.

4.2 MH2 Domain

Figure 3 shows an alignment of the Smad family MH2 domain. First identified as essential for Smad homo-trimer formation (Shi et al., 1997) this C-terminally located domain is now known as a versatile protein-protein interaction module essential for many Smad activities. Functions associated with the MH2 domain are: 1) formation of homo-trimers of R-Smads and Co-Smads, 2) formation of hetero-trimers containing two R-Smads and one Co-Smad, 3) interaction of R-Smads with the SARA adapter protein and TGF-β type I receptor kinases and 4) interaction of R-Smad/Co-Smad hetero-trimers with transcriptional activators and repressors (see Moustakas and Heldin, 2002, for a review). Phylogenetically and functionally, the MH2 domain is the core of the Smad family and is present in all members.

Note that the C-terminal receptor phosphorylation region was included in many previous analyses of the MH2 domain (including our own; Newfeld et al., 1999). However, we exclude the receptor phosphorylation region from this analysis based on structural data showing that prior to phosphorylation this C-terminal region extrudes from the MH2 domain and is not involved in homo-trimer formation (*e.g.* Wu et al., 2001). This distinction should be kept in mind when comparing data reported here with previous studies.

An examination of the MH2 alignment reveals that 24% of the amino acid residues are extremely well conserved (at least 17 of the 19 sequences have an identical or similar amino acid at a particular position). Eleven of the 47 highly conserved residues are identical in all sequences and 13 are very well conserved (a similar amino acid residue in all sequences). Many of the highly conserved residues are contained in six small regions. The largest of these regions (166-193) corresponds to the L3 loop near the C-terminus of the MH2 domain. Here 10 residues are similar or identical in all Smads and 7 are well conserved. As discussed below the L3 loop is involved in two well-documented protein-protein interactions.

The overall structure of an hSmad MH2 domain homo-trimer reveals three subdomains. There is a central β-sandwich, a loop-helix region near the amino-terminus and a helix-bundle region at the C-terminus that extends into the receptor phosphorylation region. In unphosphorylated R-Smad homo-trimers (hSmad3; Chacko et al., 2001) and in Co-Smad homo-trimers (hSmad4; Shi et al., 1997), Loop1 of the loop-helix region of one monomer packs with Helix5 of the helix-bundle region of the adjacent monomer. However, residues identified as essential for homo-trimer formation (*e.g.* Arg46) are not conserved in I-Smads suggesting that heteromeric interactions may involve other features.

Studies of phosphorylated hSmad2/hSmad4 hetero-trimers (Wu et al., 2001) and phosphorylated hSmad1/hSmad4 hetero-trimers (Qin et al., 2001) identified four amino acids as essential to complex formation based on their role in positioning the phosphorylated C-terminal serine residues within the trimer. These are either conserved (Lys114 in β8 of the β-sandwich region) or identical in all species (Lys172, Tyr178 and Arg181 in the L3 loop of the helix-bundle region). The extraordinary conservation suggests that complexes containing a phosphorylated R-Smad and any other Smad (Co-Smad or I-Smad) are assembled via the same mechanism. The absolute conservation of these residues in all Smads fits with the hypothesis that competition between Co-Smads and I-Smads to form complexes with R-Smads (functional and non-functional respectively) is an essential aspect of I-Smad inhibition (Hayashi et al., 1997).

The SARA adapter protein facilitates physical interactions between TGF-β/Activin subfamily signaling R-Smads and their type I receptors (Tsukazaki et al., 1998). Residues in hSmad2 and hSmad3 essential for interactions with SARA are located in the central β-sandwich region and flank Lys114 suggesting that phosphorylation by the receptor disrupts the Smad/SARA complex (Wu et al., 2000). The residues of hSmad2/3 that interact with SARA are Ile77, Phe84, Tyr104, Trp107 and Asn121. These amino acids are also present in DmSmad2 but not in any other sequence. Alternatively, the residues in these positions are identical in all DPP/BMP signaling R-Smads. This dichotomy suggests that an adaptor molecule specific to DPP/BMP signaling will be identified.

Two interactions between the MH2 domain of hSmad3 and the TGF-β/Activin type I receptor have been identified. Two amino acids in β2 near the amino-terminus of the MH2 domain (Asn12 and Gln13) and two just downstream in Helix1 of the Loop-Helix region (Arg66 and His67) together form a basic surface. This surface is attracted to an acidic loop created by phosphorylation of the type I receptor GS domain by the ligand-binding type II receptor (Qin et al., 2002). Of these basic amino acids the pair in Helix1 is better conserved. Arg66 and His67 are

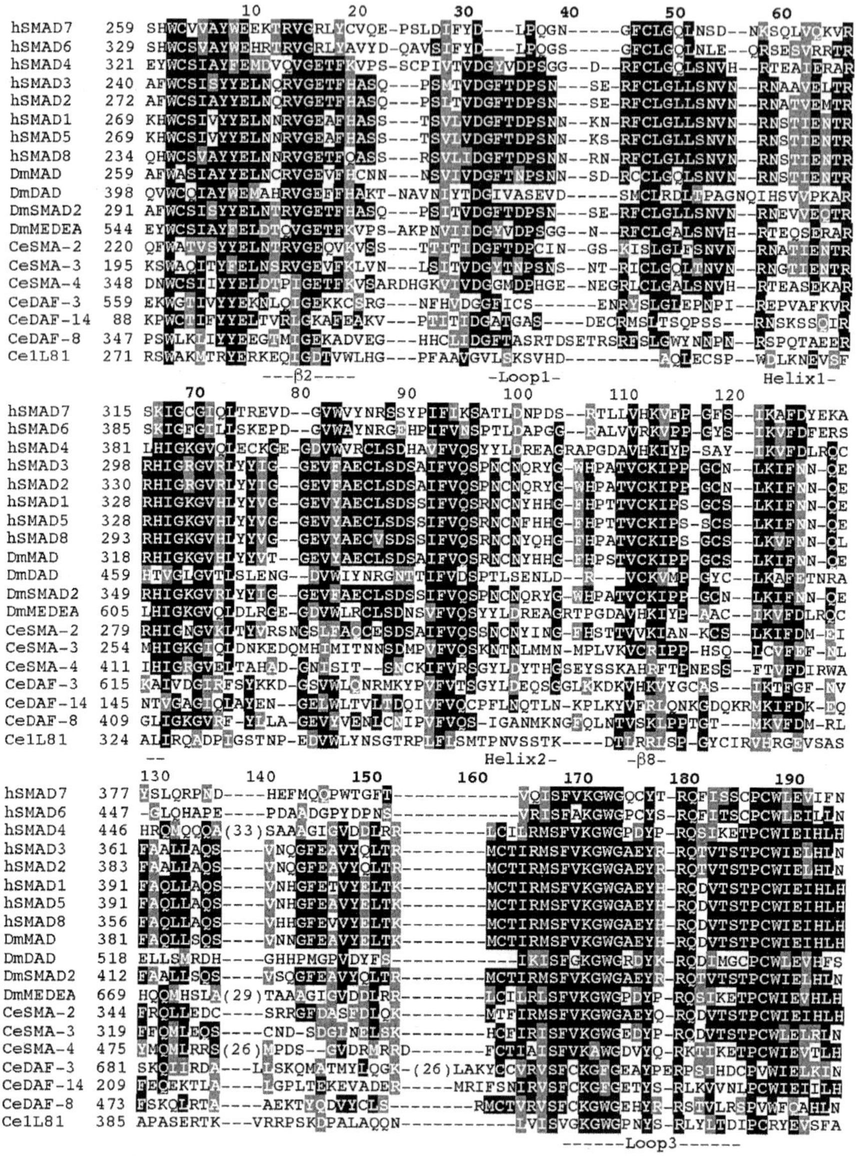

Figure 3. Smad family MH2 domain. This domain is located near the C-terminus of Smad proteins and functions in protein-protein interactions. This domain is highly conserved in all family members with DAD and Ce1L81 the most divergent. Domain extent, the representation of insertions, alignment numbering and shading are as described in Fig. 2. The evolutionally conserved portion of the MH2 domain begins at the invariant Trp261 in DmMad and ends at His431 in DmMad. The length of the alignment is 222 residues. Bold numbers 30-60 indicate the loop-helix region and 150-190 indicate the helix-bundle region. Several structural features are indicated (note that Helix1 extends two amino acids beyond the alignment break)

present in all R-Smads, except that SMA-3 has instead Met66 and His67. On the other hand, Asn12 and Gln13 are only present in hSmad2 and hSmad3 while two asparagine residues are present in hSmad1, hSmad5 and hSmad8. The distinction suggests that the basic residues in Helix1 are important for the interaction of all R-Smads with their receptors and the residues upstream mediate pathway specific interactions. There is at least one basic residue in both the upstream and Helix1 locations in all I-Smads, except Ce1L81 has two basic residues in Helix1. The presence of basic residues at these locations in I-Smads fits with the hypothesis that competition between R-Smads and I-Smads for type I receptor binding (to form functional and non-functional complexes, respectively) is a second aspect of I-Smad inhibition (Nakao et al., 1997; Hayashi et al., 1997).

A pathway-specific interaction between the MH2 domain of R-Smads and their cognate type I receptors has been identified in a study of hSmad1 and hSmad2 (Chen et al., 1998). Two residues in the L3 loop (Arg179 and Thr183) of hSmad2 interact with the L45 loop of TGF-β/Activin type I receptors but not DPP/BMP receptors. Alternatively, His179 and Asp183 in this region of hSmad1 interact only with DPP/BMP receptors. Conservation of the hSmad1 configuration in all BMP signaling R-Smads and the hSmad2 configuration in all TGF-β/Activin signaling R-Smads (plus DAF-8) supports these results. This pair of pathway-specific residues is sandwiched between the invariant Tyr178 and Arg181 involved in positioning the R-Smad phospho-serine residues in the R-Smad/Co-Smad hetero-trimer. Perhaps in addition to their role in hetero-trimer formation the tyrosine and arginine residues also act as signposts for type I receptors in their quest to identify the correct R-Smad to phosphorylate.

To date only a few of the many interactions between Smads and their transcriptional partners (activators or repressors) have been mapped. Pathway specific interactions between hSmad2/hSmad4 complexes and the transcriptional activator FAST-1 were localized to the hSmad2 MH2 domain (Chen et al., 1998). Specifically, six residues in Helix2 of the β-sandwich region that are not shared with hSmad1 are responsible for insuring that FAST-1 only interacts with hSmad2. At these positions, five are distinct between DPP/BMP signaling R-Smads and TGF-β/Activin signaling R-Smads supporting their results (for hSmad2 the residues are Pro98, Gln102, Arg103, Tyr104 and Trp107). Tyr104 and Trp107 are also essential for pathway specific interactions between TGF-β/Activin signaling R-Smads and SARA.

Pathway specific interactions between hSmad3/hSmad4 complexes and the transcriptional repressor Ski were localized to the hSmad3 MH2 domain (Qin et al., 2002). Specifically, several of the residues involved in pathway specific interactions between TGF-β/Activin signaling R-Smads and SARA and pathway specific interactions with FAST-1 also bind Ski. These include (for hSmad2) Phe84, Tyr104, Trp107. These residues are not present in any BMP signaling R-Smad but at these positions all DPP/BMP signaling R-Smads have the same amino acid. This dichotomy suggests that these residues in DPP/BMP signaling R-Smads mediate interactions with their transcriptional partners.

Near the C-terminal end of the MH2 domain, two insertions are present in the alignment. One insert is unique to DAF-3, but the other is found in all Co-Smads (hSmad4, MED and SMA-4). The insertion in the Co-Smads contains a run of alanine and glutamine residues of differing length encoded by CAG tri-nucleotide repeats (data not shown). CAG repeats are frequently found in transcription factors and are thought to encode evolutionarily variable, unstructured spacer regions (Newfeld et al., 1993, 1994). The role of this region in Co-Smads is unknown.

To explicitly examine the relationship between Smad MH2 domains we generated a phylogenetic tree (not shown) from the MH2 domain alignment in Fig. 3. The clusters of R-Smads and Co-Smads are identical in the MH2 tree and the full-length tree shown in Fig. 1. This further supports the hypothesis that the MH2 domain is the fundamental feature of Smad family proteins. We noted two differences between the trees. First, the subfamily containing only DAF-7 signaling pathway components breaks up with DAF-14 becoming a unique subfamily located between the Co-Smads and the clustered DAF-3 and DAF-8. Second, the I-Smad subfamily also breaks up with DAD and Ce1L81 forming highly divergent unique subfamilies.

Overall, the MH2 domain analysis provides further evidence against the existence of an Ecdysozoan phylum. Residues underlying pathway-specific interactions (with SARA, with receptors and with transcription factors) are always identical for human and fly members within each R-Smad subfamily but are rarely conserved in nematode Smads.

One issue concerning the similarity of the MH2 domain to domains in other proteins should be addressed here. Structural similarities between the MH2 domain and a forkhead-associated domain have been reported (Durocher et al., 2000; Lee et al., 2003). In addition, the structure of an autoinhibitory domain in interferon regulatory factors (IRFs) has similarities to the MH2 domain (Qin et al., 2003; Takahashi et al., 2003). It should be noted that the amino acid residues in these three domains are completely dissimilar but they are all capable of binding phosphoserine or phospho-threonine (two structurally very similar amino acids). Some, but not all, of these investigators clearly point out that structural similarities between the domains derive not from evolutionary conservation of amino acid sequences present in a common ancestor (homology) but from the fact that they perform the same function (convergence). In other words, the relationship between these domains is the same as the relationship between the fin of a fish and the fin of a dolphin – two structures with completely different origins that have evolved for highly efficient swimming. To date, unlike the case for the MH1 domain, there are no domains in other proteins that are considered homologous to the Smad MH2 domain.

4.3 Linker Region

Figure 4 shows alignments of Smad subfamily linker regions. As described above, this domain has received scant attention. It is not possible to generate a meaningful alignment of the entire Smad family and there are just a few abbreviated alignments of this domain in mammalian R-Smads in the literature. Here we discuss alignments containing only Smad subfamily members that identify considerable amino acid conservation.

For the R-Smad subfamily that signals for DPP/BMP proteins (Smad1/Mad subfamily – Fig. 4A) this region is well conserved in humans and flies along its entire length. For the R-Smad subfamily that signals for TGF-β/Activin proteins (Smad2/3 subfamily – Fig. 4B) this domain shows the same level of conservation. However, this domain is not alignable between these two subfamilies (nor can SMA-2 or SMA-3 be aligned with either subfamily). For the Co-Smad subfamily (Fig. 4C) two small regions are well conserved in humans and flies. One region adjoins the MH1 domain and the other adjoins the MH2 domain. Other regions (data not shown) in the Co-Smad alignment are well conserved either within vertebrates or within insects. No regions in this subfamily are conserved with SMA-4. For I-Smads one conserved region was identified that adjoins the MH1 domain. This domain is well conserved in vertebrates, moderately conserved between vertebrates and insects and only very weakly conserved in Ce1L81.

There is a biochemical interaction associated with developmental functions of vertebrate Smad family members that has been mapped to this domain. Erk kinases belong to the Mitogen-Activated Kinase (MAP) kinase family of Ser/Thr kinases. Four consensus Erk phosphorylation sites PX(S/T)P were identified in the linker region of hSmad1. Subsequently, two Erk sites were identified in hSmad3 and one of these sites is present in hSmad2. All Erk sites are phosphorylated in mammalian cells (Kretzschmar et al., 1997, 1999). Recent studies of Smad1 proteins with mutations in these sites revealed a developmental function for Erk phosphorylation in neural induction in *Xenopus* (Pera et al., 2003) and germ cell development in mice (Aubin et al., 2004).

Examination of our alignments shows that conservation of the four Erk sites in hSmad1 is highly variable. The first (in hSmad1 beginning with Pro54) is present in vertebrates and bees, the second (Pro70) is present only in vertebrates, the third (Pro84) is present in all sequences and the fourth (Pro94) is present in all except flies. Given this pattern, the most parsimonious explanation is that all four Erk sites were present in the common ancestor of human and insect Smad1/Mad and that individual sites were lost at various times in insect lineages after the divergence from vertebrates. There is even less conservation of the Erk sites in hSmad2 and hSmad3. The common site (in hSmad2 beginning at Pro57) is not present in insects and the unique site in hSmad3 (Pro96) is only present in zebrafish (DrSmad3). No Erk sites are present in the linker region of any other Smad family member. Overall, Erk phosphorylation of R-Smad function may be relevant outside vertebrates but this cannot be assumed based on the pattern of conservation.

The presence of one fully conserved Erk site in the Smad1/Mad subfamily led us to examine the linker region for other conserved phosphorylation sites. We discovered numerous consensus sites (S/T)XXX(S/T) for the glycogen synthase kinase3β (GSK3β) Ser/Thr kinase (Fiol et al., 1987) in this subfamily and in SMA-3. Further, two of these sites are conserved between vertebrates and insects. The first

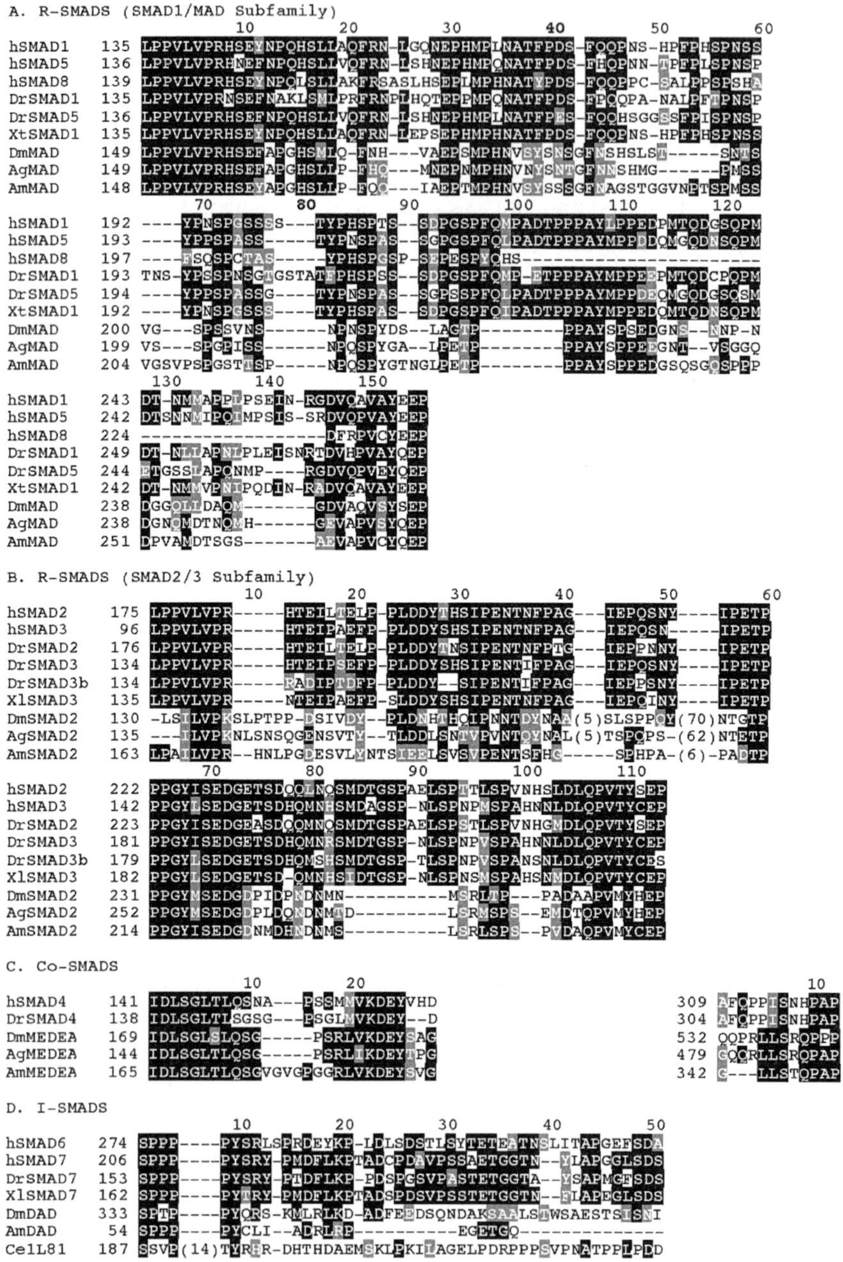

Figure 4. Smad subfamily linker region. This region encompasses all residues between the MH1 and MH2 domains. Conserved stretches were identified in alignments for each of the four major subfamilies. A) R-Smads involved in signaling by DPP/BMP proteins – the Smad1/Mad subfamily. Residue number 40 in bold indicates the location of the fully conserved GSK3β site (TFPDS in hSmad1) and number 80 the fully conserved Erk site (PHSP in hSmad1). B) R-Smads involved in signaling by

(beginning at hSmad1 Thr38) is present in all sequences and the second (Ser72) is present in all vertebrate Smad1/Smad5 sequences and in Mad. There are two sites in SMA-3 separated by the same number of amino acid residues but the region surrounding these sites is to degenerate to say with confidence that the Thr38 and Ser72 sites are conserved. No GSK3β sites are present in the linker region of any other Smad family member.

The presence of a fully conserved GSK3β site in the Smad1/Mad subfamily is intriguing because GSK3β (and its *Drosophila* homolog Zeste White3) are antagonists of Wnt family growth factor signaling. In vertebrate and insect systems TGF-β and Wnt pathways interact frequently to influence developmental processes (*e.g.* Takaesu et al., 2005) but the mechanism underlying many of these interactions is unknown. The conservation of a GSK3β site suggests that phosphorylation of Smad1/Mad subfamily members may be a mechanism utilized for growth factor "crosstalk". This hypothesis awaits experimental verification and in such experiments it is important to remember that GSK3β phosphorylation is typically a secondary event; it occurs when different serine residues in the target protein have been phosphorylated by another Ser/Thr kinase. For example, CREB is phosphorylated first at Ser133 by cAMP-dependent kinase and then at Ser129 by GSK3β (Fiol et al., 1994).

Overall, conservation of the Linker domain within but not between R-Smad subfamilies suggests that pathway-specific functions likely involve amino acids in this region. In addition, sequence conservation in this domain follows the pattern noted previously: humans and flies are similar or identical with nematodes highly divergent.

4.4 Receptor Phosphorylation Region

Figure 5 shows an alignment of the Smad family receptor phosphorylation region. At the N-terminus of the region is a stretch of amino acid residues that is well conserved in human and fly R-Smads that may function with the MH2 domain in protein-protein interactions. At the C-terminus of the region is the SSXS motif in human and fly R-Smads, the most C-terminal two serine residues of which are phosphorylated by the type I receptor to stimulate signal transduction. An examination of the alignment reveals that the second amino acid from the C-terminus in this motif is either a valine or a methionine in all human and fly R-Smads, or a conserved isoleucine in SMA-3. In addition, DAF-8 has the sequence SSRT at its terminus

Figure 4. (Continued) TGF-β/Activin proteins – the Smad2/3 subfamily. C) Co-Smads involved in signaling by DPP/BMP and TGF-β/Activin proteins. Left side – adjoins the MH1 domain. Right side – adjacent to the MH2 domain. The DrSmad4 sequence was assembled from two partial sequences to generate a contiguous sequence with the greatest agreement to the linker region of hSmad4. D) Inhibitory Smads. One conserved region that begins three amino acid residues downstream of the MH1 domain was identified. Insertions, alignment numbering and amino acid shading are as described in Fig. 2. Additional species are included in each alignment to document the extent of conservation

```
                       10        20        30
hSMAD7     425  SR---------------------------------
hSMAD6     494  NPR--------------------------------
hSMAD4     531  RALQLLDEVLHTMPIADPQPLD-------------
hSMAD3     402  GPLQWLDKVLTQMGSP-SIRCSSVS----------
hSMAD2     444  GPLQWLDKVLTQMGSP-SVRCSSMS----------
hSMAD1     442  GPLQWLDKVLTQMGSP-HNPISSVS----------
hSMAD5     442  GPLQWLDKVLTQMGSP-LNPISSVS----------
hSMAD8     407  GPLQWLDKVLTQMGSP-HNPISSVS----------
DmMAD      432  GPLQWLDKVLTQMGSP-HNAISSVS----------
DmDAD      566  HLR--------------------------------
DmSMAD2    463  GPLQWLDRVLTQMGSP-RLPCSSMS----------
DmMEDEA    750  RALQLLDEVLHAMPIDGPRAAA-------------
CeSMA-2    395  APLAWLDRVLSTMGPT-PRPISSIS----------
CeSMA-3    369  VPLAYIDQKMKQTPRTNLMEPNSMT----------
CeSMA-4    547  RPLQILDQLLKNSSQFGSS----------------
CeDAF-3    858  IAYDFMDSICQYITNCFEPLGMEDFAKLGINVSDD
CeDAF-14   265  EPADEYDTVFRINNERPEIGSRS------------
CeDAF-8    523  NPMHWVDSVLTCMGAPPRI-CSSRT----------
Ce1L81       0  -----------------------------------
```

Figure 5. Smad family receptor phosphorylation region. The region begins immediately after the end of the MH2 domain (Gly432 in DmMad) and ends at the C-terminal amino acid residue (Ser455 in DmMad). This domain is highly conserved in R-Smads and weakly conserved in Co-Smads. Alignment numbering and amino acid shading are as described in Fig. 2

indicating it could possibly be recognized by a type I receptor and phosphorylated like an R-Smad. In I-Smads, this motif is essentially absent, whereas Co-Smads show some conservation of this region. While hSmad4 and Med have no C-terminal serine residues, SMA-4 has two. Interestingly, DAF-14 has two and SMA-3 has a serine and a threonine.

As mentioned above, the presence or absence of an MH1 domain is not an accurate predictor of Smad function for nematode sequences (the DNA-binding antagonist DAF-3 has an MH1 domain and no C-terminal serine while the positively signaling DAF-8 has a nearly unrecognizable MH1 domain and three C-terminal serines). Given the conservation pattern, perhaps the number of C-terminal serine residues is a better predictor of function. From this perspective, the nematode specific subfamily composed of DAF-7 signaling pathway components (DAF-3, DAF-8 and DAF-14) can be assigned the following roles: DAF-3 is an I-Smad (no serine), DAF-14 is a Co-Smad (2 serines, like SMA-4) and DAF-8 is an R-Smad (SSRT). Given these roles, it appears that the members of this unique subfamily, a pathway unto themselves, are co-evolving to maintain their ability to interact. If this prediction is validated by experiments, this nematode subfamily is a truly unique example of developmental pathway evolution.

5. FUTURE PERSPECTIVES

One important area for future research is to investigate the diversification of the Smad family into its four major subfamilies. At present, we know that the ancestral R-Smad split into two R-Smad subfamilies after the divergence of diploblasts (cnidarians) and acoelomates (*Schistosoma*). However, the origin of Co-Smads and I-Smads is still unknown as both subfamilies are already present in nematodes (a pseudocoelomate). Additional acoelomate and pseudocoelomate species need to be surveyed to fill this gap.

A second area for future research is to test our hypotheses about the subfamily specific to the DAF-7 signaling pathway of *C. elegans*. For example, what are the biochemical interactions that underlie the inhibition of DAF-3 activity by DAF-8 and DAF-14? How do these unusual Smads interact with the equally unusual TGF-β receptor DAF-1?

In summary, our phylogenetic analysis of Smad family proteins has provided hypotheses for experimental testing and also provided explanations for experimental results that were previously difficult to interpret. In our view there is no impediment to extending positive feedback between experimental and phylogenetic studies to other signaling pathways. In fact, in addition to continuing our studies of the Smad family we have begun a phylogenetic analysis of families that participate in the Wnt signaling pathway.

ACKNOWLEDGEMENTS

We thank Peter ten Dijke and Sudhir Kumar for valuable discussions. Research in the Newfeld lab is supported by the U.S. National Institutes of Health (NCI and NHGRI).

REFERENCES

Aguinaldo, A., Turbeville, J., Linford, L., Rivera, M., Garey, J., Raff, R., and Lake, J., 1997, Evidence for a clade of nematodes, arthropods and other moulting animals. *Nature* 387: 489-493.

Aubin, J., Davy, A., and Soriano, P., 2004, In vivo convergence of BMP and MAPK signaling pathways: impact of differential Smad1 phosphorylation on development and homeostasis. *Genes Dev* 18: 1482-1494.

Beall, M., McGonigle, S., and Pearce, E., 2000, Functional conservation of *Schistosoma mansoni* Smads in TGF-β signaling. *Mol Biochem Parasitol* 111: 131-142.

Cassada, R., and Russell, R., 1975, The dauer larva, a post-embryonic developmental variant of the nematode *C. elegans*. *Dev Biol* 46: 326-342.

Chacko, B., Qin, B., Correia, J., Lam, S., de Caestecker, M., and Lin, K. 2001, The L3 loop and C-terminal phosphorylation define Smad protein trimerization. *Nat Struct Biol* 8: 248-253.

Chai, J., Wu, J., Yan, N., Massagué, J., Pavletich, N., and Shi, Y., 2003, Features of Smad3 MH1-DNA complex: roles of water and zinc in DNA binding. *J Biol Chem* 278: 20327-20331.

Chen, Y., Hata, A., Lo, R., Wotton, D., Shi, Y., Pavletich, N., and Massagué, J., 1998, Determinants of specificity in TGF-β signal transduction. *Genes Dev* 12: 2144-2152.

Dunn, N., Koonce, C., Anderson, D., Islam, A., Bikoff, E., and Robertson, E., 2005, Mice exclusively expressing the short isoform of Smad2 develop normally and are viable and fertile. *Genes Dev* 19: 152-163.

Durocher, D., Taylor, I., Sarbassova, D., Haire, L., Westcott, S., Jackson, S., Smerdon, S., and Yaffe, M., 2000, The molecular basis of FHA domain: phosphopeptide binding specificity and implications for phospho-dependent signaling mechanisms. *Mol Cell* 5: 1169-1182.

Felsenstein, J., 1985, Confidence limits on phylogenies: an approach using the bootstrap. *Evolution* 39: 783-791.

Fiol, C., Mahrenholz, A., Wang, Y., Roeske, R., and Roach, P., 1987, Formation of protein kinase recognition sites by covalent modification of the substrate: molecular mechanism for the synergistic action of casein kinase II and glycogen synthase kinase 3. *J Biol Chem* 262: 14042-14048.

Fiol, C., Williams, J., Chou, C., Wang, Q., Roach, P., and Andrisani, O., 1994, A secondary phosphorylation of CREB at Ser129 is required for the cAMP-mediated control of gene expression: a role for glycogen synthase kinase-3 in the control of gene expression. *J Biol. Chem.* 269: 32187-32193.

Hayashi, H., Abdollah, S., Qiu, Y., Cai, J., Xu, Y., Grinnell, B., Richardson, M., Topper, J., Gimbrone, M., Wrana, J., and Falb, D., 1997, The MAD-related protein Smad7 associates with the TGF-β receptor and functions as an antagonist of TGF-β signaling. *Cell* 89: 1165-1173.

Hedges, S., and Kumar, S., 2003, Genomic clocks and evolutionary timescales. *Trends Genetics* 19: 200-206.

Herpin, A., Lelong, C., and Favrel, P., 2004, TGF-β-related proteins: an ancestral and widespread superfamily of cytokines in metazoans. *Dev Comp Immunol* 28: 461-485.

Hillis, D., and Bull, J., 1993, An empirical test of bootstrapping as a method for assessing confidence in phylogenetic analysis. *Syst Biol* 42: 182-192.

Inoue, T., and Thomas, J., 2000, Targets of TGF-β signaling in *C. elegans* dauer formation. *Dev Biol* 217: 192-204.

Kretzschmar, M., Doody, J., and Massagué, J., 1997, Opposing BMP and EGF signaling pathways converge on the TGF-β family mediator Smad1. *Nature* 389: 618-622.

Kretzschmar, M., Doody, J., Timokhina, I., and Massagué, J., 1999, A mechanism of repression of TGF-β/ Smad signaling by oncogenic Ras. *Genes Dev* 13: 804-816.

Kumar, S., Tamura, K., Jakobsen, I., and Nei, M., 2001, MEGA2: molecular evolutionary genetics analysis software. *Bioinformatics* 17: 1244-1245.

Lagna, G., Hata, A., Hemmati-Brivanlou, A., and Massagué, J., 1996, Partnership between DPC4 and Smad proteins in TGF-β signaling pathways. *Nature* 383: 832-836.

Lee, G., Ding, Z., Walker, J., and Van Doren, S., 2003, NMR structure of the forkhead-associated domain from the Arabidopsis receptor kinase-associated protein phosphatase. *Proc Natl Acad Sci U S A* 100: 11261-11266.

Liu, F., Hata, A., Baker, J., Doody, J., Carcamo, J., Harland, R., and Massagué, J., 1996, A human MAD protein acting as a BMP-regulated transcriptional activator. *Nature* 381: 620-623.

Liu, F., Pouponnot, C., and Massagué, J., 1997, Dual role of the Smad4/DPC4 tumor suppressor in TGF-β-inducible transcriptional complexes. *Genes Dev* 11: 3157-3167.

Marquez, R., Singer, M., Takaesu, N., Waldrip, W., Kraytsberg, Y., and Newfeld, S., 2001, Transgenic analysis of the Smad family of TGF-β signal transducers in *Drosophila* suggests new roles and interactions between family members. *Genetics* 157: 1639-1648.

Massagué, J., Blain, S., and Lo, R., 2000, TGF-β signaling in growth control, cancer and heritable disorders. *Cell* 103: 295-309.

Moustakas, A., and Heldin, C.-H., 2002, From mono- to oligo-Smads: the heart of the matter in TGF-β signal transduction. *Genes Dev* 16: 1867-1871.

Nakao, A., Afrakhte, M., Morén, A., Nakayama, T., Christian, J., Heuchel, R., Itoh, S., Kawabata, M., Heldin, N.-E., Heldin, C.-H., and ten Dijke, P., 1997, Identification of Smad7, a TGF-β-inducible antagonist of TGF-β signaling. *Nature* 389: 631-635.

Newfeld, S., Schmid, A., and Yedvobnick, B., 1993, Homopolymer length variation in the *Drosophila* gene *mastermind. J Mol Evol* 37: 483-495.

Newfeld, S., Tachida, H., and Yedvobnick, B., 1994, Drive-selection equilibrium: homopolymer evolution in the *Drosophila* gene *mastermind. J Mol Evol* 38: 637-641.

Newfeld, S., Wisotzkey, R., and Kumar, S., 1999, Molecular evolution of a developmental pathway: phylogenetic analyses of TGF-β family ligands, receptors and Smad signal transducers. *Genetics* 152: 783-795.

Padgett, R., Wozney, J., and Gelbart, W., 1993, Human BMP sequences confer normal dorsal ventral patterning in the *Drosophila* embryo. *Proc Natl Acad Sci U S A* 90: 2905-2909.

Pera, E., Ikeda, A., Eivers, E., and De Robertis, E., 2003, Integration of IGF, FGF, and anti-BMP signals via Smad1 phosphorylation in neural induction. *Genes Dev* 17: 3023-3028.

Qin, B., Chacko, B., Lam, S., de Caestecker, M., Correia, J., and Lin, K., 2001, Structural basis of Smad1 activation by receptor kinase phosphorylation. *Mol Cell* 8: 1303-1312.

Qin, B., Lam, S., Correia, J., and Lin, K., 2002, Smad3 allostery links TGF-β receptor kinase activation to transcriptional control. *Genes Dev* 16: 1950-1963.

Qin, B., Liu, C., Lam, S., Srinath, H., Delston, R., Correia, J., Derynck, R., and Lin, K., 2003, Crystal structure of IRF-3 reveals mechanism of autoinhibition and virus-induced phosphoactivation. *Nat Struct Biol* 10: 913-921.

Raff, R., 1996, The Shape of Life: Genes, Development, and the Evolution of Animal Form. Univ. Chicago Press, Chicago, IL, USA.

Raftery, L., and Sutherland, D., 1999, TGF-β family signal transduction in *Drosophila* development: from MAD to Smads. *Dev Biol* 210: 251-68.

Sadreyev, R., and Grishin, N., 2003, COMPASS: a tool for comparison of multiple protein alignments with assessment of statistical significance. *J Mol Biol* 326: 317-336.

Saitou, N., and Nei, M., 1987, The neighbor-joining method: a new method for reconstructing phylogenetic trees. *Mol Biol Evol* 4: 406-425.

Samuel, G., Miller, D., and Saint, R., 2001, Conservation of a DPP/BMP signaling pathway in the nonbilateral cnidarian. *Acropora millepora. Evol Dev* 3: 241-250.

Sekelsky, J., Newfeld, S., Raftery, L., Chartoff, E., and Gelbart, W., 1995, Genetic characterization and cloning of *Mothers against dpp*: a gene required for *decapentaplegic* function in *Drosophila. Genetics* 139: 1347-1358.

Shi, Y., Hata, A., Lo, R., Massagué, J., and Pavletich, N., 1997, A structural basis for mutational inactivation of the tumor suppressor Smad4. *Nature* 388: 87-93.

Shi, Y., Wang, Y., Jayaraman, L., Yang, H., Massagué, J., and Pavletich, N., 1998, Crystal structure of a Smad MH1 domain bound to DNA: insights on DNA binding in TGF-β signaling. *Cell* 94: 585-594.

Stefancsik, R., and Sarkar, S., 2003, Relationship between the DNA binding domains of Smad and NFI/CTF transcription factors defines a new superfamily. *DNA Seq* 14: 233-239.

Suga, H., Ono, K., and Miyata, T., 1999, Multiple TGF-β receptor related genes in sponge and ancient gene duplications before the parazoan-eumetazoan split. *FEBS Lett* 453: 346-350.

Takaesu, N., Herbig, E., Zhitomersky, D., O'Connor, M., and Newfeld, S., 2005, DNA-binding domain mutations in Smad genes yield dominant negative proteins or a neomorphic protein that can activate Wg target genes in *Drosophila. Development* 132: 4883-4894.

Takahasi, K., Suzuki, N., Horiuchi, M., Mori, M., Suhara, W., Okabe, Y., Fukuhara, Y., Terasawa, H., Akira, S., Fujita, T., and Inagaki, F., 2003, X-ray crystal structure of IRF-3 and its functional implications. *Nat Struct Biol* 10: 922-927.

Thatcher, J., Haun, C., and Okkema, P., 1999, The DAF-3 Smad binds DNA and represses gene expression in the *C. elegans* pharynx. *Development* 126: 97-107.

Thompson, J., Gibson, T., Plewniak, F., Jeanmougin, F., and Higgins, D., 1997, The CLUSTAL-X windows interface: flexible strategies for multiple sequence alignment aided by quality analysis tools. *Nucleic Acids Res* 25: 4876-4882.

Tsukazaki, T., Chiang, T., Davison, A., Attisano, L., and Wrana, J., 1998, SARA, a FYVE domain protein that recruits Smad2 to the TGF-β receptor. *Cell* 95: 779-791.

Wu, G., Chen, Y., Ozdamar, B., Gyuricza, C., Chong, P., Wrana, J., Massagué, J., and Shi, Y., 2000, Structural basis of Smad2 recognition by SARA. *Science* 287: 92-97.

Wu, J., Hu, M., Chai, J., Seoane, J., Huse, M., Li, C., Rigotti, D., Kyin, S., Muir, T., Fairman, R., Massagué, J., and Shi, Y., 2001, Crystal structure of a phosphorylated Smad2: recognition of phosphoSerine by the MH2 domain and insights on Smad function in TGF-β signaling. *Mol Cell* 8: 1277-1289.

CHAPTER 2

C. ELEGANS TGF-β SIGNALING PATHWAYS

RICHARD W. PADGETT[1,2,3] AND GARTH I. PATTERSON[2,3]

[1] *Waksman Institute*

[2] *Department of Molecular Biology and Biochemistry, Rutgers University, Piscataway, NJ 08854-8020*

[3] *Cancer Institute of New Jersey, UMDNJ-Robert Wood Johnson Medical School, New Brunswick, NJ 08903*

Abstract: The basic building blocks of the TGF-β pathway are ubiquitous in animals, but given that it is a regulatory pathway, the functions of the pathways vary depending on the organism and cell type. *C. elegans* has two TGF-β superfamily signaling pathways that incorporate ligands, receptors, and Smads. There is an interesting ligand, *unc-129*, that does not appear to require conventional TGF-β superfamily receptors and Smads. In addition, two other TGF-β superfamily ligands have been identified. In some cases, aspects of TGF-β signaling have been modified in *C. elegans*

Keywords: *C. elegans*; Smads; TGF-β; BMP; dauer; Sma/Mab

1. INTRODUCTION

Although there are five TGF-β superfamily ligands in *C. elegans*, there are two defined signaling pathways that have the canonical components – ligands, type I and type II serine/threonine kinase receptors, and Smads. One pathway (Sma/Mab) is a conventional TGF-β pathway and controls cell size, male tail development, and immunity (Kurz and Tan, 2004; Mallo et al., 2002; Nicholas and Hodgkin, 2004; Patterson and Padgett, 2000). A second, unconventional pathway (dauer pathway) regulates entry into dauer diapause in response to harsh environmental conditions (Riddle and Albert, 1997) as well as feeding behavior, fat metabolism, egg laying, and thermotolerance. The UNC-129 ligand controls axonal migration, but apparently does not utilize canonical TGF-β superfamily components.

P. ten Dijke and C.-H. Heldin (eds.), Smad Signal Transduction, 37–53.

1.1 Identification of *C. elegans* TGF-β Pathways

The dauer pathway was one of the first pathways in *C. elegans* that was described in genetic detail. In the 1990s, molecular studies of some of the genes began to yield insights into how dauer signals are transmitted, and eventually it became clear that TGF-β governs the dauer decision. The first cloned gene in the dauer pathway was *daf-1*, which encodes a TGF-β superfamily receptor (Georgi et al., 1990). At the time, no TGF-β superfamily receptors had been cloned in any organism, so it appeared to be an orphan receptor. Soon after the cloning of *daf-1*, the Activin receptor was cloned (Mathews and Vale, 1991) and DAF-1 was recognized as a member of this unique family of serine/threonine kinase receptors. Subsequent to this, *daf-4* was cloned and shown to encode a type II receptor. The molecular cloning and the extensive genetic analysis of the dauer pathway led to the current model in which DAF-1 is the type I partner for DAF-4.

The cloning of *daf-4* allowed the connection of several important pieces of information. In addition to causing defects in dauer formation, *daf-4* mutants also produce a small body size (Sma) and abnormalities in the male tail (Mab, male abnormal). None of the other dauer mutations are Sma or Mab. This led to the hypothesis that the DAF-4 type II receptor might send two independent signals. Further investigation showed that the genes *sma-2, sma-3*, and *sma-4* encode Smads (Savage et al., 1996), and these Smads and *Drosophila* MAD (Sekelsky et al., 1995), are the eponymous founding members of the Smad family (Derynck et al., 1996). DAF-4 functions as the type II receptor in two pathways, one of which regulates dauer, and the other body size. These two pathways intersect only at DAF-4; the type I receptors and Smads are specific to each individual pathway.

1.2 Summary of the *C. elegans* Smad Genes

Alignments of all the *C. elegans* and *C. briggsae* Smads were done to assess their relationships with known classes of Smads. Using the MH2 domains, dendrograms were constructed (Fig. 1A).

DAF-4 is a type II receptor that is shared between the Sma / Mab and the dauer pathways. Specificity is imparted by using distinct type I receptors for each pathway that presumably form heterodimers with DAF-4. This formation is triggered by unique ligands, which only bind to specific sets of receptors. Once a unique set of heterodimers form, they interact with a specific set of Smads. In the case of the Sma /Mab pathway, there are two R-Smads, encoded by *sma-2* and *sma-3*. Various models have been entertained to account for two R-Smads in a single pathway, such as redundancy of the two Smads or a model in which each Smad acts on a subset of downstream targets. However, mutations in either of these Smad genes totally blocks the pathway, which indicates that these two Smads are not redundant with each other and that they do not function with a unique set of target genes – they are both required for the same output.

Both SMA-2 and SMA-3 are closely related to their *C. briggae* counterparts and fall within the cluster of R-Smads (Fig. 1A). Further, they contain conserved

C-terminal residues found in other R-Smads as well as the terminal serine residues that undergo phosophorylation. The rate of divergence between these genes in these two species is slower for the Sma/Mab Smads than for the dauer Smads, which suggests that the Sma/Mab pathway is under more stringent selection. Interestingly, changes accrue more rapidly in SMA-3 than SMA-2.

SMA-4 clearly falls in the Co-Smad cluster by sequence. It is most closely related to the *Drosophila Medea* gene and Smad4. The divergence of SMA-4 between *C. elegans* and *C. briggsae* is similar to the divergence of SMA-3. However, due to the fast molecular clock in *C. elegans*, SMA-4 is evolving about twice as fast as MEDEA and Smad4.

DAF-1 and DAF-4 are TGF-β superfamily receptors that regulate dauer formation. Subsequent molecular cloning of genes in the dauer pathway identified a ligand and three other Smads. As described above, the type II receptor, DAF-4, is shared with the *C. elegans* Sma/Mab pathway. But several other dauer components have mutant phenotypes that suggest that they have no function in the Sma/Mab pathway. All of the genes unique to the dauer pathway are highly diverged from their counterparts in *Drosophila* and vertebrates. In particular, the Smads DAF-8 and DAF-14, while clearly Smads, do not easily fall into the R-Smad, Co-Smad, or I-Smad groups based on primary sequence alone. DAF-14 is so far unique in that it has no MH1 domain, and DAF-8 has a highly diverged MH1 domain. The MH2 domains of DAF-8 and DAF-14 are highly diverged. In the comparison shown (Fig. 1A), DAF-8 is more similar to R-Smads; however, this placement is tenuous, in that changes in the algorithm used to construct the table can result in DAF-8 being placed outside of all of the known families. The MH2 domain of DAF-14 does not easily fall into the R-family based solely upon sequence. However, However, both DAF-8 and DAF-14 have a sequence at their C-termini that is similar to the SSXS phosphorylation target that is found in R-Smads, but not Co-Smads (Fig. 1B). For this reason, DAF-8 and DAF-14 have been proposed to be phosphorylated by the receptors and likely are derived from ancestral R-Smads.

The sequence of DAF-3 is more similar to Co-Smads than to other types of Smads. The DAF-3 MH2 domain aligns with other Co-Smads (Fig. 1A), and DAF-3 is missing the conserved phosphorylation site that is found in R-Smads. Unlike I-Smads, the MH1 domain of DAF-3 is a *bona fide* DNA binding domain (Thatcher et al., 1999). In addition, the MH1 domain of DAF-3 is more similar to MH1 domains of Co-Smads than to other Smads.

The Smads in the dauer pathway have evolved rapidly in *Caenorhabditis*. *C. briggsae* is closely related to *C. elegans*; morphologically, the two species are almost indistinguishable. On the DNA level, the difference between the two species is higher; the level of sequence divergence in protein coding regions is comparable to that of humans and mice. Strikingly, when the *C. elegans* Smads are compared to their *C. briggsae* orthologs, the divergence is astonishingly high. As can be seen in Fig. 1A from the length of the branches, the difference between the *C. elegans* and *C. briggsae* DAF-8 orthologs is greater than the divergence between *Drosophila* MAD and its human ortholog, Smad1. The divergence between

A

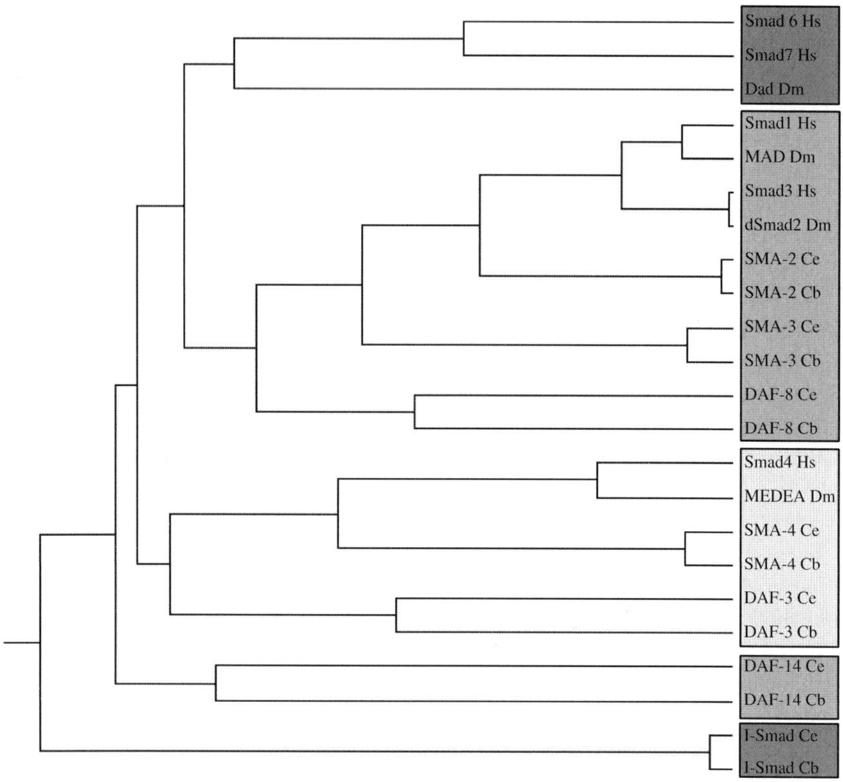
Smad 6 Hs
Smad7 Hs
Dad Dm
Smad1 Hs
MAD Dm
Smad3 Hs
dSmad2 Dm
SMA-2 Ce
SMA-2 Cb
SMA-3 Ce
SMA-3 Cb
DAF-8 Ce
DAF-8 Cb
Smad4 Hs
MEDEA Dm
SMA-4 Ce
SMA-4 Cb
DAF-3 Ce
DAF-3 Cb
DAF-14 Ce
DAF-14 Cb
I-Smad Ce
I-Smad Cb

B

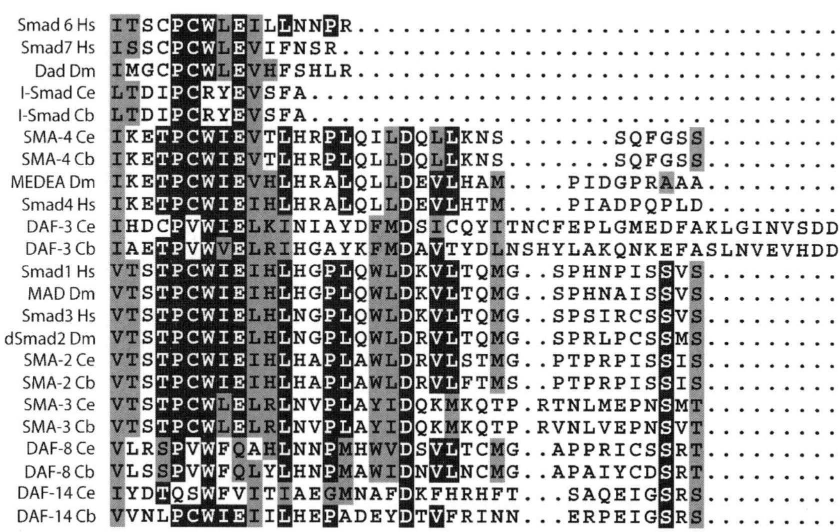
Smad 6 Hs ITSCPCWLEILLNNPR
Smad7 Hs ISSCPCWLEVIFNSR
Dad Dm IMGCPCWLEVHFSHLR
I-Smad Ce LTDIPCRYEVSFA
I-Smad Cb LTDIPCRYEVSFA
SMA-4 Ce IKETPCWIEVTLHRPLQILDQLLKNS.......SQFGSS
SMA-4 Cb IKETPCWIEVTLHRPLQLLDQLLKNS.......SQFGSS
MEDEA Dm IKETPCWIEVHLHRALQLLDEVLHAM....PIDGPRAAA
Smad4 Hs IKETPCWIEIHLHRALQLLDEVLHTM....PIADPQPLD
DAF-3 Ce IHDCPVWIELKINIAYDFMDSICQYITNCFEPLGMEDFAKLGINVSDD
DAF-3 Cb IAETPVWVELRIHGAYKFMDAVTYDLNSHYLAKQNKEFASLNVEVHDD
Smad1 Hs VTSTPCWIEIHLHGPLQWLDKVLTQMG..SPHNPISSVS
MAD Dm VTSTPCWIEIHLHGPLQWLDKVLTQMG..SPHNAISSVS
Smad3 Hs VTSTPCWIELHLNGPLQWLDKVLTQMG..SPSIRCSSVS
dSmad2 Dm VTSTPCWIELHLNGPLQWLDRVLTQMG..SPRLPCSSMS
SMA-2 Ce VTSTPCWIEIHLHAPLAWLDRVLSTMG..PTPRPISSIS
SMA-2 Cb VTSTPCWIEIHLHAPLAWLDRVLFTMS..PTPRPISSIS
SMA-3 Ce VTSTPCWLELRLNVPLAYIDQKMKQTP.RTNLMEPNSMT
SMA-3 Cb VTSTPCWLELRLNVPLAYIDQKMKQTP.RVNLVEPNSVT
DAF-8 Ce VLRSPVWFQAHLNNPMHWVDSVLTCMG..APPRICSSRT
DAF-8 Cb VLSSPVWFQLYLHNPMAWIDNVLNCMG..APAIYCDSRT
DAF-14 Ce IYDTQSWFVITIAEGMNAFDKFHRHFT...SAQEIGSRS
DAF-14 Cb VVNLPCWIEIILHEPADEYDTVFRINN...ERPEIGSRS

BMP and Activin pathway Smads is ancient, predating the divergence between arthropods and vertebrates, but the sequence differences between *C. elegans* and *C. briggsae* DAF-8 orthologs dwarfs that of Activin and BMP pathway Smads (compare the depth of the branches connecting DAF-8 orthologs to that connecting Smad1/MAD to Smad2/dSmad2). DAF-14 and DAF-3 are even more divergent. This is not a general feature of *Caenorhabditis* Smads; SMA-2, SMA-3 and SMA-4 show a level of divergence that is typical for *C. elegans* and *C. briggsae* protein comparisons.

The genome-sequencing consortium has identified one additional Smad gene in *C. elegans* with an unknown function (*tag-68*, F37D6.6). It is highly divergent from the other Smads, but has more similarity to I-Smads. It contains some conserved residues and structural features that suggest it could be an I-Smad. It contains a CCN motif in the MH1 domain, which is present in all I-Smads, and the protein terminates immediately after the MH2 domain, eliminating a conserved domain contained in Co-Smads and R-Smads (Fig. 1B). In addition, it does not have the characteristic C-terminal R-Smad serine residues, which precludes it from being an R-Smad. Taken together, this data suggests it may be a divergent I-Smad. However, biochemical studies of this gene product failed to show it functions like an I-Smad when used with mammalian receptors (K. Miyazono and R.W.P., unpublished results). Its biological function remains unknown, since a deletion of the gene fails to show any significant body size phenotypes (R.W.P. and C. Zimmerman, unpublished data). Further experiments will be needed to determine its role, but interestingly, it is conserved in *C. briggsae*, suggesting it does function.

2. THE SMA/MAB PATHWAY

Mutants of the Sma/Mab pathway genes grow more slowly than wild-type animals from hatching to adulthood. In cases where it has been examined, tissues contain the same number of cells, but the cells are smaller than wild-type. These mutants also have male tail defects (Mab), including mis-specification of sensory ray cells

Figure 1. *C. elegans* Smads. A) Smad MH2 domains. This tree shows the relationship of *C. elegans* and *C. briggsae* Smads with representative human and *Drosophila* Smads. MH2 domains and alignments were obtained from the Pfam database (Bateman et al., 2004) and refined by hand. The alignment was used to build a tree using GeneBee (Brodskii et al., 1995). B) Smad carboxy termini. The end of the MH2 domain and the remaining part of the carboxy termini were aligned. Accession numbers and abbreviations: Hs, *Homo sapiens*; Dm, *Drosophila melanogaster*; Ce, *Caenorhabditis elegans*; Cb, *Caenorhabditis briggsae*. Accession numbers, all Swiss-Prot: Smad6 Hs, O43541; Smad7 Hs, O15105; DAD Dm, O15968; Smad1 Hs, Q13485; MAD Dm, P42003; Smad3 Hs, P84022; dSmad2 Dm, Q9TZQ2; SMA-2 Ce, Q02330; SMA-2 Cb, Q61QQ6; SMA-3 Ce, P45896; SMA-3 Cb, Q612W1; DAF-8 Ce, Q21733; DAF-8 Cb, Q61DE5; Smad4 Hs, Q13485; MEDEA Dm, Q61458; SMA-4 Ce, P45897; SMA-4 Cb, Q61L35; DAF-3 Ce, Q17532; DAF-3 Cb, Q5WN84; DAF-14 Ce, Q17760; DAF-14 Cb, Q61WH4; I-Smad Ce. The Swiss-Prot entry for *C. briggsae* DAF-14, which was taken from the genome sequencing consortium gene prediction, has an error in one splice site. Correction of that error allowed the alignment with *C. elegans* DAF-4 to continue to the end of the MH2 domain, and to the C terminus of the protein, as shown in part B of this figure (A color version of this figure is freely accessible via the website of the book: http://www.springer.com/1-4020-4542-5)

and defects in spicule morphogenesis (Savage et al., 1996). The male tail has a fan comprised of nine pairs of sensory organs, called rays, which are used to locate the vulva, along with spicules, which are copulatory organs that aid in the transfer of sperm. The Sma and Mab abnormalities are identical in *sma-2, sma-3*, and *sma-4*, and are the signature of the Sma/Mab signaling pathway. The Sma genes were originally identified in the large mutant screens of Brenner in the 1970s when he was developing the nematode as a model organism for study (Brenner, 1974), but they received no attention until their link with the TGF-β pathway.

2.1 Genetic Screens have Elucidated the Signaling Pathway

The molecular identity and mutant phenotype of *daf-4* and the Sma genes suggested that a genetic screen for additional Sma mutants would identify genes in the TGF-β pathway. A clonal screen of about 17,000 genomes was carried out (Savage-Dunn et al., 2003). From this screen mutations in *dbl-1, sma-2, sma-3, sma-4, daf-4, sma-6, sma-9(schnurri)* and at least eleven other complementation groups were obtained. All the core-signaling components were identified from this screen, and several loci that modify the activity of the pathway.

2.2 The Ligand for the Sma/Mab Pathway

DBL-1 is the ligand for the Sma/Mab pathway. It was molecularly identified in a PCR screen (based upon *decapentaplegic (dpp)* and BMPs) and genetically identified from mutations that were obtained from forward and reverse genetic screens (Morita et al., 1999; Savage-Dunn et al., 2003; Suzuki et al., 1999). At the time, the two characterized TGF-β superfamily ligand genes, *daf-7* and *unc-129*, did not appear likely to function in the Sma/Mab pathway (Colavita et al., 1998; Ren et al., 1996). *daf-7* functions in the dauer pathway and shares mutant phenotypes with *daf-4*, but not *sma-2*. *unc-129* was shown to be involved in axonal guidance, a function distinct from either the dauer or Sma/Mab pathway. Therefore, an additional ligand member was hypothesized to function in the Sma/Mab pathway – it was eventually identified as *dbl-1* (Morita et al., 1999; Suzuki et al., 1999). *dbl-1* is most closely related to *Drosophila dpp* and vertebrate BMPs.

The expression pattern of *dbl-1* offered a surprise. Although the receptors and Smads are required in the hypodermis to control body size (Inoue and Thomas, 2000; Savage-Dunn et al., 2000), a *dbl-1::gfp* fusion is primarily expressed in the amphid neurons, neurons of the ventral nerve cord, and neurons and glial cells in the tail. These cells must serve as the source of the signal for all of the three developmental processes in which it is involved (Morita et al., 1999; Suzuki et al., 1999). This suggests that the ligand is actively or passively transported to receptors in the hypodermis to control body size. The phenotypic effects of *dbl-1* are dosage dependent – over-expressing the ligand generates long animals (Morita et al., 2002; Suzuki et al., 1999) as is true for over-expression of other signaling components.

2.3 *sma-6* Encodes a Type I Receptor

As the organization of the Sma/Mab pathway became better understood, it became clear that DAF-1 could not function as a type I receptor for this pathway. This led to the hypothesis that another type I receptor would function with the Sma genes to control body size. PCR screens identified a new type I receptor in the *C. elegans* genome, which mapped near *sma-6* (Krishna et al., 1999). Since the one existing allele of *sma-6* does not show spicule and male tail ray defects, it was not an ideal candidate for the new type I receptor. However, additional alleles were generated through a combination of non-complementing screens and through alleles generated from the small body size screen (Krishna et al., 1999; Savage-Dunn et al., 2003). These new alleles showed mutant tail phenotypes in addition to body size phenotypes. Subsequent molecular studies showed that *sma-6* encodes a type I receptor (Krishna et al., 1999). This work also aided in establishing how specificity of signaling is achieved – a distinct type I receptor interacts with distinct sets of Smads and regulates a set of genes distinct from those of other Smads.

2.4 Worm *schnurri* Functions with the Smads

Smads weakly bind to DNA sequences and therefore require co-factors with stronger binding affinities to aid in regulation of gene transcription. This observation has led to the discovery of many transcription factors that function with the Smads. In *Drosophila*, SCHNURRI is a zinc finger protein that functions in the DPP pathway at all stages, except the earliest embryonic stage (Arora et al., 1995; Grieder et al., 1995; Staehling-Hampton et al., 1995). In *C. elegans*, the genome consortium identified a *schnurri* homolog on the X chromosome. *sma-9* mutants from the Sma/Mab genetic screen map closely to the region containing the nematode *schnurri* homolog and they were subsequently shown to encode the *C. elegans* SCHNURRI (Liang et al., 2003). This established that *schnurri* was a *bona fide* member of the TGF-β signaling pathway, having roles in *Drosophila* and *C. elegans*.

Interestingly, *sma-9* affects the growth of animals differently than the other Sma/Mab mutants. Growth is slowed in early larval stages but is apparently normal in later larval stages (Liang et al., 2003). In addition to growth defects, *sma-9* mutants show gross male ray morphological defects, particularly rays 8-9, but never rays 4-5 or 6-7 as seen in other Sma/Mab pathway mutants. However, *sma-9* is required for specification of neurons within rays 5, 7, and 9, as are other Sma/Mab pathway components. This suggests that *sma-9* effects are more restricted than those of the other signaling components.

2.5 Regulation of Body Size by TGF-β

The primary mutant phenotype of the Sma/Mab pathway is the regulation of body size. Sma mutant animals start out with normal lengths when they are born, but they grow slower than their wild-type siblings and reach about 60% the size of normal adults, with reductions in body length and volume (Savage et al., 1996; Suzuki et al., 1999). As shown by growth curves of Sma mutants, (Savage-Dunn et al., 2000) the rate of growth is lower in TGF-β mutants at each larval stage. This implies there is no developmental switch that affects their growth rate at any particular life stage. Examination of the number of adult nuclei does not show any differences in number from wild-type, suggesting that the small body size is not caused by a decrease in cell number but rather by cell size (Flemming et al., 2000; Savage et al., 1996; Suzuki et al., 1999). Different tissues show different amounts of reduction, but the hypodermis and seam cells show the most reduction (Wang et al., 2002). Regulation of body size regulation by this mechanism is in contrast to body size regulation in *Drosophila*, where the insulin signaling pathway affects body size (Oldham et al., 2002). In *C. elegans*, the insulin pathway does not affect body size, but does have a role in dauer formation (Jia et al., 2004; Kimura et al., 1997).

The DBL-1 ligand for the Sma/Mab pathway is expressed in a subset of neurons, as described above. However, most tissues in the animals are smaller than wild-type, suggesting that the body size signal is spread throughout the body. Although their expression patterns are complex, *sma-6*, the three Smads, and *daf-4* have overlapping expression patterns in the hypodermis, intestine, and pharynx. The hypodermis has been implicated as the focus of body size (Gunther et al., 2000; Krishna et al., 1999; Patterson et al., 1997; Savage-Dunn et al., 2000). Most of the hypodermis in *C. elegans* originates from hyp7, a syncytium of about 65 nuclei that are endoreduplicated (Flemming et al., 2000). The identification of the tissue that controls body size focus was determined by using promoter fusions to drive the expression of pathway components in specific tissues and observe the resulting body size changes. Not surprisingly, two hypodermal promoters, *rol-6* and *elt-3*, driving expression of Sma/Mab signaling components, can rescue body size. Further, neither an intestinal nor a neuronal promoter are able to rescue the body size of Sma animals (Inoue and Thomas, 2000; Maduzia et al., 2002; Yoshida et al., 2001). This data indicates that the hypodermis is the focus of body size and it indicates that another signal emanating from the hypodermis affects the size of other tissues. The expression in the intestine may be responsible for the nematode innate immunity. The Sma/Mab pathway induces many downstream genes that have roles in the immune response in *C. elegans* (Mallo et al., 2002).

How does the Sma/Mab pathway regulate body size? One possibility is that *C. elegans* TGF-β mutants uncouple cell growth and mitosis. If this were true, then mutants should affect aspects of cell cycle regulation. Evidence that endoreduplication is altered in Sma/Mab mutants has been found (Flemming et al., 2000; Nystrom et al., 2002). Wild-type adult animals have a hypodermal ploidy of about 10.7. The ploidy of three small mutants, *dpy-2, daf-4* and *sma-2*, was determined. *dpy-2* is not part of the TGF-β pathway, but results in mutants roughly the same size as mutations in *daf-4* or *sma-2*. The ploidy of *daf-4* and *sma-2* is 5.8 and 7.0 respectively. However, *dpy-2* has a normal ploidy, suggesting that low ploidy is not a consequence of small body size, and that a low ploidy is associated with TGF-β mutants. At present, this data is still correlative, and it is not known if ploidy is the driving factor for cell size. Whatever the mechanism, it is not known how the signals in the hypodermis and the seam cells contribute to size differences in other tissues.

2.6 Regulation of Male Tail Development

Null mutations in the Sma/Mab pathway result in transformations of male tail sensory rays and often fusions between adjacent rays. The rays that are transformed in Sma/Mab mutants are mostly dorsal rays 5, 7, and 9, which adopt the fate of their anterior neighbor and often fuse with it, creating a fatter ray (Krishna et al., 1999; Savage et al., 1996; Suzuki et al., 1999). The basis for these phenotypes appears to be an improper migration of the cells that comprise the ray. Mutations in the Sma/Mab genes also disrupt a cellular migration necessary to form the spicule, a copulatory structure necessary for mating (Baird and Ellazar, 1999; Krishna et al., 1999; Savage et al., 1996; Suzuki et al., 1999). All Sma/Mab core signaling components affect male tail development, and *schnurri* mutants show an altered phenotype as described above.

3. THE DAUER TGF-β-LIKE PATHWAY

3.1 Dauer Pathway Receptors

The receptors in the *C. elegans* pathway function similar to receptors from other organisms. However, some unique modes of signaling and regulation have been identified. The sequence of DAF-1 is equally similar to type I and type II receptors from other systems, and the identity of DAF-1 as a type I receptor was originally in doubt. Later, the GS domain was identified in type I receptors as a phosphorylation target that allows a type I receptor to be activated by a type II receptor. This domain has been shown to be a feature of type I receptors (Wieser et al., 1995), and DAF-1 has a sequence similar to the GS domain. However, DAF-1 is unique among type I receptors in that there is evidence that it can function independently of the type II receptor, at least in some circumstances (Gunther et al., 2000). DAF-4 is the only type II receptor in *C. elegans*, but the phenotype of a *daf-4* single null mutant is less severe than that of a *daf-1; daf-4* double mutant, implying that DAF-1 can signal even when DAF-4 is absent. Furthermore, over-expression of DAF-1 can partially suppress the phenotype of the *daf-4* single mutant, providing further evidence for independent function. The type II receptor DAF-4 clearly has an indispensable role in normal signaling, but the type I receptor DAF-1 may function independently of DAF-4 to allow more precise control of downstream events in various environments. Independent functions for type I receptors in other organisms may exist, but have not been described.

The DAF-4 receptor has a novel mode of regulation. Alternate polyadenylation produces a transcript that is predicted to encode only the extracellular ligand-binding domain of DAF-4, but not the transmembrane or cytoplasmic kinase domain

(Gunther and Riddle, 2004). Transgenic expression of a construct that encodes only this alternate truncated product antagonizes the function of the full-length DAF-4, suggesting that this product negatively regulates signaling in the pathway. In addition, the level of the truncated mRNA is dramatically increased when the animals enter dauer, which is a developmental event that requires a lack of signaling in the pathway; this regulation suggests that the truncated product acts to inhibit signaling of the pathway in a developmentally appropriate manner.

3.2 The Dauer Co-Smad has Unique Properties

The *daf-3* gene encodes a third dauer Smad, and is slightly more similar to Co-Smads than R-Smads (Patterson et al., 1997). However, the dauer-defective mutant phenotype of loss of function of *daf-3* is the opposite of the dauer-constitutive phenotype of the receptors, ligand and putative R-Smads. *daf-3* is epistatic to the dauer-constitutive phenotype of the ligand, receptor and R-Smad genes (Thomas et al., 1993), which places it downstream of the R-Smads. This genetic interaction suggests that DAF-3 is negatively regulated by receptors and receptor regulated Smads. This result is unusual: in the *Drosophila* DPP pathway and the *C. elegans* Sma/Mab pathway, the genes for ligands, receptors, and R-Smads and Co-Smads all have very similar phenotypes. Furthermore, biochemical, molecular and genetic experiments with vertebrate pathways indicate a similar picture; R-Smads and Co-Smads are activated, not repressed by the receptors, and R-Smads and Co-Smads cooperate rather than interfere with each other. DAF-3 is not an I-Smad; these Smads negatively regulate receptors, R-Smads and Co-Smads. DAF-3, on the other hand, is negatively regulated by receptors, R-Smads and Co-Smads. DAF-8 and DAF-14 are not I-Smads either. In addition to the previously cited evidence that suggests they are R-Smads, there is the fact that *daf-8* and *daf-14* mutants have similar phenotypes to receptor and ligand mutants, suggesting cooperation, whereas genetic and molecular experiments with I-Smads demonstrate that these antagonize the function of the receptors.

3.3 Sno/Ski Gene Functions in the Dauer Pathway

The *daf-5* gene is related to the Sno/Ski family of oncogenes, and therefore, is similar to genes that function in other TGF-β pathways, but, like *daf-3*, has features that suggest that it has a unique regulatory function. The mutant phenotype of *daf-5* is very similar to the mutant phenotype of *daf-3*. Loss-of-function mutants of *daf-5* are dauer-defective, and epistatic to mutations in the ligand, receptors and R-Smads (Thomas et al., 1993). This result implies that DAF-3 and DAF-5 are both antagonized by the receptors and R-Smads, and that DAF-3 and DAF-5 may be cofactors. DAF-5 has weak sequence similarities to the Sno-Ski family of proto-oncogenes (da Graca et al., 2004; Tewari et al., 2004). Sno and Ski have been shown to be antagonists of TGF-β signaling and have been proposed to act as negative feedback regulators to limit TGF-β activation of gene expression (Luo, 2004).

The similarity of DAF-5 to Sno/Ski is weak, but convincing. A blast search with DAF-5 finds similarity to a 30 amino acid stretch of Sno and Ski, which does not provide convincing evidence of relatedness. Additional evidence suggests that the relationship is real and important. First, Sno/Ski function with Smads, and DAF-5 functions with DAF-3; it is hard to imagine that the similarity of DAF-5 to Sno/Ski, however weak, is not meaningful. Second, several of the residues identified by BLAST searches as conserved between DAF-5 and Sno/Ski have been shown to be critically important in Sno/Ski function (see da Graca et al., 2004, for an alignment of this domain). In fact, these residues are found in a region of Sno/Ski that binds to Smads, and are residues that actually contact Smads or that are zinc-chelating residues in a novel zinc finger that is a critical structural component of the Smad binding domain (Wu et al., 2002). Fig. 2 shows the relationship between DAF-5 and Sno/Ski homologs. An ancient family of proteins, which has been dubbed Ice/Skate, shows substantial similarity to the Sno/Ski family. When the entire protein is considered, Sno/Ski is more similar to Ice/Skate than to DAF-5; however, in the Smad binding domain, Sno and Ski are more similar to DAF-5 than either is to Ice/Skate (Fig. 2). Third, DAF-5 binds to DAF-3 in yeast two-hybrid experiments, implying functional similarity to Sno/Ski (Tewari et al., 2004). Fourth, DAF-5 shares additional similarity to Sno/Ski in the 'Dachbox' domain. This similarity

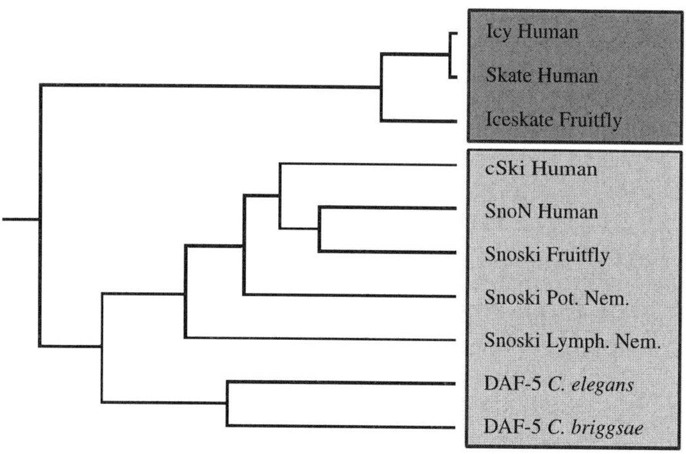

Figure 2. Sno/Ski family SDS domains. The SDS domains (defined in text) of DAF-5 and representative members of the Sno/Ski family were aligned with ClustalW and refined by hand. The alignment was used to build a tree using GeneBee (Brodskii et al., 1995). Accession Numbers and abbreviations: cSki Human, NP_003027; SnoN Human, NP_005405; Icy Human, XP_292349; Skate Human, XP_064560 Iceskate Fruitfly, NP_651946; Snowski Fruitfly, NP_609166; Snoski Pot. Nem., Potato Cyst nematode, *Globodera rostochiensis*, AY389814; Snoski Lymph Nem, *Brugia malayi*, AY389813; DAF-5 *C. briggsae*, CBG20832; DAF-5 *C. elegans*, NM_064540 (A color version of this figure is freely accessible via the website of the book: http://www.springer.com/1-4020-4542-5)

is too weak to be picked up by BLAST, but can be found by visual examination. The region of highest similarity between DAF-5 and Sno/Ski in the dachbox is a mutation hotspot for *daf-5*, implying a high degree of functional importance (da Graca et al., 2004).

4. ORPHAN TGF-β COMPONENTS

There are five TGF-β superfamily ligands in *C. elegans*. As described above, DBL-1 and DAF-7 are ligands in the two major pathways, Sma/Mab and dauer. UNC-129, which has characteristics of a BMP-like ligand, was identified from genome sequencing and from screens for mutations in axonal guidance (Colavita et al., 1998). *unc-129* mutants are weakly uncoordinated due to defects in dorsal trajectories of motor axons, similar to defects seen in *unc-5, unc-6*, and *unc-40* mutants, which are related to netrin and its receptor. *unc-129* mutants do not have Daf or Sma/Mab phenotypes and components of the Sma/Mab or dauer pathway do not have uncoordinated mutant phenotypes. This data presents a puzzle, since the nematode genome does not contain another set of canonical signaling components that might function with UNC-129. Given that, it suggests that UNC-129 may function in a non-canonical way, using a different receptor to signal. Further genetic and molecular studies will be needed to unravel this puzzle.

Two other orphan ligands of unknown function, TIG-2 and TIG-3, were identified by sequence similarity to TGF-β superfamily ligands. Mutations in *tig-2* have been generated (L. Maduzia and R.W.P., unpublished data), but no striking mutant phenotypes have been observed. Mutations are able to weakly enhance dauer mutations, and *tig-2* is strongly expressed in two unidentified head neurons. Given that the gene is conserved and intact, it is reasonable to assume it functions under unusual environmental conditions or stress. Since there are no additional TGF-β superfamily signaling components in the nematode, which could function with TIG-2, it could function as a heterodimer with one of the other ligands in a temporally or spatially restricted manner.

During the completion of the nematode genome, an additional TGF-β-like ligand gene was identified on a YAC, which spanned an incomplete part of the genome. It is known as *tig-3* (formerly Y46E12BL.1). It is expressed, although its transcripts are not abundant (Y. Funakoshi, T. Gumienny, and R.W.P., unpublished data). No mutants have been reported, so its function remains unknown. As in the case of the other orphan ligands, there are no additional signaling components that comprise an entirely new pathway. Therefore, its functions may be intertwined with one of the known ligands, perhaps forming heterodimers.

5. *C. ELEGANS* TGF-β SIGNALING DOES NOT CONTROL THE DV AXIS

One of the most striking differences in function of TGF-β in *C. elegans* is that it does not specify the dorsal-ventral axis. The formation of the dorsal-ventral axis in *Drosophila* and vertebrates is under the control of members of the BMP signaling family. Mutations in the signaling pathway result in major defects in specification of the dorsal-ventral axis, and lethality (Ferguson, 1996). In *C. elegans*, loss of TGF-β signaling does not cause this defect. For example, strong mutants of *daf-4*, which encodes the only type II receptor in *C. elegans*, have no known defects in dorsal-ventral patterning. Mutations have been identified for most of the ligands and Smads, and all of the type I receptors, and none disrupt the dorsal-ventral axis the way that the vertebrate and insect mutations do (Patterson and Padgett, 2000). Mutations in *unc-129* affect axon patterning in processes that go in a ventral to dorsal direction (Colavita et al., 1998), but this is a very subtle defect, and may or may not be related to the dorsal-ventral axis itself. More subtle defects in dorsal-ventral patterning may yet be found, but the basic axis is intact, and the mutants are healthy and fecund. Thus, *C. elegans* shows a dramatic difference in this fundamental role of BMP signaling. Two explanations come to mind. *C. elegans* diverged from other animals before the pathway evolved a dorsal-ventral function. Alternatively, the lack of a dorsal-ventral role could be a derived trait in *C. elegans*. Recent data on the evolution of simple animal phyla suggests that the latter is the correct explanation.

The traditional view has held that bilaterian animals lacking a true coelom, including acoelomates (*e.g.* platyhelminthes flatworms) and nemertina) and pseudocoelomates (*e.g.* nematodes, nematomorphs and rotifers) are "basal", which is to say that they diverged early from other bilaterian lineages. Recent molecular and developmental analyses have suggested that the evolution of nematodes and other simple pseudocoelomate and acoelomate phyla should be reevaluated. The rate of divergence of certain genes, notably ribosomal RNA genes, has been taken to represent a molecular clock for the time of divergence of various animals. A comparison of fossil evidence with frequency of mutations in ribosomal RNA has shown that, for some organisms, the molecular clock is valid. However, *C. elegans* is an example of an organism with a "fast clock", that is, an unusually high rate of mutation over time. When a variety of nematodes were examined, some (those with "normal clocks") clearly grouped with arthropods and other molting animals in a group called ecdyzoa (Aguinaldo et al., 1997). *C. elegans*, despite its high DNA sequence divergence, belongs with other nematodes, and therefore, is also in the ecdyzoa (Adoutte et al., 2000; Aguinaldo et al., 1997; Peterson and Eernisse, 2001; Telford et al., 2005). This view of animal evolution, in which all of the bilaterian phyla fall into ecdyzoa, lophotrochozoa, or deuterostomata (*e.g.* chordates and echinoderms) has achieved widespread, but not universal (Wolf et al., 2004), acceptance in a brief period of time.

This view of evolution suggests that the most recent common ancestor of humans and flies is also the most recent common ancestor of humans, flies and *C. elegans* (as well as mollusks, annelids, and most other animals). The homologous role of BMP in creating the dorsal-ventral axis in *Drosophila* and vertebrates was therefore shared by the common ancestor of all bilaterians, and the lack of a role for BMP signaling in dorsal-ventral patterning is therefore a derived trait of *C. elegans*.

6. FUTURE PERSPECTIVES

Based upon new molecular studies, Smads and their signaling partners exist in the simplest animals and therefore, have been conserved for hundreds of millions of years. While the basic building blocks of the pathway are widespread, the outputs of the pathways vary, depending on the organism and tissue. *C. elegans* contains two *bona fide* TGF-β superfamily signaling pathways that affect body size, tail development, immunity, and dauer development. Studies in *C. elegans* have been useful in identifying important components of the pathways, particularly the Smads. Early on it became apparent that the nematode different pathways had both shared and unique components. In *C. elegans* specificity is achieved by pathway specific components, which is now regarded as a generally applicable principle of TGF-β superfamily signaling.

There are interesting open questions regarding three of the ligands in *C. elegans*. The UNC-129 pathway does not obviously signal through canonical receptors, and may therefore signal through non-canonical receptor pathways, which are yet to be identified. The roles of TIG-2 and TIG-3 are not known, but given that all nematode TGF-β receptors are identified, their functions may be subtle or spatially restricted.

The most exciting future prospect in *C. elegans* TGF-β signaling is the elucidation of events downstream of the Smads. Microarray studies have allowed the identification of many genes that are regulated by these pathways. Given the large number of regulated genes, demonstrating function in the TGF-β pathway is important. The use of RNAi and classical genetic approaches will allow the rapid dissection of function of these putative targets.

ACKNOWLEDGEMENTS

This work was supported by grants from the NIH and the DOD Breast Cancer Program to GIP and RWP.

REFERENCES

Adoutte, A., Balavoine, G., Lartillot, N., Lespinet, O., Prud'homme, B., and de Rosa, R. 2000, The new animal phylogeny: reliability and implications. *Proc Natl Acad Sci U S A* 97: 4453-4456.

Aguinaldo, A.M., Turbeville, J.M., Linford, L.S., Rivera, M.C., Garey, J.R., Raff, R.A., and Lake, J.A., 1997, Evidence for a clade of nematodes, arthropods and other moulting animals. *Nature* 387: 489-493.

Arora, K., Dai, H., Kazuko, S.G., Jamal, J., O'Connor, M.B., Letsou, A., and Warrior, R., 1995, The *Drosophila schnurri* gene acts in the Dpp/TGF-β signaling pathway and encodes a transcription factor homologous to the human MBP family. *Cell* 81: 781-790.

Baird, S.E., and Ellazar, S.A., 1999, TGF-β-like signaling and spicule development in *Caenorhabditis elegans*. *Dev Biol* 212: 93-100.

Bateman, A., Coin, L., Durbin, R., Finn, R.D., Hollich, V., Griffiths-Jones, S., Khanna, A., Marshall, M., Moxon, S., Sonnhammer, E.L., Studholme, D.J., Yeats, C., and Eddy, S.R., 2004, The Pfam protein families database. *Nucleic Acids Res* 32: D138-D141.

Brenner, S., 1974, The genetics of *Caenorhabditis elegans*. *Genetics* 77: 71-94.

Brodskii, L.I., Ivanov, V.V., Kalaidzidis Ia, L., Leontovich, A.M., Nikolaev, V.K., Feranchuk, S.I., and Drachev, V.A., 1995, [GeneBee-NET: An Internet based server for biopolymer structure analysis]. *Biokhimiia* 60: 1221-1230.

Colavita, A., Krishna, S., Zheng, H., Padgett, R.W., and Culotti, J.G., 1998, Pioneer axon guidance by UNC-129, a *C. elegans* TGF-β. *Science* 281: 706-709.

da Graca, L.S., Zimmerman, K.K., Mitchell, M.C., Kozhan-Gorodetska, M., Sekiewicz, K., Morales, Y., and Patterson, G.I., 2004, DAF-5 is a Ski oncoprotein homolog that functions in a neuronal TGF-β pathway to regulate *C. elegans* dauer development. *Development* 131: 435-446.

Derynck, R., Gelbart, W.M., Harland, R.M., Heldin, C.-H., Kern, S.E., Massagué, J., Melton, D.A., Mlodzik, M., Padgett, R.W., Roberts, A.B., Smith, J., Thomsen, G.H., Vogelstein, B., and Wang, X.F., 1996, Nomenclature: vertebrate mediators of TGFβ family signals. *Cell* 87: 173.

Ferguson, E.L., 1996, Conservation of dorsal-ventral patterning in arthropods and chordates. *Curr Opin Genet Dev* 6: 424-431.

Flemming, A.J., Shen, Z.Z., Cunha, A., Emmons, S.W., and Leroi, A.M., 2000, Somatic polyploidization and cellular proliferation drive body size evolution in nematodes. *Proc Natl Acad Sci U S A* 97: 5285-5290.

Georgi, L.L., Albert, P.S., and Riddle, D.L., 1990, *daf-1*, a *C. elegans* gene controlling dauer larva development, encodes a novel receptor protein kinase. *Cell* 61: 635-645.

Grieder, N.C., Nellen, D., Burke, R., Basler, K., and Affolter, M., 1995, *schnurri* is required for *Drosophila* Dpp signaling and encodes a zinc finger protein similar to the mammalian transcription factor PRDII-BF1. *Cell* 81: 791-800

Gunther, C.V., Georgi, L.L., and Riddle, D.L., 2000, A *Caenorhabditis elegans* type I TGFβ receptor can function in the absence of type II kinase to promote larval development. *Development* 127: 3337-3347.

Gunther, C.V., and Riddle, D.L., 2004, Alternative polyadenylation results in a truncated DAF-7 BMP receptor that antagonizes DAF-7-mediated development in *Caenorhabditis elegans. J Biol Chem* 279: 39555-39564.

Inoue, T., and Thomas, J.H., 2000, Targets of TGF-β signaling in *Caenorhabditis elegans* dauer formation. *Dev Biol* 217: 192-204.

Jia, K., Chen, D., and Riddle, D.L., 2004, The TOR pathway interacts with the insulin signaling pathway to regulate *C. elegans* larval development, metabolism and life span. *Development* 131: 3897-3906.

Kimura, K.D., Tissenbaum, H.A., Liu, Y., and Ruvkun, G., 1997, *daf-2*, an insulin receptor-like gene that regulates longevity and diapause in *Caenorhabditis elegans*. *Science* 277: 942-946.

Krishna, S., Maduzia, L.L., and Padgett, R.W., 1999, Specificity of TGFβ signaling is conferred by distinct type I receptors and their associated SMAD proteins in *Caenorhabditis elegans. Development* 126: 251-260.

Kurz, C.L., and Tan, M.W., 2004, Regulation of aging and innate immunity in *C. elegans. Aging Cell* 3: 185-193.

Liang, J., Lints, R., Foehr, M.L., Tokarz, R., Yu, L., Emmons, S.W., Liu, J., and Savage-Dunn, C., 2003, The *Caenorhabditis elegans schnurri* homolog *sma-9* mediates stage- and cell type-specific responses to DBL-1 BMP-related signaling. *Development* 130: 6453-6464.

Luo, K., 2004, Ski and SnoN: negative regulators of TGF-β signaling. *Curr Opin Genet Dev* 14: 65-70.

Maduzia, L.L., Gumienny, T.L., Zimmerman, C.M., Wang, H., Shetgiri, P., Krishna, S., Roberts, A.B., and Padgett, R.W., 2002, *lon-1* regulates *Caenorhabditis elegans* body size downstream of the *dbl-1* TGFβ signaling pathway. *Dev Biol* 246: 418-428.

Mallo, G.V., Kurz, C.L., Couillault, C., Pujol, N., Granjeaud, S., Kohara, Y., and Ewbank, J.J., 2002, Inducible antibacterial defense system in *C. elegans. Curr Biol* 12: 1209-1214.

Mathews, L.S., and Vale, W.W., 1991, Expression cloning of an activin receptor, a predicted transmembrane serine kinase. *Cell* 65: 973-982.

Morita, K., Chow, K., and Ueno, N., 1999, Regulation of body length and male tail ray pattern formation of *Caenorhabditis elegans* by a member of the TGF-β family. *Development* 126: 1337-1347.

Morita, K., Flemming, A.J., Sugihara, Y., Mochii, M., Suzuki, Y., Yoshida, S., Wood, W.B., Kohara, Y., Leroi, A.M., and Ueno, N., 2002, A *Caenorhabditis elegans* TGF-β, DBL-1, controls the expression of LON-1, a PR-related protein, that regulates polyploidization and body length. *EMBO J* 21: 1063-1073.

Nicholas, H.R., and Hodgkin, J., 2004, Responses to infection and possible recognition strategies in the innate immune system of *Caenorhabditis elegans*. *Mol Immunol* 41: 479-493.

Nyström, J., Shen, Z.Z., Aili, M., Flemming, A.J., Leroi, A., and Tuck, S., 2002, Increased or decreased levels of *Caenorhabditis elegans lon-3*, a gene encoding a collagen, cause reciprocal changes in body length. *Genetics* 161: 83-97.

Oldham, S., Stocker, H., Laffargue, M., Wittwer, F., Wymann, M., and Hafen, E., 2002, The *Drosophila* insulin/IGF receptor controls growth and size by modulating PtdInsP(3) levels. *Development* 129: 4103-4109.

Patterson, G., and Padgett, R.W., 2000, TGFβ-related pathways. Roles in *Caenorhabditis elegans* development. Trends in Genetics 16: 27-33.

Patterson, G.I., Koweek, A., Wong, A., Liu, Y., and Ruvkun, G., 1997, The DAF-3 Smad protein antagonizes TGF-β-related receptor signaling in the *Caenorhabditis elegans* dauer pathway. *Genes Dev* 11: 2679-2690.

Peterson, K.J., and Eernisse, D.J., 2001, Animal phylogeny and the ancestry of bilaterians: inferences from morphology and 18S rDNA gene sequences. *Evol Dev* 3: 170-205.

Ren, P., Lim, C.S., Johnsen, R., Albert, P.S., Pilgrim, D., and Riddle, D.L., 1996, Control of *C. elegans* larval development by neuronal expression of a TGF-β homolog. *Science* 274: 1389-1391.

Riddle, D.L., and Albert, P.S., 1997, Genetic and environmental regulation of dauer larva development. In Riddle, D.L. et al. (eds.), *C. elegans II*, Cold Spring Harbor Press, pp. 739-768.

Savage, C., Das, P., Finelli, A., Townsend, S., Sun, C., Baird, S., and Padgett, R.W., 1996, *Caenorhabditis elegans* genes *sma-2, sma-3*, and *sma-4* define a conserved family of transforming growth factor β pathway components. *Proc Natl Acad Sci U S A* 93: 790-794.

Savage-Dunn, C., Maduzia, L.L., Zimmerman, C.M., Roberts, A.F., Cohen, S., Tokarz, R., and Padgett, R.W., 2003, Genetic screen for small body size mutants in *C. elegans* reveals many TGFβ pathway components. *Genesis* 35: 239-247.

Savage-Dunn, C., Tokarz, R., Wang, H., Cohen, S., Giannikas, C., and Padgett, R.W., 2000, SMA-3 smad has specific and critical functions in DBL-1/SMA-6 TGF-β related signaling. *Dev Biol* 223: 70-76.

Sekelsky, J.J., Newfeld, S.J., Raftery, L.A., Chartoff, E.H., and Gelbart, W.M., 1995, Genetic characterization and cloning of *mothers against dpp*, a gene required for *decapentaplegic* function in *Drosophila melanogaster*. *Genetics* 139: 1347-1358.

Staehling-Hampton, K., Laughon, A.S., and Hoffmann, F.M., 1995, A *Drosophila* protein related to the human zinc finger transcription factor PRDII/MBPI/HIV-EP1 is required for *dpp* signaling. *Development* 121: 3393-3403.

Suzuki, Y., Yandell, M.D., Roy, P.J., Krishna, S., Savage-Dunn, C., Ross, R.M., Padgett, R.W., and Wood, W.B., 1999, A BMP homolog acts as a dose-dependent regulator of body size and male tail patterning in *Caenorhabditis elegans*. *Development* 126: 241-250.

Telford, M.J., Wise, M.J., and Gowri-Shankar, V., 2005, Consideration of RNA secondary structure significantly improves likelihood-based estimates of phylogeny: examples from the bilateria. *Mol Biol Evol* 22: 1129-1136.

Tewari, M., Hu, P.J., Ahn, J.S., Ayivi-Guedehoussou, N., Vidalain, P.O., Li, S., Milstein, S., Armstrong, C.M., Boxem, M., Butler, M.D., Busiguina, S., Rual, J.F., Ibarrola, N., Chaklos, S.T., Bertin, N., Vaglio, P., Edgley, M.L., King, K.V., Albert, P.S., Vandenhaute, J., Pandey, A., Riddle, D.L., Ruvkun, G., and Vidal, M., 2004, Systematic interactome mapping and genetic perturbation analysis of a *C. elegans* TGF-β signaling network. *Mol Cell* 13: 469-482.

Thatcher, J.D., Haun, C., and Okkema, P.G., 1999, The DAF-3 Smad binds DNA and represses gene expression in the *Caenorhabditis elegans* pharynx. *Development* 126: 97-107.

Thomas, J.H., Birnby, D.A., and Vowels, J.J., 1993, Evidence for parallel processing of sensory information controlling dauer formation in *Caenorhabditis elegans*. *Genetics* 134: 1105-1117.

Wang, J., Tokarz, R., and Savage-Dunn, C., 2002, The expression of TGFβ signal transducers in the hypodermis regulates body size in *C. elegans*. *Development* 129: 4989-4998.

Wieser, R., Wrana, J.L., and Massagué, J., 1995, GS domain mutations that constitutively activate Tβ R-I, the downstream signaling component in the TGF-β receptor complex. *EMBO J* 14: 2199-2208.

Wolf, Y.I., Rogozin, I.B., and Koonin, E.V., 2004, Coelomata and not Ecdysozoa: evidence from genome-wide phylogenetic analysis. *Genome Res* 14: 29-36.

Wu, J.W., Krawitz, A.R., Chai, J., Li, W., Zhang, F., Luo, K., and Shi, Y., 2002, Structural mechanism of Smad4 recognition by the nuclear oncoprotein Ski: insights on Ski-mediated repression of TGF-β signaling. *Cell* 111: 357-367.

Yoshida, S., Morita, K., Mochii, M., and Ueno, N., 2001, Hypodermal expression of *Caenorhabditis elegans* TGF-β type I receptor SMA-6 is essential for the growth and maintenance of body length. *Dev Biol* 240: 32-45.

CHAPTER 3

SMADS IN *DROSOPHILA* – INTERPRETATION OF GRADED SIGNALS *IN VIVO*

LAUREL A. RAFTERY, SVETLANA KOROCHKINA AND JING CAO
Cutaneous Biology Research Center, Massachusetts General Hospital/Harvard Medical School, Charlestown, MA 02129 USA

Abstract: *Drosophila* has both BMP and Activin signaling pathways, but little is known about the functions of the Activin/TGF-β class of signals in this organism. BMP signaling has been intensively studied in this organism, and *Drosophila* continues to stand as a paradigm for understanding the functions of BMPs in development. *Drosophila* BMPs can induce different cell fates at different concentrations, a critical property for extracellular signals that direct spatial patterning of tissues. Although the endogenous BMP ligands are difficult to detect, spatial patterns of cells with phosphorylated R-Smad and nuclear co-Smad reveal the graded distribution of BMP activity in the embryonic ectoderm and the wing primordium. The question of how cells interpret different levels of BMP activity has been a major focus for studies of Smads in *Drosophila*. We review the general modes for Smad regulation of BMP target genes, including direct repression of gene expression, indirect induction through release from repression, and direct activation of gene expression. These studies integrate the mechanisms for gene regulation by Smads with the logic of spatial and temporal regulation of development by BMPs

Keywords: BMP; Dad; default repression; development; *Drosophila*; gene expression; Mad; Medea; morphogen gradient; Smox; spatial patterning

1. INTRODUCTION

Genetic studies in the fruit fly *Drosophila melanogaster* have driven a revolution in our understanding of developmental mechanisms. As discussed in the Preface, this model system had a direct impact on understanding TGF-β signaling, through the identification of the founding member of the Smad family, Mad. Studies of BMP signaling in *Drosophila* remain at the forefront of developmental biology, because they are a paradigm for understanding how different levels of an extracellular ligand can induce different cell fates. This chapter focuses on aspects of Smad signaling that are central to this function in *Drosophila*. After a brief overview of all *Drosophila* Smads, we will focus on Smad function in BMP signaling. The emphasis on *in*

P. ten Dijke and C.-H. Heldin (eds.), Smad Signal Transduction, 55–73.

vivo functions of graded BMP signals gives a different flavor to studies of Smad functions in *Drosophila*. However, many functions of BMPs in early embryogenesis and tissue morphogenesis are conserved from insects to mammals (Hogan, 1996; Whitman, 1998), so that the developmental themes emerging from *Drosophila* studies will apply to mammalian BMP signaling under many circumstances.

2. OVERVIEW OF *DROSOPHILA* SMAD GENES

As discussed in Chapter 1, there are four Smad genes in *Drosophila*, which have playful names, in the tradition of *Drosophila* geneticists. Three have been extensively studied because they regulate responses to the *Drosophila* ortholog of BMP2 and BMP4, Dpp (Fig. 1). Two *Drosophila* Smad genes were identified through maternal effect genetic modifier screens (reviewed by Raftery and Sutherland, 1999). These genes, *Mothers against dpp* (*Mad*) and *Medea*, encode a BMP R-Smad and a co-Smad, respectively (Fig. 1). When we refer to BMP signaling in this chapter, we refer to the composite BMP signal transduction pathway outlined in Fig. 1. BMPs are required in numerous tissues throughout *Drosophila* development, with functions ranging from dorsal-ventral patterning of the embryonic ectoderm (reviewed by Raftery and Sutherland, 2003) to retrograde signaling at neuromuscular junctions (reviewed by Marques, 2005). In most tissues, BMP signaling induces expression of tissue-specific target genes, and represses expression of others.

The *Drosophila* I-Smad gene, *daughters against dpp (Dad)*, was initially identified as a Dpp-induced gene during wing development (Fig. 1; reviewed by Raftery and Sutherland, 1999); it is a BMP-responsive gene in numerous tissues (*e.g.* Marty et al., 2000). In cultured cells, Dad can associate with the Tkv type I receptor, and decreases levels of phosphorylated Mad. *In vivo*, Dad down-regulates BMP signaling; however, the strength of this effect appears to be tissue dependent. Genetic manipulation of Dad levels in the wing primordium yielded only weak effects on expression of BMP target genes (Tsuneizumi et al., 1997), whereas overexpression of Dad in the germ line stem cells can strongly antagonize BMP signaling (Casanueva and Ferguson, 2004; Kawase et al., 2004; Song et al., 2004). It is not known whether Dad also antagonizes signaling through the Activin pathway.

The fourth Smad in *Drosophila*, called either Smox or dSmad2, was identified through sequence homology (reviewed by Raftery and Sutherland, 1999). The gene has been named *Smad on X* (*smox*), based on its chromosomal location. In cultured cells, Smox is phosphorylated by the Activin type I receptor Babo. Consistent with a requirement for Smox in Activin signaling *in vivo*, both *smox* and *babo* are required for specific aspects of neuronal remodeling during brain metamorphosis (Zheng et al., 2003). However, the specificity of Mad and Smox for BMP and Activin receptors, respectively, has not been critically tested *in vivo*. Candidate Activin target genes have been identified by microarray analyses (Yang et al., 2004) and null mutations in *smox* have been isolated (Wijayatunge, Stefancsik, Ulenga, and L.A.R., unpublished data). These tools will give us a better understanding of Smox and Babo functions in the future.

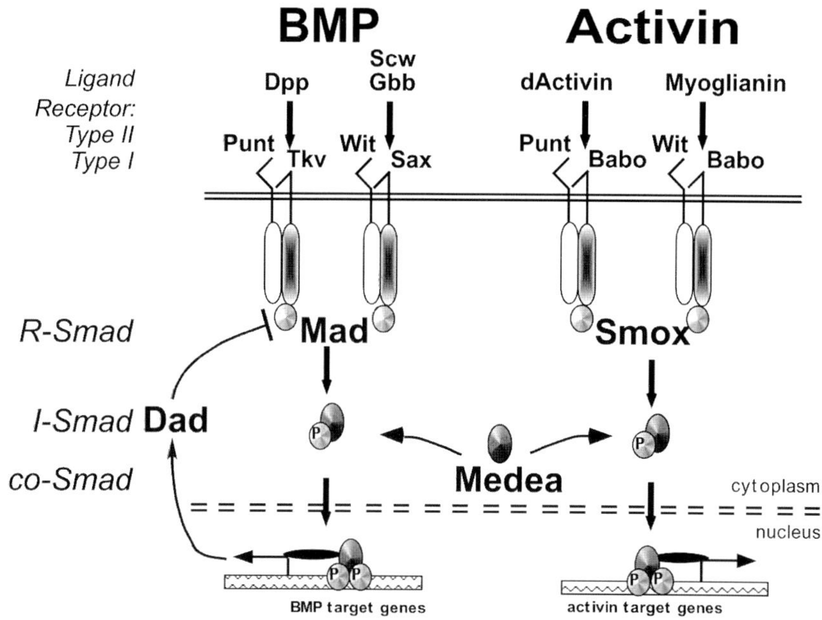

Figure 1. Drosophila TGF-β signaling pathways. Because *Drosophila* proteins are generally called by the same name as the gene, ligands, receptors, and Smads are indicated by a distinct terminology from the mammalian pathways. This diagram schematizes the Smad signal transduction pathways with the *Drosophila* names for each molecular component. The accurate stoichiometry is not depicted for molecular complexes; this is covered in the Preface. Evidence for these signal transduction pathways is reviewed by Raftery and Sutherland, except for myoglianin (Lee-Hoeflich et al., 2005) and Wit (reviewed by Marques, 2005). Genetic evidence indicates that Screw signals through Sax; association with Wit is hypothetical. The pathways for ligands encoded by *maverick* and *alp23B* have not been reported

Abbreviations: Ligands: Dpp, Decapentaplegic; Scw, Screw; Gbb, Glass-bottom boat. Receptors: Wit, Wishful thinking; Tkv, Thick veins; Sax, Saxophone; Babo, Baboon. Smads: Mad, Mothers against Dpp; Smox, Smad on X; Dad, Daughters against Dpp

3. SMADS AS INTERPRETERS OF MORPHOGEN GRADIENTS

BMP signaling aroused the interests of developmental biologists, because BMP ligands have a critical property: different levels of extracellular ligand can induce different cell fates (recently reviewed by Gurdon and Bourillot, 2001, additional references can be found there). This property led to the early identification of the *Drosophila* BMP2/4 ortholog, Dpp, as a developmental morphogen: a molecule that defines the spatial pattern of cell fates within a tissue. Widespread interest in the mechanisms that define spatial location led to an intense focus on BMP signaling during early development of the larval body plan, and later development of the adult wing. Each of these developmental systems has yielded important information about the general mechanisms employed by Smads to regulate BMP target genes. First, we briefly introduce the *Drosophila* ligands that form the BMP activity gradients and the receptors that are required for cells to respond to BMP gradients. Then, we focus on responses to gradients of BMP activity in the embryonic ectoderm and the wing primordium. The section on embryonic patterning covers two topics: nuclear accumulation of Smads as the readout of BMP activity, and the importance of derepression for Smad regulation of gene expression. The section on wing patterning also covers two topics. First, we discuss Smad-directed transcriptional repression as a major mechanism for indirect regulation of gene expression. We then briefly discuss BMP regulation of growth and proliferation in the wing primordium. Specific mechanisms for Smad transcriptional regulation will be discussed in Section 4.

3.1 The Importance of Knowing Where You Are: BMP Ligands as Morphogens

Studies of BMP signaling in *Drosophila* have focused on the mechanisms that convey information about spatial location, or pattern formation. A paradigm of developmental biology is that cells determine their spatial location from the graded distribution of a morphogen, a molecule that induces different cell fates when present at different concentrations (Fig. 2). Specific genes respond to different levels of morphogen, resulting in a nested pattern of gene expression responses. The combination of genes that are expressed determines the ultimate fate of the cell: the type of tissue it forms, its rate of proliferation, its shape, and its competence to respond to other signals. Smads provide the molecular link between BMP ligand activation of cell surface receptors and the gene expression responses that determine cell fate.

Among the first molecules demonstrated to be morphogens were two ligands in the TGF-β family, Activin in frogs, and Dpp in flies (the literature for this topic can be entered from reviews by Gurdon and Bourillot, 2001; Tabata, 2001). It has been remarkably difficult to visualize the tissue distribution of the endogenous ligands. For fly Dpp, the distribution of tagged, transgenic ligand *in vivo* parallels the endogenous gradients inferred by indirect approaches (Entchev et al., 2000; Shimmi et al., 2005; Teleman and Cohen, 2000; Wang and Ferguson, 2005). The current debate about mechanisms that create graded distributions of active BMP ligands are beyond the scope of this chapter (this literature can be entered from Gonzalez-Gaitan, 2003; Mizutani et al., 2005; Tabata and Takei, 2004).

Although most *Drosophila* studies of BMP morphogen gradients have focused specifically on Dpp, gradients of active BMPs involve additional BMP ligands (reviewed in Ashe, 2005; Podos and Ferguson, 1999). For dorsal-ventral patterning during embryogenesis, the BMP ligand Screw is required. For wing proximal-distal patterning, the BMP ligand Gbb (previously called 60A) also plays a role. Similarly, two type I receptors are required in each tissue, Tkv and Sax. The mechanism for collaboration between two BMP ligands or receptors is under debate (Shimmi et al., 2005; Wang and Ferguson, 2005). In both tissues, the Activin type II receptor, Punt, is essential for BMP signaling. In either tissue, BMP target genes are regulated by Mad and Medea through both direct and indirect mechanisms.

3.2 Dorsal-ventral Patterning of the Embryonic Ectoderm

Spatial gradients of BMP activity pattern the larval body plan along the dorsal-ventral axis during blastoderm and gastrula stages of embryogenesis, a period of about one and a half hours (reviewed by Raftery and Sutherland, 2003). This is a major transitional period for embryonic gene regulation; in dorsal-ventral patterning there is a transition from initial patterning by Toll receptor/NF-κB signal transduction to ectodermal patterning by BMP receptor/Smad signal transduction

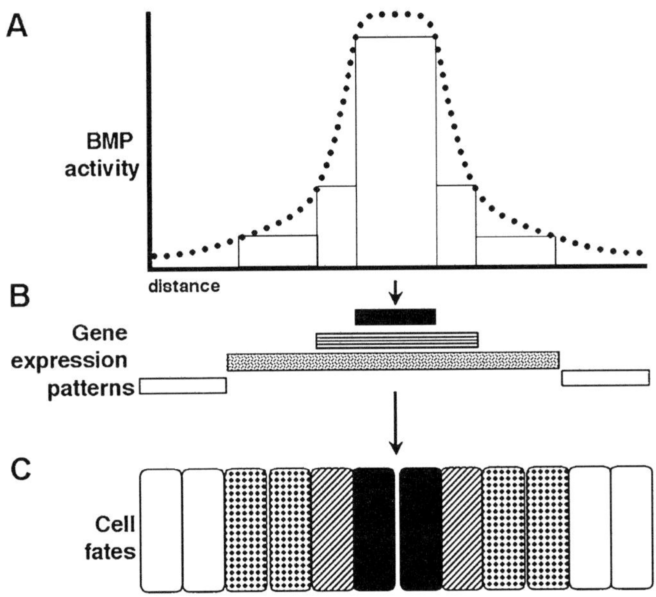

Figure 2. A hypothetical BMP morphogen gradient. *A.* A graph of BMP activity distributed across a hypothetical tissue is indicated by the dotted line. The Y-axis indicates the level of BMP activity; the X-axis indicates distance across the tissue. For this hypothetical tissue, BMP activity is highest in the middle (medial region); and lowest at the edges (lateral regions). Under the dotted line are rectangles that indicate the region of the tissue that is exposed to a specific level of BMP activity. The gradient is symmetrical, so cells on both sides are exposed to similar levels of BMP activity. *B.* Expression patterns for BMP target genes that respond to different levels of BMP activity. The black bar indicates the region of the tissue that expresses a gene induced only by very high levels of BMP activity. The striped bar and stippled bar indicate the regions that express genes induced by moderate and low levels of BMP activity, respectively. The open bars indicate the lateral regions that express a gene turned off in response to low levels of BMP activity. *C.* Cells that are exposed to different threshold levels of BMP activity adopt different fates. Cells exposed to very high BMP activity adopt the black fate. Cells exposed to moderate levels of BMP activity adopt the hatched fate. Cells exposed to low levels of BMP activity adopt the checkered fate. Cells that are not exposed to BMP activity adopt the white fate

(reviewed by Morisato and Anderson, 1995). Formation of all dorsal ectodermal tissues requires BMP signaling; thus, embryos that lack a protein required for BMP signal transduction develop only ventral ectodermal tissues. This was observed for embryos lacking all functional Mad and for embryos lacking all functional Medea (reviewed in Raftery and Sutherland, 1999).

3.2.1 In vivo *validation of signal transduction mechanism*

Initial studies of the *Mad* and *Medea* genes demonstrated that they regulate BMP target genes in every tissue tested (reviewed in Raftery and Sutherland, 1999). In cultured cell systems, Mad met the criteria for an R-Smad and Medea met the criteria for a co-Smad. *In vivo* evidence for these molecular functions required additional reagents.

Initial immunodetection studies suggested that endogenous Mad was predominantly cytoplasmic *in vivo*, even at sites where BMP signaling was known to be active (Newfeld et al., 1996). Until anti-phospho-Smad antibodies became available (Persson et al., 1998), *in vivo* nuclear accumulation of endogenous Mad or tagged transgenic Mad could only be detected with hyperstimulation of BMP signaling (Dobens et al., 2000; Newfeld et al., 1997). In contrast, anti-phospho-Smad antibody readily detects nuclear accumulation of endogenous Mad (Fig. 3; reviewed in Raftery and Sutherland, 2003). These observations suggest that only a small fraction of available Mad is phosphorylated and present in the nucleus under endogenous BMP signaling conditions. Each of the BMP ligands Dpp and Screw and each of the type I receptors Tkv and Sax are required for detectable phospho-Mad accumulation in blastoderm embryos.

For Medea, nuclear accumulation of endogenous protein can be detected under endogenous signaling conditions in blastoderm embryos (Fig. 3; Sutherland et al., 2003). It is likely that Medea shuttles through the nucleus under non-signaling conditions (Chapter 10), but this does not appear to give a strong signal for immunodetection. Detectable nuclear accumulation requires each of the ligands Dpp and Screw. The requirement for receptors has not been tested. Global expression of activated-Tkv is associated with nuclear Medea in all cells at this stage (Fig. 3). Importantly, nuclear accumulation of endogenous Medea requires Mad at this stage of development (Fig. 3). Thus, an R-Smad is necessary to achieve detectable nuclear accumulation of a co-Smad *in vivo*.

3.2.2 Nuclear Smads as the readout of BMP gradients

The Smad signal transduction mechanism led to the prediction that nuclear levels of Smads are the intracellular readout of TGF-β family morphogen gradients. Genetic manipulations in *Drosophila* fully support this model (reviewed by Raftery and Sutherland, 2003; Tabata, 2001). In *Drosophila* blastoderm embryos, levels and distribution of phospho-Mad correlate well with levels and distribution of tagged Dpp ligand (Shimmi et al., 2005; Wang and Ferguson, 2005), and distribution of nuclear Medea parallels the phospho-Mad patterns, with a slight time delay (Sutherland et al., 2003).

These studies all indicate that low levels of BMP activity are initially distributed in a shallow gradient over a broad dorsal region of mid-blastoderm embryos

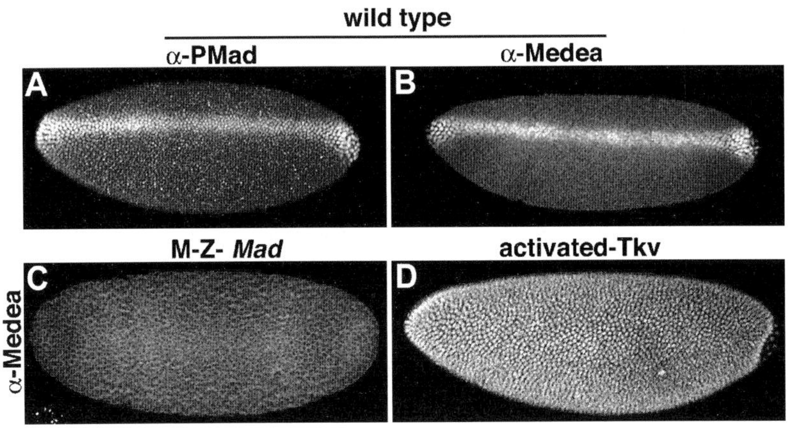

Figure 3. Phosphorylated Mad and nuclear Medea reveal a gradient of BMP activity at the dorsal midline of late blastoderm embryos. All images reprinted with permission from Sutherland et al., 2003. *A* and *B*. Anterior left, dorsal view. *C*. Anterior left, dorsal and ventral sides indistinguishable. *D*. Anterior left, side view with dorsal up. *A*. Wild-type embryo stained with antiserum that detects receptor-phosphorylated Mad (PS1, Persson et al., 1998). Strong accumulation of phosphorylated Mad is seen in a narrow band of cells that extends along the dorsal midline, wider at the termini of the embryo than in the central, segmented region. *B*. Strong accumulation of nuclear Medea occurs in a similar band of cells at the dorsal midline, detected by immunofluorescence with anti-Medea (Sutherland et al., 2003). *C*. Maternal null, zygotic null (M-Z-) *Mad* mutant embryo has cytoplasmic staining for Medea, but the strong band of nuclear Medea is absent. *D*. Global expression of constitutively activated Tkv is associated with strong nuclear accumulation of Medea throughout the embryo

(reviewed in Ashe, 2005; Raftery and Sutherland, 2003). By the end of blastoderm stages, high levels of BMP activity are concentrated at the dorsal midline, forming a steep gradient over a much narrower region (Fig. 3). The mechanisms for this redistribution of BMP activity are emerging from studies of extracellular BMP binding proteins and metalloproteases (reviewed by Ashe, 2005; Meinhardt and Roth, 2002), but are beyond the scope of this chapter. Signal transduction through Mad and Medea is required to produce the sharp peak of strong BMP activity in the second gradient, suggesting new modes of feedback to concentrate BMP activity at an appropriate location (Wang and Ferguson, 2005). The BMP target genes required for this feedback have not been identified.

3.2.3 *Gene expression responses to threshold levels of BMP activity*

The spatial gradient of BMP activity in blastoderm embryos was inferred from genetic and developmental analyses in the 1990's (reviewed by Podos and Ferguson, 1999). Patterns of expression for BMP target genes, including *hindsight, u-shaped, Race, zen, tailup, pannier*, and *dpp*, form a nested group (Ashe et al., 2000; Jazwinska et al., 1999). Expression domains for these genes all center at the dorsal midline, but extend away from the midline over different distances (hypothetical pattern in Fig. 2). Genes with broad expression domains were predicted to be highly sensitive to BMP signals, and thus expressed throughout the region exposed to BMP activity (Fig. 2B, stippled bar). Conversely, genes with narrow expression domains were predicted to be relatively insensitive to BMP signals, and thus only expressed in regions with the highest level of BMP activity (Fig. 2B, black bar). Studies of the BMP response elements from specific target genes uphold these predictions. In this section, we give an overview of the two modes of regulation that are important for induction of BMP target genes, direct transcriptional activation and derepression (Fig. 4), in Section 4, we describe the specific mechanisms for Smad regulation of gene expression.

The most broadly-expressed BMP target genes, such as *pannier* and *dpp*, are relatively insensitive to genetic reductions in BMP activity (reviewed by Raftery and Sutherland, 2003). Most *Drosophila* BMP target genes are repressed in the absence of a signal, a mechanism that has been called default repression (Barolo and Posakony, 2002). Genetic studies indicate that an important function of BMP signaling is to derepress these genes (reviewed by Affolter et al., 2001; Raftery and Sutherland, 2003). In blastoderm embryos, expression of the transcriptional repressor, Brinker, is regulated by the NF-κB pathway, independent of BMPs. Derepression of *pannier* and *dpp* appears to be very sensitive to BMP activity. Expression of *pannier* occurs in regions where phospho-Mad is only transiently detected; expression of *dpp* extends over a larger domain, including cells where phospho-Mad has not been detected. It will be important to characterize the BMP response elements that confer blastoderm expression of these genes, to understand the mechanisms for these most sensitive responses.

The *zen* gene is distinct from other blastoderm BMP target genes, because its pattern of expression changes with the changing pattern of the BMP activity gradient (reviewed by Raftery and Sutherland, 2003). The initial pattern of *zen* expression is broad; but it becomes refined to a narrow stripe at the dorsal midline. Analysis of the BMP response element from *zen* revealed that its expression is directly stimulated by BMP signaling through Mad (Rushlow et al., 2001). For *zen*, activation of expression is concomitant with derepression, because Smads compete with the repressor, Brinker, for overlapping binding sites. We will refer to genes that require both derepression and direct transcriptional activation by Smads as class II genes (Fig. 4). We return to this mechanism in section 4.2 and to the role of Zen as a co-factor for Smad transcriptional activation in section 4.4.

BMP target genes with narrow domains of expression are thought to respond only to high levels of BMP activity; their expression correlates with the location of the second, narrow gradient of BMP activity in blastoderm embryos (Fig. 3; reviewed in Raftery and Sutherland, 2003). At least one of these, *tailup*, is a class II gene. However, others, such as *Race, u-shaped* and *hindsight*, do not appear to require derepression; they appear to be regulated only by Smad transcriptional activation. We call these class III genes. The BMP response element from the class III gene *Race* has been analyzed in detail (Wharton et al., 2004; Xu et al., 2005); we discuss the mechanisms for its regulation in sections 4.2 and 4.3.

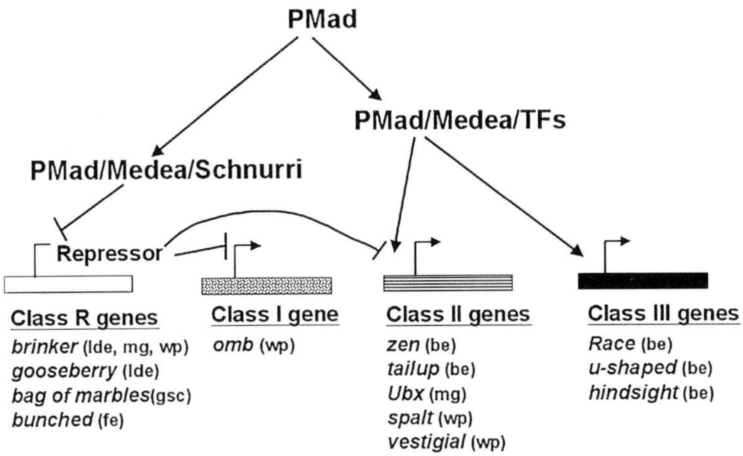

Figure 4. Modes of gene regulation in response to BMP signaling in *Drosophila*. Phosphorylated Mad (PMad) can participate in transcriptional repressor complexes with Medea and Schnurri, or in transcriptional activator complexes with other transcription factors. The Smad composition of transcriptional activator complexes has not been investigated biochemically in *Drosophila*; often the requirement for either Mad or Medea is tested genetically, and DNA binding is demonstrated for one or the other. The gene silencer complex includes a 2:1 ratio of Mad and Medea, and requires Schnurri (Gao et al., 2005; Pyrowolakis et al., 2004). Genes that are repressed by PMad/Medea/Schnurri complexes are here called class R genes (white bar). Note that *brinker* is not regulated by BMP signaling during cellular blastoderm; BMP-directed repression begins after gastrulation. Genes that are induced by BMP signaling can be indirectly or directly activated. The one characterized class I gene (stippled bar) is indirectly activated; it is released from repression when the PMad/Medea/Schnurri complex blocks expression of a repressor. Class II genes (striped bar) require both indirect and direct mechanisms for normal levels of expression; they are released from repression and induced by Smad binding to the BMP response element. Class III genes (black bar) are directly activated by Smads, with no evidence for indirect regulation by release from repression. Genes for each class are listed (Gao et al., 2005; Kawase et al., 2004; Kirkpatrick et al., 2001; Pyrowolakis et al., 2004; Saller et al., 2002; Song et al., 2004; also reviewed in Affolter et al., 2001; Raftery and Sutherland, 2003). Many of the genes are tissue specific targets: be, blastoderm embryo; fe, follicular epithelium; gsc, germline stem cells; lde, later dorsal ectoderm; mg, midgut; wp, wing primordium

3.3 Organizing Growth and Pattern in Wing Primordia

BMP signaling is required for both spatial patterning and growth of the wing primordium. A medial BMP activity gradient, centered at the anterior-posterior compartment boundary, was inferred from genetic and developmental analyses during the 1990's (reviewed by Tabata, 2001). The BMP activity gradient appears to be stable for hours, perhaps days, in this tissue. Patterns of expression for the BMP target genes *vestigial*, *omb*, *dad*, and *spalt* form a nested set, consistent with differential sensitivity to levels of BMP activity. Immunodetection of phospho-Mad indicates a shallow, asymmetrical gradient of BMP activity (Fig. 5). Detection of nuclear Medea has not been reported. The boundaries for expression patterns of the BMP-responsive genes do not correlate to obvious changes in the levels of phospho-Mad.

The shallow gradient of phospho-Mad suggests that BMP target genes in this tissue are sensitive to subtle variations in BMP activity. At present, only class I and class II target genes have been identified for the wing primordium. Thus, regulation by derepression is a central mechanism for BMP responses in this tissue. These studies have highlighted the importance of Smad transcriptional repression for cell fate determination.

3.3.1 Default repression and induction by derepression

It has been proposed that the potent developmental signals – BMP, Notch, EGF, Wnt, and Hh – employ a common logic for gene regulation (Barolo and Posakony, 2002). Default repression is one of three rules in this common logic: genes regulated by developmental signaling are repressed in the absence of a stimulus. For

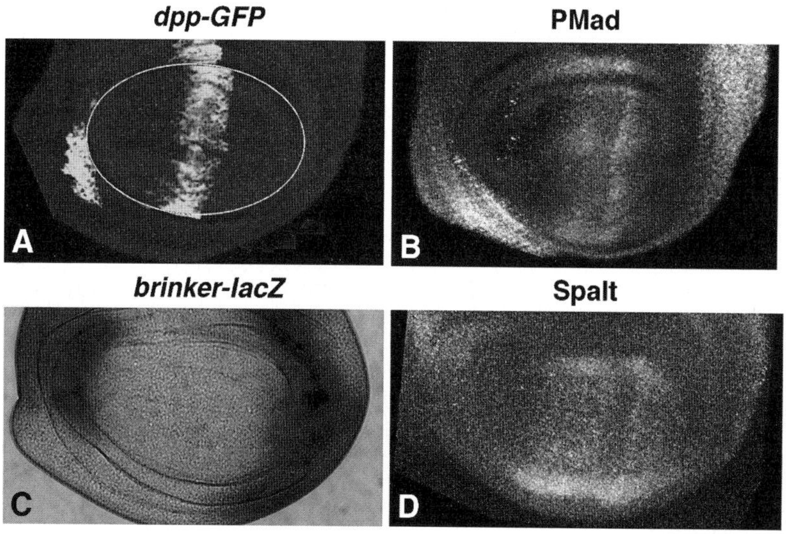

Figure 5. Phosphorylated Mad and BMP target genes in the wing primordium. In all panels, anterior is left and dorsal is up. Region of wing primordium is in the center of each image, and is outlined by a white oval in *A*. Images in *A,B,D* show confocal fluorescence image of immunostaining in white. Image in *C* shows bright field image of histochemical staining in black. *A*. Dpp is produced in a medial stripe of cells, along anterior side of the anterior-posterior compartment boundary. Here it is indicated by a binary reporter expression system, *dpp[blk]-Gal4; UAS-GFP*. *B*. Pattern of phosphorylated Mad (PMad) accumulation. Note that the levels of PMad form two peaks at the center of the wing primordium; these peaks are on either side of the *dpp* expressing cells. PMad levels are lower in the *dpp*-expressing cells, due to low levels of Tkv type I receptor (reviewed by Tabata, 2001). *C*. BMP signaling represses expression of *brinker*. Thus, *brinker* is only expressed in lateral regions of the wing primordium, shown here by β-galactosidase activity staining to detect the BMP-responsive reporter *brinker-lacZB14* (Müller et al., 2003). *D*. BMP signaling induces *spalt* expression, detected here with affinity-purified anti-Spalt (Domingos et al., 2004). Note that Spalt levels co-vary with PMad in the medial region of the wing primordium, so that expression is lower in *dpp*-expressing cells than in flanking cells. Spalt is a class II gene; its expression is repressed by Brinker in lateral regions and elevated by BMP signaling in the center of the primordium (Barrio and de Celis, 2004)

the Notch, Wnt, and Hh signal transduction pathways, the signal transducing transcription factors are the default repressors, Su(H), TCF, and Ci, respectively. Default repression by Smad proteins has not been observed, except in the nematode *C. elegans* (see Chapters 1 and 2).

The importance of default repression in BMP signaling was established from studies of the transcriptional repressor Brinker. Brinker is the repressor for class I and class II genes in embryos, wing primordia, and other tissues (reviwed by Affolter et al., 2001). Brinker DNA binding sites were identified from the minimal BMP response element for *omb*, a class I gene (Sivasankaran et al., 2000). BMP signaling is not necessary for *omb* expression in mutant cells that lack Brinker. Thus, BMP signaling regulates *omb* indirectly, by silencing *brinker* gene expression; the Smad silencer complex is described in section 4.1. Brinker repression of BMP target genes has a major role in defining the spatial patterns for their expression in the wing primordium.

In normal wing primordia, BMP signaling silences *brinker* expression in a broad, medial domain, so that Brinker only accumulates in cells at the lateral edges of the primordium (Fig. 5; reviewed by Tabata, 2001). The phospho-Mad gradient leads to inverse gradients of Brinker accumulation at each edge. In these lateral domains, Brinker is required to repress expression of the BMP target genes *omb, spalt, dad*, and *vestigial*. Genes that are more broadly expressed are thought to require higher levels of Brinker for full repression than genes that are expressed in narrow, medial domains. Consistent with this, increasing the number of Brinker binding sites in a reporter transgene gives a narrower expression pattern in the wing primordium (Müller et al., 2003). For class I genes, the sensitivity of BMP responses is thought to depend on the Brinker binding sites. For class II genes, the sensitivity of BMP responses is determined by the combination of Brinker and Smad binding sites, as we discuss in section 4.2.

3.3.2 BMP regulation of primordial wing cell growth and survival

The BMP R-Smad Mad and the BMP type I receptor Tkv are required in every cell of the wing primordium for normal rates of proliferation (Martin-Castellanos and Edgar, 2002; Moreno et al., 2002). This is the most sensitive response to BMP activity in this tissue. However, BMP regulation of proliferation does not appear to involve direct regulation of cell cycle progression (Martin-Castellanos and Edgar, 2002). The critical target genes have yet to be identified, although some evidence suggests that repression of Brinker is important for cell survival in the medial region of the wing primordium (Moreno et al., 2002).

The mechanisms that regulate proliferation and survival of cells in the wing primordium are complex, and appear to involve competition of neighboring cells for a survival factor (reviewed in Hipfner and Cohen, 2004; Johnston and Gallant 2002). Proliferation rates are influenced by the rate at which the cells increase in size, or the rate of cellular growth (reviewed by Jorgensen and Tyers, 2004). Cells that lack the BMP type I receptor Tkv grow slowly, and often undergo apoptosis in mosaic analysis experiments (Adachi-Yamada and O'Connor, 2004; Martin-Castellanos and Edgar, 2002). Slow growth is frequently associated with apoptosis in the wing primordium, and the presence of neighboring wild-type cells is a key factor in apoptosis of slowly growing mutant cells. It is thought that rapidly growing cells consume a limiting survival factor, and that slowly growing cells fail to compete for this factor. Moreno et al. (2002) proposed that Dpp is the survival factor for which cells compete. However, one group reports that wing primordium cells survive when they cannot receive a Dpp signal; such cells extrude from the epithelium to form a cyst that persists through development (Gibson and Perrimon, 2005). The mechanisms for growth regulation and cell competition are major outstanding questions in *Drosophila* wing development (reviewed in Hipfner and Cohen, 2004; Johnston and Gallant, 2002); ongoing research in this area will resolve the controversy over Dpp as a survival factor.

4. SMAD REGULATION OF GENE EXPRESSION

Studies of vertebrate gene regulation have identified numerous Smad interacting proteins that are required for general or tissue-specific gene regulation (Chapters 11, 14, 15). In parallel, studies of response elements from tissue-specific BMP target genes have been performed in *Drosophila* (reviewed in Affolter et al., 2001; Raftery and Sutherland, 1999). For this chapter, we will consider the mechanisms that generate different gene expression responses to different levels of BMP activity (Fig. 4). First, we will consider Smad silencing of gene expression, which is essential for indirect induction of class I BMP response genes. We then discuss three mechanisms that modulate sensitivity to Smad transcriptional activation: direct competition with a default repressor, Smad binding site affinity, and feed-forward regulation of tissue specific transcription co-factors.

4.1 Gene Silencing by the Mad/Medea/Schnurri Complex

The ability of Smads to act as transcriptional repressors has emerged as a central function for BMP-directed patterning of *Drosophila* tissues. The importance of Smad-mediated gene repression was first described for regulation of *brinker* expression in the wing primordium and the midgut, and in post-gastrulation patterning of dorsal ectoderm (reviewed by Affolter et al., 2001). Characterization of the BMP response element that confers repression of *brinker* identified the binding site for Mad/Medea/Schnurri complexes (Pyrowolakis et al., 2004). This binding site was called a transcriptional silencer, because it mediates BMP-responsive silencing of reporter expression from other promoters and transcriptional activators.

schnurri encodes a zinc finger transcription factor, and has mutant phenotypes similar to those of weak *dpp* alleles (reviewed in Affolter et al., 2001). Mammalian proteins that are most similar to Schnurri are sequence-specific DNA binding proteins, and Schnurri binds the same sites (Dai et al., 2000). Initial studies of the function of Schnurri indicated that it was necessary to induce many BMP target genes, but was not necessary for expression of target genes during blastoderm stages (reviewed by Affolter et al., 2001). The reason for this partial requirement was resolved when it was demonstrated that Schnurri is necessary for BMP-directed repression of *brinker*; BMP signaling does not regulate *brinker* expression in blastoderm embryos. Thus, Schnurri is necessary to derepress class I and class II genes in most tissues.

The minimal sequence for a Mad/Medea/Schnurri silencer element is a composite of a GC-rich Mad binding site separated by five nucleotides from a Medea GTCT binding site (Gao et al., 2005; Pyrowolakis et al., 2004). The silencer protein complex appears to include two molecules of Mad and one of Medea; all three MH1 domains contact the DNA (Gao et al., 2005). The two Smad binding sites must be precisely spaced to recruit the Mad/Medea/Schnurri silencer complex; if the spacing is changed, the element drives activation of gene expression. It is thought that Schnurri does not directly contact DNA at these silencer elements, because a specific sequence does not appear to be required for Schnurri binding.

The simple sequence composition of the BMP silencer element allowed rapid identification of additional genes that are silenced by the Mad/Medea/Schnurri complex: *bag of marbles*, *gooseberry*, and *bunched* (Pyrowolakis et al., 2004). Two of these genes were previously identified as important targets for BMP regulation of cell fate. BMP-directed repression of *bag of marbles* is critical to maintain germ line stem cells (Casanueva and Ferguson, 2004; Kawase et al., 2004; Song et al., 2004), and BMP-directed repression of *bunched* is critical for anterior patterning of the follicular epithelium (Dobens et al., 2000). *bunched* encodes a transcription factor in the TSC-22/DIP/BUN family. At present, it is not known whether Bunched collaborates with Brinker to generate tissue specific responses, or whether other transcriptional repressors can take the place of Brinker in default repression of BMP target genes.

4.2 Competition between Brinker Repression and Smad Activation

Class II genes are regulated both through repression by Brinker and through direct activation by Smads. This mechanism of gene regulation has been demonstrated for target genes from three tissues where BMP signaling has been examined in depth: *zen* from the embryonic dorsal ectoderm, *Ubx* from the embryonic midgut, and *spalt* and *vg* from the wing primordium. For *spalt*, Brinker repression and Smad activation are mediated by physically separate regions of the large *cis*-regulatory region (Barrio and de Celis, 2004). For *zen, Ubx*, and *vg*, the minimal BMP response element contains overlapping Brinker and Mad binding sites (Kirkpatrick et al., 2001; Rushlow et al., 2001; Saller and Bienz, 2001). For *zen* and for *Ubx*, *in vivo* experiments with reporter transgenes confirm that these overlapping binding sites mediate the competition between activation by Mad and repression by Brinker (Rushlow et al., 2001; Saller and Bienz, 2001). Brinker repression does not simply occur by obstruction of Smad binding sites; it requires CtBP and/or Groucho as co-repressors (Winter and Campbell, 2004).

Discussions of Brinker function have been framed by the concept that Brinker is a default repressor for BMP signaling (*e.g.* Pyrowolakis et al., 2004). However, it is noteworthy that Brinker function is intertwined with Wnt signaling in the *Drosophila* midgut (Saller et al., 2002). Wnt and BMP signaling interact extensively to pattern this tissue; Brinker antagonizes both BMP and Wnt signaling to repress *Ubx* gene expression. Brinker directly competes with Mad for an overlapping binding site (Saller and Bienz, 2001). When bound to this site, Brinker repression prevents Wnt induction of expression from a nearby TCF binding site (Saller et al., 2002). Surprisingly, expression of a Brinker co-factor, Teashirt, is induced by high levels of Wnt signaling in this tissue. It is not known whether Teashirt is an obligate, or a tissue-specific, co-factor for Brinker repression. BMP and Wnt signaling coordinately regulate patterning and growth of the wing primordium (reviewed by Curtiss et al., 2002), where Teashirt is expressed in an overlapping pattern with Brinker. The relationship between Brinker, Teashirt and Wnt signaling has not been investigated in this tissue.

4.3 Smad Binding Site Affinity and Sensitivity to BMP Levels

Class II and class III genes are directly induced by Smads. Even though all class III genes are directly induced by BMP signaling, different class III genes are induced by different levels of BMP activity within the same tissue (reviewed by Raftery and Sutherland, 2003). In blastoderm embryos, the gene *Race* shows a relatively insensitive response, for it is expressed only in a narrow band of cells that experience the strongest BMP activity. One study indicates that the BMP response element from *Race* becomes sensitive to lower levels of BMP activity when the Smad binding sites are changed to high affinity sequences (Wharton et al., 2004).

The BMP response element from *Race* has three Smad binding sites that can bind either Mad or Medea (Wharton et al., 2004). Two of the Smad binding sites in this element are low affinity, one is high affinity; all are important for expression of a reporter transgene *in vivo*. The low affinity sites include four base-pair consensus mammalian Smad binding sites, but the high affinity site is quite similar to the *Drosophila* consensus GC-rich Mad binding site, GCCGC[C/G]G[C/A]. When the low affinity sites are changed to the full GC-rich sequence of the high affinity Smad binding site, the reporter is expressed in dorsal-lateral cells, which only receive low levels of BMP activity. For other genes, the GC-rich sites and the four base-pair GTCT sites were defined experimentally as distinct Mad and Medea binding sites, respectively (Gao et al., 2005; Xu et al., 1998). The specific contributions of Mad versus Medea in direct transcriptional activation of *Drosophila* genes have not been examined systematically.

As discussed above, several class II genes are regulated by direct competition between binding of the Brinker repressor and Smad activators. In contrast, Brinker repression of the *spalt* gene involves a physically separate region from the sequences required for Medea activation (Barrio and de Celis, 2004). The separation of Medea activation and Brinker repression sites may give greater flexibility in the range of gene expression, so that Smad and Brinker binding sites can be optimized independently (see also Winter and Campbell, 2004). Expression of *spalt* appears to be sensitive both to Brinker levels, based on its narrow, medial expression pattern, and to variations in phospho-Mad levels, based on patterns of immunostaining (Fig. 5). In summary, the particular combination of binding sites for indirect derepression and direct transcriptional activation gives each target gene a distinct pattern of expression in response to a spatial gradient of BMP activity.

4.4 Feed-forward Regulation of Co-activator Genes

One general question for signal transduction by Smads is how to generate different responses in different tissues. This question has been addressed in multiple tissues in multiple organisms (see Chapters 14 and 15); it is clear that a wide variety of sequence-specific transcription factors act as co-factors to generate tissue-specific gene expression responses. Two general themes have emerged from studies of BMP target genes in *Drosophila*. One is that tissue-specific gene expression responses can emerge from combinatorial, but independent, action of multiple transcriptional regulators. For example, a 312 base pair regulatory element drives *eve* gene expression specifically in *Drosophila* cardiac progenitor cells (Halfon et al., 2000). Cell-type specific expression from this regulatory element requires independent DNA binding by three signal transducers, Mad (BMP signaling), dTCF (Wnt signaling), and Ptd (EGF signaling), and two tissue-specific transcription factors, Tinman and Twist.

The second theme is that cell fate determination by BMP signaling involves an amplification mechanism, in which BMP signaling increases and maintains expression of an essential, collaborating tissue-specific transcription factor. This BMP-dependent amplification mechanism has been identified during specification of three tissues: amnioserosa, dorsal mesoderm, and midgut endoderm. The *zen* gene is a class II BMP target in blastoderm and gastrula embryos; BMP signaling is necessary to maintain high levels of *zen* expression in the amnioserosa primordium (Rushlow et al., 2001). Zen is a homeodomain transcription factor required to specify the amnioserosa fate. *Race* is a class III BMP target gene that is expressed in the amnioserosa primordium. The 533 base-pair BMP response element from *Race* contains both Mad and Zen binding sites, and both are required for expression *in vivo* (Xu et al., 2005). The precise pattern of *Race* expression depends on the pattern of Zen expression, and also on the level of BMP activity, as we discussed above (Wharton et al., 2004).

For two other homeodomain transcription factors, amplification involves autoregulation. We mentioned above that *eve* expression in the cardiac mesoderm requires both Mad and Tinman, an Nkx homeodomain transcription factor. Strong expression of *tinman* in the dorsal mesoderm is maintained by BMP signaling, mediated by Mad and Medea binding sites within the 349 base-pair dorsal mesoderm regulatory element (Xu et al., 1998). Tinman is necessary to maintain dorsal mesoderm expression of its own gene through binding sites in this same regulatory element. Both Tinman and Smad binding sites are required *in vivo* to drive expression of *lacZ* from this regulatory element. Similar autoregulation is required for BMP-induced expression of the homeobox gene *labial* during midgut development (reviewed by

Raftery and Sutherland, 1999). This feedback amplification of co-factor expression provides a mechanism to sharply define the spatial domain of gene expression within an embryonic region that is exposed to a gradient of BMP activity. Furthermore, BMP-dependent maintenance of a tissue-specific co-factor gives the spatial pattern for sequential restriction of early cell fates, such as the restriction of general mesoderm to the dorsal mesoderm fates of cardiac and visceral mesoderm or the restriction of dorsal ectoderm to the amnioserosa fate.

5. SUMMARY AND PROSPECTS

The most widely recognized contribution of *Drosophila* developmental genetics has been gene discovery. We anticipate that *Drosophila* research on Activin-like signaling will provide an avenue for discovery of additional Smad modulators. In addition, *Drosophila* developmental genetics is an outstanding system for assembling regulatory pathways and networks; unexpected intersections with other molecular systems may be uncovered through genetic modifier screens ongoing in many laboratories.

The challenge for this next decade of research in Smad signaling is to move beyond proteins and genes to cellular behaviors and organ function. As we have illustrated in this chapter, current studies in *Drosophila* are taking our simple understanding of genetic pathways and regulatory networks into four dimensions, the three spatial dimensions of a developing tissue in combination with the fourth dimension of time. By putting each regulatory interaction into a spatial and temporal framework, *Drosophila* studies are leading to models for the regulatory systems that determine where and when a BMP signal will strengthen, how far it will travel, how it will be quenched, and how it interacts with other extracellular signals to establish tissue architecture and organization of cell types. Studies of Smad function will be central to understanding feedback/feedforward controls of BMP signaling (as indicated by results of Wang and Ferguson, 2005; Xu et al., 2005). Similarly, such studies will define the mechanisms that ensure different responses each time a TGF-β family signal is used during sequential steps, as multi-potent precursor cells become restricted to specific cell fates. Studies in *Drosophila* will continue to be the paradigm for understanding TGF-β signaling in the four dimensions of a living organism, because of the wealth of information available about the fly and the ease of *in vivo* analyses in this model system.

ACKNOWLEDGEMENTS

We thank Hilary Ashe, Marjorie Guitard, Allen Laughon, and Xiaodong Wu for comments and discussion, Gerard Campbell for *brk-lacZ*, Bertrand Mollereau for anti-Spalt, and Peter ten Dijke for PS1. We apologize for the numerous excellent primary research papers that have not been cited, due to space restrictions. Work on TGF-β signaling in the Raftery laboratory is supported by a grant from NIGMS.

REFERENCES

Adachi-Yamada, T., and O'Connor, M.B., 2004, Mechanisms for removal of developmentally abnormal cells: cell competition and morphogenetic apoptosis. *J Biochem (Tokyo)* 136: 13-17.

Affolter, M., Marty, T., Vigano, M.A., and Jazwinska, A., 2001, Nuclear interpretation of Dpp signaling in *Drosophila*. *EMBO J* 20: 3298-3305.

Ashe, H.L., 2005, BMP signalling: synergy and feedback create a step gradient. *Curr Biol* 15: R375-377.

Ashe, H.L., Mannervik, M., and Levine, M., 2000, Dpp signaling thresholds in the dorsal ectoderm of the *Drosophila* embryo. *Development* 127: 3305-3312.

Barolo, S., and Posakony, J., 2002, Three habits of highly effective signaling pathways: principles of transcriptional control by developmental cell signaling. *Genes Dev* 16: 1167-1181.

Barrio, R., and de Celis, J.F., 2004, Regulation of spalt expression in the *Drosophila* wing blade in response to the Decapentaplegic signaling pathway. *Proc Natl Acad Sci U S A* 101: 6021-6026.

Casanueva, M.O., and Ferguson, E.L., 2004, Germline stem cell number in the *Drosophila* ovary is regulated by redundant mechanisms that control Dpp signaling. *Development* 131: 1881-1890.

Curtiss, J., Halder, G., and Mlodzik, M., 2002, Selector and signalling molecules cooperate in organ patterning. *Nat Cell Biol* 4: E48-51.

Dai, H., Hogan, C., Gopalakrishnan, B., Torres-Vazquez, J., Nguyen, M., Park, S., Raftery, L.A., Warrior, R., and Arora, K., 2000, The zinc finger protein schnurri acts as a Smad partner in mediating the transcriptional response to decapentaplegic. *Dev Biol* 227: 373-387.

Dobens, L.L., Peterson, J., Treisman, J., and Raftery, L.A., 2000, *Drosophila bunched* integrates opposing DPP and EGF signals to set the operculum boundary. *Development* 127: 745-754.

Domingos, P.M., Brown, S., Barrio, R., Ratnakumar, K., Frankfort, B.J., Mardon, G., Steller, H., and Mollereau, B., 2004, Regulation of R7 and R8 differentiation by the spalt genes. *Dev Biol* 273: 121-133.

Entchev, E.V., Schwabedissen, A., and Gonzalez-Gaitan, M., 2000, Gradient formation of the TGF-β homolog Dpp. *Cell* 103: 981-991.

Gao, S., Steffen, J., and Laughon, A., 2005, Dpp-responsive silencers are bound by a trimeric Mad-Medea complex. *J Biol Chem* 280: 36158-36164.

Gibson, M.C., and Perrimon, N., 2005, Extrusion and death of DPP/BMP-compromised epithelial cells in the developing *Drosophila* wing. *Science* 307: 1785-1789.

Gonzalez-Gaitan, M., 2003, Signal dispersal and transduction through the endocytic pathway. *Nat Rev Mol Cell Biol* 4: 213-224.

Gurdon, J.B., and Bourillot, P.Y., 2001, Morphogen gradient interpretation. *Nature* 413: 797-803.

Halfon, M.S., Carmena, A., Gisselbrecht, S., Sackerson, C.M., Jimenez, F., Baylies, M.K., and Michelson, A.M., 2000, Ras pathway specificity is determined by the integration of multiple signal-activated and tissue-restricted transcription factors. *Cell* 103: 63-74.

Hipfner, D.R., and Cohen, S.M., 2004, Connecting proliferation and apoptosis in development and disease. *Nat Rev Mol Cell Biol* 5: 805-815.

Hogan, B.L.M., 1996, Bone morphogenetic proteins: multifunctional regulators of vertebrate development. *Genes Dev* 10: 1580-1594.

Jazwinska, A., Rushlow, C., and Roth, S., 1999, The role of *brinker* in mediating the graded response to Dpp in early *Drosophila* embryos. *Development* 126: 3323-3334.

Johnston, L.A., and Gallant, P., 2002, Control of growth and organ size in *Drosophila*. *Bioessays* 24: 54-64.

Jorgensen, P., and Tyers, M., 2004, How cells coordinate growth and division. *Curr Biol* 14: R1014-1027.

Kawase, E., Wong, M.D., Ding, B.C., and Xie, T., 2004, Gbb/Bmp signaling is essential for maintaining germline stem cells and for repressing *bam* transcription in the *Drosophila* testis. *Development* 131: 1365-1375.

Kirkpatrick, H., Johnson, K., and Laughon, A., 2001, Repression of dpp targets by binding of brinker to mad sites. *J Biol Chem* 276: 18216-18222.

Lee-Hoeflich, S.T., Zhao, X., Mehra, A., and Attisano, L., 2005, The *Drosophila* type II receptor, Wishful thinking, binds BMP and myoglianin to activate multiple TGFβ family signaling pathways. *FEBS Lett* 579: 4615-4621.

Marques, G., 2005, Morphogens and synaptogenesis in *Drosophila*. *J Neurobiol* 64: 417-434.

Martin-Castellanos, C., and Edgar, B.A., 2002, A characterization of the effects of Dpp signaling on cell growth and proliferation in the *Drosophila* wing. *Development* 129: 1003-1013.

Marty, T., Muller, B., Basler, K., and Affolter, M., 2000, Schnurri mediates dpp-dependent repression of *brinker* transcription. *Nat Cell Biol* 2: 745-749.

Meinhardt, H., and Roth, S., 2002, Developmental biology: sharp peaks from shallow sources. *Nature* 419: 261-262.

Mizutani, C.M., Nie, Q., Wan, F.Y., Zhang, Y.T., Vilmos, P., Sousa-Neves, R., Bier, E., Marsh, J.L., and Lander, A.D., 2005, Formation of the BMP activity gradient in the *Drosophila* embryo. *Dev Cell* 8: 915-924.

Moreno, E., Basler, K., and Morata, G., 2002, Cells compete for Decapentaplegic survival factor to prevent apoptosis in *Drosophila* wing development. *Nature* 416: 755-759.

Morisato, D., and Anderson, K.V., 1995, Signaling pathways that establish the dorsal-ventral pattern of the *Drosophila* embryo. *Annu Rev Genet* 29: 371-399.

Müller, B., Hartmann, B., Pyrowolakis, G., Affolter, M., and Basler, K., 2003, Conversion of an extracellular Dpp/BMP morphogen gradient into an inverse transcriptional gradient. *Cell* 113: 221-233.

Newfeld, S.J., Chartoff, E.H., Graff, J.M., Melton, D.A., and Gelbart, W.M., 1996, *Mothers against dpp* encodes a conserved cytoplasmic protein required in DPP/TGF-β responsive cells. *Development* 122: 2099-2108.

Newfeld, S.J., Mehra, A., Singer, M.A., Wrana, J.L., Attisano, L., and Gelbart, W.M., 1997, *Mothers against dpp* participates in a DPP/TGF-β responsive serine-threonine kinase signal transduction cascade. *Development* 124: 3167-3176.

Persson, U., Izumi, H., Souchelnytskyi, S., Itoh, S., Grimsby, S., Engström, U., Heldin, C.-H., Funa, K., and ten Dijke, P., 1998, The L45 loop in type I receptors for TGF-β family members is a critical determinant in specifying Smad isoform activation. *FEBS Lett* 434: 83-87.

Podos, S.D., and Ferguson, E.L., 1999, Morphogen gradients: new insights from DPP. *Trends Genet* 15: 396-402.

Pyrowolakis, G., Hartmann, B., Muller, B., Basler, K., and Affolter, M., 2004, A simple molecular complex mediates widespread BMP-induced repression during *Drosophila* development. *Dev Cell* 7: 229-240.

Raftery, L., and Sutherland, D., 1999, TGF-β family signal transduction in *Drosophila*: from *Mad* to Smads. *Dev Biol* 210: 251-268.

Raftery, L.A., and Sutherland, D.J., 2003, Gradients and thresholds: BMP response gradients unveiled in *Drosophila* embryos. *Trends Genet* 19: 701-708.

Rushlow, C., Colosimo, P.F., Lin, M., Xu, M., and Kirov, N., 2001, Transcriptional regulation of the *Drosophila* gene zen by competing Smad and Brinker inputs. *Genes Dev* 15: 340-351.

Saller, E., and Bienz, M., 2001, Direct competition between Brinker and *Drosophila* Mad in Dpp target gene transcription. *EMBO Rep* 2: 298-305.

Saller, E., Kelley, A., and Bienz, M., 2002, The transcriptional repressor Brinker antagonizes Wingless signaling. *Genes Dev* 16: 1828-1838.

Shimmi, O., Umulis, D., Othmer, H., and O'Connor, M.B., 2005, Facilitated transport of a Dpp/Scw heterodimer by Sog/Tsg leads to robust patterning of the *Drosophila* blastoderm embryo. *Cell* 120: 873-886.

Sivasankaran, R., Vigano, M.A., Muller, B., Affolter, M., and Basler, K., 2000, Direct transcriptional control of the dpp target omb by the DNA binding protein brinker. *EMBO J* 19: 6162-6172.

Song, X., Wong, M.D., Kawase, E., Xi, R., Ding, B.C., McCarthy, J.J., and Xie, T., 2004, Bmp signals from niche cells directly repress transcription of a differentiation-promoting gene, bag of marbles, in germline stem cells in the *Drosophila* ovary. *Development* 131: 1353-1364.

Sutherland, D.J., Li, M., Liu, X.Q., Stefancsik, R., and Raftery, L.A., 2003, Stepwise formation of a SMAD activity gradient during dorsal-ventral patterning of the *Drosophila* embryo. *Development* 130: 5705-5716.

Tabata, T., 2001, Genetics of morphogen gradients. *Nat Rev Genet* 2: 620-630.

Tabata, T., and Takei, Y., 2004, Morphogens, their identification and regulation. *Development* 131: 703-712.

Teleman, A.A., and Cohen, S.M., 2000, Dpp gradient formation in the *Drosophila* wing imaginal disc. *Cell* 103: 971-980.

Tsuneizumi, K., Nakayama, T., Kamoshida, Y., Kornberg, T., Christain, J., and Tabata, T., 1997, *Daughters against dpp* modulates *dpp* organizing activity in *Drosophila* wing development. *Nature* 389: 627-630.

Wang, Y.C., and Ferguson, E.L., 2005, Spatial bistability of Dpp-receptor interactions during *Drosophila* dorsal-ventral patterning. *Nature* 434: 229-234.

Wharton, S.J., Basu, S.P., and Ashe, H.L., 2004, Smad affinity can direct distinct readouts of the embryonic extracellular Dpp gradient in *Drosophila*. *Curr Biol* 14: 1550-1558.

Whitman, M., 1998, Smads and early developmental signaling by the TGFβ superfamily. *Genes Dev* 12: 2445-2462.

Winter, S.E, and Campbell, G., 2004, Repression of Dpp targets in the *Drosophila* wing by Brinker. *Development* 131: 6071-6081.

Xu, M., Kirov, N., and Rushlow, C., 2005, Peak levels of BMP in the *Drosophila* embryo control target genes by a feed-forward mechanism. *Development* 132: 1637-1647.

Xu, X., Yin, Z., Hudson, J., Ferguson, E., and Frasch, M., 1998, Smad proteins act in combination with synergistic and antagonistic regulators to target Dpp responses to the *Drosophila* mesoderm. *Genes Dev* 12: 2354-2370.

Yang, M., Nelson, D., Funakoshi, Y., and Padgett, R.W., 2004, Genome-wide microarray analysis of TGFβ signaling in the *Drosophila* brain. *BMC Dev Biol* 4: 14.

Zheng, X., Wang, J., Haerry, T.E., Wu, A.Y., Martin, J., O'Connor, M.B., Lee, C.H., and Lee, T., 2003, TGF-β signaling activates steroid hormone receptor expression during neuronal remodeling in the *Drosophila* brain. *Cell* 112: 303-315.

CHAPTER 4

DELINEATING THE TGF-β/SMAD-INDUCED CYTOSTATIC RESPONSE

FANG LIU[1,2,3]

[1] *Center for Advanced Biotechnology and Medicine*
[2] *Susan Lehman Cullman Laboratory for Cancer Research, Department of Chemical Biology, Ernest Mario School of Pharmacy, Rutgers, The State University of New Jersey*
[3] *Cancer Institute of New Jersey*
679 Hoes Lane, Piscataway, NJ 08854, USA

Abstract: TGF-β potently inhibits cell proliferation by causing cell cycle arrest at the G1 phase. It exerts the antiproliferative response through inhibition of G1 CDK activities. Smads play key roles in mediating the TGF-β growth-inhibitory response. Smads downregulate the expression of the protooncogene product c-Myc and inhibitor of differentiation Id (Id1, Id2, and Id3) and upregulate the expression of CDK inhibitors p15 and p21. The mechanisms for TGF-β/Smad antiproliferative response vary considerably from one cell type to another. The intensity of the TGF-β cytostatic effect is dependent on a combinatorial nature. The cytostatic effect of TGF-β/Smad is a crucial guardian for tissue homeostasis. Accordingly, TGF-β and Smads are tumor suppressors during early stages of tumorigenesis. Understanding of the TGF-β/Smad cytostatic program is important for better preventing and treating cancers

Keywords: antiproliferative response; cancer; CDK; cell cycle; G1 arrest; G1 cell cycle control; TGF-β

1. INTRODUCTION

In the canonical view of the cell cycle, periodic expression of distinct cyclins activates different cyclin-dependent kinases (CDKs) at specific phases of the cell cycle. During the G1 phase, cells sense the environmental cues and transmit them to the cell cycle machinery (Sherr and Roberts, 1999). Cell cycle progression from the G1 to S phase is controlled by G1 cyclins and their associated CDKs (Sherr and Roberts, 1999). Two classes of G1 cyclins are present: D-type cyclins (cyclins D1, D2, and D3) and E-type cyclins (cyclins E1 and E2). Cyclin D,

P. ten Dijke and C.-H. Heldin (eds.), Smad Signal Transduction, 75–91.

which senses extracellular mitogenic stimulation, binds to and activates CDK4 and the homologous CDK6 in early-mid G1 phase. CDK2 is activated by cyclin E at late G1 phase and by cyclin A at S phase. Activation of CDK2 or CDK4 also requires phosphorylation by CDK activating kinase (CAK) and removal of inhibitory phosphorylation on a threonine and a tyrosine by the dual specificity phosphatase CDC25A. CDK activities are constrained by two classes of CDK inhibitors (Sherr and Roberts, 1999). The first class includes four members of the INK4 family, p16, p15, p18, and p19, which specifically inhibit CDK4 and CDK6. The second class includes three members of the CIP/KIP family, p21, p27, and p57, which inhibit cyclin E-CDK2 and cyclin A-CDK2 activities. While p27 inhibits cyclin E-CDK2, it is an assembly factor for cyclin D-CDK4/6 complexes. Induction of cyclin D-CDK4/6 complexes enables p27 to be redistributed from cyclin E-CDK2 to cyclin D-CDK4/6, thus stabilizing cyclin D-CDK4/6 complexes as well as triggering cyclin E-CDK2 activity. p27 redistribution represents another mechanism for coupling the cell cycle machinery with growth factor stimulation, via cyclin D.

Both cyclin D-CDK4/6 and cyclin E-CDK2 target Rb for phosphorylation (Sherr and Roberts, 1999). The generally accepted view is that cyclin D-CDK4/6 initiates Rb phosphorylation in mid-G1 phase after which cyclin E-CDK2 becomes active and completes this process by phosphorylating Rb on additional sites. The prior phosphorylation by CDK4/6 is necessary for Rb phosphorylation by CDK2. Hypophosphorylated Rb binds to and inhibits the E2F transactivation domain. Rb phosphorylation releases the active E2F to allow the expression of genes that mediate the entry into S phase. Rb and the related p107 and p130 were the only substrates demonstrated for CDK4. Recent studies have shown that both CDK4 and CDK2 also phosphorylate Smad3 and inhibit its transcriptional activity and antiproliferative function (Matsuura et al., 2004), thus adding another level of control to the cell cycle machinery. The Rb pathway in cancer and the implications for Smad3 phosphorylation by CDK in tumorigenesis and TGF-β resistance is discussed in Sections 4 and 5.

2. TARGETS OF THE TGF-β/SMAD CYTOSTATIC PROGRAM

TGF-β potently inhibits epithelial, neuronal, and hematopoietic cell proliferation by causing cell cycle arrest at the G1 phase (Siegel and Massagué, 2003). It exerts the cytostatic effect often through inhibition of G1 CDK activity. This is achieved by multiple mechanisms, the most studied ones through transcriptional regulation by Smads. The important targets are discussed below. Smads downregulate the expression of c-Myc and Id (Id1, Id2, Id3) and upregulate the expression of p15 and p21, leading to cell cycle arrest at the G1 phase (Fig. 1). These events occur in several cell types examined. The TGF-β/Smad antiproliferative response displays cell type specificity. A Smad target can be induced to a much greater extent in one cell type than in another cell type. The extent of the TGF-β cytostatic effect is orchestrated on a combinatorial basis.

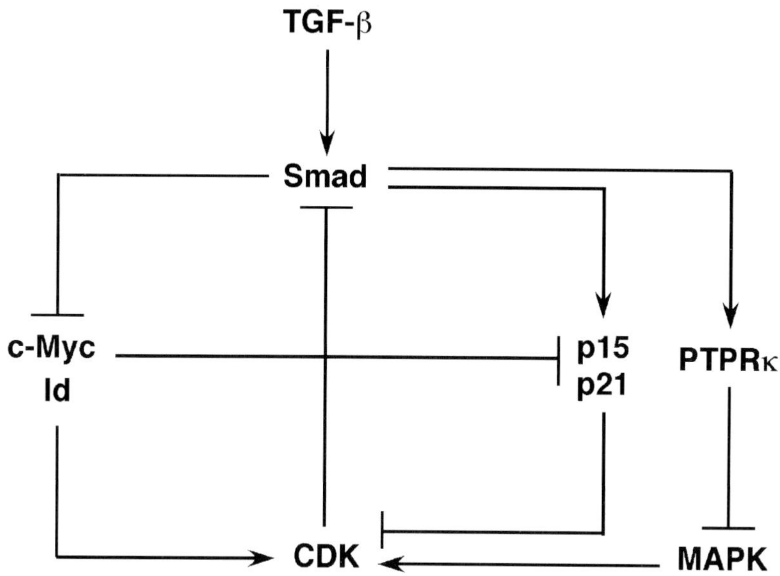

Figure 1. The TGF-β/Smad cytostatic program through transcriptional regulation. Smads are activated in response to TGF-β treatment. Smads downregulate the expression of c-Myc and Id, and upregulate the expression of p15 and p21. Downregulation of c-Myc is also necessary for Smads to activate the expression of p15 and p21 in cooperation with cellular factors. Smads also activate the expression of PTPRκ, which leads to the inhibition of MAPK activity. All these events result in the inhibition of G1 CDK activities, thus causing cell cycle arrest at the G1 phase. Smad3 is a physiological substrate of CDK. CDK phosphorylation of Smad3 inhibits its transcriptional activity and antiproliferative function

2.1 Downregulation of *c-Myc*

TGF-β downregulation of the expression of *c-Myc* was discovered over 15 years ago (Pietenpol et al., 1990). Downregulation of *c-Myc* is a key event in TGF-β-mediated cytostatic program in multiple cell types analyzed, and is defective in breast cancer cells (Chen et al., 2001). c-Myc promotes cell proliferation and cell growth. c-Myc can activate or repress transcription depending on the target gene. Both c-Myc mRNA and protein are short-lived. Their levels are rapidly reduced in response to TGF-β treatment.

TGF-β induces a repressor complex containing Smad3, Smad4, E2F4 (or E2F5) and the Rb family member p107 to bind to the TGF-β inhibitory element (TIE) of the *c-Myc* promoter and repress its transcription (Chen et al., 2002; Yagi et al., 2002). It was demonstrated that the Smad3 MH2 domain binds to E2F4 and p107, allowing the formation of the multimeric protein complex that is recruited to the TIE of the *c-Myc* promoter upon TGF-β treatment, and that E2F4 and p107 are functionally required for TGF-β-mediated repression of *c-Myc* (Chen et al., 2002). The *c-Myc* TIE is a composite element comprised of a Smad binding element and an E2F binding site (Chen et al., 2002; Frederick et al., 2004; Yagi et al., 2002). The sequence of the TIE is -92 5'-TTCTCAGAGGCTTGGCGGGAAAAG-AACGG-3' -63, which contains a consensus E2F binding site, 5'-GGCGGGAAA-3'. It was initially shown that Smad3 binds to the 5'-GGCT-3' sequence of the TIE (Chen et al., 2002), which fits the previous findings that the second position of the SBE box 5'-GTCT-3' can tolerate substitutions (see Chapter 11). A subsequent study with methylation interference footprinting as well as thorough mutagenesis, however, provided persuasive evidence that the 5'-GGCT-3' sequence is not essential for either Smad3 binding or TGF-β-mediated repression of the *c-Myc* reporter gene activity (Frederick et al., 2004). Instead, based on the gel mobility shift assay, DNA affinity precipitation as well as methylation interference analysis, it was concluded that Smad3 binds to a repressive Smad binding element that is maximally comprised of 5'-TTGGCGGGAA-3' in the TIE (Frederick et al., 2004). It is proposed that in the endogenous context, the Smad3 MH1 domain binds to the repressive Smad binding element and the E2F4 DNA binding domain simultaneously contacts the overlapping E2F site of the TIE in the *c-Myc* promoter. The *c-Myc* downregulation is fascinating at the Smad3 and E2F4 DNA binding level. A thorough understanding of this process may also shed light into mechanisms of TGF-β/Smad-mediated repression of other genes.

2.2 Upregulation of p15 and p21

c-Myc suppresses the expression of CDK inhibitors p15 and p21. Miz-1 (Myc-interacting zinc-finger 1) binds to the p15 initiator and the proximal promoter region of p21 and activates their transcription. By binding to Miz-1, c-Myc inhibits Miz1-mediated transcriptional activation of p15 and p21 (Seoane et al., 2001; Staller et al., 2001). c-Myc also inhibits TGF-β induction of p15 and p21 in the initial one-to-two hours of TGF-β treatment through binding to Smad2 and Smad3 and suppressing their function (Feng et al., 2002). Conditional overexpression of low levels of c-Myc in Mv1Lu mink lung epithelial cells inhibits p15 induction by TGF-β (Warner et al., 1999). Similarly, expression of c-Myc blocks p21 induction by TGF-β in HaCaT human keratinocytes (Claassen and Hann, 2000). Thus, downregulation of c-Myc is necessary for the subsequent induction of p15 and p21. After the c-Myc level is decreased, Smads cooperate with cellular factors, such as Sp1 and FoxO, to activate p15 and p21 expression (Feng et al., 2000; Moustakas and Kardassis 1998; Pardali et al., 2000; Seoane et al., 2001; Seoane et al., 2004). The induced p15 binds to CDK4/6 and prevents their interaction with cyclin D (Hannon and Beach, 1994; Siegel and Massagué, 2003). As a result, the CDK inhibitor p27 is released from the cyclin D-CDK4 complex, and redistributes to bind to the cyclin E-CDK2 complex to inhibit its activity. p21 also inhibits cyclin E-CDK2 activity. The coordinated inhibition of CDK4/6 and CDK2 activities leads to cell cycle arrest induced by TGF-β (Siegel and Massagué, 2003).

FoxO factors belong to the Forkhead transcription factor family. FoxO includes FoxO1, FoxO3, and FoxO4. In response to TGF-β, FoxO interacts with Smad3 and Smad4 but not Smad2 (Seoane et al., 2004). FoxO binds to a consensus Forkhead binding element present immediately upstream of the Smad binding elements in the distal part of the p21 promoter and cooperates with Smads to activate p21 expression (Seoane et al., 2004). In addition to p21, ongoing studies by transcriptional profiling of FoxO-defective cells have identified a set of genes, including p15, whose induction by TGF-β depends on FoxO function (J. Massagué, personal communication). It will be interesting to determine the similar as well as distinct properties of the FoxO-Smad complexes in regulation of the various targets.

FoxO is negatively regulated by the PI3K/Akt pathway, which is activated by growth and survival factors such as insulin-like growth factor-1. Akt phosphorylates FoxO. The phosphorylation can force FoxO out of the nucleus (Brunet et al., 1999). Exclusion of FoxO from the nucleus, thus unable to activate p21 and p15, may contribute to the evasion of TGF-β cytostatic effect in certain cancer cells.

FoxO is also under the negative control of FoxG1, a transcriptional repressor of the Forkhead transcription factor family. FoxG1 protects neuroepithelial progenitor cells from cytostatic and differentiative signals. FoxG1 directly targets FoxO. FoxG1 binds to FoxO at basal state and associates with FoxO-Smad3-Smad4 in the presence of TGF-β, inhibiting the FoxO-Smad complex to activate the p21 expression, thus protecting the telencephalic neuroepithelium from the cytostatic effect of TGF-β (Seoane et al., 2004).

Thus, FoxO integrates the three signals TGF-β, PI3K/Akt, and FoxG1 to tightly regulate the expression of p21 and p15. The Smad/FoxO/Akt/FoxG1 network contributes to the TGF-β resistance during telencephalic neuroepithelium development as well as in glioblastoma progression (Seoane et al., 2004).

While FoxO plays an important role in p21 induction in keratinocytes, neuroepithelial cells, and possibly some other cell types, FoxO does not appear to play a significant role in p21 induction in stomach epithelial cells (Chi et al., 2005). Instead, Runx3, a transcription factor of the Runx family, is essential for p21 induction in stomach epithelial cells (Chi et al., 2005). Runx3 is a tumor suppressor for gastric cancer. Gastric mucosa from Runx3 null mouse exhibits hyperplasia due to increased proliferation and suppressed apoptosis in epithelial cells, and the cells have diminished responses to the growth inhibitory and apoptosis-inducing effects of TGF-β. Approximately 50% gastric cancers lack the expression of Runx3 due to hemizygous deletion and hypermethylation of the promoter region. The p21 promoter contains five Runt-binding sites. Mutational analysis suggests that four or all five sites are important for Runx3-mediated p21 expression. Runx3 interacts with Smad2, Smad3 and Smad4, and synergize with Smads for TGF-β-dependent induction of p21, accompanied by inhibition of cell proliferation. In contrast, Runx3 (R122C), a mutation identified from a human gastric patient, disrupts the ability of Runx3 to activate p21 expression. In addition, Runx3 and p21 are colocalized in human and mouse gastric epithelium (Chi et al., 2005). These observations provide strong evidence that Runx3 is a gastric-specific partner of Smads for p21 expression. Smad3 has also a tumor suppressive role in human gastric cancer (Han et al., 2004). Smad3 null mice display squamous hyperplasia in the stomach (C. Deng, personal communication). In addition, Smad4 heterozygous mice develop gastric polyposis and cancer (Xu et al., 2000). Thus, the TGF-β cytostatic effects likely play important tumor suppressive roles for gastric cancers.

2.3 Upregulation of p57

TGF-β exerts a potent growth-inhibitory response in hematopoietic cells. Induction of p15 and p21 is important for the cytostatic effect of TGF-β in epithelial cells and keratinocytes but does not occur in hematopoietic cells. Recent studies have shown that the CDK inhibitor p57 is the only CDK inhibitor induced by TGF-β in hematopoietic cells (Scandura et al., 2004). Both the p57 mRNA and proteins are induced. The induction requires transcription, and is dependent on a conserved region in the proximal part of the p57 promoter. Upregulation of p57 is essential for TGF-β-mediated cell cycle arrest in hematopoietic cells (Scandura et al., 2004). p57 is a putative tumor suppressor. Its expression is often silenced by promoter hypermethylation in hematologic malignancies. Future studies are necessary to determine whether Smads activate the expression of p57.

2.4 Downregulation of Id (Id1, Id2 and Id3)

Id proteins inhibit cell cycle exit and differentiation. Genome-wide transcriptional profiling of three human epithelial cells lines, HaCaT keratinocytes, HPL1 lung epithelial cells, and MCF10A mammary epithelial cells, was employed to identify common targets for TGF-β cytostatic response in these three cell lines. Cells were treated with TGF-β for 3 hours in this study. Only six genes were identified as direct cell cycle regulators. The list includes p15 and p21 in HaCaT and HPL1 cells and c-Myc, Id1, Id2 and Id3 in all three cell lines (Kang et al., 2003). TGF-β downregulation of Id was also identified from two other independent microarray screens (Kowanetz et al., 2004; Untergasser et al., 2005), suggesting that Id is a common target of TGF-β in several cell types.

For downregulation of Id1, a Smad3-containing transcriptional complex activates ATF3, a transcriptional repressor, which then together with Smad3 and Smad4 represses the transcription of the Id1 promoter (Kang et al., 2003). Among the Id proteins, downregulation of Id2 is most important, as Id2, but not Id1 or Id3, can bind to the Rb family members and inhibit their function. In addition, the growth-inhibitory activities of p16 and p21 can be effectively antagonized by high levels of Id2 but not by Id1 or Id3 (see references in Lasorella et al., 2000). Id2 is a transcriptional target of N-Myc and c-Myc, and it has been shown that c-Myc in conjunction with Max can bind to the E-box in the Id2 promoter to stimulate its expression (Lasorella et al., 2000). TGF-β downregulation of Id2 is a secondary effect. TGF-β downregulation of c-Myc, which is an activator of the Id2 expression, leads to the reduced Id2 expression. In addition, TGF-β also increases Mad expression, which appears to be an indirect effect of Smads (Siegel et al., 2003). Mad and Max form a repressor complex, which binds to and inhibits Id2 promoter. The increased level of Mad, and thus the enhanced repression of the Id2 promoter, also contributes to the reduced levels of Id2 in the presence of TGF-β.

Although BMP induces even higher level of p21 expression than TGF-β, BMP exerts only weak growth-inhibitory effect on epithelial cells (Pardali et al., 2005). This is due to induction of Id2 expression by BMP (Pardali et al., 2005). This finding supports the notion that downregulation of Id2 is an important event for TGF-β-mediated growth inhibition.

2.5 Downregulation of CDC25A

CDC25A promotes cell cycle progression from G1 to S phase by dephosphorylating and activating CDK2 and CDK4/6. Downregulation of CDC25A is also important for the TGF-β growth-inhibitory effect. Different cell types have evolved distinct mechanisms to downregulate CDC25A expression or inhibit CDC25A activity. Three mechanisms are described (Fig. 2). The latest one involves Skp1-cullin-β-TrCP ($SCF^{\beta\text{-}TrCP}$)-mediated ubiquitination and degradation of CDC25A. TGF-β-induces ubiquitination and degradation of CDC25A, and Smad3 plays an important role in this process (Ray et al., 2005). The stability of CDC25A is decreased in Smad3-overexpressing cells and is increased in Smad3 null cells or Smad3 siRNA-treated cells. Smad3 appears to facilitate CDC25A and $SCF^{\beta\text{-}TrCP}$ interaction

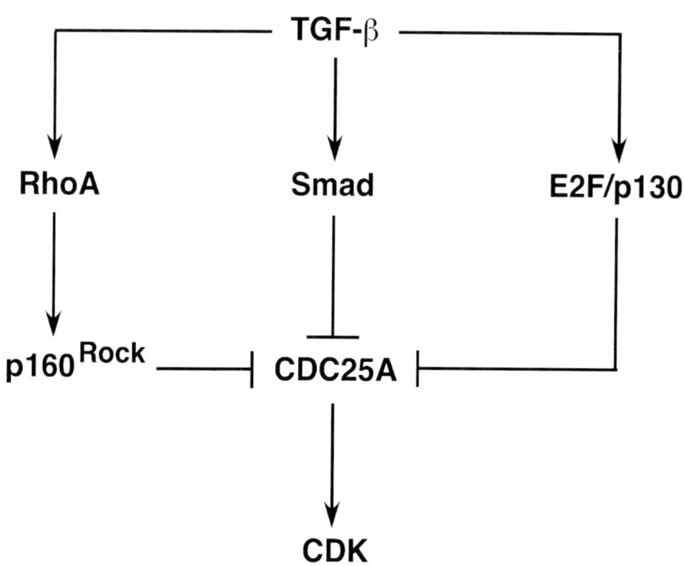

Figure 2. TGF-β downregulates CDC25A through several distinct mechanisms. TGF-β can downregulate CDC25A through ubiquitin-dependent degradation. Smad3 plays an important role in this process. TGF-β activates RhoA, which in turn, activates $p160^{Rock}$. This leads to $p160^{Rock}$ phosphorylating CDC25A and inhibiting its enzymatic activity. TGF-β can also downregulate CDC25A through the E2F/p130 repressor complex that recruits histone deacetylase

and subsequent ubiquitination and degradation (Ray et al., 2005). Another mechanism is through enzymatic inhibition of CDC25A phosphatase activity (Bhowmick et al., 2003). TGF-β signaling activates RhoA and $p160^{ROCK}$ in epithelial cells. $p160^{ROCK}$ then translocates to the nucleus, phosphorylates and inhibits the activity of CDC25A. The third mechanism, which was the first described, involves transcriptional repression of the CDC25A promoter through binding of the E2F-p130 repressor that recruits histone deacetylase (Iavarone and Massagué, 1999). This mechanism was identified for CDC25A downregulation in HaCaT keratinocytes. CDC25A downregulation in keratinocytes is a delayed event, subsequent to p15 and p21 induction. In mammary epithelial cells, however, TGF-β rapidly downregulates CDC25A mRNA levels. Since the E2F site-containing sequence in the CDC25A promoter 5'-TTTGGCGCCAA-3' (-62 to -52) is similar to the repressive Smad binding element/E2F site in the TIE of the *c-Myc* promoter 5'-TTTCCCGCCAA-3' (-71 to -81) (Frederick et al., 2004), it will be interesting to determine whether Smad3 is recruited to this site for the rapid downregulation of CDC25A in mammary epithelial cells.

2.6 Downregulation of Cyclin A

Both TGF-β and Activin downregulate the expression of cyclin A (Burdette et al., 2005; Feng et al., 1995; and references therein). Cyclin A downregulation by TGF-β occurs in multiple cell lines analyzed, and cyclin A was also identified as a TGF-β repressive target gene from a genome-wide transcriptional profiling of normal murine mammary gland epithelial cells (Xie et al., 2003). While regulation of p15, p21, c-Myc, and Id directly correlates with TGF-β-induced cell cycle arrest, downregulation of cyclin A may also contribute to the TGF-β cytostatic program. Cyclin A-CDK2 and cyclin A-CDC2 phosphorylate and inhibit Rb (Sherr and Roberts, 1999), and cyclin A-CDK2 also phosphorylates Smad3 (Matsuura et al., 2004). Overexpression of Smad3 causes downregulation of cyclin A (Burdette et al., 2005; our unpublished results). Thus, Smad3 may play an important role in cyclin A downregulation. Cyclin A is also a target gene of E2F. Its synthesis is necessary for S phase entry. Since the *c-Myc* downregulation occurs through the TIE that is a composite site for Smad3 and E2F, it will be interesting to determine whether downregulation of cyclin A occurs through a similar DNA sequence.

2.7 Upregulation of Protein Tyrosine Phosphatase Receptor Type Kappa (PTPRκ)

TGF-β induces the expression of PTPRκ in HaCaT keratinocytes and MCF10A mammary epithelial cells (Wang et al., 2005; Yang et al., 1996). The mRNA induction is resistant to cycloheximide treatment (Yang et al., 1996). Moreover, Smads bind to the promoter region of PTPRκ in chromatin immunoprecipitation assay. In fact, PTPRκ was identified as a Smad target gene in a systematic search by chromatin immunoprecipitation (Wang et al., 2005). TGF-β induced G1 arrest is abrogated in MCF10A cells that are treated with PTPRκ siRNA (Wang et al., 2005). PTPRκ interacts with EGF receptor and HER2 at endogenous protein levels, thus inhibiting the basal and ErbB ligand-induced receptor phosphorylation and MAP kinase activation. This can lead to the reduced levels of cyclin D and cyclin E (Wang et al., 2005), thus lower levels of the CDK activity (Fig. 1). Interestingly, PTPRκ is also necessary for TGF-β to promote cell migration in MCF10A cells that overexpress HER2 (Wang et al., 2005). Thus, it appears that PTPRκ can mediate both the tumor suppressor and tumor promoter functions of TGF-β/Smad.

2.8 Smad-independent Cytostatic Response

TGF-β can directly inhibit the activity of the CDK2 activating kinase (CDK2 CAK) in human HepG2 hepatocellular carcinoma cells and cause cell cycle arrest at the G1 phase (Nagahara et al., 1999). p15 gene is deleted in HepG2 cells. Other cell cycle regulators that were examined, including p21, p57 and CDC25A, were unchanged in the TGF-β treated HepG2 cells, suggesting that the G1 arrest may occur independent of Smads (Nagahara et al., 1999). TGF-β also down-regulates the CDK4 protein level in certain responsive cells, such as the Mv1Lu lung epithelial cells, a late event in the TGF-β response (Ewen et al., 1993). This downregulation occurs through inhibition of CDK4 translation. The CDK4 mRNA 5' untranslated region mediates this inhibition (Ewen et al., 1995).

In addition, TGF-β inhibits p70 S6 kinase via protein phosphatase 2A to induce G1 arrest in EpH4 mammary epithelial cells (Petritsch et al., 2000). Interestingly, TGF-β also induces growth arrest through Smads in this cell line. The two pathways are parallel, either one sufficient to induce cell cycle arrest. Disruption of both pathways, Smad-mediated transcriptional activation of CDK inhibitors and the inhibition of S6 kinase, are necessary for these cells to escape the TGF-β cytostatic response. It remains to be determined how widespread the inhibition of S6 kinase pathway operates among TGF-β-responsive cells.

3. ROLES OF SMAD2, SMAD3, AND SMAD4 IN TGF-β CYTOSTATIC PROGRAM

Smad3 plays an important role in the TGF-β cytostatic program, as indicated by its critical role in the downregulation of c-Myc, Id, CDC25A, cyclin A and upregulation of p15 and p21. Other evidence also supports this view. For instance, a variety of primary cells from Smad3 null mice, such as fibroblasts, keratinocytes, astrocytes, and T cells are largely resistant to the TGF-β cytostatic effect (Datto et al., 1999; Matsuura et al., 2004; Rich et al., 1999; Yang et al., 1999). Hyperproliferation is a component in the Smad3 null mice that develop metastatic colon cancer (Zhu et al., 1998). Knock-down of endogenous Smad3 by siRNA leads to diminished TGF-β cytostatic effect in a variety of TGF-β responsive cells analyzed (Kim et al., 2005). Interestingly, depletion of endogenous Smad2 by siRNA markedly enhanced the TGF-β cytostatic effect, accompanied by increased levels of p15 and p21 and reduced levels of c-Myc. Moreover, a single event of depletion of Smad2 by siRNA can restore TGF-β growth-inhibitory effect in cells that are resistant to TGF-β. These observations have led to the suggestion that the endogenous ratio of Smad3 and Smad2 is a determinant for the TGF-β cytostatic program (Kim et al., 2005). Genome-wide microarray analysis using Smad2 deficient cells also showed that Smad2 often negatively regulates TGF-β-responsive genes (see Chapter 17). On the other hand, an early study showed that Smad2 is necessary for p15 induction by TGF-β (Feng et al., 2000). In addition, overexpression of Smad2 in Mv1Lu cells inhibits anchorage-independent colony formation in soft agar, although this assay is complicated by the involvement of the extracellular matrix issue. Importantly, overexpression of Smad2 in Mv1Lu cells also suppresses cell proliferation and angiogenesis when injected into SCID mice (Sjöblom et al., 2004). The role of Smad2 in TGF-β cytostatic response remains to be further characterized. This is particularly important in light of the fact that Smad2 is a tumor suppressor. Smad4 is essential for the TGF-β cytostatic effect. For example, knock-down of Smad4 abolishes the growth arrest induced by TGF-β in HaCaT cells (Levy et al., 2005). Future studies are necessary to narrow down the targets of Smad4 that are necessary for the TGF-β antiproliferative response. The importance of each of the three Smads can vary in a cell-type dependent manner. For example, based on a study using an improved antisense technology, Smad3 plays a major role in mediating the TGF-β growth-inhibitory response in HaCaT cells, whereas Smad2, Smad3, and Smad4 contribute approximately equally in MCF10A cells (Kretschmer et al., 2003). This adds another level of complexity in the TGF-β/Smad cytostatic program.

4. SMADS ARE TUMOR SUPPRESSORS

The cytostatic effect of TGF-β enables it to be a potent tumor suppressor during the early stages of tumorigenesis. Accordingly, as described in Chapter 20, TGF-β receptors and Smads are tumor suppressors. Smad4 is a tumor suppressor in pancreatic and colon cancers, and is also mutated in a variety of other types of cancers at a substantially lower frequency. Smad2 is a tumor suppressor in colon cancer and small cell lung cancer. Smad3 null mice can develop metastatic colon cancer, which appears to be dependent on the genetic background (Zhu et al., 1998). Smad3 is a tumor suppressor for T cell acute lymphoblastic leukemia (ALL). Although Smad3 mRNA is expressed at normal levels, Smad3 protein is lacking in T cell ALL (Wolfraim et al., 2004). Smad3 is also a tumor suppressor in gastric cancer (Han et al., 2004). Smad3 protein is undetectable in a significant proportion of gastric cancer patients and gastric cancer cell lines and the Smad3 mRNA is simultaneously lost (Han et al., 2004). Introduction of Smad3 into Smad3 deficient gastric cancer cells restores TGF-β-dependent p15 and p21 induction and the cytostatic effect. Moreover, these Smad3 expressing cells have dramatically reduced tumorigenicity when analyzed in the nude mice injection assay. In addition, Smad3 levels decrease during carcinogenesis in some tissues. For example, nuclear Smad3 level is reduced in high grade breast cancers (Jeruss et al., 2003). During skin carcinogenesis, Smad3 level is severely lost (He et al., 2001). Unlike a typical tumor suppressor, which displays loss of heterozygocity, Smad3 seems to be often inactivated or inhibited by epigenetic-like mechanisms, such as by loss or reduction of its expression.

The Rb pathway, in which p16 inhibits CDK4/6 and prevents them from phosphorylating Rb, is disrupted in almost all human cancers (Ortega et al., 2003; Sherr, 1996). Frequent inactivation of Rb is limited to only a subset of human cancers. In contrast, inactivation of p16 or overexpression of cyclin D1 occurs frequently in human cancers, thus resulting in high levels of CDK activities (Ortega et al., 2003; Sherr, 1996). Cyclin D1 is amplified, translocated, or overexpressed in many cancers. Cyclin D1 has both CDK-dependent and CDK-independent activities, both of which contribute to tumorigenesis. Cyclin D2 and cyclin D3 are also amplified or overexpressed in several types of cancers. p16 is inactivated through deletion, mutation or promoter methylation in a wide variety of cancers. The frequency of p16 inactivation in human cancer seems to be second only to that of p53. CDK4 is amplified and overexpressed in a variety of cancer or cancer cell lines. CDK6 is amplified or translocated in certain types of cancers. Point mutations in CDK4 or CDK6, such as $CDK4^{R24C}$, disrupt the ability of p16 to bind to and inhibit CDK4 or CDK6, thus generating a dominant oncogene. When the $CDK4^{R24C}$ is introduced into mice by knock-in, it is highly oncogenic, resulting in a wide spectrum of tumors. In addition, overexpression of cyclin E or downregulation of p27 also occurs in some cancers. Since both CDK4 and CDK2 phosphorylate Smad3 and inhibit its antiproliferative function (Matsuura et al., 2004), extensive phosphorylation of Smad3 by CDK may be an important mechanism to diminish Smad3 tumor suppressive function in tumorigenesis.

5. TGF-β RESISTANCE IN CANCER

The vast majority of human cancers are resistant to the cytostatic effect of TGF-β. Most cancer cells retain normal TGF-β receptors and Smad proteins. The molecular basis for the broad loss of TGF-β sensitivity is not clear. Defective c-Myc downregulation and the Smad/FoxO/Akt/FoxG1 network contribute to the TGF-β resistance in some cancer cells as described in Sections 2.1 and 2.2. The endogenous ratio of Smad3 over Smad2 has recently been suggested as a determinant for TGF-β resistance (Kim et al., 2005). As described above, tumor cells often contain high levels of CDK activity. Smad3 is a very good substrate for both CDK4 and CDK2. Mutation of the CDK phosphorylation sites increases the ability of Smad3 to downregulate c-Myc and to upregulate p15 (Matsuura et al., 2004). Thus, inhibition of Smad3 transcriptional activity and antiproliferative function by CDK phosphorylation may contribute to tumorigenesis and TGF-β resistance in cancers. Two related findings support this hypothesins. First, it was shown that overexpression of CDK4 in Mv1Lu mink lung epithelial cells through generation of stable clones resulted in TGF-β resistance in these cells (Ewen et al., 1993). Overexpression of CDK4 led to increased levels of the cyclin D-CDK4 complex, accompanied with higher levels of the CDK4 kinase activity (Ewen et al., 1993). Second, Xuedong Liu and his coworkers at the University of Colorado, Boulder, recently searched for proteins that when overexpressed in Mv1Lu cells, render the cells resistant to the TGF-β growth-inhibitory effect. They introduced a retroviral cDNA library from mouse NIH 3T3 cells into the Mv1Lu cells, and selected for cells that are resistant to the TGF-β growth-inhibitory effect. One of the cell lines obtained encodes the full-length mouse CDK4. When this CDK4 cDNA was reintroduced into the parental Mv1Lu cells, it caused TGF-β resistance (X. Liu, personal communication). While it remains to be elucidated whether CDK phosphorylation and inhibition of Smad3 activity is a leading cause for the TGF-β resistance in these CDK4 overexpressed cells, these findings highlight the importance of CDK4 in TGF-β resistance.

6. NEW CHALLENGES TO THE CELL CYCLE AND TGF-β/SMAD CYTOSTATIC PROGRAM

Our current understanding of the TGF-β/Smad cytostatic program is based on the classic view of the cell cycle, in which cyclin D-CDK4/6 and cyclin E-CDK2 are essential for cell cycle progression from the G1 to S phase. The TGF-β cytostatic program, through Smad-dependent or Smad-independent mechanisms, directly or indirectly leads to the inhibition of both CDK4/6 and CDK2, thus causing cell cycle arrest at the G1 phase. Recent knock-out studies in mice, however, challenge the canonical view of the cell cycle. All three D-type cyclins, CDK4 and CDK6, the two E-type cyclins, or CDK2 have been deleted (Kozar et al., 2004; Malumbres et al., 2004; Ortega et al., 2003; Parisi et al., 2003). Mice with disruption of all three D-type cyclins or CDK4 and CDK6 survived most of the embryonic development. They die at ~E16 due to a major failure in hematopoiesis. Most organogenesis and tissue development seem unaffected (Kozar et al., 2004; Malumbres et al., 2004). Mice with cyclin E1 and cyclin E2 deletion are embryonic lethal at ~E11.5 due to the loss of placenta tissue. Interestingly, replacement of mutant placental cells with wild-type cells rescued some cyclins E1 and E2 double mutant embryos to birth, indicating that cyclin E is dispensable for embryonic cell cycles (Parisi et al., 2003). CDK2 knock-out mice are viable, except that both male and female are sterile due to meiotic defects (Ortega et al., 2003). It is surprising that most of the embryonic development can occur in the absence of these classically viewed master regulators of the cell cycle. It is not clear, however, to what extent compensatory mechanisms may have contributed to these phenotypes. Two other recent studies also raise questions about CDK2. One study showed that inhibition of CDK2 by p27, dominant negative CDK2, antisense or RNAi did not prevent cancer cells from proliferating, in contrast to the G1 arrest by inhibiting CDK4 (Tetsu et al., 2003). CDK4 was found to be able to phosphorylate Rb even at the CDK2 preferred sites. It was postulated that the high levels of CDK4 and E2F activities in cancer cells may compensate for the requirement of CDK2 (Tetsu et al., 2003). Another study showed that p27 and p21 can still inhibit the cell cycle and tumor progression even in the absence of CDK2 (Martin et al., 2005). While this raises a possibility that CDK2 might not mediate the inhibitory effects of p27 and p21, it is more likely that CDK2 null cells possess or have developed compensatory mechanisms to bypass the requirement for CDK2. Generation of a CDK2, CDK4 and CDK6 triple knock-out mouse, which is currently ongoing (Malumbres et al., 2004), is expected to be able to answer a number of fundamental questions. Whatever the results are, they will directly impact our view of the TGF-β/Smad cytostatic response.

7. CONCLUSIONS AND PERSPECTIVES

TGF-β's cytostatic program has attracted many investigators because of its prominent connection with cancer. Smad proteins play important roles in mediating the TGF-β growth-inhibitory response by regulating cell cycle components. Smads can downregulate c-Myc, Id1, Id2, Id3, CDC25A, and cyclin A, and upregulate p15, p21, and PTPRκ. TGF-β/Smad cytostatic responses can vary significantly among different cell types. The extent of the cytostatic effect is determined by a combinatorial basis.

It is interesting that c-Myc, cyclin A and CDC25A, the three downregulated genes, all contain an E2F binding site in their promoters. One possibility is that E2F may integrate both growth stimulation and TGF-β cytostatic program. Structural studies to determine how Smad3 binds to the newly identified repressive Smad binding element, which overlaps with the E2F site, will provide fundamental insights into this fascinating DNA binding feature, which is part of the TGF-β/Smad cytostatic program.

Genome-wide transcriptional profiling led to the identification of Id (Id1, Id2, Id3) as Smad target genes for the TGF-β growth-inhibitory effect. A chromatin immunoprecipitation screen found PTPRκ, which seems necessary for both the growth-inhibitory and promigratory functions of TGF-β. These techniques, especially when coupled with RNAi, knock-out cells, or a specific diseased state, such as a certain cancer, will continue to be very useful in identification of tissue or cell type-specific targets for the various activities of TGF-β/Smad, including the cytostatic program.

It is important to emphasize that the TGF-β cytostatic program is often analyzed in cell proliferation assays and/or by analysis of cell cycle regulators in cell culture systems. The tumor suppressive activities of Smads in living organisms are a combination of cytostatic effect, promoting apoptosis (Chapter 6), and also likely their ability to inhibit angiogenesis. In this regard, overexpressed Smad2, Smad3, or Smad4 can all potently inhibit angiogenesis. The tumor suppressive activities of Smads can be out-competed by their tumor promoter functions under certain conditions (see Chapters 7 and 20).

Lastly and importantly, our view of the TGF-β cytostatic program relies on our understanding of cell cycle control. Currently, the cell cycle field is experiencing new challenges, new models, and new speculations. The canonical view of the cell cycle is based on extensive biochemical and cell culture work for over 15 years.

Mammalian cells are more plastic than we previously envisioned. Some of the questions may not be answered soon. Nevertheless, we all look forward to new findings and new breakthroughs that can definitively resolve the issue on the roles of GI cyclins and CDKs in cell cycle control.

ACKNOWLEDGEMENTS

I thank many colleagues for stimulating discussions. I also thank the American Association for Cancer Research-National Foundation for Cancer Research, the Burroughs Wellcome Fund, the Sidney Kimmel Foundation for Cancer Research, the Pharmaceutical Research and Manufacturer of America Foundation, the Emerald Foundation, the New Jersey Commission on Cancer Research, the Department of Defense Breast Cancer Research Program, and the National Institutes of Health for support the research in my laboratory.

REFERENCES

Bhowmick, N.A., Ghiassi, M., Aakre, M., Brown, K., Singh, V., and Moses, H.L., 2003, TGF-β-induced RhoA and p160ROCK activation is involved in the inhibition of Cdc25A with resultant cell-cycle arrest. *Proc Natl Acad Sci U S A* 100: 15548-15553.

Brunet, A., Bonni, A., Zigmond, M.J., Lin, M.Z., Juo, P., Hu, L.S., Anderson, M.J., Arden, K.C., Blenis, J., and Greenberg, M.E., 1999, Akt promotes cell survival by phosphorylating and inhibiting a Forkhead transcription factor. *Cell* 96:, 857-868.

Burdette, J.E., Jeruss, J.S., Kurley, S.J., Lee, E.J., and Woodruff, T.K., 2005, Activin A mediates growth inhibition and cell cycle arrest through Smads in human breast cancer cells. *Cancer Res* 65: 7968-7975.

Chen, C.R., Kang, Y., and Massagué, J., 2001, Defective repression of c-Myc in breast cancer cells: A loss at the core of the transforming growth factor β growth arrest program. *Proc Natl Acad Sci U S A* 98: 992-999.

Chen, C.R., Kang, Y., Siegel, P.M., and Massagué, J., 2002, E2F4/5 and p107 as Smad cofactors linking the TGFβ receptor to c-Myc repression. *Cell* 110: 19-32.

Chi, X.Z., Yang, J.O., Lee, K.Y., Ito, K., Sakakura, C., Li, Q.L., Kim, H.R., Cha, E.J., Lee, Y.H., Kaneda, A., Ushijima, T., Kim, W.J., Ito, Y., and Bae, S.C., 2005, RUNX3 suppresses gastric epithelial cell growth by inducing $p21^{(WAF1/Cip1)}$ expression in cooperation with transforming growth factor β-activated SMAD. *Mol Cell Biol* 25: 8097-8107.

Claassen, G.F., and Hann, S.R., 2000, A role for transcriptional repression of p21CIP1 by c-Myc in overcoming transforming growth factor β-induced cell-cycle arrest. *Proc Natl Acad Sci U S A* 97: 9498-9503.

Datto, M.B., Frederick, J.P., Pan, L., Borton, A.J., Zhuang, Y., and Wang, X.F., 1999, Targeted disruption of Smad3 reveals an essential role in transforming growth factor β-mediated signal transduction. *Mol Cell Biol* 19: 2495-2504.

Ewen, M.E., Oliver, C.J., Sluss, H.K., Miller, S.J., and Peeper, D.S., 1995, p53-dependent repression of CDK4 translation in TGF-β-induced G1 cell-cycle arrest. *Genes Dev* 9: 204-217.

Ewen, M.E., Sluss, H.K., Whitehouse, L.L., and Livingston, D.M., 1993, TGFβ inhibition of Cdk4 synthesis is linked to cell cycle arrest. *Cell* 74: 1009-1020.

Feng, X.H., Filvaroff, E.H., and Derynck, R., 1995, Transforming growth factor-β (TGF-β)-induced down-regulation of cyclin A expression requires a functional TGF-β receptor complex. Characterization of chimeric and truncated type I and type II receptors. *J Biol Chem* 270: 24237-24245.

Feng, X.-H., Liang, Y.Y., Liang, M., Zhai, W., and Lin, X., 2002, Direct interaction of c-Myc with Smad2 and Smad3 to inhibit TGF-β-mediated induction of the CDK inhibitor p15(Ink4B). *Mol Cell* 9: 133-143.

Feng, X.-H., Lin, X., and Derynck, R., 2000, Smad2, Smad3 and Smad4 cooperate with Sp1 to induce p15(Ink4B) transcription in response to TGF-β. *EMBO J* 19: 5178-5193.

Frederick, J.P., Liberati, N.T., Waddell, D.S., Shi, Y., and Wang, X.F., 2004, Transforming growth factor β-mediated transcriptional repression of c-Myc is dependent on direct binding of Smad3 to a novel repressive Smad binding element. *Mol Cell Biol* 24: 2546-2559.

Han, S.U., Kim, H.T., Seong do, H., Kim, Y.S., Park, Y.S., Bang, Y.J., Yang, H.K., and Kim, S.J., 2004, Loss of the Smad3 expression increases susceptibility to tumorigenicity in human gastric cancer. *Oncogene* 23: 1333-1341.

Hannon, G.J., and Beach, D., 1994, p15INK4B is a potential effector of TGF-β-induced cell cycle arrest. *Nature* 371: 257-261.

He, W., Cao, T., Smith, D.A., Myers, T.E., and Wang, X.J., 2001, Smads mediate signaling of the TGFβ superfamily in normal keratinocytes but are lost during skin chemical carcinogenesis. *Oncogene* 20: 471-483.

Iavarone, A., and Massagué, J., 1999, E2F and histone deacetylase mediate transforming growth factor β repression of cdc25A during keratinocyte cell cycle arrest. *Mol Cell Biol* 19: 916-922.

Jeruss, J.S., Sturgis, C.D., Rademaker, A.W., and Woodruff, T.K., 2003, Down-regulation of activin, activin receptors, and Smads in high-grade breast cancer. *Cancer Res* 63: 3783-3790.

Kang, Y., Chen, C.R., and Massagué, J., 2003, A self-enabling TGFβ response coupled to stress signaling: Smad engages stress response factor ATF3 for Id1 repression in epithelial cells. *Mol Cell* 11: 915-926.

Kim, S.G., Kim, H.A., Jong, H.S., Park, J.H., Kim, N.K., Hong, S.H., Kim, T.Y., and Bang, Y.J., 2005, The Endogenous Ratio of Smad2 and Smad3 Influences the Cytostatic Function of Smad3. *Mol Biol Cell* 16: 4672-4683.

Kowanetz, M., Valcourt, U., Bergström, R., Heldin, C.-H., and Moustakas, A., 2004, Id2 and Id3 define the potency of cell proliferation and differentiation responses to transforming growth factor β and bone morphogenetic protein. *Mol Cell Biol* 24: 4241-4254.

Kozar, K., Ciemerych, M.A., Rebel, V.I., Shigematsu, H., Zagozdzon, A., Sicinska, E., Geng, Y., Yu, Q., Bhattacharya, S., Bronson, R.T., Akashi, K., and Sicinski, P., 2004, Mouse development and cell proliferation in the absence of D-cyclins. *Cell* 118: 477-491.

Kretschmer, A., Moepert, K., Dames, S., Sternberger, M., Kaufmann, J., and Klippel, A., 2003, Differential regulation of TGF-β signaling through Smad2, Smad3 and Smad4. *Oncogene* 22: 6748-6763.

Lasorella, A., Noseda, M., Beyna, M., Yokota, Y., and Iavarone, A., 2000, Id2 is a retinoblastoma protein target and mediates signalling by Myc oncoproteins. *Nature* 407: 592-598.

Levy, L., and Hill, C.S., 2005, Smad4 dependency defines two classes of transforming growth factor β (TGF-β) target genes and distinguishes TGF-β-induced epithelial-mesenchymal transition from its antiproliferative and migratory responses. *Mol Cell Biol* 25: 8108-8125.

Malumbres, M., Sotillo, R., Santamaria, D., Galan, J., Cerezo, A., Ortega, S., Dubus, P., and Barbacid, M., 2004, Mammalian cells cycle without the D-type cyclin-dependent kinases Cdk4 and Cdk6. *Cell* 118: 493-504.

Martin, A., Odajima, J., Hunt, S.L., Dubus, P., Ortega, S., Malumbres, M., and Barbacid, M., 2005, Cdk2 is dispensable for cell cycle inhibition and tumor suppression mediated by $p27^{(Kip1})$ and $p21^{(Cip1)}$. *Cancer Cell* 7: 591-598.

Matsuura, I., Denissova, N.G., Wang, G., He, D., Long, J., and Liu, F., 2004, Cyclin-dependent kinases regulate the antiproliferative function of Smads. *Nature* 430: 226-231.

Moustakas, A., and Kardassis, D., 1998, Regulation of the human p21/WAF1/Cip1 promoter in hepatic cells by functional interactions between Sp1 and Smad family members. *Proc Natl Acad Sci U S A* 95: 6733-6738.

Nagahara, H., Ezhevsky, S.A., Vocero-Akbani, A.M., Kaldis, P., Solomon, M.J., and Dowdy, S.F., 1999, Transforming growth factor β targeted inactivation of cyclin E:cyclin-dependent kinase 2 (Cdk2) complexes by inhibition of Cdk2 activating kinase activity. *Proc Natl Acad Sci U S A* 96: 14961-14966.

Ortega, S., Malumbres, M., and Barbacid, M., 2002, Cyclin D-dependent kinases, INK4 inhibitors and cancer. *Biochim Biophys Acta* 1602: 73-87.

Ortega, S., Prieto, I., Odajima, J., Martin, A., Dubus, P., Sotillo, R., Barbero, J.L., Malumbres, M., and Barbacid, M., 2003, Cyclin-dependent kinase 2 is essential for meiosis but not for mitotic cell division in mice. *Nat Genet* 35: 25-31.

Pardali, K., Kowanetz, M., Heldin, C.-H., and Moustakas, A., 2005, Smad pathway-specific transcriptional regulation of the cell cycle inhibitor $p21^{(WAF1/Cip1)}$. *J Cell Physiol* 204: 260-272.

Pardali, K., Kurisaki, A., Morén, A., ten Dijke, P., Kardassis, D., and Moustakas, A., 2000, Role of Smad proteins and transcription factor Sp1 in $p21^{(Waf1/Cip1)}$ regulation by transforming growth factor-β. *J Biol Chem* 275: 29244-29256.

Parisi, T., Beck, A.R., Rougier, N., McNeil, T., Lucian, L., Werb, Z., and Amati, B., 2003, Cyclins E1 and E2 are required for endoreplication in placental trophoblast giant cells. *EMBO J* 22: 4794-4803.

Petritsch, C., Beug, H., Balmain, A., and Oft, M., 2000, TGF-β inhibits p70 S6 kinase via protein phosphatase 2A to induce G(1) arrest. *Genes Dev* 14: 3093-3101.

Pietenpol, J.A., Stein, R.W., Moran, E., Yaciuk, P., Schlegel, R., Lyons, R.M., Pittelkow, M.R., Munger, K., Howley, P.M., and Moses, H.L., 1990, TGF-β1 inhibition of c-myc transcription and growth in keratinocytes is abrogated by viral transforming proteins with pRB binding domains. *Cell* 61: 777-785.

Ray, D., Terao, Y., Nimbalkar, D., Chu, L.H., Donzelli, M., Tsutsui, T., Zou, X., Ghosh, A.K., Varga, J., Draetta, G.F., and Kiyokawa, H., 2005, Transforming growth factor β facilitates β-TrCP-mediated degradation of Cdc25A in a Smad3-dependent manner. *Mol Cell Biol* 25: 3338-3347.

Rich, J.N., Zhang, M., Datto, M.B., Bigner, D.D., and Wang, X.F., 1999, Transforming growth factor-β-mediated p15(INK4B) induction and growth inhibition in astrocytes is SMAD3-dependent and a pathway prominently altered in human glioma cell lines. *J Biol Chem* 274: 35053-35058.

Scandura, J.M., Boccuni, P., Massagué, J., and Nimer, S.D., 2004, Transforming growth factor β-induced cell cycle arrest of human hematopoietic cells requires p57KIP2 up-regulation. *Proc Natl Acad Sci U S A* 101: 15231-15236.

Seoane, J., Le, H.V., Shen, L., Anderson, S.A., and Massagué, J., 2004, Integration of Smad and forkhead pathways in the control of neuroepithelial and glioblastoma cell proliferation. *Cell* 117: 211-223.

Seoane, J., Pouponnot, C., Staller, P., Schader, M., Eilers, M., and Massagué, J., 2001, TGFβ influences Myc, Miz-1 and Smad to control the CDK inhibitor p15INK4b. *Nat Cell Biol* 3: 400-408.

Sherr, C.J., 1996, Cancer cell cycles. *Science* 274: 1672-1677.

Sherr, C.J., and Roberts, J.M., 1999, CDK inhibitors: positive and negative regulators of G1-phase progression. *Genes Dev* 13: 1501-1512.

Siegel, P.M., and Massagué, J., 2003, Cytostatic and apoptotic actions of TGF-β in homeostasis and cancer. *Nat Rev Cancer* 3: 807-821.

Siegel, P.M., Shu, W., and Massagué, J., 2003, Mad upregulation and Id2 repression accompany transforming growth factor (TGF)-β-mediated epithelial cell growth suppression. *J Biol Chem* 278: 35444-35450.

Sjöblom, T., Yakymovych, I., Heldin, C.-H., Östman, A., and Souchelnytskyi, S., 2004, Smad2 suppresses the growth of Mv1Lu cells subcutaneously inoculated in mice. *Eur J Cancer* 40: 267-274.

Staller, P., Peukert, K., Kiermaier, A., Seoane, J., Lukas, J., Karsunky, H., Moroy, T., Bartek, J., Massagué, J., Hanel, F., and Eilers, M., 2001, Repression of p15INK4b expression by Myc through association with Miz-1. *Nat Cell Biol* 3: 392-399.

Tetsu, O., and McCormick, F., 2003, Proliferation of cancer cells despite CDK2 inhibition. *Cancer Cell* 3: 233-245.

Untergasser, G., Gander, R., Lilg, C., Lepperdinger, G., Plas, E., and Berger, P., 2005, Profiling molecular targets of TGF-β1 in prostate fibroblast-to-myofibroblast transdifferentiation. *Mech Ageing Dev* 126: 59-69.

Wang, S.E., Wu, F.Y., Shin, I., Qu, S., and Arteaga, C.L., 2005, Transforming growth factor β (TGF-β)-Smad target gene protein tyrosine phosphatase receptor type kappa is required for TGF-β function. *Mol Cell Biol* 25: 4703-4715.

Warner, B.J., Blain, S.W., Seoane, J., and Massagué, J., 1999, Myc downregulation by transforming growth factor β required for activation of the p15(Ink4b) G(1) arrest pathway. *Mol Cell Biol* 19: 5913-5922.

Wolfraim, L.A., Fernandez, T.M., Mamura, M., Fuller, W.L., Kumar, R., Cole, D.E., Byfield, S., Felici, A., Flanders, K.C., Walz, T.M., Roberts, A.B., Aplan, P.D., Balis, F.M., and Letterio, J.J., 2004, Loss of Smad3 in acute T-cell lymphoblastic leukemia. *N Engl J Med* 351: 552-559.

Xie, L., Law, B.K., Aakre, M.E., Edgerton, M., Shyr, Y., Bhowmick, N.A., and Moses, H.L., 2003, Transforming growth factor β-regulated gene expression in a mouse mammary gland epithelial cell line. *Breast Cancer Res* 5: R187-198.

Xu, X., Brodie, S.G., Yang, X., Im, Y.H., Parks, W.T., Chen, L., Zhou, Y.X., Weinstein, M., Kim, S.J., and Deng, C.X., 2000, Haploid loss of the tumor suppressor Smad4/Dpc4 initiates gastric polyposis and cancer in mice. *Oncogene* 19: 1868-1874.

Yagi, K., Furuhashi, M., Aoki, H., Goto, D., Kuwano, H., Sugamura, K., Miyazono, K., and Kato, M., 2002, c-myc is a downstream target of the Smad pathway. *J Biol Chem* 277: 854-861.

Yang, X., Letterio, J.J., Lechleider, R.J., Chen, L., Hayman, R., Gu, H., Roberts, A.B., and Deng, C., 1999, Targeted disruption of SMAD3 results in impaired mucosal immunity and diminished T cell responsiveness to TGF-β. *EMBO J* 18: 1280-1291.

Yang, Y., Gil, M., Byun, S.M., Choi, I., Pyun, K.H., and Ha, H., 1996, Transforming growth factor-β1 inhibits human keratinocyte proliferation by upregulation of a receptor-type tyrosine phosphatase R-PTP-κ gene expression. *Biochem Biophys Res Commun* 228: 807-812.

Zhu, Y., Richardson, J.A., Parada, L.F., and Graff, J.M., 1998, Smad3 mutant mice develop metastatic colorectal cancer. *Cell* 94: 703-714.

CHAPTER 5

SMADS IN MESENCHYMAL DIFFERENTIATION

RIK DERYNCK, LISA CHOY, AND TAMARA ALLISTON

Departments of Cell and Tissue Biology, and Anatomy, Programs in Cell Biology and Developmental Biology, University of California, San Francisco, CA 94143

Abstract: Transforming growth factor-β (TGF-β) family members play key roles in development through their regulatory roles in cell and tissue differentiation. Among the differentiation lineages, mesenchymal tissue differentiation into bone, fat, cartilage or muscle is strongly regulated by TGF-β family members. Smads have shown themselves to function as cell-intrinsic regulators of mesenchymal stem cell differentiation into osteoblasts, adipocytes, chondrocytes and myocytes. Their activities are defined by autocrine and paracrine signals from TGF-β family members, and regulate the proliferation of the mesenchymal progenitor cell pool, the selection of the lineage along which the cells will differentiate, and the progression of differentiation. At the molecular level, Smads can inhibit progression of differentiation, through functional repression of key transcription factors that drive differentiation, or alternatively activate the expression, or enhance the activities, of such transcription factors to drive the selection of a lineage and progression along a particular lineage

Keywords: osteoblast, adipocyte, chondrocyte, myocyte, TGF-β, BMP

1. INTRODUCTION

Members of the TGF-β family act as key determinants of development of multicellular metazoan organisms, from Planaria and *C. elegans* to *Drosophila* and vertebrates. They play critical roles in defining the body plan and in the formation of most if not all organs. These functions are largely dictated by their abilities to promote or inhibit tissue differentiation and to mediate tissue interactions, such as the interactions between adjacent epithelial and mesenchymal tissues that are instrumental in organ development.

The multiple roles of TGF-β family members in development are reflected in the complex temporal and spatial regulation of the different TGF-β-related proteins and receptor complexes, and the presence of Smads in most, if not all cell types.

P. ten Dijke and C.-H. Heldin (eds.), Smad Signal Transduction, 93–112.

Thus, depending on the availability and activity of specific TGF-β family ligands and receptors at the cell's surface, the Smads will integrate these signals in a time- and space-dependent manner to influence cellular activities. The signaling from the receptors to the Smads is activated through both autocrine and paracrine signaling, in response to TGF-β family ligand expression by the same cell or neighboring cells. During development, the protein levels of some TGF-β family ligands or their inhibitors can manifest themselves as gradients, thus resulting in graded cell responses to the level of available ligand (Green et al., 1992; Dosch et al., 1997). Thus, the nature of the differentiation response to TGF-β family proteins *in vivo* can differ depending on the concentration of available ligand and on the crosstalk with other pathways, as is sometimes also observed in cell culture (Wang et al., 1993). Smads then integrate TGF-β family signals with those from other signaling pathways by virtue of their interactions with transcription factors that drive cell lineage-specific differentiation. In this way, the Smads act as cell-intrinsic regulators of differentiation pathways that are controlled by autocrine and paracrine signals.

Signaling by TGF-β family factors has been shown to affect various differentiation pathways, including those of the hematopoietic and immune cell lineages, epithelial lineages and hemangioblasts, with their effects depending on the nature of the ligand and the responding heteromeric receptor complexes. Accordingly, the Smads function as cell-intrinsic regulators of these differentiation pathways. In this chapter we will discuss the roles of Smad signaling in the differentiation of mesenchymal stem cells – focusing primarily on their differentiation into osteoblasts, adipocytes, chondrocytes, and myocytes.

2. MESENCHYMAL DIFFERENTIATION

Mesenchymal cells generally derive from mesodermal origin, although a population of mesenchymal cells also derives from neural crest cells that have migrated to contribute to various mesenchymal structures primarily in the craniofacial area. Mesenchymal stem cells have long been known to differentiate into four specialized lineages, *i.e.* cartilage cells (chondrocytes), bone matrix-depositing cells (osteoblasts), muscle cells (myocytes) or fat cells (adipocytes). Recent studies suggest that mesenchymal stem cells may also have potential to differentiate into additional cell lineages, such as hepatocytes or neurons. Each of these cell types is marked by the expression of a distinct set of proteins, characteristic of the specialized functions of each differentiation lineage. Whereas chondrocytes, osteoblasts and adipocytes are all mononuclear, skeletal myocytes organize themselves in multinuclear myofibers. As the cells differentiate along these lineages, their proliferation decreases and the fully differentiated cells stop dividing.

The differentiation along each of these lineages is driven by the expression and activities of defined transcription factors that are regulated by signaling inputs and through interactions with other proteins. Thus, Runx2, a runt domain transcription factor, functions as a key driver in the selection of the osteoblast differentiation pathway, and the progression from an undifferentiated mesenchymal cell to a functional osteoblast (Harada and Rodan, 2003). In myoblast differentiation, myogenic bHLH (basic helix-loop-helix) proteins form heterodimeric complexes with E proteins and drive the progression of myogenic differentiation. Their activities and abilities to drive myogenic differentiation are largely dependent upon their interaction with MEF (myocyte enhancer factor) proteins, another set of transcription factors that serve as coactivators for myogenic bHLH proteins (Sartorelli and Caretti, 2005). Adipogenic differentiation is driven by the sequential and parallel activities of two types of transcription factors, the C/EBPs (CCAAT/enhancer binding proteins) and PPARγ (peroxisome proliferators-activated receptor γ). C/EBPs β and δ, members of a larger family of transcription factors characterized by a basic/leucine zipper, activate the expression of PPARγ in pre-adipocytes. PPARγ is a nuclear receptor that functions through heterodimerization with another nuclear receptor RXR (retinoid X receptor), and, in cooperation with C/EBPα, drives the expression of various proteins that characterize the differentiated adipocyte (Otto and Lane, 2005). Finally, chondrogenic differentiation from mesenchymal progenitors is controlled by Nkx3.2 and the Sox (Sry-type high mobility group box) transcription factors, characterized by HMG (high mobility group) box DNA binding domains. Nkx3.2 is required for Sox9 expression and represses the expression of osteogenic transcription factors. Sox9 is essential for the initiation of chondrogenic differentiation during mesenchymal condensation, yet also plays a role in slowing down or preventing terminal conversion of chondrocytes into hypertrophic chondrocytes. L-Sox5 and Sox6, two closely related Sox proteins that lack a transactivation domain are required for progression during chondrogenesis, presumably in a complex with Sox9 (Lefebvre and Smits, 2005).

In addition to these four specialized and distinct cell types, mesenchymal cells also give rise to the fibroblasts that are located in and produce the connective tissue or stroma in between many other cell types and tissues. Perhaps because these cells do not exhibit a striking phenotypic or functional progression, their differentiation is poorly characterized. The key regulatory transcription factors for the fibroblast lineage have not been identified. Nonetheless, their importance in developmental processes and cancer development and progression is starting to be appreciated. Connective tissue cells can be highly organized, whereby the cells are closely packed and aligned along the same direction. When they organize themselves in this way to form tendons that provide great mechanical strength, they are often called tenocytes, another mesenchymal cell population that is poorly characterized (Docheva et al., 2005).

3. SMADS IN MESENCHYMAL STEM CELLS

Mesenchymal stem cells are present at most if not all sites where differentiated mesenchymal cell types are found, as well as in loose connective tissue interspersed between other tissues. Thus, besides the adipocytes themselves and the cell types associated with the high vascularization of this tissue, adipose tissue contains seemingly undifferentiated mesenchymal cells that serve as cell reservoir to generate new fully differentiated fat cells (Zuk et al., 2001). These adipose stromal cells maintain their self-renewal and are fully capable to differentiate along other lineages, depending on the environment and signals to which they are exposed. Conceptually similarly, mesenchymal cells, referred to as satellite cells, are interspersed in between the muscle fibers of differentiated muscle tissue (Dhawan and Rando, 2005). They can be activated to differentiate into myoblasts and myocytes, yet have maintained their pluripotentiality and can be stimulated to differentiate into osteoblasts and chondrocytes. In the periosteum and perichondrium, which cover the surfaces of bone and cartilage, respectively, undifferentiated mesenchymal cells can differentiate from mesenchymal stem cells into functional osteoblasts or chondrocytes that deposit the bone or cartilage matrix (de Crombrugghe et al., 2001). Finally, the stromal cells found in bone marrow normally differentiate into osteoblasts or adipocytes, while also maintaining a pool of undifferentiated stem cells. The bone marrow stroma thus represents a major source of pluripotential mesenchymal stem cells (Pittenger et al., 1999). These cells have the ability to differentiate into the four specialized mesenchymal cell types and have gained prominence for their potential to be used as a cell source for cell-based therapies in the repair of numerous tissue types (Caplan, 2005; Polak and Hench, 2005).

The selection and progression of differentiation of mesenchymal stem cells along their individual lineages depend on the location of these cells and the signals to which they are exposed, either through direct cell contact or through soluble mediators, including the TGF-β family members. Signals that stimulate proliferation, such as TGF-β, will generally inhibit differentiation, whereas growth inhibitory signals, typically triggered by cell-cell contact, may promote differentiation. Clearly, the nature of the signals and the combination of the different signals are the major determinants of the initiation and progression of a defined differentiation pathway, through their abilities to activate expression of transcription factors and by regulating their function.

The prominent role of Smad signaling in mesenchymal differentiation is apparent in the maintenance of mesenchymal stem cells in an undifferentiated state, in the selection of a differentiation lineage and the progression of differentiation along that defined lineage. Although the expression of the Smads during development and tissue differentiation has not yet been fully characterized, it appears that the receptor-activated Smads and Smad4 are expressed in both undifferentiated and differentiated mesenchymal cells, thus providing the cells with the potential to respond to signaling by the diverse TGF-β family members. Since mesenchymal cells express several TGF-β family ligands and receptors, the autocrine and paracrine regulation of the endogenous Smads is a key aspect of the regulation of mesenchymal differentiation. Thus, TGF-β family members are thought to play a major role in the selection of the differentiation lineage and in the progression of differentiation. Since TGF-β stimulates proliferation of mesenchymal cells, yet inhibits myogenic, adipogenic, osteoblast and chondrogenic differentiation, local expression of TGF-β with consequent TGF-β signaling through Smad3 may be a determinant of the maintenance and expansion of the mesenchymal stem cell population stem cell pool and the maintenance of these cells in an uncommitted state. Considering the role of BMP (bone morphogenetic protein) signaling in mesenchymal differentiation, one should also assume that regulation of the activation state of the BMP-regulated Smads is an essential part of keeping the cells undifferentiated and pluripotential, as apparent in mouse embryonic stem cells (Ying et al., 2003). Thus, even though little is as yet known about how Smads are involved in the physiology of mesenchymal stem cells, the Smad transcription factors are an integral part of the cell-intrinsic circuitry that regulates selection and progression of differentiation.

4. SMAD SIGNALING IN OSTEOBLAST DIFFERENTIATION

Consistent with the expression of the full complement of Smads and of bone morphogenetic protein (BMP) and TGF-β receptors in mesenchymal cells and osteoblasts, BMPs and TGF-βs regulate differentiation into osteoblasts (Fig. 1). BMPs, which act through Smad1, 5 and 8, stimulate both the commitment of

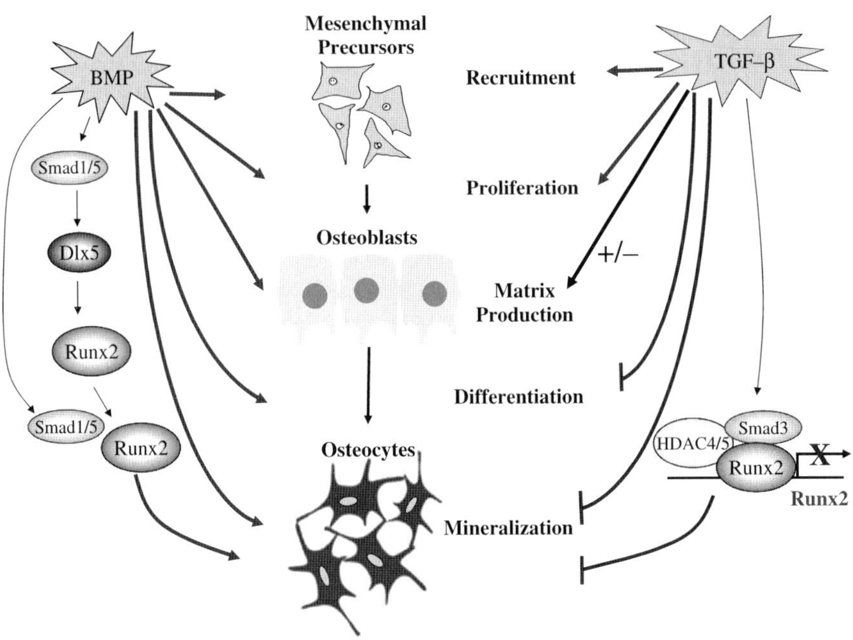

Figure 1. BMP and TGF-β regulate osteoblast differentiation. BMPs induce the expression of Dlx5, which, in turn, induces Runx2 expression. BMP-responsive Smads 1 and 5 bind Runx2 and enhance its transcription activity to promote osteogenesis. TGF-β promotes recruitment and proliferation of mesenchymal precursors at the sites of bone formation, but inhibits progression of osteoblast differentiation. TGF-β activates Smad3 to bind and repress Runx2 by recruiting class II histone deacetylases (HDAC4 and 5). In this way, TGF-β represses the expression of Runx2 target genes, including *runx2* itself

mesenchymal stem cells to the osteoblast lineage as well as the progression of osteoblast differentiation. BMPs induce the expression of osteoblast markers by C2C12 myoblasts (Katagiri et al., 1994; Lee et al., 2000) and by 3T3-F442A pre-adipocytes (Skillington et al., 2002), while inhibiting the normal myogenic or adipocytic differentiation of these cell types. These findings suggest that BMP-activated Smads not only stimulate osteoblast differentiation of undifferentiated mesenchymal cells, but may also redirect mesenchymal differentiation into the osteoblast lineage. These conclusions are consistent with the observations that local administration of BMPs *in vivo* results in ectopic bone formation at various sites, including in muscle (Wozney et al., 1988). A developmental role for BMPs, and by extension BMP-activated Smads, in osteoblast differentiation is also supported by the many studies in which the activity of the BMP pathway is induced or ablated in mice in a targeted manner (Canalis and Economides, 2003). However, the individual contributions of the different BMP-activated Smads in osteoblast differentiation and bone matrix deposition by osteoblasts are as yet largely unexplored.

In contrast to the BMPs, TGF-β, which activates Smad2 and Smad3, inhibits osteoblast differentiation in culture, and appears to do so through Smad3 (Alliston et al., 2001). However, TGF-β stimulates proliferation of mesenchymal cells and (pre-)osteoblasts, and serves as chemoattractant of mesenchymal cells (Pfeilschifter et al., 1990), thus allowing for an expansion of the mesenchymal progenitor pool that differentiates into osteoblasts (Fig. 1). This effect of TGF-β on (pre)osteoblast proliferation and recruitment explains why subperiosteal injection of TGF-β results in increased numbers of osteoblasts with consequent bone formation (Joyce et al., 1990) or why transgenic expression of TGF-β2 in preosteoblasts results in higher numbers of osteoblasts, with a concomitant increase in bone deposition by osteoblasts (Erlebacher and Derynck, 1996). Thus, even though TGF-β inhibits or slows down osteoblast differentiation, its positive effect on osteoblast cell proliferation causes an overall increase in the bone matrix deposition rate. Conversely, decreasing the TGF-β responsiveness of preosteoblasts and osteoblasts decreases the numbers of osteoblasts and rate of bone matrix deposition (Filvaroff et al., 1999). That autocrine TGF-β signaling plays an important regulatory role in osteoblast differentiation is also supported through the use of a specific inhibitor of the TβRI kinase, which normally phosphorylates and thereby activates Smad2 and Smad3. Inhibition of autocrine TGF-β signaling enhances BMP-induced osteoblast differentiation of C2C12 cells. Presumably, the TβRI kinase inhibitor enhances BMP-inducible osteoblast differentiation by preventing the TGF-β-inducible expression of the inhibitory Smads (Maeda et al., 2004).

The transcriptional mechanisms through which BMP-activated Smads promote, and TGF-β-activated Smad3 inhibits osteoblast differentiation are starting to be unraveled (Fig. 1). Even though BMP treatment induces the expression of Runx2, the "master" transcription factor in osteoblast differentiation (Ducy et al., 1997; Lee et al., 2000), it is unlikely that BMP-activated Smads directly activate the *runx2* promoter. Instead, the induction of the homeobox transcription factors Dlx-5, Msx1 and Msx2 by BMPs may be primary events in BMP's ability to induce osteoblast differentiation (Marazzi et al., 1997; Miyama et al., 1999; Brugger et al., 2004). Dlx5 in turn would then activate expression of *runx2*, which further amplifies *runx2* expression and activates other genes involved in osteoblast differentiation and function (Lee et al., 2003, 2005). The effect of BMPs on osteoblast differentiation is further explained by the physical interaction and functional cooperation of BMP-activated Smads with Runx2. Indeed, Smad1 and Smad5 have been shown to associate with Runx2 in response to BMP signaling, and to functionally cooperate with Runx2 in inducing osteoblast differentiation of C2C12 cells (Lee et al., 2000; Zhang et al., 2000). This cooperation is accompanied by, and may require, the recruitment of Smad/Runx2 complexes to defined subnuclear foci (Zaidi et al., 2002). The cooperation and association of BMP-activated Smads with Runx2 may concomitantly play a key role in suppressing the myoblast phenotype and myogenic differentiation (see below).

The inhibitory effect of TGF-β on osteoblast differentiation appears to be effected at different levels. The best characterized and likely the predominant mechanism of this inhibition relates to the direct interaction of Smad3 with Runx2 in response to TGF-β. Rather than resulting in a functional cooperation, as proposed for BMP-activated Smads, TGF-β-activated Smad3 represses Runx2 function through its physical association with Runx2 at Runx2 binding sites in the regulatory promoter sequences of Runx2 target genes. This TGF-β/Smad3-mediated repression of Runx2 function consequently confers repression of transcription from the *runx2* and *osteocalcin* genes, and most likely other Runx2 target genes that are involved in the differentiation and function of osteoblasts (Alliston et al., 2001; Fig. 1). The repression of Runx2 activity by Smad3 results from the TGF-β-induced association of class IIa histone deacetylases, *i.e.* HDAC4 and/or HDAC5, with Smad3, and consequent formation of a Smad3/Runx2/HDAC complex at the Runx2 binding DNA sequences. This recruitment confers histone deacetylation at the Runx2 binding osteocalcin promoter sequences, primarily of histone 4. Thus, the repression of Runx2 function by Smad3 is mediated by HDAC recruitment by Smad3, and this mechanism plays a key role in the inhibition of osteoblast differentiation by autocrine or paracrine TGF-β (Kang et al., 2005).

The functional interactions of Smads with Runx2, whether it is cooperation of BMP-activated Smad1 or Smad5 with Runx2, or inhibition of Runx2 function by Smad3 in response to TGF-β, illustrate the roles of Smads as cell-intrinsic regulators of osteoblast differentiation. Thus, enhancing or decreasing the levels of expression and/or activation of Smads profoundly affects the extent and kinetics of osteoblast differentiation. For example, increasing the levels of Smad3, thereby enhancing the overall Smad3 activation through autocrine TGF-β signaling, inhibits osteoblast differentiation, whereas dominant negative interference with Smad3 function can enhance osteoblast differentiation and confer resistance to the inhibition of differentiation by TGF-β (Alliston et al., 2001). By interfacing with the Smads, other pathways can impact the progression of osteoblast differentiation. Menin is one such protein. Menin can enhance the activity of BMP-activated Smad/Runx2 complexes to promote osteoblast differentiation but can also potentiate the osteoblast-inhibitory activities of TGF-β by forming complexes with Smad3 later in differentiation (Sowa et al., 2004).

Since Smad3 recruits class II HDACs to Runx2 in response to TGF-β, and such recruitment occurs in the context of autocrine TGF-β signaling, class II HDACs also function as cell intrinsic regulators of osteoblast differentiation in functional connection with the TGF-β signaling system. Accordingly, interference with class II HDAC function strongly enhances and accelerates osteoblast differentiation (Kang et al., 2005). These observations raise the possibility that Smad signaling as well as class II HDAC activity represent potential targets for therapy of metabolic bone diseases, *i.e.* by enhancing osteoblast activity and bone matrix deposition.

5. SMAD SIGNALING IN ADIPOGENIC DIFFERENTIATION

Compared to osteoblast differentiation, much less is known about the regulation of adipogenic differentiation by TGF-β and BMP-related factors, and most of our knowledge is gained from cell culture experiments. BMPs, such as BMP-2, -4 and -7, affect the adipogenic differentiation of undifferentiated mesenchymal cells, but, while some reports conclude that BMPs inhibit adipogenic differentiation, others provide evidence for stimulation by BMPs. For example, BMP-2 was shown to inhibit adipocyte differentiation of bone marrow stromal cell lines and 3T3-F442A pre-adipocytes (Gimble et al., 1995; Gori et al., 1999; Skillington et al., 2002), but BMP-2 and BMP-7 conversely stimulate adipocyte differentiation of multipotential 10T1/2 cells and 3T3-L1 preadipocytes (Ahrens et al., 1993; Rebbapragada et al., 2003). This discrepancy may relate to differences in cell systems and/or culture conditions, *e.g.* relative levels of ligand for PPARγ, the central transcription factor in the adipocyte differentiation pathway, in the culture medium (Sottile and Seuwen, 2000). The finding that BMPs can stimulate both osteoblast and adipocyte differentiation of the same cell population (Asahina et al., 1996) may suggest a common early differentiation step involving BMP-activated Smads, the identification of which would greatly enhance our understanding of the normal differentiation of bone marrow stromal cells into osteoblasts versus adipocytes. Whether osteoblast or adipocyte differentiation in response to BMPs is favored may depend on the relative ratios of BMP-activated type I receptors or Smads. Thus, it has been proposed that BMPR-IB signaling favors osteoblast differentiation, while BMPR-IA signaling favors adipogenic differentiation of 2T3 mesenchymal cells (Chen et al., 1998). It remains, however, unclear how BMPs function in adipogenic differentiation *in vivo*, or even to which BMPs differentiating adipocytes are physiologically exposed.

A BMP-related protein that may be of particular relevance for adipocyte differentiation is GDF-3 (growth and differentiation factor-3), which in mice is primarily expressed in adipose tissue (McPherron and Lee, 1993) and regulated by high fat diet feeding (Witthuhn and Bernlohr, 2001). Exposure of mice to viral vectors expressing GDF-3 results in increased body fat with prominent adipocyte hypertrophy, when these mice are fed a high fat diet. Furthermore, GDF-3 stimulates

PPARγ expression in primary adipocytes and 3T3-L1 pre-adipocytes in culture (Wang et al., 2004). These observations suggest a physiological role of GDF-3 in adipogenic differentiation and fat formation. However, although GDF-3 is structurally related to BMPs, it is unknown which receptors or Smads are activated by GDF-3 and, thus, we lack insight into the mechanism of adipogenic stimulation by GDF-3.

In contrast to the BMPs, TGF-β blocks adipogenic differentiation both *in vitro* (Ignotz and Massagué, 1985; Torti et al., 1989) and *in vivo* (Clouthier et al., 1997), while it stimulates the proliferation of preadipocytes (Jeoung et al., 1995; Choy et al., 2000). As with mesenchymal cells differentiating into osteoblasts, autocrine and paracrine TGF-β responsiveness thus allows an expansion of the progenitor population thereby allowing an increased number of cells to differentiate into adipocytes. Such scenario would be compatible with the expression of TGF-β by fat and its increased expression in obese adipose tissue (Samad et al., 1997). The inhibition of differentiation by TGF-β is mediated by Smad3. Thus, increased activity of Smad3, but not Smad2, inhibits adipogenic conversion, while interfering with Smad3 function enhances and accelerates adipose conversion in culture and confers resistance to inhibition of adipocyte differentiation by TGF-β. Increased expression of Smad6 or Smad7 strongly inhibits adipogenic differentiation, and thereby remarkably cooperates with the inhibitory activity of TGF-β, even though these "inhibitory" Smads have been characterized as inhibitors of TGF-β and BMP signaling (Choy et al., 2000). Thus, as with osteoblast differentiation, the Smad system, or at least Smad3 and the inhibitory Smads, act as a cell-intrinsic system to regulate progression of differentiation in response to external factors.

Myostatin, initially identified as GDF-8 and a regulator of muscle mass as will be discussed further below, also affects adipogenesis. Mice lacking myostatin expression due to targeted gene inactivation develop less adipose tissue (McPherron and Lee, 2002), suggesting that myostatin promotes adipose conversion of mesenchymal cells. This would be consistent with the observation that *myostatin* promotes adipogenic differentiation of 10T1/2 cells (Artaza et al., 2005). On the other hand, myostatin inhibits BMP-induced adipogenic differentiation of 10T1/2 mesenchymal cells and 3T3-L1 preadipocytes (Rebbapragada et al., 2003). As myostatin acts through two type I receptors, *i.e.* ActRIIB known to mediate Activin signaling through Smad2, and TβRI (Rebbapragada et al., 2003), the effector of TGF-β signaling through Smad2 and Smad3, this inhibition by myostatin is consistent with the inhibition of adipocyte differentiation of 3T3-L1 cells by Activin (Hirai et al., 2005) or TGF-β. How this inhibition of differentiation is reconciled with the decreased fat formation in myostatin-deficient mice is unclear, but presumably involves metabolic effects of myostatin at the organismal level.

Whereas the contribution of non-Smad mechanisms in this context has yet to be elucidated, the strong inhibition by TGF-β is primarily mediated by Smad3 as transcriptional repressor. Epistasis and functional and physical interaction analyses

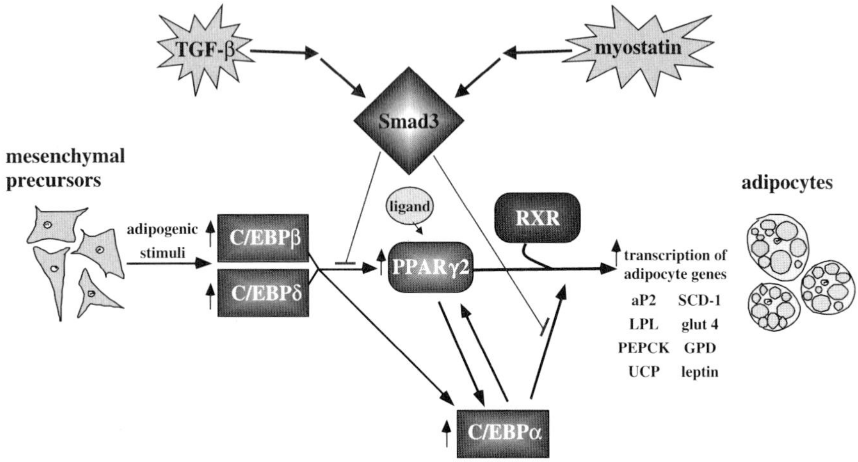

Figure 2. Transcription factor cascade in adipocyte differentiation from mesenchymal progenitor cells, and inhibition of adipogenesis by TGF-β or myostatin. Progenitor cells, exposed to adipogenic stimuli, activate the expression and function of C/EBPβ and δ, which directly induce transcription of the PPARγ gene. PPARγ2 is the adipocyte-specific isoform of PPARγ and is specifically required for activation of adipocyte genes. It heterodimerizes with RXR and requires activation by unknown endogenous ligand(s) to activate transcription of genes in the adipocyte differentiation program. C/EBPα is then induced and takes over the function of C/EBPβ and δ, maintaining the transcription of PPARγ and cooperating with PPARγ to activate the genes of the adipocyte transcription program. PPARγ and C/EBPα stimulate each other's expression to create a positive feedback loop and drive adipogenesis. TGF-β or myostatin binding to their cell-surface receptors results in activation of Smad3. Activated Smad3 physically interacts directly with C/EBPβ, δ and α, and represses their transcription function, thus inhibiting activation of the adipocyte differentiation program

revealed that Smad3 targets the C/EBPs for functional repression through physical interactions (Fig. 2; Choy and Derynck, 2003).

Thus, adipogenic conversion of NIH-3T3 cells by ectopic C/EBPα, β or δ expression is efficiently blocked by TGF-β/Smad3 signaling, which represses the transcription function of these C/EBPs without a decrease in C/EBP protein levels. As C/EBPβ and δ activate PPARγ expression during differentiation, their functional repression prevents adipogenic conversion and the expression of other C/EBP target genes. The repression of C/EBP function by TGF-β-activated Smad3 correlates with the direct association of Smad3 with the transactivation domain of the C/EBPs in response to TGF-β, resulting in repression of their transcription function. Smad3 binding does not interfere with the binding of the C/EBPs to their cognate DNA sequence in regulatory promoter regions. In contrast to the C/EBPs, Smad3 does not interact with PPARγ and does not affect its transcription function. Thus, as with osteoblast differentiation, the inhibition of adipogenic differentiation by TGF-β is mediated through transcriptional repression by Smad3 that targets critical transcription factors, whose function is essential in the progression of differentiation.

6. SMADS AS REGULATORS OF CHONDROCYTE DIFFERENTIATION

Studying the regulation of chondrocyte differentiation by TGF-β family members has been hampered by problems associated with recapitulating the *in vivo* differentiation cascade of chondrocytes in cell culture. Nevertheless, the use of organ cultures and mouse model studies have brought some insight into the roles of Smads in chondrocyte differentiation. Several TGF-β family members play critical roles in chondrogenic differentiation – including the BMPs, GDFs and TGF-βs.

The three TGF-β isoforms are expressed in the perichondrium and differentiating chondrocytes (Millan et al., 1991). TGF-β stimulates the proliferation of chondrocyte precursors and the selection of a chondrocyte fate, thus increasing the number of cells in the chondrogenic lineage (Seyedin et al., 1985; Kulyk et al., 1989). Additionally, TGF-β promotes chondrogenesis in culture of early, undifferentiated mesenchyme, but inhibits terminal differentiation in high density chondrocyte cultures or organ cultures (Serra et al., 1999). The positive effects of TGF-β on early stages of chondrogenesis are dependent on TGF-β activated Smads 2 and 3. Smad3 binds the chondrogenic transcription factor Sox9, thereby stimulating the recruitment of the coactivator CBP/p300 (Furumatsu et al., 2005). Early in chondrocyte differentiation, TGF-β activates Smads 2 and 3 to increase the expression of cartilage matrix proteins aggrecan and collagen II. The importance of Smad-meditated signaling by TGF-β decreases with chondrocyte differentiation. Later in chondrocyte differentiation, TGF-β preferentially activates non-Smad signaling pathways, which inhibit the progression of chondrocyte differentiation (Watanabe et al., 2001; Qiao et al., 2005). Smad3 is important for the inhibition of terminal hypertrophic chondrocyte differentiation. However, this important role of Smad3 occurs in the perichondrial cells, where TGF-β induces the expression of PTHrP through a Smad3 dependent pathway. PTHrP then acts directly on the hypertrophic chondrocytes to slow their terminal differentiation (Alvarez and Serra, 2004).

GDF5, a BMP-like protein, and several BMPs are expressed in the perichondrium and in developing joints (Storm et al., 1994). GDF-5 signals through BMPR-IB (Nishitoh et al., 1996), whereas other BMPs use BMPR-IA and/or BMPR-IB as their major type I receptors, both of which are expressed in developing cartilage (Yoon et al., 2005). Addition of GDF5 increases condensation of differentiating chondrocytes (Hotten et al., 1996) and chondrocyte proliferation. Increased GDF5 expression *in vivo* confers an expansion of the cartilage elements with an enlarged hypertrophic zone and decreased proliferative zone (Tsumaki et al., 1999). Inactivation of the BMPR-IB gene results in a skeletal phenotype very similar to the GDF5-/- mouse phenotype, suggesting the importance of GDF5 signaling through the BMPR-IB receptor for chondrogenesis and joint formation (Baur et al., 2000; Yi et al., 2000).

Targeted inactivation of both the BMPR-IA and BMPR-IB genes in chondrocytes severely disrupts cartilage formation, with impaired chondrocyte proliferation and differentiation, and a concomitant lack of expression of Sox9, L-Sox5 and Sox6 (Yoon et al., 2005). This phenotype is more severe than either the GDF5-/- or

BMPR-IB-/- mice – suggesting additional roles of BMPs and BMPR-IA-mediated signaling in chondrogenesis. BMPs have been shown to induce cartilage-specific Sox genes in mesenchymal cells in culture (Chimal-Monroy et al., 2003). These data indicate a stimulatory role of BMP signaling through Smad1 and Smad5 on chondrocyte differentiation. An important role of BMPs and BMP-activated receptors and Smads is also consistent with the generation of cartilage from mesenchymal stem cells following intramuscular injection of BMPs (Wozney et al., 1988).

The impaired Sox transcription factor expression occurring in the absence of BMP signaling (Yoon et al., 2005) suggests that BMP signaling is required at an early stage of chondrocyte differentiation. This is consistent with active BMP signaling at sites of mesenchymal condensation (Murtaugh et al., 1999). At the molecular level, this requirement is supported by the observation that BMP signaling represses the transcriptional activity of the Shh-induced transcription factor Nkx3.2. This repression in mesenchymal cells is required for the activation of Sox9, L-Sox5 and Sox6 expression in the progression of chondrogenesis, presumably by repressing the expression of Nkx3.2-activated "anti-chondrogenic" proteins, one of which is Runx2, the transcription factor that drives osteoblast differentiation (Kim et al., 2003; Lengner et al., 2005). The BMP-induced repression of Nkx3.2 is direct and is mediated by Smad1 and Smad4, which serve as BMP-activated transcription effectors of and form a complex in response to BMP signaling. Thus Smad1/4 stabilize the interaction of the histone deacetylase HDAC1 with Nkx3.2 at an Nkx3.2 binding DNA sequence in response to BMP. The interaction of HDAC1, which in turn recruits the Sin3A corepressor complex, results in the repression of Nkx3.2 at Nkx3.2-responsive genes, thus allowing progression of differentiation (Kim et al., 2003). The observations that BMPs additionally promote progression of chondrogenic differentiation at later stages suggest that BMP-activated Smads may also interact with Sox transcription factors to increase their activity. The critical role of Smads in chondrogenesis is highlighted by the severely disorganized growth plate and dwarfism in mice with a targeted deletion of Smad4 in chondrocytes (Zhang et al., 2005). These chondrocytes exhibit deficiencies in both TGF-β and BMP responsiveness, implicating Smad signaling by the TGF-β family as a key cell-intrinsic regulator of chondrogenesis.

7. SMADS AS REGULATORS OF MYOGENIC DIFFERENTIATION

Similarly to the other mesenchymal differentiation lineages, the generation of muscle cells from its progenitors is regulated by members of the TGF-β family. TGF-βs, myostatin and BMPs act as potent inhibitors of the progression of myogenic differentiation, although TGF-β may also play a stimulatory role in the generation of myoblasts, *i.e.* presumably at early stages of differentiation.

TGF-β is a potent inhibitor of myoblast differentiation. Adding TGF-β to myoblast cultures, *e.g.* C2C12 myoblasts, blocks the progression of differentiation and, thus, the expression of proteins known to be associated with myogenic differentiation as well as the formation of multinucleated myofibers (Olson et al., 1986). This inhibition of differentiation is mediated by Smad3; thus, increased Smad3 activation levels block myoblast differentiation in the absence of exogenously added TGF-β, presumably fully dependent on autocrine TGF-β signaling, while dominant negative interference with Smad3 signaling interferes with the inhibition of myogenic differentiation by TGF-β (Liu et al., 2001). On the other hand, several observations reveal a stimulatory role for TGF-β in myoblast generation or early differentiation. Thus, blocking TGF-β signaling by expressing a dominant negative version of TβRII, the type II TGF-β receptor, blocks myogenic differentiation (Filvaroff et al., 1994). Also, inhibition of TGF-β activity inhibits myotome induction, whereas TGF-β together with FGF induce myotome formation (Stern et al., 1997). In addition, the location of TGF-β within muscle tissues during early embryonic development is consistent with its involvement in muscle pattern formation. These and other observations led to the suggestion that the high level of TGF-β in the first wave of myofiber formation inhibits the fusion of late but not early myoblasts, and that the onset of secondary myofiber formation is triggered by a decrease in the local TGF-β levels (Cusella-De Angelis et al., 1994). These observations are also consistent with a role of TGF-β in the expansion and maintenance of the cells that give rise to muscle cells, and in the competence of the progenitors to initiate myogenic differentiation (Filvaroff et al., 1994).

Myostatin, a TGF-β-related protein initially identified as GDF-8, is expressed in the myotome layer during development and then primarily in muscle cells. Targeted gene inactivation conferred an increased muscularity in mice, associated with increased cell proliferation and muscle cell hypertrophy (McPherron et al., 1997). Consistent with this phenotype, myostatin inhibits the proliferation and differentiation of myoblasts (Langley et al., 2002; McCroskery et al., 2003; Wagner et al., 2005). In addition, myoblasts and satellite cells cultured from myostatin null mice proliferate and differentiate more rapidly (McCroskery et al., 2003; Wagner et al., 2005). Thus, myostatin should be considered as an endogenous and autocrine inhibitor of muscle generation. Myostatin exerts its biological activities by binding to ActRIIB, *i.e.* initially identified as a type II Activin receptor, in partnership with ActRIB, the Activin type I receptor, or TβRI, the major type I TGF-β receptor (Rebbapragada et al., 2003). While this would predict that both Smad2 and Smad3 would be activated, only Smad3 was shown to be activated by myostatin in myoblasts (Langley et al., 2002). The mechanisms underlying the inhibition of myogenic differentiation by myostatin are therefore likely to be identical to the Smad3-mediated mechanisms exerted by TGF-β.

At the molecular level, TGF-β-activated Smad3 inhibits myogenic differentiation through its association with myogenic transcription factors in response to TGF-β, with consequent interference with transcription complex formation at the regulatory DNA sequences (Fig. 3). In response to TGF-β, Smad3 directly represses the transcription activity of myogenic bHLH transcription factors, such as MyoD and myogenin that drive myogenic differentiation. Smad3 associates with the bHLH domain of MyoD (or other myogenic bHLH transcription factors) in response to

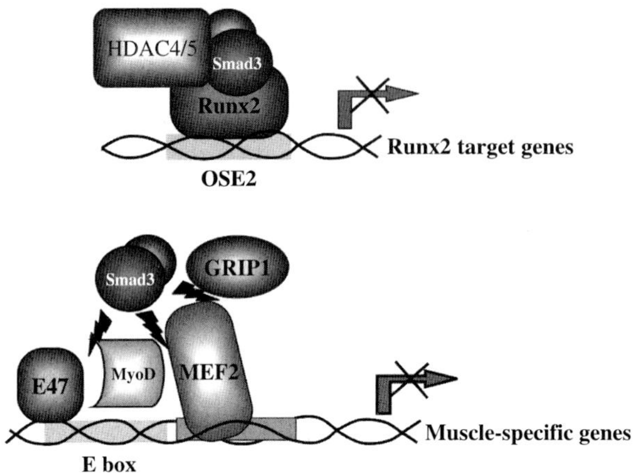

Figure 3. Mechanisms of transcription repression in response to TGF-β in osteoblast and myoblast differentiation. In osteoblast differentiation, TGF-β-activated Smad3 recruits histone deacetylases 4 and/or 5 to Runx2 at the promoter, thus forming a stable complex of Runx2, Smad3 and histone decetylase 4 or 5 at the Runx2 binding DNA sequence (OSE2). This complex confers deacetylation of histone 4 in response to TGF-β. In myogenic transcription, TGF-β-activated Smad3 binds to the HLH domain of myogenic bHLH transcription factors, such as MyoD, thus interfering with the dimerization of the bHLH protein with an E protein and consequently with DNA binding of the dimer to the E box DNA sequence. Smad3 also interferes with binding of the coactivator GRIP1 with MEF2, as well as with MEF2's association with MyoD

TGF-β. Since this domain is required for association with E-protein partners such as E12 and E47, this association interferes with its dimerization with an E protein. As the dimerization is required for efficient DNA binding of MyoD (or other myogenic bHLH proteins), this interference by Smad3 prevents efficient DNA binding of MyoD to E boxes, thus preventing transcriptional activation (Liu et al., 2001). Smad3 also represses the function of MEF2, another class of myogenic transcription factors that physically interact with myogenic bHLH transcription factors and strongly enhance their activity through transcriptional cooperation (Black and Olson, 1998). This interaction of Smad3 disrupts the association of MEF2 with GRIP1 (glucocorticoid receptor-interacting protein 1), a coactivator required for MEF2's activity in myogenic differentiation. Consistent with this physical displacement, TGF-β signaling blocks the GRIP1-induced redistribution of MEF2 to discrete, subnuclear loci and the GRIP1 recruitment to the myogenin promoter (Liu et al., 2004).

BMPs also inhibit myogenic differentiation, but do so through a mechanism that differs from TGF-β-induced inhibition of myogenesis. This is reflected in the differentiation response of the myoblasts to ligand; whereas BMPs concomitantly initiate osteoblast differentiation, TGF-β does not. A most critical difference at the molecular level is that BMPs directly activate expression of Id1 (inhibitor of differentiation/DNA binding 1), whereas TGF-β does not (Katagiri et al., 1994; Ogata et al., 1993; Nakashima et al., 2001). Id proteins have an HLH domain, but lack the basic domain that allows for DNA binding. Thus, Id proteins heterodimerize with E proteins, thereby preventing them through dominant negative interference from interacting with bHLH proteins and binding to DNA (Norton, 2000). Consequently, BMP-induced Id1 expression in myoblasts (Katagiri et al., 1994) inhibits the activity of the myogenic bHLH transcription factors, additionally leading to their accelerated degradation (Vinals and Ventura, 2004). Consistent with the direct transcriptional activation of Id1 upon BMP stimulation, the Id1 promoter contains Smad binding sites capable of binding to BMP-activated Smad1 and Smad4 (Korchynskyi and ten Dijke, 2002; Lopez-Rovira et al., 2002). The induction of Id protein expression by BMPs is not restricted to mesenchymal cells and occurs in *e.g.* embryonic stem cells (Hollnagel and Nordheim, 1999) and neural progenitor cells (Nakashima et al., 2001). In the latter cells, BMP-induced Id1 expression redirects the cells from neuronal to astroglial differentiation through inhibition of neurogenic HLH transcription factors (Nakashima et al., 2001), thus mechanistically resembling the BMP-induced conversion of myoblast cells. These findings illustrate a mechanism whereby cells that are committed to a certain lineage, through expression of differentiation-specific HLH factors, can be redirected into another differentiation pathway by BMP signaling.

8. SUMMARY AND PERSPECTIVES

The Smads have revealed themselves as cell-intrinsic regulators of mesenchymal differentiation that respond to autocrine as well as paracrine signals from the extracellular TGF-β family proteins. They help define both the selection of the differentiation lineage and the progression of differentiation along a particular pathway. The molecular mechanisms at the transcription level illustrate how Smads efficiently regulate differentiation, through interactions with key transcription factors involved in the selection or progression of mesenchymal differentiation. These results provide paradigms for how the Smad activities regulate other types of differentiation. Although the principles have been clarified for how Smads, in response to TGF-β family members, can repress or stimulate differentiation along a specific lineage, much work remains to be done to delineate the full extent of the regulation of mesenchymal differentiation by Smads at distinct levels during selection and progression of a differentiation pathway. In addition, in most cases, we do not know which ligands and receptors naturally regulate the Smad activities in the generation and differentiation of tissues and in tissue regeneration during repair. Considering the complexity of the TGF-β family and the many possible receptor combinations, defining the natural ligand/receptor combinations will continue to be an extensive and tedious undertaking. Finally, as epigenetic mechanisms define the regulation of gene expression during differentiation, possibly in a lineage-dependent manner, it will be of great interest to unravel whether and how Smads can affect, and possibly redirect, chromatin remodeling.

REFERENCES

Ahrens, M., Ankenbauer, T., Schröder, D., Hollnagel, A., Mayer, H., and Gross, G., 1993, Expression of human bone morphogenetic proteins-2 or -4 in murine mesenchymal progenitor C3H10T1/2 cells induces differentiation into distinct mesenchymal cell lineages. *DNA Cell Biol* 12: 871-880.

Alliston, T., Choy, L., Ducy, P., Karsenty, G., and Derynck, R., 2001, TGF-β-induced repression of CBFA1 by Smad3 decreases cbfa1 and osteocalcin expression and inhibits osteoblast differentiation. *EMBO J* 20: 2254-2272.

Alvarez, J., and Serra, R., 2004, Unique and redundant roles of Smad3 in TGF-β-mediated regulation of long bone development in organ culture. *Dev Dyn* 230: 685-699.

Artaza, J.N., Bhasin, S., Magee, T.R., Reisz-Porszasz, S., Shen, R., Groome, N.P., Fareez, M.M., and Gonzalez-Cadavid, N.F., 2005, Myostatin inhibits myogenesis and promotes adipogenesis in C3H 10T(1/2) mesenchymal multipotent cells. *Endocrinology* 146: 3547-3557.

Asahina, I., Sampath, T.K., and Hauschka, P.V., 1996, Human osteogenic protein-1 induces chondroblastic, osteoblastic, and/or adipocytic differentiation of clonal murine target cells. *Exp Cell Res* 222: 38-47.

Baur S.T., Mai J.J., and Dymecki, S.M., 2000, Combinatorial signaling through BMP receptor IB and GDF5: shaping of the distal mouse limb and the genetics of distal limb diversity. *Development* 127: 605-619.

Black, B.L., and Olson, E.N., 1998, Transcriptional control of muscle development by myocyte enhancer factor-2 (MEF2) proteins. *Annu Rev Cell Dev Biol* 14: 167-196.

Brugger, S.M., Merrill, A.E., Torres-Vazquez J, Wu, N., Ting, M.C., Cho, J.Y., Dobias, S.L., Yi, S.E., Lyons, K., Bell, J.R., Arora, K., Warrior, R., and Maxson, R., 2004, A phylogenetically conserved cis-regulatory module in the Msx2 promoter is sufficient for BMP-dependent transcription in murine and *Drosophila* embryos. *Development* 131: 5153-5165.

Canalis, E., Economides, A.N., and Gazzerro, E., 2003, Bone morphogenetic proteins, their antagonists, and the skeleton. *Endocr Rev* 24: 218-235.

Caplan, A.I., 2005, Mesenchymal stem cells: cell-based reconstructive therapy in orthopedics. *Tissue Eng 11*: 1198-1211.

Chen, D., Ji, X., Harris, M.A., Feng, J.Q., Karsenty, G., Celeste, A.J., Rosen, V., Mundy, G.R., and Harris, S.E., 1998, Differential roles for bone morphogenetic protein (BMP) receptor type IB and IA in differentiation and specification of mesenchymal precursor cells to osteoblast and adipocyte lineages. *J Cell Biol* 142: 295-305.

Chimal-Monroy, J., Rodriguez-Leon, J., Montero, J.A., Ganan, Y., Macias, D., Merino, R., and Hurle, J.M., 2003, Analysis of the molecular cascade responsible for mesodermal limb chondrogenesis: Sox genes and BMP signaling. *Dev Biol* 257: 292-301.

Choy, L., Skillington, J., and Derynck, R., 2000, Roles of autocrine TGF-β receptor and Smad signaling in adipocyte differentiation. *J Cell Biol* 149: 667-682.

Choy, L., and Derynck, R., 2003, Transforming growth factor-β inhibits adipocyte differentiation by Smad3 interacting with CCAAT/enhancer-binding protein (C/EBP) and repressing C/EBP transactivation function. *J Biol Chem* 278: 9609-9619.

Clouthier, D.E., Comerford, S.A., and Hammer, R.E., 1997, Hepatic fibrosis, glomerulosclerosis, and a lipodystrophy-like syndrome in PEPCK-TGF-β1 transgenic mice. *J Clin Invest* 100: 2697-2713.

Cusella-De Angelis, M.G., Molinari, S., Le Donne, A., Coletta, M., Vivarelli, E., Bouche, M., Molinaro, M., Ferrari, S., and Cossu, G., 1994, Differential response of embryonic and fetal myoblasts to TGF β: a possible regulatory mechanism of skeletal muscle histogenesis. *Development* 120: 925-933.

de Crombrugghe, B., Lefebvre, V., and Nakashima, K., 2001, Regulatory mechanisms in the pathways of cartilage and bone formation. *Curr Opin Cell Biol* 13: 721-7.

Dhawan, J., and Rando, T.A., 2005, Stem cells in postnatal myogenesis: molecular mechanisms of satellite cell quiescence, activation and replenishment. *Trends Cell Biol* 15: 666-673.

Docheva, D., Hunziker, E.B., Fassler, R., and Brandau, O., 2005, Tenomodulin is necessary for tenocyte proliferation and tendon maturation. *Mol Cell Biol* 25: 699-705.

Dosch, R., Gawantka, V., Delius, H., Blumenstock, C., and Niehrs, C., 1997, Bmp-4 acts as a morphogen in dorsoventral mesoderm patterning in *Xenopus*. *Development* 124: 2325-2334.

Ducy, P., Zhang, R., Geoffroy, V., Ridal, A.L., and Karsenty, G., 1997, Osf2/Cbfa1: a transcriptional activator of osteoblast differentiation. *Cell* 89: 747-754.

Erlebacher, A., and Derynck, R., 1996, Increased expression of TGF-β2 in osteoblasts results in an osteoporosis-like phenotype. *J Cell Biol* 132: 195-210.

Filvaroff, E.H., Ebner, R., and Derynck, R., 1994, Inhibition of myogenic differentiation in myoblasts expressing a truncated type II TGF-β receptor. *Development* 121: 185-195.

Filvaroff, E., Erlebacher, A., Ye, J., Gitelman, S.E., Lotz, J, Heillman, M., and Derynck, R., 1999, Inhibition of TGF-β receptor signaling in osteoblasts leads to decreased bone remodeling and increased trabecular bone mass. *Development* 126: 4267-4279.

Furumatsu, T., M. Tsuda, M., Yoshida, K., Taniguchi, N., Ito, T., Hashimoto, M., Ito, T., and Asahara, H., 2005, Sox9 and p300 cooperatively regulate chromatin-mediated transcription. *J Biol Chem* 280: 35203-35208.

Gimble, J.M., Morgan, C., Kelly, K., Wu, X., Dandapani, V., Wang, C.S., and Rosen, V., 1995, Bone morphogenetic proteins inhibit adipocyte differentiation by bone marrow stromal cells. *J Cell Biochem* 58: 393-402.

Green, J.B.A., New, H.V., and Smith, J.C., 1992, Responses of embryonic *Xenopus* cells to activin and FGF are separate by multiple dose thresholds and correspond to distinct axes of mesoderm. *Cell* 71: 731-739.

Gori, F., Thomas, T., Hicok, K.C., Spelsberg, T.C., and Riggs, B.L., 1999, Differentiation of human marrow stromal precursor cells: bone morphogenetic protein-2 increases OSF2/CBFA1, enhances osteoblast commitment, and inhibits late adipocyte maturation. *J Bone Miner Res* 14: 1522-1535.

Harada, S., and Rodan, G.A., 2003, Control of osteoblast function and regulation of bone mass. *Nature* 423: 349-355.

Hirai, S., Yamanaka, M., Kawachi, H., Matsui, T., and Yano, H., 2005, Activin A inhibits differentiation of 3T3-L1 preadipocyte. *Mol Cell Endocrinol* 232: 21-26.

Hollnagel, A., Oehlmann, V., Heymer, J., Ruther, U., and Nordheim, A., 1999, Id genes are direct targets of bone morphogenetic protein induction in embryonic stem cells. *J Biol Chem* 274: 19838-19845.

Hotten, G.C., Matsumoto, T., Kimura, M., Bechtold, R.F., Kron, R., Ohara, T., Tanaka, H., Satoh, Y., Okazaki, M., Shirai, T., Pan, H., Kawai, S., Pohl. J.S., and Kudo, A., 1996, Recombinant human growth/differentiation factor 5 stimulates mesenchyme aggregation and chondrogenesis responsible for the skeletal development of limbs. *Growth Factors* 13: 65-74.

Ignotz, R.A., and Massagué, J., 1985, Type β transforming growth factor controls the adipogenic differentiation of 3T3 fibroblasts. *Proc Natl Acad Sci U S A* 82: 8530-8534.

Jeoung, D.-I., Tang, B., and Sonenberg, M., 1995, Mitogenic response to TGF-β in 3T3-F442A cells. *Biochem Biophys Res Comm* 216: 964-969.

Joyce, M.E., Roberts, A.B., Sporn, M.B., and Bolander, M.E., 1990, Transforming growth factor-β and the initiation of chondrogenesis and osteogenesis in the rat femur. *J Cell Biol* 110: 2195-2207.

Kang, J.S., Alliston, T., Delston, R., and Derynck, R., 2005, Repression of Runx2 function by TGF-β through recruitment of class II histone deacetylases by Smad3. *EMBO J* 24: 2543-2555.

Katagiri, T., Yamaguchi, A., Komaki, M., Abe, E., Takahashi, N., Ikeda, T., Rosen, V., Wozney, J. M., Fujisawa-Sehara, A., and Suda, T., 1994, Bone morphogenetic protein-2 converts the differentiation pathway of C2C12 myoblasts into the osteoblast lineage. *J Cell Biol* 127, 1755-1766.

Kim, D.W., and Lassar, A.B., 2003, Smad-dependent recruitment of a histone deacetylase/Sin3A complex modulates the bone morphogenetic protein-dependent transcriptional repressor activity of Nkx3.2. *Mol Cell Biol* 23: 8704-8717.

Korchynskyi, O., and ten Dijke, P., 2002, Identification and functional characterization of distinct critically important bone morphogenetic protein-specific response elements in the Id1 promoter. *J Biol Chem* 277: 4883-4891.

Kulyk, W.M., Rodgers, B.J., Greer, K., and Kosher, R.A., 1989, Promotion of embryonic chick limb cartilage differentiation by transforming growth factor-β. *Dev Biol* 135: 424-430.

Langley, B., Thomas, M., Bishop, A., Sharma, M., Gilmour, S., and Kambadur, R., 2002, Myostatin inhibits myoblast differentiation by down-regulating MyoD expression. *J Biol Chem* 277: 49831-49840.

Lee, K.S., Kim, H.J., Li, Q. L., Chi, X. Z., Ueta, C., Komori, T., Wozney, J.M., Kim, E.G., Choi, J.Y., Ryoo, H.M., and Bae, S.C., 2000, Runx2 is a common target of transforming growth factor β1 and bone morphogenetic protein 2, and cooperation between Runx2 and Smad5 induces osteoblast-specific gene expression in the pluripotent mesenchymal precursor cell line C2C12. *Mol Cell Biol* 20: 8783-8792.

Lee, M.H., Kim, Y.J., Kim, H.J., Park, H.D., Kang, A.R., Kyung, H.M., Sung, J.H., Wozney, J.M., Kim, H.J., and Ryoo, H.M., 2003, BMP-2-induced Runx2 expression is mediated by Dlx5, and TGF-β1 opposes the BMP-2-induced osteoblast differentiation by suppression of Dlx5 expression. *J Biol Chem* 278: 34387-34394.

Lee, M.H., Kim, Y.J., Yoon, W.J., Kim, J.I., Kim, B.G., Hwang, Y.S., Wozney, J.M., Chi, X.Z., Bae, S.C., Choi, K.Y., Cho, J.Y., Choi, J.Y., and Ryoo, H.M., 2005, Dlx5 specifically regulates Runx2 type II expression by binding to homeodomain-response elements in the Runx2 distal promoter. *J Biol Chem* 280: 35579-35587.

Lefebvre, V., and Smits, P., 2005, Transcriptional control of chondrocyte fate and differentiation. *Birth Defects Res C Embryo Today* 75: 200-212.

Lengner, C.J., Hassan, M.Q., Serra, R.W., Lepper, C., van Wijnen, A.J., Stein, J.L., Lian, J.B., and Stein, G.S., 2005, Nkx3.2-mediated repression of Runx2 promotes chondrogenic differentiation. *J Biol Chem* 280: 15872-15879.

Liu, D., Black, B.L., and Derynck, R., 2001, TGF-β inhibits muscle differentiation through functional repression of myogenic transcription factors by Smad3. *Genes Dev* 15: 2950-2966.

Liu, D., Kang, J.S., and Derynck, R., 2004, TGF-β-activated Smad3 represses MEF2-dependent transcription in myogenic differentiation. *EMBO J* 23: 1557-1566.

Lopez-Rovira, T., Chalaux, E., Massagué, J., Rosa, J.L., and Ventura, F., 2002, Direct binding of Smad1 and Smad4 to two distinct motifs mediates bone morphogenetic protein-specific transcriptional activation of Id1 gene. *J Biol Chem* 277: 3176-31785.

Maeda, S., Hayashi, S., Komiya, S., Imamura, T., and Miyazono, K., 2004, Endogenous TGF-β signaling suppresses maturation of osteoblastic mesenchymal cells. *EMBO J* 23: 552-563.

Marazzi, G., Wang, Y., and Sassoon D., 1997, Msx2 is a transcriptional regulator in the BMP4-mediated programmed cell death pathway. *Dev Biol* 186: 127-138.

McCroskery, S., Thomas, M., Maxwell, L., Sharma, M., and Kambadur, R., 2003, Myostatin negatively regulates satellite cell activation and self-renewal. *J Cell Biol* 162: 1135-1147.

McPherron, A.C., and Lee, S.-J., 1993, GDF-3 and GDF-9: two new members of the transforming growth factor-β superfamily containing a novel pattern of cysteines. *J Biol Chem* 268: 3444-3449.

McPherron, A.C., Lawler, A.M., and Lee, S.-J., 1997, Regulation of skeletal muscle mass in mice by a new TGF-β superfamily member. *Nature* 387: 83-90.

McPherron, A.C., and Lee, S.-J., 2002, Suppression of body fat accumulation in myostatin-deficient mice. *J Clin Invest* 109: 595-601.

Millan, F.A., Denhez, F., Kondaiah, P., and Akhurst, R.J., 1991, Embryonic gene expression patterns of TGF β1, β2 and β3 suggest different developmental functions *in vivo*. *Development* 111: 131-43.

Miyama, K., Yamada, G., Yamamoto, T.S., Takagi, C., Miyado, K., Sakai, M., Ueno, N., and Shibuya, H., 1999, A BMP-inducible gene, dlx5, regulates osteoblast differentiation and mesoderm induction. *Dev Biol* 208: 123-133.

Murtaugh, L.C., Chyung, J.H., and Lassar, A.B., 1999, Sonic hedgehog promotes somitic chondrogenesis by altering the cellular response to BMP signaling. *Genes Dev* 13: 225-237.

Nakashima, K., Takizawa, T., Ochiai, W., Yanagisawa, M., Hisatsune, T., Nakafuku, M., Miyazono, K., Kishimoto, T., Kageyama, R., and Taga, T., 2001, BMP2-mediated alteration in the developmental pathway of fetal mouse brain cells from neurogenesis to astrocytogenesis. *Proc Natl Acad Sci U S A* 98: 5868-5873.

Nishitoh, H., Ichijo, H., Kimura, M., Matsumoto, T., Makishima, F., Yamaguchi, A., Yamashita, H., Enomoto, S., and Miyazono, K., 1996, Identification of type I and type II serine/threonine kinase receptors for growth/differentiation factor-5. *J Biol Chem* 271: 21345-21352.

Norton, J.D., 2000, Id helix-loop-helix proteins in cell growth, differentiation and tumorigenesis. *J Cell Sci* 113: 3897-3905.
Ogata, T., Wozney, J.M., Benezra, R., and Noda, M., 1993, Bone morphogenetic protein 2 transiently enhances expression of a gene, Id (inhibitor of differentiation), encoding a helix-loop-helix molecule in osteoblast-like cells. *Proc Natl Acad Sci U S A* 90: 9219-9222.
Olson, E.N., Sternberg, E., Hu, J.S., Spizz, G., and Wilcox, C., 1986, Regulation of myogenic differentiation by type β transforming growth factor. *J Cell Biol* 103: 1799-1805.
Otto, T.C., and Lane, M.D., 2005, Adipose development: from stem cell to adipocyte. *Crit Rev Biochem Mol Biol* 40: 229-242.
Pfeilschifter, J., Wolf, O., Naumann, A., Minne, H.W., Mundy, G.R., and Ziegler, R., 1990, Chemotactic response of osteoblast-like cells to transforming growth factor β. *J Bone Miner Res* 5: 825-830.
Pittenger, M.F., Mackay, A.M., Beck, S.C., Jaiswal, R.K., Douglas, R., Mosca, J.D., Moorman, M.A., Simonetti, D.W., Craig, S., and Marshak, D.R., 1999, Multilineage potential of adult human mesenchymal stem cells. *Science* 284: 143-147.
Polak, J., and Hench, L., 2005, Gene therapy progress and prospects in tissue engineering. *Gene Ther* 12: 1725-1733.
Qiao, B., Padilla, S.R., and Benya, P.D., 2005, Transforming growth factor (TGF)-β-activated kinase 1 mimics and mediates TGF-β-induced stimulation of type II collagen synthesis in chondrocytes independent of Col2a1 transcription and Smad3 signaling. *J Biol Chem* 280: 17562-17571.
Rebbapragada, A., Benchabane, H., Wrana, J.L., Celeste, A.J., and Attisano, L., 2003, Myostatin signals through a transforming growth factor β-like signaling pathway to block adipogenesis. *Mol Cell Biol* 23: 7230-7242.
Samad, F., Yamamoto, K., Pandey, M., and Loskutoff, D.J., 1997, Elevated expression of transforming growth factor-β in adipose tissue from obese mice. *Mol Med* 3: 37-48.
Sartorelli, V., and Caretti, G., 2005, Mechanisms underlying the transcriptional regulation of skeletal myogenesis. *Curr Opin Genet Dev* 15: 528-535.
Serra, R., Karaplis, A., and Sohn, P., 1999, Parathyroid hormone-related peptide (PTHrP)-dependent and -independent effects of transforming growth factor β (TGF-β) on endochondral bone formation. *J Cell Biol* 145: 783-794.
Seyedin, S.M., Thomas, T.C., Thompson, A.Y., Rosen, D.M., and Piez, K.A., 1985, Purification and characterization of two cartilage-inducing factors from bovine demineralized bone. *Proc Natl Acad Sci U S A* 82: 2267-2271.
Sowa, H., Kaji, H., Hendy, G.N., Kanaff, L., Komori, T., Sugimoto, T., and Chihara, K., 2004, Menin is required for bone morphogenetic protein 2- and transforming growth factor β-regulated osteoblastic differentiation through interaction with Smads and Runx2. *J Biol Chem* 279: 40267-40275.
Skillington, J., Choy, L., and Derynck, R., 2002, Bone morphogenetic protein and retinoic acid signaling cooperate to induce osteoblast differentiation of preadipocytes. *J Cell Biol* 159: 135-146.
Sottile, V., and Seuwen, K., 2000, Bone morphogenetic protein-2 stimulates adipogenic differentiation of mesenchymal precursor cells in synergy with BRL 49653 (rosiglitazone), *FEBS Lett* 475: 201-204.
Stern, H.M., Lin-Jones, J., and Hauschka, S.D., 1997, Synergistic interactions between bFGF and a TGF-β family member may mediate myogenic signals from the neural tube. *Development* 124: 3511-3523.
Storm, E.E., Huynh, T.V., Copeland, N.G., Jenkins, N.A., Kingsley, D.M., and Lee, S.J., 1994, Limb alterations in brachypodism mice due to mutations in a new member of the TGF β-superfamily. *Nature* 368: 639-643.
Suzuki, A., Ueno, N., and Hemmati-Brivanlou, A., 1997, *Xenopus* msx1 mediates epidermal induction and neural inhibition by BMP4. *Development* 124: 3037-3044.
Torti, S.V., Larrick, J.W., and Ringold, G.M., 1989, Modulation of adipocyte differentiation by tumor necrosis factor and transforming growth factor β. *J Cell Biol* 108: 1105-1113.
Tsumaki, N., Tanaka, K., Arikawa-Hirasawa, E., Nakase, T., Kimura, T., Thomas, J.T., Ochi, T., Luyten, F.P., and Yamada, Y., 1999, Role of CDMP-1 in skeletal morphogenesis: promotion of mesenchymal cell recruitment and chondrocyte differentiation. *J Cell Biol* 144: 161-173.

Vinals, F., and Ventura, F., 2004, Myogenin protein stability is decreased by BMP-2 through a mechanism implicating Id1. *J Biol Chem* 279: 45766-45772.

Wagner, K.R., Liu, X., Chang, X., and Allen, R.E., 2005, Muscle regeneration in the prolonged absence of myostatin. *Proc Natl Acad Sci U S A* 102: 2519-2524.

Wang, E.A., Israel, D.I., Kelly, S., and Luxenberg, D.P., 1993, Bone morphogenetic protein-2 causes commitment and differentiation in C3H10T1/2 and 3T3 cells. *Growth Factors* 9: 57-71.

Wang, W., Yang, Y., Meng, Y., and Shi, Y., 2004, GDF-3 is an adipogenic cytokine under high fat dietary condition. *Biochem Biophys Res Commun* 321: 1024-1031.

Watanabe, H., de Caestecker, M.P., and Yamada, Y., 2001, Transcriptional cross-talk between Smad, ERK1/2, and p38 mitogen-activated protein kinase pathways regulates transforming growth factor-β-induced aggrecan gene expression in chondrogenic ATDC5 cells. *J Biol Chem* 276: 14466-14473.

Witthuhn, B.A., and Bernlohr, D.A., 2001, Upregulation of bone morphogenetic protein GDF-3/Vgr-2 expression in adipose tissue of FABP4/aP2 null mice. *Cytokine* 14: 129-135.

Wozney, J.M., Rosen, V., Celeste, A.J., Mitsock, L.M., Whitters, M.J., Kriz, R.W., Hewick, R.M., and Wang, E.A., 1988, Novel regulators of bone formation: molecular clones and activities. *Science* 242: 1528-1534.

Yi, S.E., Daluiski, A., Pederson, R., Rosen, V., and Lyons, K.M., 2000, The type I BMP receptor BMPR-IB is required for chondrogenesis in the mouse limb. *Development* 127: 621-630.

Ying, Q.L., Nichols, J., Chambers, I., and Smith, A., 2003, BMP induction of Id proteins suppresses differentiation and sustains embryonic stem cell self-renewal in collaboration with STAT3. *Cell* 115: 281-292

Yoon, B.S., Ovchinnikov, D.A., Yoshii, I., Mishina, Y., Behringer, R.R., and Lyons, K.M., 2005, Bmpr1a and Bmpr1b have overlapping functions and are essential for chondrogenesis *in vivo*. *Proc Natl Acad Sci U S A* 102: 5062-5067.

Zaidi, S.K., Sullivan, A.J., van Wijnen, A.J., Stein, J.L., Stein, G.S., and Lian, J.B., 2002, Integration of Runx and Smad regulatory signals at transcriptionally active subnuclear sites. *Proc Natl Acad Sci U S A* 99: 8048-8053.

Zhang, Y.W., Yasui, N., Ito, K., Huang, G., Fujii, M., Hanai, J., Nogami, H., Ochi, T., Miyazono, K., and Ito, Y., 2000, A RUNX2/PEBP2αA/CBFA1 mutation displaying impaired transactivation and Smad interaction in cleidocranial dysplasia. *Proc Natl Acad Sci U S A* 97: 10549-10554.

Zhang, J., Tan, X., Li, W., Wang, Y., Wang, J., Cheng, X., and Yang, X., 2005, Smad4 is required for the normal organization of the cartilage growth plate. *Dev Biol* 284: 311-322.

Zuk, P.A., Zhu, M., Mizuno, H., Huang, J., Futrell, J.W., Katz, A.J., Benhaim, P., Lorenz, H.P., and Hedrick, M.H., 2001, Multilineage cells from human adipose tissue: implications for cell-based therapies. *Tissue Eng* 7: 211-228.

CHAPTER 6

SMAD PROTEINS IN APOPTOTIC AND SURVIVAL SIGNALING

ANDREW R. CONERY AND KUNXIN LUO
University of California, Berkeley, and Lawrence Berkeley National Laboratory

Abstract: Perhaps the most critical decision to be made during the life of a cell is whether to live or to die. Signaling events initiated by TGF-β superfamily cytokines play a critical role in specifying cell death and survival responses. Of particular importance in this signaling process are the Smad proteins. By interacting with general and tissue-specific transcription factors, R-Smad proteins regulate the expression of key target genes in apoptotic and survival signaling. The inhibitory Smad7 functions both as an antagonist of TGF-β signaling and as an integrator of multiple cellular pathways. Smad proteins can cooperate with other signaling pathways to regulate cell survival and can also function as the site of regulation by other pathways. Knowledge of how the Smads function in the generation of cell death and survival aids in our understanding of developmental disorders and cancer in which defective cell death responses have been implicated

Keywords: TGF-β; BMP; apoptosis; survival; caspase; JNK; p38; Akt

1. INTRODUCTION

Apoptosis, or programmed cell death, plays a central role in the selection of appropriately differentiated cells during embryonic development and in the elimination of damaged, abnormal, or aging cells during post-embryonic life. Cytokines of the transforming growth factor-β (TGF-β) superfamily are essential mediators of apoptosis and survival decisions. Depending upon the cellular context and the differentiation status of the target cell, TGF-β family cytokines have the capability of promoting either cell survival or cell death.

Vertebrate limb development is a well characterized example of the opposing function of TGF-β family members in the context of cell death (see Chapter 5). As revealed in both chicken and mouse, BMP signaling is responsible for the elimination of undifferentiated mesenchymal cells between differentiating cartilage

P. ten Dijke and C.-H. Heldin (eds.), Smad Signal Transduction, 113–129.

and also in the web between digits (Merino et al., 1999; Zuzarte-Luis and Hurle, 2005). Within the digits, however, BMP promotes the survival of prechondrocytes and further bone development (Merino et al., 1999).

TGF-β1 promotes either apoptosis or survival of T cells depending on their differentiation status. As suggested by the inflammatory phenotype of homozygous null mice, TGF-β1 plays a significant role in regulating T cell homeostasis by promoting apoptosis of certain populations and survival of others (Wahl et al., 2000). Further, while TGF-β1 induces apoptosis in a number of epithelial and myeloid cells, in several cell types such as osteoblasts, hippocampal neurons, and microglia, TGF-β1 promotes survival (Sanchez-Capelo, 2005).

These observations imply that the response of a target cell to stimulation with a TGF-β family cytokine cannot be predicted solely by the identity of the cytokine. Rather, whether a cell lives or dies in response to stimulation with a TGF-β family cytokine depends on downstream signaling events. While Smad-independent pathways play a role in the generation of these responses, the apoptotic or survival programs are mediated to a large extent by Smad-dependent signaling.

1.1 Apoptotic and Survival Signaling Mechanisms

The end result of cellular apoptosis pathways is the activation of a cascade of cysteine proteases known as caspases that proteolyze specific targets to bring about the changes in cell physiology necessary for a dead cell to be phagocytosed. There exist two mechanisms by which a cell can activate the caspase cascade (Zimmermann et al., 2001). In the first, signals from the nucleus such as transcription of specific genes or DNA damage are transduced to the mitochondria. Pro-apoptotic members of the Bcl-2 family of proteins facilitate the release of mitochondrial cytochrome c and other factors, which act with an initiator caspase (caspase-9) to activate the executioner caspase (caspase-3) and initiate the caspase cascade. In the second apoptotic mechanism, ligands outside of the cell, either soluble or attached to the surface of a neighboring cell, signal through a death receptor complex to activate another initiator caspase (caspase-8) and the downstream caspase cascade. The two pathways are not mutually exclusive and often cross-talk to amplify the apoptotic signal. Survival signals can be generated by the inhibition of either pathway of cell death, as through the downregulation of pro-apoptotic members of the Bcl-2 family of proteins or the Fas death receptor ligand FasL.

2. SMAD-MEDIATED APOPTOSIS AND SURVIVAL IN A PHYSIOLOGICAL CONTEXT

As prominent players in TGF-β superfamily signal transduction, the R-Smads are intimately involved in the generation of cell death and survival responses. Evidence suggests that the role of the inhibitory Smad7 is more complex than would be surmised from its known role as an antagonist of R-Smad signaling. The involvement of the Smads in promoting cell death and survival has been established to a large extent through studies of normal development and phenotypic examination of transgenic mice, with validation through cell culture studies. The tissues or cells in which Smad signaling plays a role in generating cell death or survival are discussed below and summarized in Fig. 1.

2.1 BMP-specific Smads

As discussed above, BMP can induce either apoptosis or survival depending on the target cell. Consistent with this observation, the BMP-specific R-Smads (Smad1, Smad5, and Smad8) are involved in generating both responses in target cells. Smad1 transduces the apoptotic signal initiated by high doses of BMP-7 in tubule cells (Piscione et al., 2001), by BMP-2 in mature colon epithelial cells (Hardwick et al., 2004) and by BMP-2 or BMP-7 in human pulmonary vascular smooth muscle cells (Zhang et al., 2003). Both Smad1 and Smad8 are activated during the course of BMP-5-induced interdigital apoptosis during limb development (Zuzarte-Luis et al., 2004).

BMP signaling plays a significant role in selecting the hair follicle keratinocytes that survive the catagen phase of a hair cycle. Under these circumstances, BMP signaling through Smad1 promotes keratinoycte survival, while BMP-initiated MAP kinase signaling promotes apoptosis (Botchkarev and Sharov, 2004). Similarly, BMP-2 prevents serum starvation-induced apoptosis of rat cardiac myocytes in a Smad1-dependent manner (Izumi et al., 2001). Mice lacking Smad5 die at midgestation and display greatly reduced numbers of vascular smooth muscle cells and excessive apoptosis of mesenchymal cells, pointing to a role for Smad5 in promoting survival during angiogenesis (Weinstein et al., 2000) (see Chapter 8). Further investigation of Smad5 knock-out mice also reveals a survival role for Smad5 in cardiac myocytes during heart development (Sun et al., 2005). *Xenopus* embryos lacking the expression of maternal Smad8/xSmad11 fail to undergo gastrulation and instead display pervasive apoptosis, suggesting a maternal BMP/Smad8-dependent survival signaling pathway (Miyanaga et al., 2002).

2.2 TGF-β/Activin-specific Smads

Evidence suggests that although TGF-β signals through both Smad2 and Smad3, the latter plays a more significant role in the induction of TGF-β-induced apoptosis. A comparison of the phenotypes of disrupting Smad2 and Smad3 expression in mice (see Chapter 8) provides the first hint of the divergent roles of the two proteins in apoptotic signaling. In contrast to the gross developmental defects of mice lacking expression of Smad2, mice lacking Smad3 have no obvious phenotype at birth (Weinstein et al., 2000). However, in a specific genetic background these mice die within several months of birth due to a wasting syndrome that is associated with excessive abcesses and inflammation. As with the TGF-β1 knock-out mouse,

BMP R-Smads			
Apoptosis		*Survival*	
Smad1	tubule cells mature colon epithelial cells pulmonary vascular smooth muscle cells	**Smad1**	hair follicle keratinocytes cardiac myocytes
		Smad5	vascular smooth muscle cells mesenchymal cells cardiac myocytes
Smad1/ Smad8	interdigital mesenchymal cells	**Smad8/ xSmad11**	gastrulating *Xenopus* embryo
TGF-β/Activin R-Smads			
Apoptosis		*Survival*	
Smad2	rat fibroblast cells (synergy with c-myc)	**Smad3**	ovarian follicles CD 8+ T cells neurons osteoblasts
Smad3	developing T cells kidney (ureteral obstruction) involuting mammary gland gastric carcinoma cells lung, intestinal epithelial cells rat, human hepatoma cells B lymphocytes, myeloid cell line		
Smad2/3	human hepatoma cells ovarian cancer cells		
Common Smad			
Apoptosis		*Survival*	
Smad4	canine kidney cells pancreatic cancer cells (anoikis)		
Inhibitory Smads			
Apoptosis		*Survival*	
Smad7	rat prostatic epithelial cells prostatic carcinoma cells mesangial cells podocytes	**Smad6/ Smad7**	mouse B-cell hybridomas
		Smad7	developing tooth B lymphocytes human hepatoma cells

Figure 1. Role of specific Smad proteins in generating apoptosis and survival responses

this may be due to defective T cell homeostasis resulting from abrogation of the apoptotic response. Further, these mice show reduced levels of apoptosis in the kidney after unilateral ureteral obstruction as well as a decrease in apoptotic cells in the involuting mammary gland (Inazaki et al., 2004; Yang et al., 2002). The functional difference between Smad2 and Smad3 in the context of apoptosis is also clearly demonstrated by *in vitro* cell culture experiments using epithelial and myeloid cell lines. Overexpression of Smad3, but not Smad2, greatly potentiates the induction of apoptosis in response to TGF-β, while expression of a dominant negative Smad3 or antisense Smad3 inhibits TGF-β-induced apoptosis (Kim et al., 2002; Yamamura et al., 2000).

Despite the clear difference in the roles of Smad2 and Smad3 in TGF-β-induced apoptosis, Smad2 can act independently or in concert with Smad3 to promote apoptosis under certain circumstances. In a rat fibroblast cell line, Smad2 promotes c-Myc-induced apoptosis by antagonizing the protective effects of TGF-β (Mazars et al., 2000). Activin A-induced apoptosis of human hepatoma cells is mediated by both Smad2 and Smad3 (Kanamaru et al., 2002). Dominant negative mutants of either Smad2 or Smad3 are sufficient to block ovarian cancer cell apoptosis induced by Nodal (Xu et al., 2004). It is unclear why TGF-β-induced apoptosis is mediated by Smad3 while Activin- and Nodal-induced apoptosis can be mediated by either Smad2 or Smad3. One possibility is that the promoters of TGF-β-responsive apoptotic target genes favor Smad3 or Smad3/Smad4 binding over Smad2/Smad4 complexes. Alternatively, the tissue-specific transcription factors that target the Smads to TGF-β-responsive apoptotic target genes (see below) may preferentially bind to Smad3 over Smad2.

In certain cells, Smad3 participates in anti-apoptotic signaling. In osteoblasts, parathyroid hormone-induced Smad3 promotes survival in the presence of etoposide or dexamethasone (Sowa et al., 2003). Mice lacking Smad3 display reduced fertility compared to wild-type mice, and this may be due to defects in survival during ovarian folliculogenesis (Tomic et al., 2004). In the maintenance of T cell homeostasis, TGF-β1 signals through Smad3 to selectively induce the survival of activated CD8+ T cells and to negatively regulate the proliferation of CD4+ T cells (McKarns and Schwartz, 2005). Activin A acts as a neuron survival factor in cooperation with basic fibroblast growth factor by signaling through Smad3 (Bao et al., 2005).

2.3 Common Smad

Aside from its involvement in TGF-β family signal transduction as a partner of the R-Smads, there are limited examples of Smad4 being uniquely required for apoptotic or survival signaling. Smad4 overexpression can induce apoptosis in certain cell lines, but this may be due to forced activation of R-Smad-mediated transcription of apoptotic target genes (Atfi et al., 1997). In contrast, re-expression of Smad4 in a Smad4 null cancer cell line results not in apoptosis but in increased sensitivity to TGF-β-induced anoikis, or apoptosis resulting from detachment from the appropriate growth surface (Ramachandra et al., 2002). This suggests that Smad4 may have some role in preventing the detachment of nascent metastatic cells and tumor invasion.

2.4 Inhibitory Smads

Both Smad6 and Smad7 have been described as having antagonistic effects on apoptotic signaling downstream of BMP and TGF-β/Activin, respectively, as would be expected from their established mechanisms of action (see Chapter 19). In mouse

B cells, Smad7 is induced by Activin and acts as an inhibitor of Activin-induced apoptosis, while both Smad6 and Smad7 are induced by BMP and inhibit BMP-induced apoptosis (Ishisaki et al., 1999). Reduction of Smad7 expression during tooth development results in excessive apoptosis, while reduction of Smad2 expression has the opposite effect (Ito et al., 2001). CD40-induced Smad7 serves to protect B-lymphocytes from apoptosis induced by TGF-β (Patil et al., 2000). Overexpression of Smad7 in hepatoma cells is sufficient to block TGF-β-induced apoptosis (Yamamura et al., 2000).

The role of Smad7 in cell death and survival signaling is not only to antagonize signals transduced by the cytokine receptors and R-Smads. As a transcriptional target for TGF-β/R-Smad signaling, Smad7 acts as a downstream transducer of the apoptotic signal. In a number of cell lines Smad7 is induced by TGF-β as part of the apoptotic program (Sanchez-Capelo, 2005). Overexpression of Smad7 is sufficient to induce apoptosis in these cell lines, and downregulation of Smad7 expression results in decreased sensitivity to TGF-β-induced apoptosis.

The role of Smad7 as either a survival or apoptotic factor may extend beyond the antagonism of TGF-β signaling or the transduction of TGF-β apoptotic signals. In a pancreatic cancer cell line, Smad7 functions to promote survival in the presence of a DNA damage-inducing chemotherapeutic agent (Arnold et al., 2004). Perhaps independently of TGF-β signaling, Smad7 is required for chemotherapeutic induction of apoptosis in prostate cancer cells (Davoodpour and Landström, 2005). From these observations it is clear that the role of Smad7 in survival and apoptotic signaling is much more complex than the roles that may be surmised from its known biochemical functions in the TGF-β signaling network.

3. SIGNALING EVENTS DOWNSTREAM OF THE SMADS

Regardless of the response of a target cell to stimulation with a TGF-β superfamily cytokine, the signaling events leading to the nuclear translocation of receptor-phosphorylated Smads do not vary appreciably. What generates the diverse responses to TGF-β family stimulation are the signaling events downstream of the Smads. In order to induce cell death or promote survival, the Smads must interact with the appropriate general and tissue-specific transcription factors to activate the transcription of the genes that will generate the proper response. While many of the cofactors and targets of Smad-mediated apoptosis and survival remain unclassified, the details that are known allow a general framework of events downstream of the Smads to be constructed (Fig. 2).

3.1 R-Smad-binding Transcription Factors in Apoptotic Signaling

As with any transcriptional response to TGF-β family cytokines, the apoptotic program is initiated by cooperation between the nuclear R-Smad proteins and interacting proteins with DNA binding and transactivation functions. However, while it is clear that general transcription factors play a vital role in activating the transcription of diverse TGF-β superfamily target genes, evidence for the definitive involvement of these transcription factors in the apoptotic program is limited (Derynck and Zhang, 2003). In human hepatoma cells, TGF-β-induced apoptosis is correlated with an increase in the transactivating activity of the AP-1 complex (Yamamura et al., 2000). Further, TGF-β enhances the formation of a JunD-FosB AP-1 heterodimer, and overexpression of FosB enhances TGF-β-induced apoptosis. Although there is little conclusive evidence for the involvement of these general transcription factors or the chromatin remodeling proteins p300/CBP as binding partners for Smads during apoptotic signaling, their involvement in other Smad-mediated transcriptional responses suggests they are likely to be prominently involved.

Specificity of the response to TGF-β stimulation is due in large part to the expression of Smad binding transcriptional cofactors that dictate the response to Smad signaling in the target cell. The Runx3 protein may represent a tissue-specific pro-apoptotic transcription factor for Smad-mediated apoptosis. Runx3 is a tumor suppressor whose loss is associated with gastric cancer, and perhaps other types of cancer (Ito and Miyazono, 2003). Loss of Runx3 expression in an esophageal adenocarcinoma cell line results in resistance to TGF-β-induced apoptosis, while restoration of Runx3 expression results in increased sensitivity to TGF-β-induced apoptosis (Torquati et al., 2004). This complex further interacts with general transcription factors to activate the transcription of BMP or TGF-β target genes (Fig. 2).

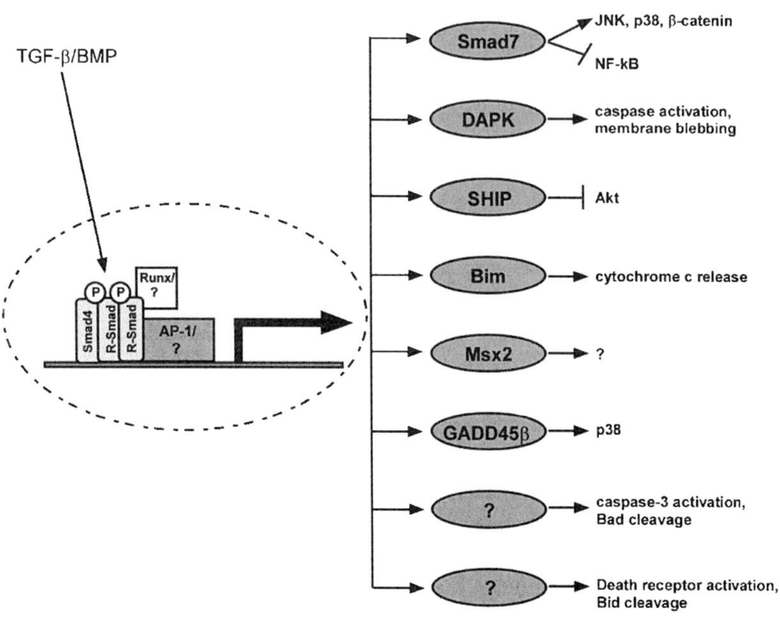

Figure 2. R-Smad-dependent downstream events in apoptotic signaling

3.2 Transcriptional Targets and Downstream Events in R-Smad-mediated Apoptosis

Although a number of pro-apoptotic transcriptional targets of TGF-β family cytokines have been identified, the potential involvement of Smad-independent apoptotic pathways complicates the assignment of many of these genes as effectors of Smad-mediated apoptosis. In TGF-β-induced apoptosis, several Smad-dependent transcriptional targets have been characterized (Fig. 2). As discussed above, in certain cell types Smad7 is induced by upstream R-Smad signaling and can activate downstream apoptotic pathways. Another transcriptional target of TGF-β whose upregulation is Smad-dependent is death-associated protein kinase, or DAP kinase (Jang et al., 2002). DAP kinase acts to promote mitochondrial apoptosis and may also facilitate membrane blebbing and the formation of autophagic vesicles that are hallmark phenotypes of apoptosis (Inbal et al., 2002). In hematopoietic cells, Activin and TGF-β upregulate SHIP, a lipid phosphatase that promotes apoptosis by decreasing the activity of the anti-apoptotic kinase Akt (Sanchez-Capelo, 2005). GADD45β is a TGF-β immediate early response gene whose expression is induced by Smad2, Smad3, and Smad4. GADD45β functions by activating the p38 MAP kinase, which can in turn initiate an apoptotic program (Sanchez-Capelo, 2005). In B lymphocytes, TGF-β induces the expression of the BH3-only protein Bim in a Smad3-dependent manner (Wildey et al., 2003). Bim acts to promote apoptosis by facilitating the release of mitochondrial cytochrome c and the initiation of the caspase cascade (Zong et al., 2001).

The Msx2 homeobox–containing transcription factor represents a Smad-dependent transcriptional target of BMP-induced developmental apoptosis (Fig. 2). BMP4 induces the expression of Msx2 in cells of the developing limb and the cephalic neural crest that are destined to undergo apoptosis, and overexpression of Msx2 is sufficient to induce cell death (Chen and Zhao, 1998). Expression of Msx2 in response to BMP requires the two Smad binding elements present in the promoter, although the Smad binding sites are not sufficient for BMP-induced transcriptional activation (see below) (Brugger et al., 2004). Events downstream of Msx2 are uncharacterized, but its role as a transcriptional repressor suggests it may function to downregulate the expression of anti-apoptotic genes such as those of the Bcl-2 family (Catron et al., 1996). Msx2 is also able to upregulate the expression of BMP-4, suggesting an amplification of the cell death signal (Chen and Zhao, 1998).

In several cases downstream events in the apoptotic signaling cascade have been found to be Smad-dependent, although the direct transcriptional targets of the Smad signaling are unknown (Fig. 2). In hepatoma cells, TGF-β-dependent apoptosis proceeds via Smad3-dependent activation of caspase-3 (the executioner caspase). Caspase-3 in turn cleaves the pro-apoptotic Bad protein, resulting in an amplification of the apoptotic signal (Sanchez-Capelo, 2005). In human gastric carcinoma cells, Smad3 participates in TGF-β-induced apoptosis by activating the Fas death receptor pathway in a mechanism that is independent of the Fas ligand (Sanchez-Capelo, 2005). The death receptor signal is further amplified by caspase-8-mediated cleavage of the Bid protein, which induces mitochondrial cytochrome c release. Although transcriptional upregulation of the Fas receptor is observed, the delayed time course of induction and the requirement for new protein synthesis suggest that this is not the direct target of Smad3. Rather than the standard activation of transcription, these downstream events may be due to direct effects of the Smad3 protein on some unknown target. In support of this idea, TGF-β promotes DNA damage-induced apoptosis by promoting a Smad3-BRCA1 complex that results in reduced DNA repair (Dubrovska et al., 2005).

3.3 R-Smad-mediated Survival Signaling

Events downstream of the Smads in anti-apoptotic signaling are not well characterized but are presumably analogous to pro-apoptotic signaling events in the activation of target gene transcription by interactions with specific transcription factors. Notably, evidence from work in astrocytes shows that adenine nucleotide translocator 1 functions as a Smad- and Sp1-inducible survival factor (Law et al., 2004). Smad-mediated survival signaling can also involve the transcriptional repression of pro-apoptotic genes. Myocytes differentiated from Smad5 -/- ES cells express significantly higher levels of p53, leading to a disruption of mitochondrial membrane potential (Sun et al., 2005). This suggests that BMP-regulated Smad5 may promote survival by downregulating p53 and the mitochondrial apoptotic pathway.

3.4 Smad7-mediated Apoptotic Signaling

As discussed above, the function of Smad7 is more complicated than as a simple antagonist of TGF-β/R-Smad signaling. As a transducer of the TGF-β apoptotic signal in certain cell types, Smad7 acts to link TGF-β signaling with other cellular signaling pathways (Fig. 2). Of particular note is the downstream activation of MAP kinase pathways by Smad7. Smad7 is able to activate both JNK and p38 MAP kinase pathways, and may serve as a scaffold protein to link the kinases of this cascade (Sanchez-Capelo, 2005; Schuster and Krieglstein, 2002). Notably, the ability of Smad7 to activate downstream MAPK signaling is independent of its ability to inhibit TGF-β signaling. Through an inhibition of the pro-survival activity of NF-κB, Smad7 functions to promote apoptosis of epithelial cells treated with a variety of apoptosis-inducing agents including TGF-β (Sanchez-Capelo, 2005). Smad7 may also serve as a functional link with the Wnt/β-catenin signaling pathway. Smad7 directly interacts with β-catenin and LEF1 in a TGF-β-dependent manner, promoting the nuclear accumulation of β-catenin and TGF-β-induced apoptosis (Edlund et al., 2005). Through all of these mechanisms, Smad7 may act as a pro-apoptotic factor both downstream and independently of TGF-β signaling.

3.5 Smad7-mediated Survival Signaling

In contrast to the variety of mechanisms by which Smad7 promotes apoptosis, there is limited evidence of survival pathways downstream of Smad7 independent of its function as a TGF-β antagonist. In pancreatic cancer cells, Smad7 promotes survival in the presence of a DNA damage-inducing agent by upregulation of thioredoxin (Arnold et al., 2004). Survival is in part due to inhibition of apoptosis signal-regulating kinase by thioredoxin and an upregulation of NF-κB activity. The diverse and seemingly antagonistic functions of Smad7 point to the complicated nature of this protein in survival and apoptotic signaling.

4. CROSSTALK WITH OTHER SIGNALING PATHWAYS

The ultimate response of a cell to stimulation with a TGF-β family cytokine represents the integration of all signaling pathways active in the cell at the time of stimulation. Due to their critical position as a link between the cytokine and downstream effectors, the R-Smads are often the site of crosstalk between signaling pathways (see Chapters 15 and 16). Based on the temporal relationship of these signaling pathways with the Smads, the interacting pathway can be classified as downstream, convergent, or upstream (Fig. 3). Note that several pathways crosstalk with the Smads at multiple steps.

4.1 Activation of Downstream Pathways by Smad Signaling

A pathway may be considered downstream of the Smads if it is activated through Smad-dependent transcriptional events during the generation of the response. As mentioned above R-Smad-induced Smad7 is required in a number of cells to activate downstream JNK or p38 pathways, which in turn promote apoptosis. In other cases, activation of downstream MAP kinase pathways requires Smad-dependent transcriptional activation of GADD45β (Sanchez-Capelo, 2005).

4.2 Convergence with Other Pathways

Smad pathways can converge with other cellular pathways at the level of transcriptional activation to produce effects on apoptosis and survival. Signals from the JNK and Smad pathways converge through cooperation between AP-1 and the Smads to promote apoptosis of hepatoma and myeloid cells (Yamamura et al., 2000). One mechanism by which NF-κB can inhibit TGF-β-induced apoptosis is by disrupting this cooperation through an inhibition of AP-1 transcriptional activity (Arsura et al., 2003). BMP-induced Smad1 and Smad4 cooperate with Wnt/β-catenin-induced LEF1 to induce the expression of Msx2, which may be involved in neural crest apoptosis (see above) (Hussein et al., 2003). In neurons, signals from Activin A-induced Smad3 converge with basic fibroblast growth factor-induced Erk to induce the expression of tyrosine hydroxylase to promote survival (Bao et al., 2005). Note that the convergence of pathways at the level of transcription requires the existence of multiple promoter elements and thus contributes to the generation of specificity in addition to allowing for integration of multiple signals.

4.3 Upstream Effects

Upstream effects of other pathways on Smads may be direct, as in the phosphorylation in the R-Smad linker region (see Chapter 12), or indirect, as in the upregulation of Smad7 to inhibit R-Smad phosphorylation. Phosphorylation of the R-Smads can have either positive or negative effects on Smad activity. The MEKK-1/JNK pathway promotes Smad activity by direct phosphorylation in response to stress (Derynck and Zhang, 2003). Ras- or EGF-induced Erk inhibits

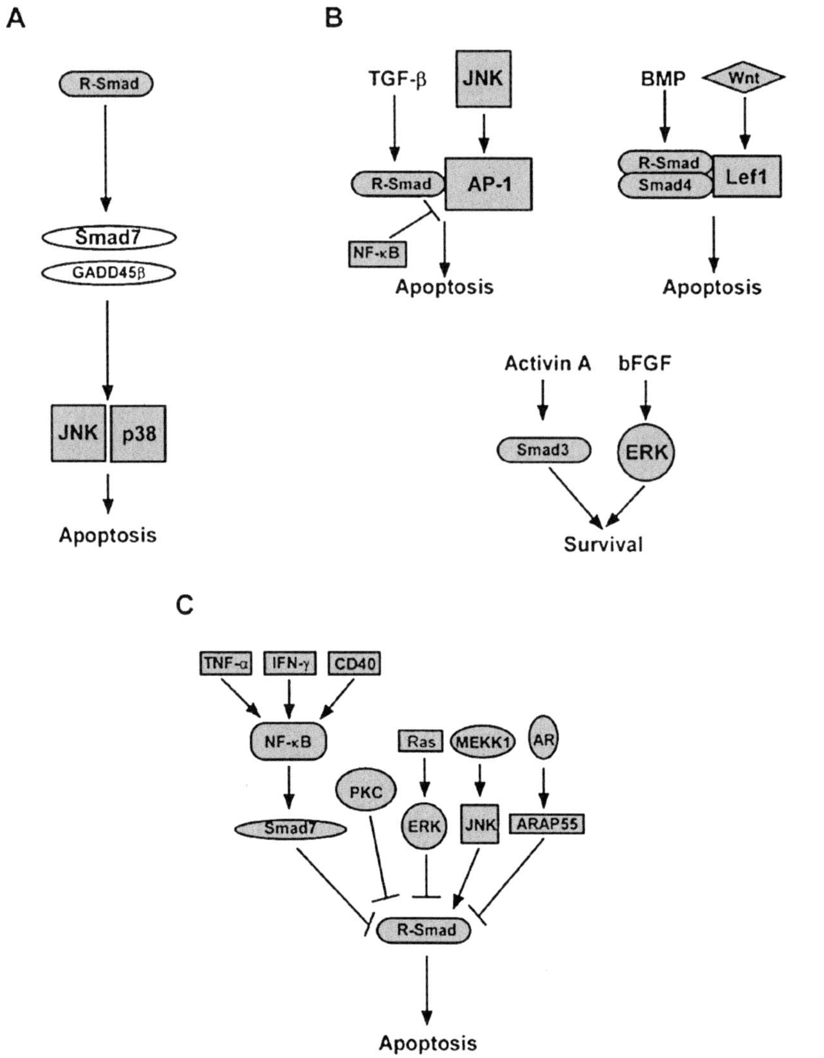

Figure 3. Crosstalk with other pathways. A, Downstream activation of MAPK pathways. B, Convergent signaling. C, Upstream effects of other pathways

the nuclear translocation of R-Smad proteins by phosphorylation in the linker region, while protein kinase C inhibits the DNA binding ability of Smad3 by direct phosphorylation in the MH1 domain (Derynck and Zhang, 2003). Through direct binding of both androgen receptor and androgen receptor-associated protein 55, the androgen receptor pathway negatively regulates Smad3 transcriptional activity (Chipuk et al., 2002; Wang et al., 2005). CD40, IFN-γ, and TNF-α negatively regulate TGF-β/Smad signaling through NF-κB-dependent expression of Smad7 (Bitzer et al., 2000; Patil et al., 2000; Ulloa et al., 1999). Although there is limited evidence that these mechanisms directly affect Smad-mediated apoptosis, modulation of Smad-mediated apoptosis would be consistent with the known roles of these pathways in promoting apoptosis or survival.

4.4 Smad Regulation and the Determination of Sensitivity to TGF-β-induced Apoptosis

An important question in TGF-β biology is why some cells undergo apoptosis in response to TGF-β while other cells undergo only growth arrest (see Chapter 4). To answer this question, one can draw hints from physiological events in which tight regulation of TGF-β-induced apoptosis is critical. Perhaps the best studied manifestation of TGF-β-induced apoptosis in a physiological context is the process of liver regeneration. After natural or experimentally-induced liver injury, the liver undergoes rapid regeneration for several days until it regains its original size. TGF-β plays a significant role in halting liver regeneration through induction of apoptosis and growth arrest in hepatocytes (Michalopoulos and DeFrances, 1997; Oberhammer et al., 1992). The negative effects of TGF-β are antagonized by factors such as insulin, interleukin-6, and hepatocyte growth factor, all of which are capable of inhibiting TGF-β-induced apoptosis both *in vitro* and *in vivo* (Chen et al., 1998; Hong et al., 2000).

Mechanistically, the inhibition of TGF-β-induced apoptosis by these factors requires the phosphatidylinositol 3-kinase (PI3-kinase) pathway (Chen et al., 1998). Activation of PI3-kinase by growth factors and hormones generates a strong survival signal via the stimulation of the serine/threonine kinase Akt/protein kinase B (Vincent and Feldman, 2002). Active PI3-kinase generates 3'-OH-phosphorylated phosphatidylinositols, which recruit Akt to the plasma membrane where it is phosphorylated at two sites (Brazil et al., 2004). Active Akt phosphorylates a number of pro-apoptotic substrates, including Bad, IκB kinase-β, and FOXO family transcription factors, to promote survival (Brazil et al., 2004). In mesenchymal cells and in cells undergoing epithelial-mesenchymal transdifferentiation, TGF-β promotes survival by activating Akt through what is likely a Smad-independent mechanism (Sanchez-Capelo, 2005).

In part because of its general anti-apoptotic function, expression of a constitutively active Akt is sufficient to inhibit TGF-β-induced apoptosis (Fig. 4A) (Chen et al., 1998; Conery et al., 2004; Remy et al., 2004). However, it is not the activation state of Akt per se that determines whether a cell undergoes TGF-β-induced

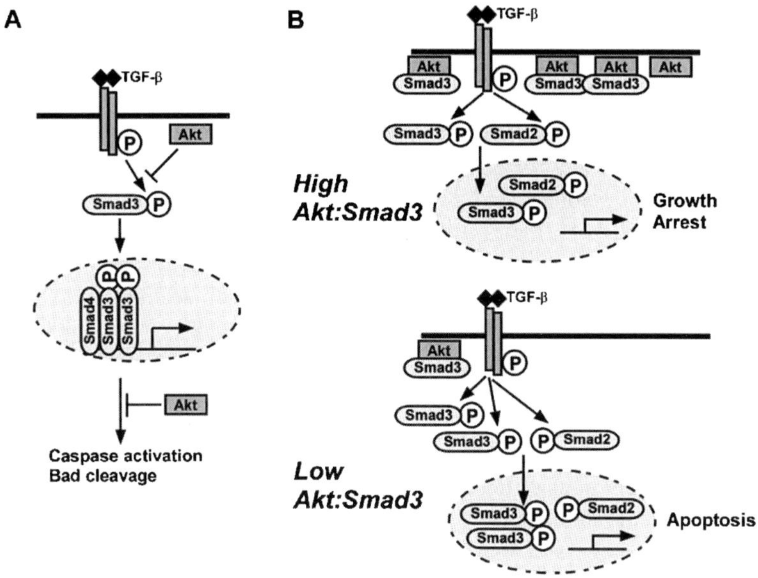

Figure 4. Regulation of TGF-β-induced apoptosis by Akt. A, Inhibition of Smad3 and downstream events by Akt. B, Modulation of the response to TGF-β by Akt

apoptosis. After PI3-kinase-stimulated membrane recruitment and phosphorylation, Akt modulates the response of cells to TGF-β independently of its kinase activity by targeting the unique apoptotic function of Smad3. Phosphorylated, membrane-associated Akt directly interacts with Smad3, but not Smad2, and inhibits its phosphorylation, oligomerization with Smad4, and nuclear translocation by physical sequestration (Conery et al., 2004; Remy et al., 2004). In this way Akt inhibits Smad3-mediated transcription as well as the TGF-β/Smad3-mediated apoptotic response. Since Smad2 and Smad-independent signaling remain intact, Akt inhibits TGF-β-induced apoptosis while keeping growth arrest and other responses intact. The ratio of Akt to Smad3 in a cell can thus dictate the response of a cell to TGF-β: a cell with a high level of Akt relative to Smad3 will undergo growth arrest or some other response, while a cell with a low level of Akt relative to Smad3 will undergo apoptosis (Fig. 4B). Further, the unique role of Smad3 in TGF-β apoptotic signaling suggests a mechanism by which tumor cells can upregulate the expression or phosporylation status of Akt to inhibit TGF-β-induced apoptosis while maintaining responsiveness to tumor-promoting effects of TGF-β.

5. SUMMARY AND PERSPECTIVES

The Smad proteins play a critical role in apoptosis and survival signaling both as signaling intermediates of TGF-β superfamily cytokines and as sites of integration of multiple cellular pathways. Receptor-activated R-Smads interact with general and tissue-specific transcription factors to activate the transcription of a number of genes involved in downstream apoptotic and survival signaling. Smad7 has multiple functions: antagonist of TGF-β-induced apoptosis, transcriptional target and effector of TGF-β-induced apoptosis, and inducer of the activity of MAPK and other pathways. Cross-talk of Smads with other pathways may involve Smad-mediated activation of MAPK pathways, cooperation with other pathways at the level of transcription, or modulation of Smad activity by upstream signaling events. By modulating Smad3 activity, the PI3-kinase/Akt signaling pathway plays a unique role in determining the sensitivity of cells to TGF-β-induced apoptosis.

Despite the large body of research on the role of Smads in cell death and survival, there are many unanswered questions that drive future research. While it is clear that TGF-β family members are important in generating cell death and survival responses during development and tissue homeostasis, the temporal and spatial role of the Smads is largely unknown. This may be addressed by using tissue-specific or conditional knock-out mice. The transcriptional cofactors of the Smads in this context are largely unclassified, as are many transcriptional targets. Another unresolved question is how to reconcile the role of Smad7 as an inhibitor of TGF-β signaling with its more general role as an inducer of multiple cellular apoptotic and survival pathways. It is possible that there are unknown binding partners or post-translational modifications of Smad7 that switch it from an antagonist of TGF-β signaling to a more general role in integrating cellular pathways. Finally, what has become clear from the research into the role of Smads in cell death and survival is that none of the Smad-mediated responses could occur in the absence of other signaling pathways. Future work will no doubt shed more light on how the Smads fit into the complex signaling networks that control whether a cell lives or dies.

ACKNOWLEDGEMENTS

Research on TGF-β/Smad signaling in our laboratory is supported by grants from the American Cancer Society, DOE-OBER and NIH R01s to K.L.

REFERENCES

Arnold, N.B., Ketterer, K., Kleeff, J., Friess, H., Buchler, M.W., and Korc, M., 2004, Thioredoxin is downstream of Smad7 in a pathway that promotes growth and suppresses cisplatin-induced apoptosis in pancreatic cancer. *Cancer Res* 64: 3599-3606.

Arsura, M., Panta, G.R., Bilyeu, J.D., Cavin, L.G., Sovak, M.A., Oliver, A.A., Factor, V., Heuchel, R., Mercurio, F., Thorgeirsson, S.S., and Sonenshein, G.E., 2003, Transient activation of NF-κB through a TAK1/IKK kinase pathway by TGF-β1 inhibits AP-1/SMAD signaling and apoptosis: implications in liver tumor formation. *Oncogene* 22: 412-425.

Atfi, A., Buisine, M., Mazars, A., and Gespach, C., 1997, Induction of apoptosis by DPC4, a transcriptional factor regulated by transforming growth factor-β through stress-activated protein kinase/c-Jun N-terminal kinase (SAPK/JNK) signaling pathway. *J Biol Chem* 272: 24731-24734.

Bao, Y.L., Tsuchida, K., Liu, B., Kurisaki, A., Matsuzaki, T., and Sugino, H., 2005, Synergistic activity of activin A and basic fibroblast growth factor on tyrosine hydroxylase expression through Smad3 and ERK1/ERK2 MAPK signaling pathways. *J Endocrinol* 184: 493-504.

Bitzer, M., von Gersdorff, G., Liang, D., Dominguez-Rosales, A., Beg, A.A., Rojkind, M., and Böttinger, E.P., 2000, A mechanism of suppression of TGF-β/SMAD signaling by NF-κ B/RelA. *Genes Dev* 14: 187-197.

Botchkarev, V.A., and Sharov, A.A., 2004, BMP signaling in the control of skin development and hair follicle growth. *Differentiation* 72: 512-526.

Brazil, D.P., Yang, Z.Z., and Hemmings, B.A., 2004, Advances in protein kinase B signalling: AKTion on multiple fronts. *Trends Biochem Sci* 29: 233-242.

Brugger, S.M., Merrill, A.E., Torres-Vazquez, J., Wu, N., Ting, M.C., Cho, J.Y., Dobias, S.L., Yi, S.E., Lyons, K., Bell, J.R., Arora, K., Warrior, R., and Maxson, R., 2004, A phylogenetically conserved cis-regulatory module in the *Msx2* promoter is sufficient for BMP-dependent transcription in murine and *Drosophila* embryos. *Development* 131: 5153-5165.

Catron, K.M., Wang, H., Hu, G., Shen, M.M., and Abate-Shen, C., 1996, Comparison of MSX-1 and MSX-2 suggests a molecular basis for functional redundancy. *Mech Dev* 55: 185-199.

Chen, R.H., Su, Y.H., Chuang, R.L., and Chang, T.Y., 1998, Suppression of transforming growth factor-β-induced apoptosis through a phosphatidylinositol 3-kinase/Akt-dependent pathway. *Oncogene* 17: 1959-1968.

Chen, Y., and Zhao, X., 1998, Shaping limbs by apoptosis. *J Exp Zool* 282: 691-702.

Chipuk, J.E., Cornelius, S.C., Pultz, N.J., Jorgensen, J.S., Bonham, M.J., Kim, S.J., and Danielpour, D., 2002, The androgen receptor represses transforming growth factor-β signaling through interaction with Smad3. *J Biol Chem* 277: 1240-1248.

Conery, A.R., Cao, Y., Thompson, E.A., Townsend, C.M., Jr., Ko, T.C., and Luo, K., 2004, Akt interacts directly with Smad3 to regulate the sensitivity to TGF-β induced apoptosis. *Nat Cell Biol* 6: 366-372.

Davoodpour, P., and Landström, M., 2005, 2-Methoxyestradiol-induced apoptosis in prostate cancer cells requires Smad7. *J Biol Chem* 280: 14773-14779.

Derynck, R., and Zhang, Y.E., 2003, Smad-dependent and Smad-independent pathways in TGF-β family signalling. *Nature* 425: 577-584.

Dubrovska, A., Kanamoto, T., Lomnytska, M., Heldin, C.-H., Volodko, N., and Souchelnytskyi, S., 2005, TGFβ1/Smad3 counteracts BRCA1-dependent repair of DNA damage. *Oncogene* 24: 2289-2297.

Edlund, S., Lee, S.Y., Grimsby, S., Zhang, S., Aspenström, P., Heldin, C.-H., and Landström, M., 2005, Interaction between Smad7 and β-catenin: importance for transforming growth factor β-induced apoptosis. *Mol Cell Biol* 25: 1475-1488.

Hardwick, J.C., Van Den Brink, G.R., Bleuming, S.A., Ballester, I., Van Den Brande, J.M., Keller, J.J., Offerhaus, G.J., Van Deventer, S.J., and Peppelenbosch, M.P., 2004, Bone morphogenetic protein 2 is expressed by, and acts upon, mature epithelial cells in the colon. *Gastroenterology* 126: 111-121.

Hong, F., Nguyen, V.A., Shen, X., Kunos, G., and Gao, B., 2000, Rapid activation of protein kinase B/Akt has a key role in antiapoptotic signaling during liver regeneration. *Biochem Biophys Res Commun* 279: 974-979.

Hussein, S.M., Duff, E.K., and Sirard, C., 2003, Smad4 and β-catenin co-activators functionally interact with lymphoid-enhancing factor to regulate graded expression of Msx2. *J Biol Chem* 278: 48805-48814.

Inazaki, K., Kanamaru, Y., Kojima, Y., Sueyoshi, N., Okumura, K., Kaneko, K., Yamashiro, Y., Ogawa, H., and Nakao, A., 2004, Smad3 deficiency attenuates renal fibrosis, inflammation, and apoptosis after unilateral ureteral obstruction. *Kidney Int* 66: 597-604.

Inbal, B., Bialik, S., Sabanay, I., Shani, G., and Kimchi, A., 2002, DAP kinase and DRP-1 mediate membrane blebbing and the formation of autophagic vesicles during programmed cell death. *J Cell Biol* 157: 455-468.

Ishisaki, A., Yamato, K., Hashimoto, S., Nakao, A., Tamaki, K., Nonaka, K., ten Dijke, P., Sugino, H., and Nishihara, T., 1999, Differential inhibition of Smad6 and Smad7 on bone morphogenetic protein- and activin-mediated growth arrest and apoptosis in B cells. *J Biol Chem* 274: 13637-13642.

Ito, Y., and Miyazono, K., 2003, RUNX transcription factors as key targets of TGF-β superfamily signaling. *Curr Opin Genet Dev* 13: 43-47.

Ito, Y., Zhao, J., Mogharei, A., Shuler, C.F., Weinstein, M., Deng, C., and Chai, Y., 2001, Antagonistic effects of Smad2 versus Smad7 are sensitive to their expression level during tooth development. *J Biol Chem* 276: 44163-44172.

Izumi, M., Fujio, Y., Kunisada, K., Negoro, S., Tone, E., Funamoto, M., Osugi, T., Oshima, Y., Nakaoka, Y., Kishimoto, T., Yamauchi-Takihara, K., and Hirota, H., 2001, Bone morphogenetic protein-2 inhibits serum deprivation-induced apoptosis of neonatal cardiac myocytes through activation of the Smad1 pathway. *J Biol Chem* 276: 31133-31141.

Jang, C.W., Chen, C.H., Chen, C.C., Chen, J.Y., Su, Y.H., and Chen, R.H., 2002, TGF-β induces apoptosis through Smad-mediated expression of DAP-kinase. *Nat Cell Biol* 4: 51-58.

Kanamaru, C., Yasuda, H., and Fujita, T., 2002, Involvement of Smad proteins in TGF-β and activin A-induced apoptosis and growth inhibition of liver cells. *Hepatol Res* 23: 211-219.

Kim, B.C., Mamura, M., Choi, K.S., Calabretta, B., and Kim, S.J., 2002, Transforming growth factor β1 induces apoptosis through cleavage of BAD in a Smad3-dependent mechanism in FaO hepatoma cells. *Mol Cell Biol* 22: 1369-1378.

Law, A.K., Gupta, D., Levy, S., Wallace, D.C., McKeon, R.J., and Buck, C.R., 2004, TGF-β1 induction of the adenine nucleotide translocator 1 in astrocytes occurs through Smads and Sp1 transcription factors. *BMC Neurosci* 5: 1.

Mazars, A., Tournigand, C., Mollat, P., Prunier, C., Ferrand, N., Bourgeade, M.F., Gespach, C., and Atfi, A., 2000, Differential roles of JNK and Smad2 signaling pathways in the inhibition of c-Myc-induced cell death by TGF-β. *Oncogene* 19: 1277-1287.

McKarns, S.C., and Schwartz, R.H., 2005, Distinct effects of TGF-β1 on CD4+ and CD8+ T cell survival, division, and IL-2 production: a role for T cell intrinsic Smad3. *J Immunol* 174: 2071-2083.

Merino, R., Ganan, Y., Macias, D., Rodriguez-Leon, J., and Hurle, J.M., 1999, Bone morphogenetic proteins regulate interdigital cell death in the avian embryo. *Ann N Y Acad Sci* 887: 120-132.

Michalopoulos, G.K., and DeFrances, M.C., 1997, Liver Regeneration. *Science* 276: 60-66.

Miyanaga, Y., Torregroza, I., and Evans, T., 2002, A maternal Smad protein regulates early embryonic apoptosis in *Xenopus laevis*. *Mol Cell Biol* 22: 1317-1328.

Oberhammer, F.A., Pavelka, M., Sharma, S., Tiefenbacher, R., Purchio, A.F., Bursch, W., and Schulte-Hermann, R., 1992, Induction of apoptosis in cultured hepatocytes and in regressing liver by transforming growth factor β1. *Proc Natl Acad Sci U S A* 89: 5408-5412.

Patil, S., Wildey, G.M., Brown, T.L., Choy, L., Derynck, R., and Howe, P.H., 2000, Smad7 is induced by CD40 and protects WEHI 231 B-lymphocytes from transforming growth factor-β-induced growth inhibition and apoptosis. *J Biol Chem* 275: 38363-38370.

Piscione, T.D., Phan, T., and Rosenblum, N.D., 2001, BMP7 controls collecting tubule cell proliferation and apoptosis via Smad1-dependent and -independent pathways. *Am J Physiol Renal Physiol* 280: F19-33.

Ramachandra, M., Atencio, I., Rahman, A., Vaillancourt, M., Zou, A., Avanzini, J., Wills, K., Bookstein, R., and Shabram, P., 2002, Restoration of transforming growth factor β signaling by functional expression of smad4 induces anoikis. *Cancer Res* 62: 6045-6051.

Remy, I., Montmarquette, A., and Michnick, S.W., 2004, PKB/Akt modulates TGF-β signalling through a direct interaction with Smad3. *Nat Cell Biol* 6: 358-365.

Sanchez-Capelo, A., 2005, Dual role for TGF-β1 in apoptosis. *Cytokine Growth Factor Rev* 16: 15-34.

Schuster, N., and Krieglstein, K., 2002, Mechanisms of TGF-β-mediated apoptosis. *Cell Tissue Res* 307: 1-14.

Sowa, H., Kaji, H., Iu, M.F., Tsukamoto, T., Sugimoto, T., and Chihara, K., 2003, Parathyroid hormone-Smad3 axis exerts anti-apoptotic action and augments anabolic action of transforming growth factor β in osteoblasts. *J Biol Chem* 278: 52240-52252.

Sun, Y., Zhou, J., Liao, X., Lu, Y., Deng, C., Huang, P., Chen, Q., and Yang, X., 2005, Disruption of Smad5 gene induces mitochondria-dependent apoptosis in cardiomyocytes. *Exp Cell Res* 306: 85-93.

Tomic, D., Miller, K.P., Kenny, H.A., Woodruff, T.K., Hoyer, P., and Flaws, J.A., 2004, Ovarian follicle development requires Smad3. *Mol Endocrinol* 18: 2224-2240.

Torquati, A., O'Rear, L., Longobardi, L., Spagnoli, A., Richards, W.O., and Daniel Beauchamp, R., 2004, RUNX3 inhibits cell proliferation and induces apoptosis by reinstating transforming growth factor β responsiveness in esophageal adenocarcinoma cells. *Surgery* 136: 310-316.

Ulloa, L., Doody, J., and Massagué, J., 1999, Inhibition of transforming growth factor-β/SMAD signalling by the interferon-gamma/STAT pathway. *Nature* 397: 710-713.

Wahl, S.M., Orenstein, J.M., and Chen, W., 2000, TGF-β influences the life and death decisions of T lymphocytes. *Cytokine Growth Factor Rev* 11: 71-79.

Wang, H., Song, K., Sponseller, T.L., and Danielpour, D., 2005, Novel function of androgen receptor-associated protein 55/Hic-5 as a negative regulator of Smad3 signaling. *J Biol Chem* 280: 5154-5162.

Weinstein, M., Yang, X., and Deng, C., 2000, Functions of mammalian Smad genes as revealed by targeted gene disruption in mice. *Cytokine Growth Factor Rev* 11: 49-58.

Wildey, G.M., Patil, S., and Howe, P.H., 2003, Smad3 potentiates transforming growth factor β (β)-induced apoptosis and expression of the BH3-only protein Bim in WEHI 231 B lymphocytes. *J Biol Chem.*

Vincent, A.M., and Feldman, E.L., 2002, Control of cell survival by IGF signaling pathways. *Growth Horm IGF Res* 12: 193-197.

Xu, G., Zhong, Y., Munir, S., Yang, B.B., Tsang, B.K., and Peng, C., 2004, Nodal Induces Apoptosis and Inhibits Proliferation in Human Epithelial Ovarian Cancer Cells via Activin Receptor-Like Kinase 7. *J Clin Endocrinol Metab* 89: 5523-5534.

Yamamura, Y., Hua, X., Bergelson, S., and Lodish, H.F., 2000, Critical role of Smads and AP-1 complex in transforming growth factor-β-dependent apoptosis. *J Biol Chem* 275: 36295-36302.

Yang, Y.A., Tang, B., Robinson, G., Hennighausen, L., Brodie, S.G., Deng, C.X., and Wakefield, L.M., 2002, Smad3 in the mammary epithelium has a nonredundant role in the induction of apoptosis, but not in the regulation of proliferation or differentiation by transforming growth factor-β. *Cell Growth Differ* 13: 123-130.

Zhang, S., Fantozzi, I., Tigno, D.D., Yi, E.S., Platoshyn, O., Thistlethwaite, P.A., Kriett, J.M., Yung, G., Rubin, L.J., and Yuan, J.X., 2003, Bone morphogenetic proteins induce apoptosis in human pulmonary vascular smooth muscle cells. *Am J Physiol Lung Cell Mol Physiol* 285: L740-754.

Zimmermann, K.C., Bonzon, C., and Green, D.R., 2001, The machinery of programmed cell death. *Pharmacol Ther* 92: 57-70.

Zong, W.X., Lindsten, T., Ross, A.J., MacGregor, G.R., and Thompson, C.B., 2001, BH3-only proteins that bind pro-survival Bcl-2 family members fail to induce apoptosis in the absence of Bax and Bak. *Genes Dev* 15: 1481-1486.

Zuzarte-Luis, V., and Hurle, J.M., 2005, Programmed cell death in the embryonic vertebrate limb. *Semin Cell Dev Biol* 16: 261-269.

Zuzarte-Luis, V., Montero, J.A., Rodriguez-Leon, J., Merino, R., Rodriguez-Rey, J.C., and Hurle, J.M., 2004, A new role for BMP5 during limb development acting through the synergic activation of Smad and MAPK pathways. *Dev Biol* 272: 39-52.

CHAPTER 7

TGF-β/SMAD SIGNALING IN EPITHELIAL TO MESENCHYMAL TRANSITION

ARISTIDIS MOUSTAKAS, MARCIN KOWANETZ, AND SYLVIE THUAULT

Ludwig Institute for Cancer Research, Biomedical Center, Uppsala University, Uppsala, Sweden

Abstract: Epithelial-mesenchymal transition (EMT) allows polarized epithelial layers to locally dissolve and create fibroblast-like or myofibroblastic cell derivatives. EMT resembles other processes of transdifferentiation from one differentiated cell type to another and is critical for proper embryonic development. Furthermore, EMT describes similar processes between endothelial and myoepithelial cells occurring during blood vessel remodeling. All such processes are triggered and controlled by the concerted action of multiple extracellular growth/morphogenetic factors, TGF-β being one of them. The mechanism by which TGF-β elicits EMT has been a major topic of current research. This is due to the relevance EMT has to the processes of tumor cell invasiveness and metastasis on the one hand, and to fibrotic conditions on the other. Here we emphasize on mechanisms of signal transduction and specific gene targets of the TGF-β pathway that are critical effectors of EMT

Keywords: adherens junction; cadherin; EMT; Id; metastasis; polarity; tight junction; transcription factor

1. INTRODUCTION

Epithelial-mesenchymal transition (EMT) is a global change in epithelial differentiation that converts fully polarized epithelia to mesenchymal, migratory cells (Hay, 2005). During EMT, cell-cell and cell-matrix adhesion is disrupted, surrounding matrix is degraded and cells become more motile and invasive (Fig. 1A). EMT and the equivalent phenomenon of endothelial-myofibroblast transition represent basic morphogenetic processes that support normal development and disease pathogenesis, including carcinoma invasiveness and tissue fibrosis (Hay, 2005; Thiery, 2003; Zeisberg and Kalluri, 2004). On the other hand, certain skepticism is still expressed, especially from tumor pathologists, regarding the importance EMT has during tumor progression and metastasis (Tarin et al., 2005). EMT can be triggered

P. ten Dijke and C.-H. Heldin (eds.), Smad Signal Transduction, 131–150.

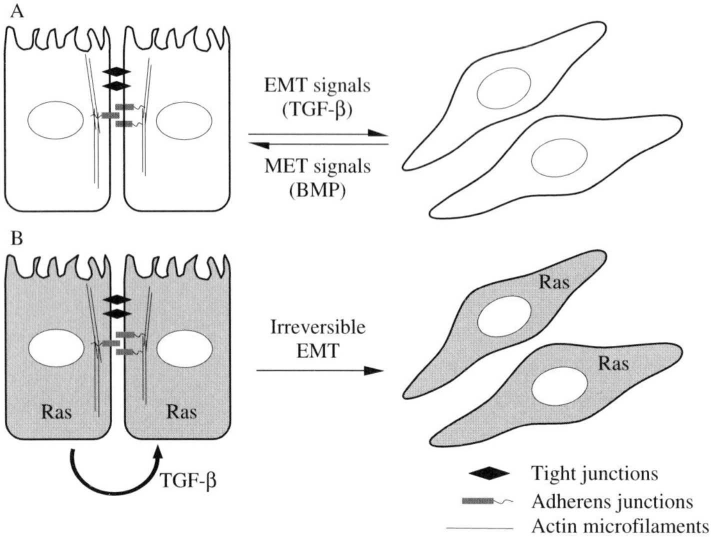

Figure 1. EMT in normal and malignant cells. (A) A pair of polarized epithelial cells is drawn, with tight (black diamonds) and adherens junctions (transmembrane grey cadherins tethered to actin microfilaments in the cytoplasm). Upon exposure to TGF-β, EMT generates mesenchymal cells. EMT is reversible and MET can be stimulated by BMPs. (B) Epithelial cells transformed by Ras (grey cytoplasm) oversecrete TGF-β, which acts in an auto- or paracrine manner (curved arrow), leading to irreversible EMT

by various signaling pathways such as hepatocyte growth factor/scatter factor or fibroblast growth factor (Thiery, 2003). Here, we focus exclusively on TGF-β superfamily pathways.

2. TGF-β SUPERFAMILY SIGNALING AND EMT

2.1 *In vitro* Cell Models of EMT

The molecular features of EMT are easier to dissect in cell models under *in vitro* culture conditions, including 3D cultures where epithelial cells grow within a matrix (*e.g.* collagen) and thus can invade the matrix, differentiate into tubular epithelia and even exhibit branching morphogenesis. *In vitro* systems provided the first clues about the role of TGF-β superfamily members in modulating EMT. Under *in vitro* conditions, primary human mammary epithelial cells (HMEC) (Valcourt et al., 2005), immortalized normal mouse and human mammary epithelial cells (NMuMG, HMLE and MCF-10A) (Brown et al., 2004; Miettinen et al., 1994; Valcourt et al., 2005), primary and established mouse (MCT) or human (HKC) renal proximal tubular epithelial cells (Brown et al., 2004; Li et al., 2003b; Sato et al., 2003), primary and immortalized human epidermal keratinocytes (NHEK and HaCaT) (Valcourt et al., 2005; Zavadil et al., 2001), primary rat alveolar epithelial cells and immortalized normal human lung epithelial cells (AEC, HPL1) (Valcourt et al., 2005; Willis et al., 2005), mouse lens epithelial cells (Kowanetz et al., 2004; Saika et al., 2004) and fetal rat hepatocytes (Valdes et al., 2002), undergo reversible EMT in response to TGF-β1 (Fig. 1A), based on the loss of epithelial markers (E-cadherin, specific keratins, ZO-1) and the acquisition of mesenchymal markers (fibronectin, Fsp1, α-smooth muscle actin, vimentin). In addition, under 2D culture conditions, the above cells reorganize their actin cytoskeleton from a cortical arrangement surrounding the polarized plasma membrane to the typical stress-fiber arrangement seen in fibroblasts when cultured on plastic. However, the relevance of actin reorganization on the development of EMT is disputed (Hay, 2005). The reversible *in vitro* EMT response of normal epithelial cells of diverse species and tissue origins is thought to mimic the ability of normal cells to undergo EMT *in vivo* during histo- or organogenesis.

In addition to normal cells, tumor and fibrotic cells of epithelial origin (colonic, kidney, liver, lung, mammary gland, skin, etc.) undergo EMT *in vitro* (Bates and Mercurio, 2005; Gotzmann et al., 2002; Han et al., 2005; Oft et al., 1998; Portella et al., 1998). *In vitro*, EMT of carcinoma cells appears to be irreversible (Fig. 1B), suggesting that transformed cells cannot undergo mesenchymal-epithelial transition (MET) (Fig. 1). Based on experiments with mouse mammary carcinomas oncogenically transformed by Ras or Raf, spindle cell carcinomas derived from mice exposed to chemical carcinogens, colon and hepatocarcinomas, a prominent model has been developed whereby the ability of oncogenes such as *Ras* or *Raf* to induce EMT and tumor cell invasiveness *in vitro* and *in vivo* depends on the activation of the TGF-β pathway (Gotzmann et al., 2002; Janda et al., 2002; Lehmann et al., 2000; Oft et al., 1998; Portella et al., 1998). In the absence of an overactive TGF-β pathway, Ras alone is incapable of driving tumor cell invasiveness and metastasis (Oft et al., 1998). In all these cases, carcinomas oversecrete TGF-β, are sensitized to TGF-β signaling, and while they lose their ability to be growth inhibited or undergo apoptosis in response to TGF-β, they exhibit characteristic EMT both in 2D and 3D matrix culture conditions (Fig. 1B). This stands in contrast to the EMT response of normal epithelial cells which is linked to their ability to be growth arrested or undergo apoptosis in response to TGF-β (Valcourt et al., 2005). We propose that the difference between reversible EMT of normal epithelial cells and irreversible EMT of oncogenically transformed cells, must be linked to the mechanism that controls cell cycle progression and which is differentially regulated in these two distinct cases.

2.2 *In vivo* Models of EMT

Due to the reversible nature of EMT, that is often followed by MET, its manifestation *in vivo* is technically difficult to assess. Despite that, advanced imaging technology (see last section) and transgenic mouse models seem to bypass this limitation. The best examples of EMT in response to TGF-β *in vivo* come from studies of normal organogenesis. Studies of palate development have shown a clear role of TGF-β3 in the formation of the connective tissue across the palate from epithelial cells via EMT (Nawshad et al., 2004). This explains the dramatic cleft palate phenotype seen in knock-out mice for TGF-β3. In the heart, valve formation depends on EMT, except that the cell type of origin for the resulting fibroblasts in the heart valves and septa, are endothelial cells (Person et al., 2005). In this organ, TGF-β2 induces local dissolution of the endothelial layer for which the activity of the type III receptor/betaglycan is critical, and later, TGF-β3 induces EMT and formation of mesenchymal progenitors in both chicken and mice. Interestingly, this *in vivo* endothelial cell transformation depends on the action of the classical TGF-β type II receptor and of the type I receptor named Activin receptor-like 2 (ALK2), which stimulates downstream BMP-specific Smads (Desgrosellier et al., 2005). Further molecular analysis of the mechanism of activation of specific receptors and Smads during EMT in the heart is warranted.

TGF-β also induces EMT *in vivo* during tumor progression. EMT of invasive spindle carcinomas was linked to the ability of TGF-β1 to enhance tumor progression in mice (Cui et al., 1996). Targeted expression of TGF-β1 in keratinocytes and chemical carcinogenesis in the skin, led to enhanced rate of aggressive and highly malignant tumors. These keratinocytes were undergoing EMT in response to the transgenic TGF-β1 during *in vivo* tumor progression, and in addition, they oversecreted TGF-β3 that was proposed to maintain stable spindle carcinoma cells (Cui et al., 1996; Portella et al., 1998). This *in vivo* scenario might well mimic alterations occurring during human malignancy in the skin, since human tumors oversecrete TGF-β1 and concomitantly show deregulated levels of the TGF-β type II receptor (Han et al., 2005). Similarly, in the mammary carcinoma model (Ras) discussed above, *in vivo* tumor invasiveness and metastasis not only depends on the continuous action of TGF-β, but more to the point, those cells exhibiting the more overt and irreversible EMT correlated with the highest rate of tumor aggressiveness and metastasis (Oft et al., 1998). These experiments led to the formation of a model of dual action of TGF-β in cancer (see Chapter 20). Accordingly, TGF-β inhibits the growth of normal epithelial cells (and induces reversible EMT) and in doing so acts as a tumor suppressor of early-stage, relatively benign adenomas; TGF-β also acts on carcinoma cells, establishing irreversible EMT that contributes to tumor aggressiveness and metastasis due to the cooperative action of oncogenes such as *Ras* with TGF-β (Akhurst and Derynck, 2001). One obvious caveat in the model of dual action of TGF-β during tumorigenesis is what governs the specific outcome of EMT in normal versus malignant epithelial cells. The most probable answer to this question must reside in the crosstalk mechanisms between cell proliferation and differentiation (EMT).

In addition to cancer, tissue fibrosis is a broad pathological condition that involves processes of EMT *in vivo* and TGF-β is a major factor controlling the progression and outcome of the fibrotic disorders (reviewed in Zeisberg and Kalluri, 2004). Prominent cases of tissue fibrosis with a clear role of TGF-β-mediated EMT have been reported in the kidney, during conditions that lead to chronic renal failure (Li et al., 2003b; Sato et al., 2003). For example, upon ureteral obstruction or diabetic nephropathy, bioactive TGF-β levels increase dramatically, leading to EMT of the tubular epithelial cells and enhanced extracellular matrix deposition, which cumulatively lead to the development of tubulointerstitial fibrosis in the kidney (Li et al., 2003b; Sato et al., 2003). EMT contributing to fibrotic disorders is of primary clinical relevance as efficacious pharmacological treatment of these disorders has been achieved in mouse models, as we discuss in the last section.

2.3 Ligand Specificity in EMT

The TGF-β superfamily includes 34 ligands that exhibit conserved but also rather distinct physiological functions (see Preface). All examples of EMT illustrated above are based on a single group of ligands, the TGF-βs. In fact, all three TGF-β isoforms, TGF-β1, -β2 and -β3 are capable of eliciting EMT *in vitro* (Valcourt et al., 2005). These are the major ligands that are constitutively secreted by carcinoma or fibrotic cells and they are also involved in EMT during development. We have scrutinized the question of ligand specificity during EMT by the TGF-β superfamily and concluded that all signaling pathways that activate Smad2 and Smad3, *i.e.* TGF-β/Activin/Nodal/Myostatin and other ligands, should be able to elicit EMT *in vitro* and *in vivo*, because their corresponding receptors, when ectopically expressed have the capacity to induce EMT (Valcourt et al., 2005). However, this prediction awaits verification from cases where non-TGF-β ligands of the superfamily will actually be identified as critical inducers of EMT. On the other hand, the second branch of ligands of the superfamily that activate Smad1, Smad5 and Smad8 signaling pathways, *i.e.* BMPs/GDFs/MIS seem to be incapable of eliciting EMT (Valcourt et al., 2005). The mechanism for this specificity is explained below.

BMPs are particularly interesting because, they not only fail to induce EMT *in vitro* (Valcourt et al., 2005), but more importantly, BMPs antagonize the TGF-βs and can revert mesenchymal cells produced by the activity of TGF-βs to epithelial cells (MET) (Fig. 1A). This principle has been observed so far in normal mammary, lens and fibrotic kidney epithelial cells *in vitro* (Kowanetz et al., 2004; Zeisberg et al., 2005), and in the fibrotic kidney *in vivo* (Zeisberg and Kalluri, 2004). The fibrotic kidney has provided the first clear evidence that BMP-7 is a true inducer of MET by acting on adult renal fibroblasts, and based on this capacity, BMP-7 leads to regeneration and healing of the injured kidney, thus acting therapeutically (Zeisberg and Kalluri, 2004). It will be clinically important to identify additional BMP/GDF members that might induce MET in other organs, during tissue fibrosis and also during tumor progression.

3. SMAD AND ALTERNATIVE SIGNALS ELICIT EMT

Growing evidence supports strongly the model that EMT is elicited by a large group of cellular mechanisms that operate either at the gene level or at various non-genomic targets by utilizing the concerted action of Smads and alternative signaling proteins. It is worth noting that despite the prevalence of the Smad pathway in the TGF-β world, most published reports on signaling mechanisms that lead to EMT downstream of TGF-β describe non-Smad proteins as the critical effectors. Here we summarize the most prominent examples of signaling mechanisms.

3.1 The Role of Smads in EMT

The role of Smad3 as a mediator of EMT in normal mammary epithelial cells (NMuMG) was established not long after the discovery of Smads (Piek et al., 1999). Despite that, the importance of Smad signaling for EMT only recently has become fully accepted due to the combination of a series of *in vitro* and *in vivo* studies in mouse models (Itoh et al., 2003; Kowanetz et al., 2004; Li et al., 2003a; Li et al., 2003b; Oft et al., 2002; Saika et al., 2004; Sato et al., 2003; Tian, 2003 #6381; Tian et al., 2004; Valcourt et al., 2005). Strong evidence of a general role of Smad signaling during EMT came out of studies using a TGF-β type I receptor bearing point mutations at its L45 loop, the Smad-binding site of the receptor, which fails to recruit and activate the Smad pathway (Itoh et al., 2003). Such mutant receptor fails to elicit EMT downstream of TGF-β. This mutant can also induce a more epithelial phenotype in cells that ectopically express it (Valcourt et al., 2005). Analysis of an L45 mutant receptor in tumor growth and metastasis assays in xenografted mice, revealed that it has two capacities: it enhances tumor growth of relatively benign mammary tumor cells, and it inhibits significantly the metastatic potential of more aggressive mammary tumor cells to the lung (Tian et al., 2004). In both cases, the L45 mutant receptor would block TGF-β signaling, which would mediate tumor suppressor functions in the former case and pro-metastatic functions in the latter.

Our original study on the role of Smad3 in NMuMG (Piek et al., 1999) was recently expanded to analyze the possible role of all Smad proteins during EMT of this *in vitro* cell model (Valcourt et al., 2005). We have demonstrated that only Smads of the TGF-β branch, *i.e.* Smad2 and Smad3 together with the Co-Smad, Smad4, are capable of contributing to EMT. We have repeatedly observed a better ability of Smad3 to elicit EMT compared to Smad2, which correlates with the primary role Smad3 has in mediating the epithelial cytostatic response and global regulation of gene expression (see Chapters 4 and 17). Using the mouse model of squamous carcinoma that depends on the cooperation of Ras and TGF-β signaling, Oft and colleagues convincingly showed that tumor cell invasiveness and metastasis required critically the activity of Smad2 downstream of TGF-β (Oft et al., 2002). Similarly, tissue-specific knock-out of Smad4 in the mammary gland resulted in the development of squamous carcinoma; explanted Smad4 knock-out carcinoma cells failed to undergo EMT in response to TGF-β, in contrast to their wild-type counterparts (Li et al., 2003a). In contrast, when RNAi was used to achieve a partial knock down of Smad4 in HaCaT keratinocytes and colon carcinoma cells, TGF-β was found competent to elicit EMT even when Smad4 levels were very low in the cell (Levy and Hill, 2005). The latter suggests that EMT may require in general a very low degree of Smad signaling, a hypothesis worth investigating further, as it has direct applications for the pharmacological treatment of aggressive or metastatic tumors. Additional studies have firmly placed Smad3 as a critical mediator of EMT *in vivo* during mammary tumor cell invasiveness, during lens epithelial fibrosis caused by injury in post-cataract surgery and during renal tubulointestinal fibrosis (Saika et al., 2004; Sato et al., 2003; Tian et al., 2003). In the latter case, knock-out mice for Smad3 are protected from kidney fibrosis induced by ureteral obstruction and explanted primary renal epithelial cells from the same mice fail to undergo EMT in response to TGF-β, in contrast to the same cells explanted from wild-type mice (Sato et al., 2003). Smad3, also plays a key regulatory role in the chronic auto-induction of TGF-β by exposure of renal cells to the same cytokine, which is required for effective establishment of the fibrotic phenotype. In a similar model of renal fibrogenesis, the Smad pathway induces expression of integrin-linked kinase (ILK) (Fig. 3), which plays a critical role in both eliciting the EMT phenotype and inducing excessive production of extracellular matrix and of metalloproteases by kidney epithelial cells that have undergone EMT (Li et al., 2003b).

Finally, transcriptional repressors such as the nuclear proto-oncogene *c-Ski* and the multifunctional transcription factor YY1 block efficiently TGF-β-driven EMT at least in mammary NMuMG cells (Kurisaki et al., 2003; Takeda et al., 2004). Ski and YY1 bind to Smad proteins and respectively repress either Smad transcriptional activity or Smad DNA-binding.

Overall, the current evidence places the Smad pathway downstream of TGF-β as critical mediator of the EMT and pro-metastatic response of epithelial cells. This emphasizes the role of gene targets of the Smad pathway in EMT (as we discuss later) and is compatible with the notion that EMT represents a switch in epithelial cell differentiation, which should involve changes in the repertoire of gene expression of the cell (Hay, 2005).

3.2 Non-Smad Signal Transducers and EMT under the Control of TGF-β

The role of non-Smad effectors during EMT induced by TGF-β superfamily members gains increasing acceptance in the field. In the case of carcinoma cells with an overactive *Ras* oncogene, EMT depends on mitogen activated protein kinases (MAPK), *e.g.* extracellular regulated kinase (Erk), and on phospho-inositol 3′ kinase (PI3K) activities (Gotzmann et al., 2002; Janda et al., 2002). A similar role of Erk has been established in immortalized HaCaT keratinocytes (Zavadil et al., 2001). In addition, EMT of the breast cancer model that depends on the cooperation of Ras with TGF-β, and metastasis of such tumor cells in mice, critically depend on the endogenous NF-κB pathway (Huber et al., 2004). In this model, TGF-β, in parallel to the Smad pathway, activates the inhibitor of NF-κB (IκB) kinase 2 (IKK-2), which phosphorylates and leads to degradation of IκBα, thus releasing NF-κB in its active form. Presently, it remains unclear how TGF-β activates IKK-2 to elicit the transcriptional activity of NF-κB (Huber et al., 2004). Possible mechanisms might involve the TGF-β-activated kinase 1 (TAK1) that directly phosphorylates IKKs, the cooperation of Smads with NF-κB at the transcriptional level, or the ability of IKK-2 to phosphorylate and inactivate transcription factors of the FoxO family, which interact and cooperate with Smads in regulation of critical gene targets (reviewed in (Moustakas and Heldin, 2005).

Work done using another prominent cell model of *in vitro* EMT, NMuMG, provided a plethora of non-Smad signaling proteins as positive mediators of EMT downstream of TGF-β. These include MAPKs such as Erk, p38, the phospholipid kinase PI3K, the small GTPase RhoA and its downstream kinase ROCK1, which are best known for their mechanism of regulation of actin cytoskeleton dynamics and epithelial cell polarity (Bakin et al., 2002; Bhowmick et al., 2001a; Xie et al., 2004). Currently, none of the above non-Smad effectors have been linked directly to the TGF-β receptors during EMT. Efforts to achieve this, have utilized the L45 mutant type I receptor that cannot activate the Smads, while it can activate endogenous p38 or Jun N-terminal kinase (JNK) pathways (Itoh et al., 2003; Yu et al., 2002), as discussed above. At least in one report (Yu et al., 2002), evidence was provided in favor of p38 as a mediator of EMT. In addition, ILK, which mediates EMT downstream of TGF-β, was thought to represent another critical non-Smad effector (Lee et al., 2004). However, it has been recently demonstrated that ILK expression is induced at the gene level by the Smad pathway and then cooperates with the PI3K-Akt pathway to promote renal cell survival and EMT (Li et al., 2003b). A similar scenario operates under the control of BMP-7, which in the same organ, stimulates ureteric bud formation, a process that involves morphogenesis from renal epithelial cells, and which requires the activity of ILK (Leung-Hagesteijn et al., 2005). In addition to ILK, integrins themselves have been implicated in the regulation of EMT by TGF-β (Bates et al., 2005; Bhowmick et al., 2001b). In NMuMG cells, integrin β1 function is required for TGF-β to activate the endogenous p38 MAPK pathway and elicit EMT (Bhowmick et al., 2001b). Inhibition of integrin β1 function was effective in blocking EMT in this normal mammary epithelial cell model. Similarly, in human colon carcinoma cells that undergo EMT as they become more invasive and migratory, integrin β6 expression is induced, leading to activation of autocrine TGF-β, which enhances the transition of these colon cells to fibroblast-like cells and the secretion of extracellular matrix (Bates et al., 2005). In the latter case, the role of integrin β6 appears to be upstream of TGF-β, as a regulator of activation of this cytokine, instead of acting as a direct transmitter of TGF-β signals towards EMT.

The most compelling evidence for the role of a non-Smad effector downstream of TGF-β during EMT of NMuMG cells is based on a direct link between the TGF-β receptors and a polarity complex that regulates the Rho GTPase pathway (Ozdamar et al., 2005). In polarized cells, the TGF-β type I receptor segregates to tight junctions by interacting with the integral membrane protein occludin, where it also interacts with the polarity protein Par6 (Fig. 2). In another cell model, proximal tubular epithelial HK-2 cells that also undergo EMT in response to TGF-β, the type II receptor segregates in adherens junctions by forming complexes with E-cadherin and β-catenin (Tian and Phillips, 2002). Whether the latter occurs in NMuMG cells where the Par6 mechanism was elucidated remains to be examined

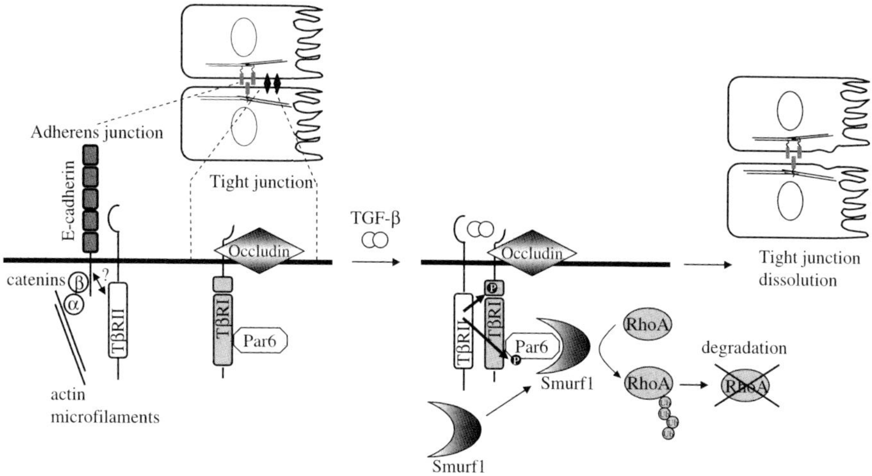

Figure 2. TGF-β receptor-mediated tight junction dissolution. A detail of a polarized epithelial cell focuses (dotted lines) on tight and adherens junction complexes. In tight junctions resides the integral membrane protein occludin that interacts with the TGF-β type I receptor (TβRI), which also makes complexes with the cytoplasmic protein Par6. The type II receptor (TβRII) possibly (arrow with question mark) resides in adherens junctions tethered to E-cadherin, which also forms complexes with β- and α-catenins and actin microfilaments. Upon exposure of cells to TGF-β, the type II receptor is recruited to the occludin complex and phosphorylates (straight thick arrows) the type I receptor and Par6. This leads to recruitment of the Smurf1 ubiquitin ligase which ubiquitylates (small circles labeled Ub) and degrades the small GTPase RhoA. The latter event contributes to local dissolution of tight junctions. The model is adapted from (Ozdamar et al., 2005)

(Fig. 2). Upon TGF-β signaling, the type II receptor is recruited to tight junctions and phosphorylates not only the type I receptor but also the type I receptor-tethered Par6, leading to recruitment of the ubiquitin ligase Smurf1 (see Chapter 13), and subsequent ubiquitylation and degradation of RhoA (Fig. 2) (Ozdamar et al., 2005). This leads to local disassembly of the actin cytoskeleton and dissolution of tight junctions, one of the hallmarks of EMT.

Direct phosphorylation of Par6 by the type II receptor kinase is the first example of a non-TGF-β receptor protein substrate for this receptor kinase. This opens the exciting possibility that, in addition to the type I receptor of TGF-β, other signaling proteins may become phosphorylated by the type II receptor serine/threonine kinase, and thus promises that the above listed plethora of non-Smad effectors may eventually become linked to the activity of the TGF-β receptor complex.

In conclusion, we currently consider a complex signaling network downstream of TGF-β as the mediator of EMT responses. Non-Smad signals may regulate the direct dissolution of epithelial cell adhesion, a process that appears also dependent on Smad-dependent regulation of target genes (see below). In addition, Smad signals regulate gene expression that is critical for the change in differentiation and the generation of the mesenchymal cell.

4. CRITICAL GENE TARGETS OF TGF-β FOR EMT

Upon delineating signaling pathways that mediate EMT under the control of TGF-β members, an important task has been the identification of target genes of these pathways that play direct effector roles for EMT, or that alternatively are required for the maintenance and full deployment of EMT. This task is crucial since we still lack clear markers of the EMT process that could be used in screening protocols in the laboratory and the clinic in order to define unequivocally the importance of EMT in various pathophysiological processes (Hay, 2005). Over the past 4 years several reports of large scale transcriptomic analyses of the EMT process per se under *in vitro* and *in vivo* conditions have appeared (LaGamba et al., 2005; Valcourt et al., 2005; Xie et al., 2003; Zavadil et al., 2001). In addition, microarray screens performed to identify pro-invasiveness and metastasis genes are informative for the process of EMT and some of these reports have made direct implications for the role of TGF-β in supporting tumor cell invasiveness (Jechlinger et al., 2003; Kang et al., 2003). Most of these reports offer interesting gene lists that correlate with the processes of EMT, tumor cell invasiveness or metastasis. In some cases, short gene lists, commonly called gene signatures, have been derived. Despite the existence of such large scale gene expression data, the panel of genes whose function is linked to EMT remains short (Fig. 3). We will focus here on those specific gene targets of TGF-β that provide a deeper understanding of the mechanism of EMT.

A transcriptomic screen in immortalized HaCaT keratinocytes led to the discovery that TGF-β, via the Smad pathway, induces expression of the ligand of Notch signaling, Jagged1 (Zavadil et al., 2001). In addition, TGF-β cooperates with Notch signaling in regulating expression of a well-established target gene of Notch, the *Hey1* transcriptional repressor (Zavadil et al., 2004). Such coordinate regulation of *Hey1* also occurs in mammary and kidney epithelial cells providing an integrated mechanism of EMT that is based on crosstalk between TGF-β/Smad3 and Jagged1/Notch signaling.

Other microarray screens led to the identification of two important regulators of actin dynamics. TGF-β, via Smad3, induces expression of the guanine exchange factor NET1, which accumulates and contributes to the sustained and long term activation of Rho GTPases, thus leading to robust changes in actin dynamics (Shen et al., 2001). A clear role of NET1 during TGF-β-induced EMT has not yet been elucidated. In addition, TGF-β, via the coordinate action of Smads and the p38 MAPK induces expression of several *tropomyosin* genes (Bakin et al., 2004). The increased levels of tropomyosins contribute to contractility of the actin cytoskeleton and cell motility of metastatic carcinoma cells. However, tropomyosin regulation may not play direct effector roles on EMT, but rather is utilized by cells whose differentiation has already been switched to the mesenchymal fate. Another factor acting in a similar manner is the homeobox transcription factor CUTL1, whose expression is directly induced by TGF-β and which then activates expression of many other genes that are important for cell motility, tumor cell invasiveness and extracellular matrix deposition (Michl et al., 2005). CUTL1 is as a poor prognosis marker for metastatic carcinoma of the breast and its role during mammary epithelial

EMT remains unchallenged. Finally, the gene for the adaptor protein disabled-2 (Dab2) that participates in clathrin-mediated endocytosis is a target of TGF-β signaling (Prunier and Howe, 2005). Dab2 is required for TGF-β-induced EMT at least in NMuMG cells and, in its presence, epithelial cells are protected from apoptosis so that they can switch differentiation profile. It is most certain that additional gene targets of the TGF-β/Smad pathway will be soon uncovered with the hope that the most immediate and direct effectors of the EMT process downstream of TGF-β may become known.

4.1 Regulators of *E-cadherin* Gene Expression

Another hallmark of EMT is the dissolution of adherens junctions between neighboring epithelial cells, thus leading to the ability of cells to detach from each other. A critical component of adherens junctions is the cell adhesion and integral protein E-cadherin (Peinado et al., 2004). During EMT, *E-cadherin* mRNA and protein levels decrease and the protein delocalizes from adherens junctions leading to their dissolution. An intimate protein partner of E-cadherin is the intracellular protein β-catenin. which participates as a signal transducer downstream of Wnt to activate gene expression, by modulating the activity of the high mobility group transcription factors, lymphoid enhancer factor/T cell factor (LEF/TCF) (Nawshad et al., 2004). Similar to other pathways that mediate EMT, the mechanism of *E-cadherin* gene repression by TGF-β has been of primary interest (Peinado et al., 2004). For this reason, many investigators have focused on the analysis of transcriptional repressors of the *E-cadherin* gene, such as members of the Snail family of zinc finger proteins (Snail, Slug), two-handed zinc finger/homeodomain proteins (ZEB1, ZEB2), basic helix-loop-helix (bHLH) proteins (E12/E47, Twist) and high mobility group proteins (LEF-1) (Fig. 3) (Peinado et al., 2004). All these repressors bind to specific sequences, named E-boxes, in the proximal promoter region of the *E-cadherin* gene, and recruit transcriptional co-repressors and histone deacetylases to establish stable repression of the gene. A number of these transcriptional repressors (*e.g.* Snail, Slug, LEF-1) have been recently recognized in the transcriptomic analyses of epithelial cells undergoing TGF-β-induced EMT (LaGamba et al., 2005; Zavadil et al., 2001).

The current model of how TGF-β mediates repression of *E-cadherin* gene expression during EMT can be summarized as follows (Fig. 3): a) TGF-β and oncogenic Ras activate the Erk and PI3K pathways leading to induction of Snail expression (Peinado et al., 2003). Smad3 is directly involved in the induction of *Snail* gene expression by TGF-β since renal epithelial cells derived from knock-out mice for Smad3 fail to show responsiveness of their *Snail* gene levels upon stimulation with TGF-β *in vitro* and *in vivo* (Sato et al., 2003). Interestingly, the cleft palate phenotype associated with the TGF-β3 knock-out mouse discussed above is linked to misregulation of Snail expression by TGF-β (Martinez-Alvarez et al., 2004). b) At least in the chicken, the role of Snail seems to be taken by the related transcriptional repressor Slug, which mediates EMT during normal vulval development in the heart under the instruction of TGF-β2 (Martinez-Alvarez et al., 2004;

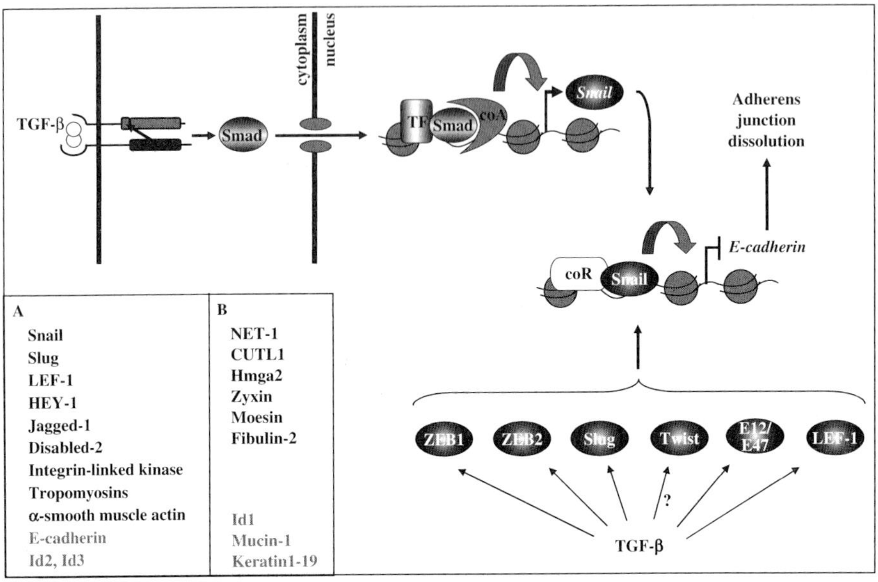

Figure 3. E-cadherin gene repression and other gene targets of TGF-β. TGF-β receptors activate cytoplasmic Smads that move to the nucleus and bind to transcription factors (TF) and co-activators (coA) to induce transcription of *Snail*. Snail protein accumulates and binds to the E-boxes of the *E-cadherin* promoter, thus recruiting corepressors (coR) and inhibiting expression of this gene. Reduction of E-cadherin protein levels leads to dissolution of adherens junctions. A similar mechanism of *E-cadherin* repression can be mediated by other transcriptional repressors, ZEB1, ZEB2, Slug, Twist, E12/E47 and LEF-1. Question mark indicates not yet described links between TGF-β and a repressor. (A) Genes regulated by TGF-β during EMT. The functional role of these genes in EMT has been demonstrated experimentally. (B) Genes regulated by TGF-β whose role on EMT is only suggested based on their known functional properties and their demonstrated involvement in the regulation of epithelial polarity, actin dynamics or metastasis. In this table, all genes not discussed in the text are based on our recent transcriptomic analysis of NMuMG cells (Valcourt et al., 2005). In tables A and B genes in black are upregulated and genes in grey are downregulated

Romano and Runyan, 2000). c) TGF-β-activated Smads inhibit expression of Id proteins, which inhibit DNA binding of bHLH transcription factors such as E12 and E47 (Kondo et al., 2004; Kowanetz et al., 2004). When Id levels are reduced, E12/E47 are relieved from inhibition and repress *E-cadherin* expression (Kondo et al., 2004). d) TGF-β-activated Smads directly interact with ZEB1 and ZEB2 and the complexes participate in repression of *E-cadherin* gene expression mediated by the same E-box sequences to which Snail, Slug and E47 bind (Comijn et al., 2001; Peinado et al., 2004). e) During normal palate development, TGF-β3 induces expression of the high mobility group factor LEF-1 via the Smad pathway (Nawshad et al., 2004). Upon induction of high LEF-1 expression, this factor interacts with Smad proteins (instead of its canonical transactivator β-catenin), which together regulate expression of several genes involved in EMT, including *E-cadherin* (LaGamba et al., 2005). A similar mechanism has been reported during

EMT of mammary epithelial cells that are oncogenicaly transformed by a synthetic *Fos-estrogen receptor* fusion oncogene, which activates autocrine TGF-β signaling via Smads and β-catenin/LEF-1 to elicit *E-cadherin* repression (Eger et al., 2004). Finally, a novel and critical repressor of *E-cadherin* expression that is involved in tumor cell invasiveness and metastasis, the bHLH factor Twist (Yang et al., 2004), has not yet been linked to TGF-β-induced EMT (Fig. 3).

An important open question is whether these distinct transcriptional mechanisms of *E-cadherin* repression act in concert downstream of TGF-β, or whether cells of different tissue origin select a subset of these repressors to mediate shut down of *E-cadherin* mRNA synthesis. In addition, based on the Par6 mechanism whereby the TGF-β receptor activates proteasomal degradation of RhoA and tight junction dissolution, we hypothesize that TGF-β might additionally induce local destruction of E-cadherin protein at adherens junctions. Finally, despite the plethora of studies around the mechanism of transcriptional repression of *E-cadherin*, very little is known about the transcriptional induction of genes that will establish the new mesenchymal phenotype during EMT.

4.2 A Central Role for Id Proteins in the Control of EMT by TGF-β

The recent finding that transcriptional repression of *Id* (*Id2*, *Id3*) gene expression by TGF-β is linked to the process of EMT, has provided new mechanistic insight into the signaling pathways that mediate this process (Fig. 4) (Kondo et al., 2004; Kowanetz et al., 2004). Although all three *Id* genes (*Id1-3*) are downregulated by TGF-β, only *Id2* and *Id3* have been rigorously analyzed. The TGF-β Smad pathway leads to transcriptional repression of *Id* genes, whereas the BMP Smads do the opposite and induce robust levels of Ids in epithelial cells. Downregulation of Id2 and Id3 is important for TGF-β to mediate E-cadherin, ZO-1 downregulation and eventually EMT (Kowanetz et al., 2004). In the case of *E-cadherin* gene repression, a decrease in Id2 levels is critical for the E12/E47 bHLH factors to be able to bind to the *E-cadherin* promoter and elicit the repression (Kondo et al., 2004). In contrast, BMPs induce high levels of all Ids that enforce global inhibition of various bHLH proteins, and failure to induce EMT. When endogenous Id levels are knocked down by RNA interference (RNAi), BMPs are capable of inducing EMT in mammary or lens epithelial cells (Kowanetz et al., 2004). *Id* gene regulation by TGF-β explains the aforementioned antagonism between TGF-β and BMP, whereby, BMP dominantly antagonizes TGF-β-induced EMT and promotes MET (Kowanetz et al., 2004; Zeisberg and Kalluri, 2004; Zeisberg et al., 2005).

Interestingly, the role of Ids in defining how epithelial cells will respond to TGF-βs versus BMPs is not only pertinent to the process of EMT but also to the process of growth arrest (Kowanetz et al., 2004). This provides a molecular mechanism by which the two processes, EMT and growth arrest, become intimately linked at least in normal epithelial cells. TGF-β downregulates Ids together with the proto-oncogene *myc* and induces expression of cell cycle inhibitors such as *p21*,

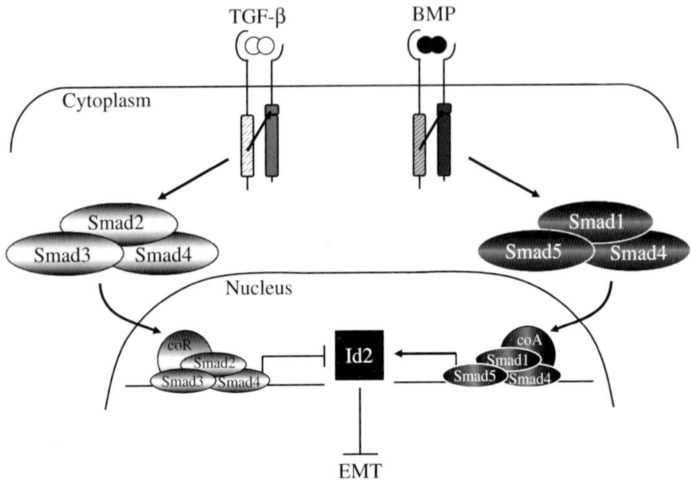

Figure 4. Id2 plays central roles in the regulation of EMT by TGF-β superfamily members. TGF-β receptors activate Smad2 and Smad3 to repress expression of Id2, whereas BMP receptors activate Smad1 and Smad5 to induce expression of Id2. The exact stoichiometry of the Smad complexes on the Id2 enhancer/silencer and the specific R-Smads that mediate *Id2* gene regulation *in vivo* are not yet known. High Id2 levels inhibit EMT and preserve epithelial differentiation. TGF-β downregulates Id2 expression in order to elicit EMT. BMP fails to elicit EMT and promotes MET due to the upregulation of Id2. Co-repressor (coR) and co-activator (coA) complexes are illustrated on the *Id2* gene in complex with Smads

which are hallmark events during epithelial cytostasis (see Chapter 4). BMPs also repress *c-myc* and induce robust *p21* expression, but, in addition, BMPs induce high Id levels. Thus, Ids play dominant roles over c-myc and p21, and in the presence of high Id levels, BMPs fail to establish an efficient cytostatic response in epithelial cells (Pardali et al., 2005). This implies that there must be additional genes that govern the cytostatic and EMT responses to TGF-β superfamily members, and expression of such genes must be regulated by Id proteins. These genes may include novel regulators of these two important processes, and understanding how their function might become subverted during tumorigenesis or in fibrotic diseases has clinical relevance.

With respect to the role of EMT in tumor progression, the Id model (Fig. 4) suggests that at early stages of carcinoma invasiveness when EMT occurs, Id levels must be reduced. However, upon establishment of tumor growth in a metastatic site, Id levels must gradually increase as the transitory mesenchymal cells undergo MET and establish a new (semi-) differentiated tumor. The latter is actually observed in the case of Id2, whose levels are abnormally high in carcinomas that have metastasized to bone or lung (Perk et al., 2005). The challenging task for the future remains to observe the transitory reduction of Id expression in carcinomas that prepare for metastasis *in vivo*.

5. CLINICAL RELEVANCE AND FUTURE PERSPECTIVES

Despite the early times in understanding every detail of TGF-β-driven EMT *in vitro* and *in vivo*, promising lead discoveries have already offered new therapeutic approaches towards cancer and fibrosis (see Chapters 21 and 22). Based on the central role of Smad3 as mediator of EMT, drugs that would target the function of this protein might be desirable (Sato et al., 2003). However, such Smad3-targeted therapy might affect many processes downstream of TGF-β signaling, as Smad3 seems to be involved in all aspects of TGF-β physiology at least in adult cells of all tissue types examined so far. Design of EMT-specific drugs awaits the discovery of novel mediators of the process that are amenable to pharmaceutical intervention and play critical roles during disease. The emerging technologies of single cell micro-dissection coupled to bioluminescence analysis of tagged cell types and large scale gene expression analysis is a major promise in this field, and has already provided exciting new discoveries in the realm of metastatic tumors (Wang et al., 2005b).

The understanding of non-Smad signaling pathways that mediate EMT, at least during tissue fibrosis, has proven influential, in the sense that the tyrosine kinase-specific drug, imatinib, has been used successfully to block lung and kidney fibrosis in experimental mouse and rat models (Wang et al., 2005a). The rationale behind the use of imatinib was based on the understanding that TGF-β utilizes the c-Abl kinase in mediating its signals at least in fibroblasts. In addition to tyrosine kinases as therapeutic targets of renal fibrosis that involves TGF-β-driven EMT, ILK represents another attractive target for intervention as exemplified by proof-of-prinicple experiments in mouse models using dominant-negative mutants of ILK (Li et al., 2003b). In the case of EMT that links to breast cancer invasiveness and metastasis, the NF-κB pathway presents another attractive site for intervention based on the critical role IKK-2 plays in the Ras-TGF-β cooperative tumor model described above (Huber et al., 2004). Proof-of-principle experiments in mouse models of breast cancer metastasis to lung, convincingly demonstrated that inhibition of NF-κB activity is sufficient to abrogate dramatically the metastatic protential of the breast cancer cells (Huber et al., 2004). This raises the important point that combinatorial pharamacologic regimes that target c-Abl, ILK and IKKs might present a more efficient way to attack EMT against tissue fibrosis and cancer. Identification of additional signaling effectors of TGF-β that contribute to EMT may allow the development of even more sophisticated pharmacological approaches in the battle against these major human diseases.

Wherever possible in this chapter we attempted to present relatively complete models of signal transduction and gene regulation downstream of TGF-β ligands that describe the onset and establishment of EMT. However, we are far away from being able to describe comprehensive mechanisms. The real challenge remains for the future to fill in all the important gaps and understand the complex signaling networks operating *in vivo*. The complexity of Smad and non-Smad proteins that orchestrate the EMT response promises several additional important players to be discovered. Influenced from the example of the TGF-β receptor/occludin/Par6 mechanism (Ozdamar et al., 2005), we suggest that focusing on studies of receptor localization and on the identification of novel receptor substrates during EMT should uncover novel mechanistic details of EMT. The direct role of Smads in such processes also deserves further attention.

In addition to the development of novel therapeutic protocols against EMT that leads to tissue fibrosis or tumor metastasis, the hunt for new molecular markers that can be used either prognostically or diagnostically to define EMT in human patients is another major task of this field of research (Huber et al., 2005). It is worth summarizing that among the various genes and proteins discussed in this chapter at least two of them have already been analyzed in enough detail to consider them as important novel indicators of cancer progression. Integrin β6 expression levels mark the aggressiveness of colon carcinomas in humans and this protein can serve as a prognostic marker for patient life expectancy (Bates et al., 2005). Expression levels of the transcriptional repressor Snail can also serve as valid indicators of breast cancer recurrence in women, after primary tumor surgery (Moody et al., 2005). High Snail expression predicts a significantly lower chance for relapse-free survival in women with breast cancer. As outlined above, the gene targets of the TGF-β pathway that are true effectors of EMT and which contribute to the establishment of this process remain poorly understood. Transcriptomic approaches are promising and need to be coupled to extensive functional dissection of regulated genes. Alternatively, approaches based on RNAi, such as the one that linked CUTL1 to metastasis (Michl et al., 2005), warrant future use in the EMT field. One last promise of such research is the identification of new and more reliable markers of the mesenchymal cell phenotype and of the EMT process overall.

ACKNOWLEDGEMENTS

Owing to space limitations, selected literature is cited in this chapter. Funding of the authors' work is provided by the Ludwig Institute for Cancer Research, the Swedish Cancer Society, the Swedish Research Council and the European Commission FP6 programme. We thank our colleagues A. Gaal, A. Kurisaki, K. Pardali and U. Valcourt for their contributions to the scientific work emanating from our laboratory. Special thanks to Carl-Henrik Heldin (Ludwig Institute for Cancer Research, Uppsala) and P. ten Dijke (Leiden University Medical Center) for continuous encouragement and collaborations.

REFERENCES

Akhurst, R.J., and Derynck, R., 2001, TGF-β signaling in cancer – a double-edged sword. *Trends Cell Biol* 11: S44-51.

Bakin, A.V., Rinehart, C., Tomlinson, A.K., and Arteaga, C.L., 2002, p38 mitogen-activated protein kinase is required for TGFβ-mediated fibroblastic transdifferentiation and cell migration. *J Cell Sci* 115: 3193-3206.

Bakin, A.V., Safina, A., Rinehart, C., Daroqui, C., Darbary, H., and Helfman, D.M., 2004, A critical role of tropomyosins in TGF-β regulation of the actin cytoskeleton and cell motility in epithelial cells. *Mol Biol Cell* 15: 4682-4694.

Bates, R.C., Bellovin, D.I., Brown, C., Maynard, E., Wu, B., Kawakatsu, H., Sheppard, D., Oettgen, P., and Mercurio, A.M., 2005, Transcriptional activation of integrin β6 during the epithelial-mesenchymal transition defines a novel prognostic indicator of aggressive colon carcinoma. *J Clin Invest* 115: 339-347.

Bates, R.C., and Mercurio, A.M., 2005, The Epithelial-Mesenchymal Transition (EMT) and Colorectal Cancer Progression. *Cancer Biol Ther* 4: 365-370.

Bhowmick, N.A., Ghiassi, M., Bakin, A., Aakre, M., Lundquist, C.A., Engel, M.E., Arteaga, C.L., and Moses, H.L., 2001a, Transforming growth factor-β1 mediates epithelial to mesenchymal transdifferentiation through a RhoA-dependent mechanism. *Mol Biol Cell* 12: 27-36.

Bhowmick, N.A., Zent, R., Ghiassi, M., McDonnell, M., and Moses, H.L., 2001b, Integrin β1 signaling is necessary for transforming growth factor-β activation of p38 MAPK and epithelial plasticity. *J Biol Chem* 276: 46707-46713.

Brown, K.A., Aakre, M.E., Gorska, A.E., Price, J.O., Eltom, S.E., Pietenpol, J.A., and Moses, H.L., 2004, Induction by transforming growth factor-β1 of epithelial to mesenchymal transition is a rare event in vitro. *Breast Cancer Res* 6: R215-231.

Comijn, J., Berx, G., Vermassen, P., Verschueren, K., van Grunsven, L., Bruyneel, E., Mareel, M., Huylebroeck, D., and van Roy, F., 2001, The two-handed E box binding zinc finger protein SIP1 downregulates E-cadherin and induces invasion. *Mol Cell* 7: 1267-1278.

Cui, W., Fowlis, D.J., Bryson, S., Duffie, E., Ireland, H., Balmain, A., and Akhurst, R.J., 1996, TGFβ1 inhibits the formation of benign skin tumors, but enhances progression to invasive spindle carcinomas in transgenic mice. *Cell* 86: 531-542.

Desgrosellier, J.S., Mundell, N.A., McDonnell, M.A., Moses, H.L., and Barnett, J.V., 2005, Activin receptor-like kinase 2 and Smad6 regulate epithelial-mesenchymal transformation during cardiac valve formation. *Dev Biol* 280: 201-210.

Eger, A., Stockinger, A., Park, J., Langkopf, E., Mikula, M., Gotzmann, J., Mikulits, W., Beug, H., and Foisner, R., 2004, β-Catenin and TGFβ signalling cooperate to maintain a mesenchymal phenotype after FosER-induced epithelial to mesenchymal transition. *Oncogene* 23: 2672-2680.

Gotzmann, J., Huber, H., Thallinger, C., Wolschek, M., Jansen, B., Schulte-Hermann, R., Beug, H., and Mikulits, W., 2002, Hepatocytes convert to a fibroblastoid phenotype through the cooperation of TGF-β1 and Ha-Ras: steps towards invasiveness. *J Cell Sci* 115: 1189-1202.

Han, G., Lu, S.L., Li, A.G., He, W., Corless, C.L., Kulesz-Martin, M., and Wang, X.-J., 2005, Distinct mechanisms of TGF-β1-mediated epithelial-to-mesenchymal transition and metastasis during skin carcinogenesis. *J Clin Invest* 115: 1714-1723.

Hay, E.D., 2005, The mesenchymal cell, its role in the embryo, and the remarkable signaling mechanisms that create it. *Dev Dyn* 233: 706-720.

Huber, M.A., Azoitei, N., Baumann, B., Grünert, S., Sommer, A., Pehamberger, H., Kraut, N., Beug, H., and Wirth, T., 2004, NF-κB is essential for epithelial-mesenchymal transition and metastasis in a model of breast cancer progression. *J Clin Invest* 114: 569-581.

Huber, M.A., Kraut, N., and Beug, H., 2005, Molecular requirements for epithelial-mesenchymal transition during tumor progression. *Curr Opin Cell Biol* 17: 548-558.

Itoh, S., Thorikay, M., Kowanetz, M., Moustakas, A., Itoh, F., Heldin, C.-H., and ten Dijke, P., 2003, Elucidation of Smad requirement in transforming growth factor-β type receptor-induced responses. *J Biol Chem* 278: 3751-3761.

Janda, E., Lehmann, K., Killisch, I., Jechlinger, M., Herzig, M., Downward, J., Beug, H., and Grünert, S., 2002, Ras and TGFβ cooperatively regulate epithelial cell plasticity and metastasis: dissection of Ras signaling pathways. *J Cell Biol* 156: 299-313.

Jechlinger, M., Grunert, S., Tamir, I.H., Janda, E., Ludemann, S., Waerner, T., Seither, P., Weith, A., Beug, H., and Kraut, N., 2003, Expression profiling of epithelial plasticity in tumor progression. *Oncogene* 22: 7155-7169.

Kang, Y., Siegel, P.M., Shu, W., Drobnjak, M., Kakonen, S.M., Cordon-Cardo, C., Guise, T.A., and Massagué, J., 2003, A multigenic program mediating breast cancer metastasis to bone. *Cancer Cell* 3: 537-549.

Kondo, M., Cubillo, E., Tobiume, K., Shirakihara, T., Fukuda, N., Suzuki, H., Shimizu, K., Takehara, K., Cano, A., Saitoh, M., and Miyazono, K., 2004, A role for Id in the regulation of TGF-β-induced epithelial-mesenchymal transdifferentiation. *Cell Death Differ* 11: 1092-1101.

Kowanetz, M., Valcourt, U., Bergström, R., Heldin, C.-H., and Moustakas, A., 2004, Id2 and Id3 define the potency of cell proliferation and differentiation responses to tranforming growth factor β and bone morphogenetic protein. *Mol Cell Biol* 24: 4241-4254.

Kurisaki, K., Kurisaki, A., Valcourt, U., Terentiev, A.A., Pardali, K., ten Dijke, P., Heldin, C.-H., Ericsson, J., and Moustakas, A., 2003, Nuclear factor YY1 inhibits transforming growth factor β- and bone morphogenetic protein-induced cell differentiation. *Mol Cell Biol* 23: 4494-4510.

LaGamba, D., Nawshad, A., and Hay, E.D., 2005, Microarray analysis of gene expression during epithelial-mesenchymal transformation. *Dev Dyn* 234: 132-142.

Lee, Y.I., Kwon, Y.J., and Joo, C.K., 2004, Integrin-linked kinase function is required for transforming growth factor β-mediated epithelial to mesenchymal transition. *Biochem Biophys Res Commun* 316: 997-1001.

Lehmann, K., Janda, E., Pierreux, C.E., Rytomaa, M., Schulze, A., McMahon, M., Hill, C.S., Beug, H., and Downward, J., 2000, Raf induces TGFβ production while blocking its apoptotic but not invasive responses: a mechanism leading to increased malignancy in epithelial cells. *Genes Dev* 14: 2610-2622.

Leung-Hagesteijn, C., Hu, M.C., Mahendra, A.S., Hartwig, S., Klamut, H.J., Rosenblum, N.D., and Hannigan, G.E., 2005, Integrin-linked kinase mediates bone morphogenetic protein 7-dependent renal epithelial cell morphogenesis. *Mol Cell Biol* 25: 3648-3657.

Levy, L., and Hill, C.S., 2005, Smad4 dependency defines two classes of transforming growth factor β (TGF-β) target genes and distinguishes TGF-β-induced epithelial-mesenchymal transition from its antiproliferative and migratory responses. *Mol Cell Biol* 25: 8108-8125.

Li, W., Qiao, W., Chen, L., Xu, X., Yang, X., Li, D., Li, C., Brodie, S.G., Meguid, M.M., Hennighausen, L., and Deng, C.-X., 2003a, Squamous cell carcinoma and mammary abscess formation through squamous metaplasia in Smad4/Dpc4 conditional knockout mice. *Development* 130: 6143-6153.

Li, Y., Yang, J., Dai, C., Wu, C., and Liu, Y., 2003b, Role for integrin-linked kinase in mediating tubular epithelial to mesenchymal transition and renal interstitial fibrogenesis. *J Clin Invest* 112: 503-516.

Martinez-Alvarez, C., Blanco, M.J., Perez, R., Rabadan, M.A., Aparicio, M., Resel, E., Martinez, T., and Nieto, M.A., 2004, Snail family members and cell survival in physiological and pathological cleft palates. *Dev Biol* 265: 207-218.

Michl, P., Ramjaun, A.R., Pardo, O.E., Warne, P.H., Wagner, M., Poulsom, R., D'Arrigo, C., Ryder, K., Menke, A., Gress, T., and Downward, J., 2005, CUTL1 is a target of TGFβ signaling that enhances cancer cell motility and invasiveness. *Cancer Cell* 7: 521-532.

Miettinen, P.J., Ebner, R., Lopez, A.R., and Derynck, R., 1994, TGF-β induced transdifferentiation of mammary epithelial cells to mesenchymal cells: involvement of type I receptors. *J Cell Biol* 127: 2021-2036.

Moody, S.E., Perez, D., Pan, T.C., Sarkisian, C.J., Portocarrero, C.P., Sterner, C.J., Notorfrancesco, K.L., Cardiff, R.D., and Chodosh, L.A., 2005, The transcriptional repressor Snail promotes mammary tumor recurrence. *Cancer Cell* 8: 197-209.

Moustakas, A., and Heldin, C.-H., 2005, Non-Smad TGF-β signals. *J Cell Sci* 118: 3573-3584.

Nawshad, A., LaGamba, D., and Hay, E.D., 2004, Transforming growth factor β (TGFβ) signalling in palatal growth, apoptosis and epithelial mesenchymal transformation (EMT). *Arch Oral Biol* 49: 675-689.

Oft, M., Akhurst, R.J., and Balmain, A., 2002, Metastasis is driven by sequential elevation of H-ras and Smad2 levels. *Nat Cell Biol* 4: 487-494.

Oft, M., Heider, K.H., and Beug, H., 1998, TGFβ signaling is necessary for carcinoma cell invasiveness and metastasis. *Curr Biol* 8: 1243-1252.

Ozdamar, B., Bose, R., Barrios-Rodiles, M., Wang, H.R., Zhang, Y., and Wrana, J.L., 2005, Regulation of the polarity protein Par6 by TGFβ receptors controls epithelial cell plasticity. *Science* 307: 1603-1609.

Pardali, K., Kowanetz, M., Heldin, C.-H., and Moustakas, A., 2005, Smad pathway-specific transcriptional regulation of the cell cycle inhibitor $p21^{WAF1/Cip1}$. *J Cell Physiol* 204: 260-272.

Peinado, H., Portillo, F., and Cano, A., 2004, Transcriptional regulation of cadherins during development and carcinogenesis. *Int J Dev Biol* 48: 365-375.

Peinado, H., Quintanilla, M., and Cano, A., 2003, Transforming growth factor β-1 induces snail transcription factor in epithelial cell lines: mechanisms for epithelial mesenchymal transitions. *J Biol Chem* 278: 21113-21123.

Perk, J., Iavarone, A., and Benezra, R., 2005, Id family of helix-loop-helix proteins in cancer. *Nat Rev Cancer* 5: 603-614.

Person, A.D., Klewer, S.E., and Runyan, R.B., 2005, Cell biology of cardiac cushion development. *Int Rev Cytol* 243: 287-335.

Piek, E., Moustakas, A., Kurisaki, A., Heldin, C.-H., and ten Dijke, P., 1999, TGF-β type I receptor/ALK-5 and Smad proteins mediate epithelial to mesenchymal transdifferentiation in NMuMG breast epithelial cells. *J Cell Sci* 112: 4557-4568.

Portella, G., Cumming, S.A., Liddell, J., Cui, W., Ireland, H., Akhurst, R.J., and Balmain, A., 1998, Transforming growth factor β is essential for spindle cell conversion of mouse skin carcinoma *in vivo*: implications for tumor invasion. *Cell Growth Differ* 9: 393-404.

Prunier, C., and Howe, P.H., 2005, Disabled-2 (Dab2) is required for transforming growth factor β-induced epithelial to mesenchymal transition (EMT). *J Biol Chem* 280: 17540-17548.

Romano, L.A., and Runyan, R.B., 2000, Slug is an essential target of TGFβ2 signaling in the developing chicken heart. *Dev Biol* 223: 91-102.

Saika, S., Kono-Saika, S., Ohnishi, Y., Sato, M., Muragaki, Y., Ooshima, A., Flanders, K.C., Yoo, J., Anzano, M., Liu, C.Y., Kao, W.W., and Roberts, A.B., 2004, Smad3 signaling is required for epithelial-mesenchymal transition of lens epithelium after injury. *Am J Pathol* 164: 651-663.

Sato, M., Muragaki, Y., Saika, S., Roberts, A.B., and Ooshima, A., 2003, Targeted disruption of TGF-β1/Smad3 signaling protects against renal tubulointerstitial fibrosis induced by unilateral ureteral obstruction. *J Clin Invest* 112: 1486-1494.

Shen, X., Li, J., Hu, P.P., Waddell, D., Zhang, J., and Wang, X.-F., 2001, The activity of guanine exchange factor NET1 is essential for transforming growth factor-β-mediated stress fiber formation. *J Biol Chem* 276: 15362-15368.

Takeda, M., Mizuide, M., Oka, M., Watabe, T., Inoue, H., Suzuki, H., Fujita, T., Imamura, T., Miyazono, K., and Miyazawa, K., 2004, Interaction with Smad4 is indispensable for suppression of BMP signaling by c-Ski. *Mol Biol Cell* 15: 963-972.

Tarin, D., Thompson, E.W., and Newgreen, D.F., 2005, The fallacy of epithelial mesenchymal transition in neoplasia. *Cancer Res* 65: 5996-6000; discussion 6000-6001.

Thiery, J.P., 2003, Epithelial-mesenchymal transitions in development and pathologies. *Curr Opin Cell Biol* 15: 740-746.

Tian, F., Byfield, S.D., Parks, W.T., Stuelten, C.H., Nemani, D., Zhang, Y.E., and Roberts, A.B., 2004, Smad-binding defective mutant of transforming growth factor β type I receptor enhances tumorigenesis but suppresses metastasis of breast cancer cell lines. *Cancer Res* 64: 4523-4530.

Tian, F., DaCosta Byfield, S., Parks, W.T., Yoo, S., Felici, A., Tang, B., Piek, E., Wakefield, L.M., and Roberts, A.B., 2003, Reduction in Smad2/3 signaling enhances tumorigenesis but suppresses metastasis of breast cancer cell lines. *Cancer Res* 63: 8284-8292.

Tian, Y.C., and Phillips, A.O., 2002, Interaction between the transforming growth factor-β type II receptor/Smad pathway and β-catenin during transforming growth factor-β1-mediated adherens junction disassembly. *Am J Pathol* 160: 1619-1628.

Valcourt, U., Kowanetz, M., Niimi, H., Heldin, C.-H., and Moustakas, A., 2005, TGF-β and the Smad signaling pathway support transcriptomic reprogramming during epithelial-mesenchymal cell transition. *Mol Biol Cell* 16: 1987-2002.

Valdes, F., Alvarez, A.M., Locascio, A., Vega, S., Herrera, B., Fernandez, M., Benito, M., Nieto, M.A., and Fabregat, I., 2002, The epithelial mesenchymal transition confers resistance to the apoptotic effects of transforming growth factor β in fetal rat hepatocytes. *Mol Cancer Res* 1: 68-78.

Wang, S., Wilkes, M.C., Leof, E.B., and Hirschberg, R., 2005a, Imatinib mesylate blocks a non-Smad TGF-β pathway and reduces renal fibrogenesis in vivo. *FASEB J* 19: 1-11.

Wang, W., Goswami, S., Sahai, E., Wyckoff, J.B., Segall, J.E., and Condeelis, J.S., 2005b, Tumor cells caught in the act of invading: their strategy for enhanced cell motility. *Trends Cell Biol* 15: 138-145.

Willis, B.C., Liebler, J.M., Luby-Phelps, K., Nicholson, A.G., Crandall, E.D., du Bois, R.M., and Borok, Z., 2005, Induction of epithelial-mesenchymal transition in alveolar epithelial cells by transforming growth factor-β1: potential role in idiopathic pulmonary fibrosis. *Am J Pathol* 166: 1321-1332.

Xie, L., Law, B.K., Aakre, M.E., Edgerton, M., Shyr, Y., Bhowmick, N.A., and Moses, H.L., 2003, Transforming growth factor β-regulated gene expression in a mouse mammary gland epithelial cell line. *Breast Cancer Res* 5: R187-198.

Xie, L., Law, B.K., Chytil, A.M., Brown, K.A., Aakre, M.E., and Moses, H.L., 2004, Activation of the Erk pathway is required for TGF-β1-induced EMT in vitro. *Neoplasia* 6: 603-610.

Yang, J., Mani, S.A., Donaher, J.L., Ramaswamy, S., Itzykson, R.A., Come, C., Savagner, P., Gitelman, I., Richardson, A., and Weinberg, R.A., 2004, Twist, a master regulator of morphogenesis, plays an essential role in tumor metastasis. *Cell* 117: 927-939.

Yu, L., Hebert, M.C., and Zhang, Y.E., 2002, TGF-β receptor-activated p38 MAP kinase mediates Smad-independent TGF-β responses. *EMBO J* 21: 3749-3759.

Zavadil, J., Bitzer, M., Liang, D., Yang, Y.C., Massimi, A., Kneitz, S., Piek, E., and Böttinger, E.P., 2001, Genetic programs of epithelial cell plasticity directed by transforming growth factor-β. *Proc Natl Acad Sci U S A* 98: 6686-6691.

Zavadil, J., Cermak, L., Soto-Nieves, N., and Böttinger, E.P., 2004, Integration of TGFβ/Smad and Jagged1/Notch signalling in epithelial-to-mesenchymal transition. *EMBO J* 23: 1155-1165.

Zeisberg, M., and Kalluri, R., 2004, The role of epithelial-to-mesenchymal transition in renal fibrosis. *J Mol Med* 82: 175-181.

Zeisberg, M., Shah, A.A., and Kalluri, R., 2005, Bone morphogenic protein-7 induces mesenchymal to epithelial transition in adult renal fibroblasts and facilitates regeneration of injured kidney. *J Biol Chem* 280: 8094-8100.

CHAPTER 8

GENETIC DISRUPTIONS WITHIN THE MURINE GENOME REVEAL NUMEROUS ROLES OF THE SMAD GENE FAMILY IN DEVELOPMENT, DISEASE, AND CANCER

MICHAEL WEINSTEIN[1] AND CHU-XIA DENG[2]

[1] *Department of Molecular Genetics, Ohio State University, Columbus, OH 43210 USA*

[2] *Mammalian Genetics Section, Genetics of Development and Disease Branch, Digestive and Kidney Diseases, National Institutes of Health, 10/9N105, 10 Center Drive, Bethesda, MD 20892, USA*

Abstract: Mammalian Smads consist of a gene family of 8 members that serve as intracellular mediators of TGF-β signaling. Functions of Smads have been elucidated through extensive research using gene targeting technologies to introduce various types of mutations into these genes. Analysis of mutant mice has uncovered important functions of Smads in numerous biological processes, including gastrulation, mesoderm induction and patterning, angiogenesis, cell proliferation and differentiation, organogenesis, immunological response, wound healing, and tumorigenesis. Here we summarize these studies and discuss possible mechanisms underlying phenotypic alterations associated with the deficiency of each Smad

Keywords: angiogenesis; cancer; embryo; gene targeting; mouse; organogenesis

1. INTRODUCTION

Members of the TGF-β superfamily transduce their signals through sequentially activating membrane bound receptors and intracellular mediators, the Smad proteins (reviewed in Heldin et al., 1997; Massagué, 1998; ten Dijke and Hill, 2004). Smads 1, 5, and 8 mediate signals of the BMP ligand subfamily, while Smads 2 and 3 mediate signals from TGF-β, Activins, Nodal, and other TGF-β ligands. Smad4 serves as a common mediator and forms heterotrimers with all of these Smads, which are translocated to the nucleus and function directly to induce or repress expression of TGF-β target genes. Smads 6 and 7 play important roles in inhibiting

P. ten Dijke and C.-H. Heldin (eds.), Smad Signal Transduction, 151–176.

TGF-β signals once the pathway is activated. Smads are expressed in virtually all tissues at varying levels (Flanders et al., 2001; Yang et al., 1999b). Currently, all of the *Smad* genes have been disrupted in mice by gene targeting. Phenotypes of some of these mice have been reviewed previously (Weinstein et al., 2000). Briefly, mice lacking Smads 1, 2, 4 or 5 exhibit embryonic lethality, while mice carrying targeted mutations of Smad3, 6 or 8 survive to adulthood displaying abnormalities in multiple organs/tissues. Phenotypes of these mutant mice include pregastrulation lethality (*Smad2*$^{-/-}$ and *Smad4*$^{-/-}$ mutants), failure of placenta and chorioallantoic fusion (*Smad1*$^{-/-}$), angiogenesis defects (*Smad5*$^{-/-}$), heart abnormalities (*Smad6*$^{-/-}$), defective immunologic response (*Smad3*$^{-/-}$), and tumor formation in a number of organs (*Smad3*$^{-/-}$) and (*Smad4*$^{-/-}$). Analysis of these animals revealed numerous pivotal roles for TGF-β signaling in many biologic processes, which are summarized in Table 1 and discussed in greater detail below.

Table 1. Phenotypes of Smad mutant mice revealed by conventional gene targeting

Genes	Alleles	Phenotypes	References
Smad1	null	Allantois, PGC	(Lechleider et al., 2001; Tremblay et al., 2001)
	heterozygote	Placental dysfunction, neuroectodermal defect	(Hester et al., 2005)
	MAP kinase phosph. sites	PGC, gastric, reproductive phenotypes	(Aubin et al., 2004)
	C-terminal phosph. sites	PGC, chorioallantoic fusion defect, posterior truncations	(Aubin et al., 2004)
Smad2	null	Gastrulation, extraembryonic membranes	(Hamamoto et al., 2002; Nomura and Li, 1998; Weinstein et al., 1998)
	heterozygote	gastrulation	(Liu et al., 2004a; Nomura and Li, 1998)
	Smad2$^{\Delta exon2}$	Anterior/posterior axis defect	(Waldrip et al., 1998)
	Smad2^{S276L}	Anterior/posterior axis defect, axial, anterior and cardiac abnormalities	(Vivian et al., 2002)
	Smad2^{3loxP}	Anterior/posterior axis defect, axial, anterior and cardiac abnormalities	(Liu et al., 2004a)
Smad3	null	Immune, bone, mammary, ovarian defects	(Datto et al., 1999; Yang et al., 1999b)
		Metastatic Colon cancer	(Zhu et al., 1998)
Smad4	null	Extraembrynic membrane and epiblast proliferation defects	(Sirard et al., 1998; Yang et al., 1998)
	heterozygote	Gastric polyposis and cancer	(Taketo and Takaku, 2000; Xu et al., 2000)
Smad5	null	Angiogenesis and cardiac defects, mesenchymal apoptosis, posterior truncations	(Chang et al., 1999; Yang et al., 1999a)
		Left/right asymmetry	(Chang et al., 2000)
		PGC	(Chang and Matzuk, 2001)

(Continued)

Table 1. (Continued)

Genes	Alleles	Phenotypes	References
Smad6	MH2:lacZ	Cardiac outflow tract, endocardial cushions	(Galvin et al., 2000)
Smad7	Exon1 deletion	Altered B cell responses	R. Heuchel, pers. com.
Smad8	*Smad8*^{3loxP}	mid and hindbrain reductions	(Hester et al., 2005)
Double KO			
Smad2$^{+/-}$;*Nodal*$^{+/-}$		Abnormal gastrulation, holoprosencephaly, left/right isomerism	(Nomura and Li, 1998)
Smad2$^{+/-}$;*Smad3*$^{+/-}$		Hypoplasia of the liver, craniofacial defects, embryonic lethal before E16.5	(Liu et al., 2004a)
Smad2$^{\Delta exon2/+}$;*Smad3*$^{-/-}$ *Smad2*$^{\Delta exon2/\Delta exon2}$;*Smad3*$^{+/-}$ *Smad2*$^{\Delta exon2/\Delta exon2}$;*Smad3*$^{-/-}$		Progressively diminished anterior migration of primitive streak derivatives leading to anterior patterning defects	(Dunn et al., 2004)
Smad2$^{3loxP/3loxP}$;*Smad3*$^{+/-}$ *Smad2*$^{3loxP/3loxP}$;*Smad3*$^{-/-}$		Progressively more severe anterior and hepatocytic defects	A. Chow and M. Weinstein, unpub. results
Smad2$^{+/-}$;*Apc*$^{+/-}$		Increased incidence and invasiveness of colorectal cancer	(Hamamoto et al., 2002)
Smad2$^{\Delta exon2/+}$;*Apc*$^{+/-}$		No difference from *Smad2*$^{+/+}$;*Apc*$^{+/-}$	(Takaku et al., 2002)
Smad4$^{+/-}$;*Apc*$^{+/-}$		Increased incidence and invasiveness of colorectal cancer	(Takaku et al., 1998)

2. FUNCTIONAL ANALYSIS OF MUTANT MICE LACKING SPECIFIC SMADS

2.1 Smad1

Smad1 homozygous mutants showed a disruption of gastrulation, including abnormal epiblastic twisting and delayed mesodermal differentiation, although patterning in the embryonic anterior/posterior axis was largely normal. Formation of the allantois and chorion, extraembryonic tissues that contribute to the development of the placenta, was morphologically abnormal. These abnormalities correlated with, and were possibly caused by, delayed mesodermal formation and posterior migration (Tremblay et al., 2001). There was an increase in a number of genes expressed in the extraembryonic membranes along with a failure of placental formation, which caused lethality by embryonic day 10.5 (E10.5) (Lechleider et al., 2001; Tremblay et al., 2001). In addition, *Smad1* mutants lacked primordial germ cells (Hayashi et al., 2002; Tremblay et al., 2001), which transit the allantois *en route* to the gonad. Interestingly, the *Smad1*$^{-/-}$ mutants exhibited a loss of the cell adhesion molecule vCAM, also needed for chorio-allantoic fusion and placentation, specifically in the allantois, and not in other tissues. Although the level of vCAM transcripts was unaffected in *Smad1*$^{-/-}$ embryos, vCAM loss was demonstrable at the protein level (Lechleider et al., 2001). The defects seen in the *Smad1* null embryos are likely not due to functions of Smad1 carried out in the visceral endoderm, as wild-type ES cells failed to rescue Smad1-dependent gastrulation defects when implanted within *Smad1* mutant embryos (Tremblay et al., 2001).

Smad1 translocates directly from the BMP receptors to the cell nucleus, making the level of BMP signaling directly proportional to the protein level of Smad1 present in the cell. Consequently, *Smad1* heterozygous mice exhibited haploinsufficiency phenotypes. A proportion of these exhibited placentation defects in which the labyrinthine layer of the placenta was reduced, and died before E10.5 (M. Hester and M. Weinstein, unpub. observations). Others displayed a reduction in the midbrain and hindbrain at E11.5 (Hester et al., 2005). The latter was accompanied by hypercellularity and increased cell proliferation in the dorsal aspect of the developing brain. BMPs have been known to function in dorsal/ventral (D/V) patterning of the spinal chord (reviewed in Altmann and Brivanlou, 2001), but the defect observed for the *Smad1*$^{+/-}$ embryos appeared distinct from D/V patterning defects, as the expression pattern of genes restricted in the D/V axis of the brain was normal. Interestingly, there appeared to be an increase in the level of *Pax3* and *Pax6* expression in the brains of phenotypic *Smad1* heterozygotes. The level of *Pax3* was also increased in the spinal chord, although there the expression domain of *Pax3* was expanded ventrally, suggesting a disruption of Smad1-dependent D/V functions in the spinal chord in some *Smad1* heterozygous embryos (Hester et al., 2005).

Biochemical data has suggested that Smad protein function is inhibited upon phosphorylation of the linker domain by Erk MAP kinases (de Caestecker et al., 1998; Grimm and Gurdon, 2002; Kretzschmar et al., 1997; Kretzschmar et al., 1999; Pera et al., 2003; Yue et al., 1999). Genetic experiments have confirmed this for Smad1 through specific disruption of the MAP kinase phosphorylation sites. The mutations were hypomorphic in nature, in that homozygotes were viable. However, homozygotes showed phenotypes in the stomach, including a decrease in the number of gastric zymogenic cells and an increased number of parietal cells, as well as in the reproductive tract, in which the testes were smaller, the epididymis abnormal, and the number of primordial germ cells decreased (Aubin et al., 2004). This data confirms that Smad1 does interact with MAP kinases *in vivo*.

2.2 Smad2

Smad2-null mutants suffer a failure of gastrulation and arrest development between E7.5 and E8.5 (Nomura and Li, 1998; Weinstein et al., 1998). At the affected stages, the embryo consists of a cup of cells, referred to as the egg cylinder, the bottom part of which, the epiblast, gives rise to embryonic tissues, while the top half gives rise to the extraembryonic membranes. Over these layers is wrapped a thin membrane, the visceral endoderm, which carries out important patterning functions of its own.

Smad2-null mutants failed to develop the extraembryonic portion of the egg cylinder, but instead an abnormal epiblast continued to proliferate that was largely unpatterned (Hamamoto et al., 2002; Nomura and Li, 1998; Weinstein et al., 1998). Mesoderm and endoderm were absent, and the mutants bore some similarity to embryos lacking the important TGF-β ligand, Nodal (Zhou et al., 1993). Indeed, when *Smad2*-null mutants were bred to mice lacking *Nodal*, doubly heterozygous animals were not viable but exhibited defects of gastrulation and craniofacial patterning (Nomura and Li, 1998) strongly suggesting that Smad2 carries out at least some aspects of Nodal signaling. Like *Smad1, Smad2* heterozygous mutants also exhibited haploinsufficiency phenotypes, including gastrulation defects and craniofacial abnormalities (Liu et al., 2004a; Nomura and Li, 1998).

2.2.1 *Smad2 hypomorphic alleles*

Smad2 functions have been further illuminated through the use of hypomorphic *Smad2* mutants. The first of these hypomorphs was a deletion of exon 2. Although first thought to be a null mutation, The $Smad2^{\Delta exon2}$ allele encoded a truncated C-terminal Smad2 protein still capable of phosphorylation and nuclear translocation (Randall et al., 2004). In addition, the phenotype associated with $Smad2^{\Delta exon2/\Delta exon2}$ mutants was reproducible with other known hypomorphic *Smad2* alleles (see below). $Smad2^{\Delta exon2/\Delta exon2}$ mutants, which exhibited lethality a day later than *Smad2*-null mutants, were capable of extraembryonic membrane formation and gastrulation, but suffered from a failure in anterior/posterior axis formation. As a result, the epiblast was transformed into mesoderm, which may be an expanded allantois, the most posterior mesodermal tissue. Interestingly, Smad2 exerts its functions in the visceral endoderm, as introducing wild-type ES stems cells into $Smad2^{-/-}$ embryos failed to rescue their phenotype (Waldrip et al., 1998).

$Smad2^{\Delta exon2/\Delta exon2}$ mutant ES cells have been made, and were able to produce mesoderm in teratoma and *in vitro* differentiation experiments (Heyer et al., 1999). They could contribute successfully to many tissues in wild-type embryos, although appeared unable to contribute to the definitive endodermal lineage (Tremblay et al., 2000). In addition, embryos composed entirely of $Smad2^{\Delta exon2/\Delta exon2}$ ES cells exhibited holoprosencephaly and defects of the left/right axis (Heyer et al., 1999).

Another hypomorphic allele of *Smad2* has been produced by an interesting high-throughput genotype-based screening system, in which ES cells were treated with N-ethyl nitrosourea and selected through an HPLC genotypic assay (Chen et al., 2000). This mutation ($Smad2^{S276L}$), which converted serine 276 to leucine, caused a variably penetrant embryonic lethal phenotype when homozygous. Some $Smad2^{S276L/S276L}$ embryos closely resembled $Smad2^{\Delta exon2/\Delta exon2}$ mutants, as did $Smad2^{\Delta exon2/S276L}$ embryos. Others exhibited defects in anterior patterning, as well as vascular and cardiac defects (Vivian et al., 2002). Similar results have been obtained with another hypomorphic allele, created through the insertion of a neomycin cassette into intron 10 of *Smad2*. This allele ($Smad2^{3loxP}$) was viable in the homozygous state, but caused lethality when placed over a null allele of *Smad2*. The phenotypes of $Smad2^{3loxP/-}$ embryos mirrored those seen in other studies, including empty yolk sacs similar to *Smad2*$^{\Delta exon2/\Delta exon2}$ mutants, as well as defects of the axial midline and anterior, similar to some *Smad2*$^{S276L/S276L}$ embryos (Liu et al., 2004b).

Another very interesting *Smad2* hypomorph has been created through the deletion of exon 3 of *Smad2*. This exon, which is alternatively spliced (Dunn et al., 2005; Yagi et al., 1999), encodes an insert in the MH1 domain of Smad2 which prevents it from binding to DNA (Dennler et al., 1999; Yagi et al., 1999), and its presence is one of the main differences between Smad2 and Smad3. However, mice lacking this exon of *Smad2* were viable and normal, despite the fact that the shorter isoform of *Smad2* is coexpressed with the full-length isoform in many tissues. However, the short isoform of *Smad2* lacking exon 3 was capable of rescuing the ability of *Smad2*$^{\Delta exon2/\Delta exon2}$ ES cells to contribute to the definitive endoderm, as could full-length *Smad2*, and *Smad3*. Indeed, replacement of *Smad2* with *Smad3* resulted in relatively normal development, although some embryos exhibited anterior truncations (Dunn et al., 2005).

2.2.2 Smad2/APC

There have been two reports on compound mutations of *Smad2* and *APC*, the latter an important tumor suppressor of colon cancer that functions in the wnt signaling pathway (reviewed in Kolligs et al., 2002). One report showed a substantial impact of Smad2 loss on the progression of APC-dependent tumors. A null allele of *Smad2* was generated by replacing exons 3 and 4 of *Smad2* with a *neo* cassette. This allele of *Smad2* was introduced to a mouse strain carrying an *Apc* mutation, and the two alleles (which are tightly linked) were introduced to the same chromosome by meiotic recombination. *Smad2*$^{+/-}$*;Apc*$^{+/-}$ *cis* heterozygotes developed more polyps than *Smad2*$^{+/+}$*;Apc*$^{+/-}$ controls, and exhibited postnatal lethality due to large tumors that blocked the ileum, a phenotype not seen in *Smad2*$^{+/+}$*;Apc*$^{+/-}$ animals. Moreover, tumors from *Smad2*$^{+/-}$*;Apc*$^{+/-}$ *cis* heterozygous mice exhibited increased invasive properties compared to *Apc*-initiated tumors generated in wild-type animals. These results suggest that while loss of *Smad2* does not initiate colon carcinogenesis, it indeed functions in tumor progression (Hamamoto et al., 2002). A similar study was conducted with *Smad2*$^{\Delta exon2}$ mutants in which no effect was seen on APC-dependent tumorigenesis (Takaku et al., 2002). It is likely that the discrepancy seen between these two reports was caused by the hypomorphic nature of the *Smad2*$^{\Delta exon2}$ allele, allowing sufficient Smad2 signaling to suppress intestinal carcinogenesis.

2.3 Smad3

2.3.1 Phenotypes of Smad3 deficient mice

Mice lacking *Smad3* were viable but exhibited reduced growth that was first apparent following birth and progressed after weaning into a wasting syndrome (Datto et al., 1999; Yang et al., 1999b). This syndrome was, in part, the result of a defect in mucosal immunity, as revealed by the presence of multiple abscesses in tissues adjacent to or underlying the mucus membranes, leading to lethality of majority of *Smad3*$^{-/-}$ mice at 1 to 8 months.

Smad3$^{-/-}$ mice also displayed skeletal abnormalities, including inwardly turned paws and kyphosis of the spine (Datto et al., 1999). Further investigation revealed that *Smad3*$^{-/-}$ mice developed degenerative joint disease resembling human osteoarthritis, as characterized by progressive loss of articular cartilage, formation of large osteophytes, decreased production of proteoglycans, and an abnormally increased number of type X collagen-expressing chondrocytes in synovial joints. In an *in vitro* embryonic metatarsal rudiment culture system, TGF-β significantly inhibited chondrocyte differentiation of wild-type metatarsal rudiments. However, this inhibition is diminished in metatarsal bones isolated from *Smad3*$^{-/-}$ mice. These data suggest that TGF-β/Smad3 signals inhibit terminal hypertrophic differentiation of chondrocyte and are essential for maintaining articular cartilage (Yang et al., 2001).

Smad3$^{-/-}$ mutants showed reproduction defects, including a small but reproducible reduction in the level of apoptotic cell death during mammary gland involution (Yang et al., 2002b). In addition, *Smad3*$^{-/-}$ females exhibited reduced fertility due to effects on hormonal levels, delayed follicular development, and a lack of corpora lutea. Ovaries lacking Smad3 fail to ovulate even when stimulated (Tomic et al., 2004).

Smad3$^{-/-}$ mice also exhibited respiratory defects, including delayed alveogenesis in younger animals, while older mutants suffered centrilobular emphysema (Bonniaud et al., 2004; Chen et al., 2005). In addition, *Smad3*$^{-/-}$ mutants were protected from TGF-β1 induced fibrosis in the lung (Bonniaud et al., 2004) and the kidney (Inazaki et al., 2004; Sato et al., 2003). In the latter, Smad3 increases the expression of TGFβ1 following ureteral blockage. The TGFβ1 can then cause epithelial-mesenchymal transformation (EMT) (Sato et al., 2003). In *Smad3*$^{-/-}$ mice, the amount of extracellular matrix (ECM) deposition was lowered, as was the rate of inflammatory cell invasion and apoptosis upon unilateral ureteric obstruction (Inazaki et al., 2004). Similar results have been recorded in the eye, where retinal detachment causes EMT of the retinal epithelial cells, also caused by Smad3-dependent upregulation of TGF-β1 (Saika et al., 2004). *Smad3*$^{-/-}$ retinal pigment epithelial cells fail to undergo EMT upon injury, exhibit lowered levels of ECM deposition, and a reduced fibrotic response (Saika et al., 2004). Thus loss of Smad3 protects against injury-induced fibrosis in a number of organs.

Interestingly, *Smad3* mutant mice showed an ability to heal wounds more rapidly than wild-type animals, with homozygotes exhibiting a more pronounced phenotype than heterozygous animals (Ashcroft et al., 1999). There was less extracellular matrix in the wound area, probably due to a decrease in the population of monocytes. Both of these defects contributed directly to a reduction in fibrosis, as the normal induction of matrix proteins could be restored either by administration of exogenous TGF-β to the wound area or the introduction of wild-type monocytes to the wound. However, *Smad3* null monocytes were unable to migrate into the wounds of the *Smad3*$^{-/-}$ animals, and this defect was not remediated by the application of TGF-β, suggesting that the Smad3 null mutant moncytes were refractory to the TGF-β signal. These results clearly demonstrated the importance of TGF-β and Smad3 in monocytic chemoattraction. In addition to a lower number of inflammatory cells, the wounds of the *Smad3* mutant mice exhibited an increased rate of re-epithelialization; the result of an increased rate of keratinocyte proliferation (Ashcroft et al., 1999).

2.3.2 Functions of Smad3 in tumorigenesis

A number of studies have indicated that TGF-β signaling performs an important tumor-suppressor function. It is therefore not surprising that *Smad3* mutations have been specifically implicated in tumorigenesis in mice and humans. One of the three laboratories that performed the targeted disruption of this gene reported that animals lacking *Smad3* developed metastatic colon cancer at a high frequency. The incidence of cancer was somewhat dependent on the background, as 100% of the 129 mice developed cancer, but the incidence dropped to 30% when the stain was outbred on a C57/Black 6 background (Zhu et al., 1998). Other groups have not found colon cancer in their mice (Datto et al., 1999; Yang et al., 1999b), despite extensive search and rederivation of the animals on a pure 129 background (C.D unpublished results). The cause of this discrepancy remains to be elusive, although it was initially suspected that many factors, such as hypomorphic mutation or bacterial infection might be involved (Yang et al., 1999b).

Smad3$^{-/-}$ mice exhibited reduced size of the thymus and spleen with concomitant increases in the lymph nodes due to the expansion of activated T cell populations. *Smad3*$^{-/-}$ T cells showed increased rates of proliferation that could not be inhibited by TGF-β treatment in culture, although the same dose of TGF-β treatment could inhibit B cells efficiently (Yang et al., 1999b). Despite this abnormality, T cell tumors failed to develop in the *Smad3*$^{-/-}$ mice, suggesting that *Smad3* deficiency alone is not sufficient to cause malignant transformation, and additional factors are required.

To further investigate the role of *Smad3* in the pathogenesis of lymphoid neoplasia, *Smad3*$^{+/-}$ and *Smad3*$^{-/-}$ mice were crossed with mice in which both alleles of the tumor suppressor *p27*^{Kip1} were deleted (Wolfraim et al., 2004). This study indicated that the loss of one functional *Smad3* allele impaired the inhibitory effect of TGF-β on the proliferation of normal T cells and worked in tandem with the homozygous inactivation of *p27*^{Kip1} to promote T-cell leukemogenesis in mice. Interestingly, *Smad3* mRNA and protein were either down regulated or absent in leukemia cells obtained at diagnosis from 19 children with acute leukemia, including 10 with T-cell acute lymphoblastic leukemia, 7 with pre-B-cell ALL, and 2 with acute nonlymphoblastic leukemia, although no mutations at the genomic level were found (Wolfraim et al., 2004). These observations establish an inhibitory role of Smad3 in lymphoblastic leukemia in both mice and humans.

In striking contrast, two studies showed that *Smad3*$^{-/-}$ mutant mice exhibited resistance, rather than sensitivity, to chemically induced skin carcinogenesis. Li et al. found that after treatment with 12-O-tetradecanoylphorbol-13-acetate (TPA), *Smad3*$^{-/-}$ mice exhibited reduced papilloma formation in comparison to

Smad3$^{+/+}$ mice and failed to develop squamous cell carcinomas. Further analysis revealed that *Smad3* knock-out mice were resistant to TPA-induced epidermal hyperproliferation. Although there was a decrease in tumor formation, increases were seen both in the incidence of papillomas, and in the rate of apoptosis in TPA-treated *Smad3*$^{-/-}$ skin. Expression levels of activator protein-1 family members (c-Jun, JunB, JunD, and c-Fos) and transforming growth factor (TGF)-α were significantly lower in TPA-treated *Smad3*$^{-/-}$ skin, cultured keratinocytes, and papillomas, as compared with *Smad3*$^{+/+}$ controls. All of these molecular and cellular alterations also occurred to a lesser extent in *Smad3*$^{+/-}$ mice compared with *Smad3*$^{+/+}$ mice (Li et al., 2004). In the second study, Tannehill et al. directly compared *Smad3*$^{+/-}$ and wild-type mice after they were treated with topical dimethylbenzanthracene. Their data indicated that *Smad3*$^{+/-}$ mice developed significantly fewer tumors than the wild-type group (Tannehill-Gregg et al., 2004). Previous investigations indicated that TGF-β signaling plays a complex role in carcinogenesis, acting as a suppressor early in tumor development but later as a promoter of tumor progression and metastasis (Roberts and Wakefield, 2003; Wakefield and Roberts, 2002). These studies suggest that TGF-β signaling through Smad3 serves as a promoter that enhances chemical induced skin carcinoma formation in a dosage dependent manner.

2.3.3 Smad2/3

A number of insights have been gained by examining the impact of combinations of both *Smad2* and *Smad3* mutations. A null allele of *Smad2*, when combined with a *Smad3* mutation, resulted in lethality in double heterozygous mice. These animals exhibited defects in hepatogenesis, in which the liver was formed, but failed to proliferate. Interestingly, this defect could be rescued *in vitro* by treatment with hepatocyte growth factor (HGF) (Weinstein et al., 2001). Some, but not all *Smad2*$^{+/-}$;*Smad3*$^{+/-}$ embryos exhibited abnormalities in craniofacial and axial midline patterning. The hepatogenesis, anterior, and axial defects seen in these embryos were thought to be due to an early defect of the endoderm. The definitive endoderm was formed, as well as a normal foregut. However, the endoderm was clearly abnormal, as there was disruption of gene expression. In addition, the endoderm failed to migrate properly, leaving visceral endoderm in an ectopic location adjacent to the developing heart. It was thought that this abnormal location of visceral endoderm interfered with communication between cardiac mesoderm and definitive endoderm, known to be necessary for hepatogenesis (reviewed in Zaret, 2002). As a result, genes needed for the hepatogenic developmental program, such as *Hex* and *Hnf4*, were extremely reduced in their expression, resulting in hepatogenic abnormalities (Liu et al., 2004a).

The *Smad2*$^{\Delta exon2}$ allele has also been combined with *Smad3* mutants with interesting results. *Smad2*$^{\Delta exon2/+}$;*Smad3*$^{+/-}$ animals were viable, although *Smad2*$^{\Delta exon2/+}$;*Smad3*$^{-/-}$ animals exhibited lethality due to defects of anterior and midline patterning. Marker gene analysis confirmed that tissues made from the anterior primitive streak were missing in these embryos, such that there was insufficient mesoderm and definitive endoderm to fully contribute to the anterior.

Of note, further reductions in the level of *Smad2*, using a conditional version of the *Smad2*$^{\Delta exon2}$ allele, exacerbated these phenotypes considerably, to the point where tissues from the middle of the primitive streak were affected. Indeed, further lowering *Smad2* and *Smad3* levels, by combinations of the *Smad2*$^{\Delta exon2}$ and *Smad3* mutations, resulted in increasing severity, until the doubly homozygous embryos closely resembled *Smad4* null mutants (Dunn et al., 2004).

Additional experiments introduced another hypormorphic *Smad2* allele, *Smad2*^{3loxP}, onto the *Smad3*$^{+/-}$ strain. Our recent data suggests *Smad2*$^{3loxP/3loxP}$ embryos produce roughly half as much *Smad2* as wild-type embryos, and *Smad2*$^{3loxP/+}$;*Smad3*$^{+/-}$, *Smad2*$^{3loxP/+}$;*Smad3*$^{-/-}$, and *Smad2*$^{3loxP/3loxP}$;*Smad3*$^{+/-}$ animals are all viable. However, thirty percent of *Smad2*$^{3loxP/3loxP}$;*Smad3*$^{+/-}$ embryos exhibit holoprosencephalic phenotypes and hepatic dysgenesis, similar to *Smad2*$^{+/-}$;*Smad3*$^{+/-}$ embryos. *Smad2*$^{3loxP/3loxP}$;*Smad3*$^{-/-}$ animals show axial patterning defects resulting in lethality at E9.5, similar to many *Smad2*$^{3loxP/-}$ embryos (A. Chow and M. Weinstein, unpub. data). It is interesting that despite the wide range of Smad2 levels in the many mutants, similar phenotypes result.

2.4 Smad4

2.4.1 Phenotypes of Smad4-deficient embryos

Smad4 serves as a common mediator of TGF-β signaling. As expected, Smad4-deficient embryos exhibited a phenotype that was more severe than embryos carrying targeted mutations of any other Smads (Sirard et al., 1998; Yang et al., 1998). *Smad4*$^{-/-}$ embryos died shortly after implantation, exhibiting no signs of gastrulation and mesoderm induction (Sirard et al., 1998; Yang et al., 1998). Several lines of evidence indicated that the primary defect caused by *Smad4* deficiency was restricted to the visceral endoderm, a layer of cells that tightly opposes the epiblast and extraembryonic ectoderm. First, *Smad4*$^{-/-}$ embryos exhibited reduced staining for *Hnf4*, a gene expressed in the visceral endoderm. Second, *Smad4*$^{-/-}$ ES cells failed to make a normal visceral endoderm in embryoid bodies differentiated *in vitro*. Third, *Smad4*$^{-/-}$ embryos were capable of forming mesoderm when supplied with wild-type visceral endoderm (Sirard et al., 1998). Consistently, embryos, which carried a targeted deletion of *Smad4* specifically in epiblast cells using a Cre-loxP approach, formed numerous mesodermal derivatives, including somites, heart, allantois and lateral plate mesoderm (Chu et al., 2004). These observations suggest that cells lacking signaling of TGF-β superfamily can still adopt mesodermal fates, or alternatively that the absence of Smad4 does not completely block TGF-β signals. Of note, *Smad2*$^{-/-}$;*Smad3*$^{-/-}$ double mutant embryos showed a phenotype more severe than *Smad4*-deficient epiblast (Dunn et al., 2004). This observation suggests that Smad2 and Smad3 function in a Smad4-independent manner in mediating signaling of Nodal and TGF-β subfamily in mesoderm formation and patterning, while Smad4 does not play an indispensable function in these processes.

2.4.2 *Tumor formation in Smad4$^{+/-}$ mice*

Human *Smad4* is located in chromosome 18q21.1, where loss of heterozygosity (LOH) is frequently observed in a wide variety of tumors. Mutations in *Smad4* have been identified in about 50% of pancreatic tumors, 30% of colon carcinomas, and in less than 10% of other cancers (Friedl et al., 1999; Hahn et al., 1998; Hahn et al., 1996; Takagi et al., 1996). Germline mutation of *Smad4* contributes to familial juvenile polyposis, an autosomal dominant disorder characterized by predisposition to hamartomatous polyps and gastrointestinal cancer (Howe et al., 1998).

Because of the early lethality of *Smad4*-deficient embryos, the role of Smad4 in tumorigenesis cannot be directly studied. Takagi et al. crossed their *Smad4*$^{+/-}$ mice generated into a strain in which the tumor suppressor *Apc* was inactivated (Takagi et al., 1996). Mice lacking one copy of *Apc* develop numerous intestinal and colonic polyps of small size and limited invasiveness (Oshima et al., 1995). However, *Smad4*$^{+/-}$*;Apc*$^{+/-}$ mice exhibited faster growing, more invasive polyps and gastrointestinal tumors. These polyps exhibited LOH of both *Smad4* and *Apc*. While the polyps formed in the *Apc* heterozygotes were benign and unable to form tumors in nude mice, those found in *Smad4*$^{+/-}$*;Apc*$^{+/-}$ mice were malignant and readily formed tumors in nude mice.

Indeed, *Smad4* heterozygous mice themselves developed gastric polyps when the animals were 6-12 months of age (Redman et al., 2005; Takaku et al., 1999; Xu et al., 2000). With increasing age, polyps in the antrum developed into tumors. Although some tumors exhibited loss of LOH of the *Smad4* gene, half of those analyzed maintained a normal *Smad4* allele, suggesting that LOH at the *Smad4* locus was not an obligate event in *Smad4* dependent tumorigenesis. Smad4 appears to function in a similar fashion as p53, in that haploinsufficiency at the *p53* locus is sufficient for tumorigenesis without LOH (Takagi et al., 1996). These results clearly establish *Smad4* as a suppressor of gastric cancer.

2.5 Smad5

Disruption of *Smad5* resulted in lethality between E9.5 and E11.5, with a constellation of complex phenotypes. *Smad5*$^{-/-}$ embryos exhibited a lack of blood vessels in the yolk sac and abnormal blood vessels in the embryo proper (Chang et al., 1999; Yang et al., 1999a), suggesting a defect in angiogenesis similar to embryos homozygous for mutations in *Endoglin* (Li et al., 1999), *TGF-β1* (Martin et al., 1995), *TGF-β type II receptor* (Martin et al., 1995), and *ALK1* (Oh et al., 2000). Indeed, TGF-β has been shown to promote endothelial cell proliferation and migration by signaling through ALK1 to Smad5 (Goumans et al., 2002). *Smad5* mutant yolk sacs contained increased numbers of hematopoietic precursors of the myeloid lineage (Yang et al., 1999a). Similar results were observed when *Smad5*$^{-/-}$ ES cells were differentiated into the hematopoietic lineage *in vitro* (Liu et al., 2003). *Smad5*$^{-/-}$ embryoid bodies produced increased numbers of blast colony producing cells and high proliferative potential colony-forming cells, although the proliferation of erythroid progenitors was diminished (Liu et al., 2003).

Smad5$^{-/-}$ embryos also exhibited increased levels of apoptosis in the embryonic mesenchyme, leading to a lack of stromal cells in much of the embryo. In most *Smad5*$^{-/-}$ embryos this led to defects in craniofacial patterning and anterior deletions, where the apoptosis was more severe (Yang et al., 1999a). Another likely consequence was a lack of vascular smooth muscle cells, which differentiate from mesencyhmal precursors under the direction of TGF-β signals (Goumans et al., 2003; Lebrin et al., 2005). The lack of vascular support cells correlated with increased vascular dilation in the *Smad5* mutant embryos. Mutants also exhibited cardiac defects in which the heart was exteriorized beyond the coelomic cavity, delayed formation of the foregut, and failure of the posterior gut to close. In some *Smad5* mutants, posterior truncations were also observed (Chang et al., 1999; Yang et al., 1999a). Interestingly, *Smad1* mutants carrying missense mutations in the C-terminal phosphorylation sites also exhibited posterior truncations, which were not seen in *Smad1* null embryos, suggesting that such a *Smad1* missense mutant may function in a dominant negative manner towards Smad5 (Aubin et al., 2004).

Like Smad1-null mutants, Smad5-deficient embryos exhibited abnormalities in the allantois (Chang et al., 1999), and also lacked primordial germ cells, with heterozygotes exhibiting a diminished level as compared to wild-type animals, and homozygotes exhibiting a nearly complete deficiency (Chang and Matzuk, 2001). *Smad5* mutants also exhibited defects in left/right (L/R) asymmetry, with absent or bilateral expression of several genes important for establishing the L/R axis, including *Nodal, Lefty-1, Lefty-2*, and *Pitx2* (Chang et al., 2000). This is quite interesting, as L/R patterning defects are also associated with a lack of Nodal signaling (Brennan et al., 2002; Gaio et al., 1999; Nomura and Li, 1998; reviewed in Hackett, 2002; Yost, 2001).

2.6 Smad6

Smad6 has been disrupted by targeting a lacZ reporter into the Smad6 MH2 domain, such that a Smad6:lacZ fusion protein is produced. LacZ staining demonstrated that *Smad6* expression could be found in the heart and was largely restricted to the outflow tract and endocardial cushion (Galvin et al., 2000), similar to what has been shown in chick development, where *Smad6* expression is controlled by BMP signaling (Yamada et al., 1999). Most homozygous *Smad6* mutants exhibit postnatal lethality due to cardiovascular abnormalities, including hyperplasia of the cardiac valves and outflow tract septation defects, suggesting a function for Smad6 in the regulation of endocardial cushion transformation. BMP signaling is intimately associated with osteogenesis (hence the name bone morphogenic proteins) and ectopic ossification of the cardiac outflow tracts is seen in *Smad6* mutants. Those *Smad6* mutants that survive to adulthood also exhibit elevated blood pressure (Galvin et al., 2000). These defects highlight the importance of Smad6 in the tissue-specific modulation of TGF-β superfamily signaling pathways *in vivo*.

2.7 Smad8

Smad8 has been shown to function in the BMP signaling pathway, and has been disrupted by a deletion of exon 3. Unfortunately, the allele is hypomorphic, as shown by injection of the allele into *Xenopus laevis*, and no phenotype has been seen. However, by inserting a *neo* cassette into intron 4 (*Smad8*^{3loxP}), it was shown that the *Smad8* expression level was lowered considerably due to cryptic splicing from the *neo* cassette. Ten percent of the homozygotes exhibited a phenotype extremely similar to that seen in *Smad1* heterozygotes, in which the mid- and hindbrain were reduced in size and hypercellular. Gene expression was also similar between the *Smad1*$^{+/-}$ and *Smad8*$^{3loxP/3loxP}$ mutants (Hester et al., 2005). These results suggest that *Smad8* functions similarly to *Smad1* in BMP signaling.

3. FUNCTIONAL ANALYSIS OF MUTANT MICE CARRYING TISSUE SPECIFIC MUTATIONS OF SMADS

Embryonic lethality of Smad1, 2, 4 and 5 makes it difficulty to study functions of these genes during postnatal development. Therefore, conditional alleles for these genes have been generated to overcome embryonic lethality (Table 2). Tissue specific disruption using the Cre-loxP approach has been performed for Smad2 and Smad4 in multiple tissues/organs. We have summarized major phenotypes caused by these conditional mutations in Table 2, and chose to discuss in details about phenotypes associated with Smad4 disruption in mammary epithelium (Li et al., 2003) and keratinocytes (Qiao et al., 2005).

3.1 Interplay between Smad4 and β-catenin in Transdifferentiation of Mammary Epithelial Cells and Tumor Formation

Smad4 is expressed in the mammary gland throughout development (Fig. 1A). To study a role of Smad4 in mammary gland development, we crossed Smad4 conditional mice (Yang et al., 2002a) with WAP-Cre or MMTV-Cre transgenic mice that express Cre recombinase in the mammary gland (Wagner et al., 1997). Northern blot analysis demonstrated that *Smad4* transcripts were reduced to about 10% of wild-type level in mammary tissue at day 16 of pregnancy and to about 20% at lactation day 10 in *Smad4*$^{Co/Co}$*WAP-Cre* mice (Fig. 1B). Immunohistochemical staining using an antibody to Smad4 in mutant mammary glands isolated from different developmental stages confirmed that over 90% of mammary epithelial cells were negative for the staining (Fig. 1C,D). Despite the high efficiency of Smad4 deletion in mammary epithelium, *Smad4*$^{Co/Co}$*WAP-Cre* mice showed no obvious abnormalities in their mammary gland development during the first 5 months of age, suggesting that Smad4 is dispensable for normal development of this gland in mouse.

Some *Smad4*$^{Co/Co}$*WAP-Cre* mice started to developed visible tumor masses in their mammary glands at 5 months of age, and by 16 months of age, all *Smad4*$^{Co/Co}$*WAP-Cre* mice had developed mammary tumor masses (Fig. 2A).

Table 2. Summary of phenotypes of conditional mutant mice

Genes/Exons mutated	Tissue specific disruption	Phenotypes	References
Smad1/Exon2	NA	NA	(Huang et al., 2002)
Smad2/Exon1	Epiblast	Mutant embryos fail to correctly specify the anterior definitive endoderm and prechordal plate progenitors	(Vincent et al., 2003)
Smad2/Exons9-10	Germline deletion	Mutant embryos exhibit phenotypes similar to null mutation	(Liu et al., 2004b)
Smad4/Exon1	Epiblast	Mutant embryos fail to pattern derivatives of the anterior primitive streak, such as prechordal plate, node, notochord and definitive endoderm	(Chu et al., 2004)
Smad4/Exon8	Mammary epithelium	Transdifferentiation of mammary epithelial cells into squamous epithelial cells, which is associated with β-catenin accumulation. Mutant mice developed squamous cell carcinoma and/or mammary abscesses between 5 and 16 months of age.	(Li et al., 2003)
	Hippocampus and Purkinje cells	Decreased number of cerebellar Purkinje cells and parvalbumin-positive interneurons, and significantly increased vertical activity.	(Zhou et al., 2003)
	Chondrocytes	Mutant mice exhibit dwarfism with a severely disorganized growth plates. Mutant chondrocytes exhibit decreased expression of Ihh/PTHrP, leading to the increased hypertrophic differentiation of chondrocytes.	(Zhang et al., 2005)
	Keratinocytes	Progressive hair loss due to failure of hair follicle differentiation and cycling. Mutant mice developed spontaneous squamous carcinoma from 3 months to 13 months of age. Inactivation of Pten, activation of Akt, fast proliferation and nuclear accumulation of cyclin D1 are observed in tumor cells.	(Qiao et al., 2005) (Yang et al., 2005)
	Liver	Iron accumulation in multiple organs, including liver, pancreas, and kidney.	C.D. Unpub. observation
Smad5/Exon2	Germline deletion	Mutant embryos exhibit phenotypes similar to null mutation	(Umans et al., 2003)
Smad8/Exon3	Germline deletion	Smad8 lacking exon3 is a hypomorphic allele. The homozygous mutant mice do not exhibit an obvious phenotype.	(Hester et al., 2005)

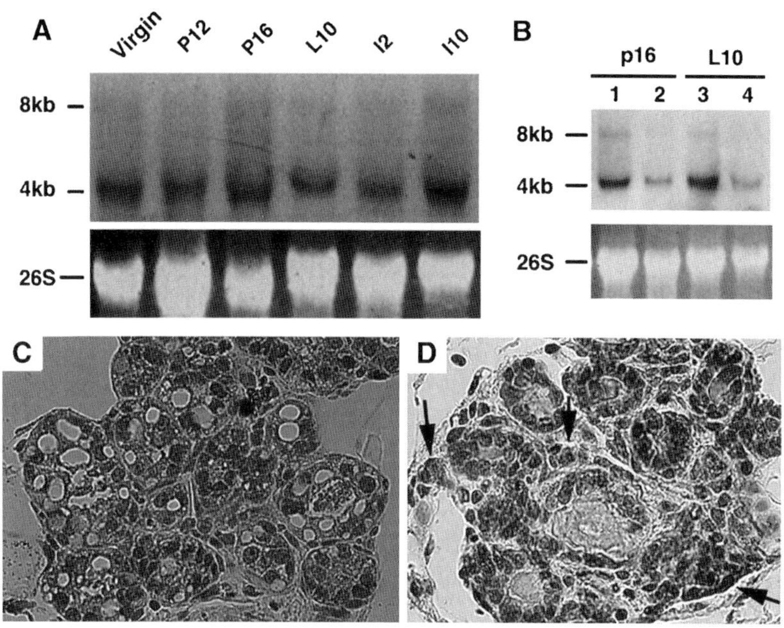

Figure 1. Expression of *Smad4* in mammary glands. (A) Northern blot analysis of *Smad4* expression in the mammary gland of wild-type at different stages, including virgin, pregnant (P) days 12 and 16, lactation (L) day 10; and involution (I) days 2 and 10. (B) Northern blot analysis of RNA isolated from p16 and L10 wild-type (lanes 1 and 3) and *Smad4*$^{Co/Co}$*WAP-Cre* (lanes 2 and 4) mice. (C,D) Immunohistochemical staining of L10 mammary glands isolated from wild-type (C) and mutant (D) using an antibody to Smad4. Mutant gland shows very few cells that are still positive for Smad4 (arrowheads) (A color version of this figure is freely accessible via the website of the book: http://www.springer.com/1-4020-4542-5)

Histological analysis revealed that majority of them (>90%) contained both squamous cell carcinoma, which frequently exhibited cancer-pearl-like keratinization and onion-skin-like structures (arrows, Fig. 2C), and mammary abscesses with varying sizes that were filled with keratin (Fig. 2D). Studying the process of abscess formation, we found that BrdU positive labeled cells were presented at the edge of the initiating abscesses (Fig. 2E,F), while cells in the center were enlarged and underwent transdifferentiation (arrow, Fig. 2E). Immunohistochemical staining for K14, a marker for basal layer keratinocytes of skin, confirmed that these cells had undergone fate changes from epithelial cells to keratinocytes (Fig. 2E,G). Thus, the BrdU-positive cells in the periphery were responsible for the continuous proliferation leading to increased volume of the abscess, while the cells in the center were responsible for the continuous keratinization (Fig. 2H,I). The keratin secreted by differentiating keratinocytes blocks the mammary ducts and eventually results in the formation of mammary abscesses. Our study also revealed that the wall of well-developed abscesses showed many features very similar to skin (Fig. 2J), suggesting that absence of Smad4 in mammary epithelium recaptures a process resembling the formation of skin epidermis.

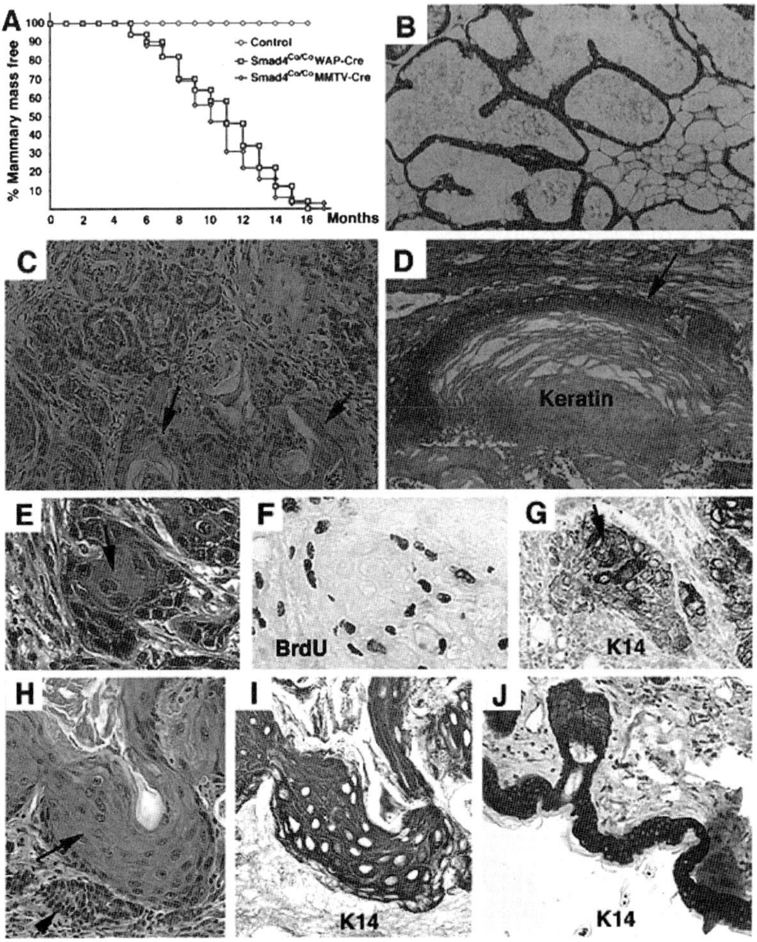

Figure 2. Squamous cell carcinoma and mammary abscess formation through squamous metaplasia in *Smad4* conditional knock-out mice. (A) A Kaplan–Meier survival curve showing percent of mammary mass *Smad4*$^{Co/Co}$*WAP-Cre* (n = 36), *Smad4*$^{Co/Co}$*MMTV-Cre* (n = 34) and control mice (n = 70). The controls include *Smad4*$^{Co/+}$*WAP-Cre*, *WAP-Cre*, and *Smad4*$^{Co/Co}$ mice. (B) A section of a L10 gland from a 6-month old control mouse. (C,D) Squamous cell carcinoma (C) and an abscess (D), showing cancer-pearl-like keratinization and onion-skin-like structures (arrows in C). (E-J). Transdifferentiation from mammary epithelial cells to keratinocytes due to the loss of Smad4. (E-G) Images of early stages of transdifferentiation revealed by H&E (E), BrdU (F), and K14 immunohistochemical staining (G). Arrows point to the center of the lesion. (H,I) Images of an abscess at later stages detected by H&E (H) and K14 immunohistochemical staining (I). Arrow points to the wall of the abscess, which is K14 positive, and arrowhead points at an alveoli, which is K14 negative. A normal, K14 stained skin sample isolated from *Smad4*$^{Co/Co}$*WAP-Cre* mice was shown in (J) (A color version of this figure is freely accessible via the website of the book: http://www.springer.com/1-4020-4542-5)

Interestingly, β-catenin was expressed in all abscesses at high levels, irrespective of their size or developmental stage (Fig. 3A). Alterations of β-catenin has been shown to cause multiple types of human cancers (Akiyama, 2000; Karayiannakis et al., 2001; Polakis, 2001). Moreover, a similar phenotype was observed in the mammary gland upon over-expression of a constitutive activated form of β-catenin (Miyoshi et al., 2002). Therefore, we suspected that the transdifferentiation of mammary alveolar epithelium into epidermal structures and squamous cell carcinoma formation in Smad4 mutant glands was, at least in part, caused by the increased expression of β-catenin. To understand the relationship between TGF-β/Smad4 and β-catenin, which serves as a major component of the Wnt signaling pathway, we tested whether β-catenin expression could be regulated by TGF-β/Smad4 signals. For this experiment, Smad4-null and wild-type cell lines were subjected to TGF-β1 treatment. Our data indicated that TGF-β1 treatment significantly decreased β-catenin in *Smad4*$^{+/+}$ cells but not in Smad4-null cells, as assayed by both western blot (Fig. 3B) and immunofluorescence (Fig. 3C). These observations suggest that TGF-β1 treatment results in β-catenin degradation, which is mediated by Smad4.

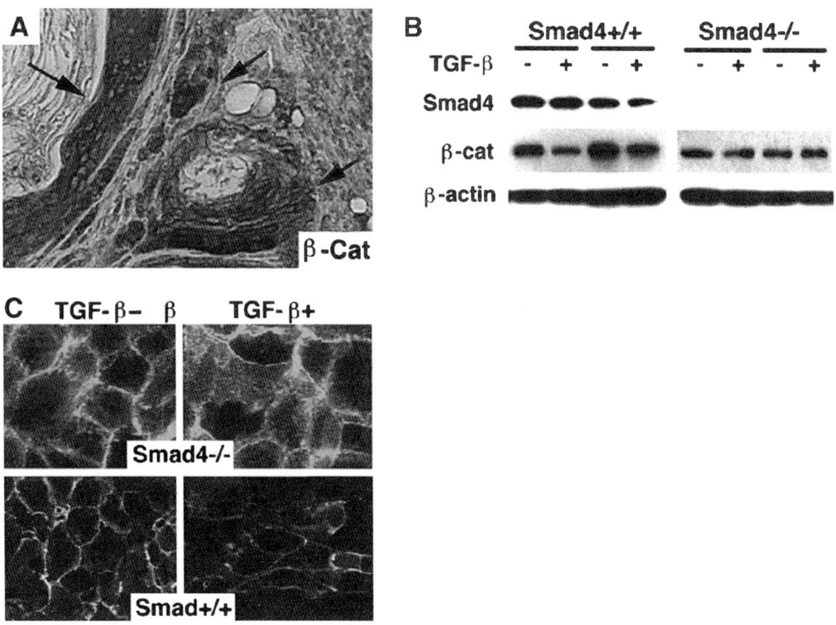

Figure 3. Loss of TGF-β-mediated β-catenin degradation and TGF-βresponsiveness in *Smad4*$^{-/-}$ cells. (A) A section from a 1-year old *Smad4*$^{Co/Co}$*WAP-Cre* mouse, showing increased levels of β-catenin. (B) Western blot showing β-catenin in *Smad4*$^{+/+}$ and *Smad4*$^{-/-}$ mammary epithelial cells after they were treated with 2 ng/ml of TGF-β for 24 hours. Treatment of TGF-β led to a decrease in β-catenin in all *Smad4*$^{+/+}$ cell lines while no changes were detected in both independently derived *Smad4*$^{-/-}$ cells. (C) EMT and decreased expression of β-catenin upon TGF-β treatment observed in wild-type, but not in *Smad4*$^{-/-}$, cells (A color version of this figure is freely accessible via the website of the book: http://www.springer.com/1-4020-4542-5)

TGF-β has been known as a potent factor that promotes epithelial-mesenchymal transition (EMT) of cultured mammary epithelial cells (Piek et al., 1999). The existence of *Smad4*-null epithelial cells provides an excellent opportunity to study the role of Smad4 in this process. Thus, we treated both *Smad4*$^{-/-}$ and wild-type cells with TGF-β1. We found that *Smad4*-null cells lost the TGF-β responsiveness and showed no signs of EMT, while EMT was readily observed in *Smad4*$^{+/+}$ cells (Fig. 3C), indicating that Smad4 is essential for TGF-β induced cell fate transformation.

Mammary epithelium and skin epidermis share a common origin as they are both derived from the ectoderm. They undergo distinct developmental outcomes as the mammary bud forms ductal branches during embryogenesis (Cardiff et al., 2000). Because the loss of Smad4 results in the epidermalization of mammary epithelia we suggest that Smad4 is required to maintain mammary epithelia and prevent them from undergoing transdifferentiation. Thus, our findings may suggest that the correct dose of TGF-β/Smad4 signals is essential in maintaining normal development of mammary epithelial cells. Therefore, we propose that TGF-β signals act through Smad4 to inhibit tumor initiation through their ability to inhibit epithelial cell proliferation. When these inhibition signals are absent due to the lack of Smad4, mammary epithelial cells increase proliferation leading to the hyperplasia and tumor initiation. Meanwhile, Smad4 also plays a potent role in determining the fate of the cells. Its absence unavoidably triggers transdifferentiation of mammary epithelial cells. A loss of these functions causes *Smad4*-null mammary epithelial cells to undergo both tumorigenesis and continuous transdifferentiation. This results in the conversion of *Smad4*-null tumor cells into highly differentiated, yet less malignant cells, leading to the mammary abscess formation.

3.2 Genetic Interactions between Smad4 and Pten/Akt Pathway during Skin Cancer Formation

During our study of the role of Smad4 in mammary glands (Li et al., 2003), we found that *Smad4*$^{Co/Co}$*;MMTV-Cre* mice, but not *Smad4*$^{Co/Co}$*;Wap-Cre* mice also developed abnormalities in skin. Using a Rosa-26 reporter strain (Soriano, 1999), we detected Cre activity in both the epidermis and hair follicles (Fig. 4A). The Cre activity largely overlaps with the Smad4 expression domain in the skin (Fig. 4B). Therefore, we investigated the role of Smad4 in skin development and tumor formation using the *Smad4*$^{Co/Co}$*;MMTV-Cre* mice.

We showed that the absence of Smad4 resulted in a block of both TGF-β and BMP signaling pathways, as demonstrated by the downregulation of *Smad6* and *7* (Fig. 4B), which are immediate downstream targets of BMP and TGF-β signaling (Ishida et al., 2000; Nagarajan et al., 1999). Mutant skin also showed a dramatically reduced expression of a number of genes that may play important functions in hair follicle formation differentiation, and/or growth. This including *Lef1*, a HMG box transcription factor that is expressed in the hair matrix and the precursor cells of the hair shaft (DasGupta and Fuchs, 1999; Kratochwil et al., 1996), and p21, a well-known cyclin-dependent kinase inhibitor (Harper et al., 1993). Consequently, Smad4 deficiency blocked hair follicle differentiation and cycling, leading to the absence of the hair shaft in the mutant hair follicles. This resulted in a progressive hair loss of mutant mice. Additionally, skin of mutant mice exhibited increased proliferation of basal keratinocytes, epidermal hyperplasia, and expansion of the outer route sheath of hair follicles, which eventually results in the spontaneous

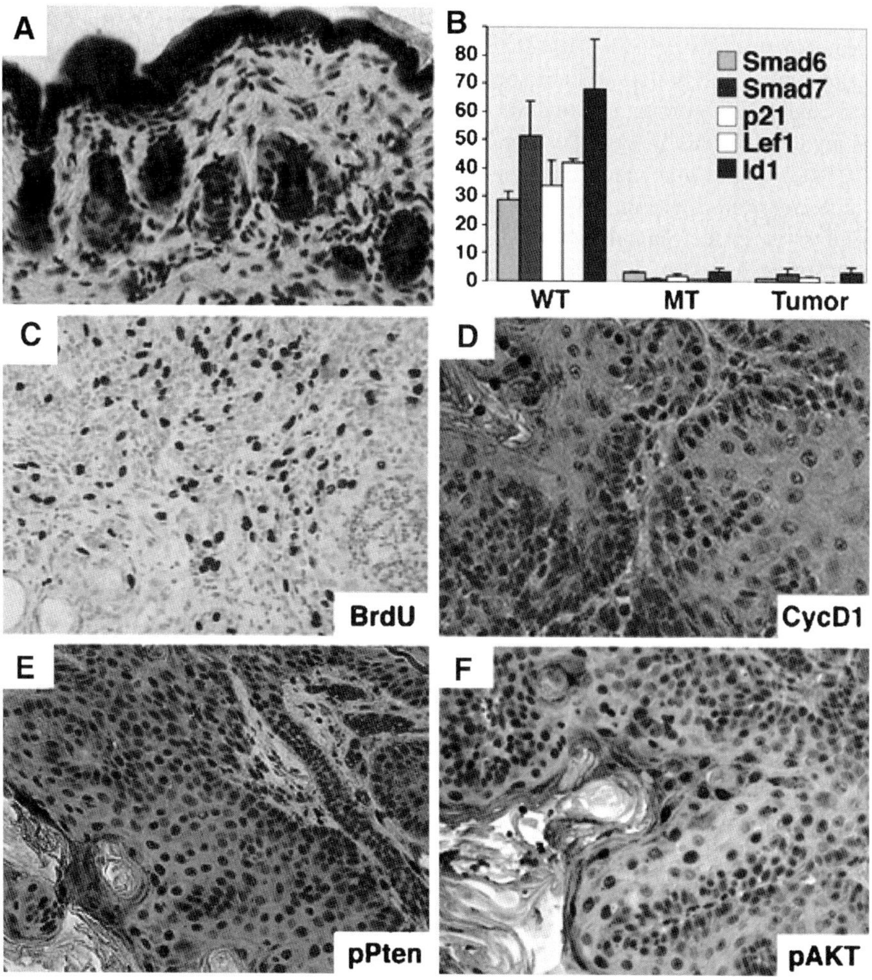

Figure 4. The absence of Smad4 in keratinocytes results in tumor formation accompanied with activation of Akt signaling. (A) Activity of *MMTV-Cre* revealed by Rosa-26 reporter mice. (B) Analysis of gene expression in wild-type (WT), mutant (MT) skin and tumors by quantitative PCR using primers as indicated. All qRT-PCR results shown are averaged from 3–5 samples in each group. (C-F) Increased cell proliferation and activation of Akt in the tumors. Immunohistochemical staining of skin tumors using antibodies as indicated (A color version of this figure is freely accessible via the website of the book: http://www.springer.com/1-4020-4542-5)

malignant skin tumors from 3 months to 13 months of age. The majority of the tumors (>80%) are poorly differentiated malignant squamous cell carcinomas in addition to other types of tumors, such as sebaceous adenoma, basal cell carcinomas, and tricoepithelomas.

We have showed earlier that inactivation of *Smad4* in mammary epithelial cells resulted in increased expression of β–catenin, and conversely, TGF-β treatment inhibited transcription of β-catenin in the wild-type mammary epithelial cells but not in *Smad4*$^{-/-}$ cells, establishing that TGF-β treatment inhibited transcription of β-catenin through Smad4 (Fig. 3). However, our expression study in the mutant keratinocytes and tumors did not show increased expression or nuclear localization of β-catenin, suggesting a distinct mechanism, rather than β-catenin activation, may be involved in this model. To this end, we observed increased cell proliferation (Fig. 4C) that is associated with an increased expression of cyclin D1 (Fig. 4D) and decrease in p21 (Fig. 4B), which may have a significant contribution of tumor progression in the Smad4 mutant skin. While both cyclin D1 and p21 are known downstream targets of TGF-β/Smad4 signaling, they are also subjected to regulation of multiple other factors.

Next, we investigated whether Smad4 deficiency could affect the signaling pathway mediated by the tumor suppressor Pten (phosphatase and tensin homolog deleted on chromosome 10). It has been shown that Pten induces G1 arrest by decreasing the level and nuclear localization of cyclin D1 (Radu et al., 2003; Weng et al., 2001). Using an antibody that detects phosphorylated Pten, an inactive form of the protein, we detected significantly increased levels of Pten phosphorylation in all the tumors (Fig. 4E). Comparable elevated levels of pAkt (activated form of Akt) were also observed in all of the tumors examined (Fig. 4F). The synergistic effect of these alterations with Smad4 deficiency may play a causal role in skin tumor formation through increasing nuclear accumulation of cyclin D1 and other unidentified factors in *Smad4*$^{Co/Co}$*MMTV-Cre* keratinocytes. Thus, different from mammary and intestinal epithelial cells, where targeted ablation of Smad4 (Li et al., 2003) or BMPR1A (He et al., 2004) results in activation of β-catenin, the absence of Smad4 in skin causes tumor formation that is accompanied by activation of Akt pathway.

4. CONCLUSION AND FUTURE DIRECTIONS

We have provided a comprehensive coverage of mutant mice carrying targeted disruptions of Smad genes. Studies of these mice have provided valuable information about functions of Smads in many biological processes in multiple tissues/organs. Conditional alleles of *Smad1, 2, 3* and 5 have also been created to overcome embryonic lethality associated with deficiency of these genes. Tissue specific disruption has been performed for Smad4 in the epiblast (Chu et al., 2004), mammary gland (Li et al., 2003), the brain (Zhou et al., 2003), chondrocyte (Zhang et al., 2005) and the skin (Qiao et al., 2005), and for Smad2 in the epiblast (Vincent et al., 2003). Conditional disruption of all these genes in more tissues/organs can be expected in the near future owing to the existence of a large collection of Cre transgenic lines (hhttp://www.mshri.on.ca/nagy/cre-pub.html). Functions of Smads have also been addressed in mice carrying double mutations of Smad2 and 3, and Smads with a number of other genes (Table 1). More than a dozen of mutant strains (null, hypomorphic, heterozygous and conditional stains) created in these gene families of 8 could potentially allow numerous combinational crosses to address synergistic and redundant functions among different Smads. TGF-β signaling plays a complex role in is involved in many different types of human cancers (Roberts and Wakefield, 2003; Wakefield and Roberts, 2002). The future efforts using this collection of mutant mice will be directed at further understanding of oncogenic and tumor suppressing pathways and mechanisms underlying these processes in wide variety of tissues/organs in order to provide better therapeutic approaches for the treatment of cancers associated with TGF-β signaling.

ACKNOWLEDGEMENTS

This research was supported for C.D by the Intramural Research Program of the Institute of Digestive and Kidney Diseases, National Institutes of Health, USA, and for M.W by funding from the National Institute of Child Health and Human Development, National Institutes of Health.

REFERENCES

Akiyama, T., 2000, Wnt/β-catenin signaling. *Cytokine Growth Factor Rev* 11: 273-282.

Altmann, C.R., and Brivanlou, A.H., 2001, Neural patterning in the vertebrate embryo. *Int Rev Cytol* 203: 447-482.

Ashcroft, G.S., Yang, X., Glick, A., Weinstein, M., Letterio, J.J., Mizel, D.E., Anzano, M., Greenwell-Wild, T., Wahl, S.M., Deng, C., and Roberts, A.B., 1999, Mice lacking SMAD 3 show accelerated wound healing and an impaired local inflammatory response. *Nature Cell Biology* 1: 260-266.

Aubin, J., Davy, A., and Soriano, P., 2004, In vivo convergence of BMP and MAPK signaling pathways: impact of differential Smad1 phosphorylation on development and homeostasis. *Genes Dev* 18: 1482-1494.

Bonniaud, P., Kolb, M., Galt, T., Robertson, J., Robbins, C., Stampfli, M., Lavery, C., Margetts, P.J., Roberts, A.B., and Gauldie, J., 2004, Smad3 null mice develop airspace enlargement and are resistant to TGF-β-mediated pulmonary fibrosis. *J Immunol* 173: 2099-2108.

Brennan, J., Norris, D.P., and Robertson, E.J., 2002, Nodal activity in the node governs left-right asymmetry. *Genes Dev* 16: 2339-2344.

Cardiff, R.D., Anver, M.R., Gusterson, B.A., Hennighausen, L., Jensen, R.A., Merino, M.J., Rehm, S., Russo, J., Tavassoli, F.A., Wakefield, L.M., Ward, J.M., and Green, J.E., 2000, The mammary pathology of genetically engineered mice: the consensus report and recommendations from the Annapolis meeting. *Oncogene* 19: 968-988.

Chang, H., Huylebroeck, D., Verschueren, K., Guo, Q., Matzuk, M.M., and Zwijsen, A., 1999, Smad5 knockout mice die at mid-gestation due to multiple embryonic and extraembryonic defects. *Development* 126: 1631-1642.

Chang, H., and Matzuk, M.M., 2001, Smad5 is required for mouse primordial germ cell development. *Mech Dev* 104: 61-67.

Chang, H., Zwijsen, A., Vogel, H., Huylebroeck, D., and Matzuk, M.M., 2000, Smad5 is essential for left-right asymmetry in mice. *Dev Biol* 219: 71-78.

Chen, H., Sun, J., Buckley, S., Chen, C., Warburton, D., Wang, X.F., and Shi, W., 2005, Abnormal mouse lung alveolarization caused by Smad3 deficiency is a developmental antecedent of centrilobular emphysema. *Am J Physiol Lung Cell Mol Physiol* 288: L683-691.

Chen, Y., Yee, D., Dains, K., Chatterjee, A., Cavalcoli, J., Schneider, E., Om, J., Woychik, R.P., and Magnuson, T., 2000, Genotype-based screen for ENU-induced mutations in mouse embryonic stem cells. *Nat Genet* 24: 314-317.

Chu, G.C., Dunn, N.R., Anderson, D.C., Oxburgh, L., and Robertson, E.J., 2004, Differential requirements for Smad4 in TGFβ-dependent patterning of the early mouse embryo. *Development* 131: 3501-12.

DasGupta, R., and Fuchs, E., 1999, Multiple roles for activated LEF/TCF transcription complexes during hair follicle development and differentiation. *Development* 126: 4557-4568.

Datto, M.B., Frederick, J.P., Pan, L., Borton, A.J., Zhuang, Y., and Wang, X.F., 1999, Targeted disruption of Smad3 reveals an essential role in transforming growth factor β–mediated signal transduction. *Mol Cell Biol* 19: 2495-2504.

de Caestecker, M.P., Parks, W.T., Frank, C.J., Castagnino, P., Bottaro, D.P., Roberts, A.B., and Lechleider, R.J., 1998, Smad2 transduces common signals from receptor serine-threonine and tyrosine kinases. *Genes Dev* 12: 1587-1592.

Dennler, S., Huet, S., and Gauthier, J.M., 1999, A short amino-acid sequence in MH1 domain is responsible for functional differences between Smad2 and Smad3. *Oncogene* 18: 1643-1648.

Dunn, N.R., Koonce, C.H., Anderson, D.C., Islam, A., Bikoff, E.K., and Robertson, E.J., 2005, Mice exclusively expressing the short isoform of Smad2 develop normally and are viable and fertile. *Genes Dev* 19: 152-163.

Dunn, N.R., Vincent, S.D., Oxburgh, L., Robertson, E.J., and Bikoff, E.K., 2004, Combinatorial activities of Smad2 and Smad3 regulate mesoderm formation and patterning in the mouse embryo. *Development* 131: 1717-1728.

Flanders, K.C., Kim, E.S., and Roberts, A.B., 2001, Immunohistochemical expression of Smads 1-6 in the 15-day gestation mouse embryo: signaling by BMPs and TGF-βs. *Dev Dyn* 220: 141-154.

Friedl, W., Kruse, R., Uhlhaas, S., Stolte, M., Schartmann, B., Keller, K.M., Jungck, M., Stern, M., Loff, S., Back, W., Propping, P., and Jenne, D.E., 1999, Frequent 4-bp deletion in exon 9 of the SMAD4/MADH4 gene in familial juvenile polyposis patients. *Genes Chromosomes Cancer* 25: 403-406.

Gaio, U., Schweickert, A., Fischer, A., Garratt, A.N., Muller, T., Ozcelik, C., Lankes, W., Strehle, M., Britsch, S., Blum, M., and Birchmeier, C., 1999, A role of the cryptic gene in the correct establishment of the left-right axis. *Curr Biol* 9: 1339-1342.

Galvin, K.M., Donovan, M.J., Lynch, C.A., Meyer, R.I., Paul, R.J., Lorenz, J.N., Fairchild-Huntress, V., Dixon, K.L., Dunmore, J.H., Gimbrone, M.A., Jr., Falb, D., and Huszar, D., 2000, A role for smad6 in development and homeostasis of the cardiovascular system. *Nat Genet* 24: 171-174.

Goumans, M.-J., Lebrin, F., and Valdimarsdottir, G., 2003, Controlling the angiogenic switch: a balance between two distinct TGF-β receptor signaling pathways. *Trends Cardiovasc Med* 13: 301-307.

Goumans, M.-J., Valdimarsdottir, G., Itoh, S., Rosendahl, A., Sideras, P., and ten Dijke, P., 2002, Balancing the activation state of the endothelium via two distinct TGF-β type I receptors. *EMBO J* 21: 1743-1753.

Grimm, O.H., and Gurdon, J.B., 2002, Nuclear exclusion of Smad2 is a mechanism leading to loss of competence. *Nat Cell Biol* 4: 519-522.

Hackett, B.P., 2002, Formation and malformation of the vertebrate left-right axis. *Curr Mol Med* 2: 39-66.

Hahn, S.A., Bartsch, D., Schroers, A., Galehdari, H., Becker, M., Ramaswamy, A., Schwarte-Waldhoff, I., Maschek, H., and Schmiegel, W., 1998, Mutations of the DPC4/Smad4 gene in biliary tract carcinoma. *Cancer Res* 58: 1124-1126.

Hahn, S.A., Schutte, M., Hoque, A.T., Moskaluk, C.A., da Costa, L.T., Rozenblum, E., Weinstein, C.L., Fischer, A., Yeo, C.J., Hruban, R.H., and Kern, S.E., 1996, DPC4, a candidate tumor suppressor gene at human chromosome 18q21.1. *Science* 271: 350-353.

Hamamoto, T., Beppu, H., Okada, H., Kawabata, M., Kitamura, T., Miyazono, K., and Kato, M., 2002, Compound disruption of smad2 accelerates malignant progression of intestinal tumors in apc knockout mice. *Cancer Res* 62: 5955-5961.

Harper, J.W., Adami, G.R., Wei, N., Keyomarsi, K., and Elledge, S.J., 1993, The p21 Cdk-interacting protein Cip1 is a potent inhibitor of G1 cyclin-dependent kinases. *Cell* 75: 805-816.

Hayashi, K., Kobayashi, T., Umino, T., Goitsuka, R., Matsui, Y., and Kitamura, D., 2002, SMAD1 signaling is critical for initial commitment of germ cell lineage from mouse epiblast. *Mech Dev* 118: 99-109.

He, X.C., Zhang, J., Tong, W.G., Tawfik, O., Ross, J., Scoville, D.H., Tian, Q., Zeng, X., He, X., Wiedemann, L.M., Mishina, Y., and Li, L., 2004, BMP signaling inhibits intestinal stem cell self-renewal through suppression of Wnt-β-catenin signaling. *Nat Genet* 36: 1117-1121.

Heldin, C.-H., Miyazono, K., and ten Dijke, P., 1997, TGF-β signalling from cell membrane to nucleus through SMAD proteins. *Nature* 390: 465-471.

Hester, M., Thompson, J.C., Mills, J., Liu, Y., El-Hodiri, H.M., and Weinstein, M., 2005, Smad1 and Smad8 function similarly in mammalian central nervous system development. *Mol Cell Biol* 25: 4683-4692.

Heyer, J., Escalante-Alcalde, D., Lia, M., Böttinger, E., Edelmann, W., Stewart, C.L., and Kucherlapati, R., 1999, Postgastrulation Smad2-deficient embryos show defects in embryo turning and anterior morphogenesis. *Proc Natl Acad Sci U S A* 96: 12595-12600.

Howe, J.R., Ringold, J.C., Summers, R.W., Mitros, F.A., Nishimura, D.Y., and Stone, E.M., 1998, A gene for familial juvenile polyposis maps to chromosome 18q21.1. *Am J Hum Genet* 62: 1129-1136.

Huang, S., Tang, B., Usoskin, D., Lechleider, R.J., Jamin, S.P., Li, C., Anzano, M.A., Ebendal, T., Deng, C., and Roberts, A.B., 2002, Conditional knockout of the Smad1 gene. *Genesis* 32: 76-79.

Inazaki, K., Kanamaru, Y., Kojima, Y., Sueyoshi, N., Okumura, K., Kaneko, K., Yamashiro, Y., Ogawa, H., and Nakao, A., 2004, Smad3 deficiency attenuates renal fibrosis, inflammation, and apoptosis after unilateral ureteral obstruction. *Kidney Int* 66: 597-604.

Ishida, W., Hamamoto, T., Kusanagi, K., Yagi, K., Kawabata, M., Takehara, K., Sampath, T.K., Kato, M., and Miyazono, K., 2000, Smad6 is a Smad1/5-induced Smad inhibitor. Characterization of bone morphogenetic protein-responsive element in the mouse Smad6 promoter. *J Biol Chem* 275: 6075-6079.

Karayiannakis, A.J., Nakopoulou, L., Gakiopoulou, H., Keramopoulos, A., Davaris, P.S., and Pignatelli, M., 2001, Expression patterns of β-catenin in in situ and invasive breast cancer. *Eur J Surg Oncol* 27: 31-36.

Kolligs, F.T., Bommer, G., and Goke, B., 2002, Wnt/β-catenin/tcf signaling: a critical pathway in gastrointestinal tumorigenesis. *Digestion* 66: 131-144.

Kratochwil, K., Dull, M., Farinas, I., Galceran, J., and Grosschedl, R., 1996, Lef1 expression is activated by BMP-4 and regulates inductive tissue interactions in tooth and hair development. *Genes Dev* 10: 1382-1394.

Kretzschmar, M., Doody, J., and Massagué, J., 1997, Opposing BMP and EGF signalling pathways converge on the TGF-β family mediator Smad1. *Nature* 389: 618-622.

Kretzschmar, M., Doody, J., Timokhina, I., and Massagué, J., 1999, A mechanism of repression of TGFβ/Smad signaling by oncogenic Ras. *Genes Dev* 13: 804-816.

Lebrin, F., Deckers, M., Bertolino, P., and ten Dijke, P., 2005, TGF-β receptor function in the endothelium. *Cardiovasc Res* 65: 599-608.

Lechleider, R.J., Ryan, J.L., Garrett, L., Eng, C., Deng, C., Wynshaw-Boris, A., and Roberts, A.B., 2001, Targeted mutagenesis of Smad1 reveals an essential role in chorioallantoic fusion. *Dev Biol* 240: 157-167.

Li, A.G., Lu, S.L., Zhang, M.X., Deng, C., and Wang, X.J., 2004, Smad3 knockout mice exhibit a resistance to skin chemical carcinogenesis. *Cancer Res* 64: 7836-7845.

Li, D.Y., Sorensen, L.K., Brooke, B.S., Urness, L.D., Davis, E.C., Taylor, D.G., Boak, B.B., and Wendel, D.P., 1999, Defective angiogenesis in mice lacking endoglin. *Science* 284: 1534-1537.

Li, W., Qiao, W., Chen, L., Xu, X., Yang, X., Li, D., Li, C., Brodie, S.G., Meguid, M.M., Hennighausen, L., and Deng, C.X., 2003, Squamous cell carcinoma and mammary abscess formation through squamous metaplasia in Smad4/Dpc4 conditional knockout mice. *Development* 130: 6143-6153.

Liu, B., Sun, Y., Jiang, F., Zhang, S., Wu, Y., Lan, Y., Yang, X., and Mao, N., 2003, Disruption of Smad5 gene leads to enhanced proliferation of high-proliferative potential precursors during embryonic hematopoiesis. *Blood* 101: 124-133.

Liu, Y., Festing, M., Thompson, J.C., Hester, M., Rankin, S., El-Hodiri, H.M., Zorn, A.M., and Weinstein, M., 2004a, Smad2 and Smad3 coordinately regulate craniofacial and endodermal development. *Dev Biol* 270: 411-426.

Liu, Y., Festing, M.H., Hester, M., Thompson, J.C., and Weinstein, M., 2004b, Generation of novel conditional and hypomorphic alleles of the Smad2 gene. *Genesis* 40: 118-123.

Martin, J.S., Dickson, M.C., Cousins, F.M., Kulkarni, A.B., Karlsson, S., and Akhurst, R.J., 1995, Analysis of homozygous TGFβ1 null mouse embryos demonstrates defects in yolk sac vasculogenesis and hematopoiesis. *Ann N Y Acad Sci* 752: 300-308.

Massagué, J., 1998, TGF-β signal transduction. *Annu Rev Biochem* 67: 753-791.

Miyoshi, K., Shillingford, J.M., Le Provost, F., Gounari, F., Bronson, R., von Boehmer, H., Taketo, M.M., Cardiff, R.D., Hennighausen, L., and Khazaie, K., 2002, Activation of β–catenin signaling in differentiated mammary secretory cells induces transdifferentiation into epidermis and squamous metaplasias. *Proc Natl Acad Sci U S A* 99: 219-224.

Nagarajan, R.P., Zhang, J., Li, W., and Chen, Y., 1999, Regulation of Smad7 promoter by direct association with Smad3 and Smad4. *J Biol Chem* 274: 33412-33418.

Nomura, M., and Li, E., 1998, Smad2 role in mesoderm formation, left-right patterning and craniofacial development. *Nature* 393: 786-790.

Oh, S.P., Seki, T., Goss, K.A., Imamura, T., Yi, Y., Donahoe, P.K., Li, L., Miyazono, K., ten Dijke, P., Kim, S., and Li, E., 2000, Activin receptor-like kinase 1 modulates transforming growth factor-β1 signaling in the regulation of angiogenesis. *Proc Natl Acad Sci U S A* 97: 2626-2631.

Oshima, M., Oshima, H., Kitagawa, K., Kobayashi, M., Itakura, C., and Taketo, M., 1995, Loss of Apc heterozygosity and abnormal tissue building in nascent intestinal polyps in mice carrying a truncated Apc gene. *Proc Natl Acad Sci U S A* 92: 4482-4486.

Pera, E.M., Ikeda, A., Eivers, E., and De Robertis, E.M., 2003, Integration of IGF, FGF, and anti-BMP signals via Smad1 phosphorylation in neural induction. *Genes Dev* 17: 3023-3028.

Piek, E., Heldin, C.-H., and ten Dijke, P., 1999, Specificity, diversity, and regulation in TGF-β superfamily signaling. *FASEB J* 13: 2105-2124.

Polakis, P., 2001, More than one way to skin a catenin. *Cell* 105: 563-566.

Qiao, W., Li, A.G., Owens, P., Xu, X., Wang, X.J., and Deng, C.X., 2005, Hair follicle defects and squamous cell carcinoma formation in *Smad4* conditional knockout mouse skin. *Oncogene* (in press).

Radu, A., Neubauer, V., Akagi, T., Hanafusa, H., and Georgescu, M.M., 2003, PTEN induces cell cycle arrest by decreasing the level and nuclear localization of cyclin D1. *Mol Cell Biol* 23: 6139-6149.

Randall, R.A., Howell, M., Page, C.S., Daly, A., Bates, P.A., and Hill, C.S., 2004, Recognition of phosphorylated-Smad2-containing complexes by a novel Smad interaction motif. *Mol Cell Biol* 24: 1106-1121.

Redman, R.S., Katuri, V., Tang, Y., Dillner, A., Mishra, B., and Mishra, L., 2005, Orofacial and gastrointestinal hyperplasia and neoplasia in smad4+/- and elf+/-/smad4+/- mutant mice. *J Oral Pathol Med* 34: 23-29.

Roberts, A.B., and Wakefield, L.M., 2003, The two faces of transforming growth factor β in carcinogenesis. *Proc Natl Acad Sci U S A* 100: 8621-8623.

Saika, S., Kono-Saika, S., Tanaka, T., Yamanaka, O., Ohnishi, Y., Sato, M., Muragaki, Y., Ooshima, A., Yoo, J., Flanders, K.C., and Roberts, A.B., 2004, Smad3 is required for dedifferentiation of retinal pigment epithelium following retinal detachment in mice. *Lab Invest* 84: 1245-1258.

Sato, M., Muragaki, Y., Saika, S., Roberts, A.B., and Ooshima, A., 2003, Targeted disruption of TGF-β1/Smad3 signaling protects against renal tubulointerstitial fibrosis induced by unilateral ureteral obstruction. *J Clin Invest* 112: 1486-1494.

Sirard, C., de la Pompa, J.L., Elia, A., Itie, A., Mirtsos, C., Cheung, A., Hahn, S., Wakeham, A., Schwartz, L., Kern, S.E., Rossant, J., and Mak, T.W., 1998, The tumor suppressor gene Smad4/Dpc4 is required for gastrulation and later for anterior development of the mouse embryo. *Genes Dev* 12: 107-119.

Soriano, P., 1999, Generalized lacZ expression with the ROSA26 Cre reporter strain. *Nat Genet* 21: 70-71.

Takagi, Y., Kohmura, H., Futamura, M., Kida, H., Tanemura, H., Shimokawa, K., and Saji, S., 1996, Somatic alterations of the DPC4 gene in human colorectal cancers in vivo. *Gastroenterology* 111: 1369-1372.

Takaku, K., Miyoshi, H., Matsunaga, A., Oshima, M., Sasaki, N., and Taketo, M.M., 1999, Gastric and duodenal polyps in Smad4 (Dpc4) knockout mice. *Cancer Res* 59: 6113-6117.

Takaku, K., Oshima, M., Miyoshi, H., Matsui, M., Seldin, M.F., and Taketo, M.M., 1998, Intestinal tumorigenesis in compound mutant mice of both Dpc4 (Smad4) and Apc genes. *Cell* 92: 645-656.

Takaku, K., Wrana, J.L., Robertson, E.J., and Taketo, M.M., 2002, No effects of Smad2 (madh2) null mutation on malignant progression of intestinal polyps in Apc(delta716) knockout mice. *Cancer Res* 62: 4558-4561.

Taketo, M.M., and Takaku, K., 2000, Gastrointestinal tumorigenesis in Smad4 (Dpc4) mutant mice. *Hum Cell* 13: 85-95.

Tannehill-Gregg, S.H., Kusewitt, D.F., Rosol, T.J., and Weinstein, M., 2004, The roles of Smad2 and Smad3 in the development of chemically induced skin tumors in mice. *Vet Pathol* 41: 278-282.

ten Dijke, P., and Hill, C.S., 2004, New insights into TGF-β-Smad signalling. *Trends Biochem Sci* 29: 265-273.

Tomic, D., Miller, K.P., Kenny, H.A., Woodruff, T.K., Hoyer, P., and Flaws, J.A., 2004, Ovarian follicle development requires Smad3. *Mol Endocrinol* 18: 2224-2240.

Tremblay, K.D., Dunn, N.R., and Robertson, E.J., 2001, Mouse embryos lacking Smad1 signals display defects in extra-embryonic tissues and germ cell formation. *Development* 128: 3609-3621.

Tremblay, K.D., Hoodless, P.A., Bikoff, E.K., and Robertson, E.J., 2000, Formation of the definitive endoderm in mouse is a Smad2-dependent process. *Development* 127: 3079-3090.

Umans, L., Vermeire, L., Francis, A., Chang, H., Huylebroeck, D., and Zwijsen, A., 2003, Generation of a floxed allele of Smad5 for cre-mediated conditional knockout in the mouse. *Genesis* 37: 5-11.

Wagner, K.U., Wall, R.J., St-Onge, L., Gruss, P., Wynshaw-Boris, A., Garrett, L., Li, M., Furth, P.A., and Hennighausen, L., 1997, Cre-mediated gene deletion in the mammary gland. *Nucleic Acids Res* 25: 4323-4330.

Wakefield, L.M., and Roberts, A.B., 2002, TGF-β signaling: positive and negative effects on tumorigenesis. *Curr Opin Genet Dev* 12: 22-29.

Waldrip, W.R., Bikoff, E.K., Hoodless, P.A., Wrana, J.L., and Robertson, E.J., 1998, Smad2 signaling in extraembryonic tissues determines anterior-posterior polarity of the early mouse embryo. *Cell* 92: 797-808.

Weinstein, M., Monga, S.P., Liu, Y., Brodie, S.G., Tang, Y., Li, C., Mishra, L., and Deng, C.X., 2001, Smad proteins and hepatocyte growth factor control parallel regulatory pathways that converge on β1-integrin to promote normal liver development. *Mol Cell Biol* 21: 5122-5131.

Weinstein, M., Yang, X., and Deng, C., 2000, Functions of mammalian Smad genes as revealed by targeted gene disruption in mice. *Cytokine Growth Factor Rev* 11: 49-58.

Weinstein, M., Yang, X., Li, C., Xu, X., Gotay, J., and Deng, C.X., 1998, Failure of egg cylinder elongation and mesoderm induction in mouse embryos lacking the tumor suppressor smad2. *Proc Natl Acad Sci U S A* 95: 9378-9383.

Weng, L., Brown, J., and Eng, C., 2001, PTEN induces apoptosis and cell cycle arrest through phosphoinositol-3-kinase/Akt-dependent and -independent pathways. *Hum Mol Genet* 10: 237-242.

Vincent, S.D., Dunn, N.R., Hayashi, S., Norris, D.P., and Robertson, E.J., 2003, Cell fate decisions within the mouse organizer are governed by graded Nodal signals. *Genes Dev* 17: 1646-1662.

Vivian, J.L., Chen, Y., Yee, D., Schneider, E., and Magnuson, T., 2002, An allelic series of mutations in Smad2 and Smad4 identified in a genotype-based screen of N-ethyl-N-nitrosourea-mutagenized mouse embryonic stem cells. *Proc Natl Acad Sci U S A* 99: 15542-15547.

Wolfraim, L.A., Fernandez, T.M., Mamura, M., Fuller, W.L., Kumar, R., Cole, D.E., Byfield, S., Felici, A., Flanders, K.C., Walz, T.M., Roberts, A.B., Aplan, P.D., Balis, F.M., and Letterio, J.J., 2004, Loss of Smad3 in acute T-cell lymphoblastic leukemia. *N Engl J Med* 351: 552-559.

Xu, X., Brodie, S.G., Yang, X., Im, Y.H., Parks, W.T., Chen, L., Zhou, Y.X., Weinstein, M., Kim, S.J., and Deng, C.X., 2000, Haploid loss of the tumor suppressor Smad4/Dpc4 initiates gastric polyposis and cancer in mice. *Oncogene* 19: 1868-1874.

Yagi, K., Goto, D., Hamamoto, T., Takenoshita, S., Kato, M., and Miyazono, K., 1999, Alternatively spliced variant of Smad2 lacking exon 3. Comparison with wild-type Smad2 and Smad3. *J Biol Chem* 274: 703-709.

Yamada, M., Szendro, P.I., Prokscha, A., Schwartz, R.J., and Eichele, G., 1999, Evidence for a role of Smad6 in chick cardiac development. *Dev Biol* 215: 48-61.

Yang, L., Mao, C., Teng, Y., Li, W., Zhang, J., Cheng, X., Li, X., Han, X., Xia, Z., Deng, H., and Yang, X., 2005, Targeted disruption of Smad4 in mouse epidermis results in failure of hair follicle cycling and formation of skin tumors. *Cancer Res* 65: 8671-8678.

Yang, X., Castilla, L.H., Xu, X., Li, C., Gotay, J., Weinstein, M., Liu, P.P., and Deng, C.X., 1999a, Angiogenesis defects and mesenchymal apoptosis in mice lacking SMAD5. *Development* 126: 1571-1580.

Yang, X., Chen, L., Xu, X., Li, C., Huang, C., and Deng, C.X., 2001, TGF-β/Smad3 signals repress chondrocyte hypertrophic differentiation and are required for maintaining articular cartilage. *J Cell Biol* 153: 35-46.

Yang, X., Letterio, J.J., Lechleider, R.J., Chen, L., Hayman, R., Gu, H., Roberts, A.B., and Deng, C., 1999b, Targeted disruption of SMAD3 results in impaired mucosal immunity and diminished T cell responsiveness to TGF-β. *EMBO J* 18: 1280-1291.

Yang, X., Li, C., Herrera, P.L., and Deng, C.X., 2002a, Generation of Smad4/Dpc4 conditional knockout mice. *Genesis* 32: 80-81.

Yang, X., Li, C., Xu, X., and Deng, C., 1998, The tumor suppressor SMAD4/DPC4 is essential for epiblast proliferation and mesoderm induction in mice. *Proc Natl Acad Sci U S A* 95: 3667-3672.

Yang, Y.A., Tang, B., Robinson, G., Hennighausen, L., Brodie, S.G., Deng, C.X., and Wakefield, L.M., 2002b, Smad3 in the mammary epithelium has a nonredundant role in the induction of apoptosis, but not in the regulation of proliferation or differentiation by transforming growth factor-β. *Cell Growth Differ* 13: 123-130.

Yost, H.J., 2001, Establishment of left-right asymmetry. *Int Rev Cytol* 203: 357-381.

Yue, J., Frey, R.S., and Mulder, K.M., 1999, Cross-talk between the Smad1 and Ras/MEK signaling pathways for TGFβ. *Oncogene* 18: 2033-2037.

Zaret, K.S., 2002, Regulatory phases of early liver development: paradigms of organogenesis. *Nat Rev Genet* 3: 499-512.

Zhang, J., Tan, X., Li, W., Wang, Y., Wang, J., Cheng, X., and Yang, X., 2005, Smad4 is required for the normal organization of the cartilage growth plate. *Dev Biol* 384; 311-22.

Zhou, X., Sasaki, H., Lowe, L., Hogan, B.L., and Kuehn, M.R., 1993, Nodal is a novel TGF-β-like gene expressed in the mouse node during gastrulation. *Nature* 361: 543-547.

Zhou, Y.X., Zhao, M., Li, D., Shimazu, K., Sakata, K., Deng, C.X., and Lu, B., 2003, Cerebellar deficits and hyperactivity in mice lacking Smad4. *J Biol Chem* 278: 42313-42320.

Zhu, Y., Richardson, J.A., Parada, L.F., and Graff, J.M., 1998, Smad3 mutant mice develop metastatic colorectal cancer. *Cell* 94: 703-714.

CHAPTER 9

TRAFFICKING OF SERINE/THREONINE KINASE RECEPTORS AND SMAD ACTIVATION

CHRISTINE LE ROY, ROHIT BOSE, AND JEFFREY L. WRANA

Center for Systems Biology, Department of Medical Genetics and Microbiology, University of Toronto, Samuel Lunenfeld Research Institute, Mount Sinai Hospital, 600 University Avenue, Toronto M5G 1X5, Canada

Abstract: Signaling by the transforming growth factor-β (TGF-β) family of growth factors involves cell surface receptor serine/threonine kinases and downstream components such as SARA and Smurfs, which bind the receptor substrates called the Smad proteins. These components partition between two endocytic pathways, which lead to two separate functions in TGF-β signaling. On the one hand, clathrin-dependent endocytosis promotes signaling by leading the receptor to the early endosome where SARA is localized. On the other, non-clathrin pathways, which are enriched for the Smurf ubiquitin ligases, lead the receptor to degradation. However, in polarized cells, ligand addition induces TGF-β receptor trafficking into junctional regions of the cell where activation of Smad-dependent and -independent pathways mediates the process of epithelial-to-mesenchymal transition, which is critical during development and tumorigenesis. In this chapter, we will focus on how compartmentalization of TGF-β receptors and their downstream components controls TGF-β signaling and its biological functions

Keywords: caveolin; EMT; endocytosis; Par6; polarity; SARA; Smurf; TGF-β; TGF-β receptor; trafficking clathrin; lipid-rafts

1. SERINE/THREONINE KINASE RECEPTORS

Serine/threonine kinase receptors are a unique class of transmembrane receptors, first identified in *C. elegans* (Georgi et al., 1990). They all share certain structural features that include a short extracellular domain that is cysteine-rich, a single transmembrane domain and an intracellular Ser/Thr kinase (Attisano and Wrana, 1996; Massagué, 1998; Shi and Massagué, 2003). There are relatively few Ser/Thr kinases encoded by mammalian genomes and thus far all of them have been shown to function as receptors for members of the TGF-β superfamily (Manning et al., 2002). Furthermore, no evidence has yet been found that they function in any

P. ten Dijke and C.-H. Heldin (eds.), Smad Signal Transduction, 177–191.

capacity other than as receptors for the TGF-β superfamily. A wealth of biochemical, structural and genetic data has demonstrated that two classes of Ser/Thr kinase receptors exist, the so-called type II and type I receptors (Attisano and Wrana, 1996; Massagué, 1998; Shi and Massagué, 2003). These two Ser/Thr kinase receptors exist as homodimers but upon ligand addition assemble into a heterotetramer that is comprised of two type II (TβRII) and two type I (TβRI) receptor subunits (Gilboa et al., 1998; Yamashita et al., 1994). Once the ligand induces assembly of a heteromeric receptor complex, the type II receptor kinase, which is constitutively active, phosphorylates the type I receptor in a unique region at the beginning of the kinase domain, called the GS region (Wrana et al., 1994). Phosphorylation of the GS region is critical to establish Smad signal transduction and provides part of a docking site for Smads (Wrana, 2000). Thus, ligand-dependent assembly of a heteromeric Ser/Thr kinase receptor complex is critical for TGF-β signal transduction *via* the Smad pathway (Derynck and Zhang, 2003; Heldin et al., 1997; Miyazono et al., 2000; Shi and Massagué, 2003; Wrana, 2000).

Smads were one of the first targets of heteromeric Ser/Thr kinase receptors and as such have received considerable attention. The receptor-regulated class of Smads (R-Smads) are direct substrates of the type I receptor and are phosphorylated by the type I kinase on the last two serine residues of a single SSXS motif that is found at the carboxy-terminus of R-Smads (Macias-Silva et al., 1996; Massagué and Chen, 2000; Wrana, 2000). Specificity in R-Smad phosphorylation and activation are thus conferred by the interaction between the R-Smad and the type I receptor (Feng and Derynck, 2005; Shi and Massagué, 2003). This divides the R-Smads into two subclasses; Smad1, Smad5 and Smad8, which are regulated by the type I receptors ALK1, ALK2, ALK3 and ALK6 and Smad2 and Smad3, which are phosphorylated by TβRI, ActRIB and ALK7 type I receptors (Massagué, 1998; Shi and Massagué, 2003). As the SSXS motif is almost the same in all R-Smads, the specificity of R-Smad activation by the type I kinase comes not from specificity in the peptide sequence of the kinase substrate, but rather by specificity in R-Smad-receptor interaction. Thus, different R-Smads only interact with distinct type I receptors (Chen et al., 1998; Shi and Massagué, 2003). This is mediated by cooperative interactions between the phosphorylated GS domain of type I receptor, which likely contacts the phosphoserine binding pocket on R-Smads and interactions between loop 45 (L45) on the type I receptors (Feng and Derynck, 1997) and loop 3 (L3) on the MH2 domains of Smads (Lo et al., 1998). The former contacts are unlikely to confer specificity, since GS regions are well conserved between all type I receptors and it is a threshold of phosphorylation that allows signaling rather than phosphorylation on any specific serines. However, complementary mutagenesis of L45 on the receptor and L3 on the Smads has shown that this putative second contact is the critical determinant that dictates specificity between Smad and receptor interactions (Feng and Derynck, 2005; Shi and Massagué, 2003). Thus, our current model for Smad regulation invokes homologous but parallel pathways that often yield unique biological outputs by regulating distinct transcriptional programs in the nucleus (Massagué and Chen, 2000; Wrana, 2000).

2. SARA (SMAD ANCHOR FOR RECEPTOR ACTIVATION)

Identified as a Smad-interacting protein, SARA binds to unphosphorylated Smads through a large 85 amino acid Smad binding domain (SBD) (Tsukazaki et al., 1998). While the SBD is unfolded in solution, it adopts a β sheet, α helix and rigid coil structure upon docking to a large hydrophobic surface on the R-Smad MH2 domain (Chong et al., 2004; Wu et al., 2000). This is a highly cooperative process and interference with either the amino- or carboxy-terminal docking sites blocks SARA-Smad interaction. In addition to binding Smads, SARA also binds receptors *via* a larger carboxy-terminal region that interacts with the heteromeric receptor complex and type II receptors (Di Guglielmo et al., 2003; Tsukazaki et al., 1998). Thus, SARA can facilitate receptor activation of Smads by performing scaffolding functions. Importantly, the third domain of interest is a FYVE domain (Tsukazaki et al., 1998), which is a special type of double zinc finger that is modular in nature and as such can be found in a number of otherwise structurally unrelated proteins (Stenmark and Aasland, 1999). FYVE domains bind with high specificity to the phosphoinositol-3'-phosphate (PtdIns3P). Different phosphoisoforms of phosphoinositides (PtdInsPs) display restricted localization to distinct membranes in cells and *via* binding to specific lipid binding domains in proteins they can regulate the subcellular localization of proteins (Le Roy and Wrana, 2005; Lemmon, 2003). Especially, FYVE domains are found in a variety of specialized locations in the cell and can serve to regulate the appropriate localization of interacting proteins (Stenmark and Aasland, 1999). Of particular relevance, PtdIns3P is highly enriched in the early endosome and thus FYVE domains can mediate the specific targeting of proteins such as SARA and the Early Endosomal Antigen 1 protein (EEA1) to this endocytic compartment. In the case of EEA1, this requires cooperative interaction with Rab5 (Stenmark et al., 1996), a regulator of endosomal trafficking, whereas the SARA FYVE domain may mediate SARA's localization independently of protein-protein interactions (Tsukazaki et al., 1998). Thus, SARA can bind endosomal membranes enriched in PtdIns3P, as well as Smad2 or Smad3 and TGF-β receptors. By nucleating a receptor-Smad complex at the membrane, SARA can thus promote TGF-β signal transduction (Di Guglielmo et al., 2003; Tsukazaki et al., 1998).

3. INHIBITORY SMADS AND C2-WW-HECT E3 UBIQUITIN LIGASES

Apart from the R-Smads, another class of Smads, the inhibitory Smads (I-Smad6 and 7) also interact with TGF-β and BMP receptors (Hayashi et al., 1997; Imamura et al., 1997; Nakao et al., 1997). Unlike R-Smads, the I-Smads do not have a SSXS motif at their carboxy-terminus and thus form stable complexes with the activated receptor complexes (Hayashi et al., 1997; Nakao et al., 1997). The I-Smads in this case can block access of R-Smads to the receptor complex and inhibit R-Smad activation (Hayashi et al., 1997). While Smad6 appears to preferentially associate with and inhibit the BMP pathway, Smad7 interacts efficiently with all the receptors and potently inhibits both BMP and TGF-β Smad signaling. I-Smads inhibit Ser/Thr kinase receptor signaling in cooperation with E3 protein ubiquitin ligases of the C2-WW-HECT domain class (Izzi and Attisano, 2004). These E3 ubiquitin ligases can catalyze both poly- and mono-ubiquitination of their associated protein substrates. Poly-ubiquitination leads to recognition by the proteasome and subsequent degradation of the ubiquitinated substrates (Ciechanover et al., 2000), while mono-ubiquitination of cell surface receptors serves as a trafficking signal that initiates endocytosis and targeting of substrates to the lysosome for degradation (Le Roy and Wrana, 2005). This class of ubiquitin ligases all contains a lipid/Ca^{2+}-binding (C2) domain, which can mediate lipid and protein-protein interactions, as well as 2-4-conserved-tryptophan (WW) domains, which mediate interaction with Pro-Pro-X-Tyr (PY) motifs of Smads. Catalytic activity is conferred by a HECT (Homologous to E6-associated protein C-Terminus) domain, which catalyzes the addition of ubiquitin chains to ligase-bound substrates. Smurf1 and Smurf2 (for Smad ubiquitin regulatory factor) were the first members of this class of ligases that were shown to interact with R-Smads and to catalyze their ubiquitin-dependent turnover (Bonni et al., 2001; Izzi and Attisano, 2004; Lin et al., 2000; Lo and Massagué, 1999; Zhang et al., 2001; Zhu et al., 1999). For example, by decreasing R-Smad1 levels Smurf1 was shown to control the magnitude of Smad signaling and cell fate decisions during early *Xenopus* gastrulation (Zhu et al., 1999). Smurfs interact with R-Smads *via* a PY motif that is found in the linker region of all R-Smads, except Smad8. Interestingly, Smad4, which does not have a PY motif, seems to be resistant to Smurf-dependent turnover (Lin et al., 2000; Zhang et al., 2001). However, Smad4 can also be poly-ubiquitinated in presence of Smad6 or Smad7 and Smurfs (Morén et al., 2005). I-Smads also have PY motifs in an analogous position as R-Smads and also interact with Smurf WW domains (Kavsak et al., 2000; Ogunjimi et al., 2005). However, in the case of I-Smad-Smurf complexes, this allows recruitment of Smurf to the activated receptor complex to catalyze receptor turnover *via* ubiquitin-dependent degradation (Di Guglielmo et al., 2003; Ebisawa et al., 2001; Kavsak et al., 2000). Of note, I-Smad7 can also bind, through a distinct region from its PY motif, the deubiquitinating enzyme UCH37 and thus has been reported to stabilize the type I TGF-β receptor (Wicks et al., 2005).

Additional interactions between Smads and other C2-WW-HECT ubiquitin ligases have since been identified, including Atrophin 1-interacting protein 4 (AIP4) (Lallemand et al., 2005), NEDD4-2 (neural precursor cell expressed, developmentally down-regulated 4-2) (Kuratomi et al., 2005; Morén et al., 2005) and Tiul1/WWP1 (*via* the Smad2-interacting transcriptional co-repressor TGF-β-induced factor) (Morén et al., 2005; Seo et al., 2004). In addition to interaction between the Smad7 PY motif and the Smurf WW domain, there is another contact between the amino-terminal domain (NTD) of Smad7 and the HECT domain of Smurf2 (Ogunjimi et al., 2005). The Smad7 NTD also possesses a leucine-rich motif (LRM) that can bind the E2 ubiquitin conjugating protein, UbcH7. This adaptor behavior of Smad7 thus allows recruitment of the E2 to the HECT domain and activates the HECT ligase presumably by facilitating the transfer of ubiquitin to the catalytic cysteine in the HECT domain (Ogunjimi et al., 2005). Thus, Smads can directly regulate the activity of ubiquitin ligases.

The most extensively investigated C2-WW-HECT family members have been Smurf1 and Smurf2. However, R-Smads and I-Smads interact with multiple family members and thus there is considerable overlap in ligase function (Izzi and Attisano, 2004). Furthermore, these ubiquitin ligases also function in multiple molecular pathways, including regulating cell polarity as an effector of the polarity protein Par6 (Wang et al., 2003) and activation of MAPK cascades (Yamashita et al., 2005). The finding that Smurf1-deficient mice have a mild phenotype (Yamashita et al., 2005) suggests that there is considerable functional overlap in the various ligases, which are involved in TGF-β pathways. Indeed, analysis in *Drosophila*, where only one Smurf family member exists, clearly indicates a critical role for Smurf family members in the spatial regulation of BMP gradient formation prior to gastrulation and in the temporal regulation of BMP signaling in the hindgut (Izzi and Attisano, 2004; Podos et al., 2001).

4. TRAFFICKING OF CELL SURFACE RECEPTORS

Regulation of signal intensity, duration and specificity is intimately linked to the localization of signaling receptors on the cell surface and their intracellular itinerary through the endosomal system. Like other receptor systems, Ser/Thr kinase receptors are subject to regulation by endocytic trafficking.

Cell surface components endocytose through clathrin- and non-clathrin-dependent pathways (Conner and Schmid, 2003; Gruenberg and Stenmark, 2004; Le Roy and Wrana, 2005; Pelkmans and Helenius, 2002) (Fig. 1). The classical clathrin pathway involves the adaptor protein (AP) complexes that associate with clathrin and nucleate the assembly of a clathrin-coated pit, which subsequently buds from the membrane in a dynamin-dependent process (Conner and Schmid, 2003) that leads to the formation of an early endosomal vesicle. This early endosomal compartment is enriched in PtdIns3P that binds to FYVE domain containing proteins such as SARA (Tsukazaki et al., 1998) and EEA1 (Stenmark and Aasland, 1999). Entry into this EEA1-positive early endosomal compartment can send receptors back to the cell surface *via* the recycling endosome, or can lead to receptor degradation in the late endosome/multivesicular endosome and then, the lysosome (Conner and Schmid, 2003; Gruenberg and Stenmark, 2004; Le Roy and Wrana, 2005). Recent analysis of trafficking and routing to the degradative compartment has indicated a key role for mono-ubiquitination of receptors targeted for degradation (Gruenberg and Stenmark, 2004; Le Roy and Wrana, 2005).

In addition to the classical clathrin-dependent pathways, non-clathrin-dependent endocytosis has also been recently characterized (Le Roy and Wrana, 2005; Pelkmans and Helenius, 2002). These endocytic routes tend to be highly sensitive to cholesterol and can be preferentially inhibited by interfering with cholesterol using drugs such as nystatin or β-methyl cyclodextrin. Little is known of the molecular

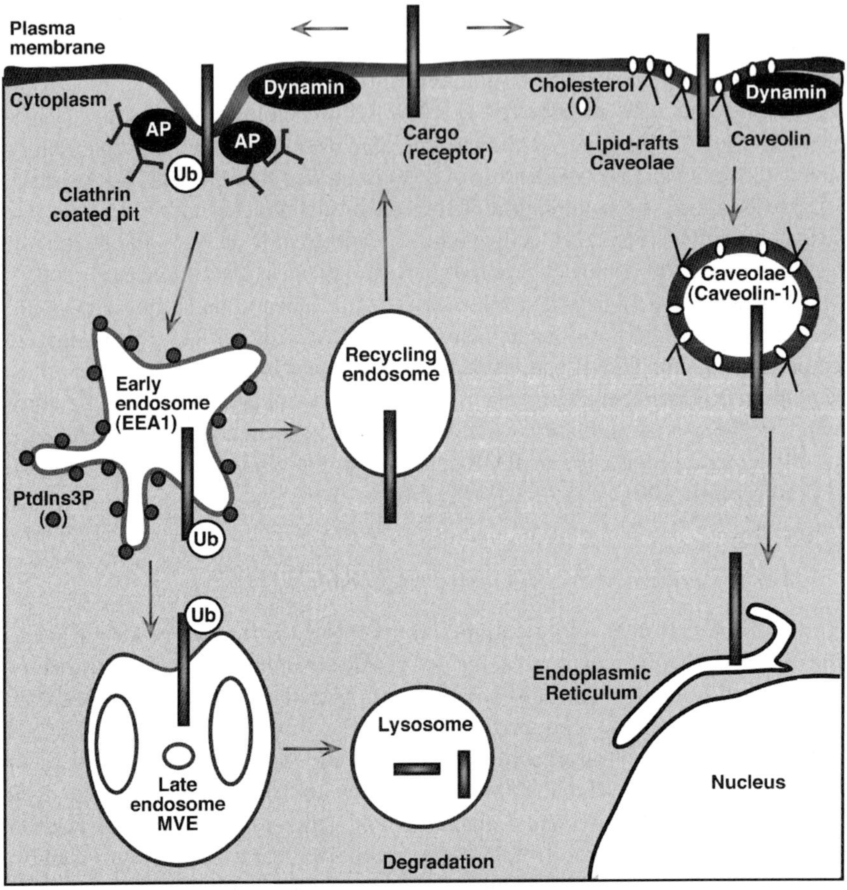

Figure 1. Clathrin- and Caveolar-Lipid-Raft-Endocytic Pathways. At the cell surface, cargo receptors are internalized through two principal endocytic pathways, the clathrin- and the caveolar-lipid-raft-routes. In the clathrin-mediated pathway, which is dependent on adaptor proteins (APs), clathrin and dynamin, receptors are internalized in clathrin-coated pits and are endocytosed into early endosomes that are enriched in phosphoinositol-3-phosphate (PtdIns3P; grey circles). From there, mono-ubiquitinated modified receptors are sorted on one hand into late/multivesicular endosomes and then lysosomes where they are degraded. On the other hand, cargo molecules are sorted from early endosomes into recycling endosomes, from where they are sent back to the cell surface. In the caveolar-lipid-raft-mediated pathway, cargo molecules are internalized into membrane microdomains, which are enriched in cholesterol (white ovals) and caveolin protein. Then, cargo receptors are endocytosed into caveolae, which are enriched in caveolin protein and reach their final destinations such as the endoplasmic reticulum

pathways that control these endocytic pathways, however they are dependent on dynamin and can lead receptors to various cellular compartments such as the endoplasmic reticulum. Furthermore, the sensitivity of these pathways to cholesterol has led to the suggestion that they are raft-dependent, although this has not been directly demonstrated. Lipid rafts are cholesterol-rich microdomains in the cell membrane, which are resistant to detergent and can be separated by floating detergent extracts of cells on sucrose cushions. One subset of these non-clathrin pathways involves the protein caveolin, which forms flask-shaped invaginations at the cell surface that can lead to internalization. Interestingly, motile caveolin-positive vesicles have been visualized and the rate of endocytosis through this pathway is controlled by a balance of caveolin-1, and the raft lipids, cholesterol and glycosphingolipids (Sharma et al., 2004) and a network of kinases (Pelkmans and Zerial, 2005).

Cell surface TGF-β receptors are subjected to internalization through clathrin- and non-clathrin-dependent pathways (Di Guglielmo et al., 2003; Le Roy and Wrana, 2005; Zwaagstra et al., 2001) (Fig. 2). Initial work on the TGF-β type II receptor indicated that it internalized *via* a dileucine motif that interacts with AP2 in the classical clathrin-dependent endocytic pathway (Ehrlich et al., 2001; Yao et al., 2002) and through its interactions with the cytosolic tail of betaglycan (an accessory receptor for TGF-β ligands), which binds the endocytic protein β-arrestin-2 (Chen et al., 2003). Once internalized, TGF-β receptors are present in EEA1-positive early endosomes where SARA can recruit the R-Smad, Smad2 (Di Guglielmo et al., 2003; Hayes et al., 2002; Itoh et al., 2002; Panopoulou et al., 2002). This recruitment is promoted by a cytoplasmic form of the promyelocytic leukaemia protein (cPML), which stimulates the enrollment of both the TGF-β receptor complex and SARA into the early endosome (Lin et al., 2004). These data suggest an important role for receptor trafficking in TGF-β signal transduction. Indeed, mislocalization of Smad2 using a dominant negative mutant of SARA interferes with TGF-β signal transduction (Tsukazaki et al., 1998), as does disrupting SARA localization using wortmannin, which inhibits PI-3' kinase and thus leads to loss of PtdIns3P (Itoh et al., 2002). Furthermore, blocking TGF-β receptor trafficking *via* clathrin coated pits, by using potassium depletion or dominant negative forms of the protein dynamin or Eps15, interferes with Smad activation (Di Guglielmo et al., 2003; Le Roy and Wrana, 2005). Consistent with this, analysis of cells mutant for Hrs/Hgs, which plays an important role in clathrin-dependent trafficking (Gruenberg and Stenmark, 2004), also have defects in TGF-β signal transduction (Miura et al., 2000). Furthermore, dissociation of Smad2 from SARA, which requires receptor-dependent phosphorylation also depends on endocytosis (Runyan et al., 2005). Thus, trafficking of TGF-β receptors *via* the clathrin pathway into the early endosome promotes Smad activation by co-localizing receptors with downstream signaling components.

While analysis of the clathrin-dependent endocytic pathway demonstrates its key role in bringing TGF-β receptors to the SARA-Smad2 complex, other work suggests that TGF-β receptors may also be present in lipid rafts (Di Guglielmo et al., 2003; Ito et al., 2004; Zwaagstra et al., 2001) and interact with caveolin (Razani et al., 2001; Schwartz et al., 2005). Furthermore, this latter pathway is inhibitory to TGF-β signal transduction (Di Guglielmo et al., 2003; Zhang et al., 2005), suggesting it may function in opposition to the clathrin-dependent pathway. In support of this, interfering with clathrin-dependent endocytosis not only blocks TGF-β signal transduction (Di Guglielmo et al., 2003; Itoh et al., 2002; Runyan et al., 2005; Zhang et al., 2005; Zwaagstra et al., 2001), but also leads

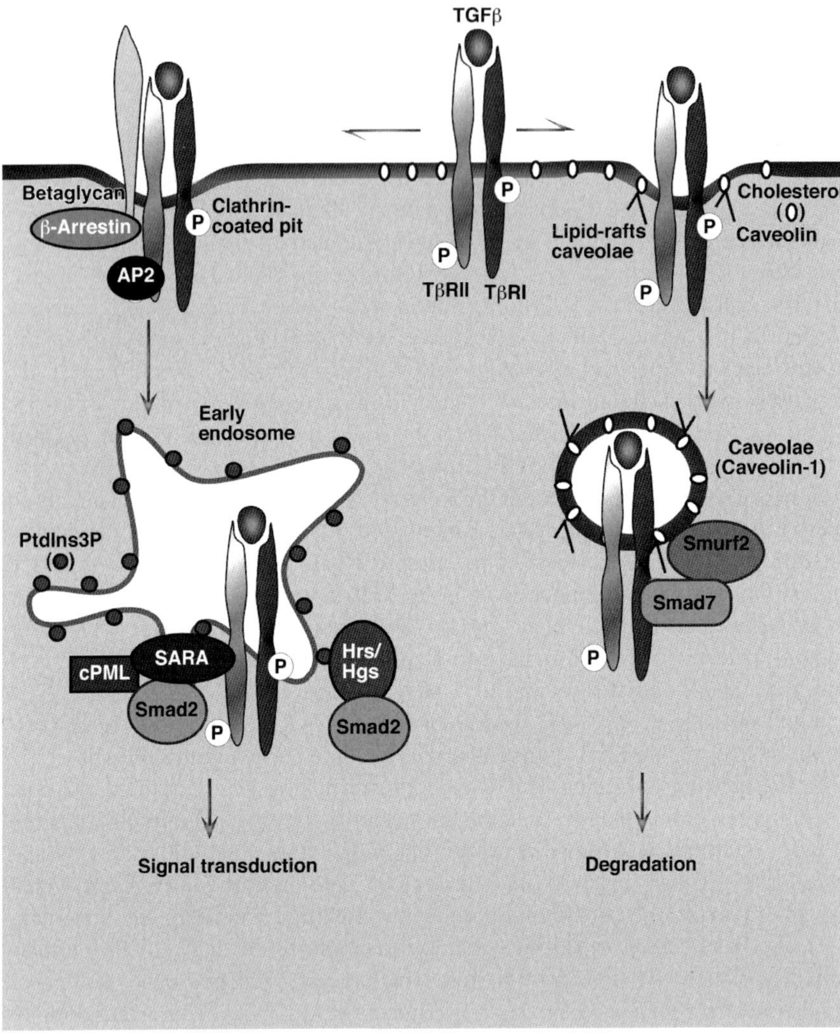

Figure 2. TGF-β Receptor Trafficking and Signaling. Upon TGF-β stimulation, cell surface ligand-bound TGF-β type II receptor (TβRII) recruits TGF-β type I receptor (TβRI) and the complex is internalized by two endocytic pathways; the clathrin- and lipid-raft-caveolae-endocytic routes. In the clathrin endocytic pathway, TβRII interacts with the betaglycan, which is associated with the β-arrestin 2 protein, and forms a complex with the Adaptor Protein 2 (AP2). Then, TGF-β receptor complex enters into early endosomes that are enriched in phosphoinositol-3-phosphate (PtdIns3P; grey circles), which are bound by the FYVE domain proteins called SARA and Hrs/Hgs. These proteins, which are also associated with R-Smads allow the phosphorylation of Smad2 by TGF-β receptor complex (event facilitated by cPML in the context of SARA/Smad2 complex) and subsequently TGF-β signal transduction. On the other hand, cell surface TGF-β receptor complex are present in the lipid-raft microdomains that are enriched in cholesterol (white ovals). They are endocytozed into caveolae, which are characterized by the presence of the caveolin-1 protein, and interact with the Smad7/Smurf2 complex that leads to their degradation

to enhanced TGF-β receptor raft localization (Di Guglielmo et al., 2003). This indicates that the clathrin endocytic route also functions to remove receptors from the inhibitory raft-caveolar pathway. In addition to their localization in the EEA1-positive compartment, TGF-β receptors were observed in caveolin-positive vesicles (Di Guglielmo et al., 2003; Schwartz et al., 2005), indicating that receptors endocytose *via* both clathrin and non-clathrin-dependent pathways. Moreover, analysis of the raft distribution and localization of Smad7-Smurf2 complexes, which target the activated receptors for degradation, shows that they reside and interact with TGF-β receptors in the raft-caveolin-positive compartment (Di Guglielmo et al., 2003; Ito et al., 2004). These data point to a complex pattern of TGF-β endocytosis in which a non-clathrin-dependent pathway mediates inhibition of TGF-β signal transduction and turnover of the receptor, whereas the clathrin-dependent pathway removes receptors from rafts and promotes signal transduction from the early endosome. These studies thus establish the early endosome as an important TGF-β signaling center and the raft-caveolin-dependent pathways as critical for receptor turnover.

The partitioning of TGF-β receptors into two distinct endocytic pathways, which leads to two separate functions in TGF-β signaling, is a dynamic process that is subjected to regulation (Di Guglielmo et al., 2003). Indeed, blockage of the clathrin endocytic pathway strongly reduces the localization of receptors in early endosomes, shifts the receptors into the lipid-raft compartment, decreases the levels of phosphorylated Smad2 and inhibits TGF-β Smad signaling. In contrast, interference with the lipid-raft endocytic route enhances receptors in the non-raft compartment and in EEA1 positive endosomes, promotes the phosphorylation of Smad2 and TGF-β-dependent signaling and blocks TGF-β receptor degradation. This suggests that TGF-β receptor compartmentalization, trafficking and signaling are subjected to a dynamic balance. Interestingly, overexpression of a wild-type form of SARA, but not a mutant form of SARA that is deleted in its FYVE domain, protects TGF-β receptor from Smad7-Smurf-dependent degradation. This protective effect, suggests that the clathrin endocytic pathway may sequester TGF-β receptors from the lipid raft compartment. Moreover, simultaneous disruption of both clathrin and lipid-raft pathways can rescue the inhibition of TGF-β signaling induced by the simple interference of clathrin endocytosis (Di Guglielmo et al., 2003) and in cells low in caveolin, clathrin-dependent internalization does not seem to be required for TGF-β signaling (Lu et al., 2002). This strongly suggests that the clathrin endocytic pathway opposes the inhibitory role of the lipid-raft compartment and that TGF-β receptor trafficking *via* the clathrin route *per se* is not critical, but may function to sequester receptors from the raft compartment.

How partitioning of TGF-β receptor trafficking into these endocytic routes is regulated is largely unknown. Indeed, substitution of the TβRII ectodomain by the GM-CSF-1 extracellular portion leads the chimeric receptor into the EEA1 compartment (Mitchell et al., 2004). Furthermore, altering the glycosylation of TβRII in Mgat5 mutant cells, accelerates TGF-β receptor endocytosis (Partridge et al., 2004). Thus, the extracellular domain of TβRII seems to be an important determinant in the partitioning of this receptor.

TGF-β receptor trafficking into distinct endocytic pathways, which fulfill different functions, may explain how TGF-β ligands act as morphogens. In developmental context, cells have to precisely and constantly assess the morphogen concentration and respond by a ratchet mechanism to the highest concentration that they experience and switch gene response when very few of their receptors are occupied by ligand. In the case of Activin, which is a morphogen that activates a TGFβ-like Smad pathway, cells respond by using such a ratchet behavior by detecting the absolute occupancy of TGF-β receptors (Dyson and Gurdon, 1998). The presence of two distinct TGF-β receptor endocytic pathways, which uncouples TGF-β signaling and turnover, could thus account for cellular ratcheting and sustained Smad responses.

5. TRAFFICKING IN POLARIZED EPITHELIAL CELLS

Early work demonstrating trafficking of TGF-β receptors was performed in cells that did not display apical-basal polarity. Interestingly, analysis of receptor localization in polarized epithelial cells has revealed that TβRI and TβRII each have distinct cell surface distributions. Regarding receptors strictly at the apical surface of polarized mouse mammary NMuMG cells, TβRII is distributed over the entire surface whereas TβRI is restricted to a discrete band around the apical periphery that colocalizes with tight junction components (Ozdamar et al., 2005). Upon TGF-β stimulation, cell-surface TβRII redistributes to the tight junctions. Furthermore, occludin, a structural component of tight junctions, interacts with TβRII only upon TGF-β stimulation, but constitutively interacts with TβRI and can modulate TβRI localization to junctions (Barrios-Rodiles et al., 2005). Many polarized epithelial cell types, such as NMuMG cells, acquire fibroblast characteristics upon TGF-β stimulation in a process known as epithelial-to-mesenchymal transition, or EMT (Valcourt et al., 2005). The distinct distribution of TGFβ receptors appears to be critical, as EMT is inhibited by overexpression of an occluding mutant that can induce TβRI mislocalization (Barrios-Rodiles et al., 2005).

TGF-β-dependent EMT is a multi-branched process that occurs during development and is also associated with the progression of carcinomas to an invasive and metastatic state (Grunert et al., 2003). It involves a change from an epithelial to a mesenchymal transcription program, a process that is known to require Smad signaling (Valcourt et al., 2005). EMT also requires that epithelial cells lose their tight junctions; this process is thought to be regulated by TGF-β receptor signaling at the tight junctions, thereby causing phosphorylation of the polarity regulator Par6, which in turn recruits Smurf1 to the tight junction regions and leads to the degradation of such targets as RhoA (Ozdamar et al., 2005). Moreover, this appears to be a Smad-independent pathway, although TGF-β signaling complexes at cell junctions may regulate both the Par6 and Smad pathways as R-Smads have been shown to colocalize with cell-cell junctions in certain cases (Aubin et al., 2004).

Basolateral cell-surface receptors may also play key roles in TGF-β regulation of epithelial plasticity. In other polarized epithelial cell types, cell-surface TβRI has been detected at the basolateral membrane (Murphy et al., 2004) and TβRII has been shown to interact with the adherens junction components E-cadherin and β-catenin, strictly in the absence of TGF-β stimulation (Tian and Phillips, 2002). These cell types exhibit greater responses to TGF-β when it is specifically applied to the basolateral surface versus the apical surface; however, the TGF-β ligand is secreted apically (Murphy et al., 2004; Tian and Phillips, 2002). Thus, it is conceivable that the relative distribution of apical versus basolateral cell-surface receptors regulates TGF-β-dependent processes such as EMT in a way analogous to the basolateral ErbB2 receptor and its apically-secreted ligand, heregulin (Vermeer et al., 2003). In this model, apically secreted TGF-β binds to apical TβRII, which re-distributes to the tight junctions, activates TβRI and begins EMT, possibly initiating access of TGF-β ligands to receptors at the basolateral surface, thereby coordinating loss of both tight and adherens junctions during EMT.

6. FUTURE PERSPECTIVE

We have only begun to scratch the surface of understanding the molecular pathways that control TGF-β receptor trafficking and its relationship to the cellular response. These studies, conducted in only a handful of model cell systems, have tended towards attempts to build and generally apply specific models that relate trafficking pathways to Smad activation. The complexity of TGF-β biology, however, promises to confound attempts to apply general models across diverse cellular systems. Therefore, understanding the molecular pathways that regulate TGF-β receptor trafficking and function is clearly an important future goal, as is understanding how these pathways contribute to TGF-β family function during development. For example, cell type or context-specific modulators may alter the activity or function of trafficking pathways and thus the qualitative and quantitative nature of TGF-β family signaling. Also of interest will be to understand how receptor trafficking intersects with intracellular pathways that control Smad inactivation and nucleocytoplasmic shuttling. Finally, one of the hallmarks of late stage cancer cells is the switch of TGF-β biological function from a tumor suppressive role to a tumor promoting role with the concomitant production of autocrine TGF-β. Therefore, understanding how cell surface receptor trafficking is altered as a function of constitutive TGF-β signaling and how this contributes cell autonomously to the aggressiveness of cancer is another major area of future investigation.

The recent discovery of the complex relationship between TGF-β receptor endocytosis and signaling has thus ushered in a dynamic view of the Ser/Thr kinase receptor system in which location of receptor activity plays a critical role in modulating the nature of the biological response to TGF-β. These findings suggest that modulating receptor trafficking may be an ideal target for therapeutic intervention. Defining the molecular mechanisms underlying this dynamic system will be critical for this goal and will undoubtedly yield many exciting surprises in the future.

ACKNOWLEDGEMENTS

Work in J.L.W.'s lab is supported by grants from the CIHR, NCIC and Genome Canada. J.L.W. is an International Scholar of the HHMI.

REFERENCES

Attisano, L., and Wrana, J.L., 1996, Signal transduction by members of the transforming growth factor-β superfamily. *Cytokine Growth Factor Rev* 7: 327-339.

Aubin, J., Davy, A., and Soriano, P., 2004, In vivo convergence of BMP and MAPK signaling pathways: impact of differential Smad1 phosphorylation on development and homeostasis. *Genes Dev* 18: 1482-1494.

Barrios-Rodiles, M., Brown, K.R., Ozdamar, B., Bose, R., Liu, Z., Donovan, R.S., Shinjo, F., Liu, Y., Dembowy, J., Taylor, I.W., Luga, V., Przulj, N., Robinson, M., Suzuki, H., Hayashizaki, Y., Jurisica, I., and Wrana, J.L., 2005, High-throughput mapping of a dynamic signaling network in mammalian cells. *Science* 307: 1621-1625.

Bonni, S., Wang, H.R., Causing, C.G., Kavsak, P., Stroschein, S.L., Luo, K., and Wrana, J.L., 2001, TGF-β induces assembly of a Smad2-Smurf2 ubiquitin ligase complex that targets SnoN for degradation. *Nat Cell Biol* 3: 587-595.

Chen, W., Kirkbride, K.C., How, T., Nelson, C.D., Mo, J., Frederick, J.P., Wang, X.F., Lefkowitz, R.J., and Blobe, G.C., 2003, β-arrestin 2 mediates endocytosis of type III TGF-β receptor and down-regulation of its signaling. *Science* 301: 1394-1397.

Chen, Y.G., Hata, A., Lo, R.S., Wotton, D., Shi, Y., Pavletich, N., and Massagué, J., 1998, Determinants of specificity in TGF-β signal transduction. *Genes Dev* 12: 2144-2152.

Chong, P.A., Ozdamar, B., Wrana, J.L., and Forman-Kay, J.D., 2004, Disorder in a target for the smad2 mad homology 2 domain and its implications for binding and specificity. *J Biol Chem* 279: 40707-40714.

Ciechanover, A., Orian, A., and Schwartz, A.L., 2000, Ubiquitin-mediated proteolysis: biological regulation via destruction. *Bioessays* 22: 442-451.

Conner, S.D., and Schmid, S.L., 2003, Regulated portals of entry into the cell. *Nature* 422: 37-44.

Derynck, R., and Zhang, Y.E., 2003, Smad-dependent and Smad-independent pathways in TGF-β family signalling. *Nature* 425: 577-584.

Di Guglielmo, G.M., Le Roy, C., Goodfellow, A.F., and Wrana, J.L., 2003, Distinct endocytic pathways regulate TGF-β receptor signalling and turnover. *Nat Cell Biol* 5: 410-421.

Dyson, S., and Gurdon, J.B., 1998, The interpretation of position in a morphogen gradient as revealed by occupancy of activin receptors. *Cell* 93: 557-568.

Ebisawa, T., Fukuchi, M., Murakami, G., Chiba, T., Tanaka, K., Imamura, T., and Miyazono, K., 2001, Smurf1 interacts with transforming growth factor-β type I receptor through Smad7 and induces receptor degradation. *J Biol Chem* 276: 12477-12480.

Ehrlich, M., Shmuely, A., and Henis, Y.I., 2001, A single internalization signal from the di-leucine family is critical for constitutive endocytosis of the type II TGF-β receptor. *J Cell Sci* 114: 1777-1786.

Feng, X.H., and Derynck, R., 1997, A kinase subdomain of transforming growth factor-β (TGF-β) type I receptor determines the TGF-β intracellular signaling specificity. *EMBO J* 16: 3912-3923.

Feng, X.H., and Derynck, R., 2005, Specificity and Versatility in TGF- Signaling Through Smads. *Annu Rev Cell Dev Biol.*

Georgi, L.L., Albert, P.S., and Riddle, D.L., 1990, daf-1, a *C. elegans* gene controlling dauer larva development, encodes a novel receptor protein kinase. *Cell* 61: 635-645.

Gilboa, L., Wells, R.G., Lodish, H.F., and Henis, Y.I., 1998, Oligomeric structure of type I and type II transforming growth factor β receptors: homodimers form in the ER and persist at the plasma membrane. *J Cell Biol* 140: 767-777.

Gruenberg, J., and Stenmark, H., 2004, The biogenesis of multivesicular endosomes. *Nat Rev Mol Cell Biol* 5: 317-323.

Grunert, S., Jechlinger, M., and Beug, H., 2003, Diverse cellular and molecular mechanisms contribute to epithelial plasticity and metastasis. *Nat Rev Mol Cell Biol* 4: 657-665.

Hayashi, H., Abdollah, S., Qiu, Y., Cai, J., Xu, Y.Y., Grinnell, B.W., Richardson, M.A., Topper, J.N., Gimbrone, M.A., Jr., Wrana, J.L., and Falb, D., 1997, The MAD-related protein Smad7 associates with the TGFβ receptor and functions as an antagonist of TGFβ signaling. *Cell* 89: 1165-1173.

Hayes, S., Chawla, A., and Corvera, S., 2002, TGF β receptor internalization into EEA1-enriched early endosomes: role in signaling to Smad2. *J Cell Biol* 158: 1239-1249.

Heldin, C.-H., Miyazono, K., and ten Dijke, P., 1997, TGF-β signalling from cell membrane to nucleus through SMAD proteins. *Nature* 390: 465-471.

Imamura, T., Takase, M., Nishihara, A., Oeda, E., Hanai, J., Kawabata, M., and Miyazono, K., 1997, Smad6 inhibits signalling by the TGF-β superfamily. *Nature* 389: 622-626.

Ito, T., Williams, J.D., Fraser, D.J., and Phillips, A.O., 2004, Hyaluronan regulates transforming growth factor-β1 receptor compartmentalization. *J Biol Chem* 279: 25326-25332.

Itoh, F., Divecha, N., Brocks, L., Oomen, L., Janssen, H., Calafat, J., Itoh, S., and ten Dijke, P., 2002, The FYVE domain in Smad anchor for receptor activation (SARA) is sufficient for localization of SARA in early endosomes and regulates TGF-β/Smad signalling. *Genes Cells* 7: 321-331.

Izzi, L., and Attisano, L., 2004, Regulation of the TGFβ signalling pathway by ubiquitin-mediated degradation. *Oncogene* 23: 2071-2078.

Kavsak, P., Rasmussen, R.K., Causing, C.G., Bonni, S., Zhu, H., Thomsen, G.H., and Wrana, J.L., 2000, Smad7 binds to Smurf2 to form an E3 ubiquitin ligase that targets the TGF β receptor for degradation. *Mol Cell* 6: 1365-1375.

Kuratomi, G., Komuro, A., Goto, K., Shinozaki, M., Miyazawa, K., Miyazono, K., and Imamura, T., 2005, NEDD4-2 (neural precursor cell expressed, developmentally down-regulated 4-2) negatively regulates TGF-β (transforming growth factor-β) signalling by inducing ubiquitin-mediated degradation of Smad2 and TGF-β type I receptor. *Biochem J* 386: 461-470.

Lallemand, F., Seo, S.R., Ferrand, N., Pessah, M., L'Hoste, S., Rawadi, G., Roman-Roman, S., Camonis, J., and Atfi, A., 2005, AIP4 restricts transforming growth factor-β signaling through a ubiquitination-independent mechanism. *J Biol Chem* 280: 27645-27653.

Le Roy, C., and Wrana, J.L., 2005, Clathrin- and non-clathrin-mediated endocytic regulation of cell signalling. *Nat Rev Mol Cell Biol* 6: 112-126.

Lemmon, M.A., 2003, Phosphoinositide recognition domains. *Traffic* 4: 201-213.

Lin, H.K., Bergmann, S., and Pandolfi, P.P., 2004, Cytoplasmic PML function in TGF-β signalling. *Nature* 431: 205-211.

Lin, X., Liang, M., and Feng, X.H., 2000, Smurf2 is a ubiquitin E3 ligase mediating proteasome-dependent degradation of Smad2 in transforming growth factor-β signaling. *J Biol Chem* 275: 36818-36822.

Lo, R.S., Chen, Y.G., Shi, Y., Pavletich, N.P., and Massagué, J., 1998, The L3 loop: a structural motif determining specific interactions between SMAD proteins and TGF-β receptors. *EMBO J* 17: 996-1005.

Lo, R.S., and Massagué, J., 1999, Ubiquitin-dependent degradation of TGF-β-activated smad2. *Nat Cell Biol* 1: 472-478.

Lu, Z., Murray, J.T., Luo, W., Li, H., Wu, X., Xu, H., Backer, J.M., and Chen, Y.G., 2002, Transforming growth factor β activates Smad2 in the absence of receptor endocytosis. *J Biol Chem* 277: 29363-29368.

Macias-Silva, M., Abdollah, S., Hoodless, P.A., Pirone, R., Attisano, L., and Wrana, J.L., 1996, MADR2 is a substrate of the TGFβ receptor and its phosphorylation is required for nuclear accumulation and signaling. *Cell* 87: 1215-1224.

Manning, G., Whyte, D.B., Martinez, R., Hunter, T., and Sudarsanam, S., 2002, The protein kinase complement of the human genome. *Science* 298: 1912-1934.

Massagué, J., 1998, TGF-β signal transduction. *Annu Rev Biochem* 67: 753-791.

Massagué, J., and Chen, Y.G., 2000, Controlling TGF-β signaling. *Genes Dev* 14: 627-644.

Mitchell, H., Choudhury, A., Pagano, R.E., and Leof, E.B., 2004, Ligand-Dependent and -Independent TGFβ Receptor Recycling Regulated by Clathrin-mediated Endocytosis and Rab11. *Mol Biol Cell.*

Miura, S., Takeshita, T., Asao, H., Kimura, Y., Murata, K., Sasaki, Y., Hanai, J.I., Beppu, H., Tsukazaki, T., Wrana, J.L., Miyazono, K., and Sugamura, K., 2000, Hgs (Hrs), a FYVE domain protein, is involved in Smad signaling through cooperation with SARA. *Mol Cell Biol* 20: 9346-9355.

Miyazono, K., ten Dijke, P., and Heldin, C.-H., 2000, TGF-β signaling by Smad proteins. *Adv Immunol* 75: 115-157.

Morén, A., Imamura, T., Miyazono, K., Heldin, C.-H., and Moustakas, A., 2005, Degradation of the tumor suppressor Smad4 by WW and HECT domain ubiquitin ligases. *J Biol Chem* 280: 22115-22123.

Murphy, S.J., Dore, J.J., Edens, M., Coffey, R.J., Barnard, J.A., Mitchell, H., Wilkes, M., and Leof, E.B., 2004, Differential trafficking of transforming growth factor-β receptors and ligand in polarized epithelial cells. *Mol Biol Cell* 15: 2853-2862.

Nakao, A., Afrakhte, M., Morén, A., Nakayama, T., Christian, J.L., Heuchel, R., Itoh, S., Kawabata, M., Heldin, N.-E., Heldin, C.-H., and ten Dijke, P., 1997, Identification of Smad7, a TGFβ-inducible antagonist of TGF-β signalling. *Nature* 389: 631-635.

Ogunjimi, A.A., Briant, D.J., Pece-Barbara, N., Le Roy, C., Di Guglielmo, G.M., Kavsak, P., Rasmussen, R.K., Seet, B.T., Sicheri, F., and Wrana, J.L., 2005, Regulation of Smurf2 ubiquitin ligase activity by anchoring the E2 to the HECT domain. *Mol Cell* 19: 297-308.

Ozdamar, B., Bose, R., Barrios-Rodiles, M., Wang, H.R., Zhang, Y., and Wrana, J.L., 2005, Regulation of the polarity protein Par6 by TGFβ receptors controls epithelial cell plasticity. *Science* 307: 1603-1609.

Panopoulou, E., Gillooly, D.J., Wrana, J.L., Zerial, M., Stenmark, H., Murphy, C., and Fotsis, T., 2002, Early endosomal regulation of Smad-dependent signaling in endothelial cells. *J Biol Chem* 277: 18046-18052.

Partridge, E.A., Le Roy, C., Di Guglielmo, G.M., Pawling, J., Cheung, P., Granovsky, M., Nabi, I.R., Wrana, J.L., and Dennis, J.W., 2004, Regulation of cytokine receptors by Golgi N-glycan processing and endocytosis. *Science* 306: 120-124.

Pelkmans, L., and Helenius, A., 2002, Endocytosis via caveolae. *Traffic* 3: 311-320.

Pelkmans, L., and Zerial, M., 2005, Kinase-regulated quantal assemblies and kiss-and-run recycling of caveolae. *Nature* 436: 128-133.

Podos, S.D., Hanson, K.K., Wang, Y.C., and Ferguson, E.L., 2001, The DSmurf ubiquitin-protein ligase restricts BMP signaling spatially and temporally during *Drosophila* embryogenesis. *Dev Cell* 1: 567-578.

Razani, B., Zhang, X.L., Bitzer, M., von Gersdorff, G., Böttinger, E.P., and Lisanti, M.P., 2001, Caveolin-1 regulates transforming growth factor (TGF)-β/SMAD signaling through an interaction with the TGF-β type I receptor. *J Biol Chem* 276: 6727-6738.

Runyan, C.E., Schnaper, H.W., and Poncelet, A.C., 2005, The role of internalization in transforming growth factor β1-induced Smad2 association with Smad anchor for receptor activation (SARA) and Smad2-dependent signaling in human mesangial cells. *J Biol Chem* 280: 8300-8308.

Schwartz, E.A., Reaven, E., Topper, J.N., and Tsao, P.S., 2005, Transforming growth factor-β receptors localize to caveolae and regulate endothelial nitric oxide synthase in normal human endothelial cells. *Biochem J* 390: 199-206.

Seo, S.R., Lallemand, F., Ferrand, N., Pessah, M., L'Hoste, S., Camonis, J., and Atfi, A., 2004, The novel E3 ubiquitin ligase Tiul1 associates with TGIF to target Smad2 for degradation. *EMBO J* 23: 3780-3792.

Sharma, D.K., Brown, J.C., Choudhury, A., Peterson, T.E., Holicky, E., Marks, D.L., Simari, R., Parton, R.G., and Pagano, R.E., 2004, Selective stimulation of caveolar endocytosis by glycosphingolipids and cholesterol. *Mol Biol Cell* 15: 3114-3122.

Shi, Y., and Massagué, J., 2003, Mechanisms of TGF-β signaling from cell membrane to the nucleus. *Cell* 113: 685-700.

Stenmark, H., and Aasland, R., 1999, FYVE-finger proteins–effectors of an inositol lipid. *J Cell Sci* 112 (Pt 23): 4175-4183.

Stenmark, H., Aasland, R., Toh, B.H., and D'Arrigo, A., 1996, Endosomal localization of the autoantigen EEA1 is mediated by a zinc-binding FYVE finger. *J Biol Chem* 271: 24048-24054.

Tian, Y.C., and Phillips, A.O., 2002, Interaction between the transforming growth factor-β type II receptor/Smad pathway and β-catenin during transforming growth factor-β1-mediated adherens junction disassembly. *Am J Pathol* 160: 1619-1628.

Tsukazaki, T., Chiang, T.A., Davison, A.F., Attisano, L., and Wrana, J.L., 1998, SARA, a FYVE domain protein that recruits Smad2 to the TGFβ receptor. *Cell* 95: 779-791.

Valcourt, U., Kowanetz, M., Niimi, H., Heldin, C.-H., and Moustakas, A., 2005, TGF-β and the Smad signaling pathway support transcriptomic reprogramming during epithelial-mesenchymal cell transition. *Mol Biol Cell* 16: 1987-2002.

Wang, H.R., Zhang, Y., Ozdamar, B., Ogunjimi, A.A., Alexandrova, E., Thomsen, G.H., and Wrana, J.L., 2003, Regulation of cell polarity and protrusion formation by targeting RhoA for degradation. *Science* 302: 1775-1779.

Vermeer, P.D., Einwalter, L.A., Moninger, T.O., Rokhlina, T., Kern, J.A., Zabner, J., and Welsh, M.J., 2003, Segregation of receptor and ligand regulates activation of epithelial growth factor receptor. *Nature* 422: 322-326.

Wicks, S.J., Haros, K., Maillard, M., Song, L., Cohen, R.E., Dijke, P.T., and Chantry, A., 2005, The deubiquitinating enzyme UCH37 interacts with Smads and regulates TGF-β signalling. *Oncogene*.

Wrana, J.L., 2000, Regulation of Smad activity. *Cell* 100: 189-192.

Wrana, J.L., Attisano, L., Wieser, R., Ventura, F., and Massagué, J., 1994, Mechanism of activation of the TGF-β receptor. *Nature* 370: 341-347.

Wu, G., Chen, Y.G., Ozdamar, B., Gyuricza, C.A., Chong, P.A., Wrana, J.L., Massagué, J., and Shi, Y., 2000, Structural basis of Smad2 recognition by the Smad anchor for receptor activation. *Science* 287: 92-97.

Yamashita, H., ten Dijke, P., Franzen, P., Miyazono, K., and Heldin, C.-H., 1994, Formation of hetero-oligomeric complexes of type I and type II receptors for transforming growth factor-β. *J Biol Chem* 269: 20172-20178.

Yamashita, M., Ying, S.X., Zhang, G.M., Li, C., Cheng, S.Y., Deng, C.X., and Zhang, Y.E., 2005, Ubiquitin ligase Smurf1 controls osteoblast activity and bone homeostasis by targeting MEKK2 for degradation. *Cell* 121: 101-113.

Yao, D., Ehrlich, M., Henis, Y.I., and Leof, E.B., 2002, Transforming growth factor-β receptors interact with AP2 by direct binding to β2 subunit. *Mol Biol Cell* 13: 4001-4012.

Zhang, X.L., Topley, N., Ito, T., and Phillips, A., 2005, Interleukin-6 regulation of transforming growth factor (TGF)-β receptor compartmentalization and turnover enhances TGF-β1 signaling. *J Biol Chem* 280: 12239-12245.

Zhang, Y., Chang, C., Gehling, D.J., Hemmati-Brivanlou, A., and Derynck, R., 2001, Regulation of Smad degradation and activity by Smurf2, an E3 ubiquitin ligase. *Proc Natl Acad Sci U S A* 98: 974-979.

Zhu, H., Kavsak, P., Abdollah, S., Wrana, J.L., and Thomsen, G.H., 1999, A SMAD ubiquitin ligase targets the BMP pathway and affects embryonic pattern formation. *Nature* 400: 687-693.

Zwaagstra, J.C., El-Alfy, M., and O'Connor-McCourt, M.D., 2001, Transforming growth factor (TGF)-β 1 internalization: modulation by ligand interaction with TGF-β receptors types I and II and a mechanism that is distinct from clathrin-mediated endocytosis. *J Biol Chem* 276: 27237-27245.

CHAPTER 10

NUCLEOCYTOPLASMIC SHUTTLING OF SMAD PROTEINS

BERNHARD SCHMIERER AND CAROLINE S. HILL

Laboratory of Developmental Signalling, Cancer Research UK London Research Institute, 44 Lincoln's Inn Fields, London, WC2A 3PX, UK

Abstract: Nuclear accumulation of Smad proteins is pivotal for transduction of TGF-β-superfamily signals from transmembrane receptors into the nucleus. It has become clear that the nucleocytoplasmic distributions of Smads, in both the absence and the presence of a TGF-β-superfamily signal, are not static, but that Smads are continuously shuttling between the nucleus and the cytoplasm under both conditions. This chapter presents the evidence for continuous nucleocytoplasmic shuttling of Smads. It goes on to review different mechanisms that have been proposed to mediate Smad nuclear import and export, and discusses the current models for the establishment of the Smads' steady state distributions in the absence and the presence of a TGF-β-superfamily signal. Finally, we address the biological relevance of continuous nucleocytoplasmic shuttling for signaling by TGF-β superfamily members

Keywords: CRM1; karyopherins; nuclear export signal; nuclear localization signal; nucleocytoplasmic transport; nucleoporins; signal sensing; TGF-β

1. INTRODUCTION

Signaling by the transforming growth factor-β (TGF-β) superfamily of ligands is involved in many different cellular processes, including cell proliferation, apoptosis, differentiation, and specification of cell fate. Signaling by this superfamily is not only important in the adult organism, but also plays a crucial role in establishing and patterning the basic body plan during embryogenesis. Many molecular players of the pathway are highly conserved from worm to man (Massagué, 1998).

The TGF-β superfamily of ligands comprises TGF-βs, Activins/Nodals, bone morphogenetic proteins (BMPs) and growth and differentiation factors (GDFs). Signaling occurs through ligand-dependent heteromerization of type II receptors

P. ten Dijke and C.-H. Heldin (eds.), Smad Signal Transduction, 193–213.

with type I receptors, both of which are transmembrane serine-threonine kinases. The constitutively active type II receptors activate the type I receptors by transphosphorylation, which in turn phosphorylate their substrates, the best characterized of which are a subset of Smad proteins (Shi and Massagué, 2003).

Eight distinct Smads have been identified and categorized into 3 functional classes. The first class are the receptor-regulated Smads or R-Smads (Smad1, 2, 3, 5, 8), which are phosphorylated by the type I receptor kinases on an SSXS motif at their extreme C-termini. As a general rule, Smad1, 5 and 8 are activated in response to BMP/GDF signals, whereas Smad2 and Smad3 are activated in response to TGF-β, Activin and Nodal signals. Such differential activation is accomplished by distinct substrate specificities of different type I receptors. In endothelial cells, TGF-β can also activate Smads 1, 5 and 8 through the tissue-specific type I receptor ALK-1 (Goumans et al., 2003). Following phosphorylation, the R-Smads form either homomeric complexes or heteromeric complexes with Smad4, which is the only member of the second functional class of Smads, the common mediator Smads or co-Smads. Both homomeric and heteromeric Smad complexes then accumulate in the nucleus and heteromeric complexes regulate the transcription of target genes in conjunction with additional co-factors (Shi and Massagué, 2003) (see Chapter 14). The role of homomeric R-Smad complexes is not yet clear. Smad4 accumulation in the nucleus is a direct consequence of R-Smad accumulation via R-Smad/Smad4 complexes. This notion is supported by the fact that Smad4 is dispensable for R-Smad accumulation in response to TGF-β, which occurs with the same efficiency in Smad4 null cells (De Bosscher et al., 2004; Liu et al., 1997; Nicolás and Hill, 2003). The third functional class of Smads is the inhibitory Smads or I-Smads (Smad6, 7) that negatively regulate the pathway by competing with R-Smads for receptor binding. Moreover, they mediate receptor inactivation by recruiting either the E3 ubiquitin ligases Smurf1/2 (Ebisawa et al., 2001; Kavsak et al., 2000) or the catalytic subunit of protein phosphatase 1 (Shi et al., 2004) to the receptors (see Chapter 19).

The R-Smads and Smad4 share a common structure (Shi and Massagué, 2003; Fig. 1). They have a highly conserved N-terminal domain, the MH1 domain (for Mad homology domain 1), which in Smad1, 3, 4, 5 and 8 is thought to directly bind DNA at the Smad binding element, GTCT. At the C-terminus is another well-conserved domain, the MH2 domain, which is responsible for receptor–Smad and Smad–Smad interactions and also for many interactions with transcriptional co-activators and co-repressors (Shi and Massagué, 2003) (see Chapter 14). Ligand-dependent phosphorylation at the C-terminal SSXS motif is critical for stabilizing Smad–Smad complexes (Wu et al., 2001). The MH1 and MH2 domains are separated by a less well conserved proline-rich linker region. Smads 6 and 7 contain a recognizable MH2 domain, but have only weak homology with the other Smads in their N-terminal regions.

Together with Smad4, the R-Smads are multifunctional in that they both transduce the signal from the cell membrane to the nucleus *and* modulate gene transcription. This dual role requires stringent control of subcellular Smad localization. The initial

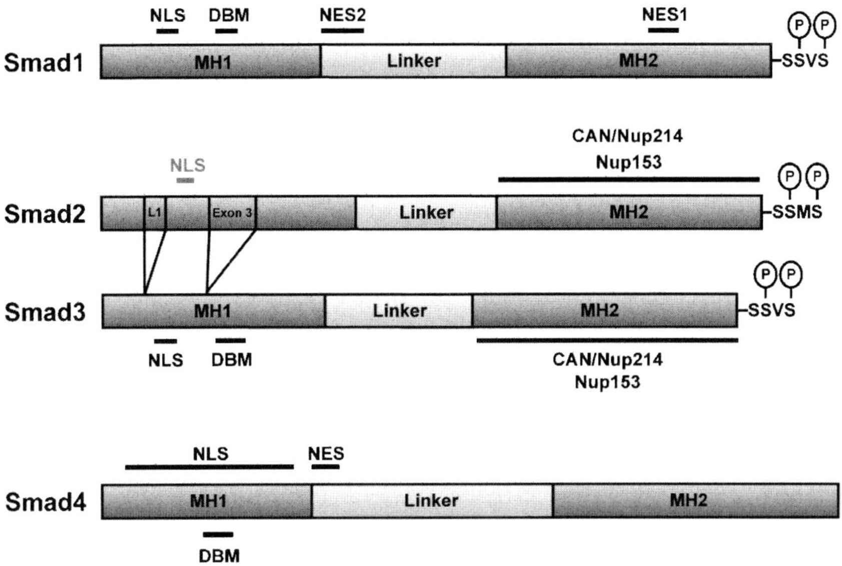

Figure 1. Smads have a conserved domain structure and harbor nuclear localization and nuclear export signals. R-Smads and Smad4 share two conserved domains, the N-terminal Mad-homology 1 (MH1) and the C-terminal MH2 domains. These domains are connected by a less well conserved, proline-rich linker region. Smad2 has two additional inserts in its MH1 domain, L1 and exon 3. In response to a TGF-β-superfamily ligand, the R-Smads, *e.g.* Smads1, 2 and 3, become phosphorylated at the C-terminal SSXS motif as indicated. The position of nuclear localization signals (NLS) and nuclear export signals (NES) as well as the DNA-binding motif (DBM) are indicated. The NLS in Smad2 is present in the sequence but has not been shown to be functional. A "hydrophobic corridor" in the MH2 domain of Smad2 and Smad3 is involved in binding to the nucleoporins CAN/Nup214 and Nup153, but full-length Smad4 seems to be required for interaction with CAN/Nup214 (not shown)

view of the pathway was that Smads would reside exclusively in the cytoplasm in the absence of a signal. Upon active signaling, they would translocate into the nucleus, regulate gene transcription and subsequently be degraded (Lo and Massagué, 1999). More recent evidence, however, challenges this simple, unidirectional view of the pathway. We know now that Smads are continuously crossing the nuclear membrane, even in unstimulated cells. Moreover, there is compelling evidence that the bulk of Smads that have accumulated in the nucleus in response to a signal are not degraded there, but rather engage in highly dynamic shuttling between the nucleus and the cytoplasm. As detailed in this chapter, such shuttling provides an elegant sensing mechanism to ensure that the concentration of active Smad in the nucleus reflects the actual signal strength at any given time.

This chapter reviews the evidence for Smad nucleocytoplasmic shuttling and discusses the different molecular transport mechanisms that have been proposed to account for nuclear import and export of different Smads. We focus on the co-Smad Smad4, on the TGF-β/Activin/Nodal-responsive R-Smads, Smad2 and Smad3, and the BMP/GDF-responsive Smad, Smad1. No studies dealing with nucleocytoplasmic transport for Smad5 and Smad8 have been done to date. We compare existing models that explain how nuclear accumulation of Smads in response to a TGF-β-superfamily ligand is achieved despite continuous Smad shuttling. Finally, the chapter explores the functional role of nucleocytoplasmic shuttling of Smads in TGF-β signaling.

2. EVIDENCE FOR SMAD NUCLEOCYTOPLASMIC SHUTTLING

2.1 Evidence for Shuttling in the Absence of Ligand Stimulation

The first evidence for nucleocytoplasmic shuttling of Smads came from studies on Smad4. In the basal state, *i.e.* in the absence of a TGF-β-superfamily ligand, Smad4 is distributed approximately equally between the nucleus and the cytoplasm; this distribution is, however, not static. Treatment of cells with leptomycin B (LMB), a specific inhibitor of the prototypic export receptor CRM1 or exportin 1 (Fornerod et al., 1997) led to rapid nuclear accumulation of Smad4 (Pierreux et al., 2000; Watanabe et al., 2000). This simple experiment shows that (1) CRM1 activity is required for nuclear export of Smad4, and (2) Smad4 must be continuously shuttling between the nucleus and the cytoplasm in the absence of a TGF-β signal. The predominantly cytoplasmic localization of the TGF-β/Activin/Nodal-responsive R-Smads remains unaffected by LMB treatment (Pierreux et al., 2000; Xu et al., 2002). Hence, there is no such simple way of determining whether R-Smads are shuttling between the nucleus and the cytoplasm, and more sophisticated approaches were required. The TGF-β/Activin/Nodal-responsive Smad2 is the best studied R-Smad in terms of nucleocytoplasmic shuttling. One study (Xu et al., 2002) took advantage of a quantitative nuclear transport reporter system (Coburn et al., 2001) as well as of heterokaryon assays to show the constitutive shuttling capability of Smad2 and to map the major export activity to the linker-MH2 domain. More recently, fluorescence perturbation experiments, *i.e.* photobleaching and photoactivation of GFP-fusions of Smad2 have provided direct visual proof of the continuous exchange between the nuclear and the cytoplasmic pools of Smad2 (Nicolás et al., 2004; Schmierer and Hill, 2005). Taken together, there is compelling evidence that Smad2 is continuously shuttling between the cytoplasm and the nucleus in the absence of a TGF-β signal.

2.2 Evidence for Shuttling in Ligandstimulated Cells

TGF-β-treatment triggers comparably slow nuclear accumulation of Smad2, Smad3 and Smad4, which reaches a maximum after about 45 minutes of signaling (Inman et al., 2002b). Once accumulated, the bulk of Smads remains maximally nuclear for at least 4 to 5 hours, after which time they start to slowly relocalize to the cytoplasm (Pierreux et al., 2000). The kinetics of Smad relocalization to the cytoplasm is thought to mirror the inactivation of the receptors and hence the slow fading of the signal. Indeed, if signaling is terminated prematurely by adding the specific type I receptor inhibitor SB-431542 (Inman et al., 2002a; Laping et al., 2002) to actively signaling cells, nuclear accumulation of Smads starts to decrease immediately, causing the bulk of Smads to relocalize to the cytoplasm within 2 hours of SB-431542 treatment (Inman et al., 2002b) (Fig. 2). This observation begs the question of how the nuclear Smads "know" that the receptors have been inactivated. The simplest explanation is continuous shuttling of Smads, even in actively signaling cells that are in a state of maximal nuclear Smad accumulation. This notion has subsequently been demonstrated by photobleaching of EGFPSmad2 and EGFPSmad4 in the cytoplasm of TGF-β-induced cells which show maximal nuclear accumulation of Smads (Nicolás et al., 2004).

Such extended cytoplasmic photobleaching causes nuclear bleaching, thus proving the continuous exchange of EGFPSmad2 and EGFPSmad4 between both compartments in the presence of a signal (Nicolás et al., 2004). In a more direct approach, Smad2 fused to photoactivatable GFP (PAGFP; Patterson and Lippincott-Schwartz, 2002) was photoactivated by multiphoton excitation in a compartment-specific manner. Within minutes, PAGFPSmad2 that had been photoactivated in the nucleus appeared in the cytoplasm, and cytoplasmically photoactivated PAGFPSmad2 appeared in the nucleus, in both TGF-β-treated and untreated cells (Schmierer and Hill, 2005). These results demonstrate that Smad2 is trickling in and out of the nucleus even in TGF-β-treated cells that have reached maximal nuclear accumulation. How is nuclear accumulation maintained for several hours under these circumstances?

Two possible mechanisms can be envisaged to explain sustained nuclear accumulation of Smad2 despite continuous exchange with the cytoplasmic pool. Cytoplasmic and nuclear pools of phosphorylated Smad2 could exist, which are in dynamic equilibrium with each other. Alternatively, sustained nuclear accumulation of Smad2 could require cycles of Smad2 phosphorylation and dephosphorylation. Several lines of evidence suggest that the latter mechanism is predominant. The Smad2 that relocalizes to the cytoplasm upon inactivation of the signal is unphosphorylated, suggesting that nuclear export is accompanied by dephosphorylation (Inman et al., 2002b; Xu et al., 2002). Moreover, if cells are treated with TGF-β for 1 hour to allow maximal nuclear accumulation of R-Smad/Smad4 complexes, and then treated with LMB to inhibit CRM1, as well as with SB-431542 to inactivate the receptor kinase, Smad4, as expected, stays trapped in the nucleus but Smad2 still redistributes to the cytoplasm (Inman et al., 2002b). These results provide strong evidence that R-Smad/Smad4 complexes dissociate in the nucleus, perhaps as a result of R-Smad dephosphorylation (Wu et al., 2001), and that monomeric R-Smads and Smad4 are then exported separately and by different mechanisms. In agreement with this idea, CRM1-dependent Smad4 export is thought to be inhibited by complex formation of Smad4 with R-Smads, which physically prevents Smad4 from interaction with CRM1 (Chen et al., 2005).

These results suggest that, in signaling cells that have reached maximal nuclear accumulation of Smads, there is continuous, if low level, R-Smad dephosphorylation. While the bulk of Smads remains complexed and nuclear, at any one time

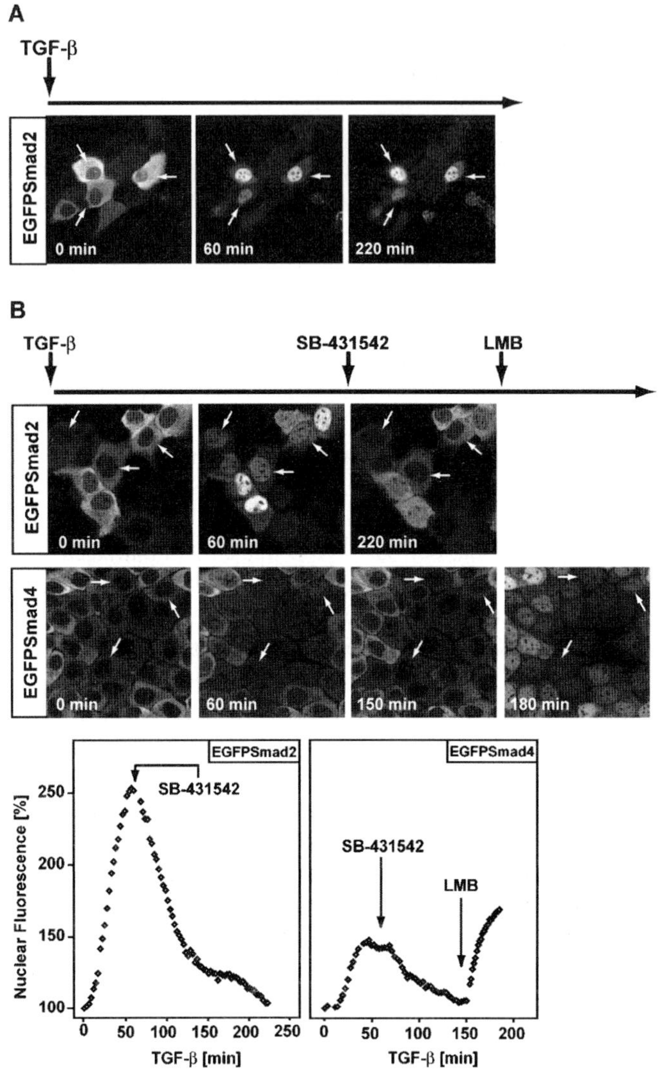

Figure 2. The Smads shuttle between the cytoplasm and nucleus in TGF-β-stimulated cells. *A.* A human keratinocyte cell line (HaCaT) stably expressing EGFPSmad2 was incubated with TGF-β for 220 min. EGFPSmad2 accumulates in the nucleus within 1 hour and remains nuclear throughout the incubation period. *B.* HaCaT cell lines stably expressing EGFPSmad2 or EGFPSmad4 were incubated with TGF-β for 1 h, followed by the type I receptor inhibitor SB-431542 for up to 160 min. In the case of the EGFPSmad4 cells, leptomycin B (LMB) was added 90 min after SB-431542 addition (150 min time point). Fluorescence images are shown at different time points after initial TGF-β treatment. Below are graphs showing quantitation of nuclear fluorescence, with fluorescence images collected every 3 min. The data show that sustained TGF-β-induced nuclear accumulation of EGFPSmads as seen in *A* is dependent upon continuous receptor activity. They also show that Smad4 is continuously shuttling in the absence of a TGF-β signal, as inhibiting nuclear export of Smad4 in SB-431542-treated cells by LMB is sufficient to induce Smad4 nuclear accumulation. Adapted from Nicolás et al., 2004, with permission

R-Smad dephosphorylation causes a small fraction of the nuclear Smad complexes to dissociate. Their monomeric subunits are then exported into the cytoplasm. If the receptors are still active, the R-Smads are rephosphorylated, form complexes and return to the nucleus. If the receptors are no longer active, the Smads slowly reaccumulate back in the cytoplasm. Hence, a constant trickle of Smads in and out of the nucleus maintains sustained nuclear accumulation of Smads that is only reversed when the receptors are inactivated (Inman et al., 2002b; Schmierer and Hill, 2005).

3. GENERAL PRINCIPLES OF NUCLEOCYTOPLASMIC TRANSPORT

Several transport mechanisms governing the translocation step of proteins across the nuclear membrane have been described. This section briefly introduces the general principles of the nucleocytoplasmic transport mechanisms that are relevant to nucleocytoplasmic shuttling of the Smads. (For detailed information on nucleocytoplasmic transport see Görlich and Kutay, 1999; Chook and Blobel, 2001; Ström and Weis, 2001; Pemberton and Paschal, 2005.)

Nucleocytoplasmic transport proceeds through the nuclear pore complex (NPC), which spans the nuclear envelope and forms an aqueous channel. In mammalian cells, NPCs are made up of multiple copies of roughly 30 different nucleoporins (Cronshaw et al., 2002). Only relatively small proteins can pass through the NPC by diffusion; proteins larger than ~20-30 kD need to be actively transported through the nuclear pore. Interaction of transport substrates with the NPC and their subsequent translocation is frequently mediated by transport receptors, termed karyopherins. These soluble proteins are able to bind cargo molecules and can interact with FG-dipeptide repeats, which are a common structural motif found in many nucleoporins. Such interactions enable the transporter-cargo complexes to pass through the nuclear pore. According to the transport direction, karyopherins can be classified into importins and exportins. However some karyopherins can mediate transport in both directions.

3.1 The RanGTP Gradient

Directionality of nucleocytoplasmic transport is established by the RanGTP gradient (Görlich and Kutay, 1999; Fig. 3). Ran is a small GTPase, which is predominantly in its GTP-bound state in the nucleus, but in its GDP-bound state in the cytoplasm. The RanGTP gradient is established and maintained because the GTPase activating protein, RanGAP, resides exclusively in the cytoplasm and depletes cytoplasmic RanGTP with the help of the cytoplasmic Ran binding protein 1 (RanBP1). Conversely, the Ran guanine nucleotide exchange factor, RCC1, is chromatin bound and hence resides exclusively in the nucleus, maintaining high nuclear levels of RanGTP. RanGDP shuttles in and out of the nucleus with the help of a specific transporter, the nuclear transport factor 2 (NTF2).

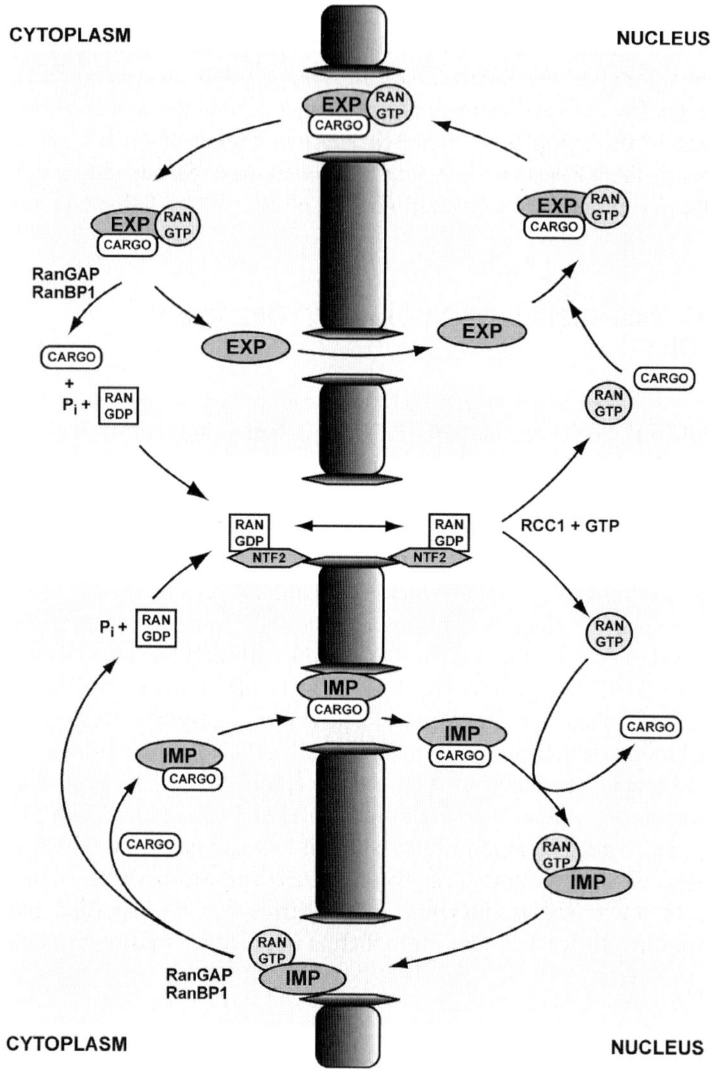

Figure 3. Nuclear import and export cycles and the involvement of the RanGTPase system. ***Upper half. Nuclear export***. RanGTP is low in the cytoplasm (left hand side) and high in the nucleus (right hand side). The exportin (EXP) binds its cargo in the nucleus together with RanGTP. This trimeric export complex then leaves the nucleus and enters the cytoplasm, where RanGAP together with RanBP1 stimulate Ran to hydrolyze its GTP. Formation of RanGDP leads to dissociation of the export complex, the export cargo is set free in the cytoplasm and the exportin translocates back into the nucleus. RanGDP is transported into the nucleus by the nuclear transport factor 2 (NTF2), where it is able to replace its GDP by GTP with the help of the chromatin-bound guanine nucleotide exchange factor, RCC1. The system is now ready for another export cycle. ***Lower half. Nuclear import***. The importin (IMP) binds the import cargo in the cytoplasm and mediates translocation of the complex into the nucleus. Binding of RanGTP to the complex triggers release of the import cargo into the nucleus. The importin-RanGTP complex translocates back into the cytoplasm, where RanGTP is converted to RanGDP and dissociates from the importin. RanGDP is transported back into the nucleus by NTF2, and the system is ready for another import cycle. Adapted from Görlich and Kutay, 1999

Importins are able to bind their cargo in the cytoplasm without the requirement for RanGTP (Fig. 3). In the classical pathway, the cargo is bound by one of several variants of importin α, which act as adaptors and recruit the actual transport receptor, importin β, forming a trimeric import complex. After transport into the nucleus, RanGTP disrupts the import complex and the cargo protein is released into the nucleoplasm. Exporters in contrast can bind their cargo only in the presence of RanGTP (*i.e.* in the nucleus) by forming a heterotrimeric complex, exporter-cargo-RanGTP. After translocation of the export complex into the cytoplasm, RanGTP hydrolyzes its GTP with the help of cytoplasmic RanGAP and RanBP1, causing dissociation and release of the exported cargo into the cytoplasm (Fig. 3).

3.2 Karyopherin-dependent Nuclear Import and Export

Recognition of cargo proteins by either importins or exportins requires certain sequence motifs. The best described motif responsible for nuclear import is the nuclear localization signal or NLS, a short basic stretch of amino acids rich in lysine and/or arginine residues, which is recognized by importin α-importin β complexes. Two subtypes of classical NLS exist (Dingwall and Laskey, 1991): those related to the SV40 large-T antigen NLS which contain a cluster of basic amino acid residues, and the bipartite NLS, which consists of two basic clusters as exemplified by the NLS in nucleoplasmin. Nuclear import is however by no means restricted to proteins exhibiting characteristic basic sequence motifs, and it is thought that less well defined features that become apparent only in the tertiary structure of the cargo protein might mediate karyopherin binding in many cases.

A variation of the classical import pathway is used by several proteins which circumvent the need for importin α by directly binding to importin β (Cingolani et al., 2002), for example ribosomal proteins (Jakel and Görlich, 1998), HIV Rev and HIV Tat (Truant and Cullen, 1999) as well as cyclin B1 (Moore et al., 1999; Takizawa et al., 1999) and Smad3 (Kurisaki et al., 2001; Xiao et al., 2000b).

Nuclear export of many proteins is mediated by the classical nuclear export signal or NES, which is defined as a short leucine/isoleucine-rich motif that is recognized by the prototypic export receptor CRM1/exportin 1 (Fornerod et al., 1997). Notably, the interaction of CRM1 with leucine-rich NESs is of low affinity, which is a requirement for efficient dissociation of the export complex (Engelsma et al., 2004).

Other export receptors include exportin 4 (Lipowsky et al., 2000), exportin 5 (Bohnsack et al., 2002) and exportin 6 (Stuven et al., 2003). In contrast to CRM1, these exportins have a very narrow substrate specificity, and to date only exportin 7 has been suggested to transport different, unrelated proteins (Mingot et al., 2004). No defined sequence motifs like the classical NES have been identified in the cargoes of these alternative exporters, and CRM1 substrates remain by far the best characterized group of export cargoes, mainly by virtue of the specific CRM1 inhibitor, LMB.

3.3 Karyopherin-independent Nuclear Transport

In addition to the karyopherin-dependent classical pathway and its variations, some proteins have been reported to enter or exit the nucleus without the need for transport receptors (Xu and Massagué, 2004). Examples of such a transport mechanism are the nuclear import of β-catenin (Fagotto et al., 1998), reviewed in (Xu and Massagué, 2004) and the nuclear import of Erk2 (Matsubayashi et al., 2001; Whitehurst et al., 2002). Proteins that employ this mode of transport are thought to directly contact the nuclear pore complex and enter the nucleus in a similar fashion to karyopherins, *i.e.* by interaction with FG-dipeptide repeats of the nucleoporins.

4. MECHANISM OF SMAD TRANSIT THROUGH THE NUCLEAR PORE

4.1 Smad Import into the Nucleus

Several studies have addressed the translocation mechanism of Smad nuclear import, and it has become apparent that, despite extensive similarities in both structure and amino acid sequence within the Smad family, results obtained for one specific Smad cannot be extended to other family members. Smad import has been studied for Smad1, Smad2, Smad3 and Smad4. There is good evidence for both karyopherin-dependent import with the major import activity residing in the MH1 domain, and karyopherin-independent import mediated by direct contacts between nucleoporins and the MH2 domain.

4.1.1 Evidence for karyopherin-dependent import of Smads

For Smad3, a karyopherin-dependent translocation into the nucleus has been found independently by two groups. One group identified a lysine rich NLS-like sequence in the MH1 domain of Smad3 (^{40}KKLKK^{44}, Fig. 1), which is conserved in all R-Smads (Xiao et al., 2000a). This motif seems to mediate karyopherin-dependent nuclear import of Smad3 by directly binding to importin β (Xiao et al., 2000b). Similarly, another study demonstrated that Smad3 interacted with importin β, and is imported into the nucleus in a Ran- and energy-dependent manner (Kurisaki et al., 2001). Both groups found that nuclear import of the isolated MH1 domain was much stronger than import of the full-length protein in an *in vitro* import assay, and that importin β interacted much more strongly with the isolated MH1 domain than with the full length protein in GST-pulldowns. TGF-β-dependent phosphorylation of Smad3 seemed to enhance its interaction with importin β and pseudo-phosphorylated Smad3 was imported more efficiently than unphosphorylated Smad3 into the nucleus *in vitro* (Kurisaki et al., 2001). The conserved NLS-like motif is non-functional in Smad2, perhaps because of an insertion encoded by exon 3 of Smad2, which is close to the NLS-like motif and might impair its accessibility for importin β binding (Fig. 1). The proposed mechanism deviates from the classical import mechanism as it does not need an importin α protein as a bridging factor between importin β and the cargo. As outlined above, several other examples of proteins that are imported in this fashion have been described.

The corresponding NLS-like motif in the MH1 domain of Smad1 (Fig. 1) has been shown to be important for nuclear import of this BMP-responsive R-Smad (Xiao et al., 2001), but no import receptor has been identified so far and it remains unclear whether the mechanism for Smad1 import is similar to the proposed mechanism for Smad3, *i.e.* if Smad1 directly interacts with importin β without the need for an importin α.

In contrast to Smad3, Smad4 seems to use the classical pathway for nuclear import. An extended, bipartite NLS has been identified in Smad4 (amino acids 45-110), which overlaps with the corresponding sequence motif responsible for Smad1 and Smad3 import, but additionally extends into the DNA-binding region of the Smad4 MH1 domain (Fig. 1). The isolated Smad4 MH1 domain interacts with importin α1 through this motif, which may be important for Smad4 import (Xiao et al., 2003b).

4.1.2 Evidence for direct nucleoporin interaction

Xu et al. (2002) have proposed a very different mechanism for the import of Smad2, which is in accordance with the finding that Smad2 cannot interact with importins due to its unique exon 3. The main import activity for Smad2 is demonstrated to reside in the MH2 domain and import seems entirely independent of karyopherins *in vitro*. Rather, Smad2 interacts directly with nuclear pore components, mainly with CAN/Nup214, which is located at the cytoplasmic side of the nuclear pore and Nup153, which is located at the nucleoplasmic side (Xu et al., 2002). These nucleoporins are thought to be the docking sites for Smad2. Additional interactions have been found with other nucleoporins, which might be important for the transit through the pore.

An identical mechanism has subsequently been proposed to be responsible for Smad3 and Smad4 import (Xu et al., 2003), in stark contrast to the previously mentioned karyopherin-dependent import mechanisms. The authors show in an *in vitro* import assay that nuclear import of recombinant Smad3 and Smad4 is independent of cytosol, and hence of transport factors. They map the main import function of Smad3 to the linker-MH2 domain, whereas import of Smad4 is only efficient in the context of the full-length protein. In addition, as for Smad2, interaction of Smad3 and Smad4 with the nuclear pore component CAN/Nup214 is important for their import. A follow up study, aimed at assessing the import mechanism by which TGF-β-induced homomeric Smad3 complexes and heteromeric Smad3-Smad4 complexes are imported, demonstrated karyopherin-independent import for such complexes in an *in vitro* import assay (Chen et al., 2005).

4.1.3 Reconciling the different Smad import mechanisms

Therefore, despite the availability of a considerable amount of data, the mechanism for Smad translocation through the nuclear pore is still a subject of debate. Many studies aiming to pinpoint the translocation mechanism have reached their main conclusions either by *in vitro* studies and/or by using isolated Smad domains. Some of the discrepancies have been attributed to the particular constructs used in the different studies (Xu et al., 2002). However, the possibility that Smads employ different modes of nuclear transport has to be taken into consideration. Determination of the relative importance of the proposed import mechanisms for the import of full-length Smads in intact cells remains a challenge for the future. To this end, additional *in vivo* studies are desirable to either reconcile the apparent contradictions or to decide which mechanism is operating for full-length Smads in living cells in which particular circumstances.

4.2 Smad Export from the Nucleus

Karyopherins involved in nuclear export are mostly very specialized proteins with a narrow range of substrates (Bohnsack et al., 2002; Lipowsky et al., 2000; Mingot et al., 2004; Stuven et al., 2003). The best studied example and the only well-characterized exporter with broad-range specificity is the prototypic export receptor CRM1 or exportin 1, which is responsible for nuclear export of a plethora of different export substrates (Fornerod et al., 1997). Identification of export substrates of CRM1 has been greatly facilitated by the availability of the drug LMB. It seems possible that at least some of the other apparently very specific exporters do in fact export many different substrates, and the development of specific inhibitors may be required to appreciate the full range of their export substrates.

Treatment of cells with LMB triggers nuclear accumulation of Smad4, indicating CRM1-dependency of Smad4 export. A canonical nuclear export signal (NES) has been identified in the N-terminal part of the linker region of Smad4 (^{142}D**L**SG**L**T**L**Q^{149}), and deletion or mutation of critical residues in this NES enhanced nuclear localization of Smad4 (Pierreux et al., 2000; Watanabe et al., 2000). However, this sequence motif is not sufficient to mediate nuclear export (Pierreux et al., 2000). Interestingly, another Smad4, XSmad4β that exists in *Xenopus laevis* and has thus far not been identified in other vertebrates, lacks the NES and is constitutively nuclear (Masuyama et al., 1999). Hence, this NES seems to be necessary to mediate nuclear export of Smad4.

The situation is somewhat more complicated for the R-Smads. A CRM1-dependent export mechanism identical to that of Smad4 has been suggested for the BMP-responsive Smad1. First, a potential NES, termed NES1, was identified in the Smad1 MH2 domain (Fig. 1). Mutation of this motif caused Smad1 to localize predominantly to the nucleus (Xiao et al., 2001). In addition, a second NES termed NES2 and located in the linker was described (Xiao et al., 2003a). NES2 partly overlaps with the functional NES found in the corresponding region of Smad4. Smad1 NES1 and NES2 are absolutely conserved between all the R-Smads. They seem to mediate CRM1-dependent export in the case of Smad1, but have no such activity in Smad2 and Smad3, whose export is clearly CRM1-independent. There is currently no explanation as to why the motifs are functional in Smad1 but not in Smad2 and Smad3. Perhaps the regions in question are buried in the protein structure in Smad2 and Smad3, thus masking the NES function. However, to date there is no structural evidence to support this hypothesis.

The export mechanisms for Smad2 and Smad3 remain elusive. For Smad2, karyopherin-independent export mediated by direct binding to nucleoporins has been suggested (Xu et al., 2002). However, the existence of an export receptor that might prove responsible for preferential export of monomeric R-Smads over complexed Smads cannot be excluded and remains a possibility (Chen et al., 2005).

Little is known about the nucleocytoplasmic transport properties of the inhibitory Smads Smad6 and Smad7. Smad7 has been reported to be nuclear in the absence of a TGF-β superfamily ligand and has been shown to be exported from the nucleus in cells transfected with a constitutively active type I receptor (Itoh et al., 1998). However, it has also been found to be cytoplasmic in other cell lines (Hanyu et al., 2001). Smad6 by contrast seems to be distributed equally between the nucleus and cytoplasm. Both Smad6 and Smad7 have been reported to be exported from the nucleus by means of interaction with the HECT type ubiquitin ligase Smurf1 (Ebisawa et al., 2001; Hanyu et al., 2001). For Smad7, this export event is CRM1 dependent and is mediated via an NES present in Smurf1 (Tajima et al., 2003). If and how this "piggyback" mechanism is regulated by TGF-β ligands is currently unknown.

5. MECHANISMS UNDERLYING TGF-β-INDUCED NUCLEAR ACCUMULATION OF SMADS

Irrespective of the exact mechanism by which Smads translocate through the nuclear pore complex, it has been of great interest to determine how the steady state distribution of Smads is established in both the resting state and in the presence of a TGF-β superfamily ligand.

5.1 Nucleocytoplasmic Smad Distribution in Uninduced Cells

It is generally accepted that, in the absence of a TGF-β signal, Smad4 is distributed approximately equally between the nucleus and the cytoplasm. Such distribution is established by the constitutive nuclear import and export functions of Smad4 which are thought to be of comparable strengths (Pierreux et al., 2000).

R-Smads reside predominantly in the cytoplasm in the absence of a signal. Importantly however, they are by no means excluded from the nucleus. Rather, a significant amount of R-Smad is nuclear in uninduced cells. Given the continuous exchange of R-Smads between the two compartments, their nucleocytoplasmic distribution directly reflects their mean residence times in both compartments. The mean residence times are determined by the translocation rates of nuclear import and nuclear export and by possible interactions with retention factors that could cause temporary sequestration of Smads from the transport machinery. Such retention factors could either act by transiently immobilizing the R-Smad or by masking its import function. The endosomal FYVE-domain protein SARA has been suggested to act as retention factor for Smad2 and Smad3 in the absence of a signal, because addition of the recombinant Smad binding domain of SARA inhibited nuclear import of Smad2 and Smad3 *in vitro* and overexpression of SARA targeted Smad2 to early endosomes *in vivo* (Xu et al., 2000, 2003). However, in cell lines expressing low levels of EGFPSmad2 no localization to early endosomes is observed and EGFPSmad2 shows a rather diffuse cytoplasmic distribution (Nicolás et al., 2004), as does endogenous Smad2 (Pierreux et al., 2000). Both findings together indicate that the amount of SARA is limiting and SARA is unlikely to bring about quantitative cytoplasmic sequestration of Smad2. However, short transient interactions, perhaps with additional factors, might be sufficient to increase the mean residence time of Smad2 in the cytoplasm to bring about the predominantly cytoplasmic localization of Smad2. Alternatively, or in addition, R-Smad import rates could simply be lower than the export rates. Quantitative kinetic studies on EGFPSmad2 in living cells suggest that the export rate is indeed two-fold higher than the import rate (Schmierer and Hill, 2005).

5.2 Nucleocytoplasmic Smad Distribution in TGF-β Induced Cells

Four distinct mechanisms could account for nuclear accumulation of a continuously shuttling, predominantly cytoplasmic protein. Growth factor signaling could either lead to a release from cytoplasmic anchoring, or to the establishment of nuclear anchoring. Alternatively, the presence of a signal could increase the transit rate through the nuclear pore during nuclear import, or could decrease the transit rate during nuclear export. All four mechanisms will increase the ratio of mean residence times in the nucleus to the mean residence time in the cytoplasm, and will cause a predominantly nuclear steady state distribution. The four mechanisms are not mutually exclusive, and more than one might operate at the same time.

Smad4 is dispensable for nuclear accumulation of R-Smads, but its own accumulation is strictly dependent on R-Smad accumulation (Chen et al., 2005; De Bosscher et al., 2004; Reguly and Wrana, 2003). Thus, the mechanism of R-Smad nuclear accumulation is of central interest. Which of the possible mechanisms outlined above is responsible for nuclear accumulation of homomeric and heteromeric Smad complexes? Does phosphorylation release R-Smads from cytoplasmic anchoring? Are Smad complexes imported into the nucleus more efficiently than the monomeric proteins? Are Smad complexes trapped in the nucleus, and if so, by what mechanism?

TGF-β-dependent increase in import has been suggested to bring about nuclear accumulation of Smad3 (Kurisaki et al., 2001). This model is based on the fact that phosphorylated Smad3 interacts more efficiently with importin β than unphosphorylated Smad3 and pseudo-phosphorylated Smad3 is imported more efficiently *in vitro* than unphosphorylated Smad3. The mechanism does not apply for Smad2, because its additional exon 3 prevents association of Smad2 with importin β.

For Smad2, an anchor-release model has been suggested based on the finding that overexpression of SARA sequesters Smad2 into clusters in the cytoplasm (Tsukazaki et al., 1998; Xu et al., 2000) and that overexpression of the Smad2-interacting nuclear protein Fast-1/FoxH1 traps Smad2 in the nucleus even in uninduced cells (Xu et al., 2002). In this model, phosphorylation-dependent disruption of the Smad2-SARA interaction (Tsukazaki et al., 1998) enables phosphorylated Smad2 to move into the nucleus, where it is retained by means of phosphorylation-dependent interactions with the transcription factor Fast-1/FoxH1 or related proteins (Xu et al., 2002). However, a Smad2 mutant (W368A) which cannot interact with the Smad-interacting motifs that are found in several transcription factors including Fast-1/FoxH1 (Randall et al., 2004), accumulates in the nucleus in response to TGF-β as efficiently as wildtype Smad2 (Schmierer and Hill, 2005). Hence, transcription factors such as Fast-1/FoxH1 are unlikely to provide nuclear retention sites for Smad2. Nevertheless, a recent *in vivo* study employing a fluorescence perturbation approach has suggested that nuclear trapping *is* an important mechanism for R-Smad accumulation in response to TGF-β (Schmierer and Hill, 2005; Fig. 4). Nuclear mobilities of EGFPSmad2 and EGFPSmad4 are strongly diminished in TGF-β-induced cells, pointing to nuclear tethering of active Smad complexes. Compartment-specific photoactivation revealed that the export rate of EGFPSmad2 is strongly decreased in TGF-β-induced cells when compared to uninduced cells, whereas the import rate is unchanged.

These data suggest that increased import does not play a predominant role in TGF-β-induced nuclear accumulation *in vivo*. Rather, selective trapping of active Smad complexes in the nucleus from a cycling pool of monomeric Smads is likely to be the most important mechanism. The nuclear trapping may be due to binding of Smad complexes to immobile nuclear components, possibly DNA, which would explain the observed decrease in mobility. Alternatively or in addition, Smad complexes might not be recognized by the export machinery, as has already been shown for Smad4 (Chen et al., 2005). Identification of the exact export mechanism of Smad2 will be essential to distinguish between these two possibilities.

6. THE BIOLOGICAL FUNCTION OF SMAD NUCLEOCYTOPLASMIC SHUTTLING

Appropriate reaction to an external stimulus requires exact timing of both the initiation and the duration of the specific response. By comparison with other signaling pathways, activation of Smads by TGF-β proceeds comparably slowly and is maintained over extended time periods in cultured cells. Maximal activation is characterized by plateauing levels of phosphorylated R-Smads and maximal nuclear accumulation of Smads after about 45 min of continuous presence of the ligand in cell culture (Inman et al., 2002b; Pierreux et al., 2000). Both features are maintained for several hours. It is not before the signal ceases, *i.e.* the actively signaling receptors are inactivated by either dephosphorylation or degradation, that the bulk of the Smads redistribute back into the cytoplasm. Highly persistent, even continuous, TGF-β signaling is observed in many cell types with autocrine TGF-β signaling *in vitro* and *in vivo*. Such steady exposure to an endogenous TGF-β-superfamily ligand establishes a steady state characterized by the constitutive presence of phosphorylated R-Smads in the nucleus. Sustained signaling is most likely required for TGF-β-induced gene transcription,

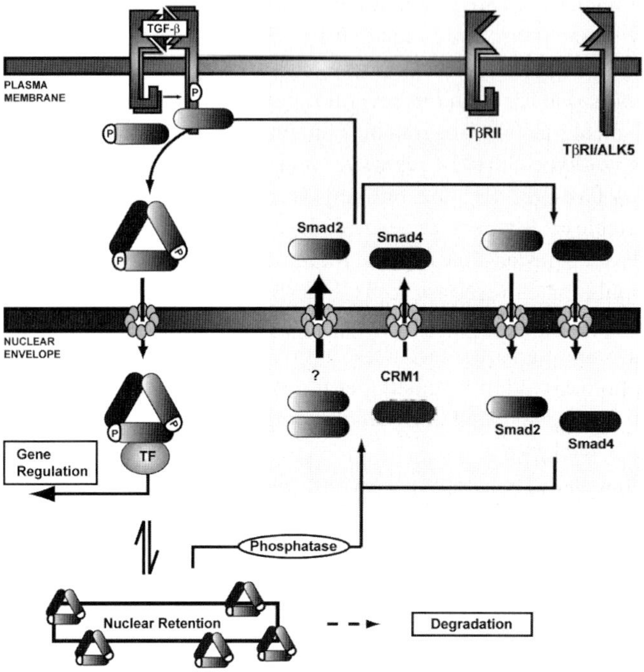

Figure 4. A model for nucleocytoplasmic shuttling of Smad2 and Smad4 in uninduced and in TGF-β-stimulated cells. In uninduced cells, cytoplasmic Smad2 is continuously imported into the nucleus, but stronger constitutive export establishes a predominantly cytoplasmic localization (shaded box). Similarly, Smad4 is also shuttling between the nucleus and cytoplasm in unstimulated cells (shaded box). Superimposed on this constitutive shuttling, receptor activation by TGF-β leads to relatively slow R-Smad phosphorylation and complex formation with Smad4. Therefore at any one time, only a small percentage of the translocating Smad2 is actually phosphorylated and complexed. However, whereas unphosphorylated, monomeric Smad2 is re-exported by an unknown transport mechanism, phosphorylated, complexed Smad2 is retained in the nucleus. Increased affinity for nuclear binding sites may be involved in this retention. TGF-β-induced Smad4 nuclear accumulation is a direct consequence of Smad2 accumulation. The pool of retained Smad complexes is in equilibrium with the presumably small fraction of Smad complexes that are actually engaged on specific target gene promoters. The model suggests that in TGF-β-treated cells, weak phosphatase activity triggers dissociation of Smad complexes, enabling monomeric Smad2 and Smad4 to be exported from the nucleus independently of each other. If the receptors are still active, Smad2 becomes rephosphorylated and re-enters the accumulation cycle. Hence, the nucleus acts as a trap, filtering phosphorylated, complexed Smad2 from the total shuttling pool. This mechanism establishes and sustains TGF-β-induced nuclear Smad accumulation by constantly monitoring the receptor activity

as premature termination of active signaling by pharmacological inhibition of TGF-β-type I receptors strongly compromises transcription from TGF-β-responsive promoters (Inman et al., 2002b). In early embryonic development by contrast, TGF-β superfamily signaling intensity can be highly dynamic. For instance, rapid activation and then inactivation of signaling by the Activin-related ligand,

Nodal, in frog embryos is an important determinant of early embryonic patterning (Lee et al., 2001).

The initial view of the pathway as unidirectional, such that the activated Smads do not exit the nucleus but rather become degraded there in a quantitative manner, does not provide a mechanism whereby the amount of phosphorylated nuclear Smads could reflect the level of activation of the receptors. Maximal accumulation would be followed by a continuous decrease in the level of active nuclear Smads, irrespective of the current level of receptor activation. Consequently it is hard to see how persistent, signal-related nuclear accumulation of Smads could be accomplished under such circumstances. The amount of nuclear Smad would depend primarily on the rates of Smad synthesis and degradation rather than being determined by the intensity of the TGF-β superfamily signal and would not reflect the activation state of the receptors. Although some nuclear degradation of Smads does occur (Lo and Massagué, 1999), it does not affect the bulk population of nuclear Smads and it will be interesting to see whether a specific subpopulation, for instance Smads directly involved in transcriptional regulation, are the preferred target for degradation.

Premature termination of active signaling by receptor inhibition causes the Smads to redistribute into the cytoplasm (Fig. 2). This key observation suggested the existence of a sensing mechanism through nucleocytoplasmic shuttling of Smads which continuously communicates the activation state of membrane-bound receptors to the Smads which have accumulated in the nucleus. The concept of Smad nucleocytoplasmic shuttling can explain the establishment, duration, and kinetics of the TGF-β response. The nucleocytoplasmic distributions of Smads in the absence and in the presence of a TGF-β superfamily signal are now seen as two distinct steady states of a highly dynamic nucleocytoplasmic shuttling equilibrium (Fig. 4). These steady states are characterized by the absence of phosphorylated R-Smads and by predominantly cytoplasmic localization of Smads in the uninduced state, and by R-Smad phosphorylation, Smad complex formation and nuclear accumulation of Smads in the induced state. The shift from one state to the other is reversible and exclusively dependent on the absence or presence of a signal. How far the equilibrium is shifted from the uninduced state towards the induced state, *i.e.* which degree of phosphorylation of R-Smads and nuclear accumulation of Smad complexes is actually achieved, is strictly dependent on the receptor activity at any given time.

Taken together, nucleocytoplasmic shuttling of Smads enables the cell to continuously monitor signal strength and provides the flexibility that is needed to react appropriately to changing signal intensities.

7. FUTURE CHALLENGES

The shuttling of the Smads between the cytoplasm and nucleus in both unstimulated cells and in cells stimulated with TGF-β has now been demonstrated by a number of different techniques. It has become clear that the nucleocytoplasmic shuttling of Smads in the presence of a TGF-β superfamily signal provides a mechanism whereby the intracellular transducers of the signal can monitor receptor activity.

This appears to be a common mechanism employed by other signaling pathways, as a very similar model has been suggested for the JAK/STAT pathway (Meyer and Vinkemeier, 2004; Vinkemeier, 2004). As is clear from this chapter, several important issues remain to be resolved. There is still some debate about the actual mechanism of Smad translocation through the nuclear pore, with R-Smad export being particularly ill-defined. The identity of the R-Smad phosphatase remains elusive and it is also not clear whether it is really nuclear as the model suggests. Although a number of models have been proposed, it is still not proven how the steady state distributions of the Smads in resting cells are established. Finally, whereas decreased export of complexed Smads seems to be important for nuclear accumulation, the nature of the trapping mechanism is unclear. Further work over the next few years will undoubtedly answer these questions.

ACKNOWLEDGEMENTS

We thank members of the Hill lab, Sara Nakielny and Giampietro Schiavo for stimulating discussions and comments on the manuscript. The work in the Hill lab is funded by Cancer Research UK, the European Commission and an Erwin Schrödinger fellowship of the Austrian Science Foundation (FWF) (# J2397-B12) to B.S.

REFERENCES

Bohnsack, M.T., Regener, K., Schwappach, B., Saffrich, R., Paraskeva, E., Hartmann, E., and Görlich, D., 2002, Exp5 exports eEF1A via tRNA from nuclei and synergizes with other transport pathways to confine translation to the cytoplasm. *EMBO J* 21: 6205-6215.

Chen, H.B., Rud, J.G., Lin, K., and Xu, L., 2005, Nuclear targeting of transforming growth factor-β-activated Smad complexes. *J Biol Chem* 280: 21329-21336.

Chook, Y.M., and Blobel, G., 2001, Karyopherins and nuclear import. *Curr Opin Struct Biol* 11: 703-715.

Cingolani, G., Bednenko, J., Gillespie, M.T., and Gerace, L., 2002, Molecular basis for the recognition of a nonclassical nuclear localization signal by importin β. *Mol Cell* 10: 1345-1353.

Coburn, G.A., Wiegand, H.L., Kang, Y., Ho, D.N., Georgiadis, M.M., and Cullen, B.R., 2001, Using viral species specificity to define a critical protein/RNA interaction surface. *Genes Dev* 15: 1194-1205.

Cronshaw, J.M., Krutchinsky, A.N., Zhang, W., Chait, B.T., and Matunis, M.J., 2002, Proteomic analysis of the mammalian nuclear pore complex. *J Cell Biol* 158: 915-927.

De Bosscher, K., Hill, C.S., and Nicolás, F.J., 2004, Molecular and functional consequences of Smad4 C-terminal missense mutations in colorectal tumour cells. *Biochem J* 379: 209-216.

Dingwall, C., and Laskey, R.A., 1991, Nuclear targeting sequences – a consensus? *Trends Biochem Sci* 16: 478-481.

Ebisawa, T., Fukuchi, M., Murakami, G., Chiba, T., Tanaka, K., Imamura, T., and Miyazono, K., 2001, Smurf1 interacts with transforming growth factor-β type I receptor through Smad7 and induces receptor degradation. *J Biol Chem* 276: 12477-12480.

Engelsma, D., Bernad, R., Calafat, J., and Fornerod, M., 2004, Supraphysiological nuclear export signals bind CRM1 independently of RanGTP and arrest at Nup358. *EMBO J* 23: 3643-3652.

Fagotto, F., Gluck, U., and Gumbiner, B.M., 1998, Nuclear localization signal-independent and importin/karyopherin-independent nuclear import of β-catenin. *Curr Biol* 8: 181-190.

Fornerod, M., Ohno, M., Yoshida, M., and Mattaj, I.W., 1997, CRM1 is an export receptor for leucine-rich nuclear export signals. *Cell* 90: 1051-1060.

Görlich, D., and Kutay, U., 1999, Transport between the cell nucleus and the cytoplasm. *Annu Rev Cell Dev Biol* 15: 607-660.

Goumans, M.-J., Valdimarsdottir, G., Itoh, S., Lebrin, F., Larsson, J., Mummery, C., Karlsson, S., and ten Dijke, P., 2003, Activin receptor-like kinase (ALK)1 is an antagonistic mediator of lateral TGFβ/ALK5 signaling. *Mol Cell* 12: 817-828.

Hanyu, A., Ishidou, Y., Ebisawa, T., Shimanuki, T., Imamura, T., and Miyazono, K., 2001, The N domain of Smad7 is essential for specific inhibition of transforming growth factor-β signaling. *J Cell Biol* 155: 1017-1027.

Inman, G.J., Nicolás, F.J., Callahan, J.F., Harling, J.D., Gaster, L.M., Reith, A.D., Laping, N.J., and Hill, C.S., 2002a, SB-431542 is a potent and specific inhibitor of Transforming Growth Factor-β superfamily type I activin receptor-like kinase receptors, ALK4, ALK5 and ALK7. *Mol Pharmacol* 62: 65-72.

Inman, G.J., Nicolás, F.J., and Hill, C.S., 2002b, Nucleocytoplasmic shuttling of Smads 2, 3 and 4 permits sensing of TGF-β receptor activity. *Mol Cell* 10: 283-294.

Itoh, S., Landström, M., Hermansson, A., Itoh, F., Heldin, C.-H., Heldin, N.-E., and ten Dijke, P., 1998, Transforming growth factor β1 induces nuclear export of inhibitory Smad7. *J Biol Chem* 273: 29195-29201.

Jakel, S., and Görlich, D., 1998, Importin β, transportin, RanBP5 and RanBP7 mediate nuclear import of ribosomal proteins in mammalian cells. *EMBO J* 17: 4491-4502.

Kavsak, P., Rasmussen, R.K., Causing, C.G., Bonni, S., Zhu, H., Thomsen, G.H., and Wrana, J.L., 2000, Smad7 binds to Smurf2 to form an E3 ubiquitin ligase that targets the TGFβ receptor for degradation. *Mol Cell* 6: 1365-1375.

Kurisaki, A., Kose, S., Yoneda, Y., Heldin, C.-H., and Moustakas, A., 2001, Transforming growth factor-β induces nuclear import of Smad3 in an importin-β1 and Ran-dependent manner. *Mol Biol Cell* 12: 1079-1091.

Laping, N.J., Grygielko, E., Mathur, A., Butter, S., Bomberger, J., Tweed, C., Fornwald, J., Lehr, R., Harling, J.D., Gaster, L.M., Callahan, J.F., and Olson, B.A., 2002, Inhibition of transforming growth factor (TGF)-β1-induced extracellular matrix with a novel inhibitor of the TGF-β type I receptor kinase activity. *Mol Pharmacol* 62: 58-64.

Lee, M.A., Heasman, J., and Whitman, M., 2001, Timing of endogenous activin-like signals and regional specification of the *Xenopus* embryo. *Development* 128: 2939-2952.

Lipowsky, G., Bischoff, F.R., Schwarzmaier, P., Kraft, R., Kostka, S., Hartmann, E., Kutay, U., and Görlich, D., 2000, Exportin 4: a mediator of a novel nuclear export pathway in higher eukaryotes. *EMBO J* 19: 4362-4371.

Liu, F., Pouponnot, C., and Massagué, J., 1997, Dual role of the Smad4/DPC4 tumor suppressor in TGFβ-inducible transcriptional complexes. *Genes Dev* 11: 3157-3167.

Lo, R.S., and Massagué, J., 1999, Ubiquitin-dependent degradation of TGF-β-activated smad2. *Nat Cell Biol* 1: 472-478.

Massagué, J., 1998, TGF-β signal transduction. *Annu Rev Biochem* 67: 753-791.

Masuyama, N., Hanafusa, H., Kusakabe, M., Shibuya, H., and Nishida, E., 1999, Identification of two Smad4 proteins in *Xenopus*. Their common and distinct properties. *J Biol Chem* 274: 12163-12170.

Matsubayashi, Y., Fukuda, M., and Nishida, E., 2001, Evidence for existence of a nuclear pore complex-mediated, cytosol-independent pathway of nuclear translocation of ERK MAP kinase in permeabilized cells. *J Biol Chem* 276: 41755-41760.

Meyer, T., and Vinkemeier, U., 2004, Nucleocytoplasmic shuttling of STAT transcription factors. *Eur J Biochem* 271: 4606-4612.

Mingot, J.M., Bohnsack, M.T., Jakle, U., and Görlich, D., 2004, Exportin 7 defines a novel general nuclear export pathway. *EMBO J* 23: 3227-3236.

Moore, J.D., Yang, J., Truant, R., and Kornbluth, S., 1999, Nuclear import of Cdk/cyclin complexes: identification of distinct mechanisms for import of Cdk2/cyclin E and Cdc2/cyclin B1. *J Cell Biol* 144: 213-224.

Nicolás, F.J., and Hill, C.S., 2003, Attenuation of the TGF-β-Smad signaling pathway in pancreatic tumor cells confers resistance to TGF-β-induced growth arrest. *Oncogene* 22: 3698-3711.

Nicolás, F.J., De Bosscher, K., Schmierer, B., and Hill, C.S., 2004, Analysis of Smad nucleocytoplasmic shuttling in living cells. *J Cell Sci* 117: 4113-4125.

Patterson, G.H., and Lippincott-Schwartz, J., 2002, A photoactivatable GFP for selective photolabeling of proteins and cells. *Science* 297: 1873-1877.

Pemberton, L.F., and Paschal, B.M., 2005, Mechanisms of receptor-mediated nuclear import and nuclear export. *Traffic* 6: 187-198.

Pierreux, C.E., Nicolás, F.J., and Hill, C.S., 2000, Transforming growth factor β-independent shuttling of Smad4 between the cytoplasm and nucleus. *Mol Cell Biol* 20: 9041-9054.

Randall, R.A., Howell, M., Page, C.S., Daly, A., Bates, P.A., and Hill. C.S., 2004, Recognition of phosphorylated-Smad2-containing complexes by a novel Smad interaction motif. *Mol Cell Biol* 24: 1106-1121.

Reguly, T., and Wrana, J.L., 2003, In or out? The dynamics of Smad nucleocytoplasmic shuttling. *Trends Cell Biol* 13: 216-220.

Schmierer, B., and Hill, C.S., 2005, Kinetic analysis of Smad nucleocytoplasmic shuttling reveals a mechanism for transforming growth factor β-dependent nuclear accumulation of Smads. *Mol Cell Biol* 25: 9845-9858.

Shi, W., Sun, C., He, B., Xiong, W., Shi, X., Yao, D., and Cao, X., 2004, GADD34-PP1c recruited by Smad7 dephosphorylates TGFβ type I receptor. *J Cell Biol* 164: 291-300.

Shi, Y., and Massagué, J., 2003, Mechanisms of TGF-β signaling from cell membrane to the nucleus. *Cell* 113: 685-700.

Ström, A.C., and Weis, K., 2001, Importin-β-like nuclear transport receptors. *Genome Biol* 2: reviews 3008.1-3008.9.

Stuven, T., Hartmann, E., and Görlich, D., 2003, Exportin 6: a novel nuclear export receptor that is specific for profilin.actin complexes. *EMBO J* 22: 5928-5940.

Tajima, Y., Goto, K., Yoshida, M., Shinomiya, K., Sekimoto, T., Yoneda, Y., Miyazono, K., and Imamura, T., 2003, Chromosomal region maintenance 1 (CRM1)-dependent nuclear export of Smad ubiquitin regulatory factor 1 (Smurf1) is essential for negative regulation of transforming growth factor-β signaling by Smad7. *J Biol Chem* 278: 10716-10721.

Takizawa, C.G., Weis, K., and Morgan, D.O., 1999, Ran-independent nuclear import of cyclin B1-Cdc2 by importin β. *Proc Natl Acad Sci U S A* 96: 7938-7943.

Truant, R., and Cullen, B.R., 1999, The arginine-rich domains present in human immunodeficiency virus type 1 Tat and Rev function as direct importin β-dependent nuclear localization signals. *Mol Cell Biol* 19: 1210-1217.

Tsukazaki, T., Chiang, T.A., Davison, A.F., Attisano, L., and Wrana, J.L., 1998, SARA, a FYVE domain protein that recruits Smad2 to the TGFβ receptor. *Cell* 95: 779-791.

Vinkemeier, U., 2004, Getting the message across, STAT! Design principles of a molecular signaling circuit. *J Cell Biol* 167: 197-201.

Watanabe, M., Masuyama, N., Fukuda, M., and Nishida, E., 2000, Regulation of intracellular dynamics of Smad4 by its leucine-rich nuclear export signal. *EMBO Rep* 1: 176-182.

Whitehurst, A.W., Wilsbacher, J.L., You, Y., Luby-Phelps, K., Moore, M.S., and Cobb, M.H., 2002, ERK2 enters the nucleus by a carrier-independent mechanism. *Proc Natl Acad Sci U S A* 99: 7496-7501.

Wu, J.W., Hu, M., Chai, J., Seoane, J., Huse, M., Li, C., Rigotti, D.J., Kyin, S., Muir, T.W., Fairman, R., Massagué, J., and Shi, Y., 2001, Crystal structure of a phosphorylated Smad2. Recognition of phosphoserine by the MH2 domain and insights on Smad function in TGF-β signaling. *Mol Cell* 8: 1277-1289.

Xiao, Z., Liu, X., Henis, Y.I., and Lodish, H.F., 2000a, A distinct nuclear localization signal in the N terminus of Smad 3 determines its ligand-induced nuclear translocation. *Proc Natl Acad Sci U S A* 97: 7853-7858.

Xiao, Z., Liu, X., and Lodish, H.F., 2000b, Importin β mediates nuclear translocation of Smad 3. *J Biol Chem* 275: 23425-23428.

Xiao, Z., Watson, N., Rodriguez, C., and Lodish, H.F., 2001, Nucleocytoplasmic shuttling of Smad1 conferred by its nuclear localization and nuclear export signals. *J Biol Chem* 276: 39404-39410.

Xiao, Z., Brownawell, A.M., Macara, I.G., and Lodish, H.F., 2003a, A novel nuclear export signal in Smad1 is essential for its signaling activity. *J Biol Chem* 278: 34245-34252.

Xiao, Z., Latek, R., and Lodish, H.F., 2003b, An extended bipartite nuclear localization signal in Smad4 is required for its nuclear import and transcriptional activity. *Oncogene* 22: 1057-1069.

Xu, L., Chen, Y.G., and Massagué, J., 2000, The nuclear import function of Smad2 is masked by SARA and unmasked by TGFβ-dependent phosphorylation. *Nat Cell Biol* 2: 559-562.

Xu, L., Kang, Y., Col, S., and Massagué, J., 2002, Smad2 nucleocytoplasmic shuttling by nucleoporins CAN/Nup214 and Nup153 feeds TGFβ signaling complexes in the cytoplasm and nucleus. *Mol Cell* 10: 271-282.

Xu, L., Alarcon, C., Col, S., and Massagué, J., 2003, Distinct domain utilization by Smad3 and Smad4 for nucleoporin interaction and nuclear import. *J Biol Chem* 278: 42569-42577.

Xu, L., and Massagué, J., 2004, Nucleocytoplasmic shuttling of signal transducers. *Nat Rev Mol Cell Biol* 5: 209-219.

CHAPTER 11

STRUCTURAL INSIGHTS INTO SMAD FUNCTION AND SPECIFICITY

YIGONG SHI

Department of Molecular Biology, Princeton University, Princeton, New Jersey, USA

Abstract: Smads contain two conserved domains, MH1 and MH2. The MH1 domain of R-Smads (except Smad2) and Smad4 adopts a compact globular fold and recognizes specific DNA sequences via a novel β-hairpin. The fact that each MH1 domain only recognizes 3–4 base pairs of DNA governs the need for Smads to cooperate among each other and with other DNA-binding proteins. The MH2 domain forms a central β-sandwich capped by a three-helix bundle on one end and a loop/helix region on the other end. Structural analyses of TGF-β receptors and the MH2 domains have revealed an essential and conserved theme in signaling. Phosphorylation of the GS region in the type I receptor results in the generation of pS/pT-X-pS motifs, which facilitate recruitment of R-Smads by binding to a positively charged and conserved surface pocket on R-Smads. After phosphorylation, the pS-X-pS motif at the C-terminus of R-Smads assists their dissociation from the receptors and aids formation of homomeric as well as heteromeric Smad complexes by binding to the same surface pocket of neighboring Smads. The MH2 domain interacts with a large number of proteins through a hydrophobic surface groove known as hydrophobic corridor

Keywords: Smad; MH1 domain; DNA-binding specificity; β-hairpin; MH2 domain; phosphoserine signaling; pS-X-pS motif; homomeric and heteromeric Smad complex

1. INTRODUCTION

Smad proteins (Smads) are the intracellular mediators of transforming growth factor β (TGF-β) signaling (See Preface). Smads transduce specific TGF-β signals from cell membrane to the nucleus, where they act as transcriptional regulators to control gene expression (Attisano and Wrana, 2002; Derynck and Zhang, 2003; Shi and Massagué, 2003; ten Dijke and Hill, 2004). Smads mediate a diverse set of cellular responses, including cell cycle regulation, recognition, differentiation, apoptosis, and specification of developmental fate, during embryogenesis as well as in mature tissues, in species ranging from flies and worms to mammals (Massagué, 1998;

P. ten Dijke and C.-H. Heldin (eds.), Smad Signal Transduction, 215–233.

Moustakas et al., 2002; Patterson and Padgett, 2000; ten Dijke et al., 2002) (see Chapters 2-8).

A TGF-β ligand initiates a cascade of signaling events by engaging specific type I and type II receptor serine/threonine kinases on the cell surface. This recognition triggers the phosphorylation and subsequent activation of the type I receptor kinase by receptor II. There are 8 distinct Smad proteins, representing three functional classes, the receptor-regulated Smad (R-Smad), the Co-mediator Smad (Co-Smad), and the inhibitory Smad (I-Smad). R-Smads (Smad1, 2, 3, 5, and 8) are directly phosphorylated and activated by the type I receptor kinases (see Chapter 12) and undergo homo-trimerization and formation of heteromeric complexes with the Co-Smad, Smad4. The activated Smad complexes are translocated into the nucleus and, in conjunction with other nuclear co-factors, regulate the transcription of target genes (see Chapters 14 and 17). The I-Smads, namely Smad6 and Smad7, inhibit TGF-β signaling by competing with R-Smads for receptor interaction and by targeting the receptors for degradation (see Chapters 13 and 19). Smad6 also specifically competes with Smad4 for binding to receptor-activated Smad1, forming an inactive Smad1-Smad6 complex (Hata et al., 1998).

Many key events in TGF-β signaling have been documented at the cellular, molecular, and mechanistic levels following the last decade of intense investigation. Cell biological investigation is complemented by biochemical and structural analysis, giving rise to an unprecedented level of clarity in all aspects of the signal transduction process. In this chapter, we focus on the structural biology of Smad proteins in TGF-β signaling. Specifically, we will discuss the structure-function relationship of Smad proteins and their cognate complexes with DNA and with other interacting proteins.

2. SEQUENCE AND STRUCTURE OF SMADS

2.1 Sequence Features of Smads

The R-Smad and Co-Smad proteins, with about 500 amino acids in length, contain two conserved structural domains, the amino-terminal MH1 (MAD homology 1) domain and the carboxy-terminal MH2 (MAD homology 2) domain (Fig. 1). The R-Smads, but not the Co-Smad, contain a characteristic Ser-Xaa-Ser (SXS) motif at their carboxy-termini. Both serine residues in the SXS motif are phosphorylated by the type I receptor during signaling (see Chapter 12). Among the R-Smads, Smad2 and Smad3 respond to signaling by the TGF-β subfamily, whereas Smad1, Smad5, and Smad8 respond to the BMP subfamily.

The MH1 domain of Smad4 and most R-Smads (except the most common splice form of Smad2) is responsible for sequence-specific DNA binding and negative regulation of the MH2 domain (Fig. 1). It also contains a nuclear localization sequence within the core domain and is followed by a nuclear export sequence at its carboxy-terminus. The amino-terminal domain of I-Smads exhibits weak sequence similarity to the MH1 domain of R-Smads, but does not bind to DNA. In contrast, the MH2 domain is highly conserved among all Smad proteins, and is responsible for interactions with receptor and SARA, formation of homomeric as well as heteromeric Smad complexes, and directly contacting the nuclear pore complex for nucleocytoplasmic shuttling (see Chapter 10). The MH2 domain also plays an important role in nuclear localization (Xu et al., 2000). Importantly, phosphorylation of the SXS motif in the MH2 domain of an R-Smad drives its activation. Both the MH1 and MH2 domains interact with a large number of proteins in the nucleus, affecting transcription (see Chapter 15).

Although the intervening sequences between the MH1 and MH2 domains are divergent among Smads, these regions contain multiple phosphorylation sites, which allow specific crosstalks with other signaling pathways (see Chapters 15 and 16), and a PY motif, which mediates specific interaction with the Smad ubiquitination regulatory factors (Smurfs). Smurf1 and Smurf2 are HECT-domain-containing E3 ubiquitin ligases that target Smads as well as Smad-associated TGF-β receptors for degradation by the 26S proteasome (see Chapter 13).

2.2 MH1 Domain: A Novel DNA-binding Motif

2.2.1 Structure of the MH1 domain

The structure of the conserved MH1 domain, as exemplified by Smad3 (Chai et al., 2003; Shi et al., 1998), adopts a novel globular fold, containing four α-helices and six short β-strands (Fig. 1A). The β-strands form two small β-sheets and one protruding β-hairpin that is directly responsible for sequence-specific DNA recognition. The β-hairpin is located next to the positively charged H2 helix (Fig. 1A), which contains the reported nuclear localization sequence.

The MH1 domain is stabilized by a tightly bound zinc atom, which is coordinated by three cysteines (Cys64, Cys109, and Cys121) and one histidine (His126) in Smad3. These four residues are invariant among members of the Smad family as well as their homologues in *Drosophila* and *C. elegans*, suggesting the conserved nature of zinc binding (Chai et al., 2003; Grishin, 2001). The structure of the Smad MH1 domain does not resemble any other sequence-specific DNA-binding protein for which structural information is available.

2.2.2 Specific DNA recognition by the MH1 domain

The primary function of the MH1 domain is to recognize specific DNA sequences. The minimal Smad binding element (SBE), initially identified as the optimal DNA-binding sequence for Smad3 and Smad4 (Zawel et al., 1998), contains only four base pairs, 5'-AGAC-3' (or 5'-GTCT-3') (Dennler et al., 1998; Yingling et al., 1997; Zawel et al., 1998), although most naturally occurring DNA sequences contains an extra base C at the 5' end. The crystal structure of the Smad3 MH1 domain bound to SBE revealed that a highly conserved β-hairpin makes specific contacts to three bases of the SBE (Fig. 2B; Chai et al., 2003; Shi et al., 1998). The side chain of Arg74 donates a pair of hydrogen bonds to the first base of **G**TCT whereas Gln76 and Lys81 make base-specific contacts to the first and second bases of **AG**AC,

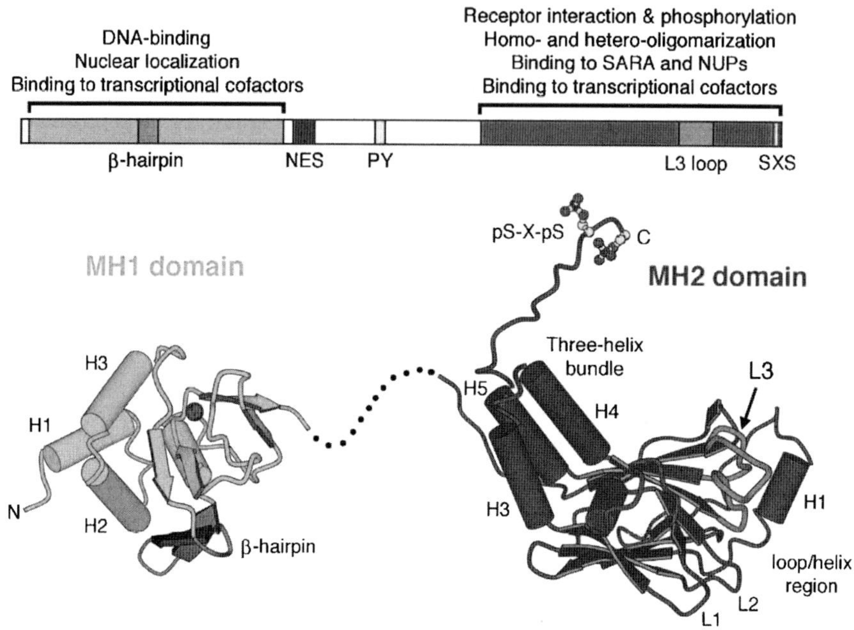

Figure 1. Conserved sequence and structure of Smad proteins. Smad proteins contain two conserved domains, MH1 and MH2, and a divergent intervening region. The MH1 domain adopts a novel zinc-containing fold, is responsible for sequence-specific DNA recognition, and contains a nuclear localization signal in the H2 helix. The β-hairpin is a novel DNA-binding motif in the MH1 domain and is highly conserved in R-Smads and Smad4. The MH2 domain adopts a cradle-like structure, with a central β-sheet as the base and the three-helix bundle and the loop/helix region as two ends of the cradle. The MH2 domain is responsible for interactions with receptor and SARA, formation of homomeric as well as heteromeric Smad complexes, and directly contacting the nuclear pore complex for nucleocytoplasmic shuttling. The R-Smads contain a characteristic SXS motif at their carboxy termini. Both the MH1 and MH2 domains interact with a large number of proteins in the nucleus. The intervening sequences contain multiple phosphorylation sites, which allow specific crosstalks with other signaling pathways, and a PY motif, which mediates specific interaction with Smurfs. Fig. 4 was prepared using GRASP (Nicholls et al., 1991) and all other figures were prepared using MOLSCRIPT (Kraulis, 1991) (A color version of this figure is freely accessible via the website of the book: http://www.springer.com/1-4020-4542-5)

Abbreviations: MH1, MAD homology 1; SARA, Smad anchor for receptor activation; Smurf, Smad ubiquitination regulatory factor

respectively (Fig. 2B). A number of well-ordered water molecules stabilize DNA recognition by the MH1 domain.

The DNA-binding β-hairpin is among the most highly conserved regions in the Smad proteins. All residues except the two at the turn of the β-hairpin in Smad4 are invariant among mammalian R-Smads and Smad4 (Fig. 2C). The conserved nature of the DNA-binding β-hairpin, as well as its surrounding sequence elements among all R-Smads, strongly suggests that other R-Smads should specifically bind to SBE as well. Compared to other Smads, the most common splice form of Smad2 contains a unique 30-residue insertion between the DNA-binding β-hairpin and the helix H2; this sequence variation resulted in the poor DNA-binding ability of Smad2, presumably due to disruption of the conformation of the DNA-binding hairpin (Shi et al., 1998). In support of this explanation, DNA binding was restored in the alternatively spliced variant of Smad2 lacking this insertion.

2.2.3 *Implications for DNA target selection*

The relatively low affinity of DNA binding by Smads has created difficulty in the identification of biologically relevant Smad-responsive DNA elements. For example, while Smad1, Smad3, and Smad4 bind specifically to SBE (Johnson et al.,

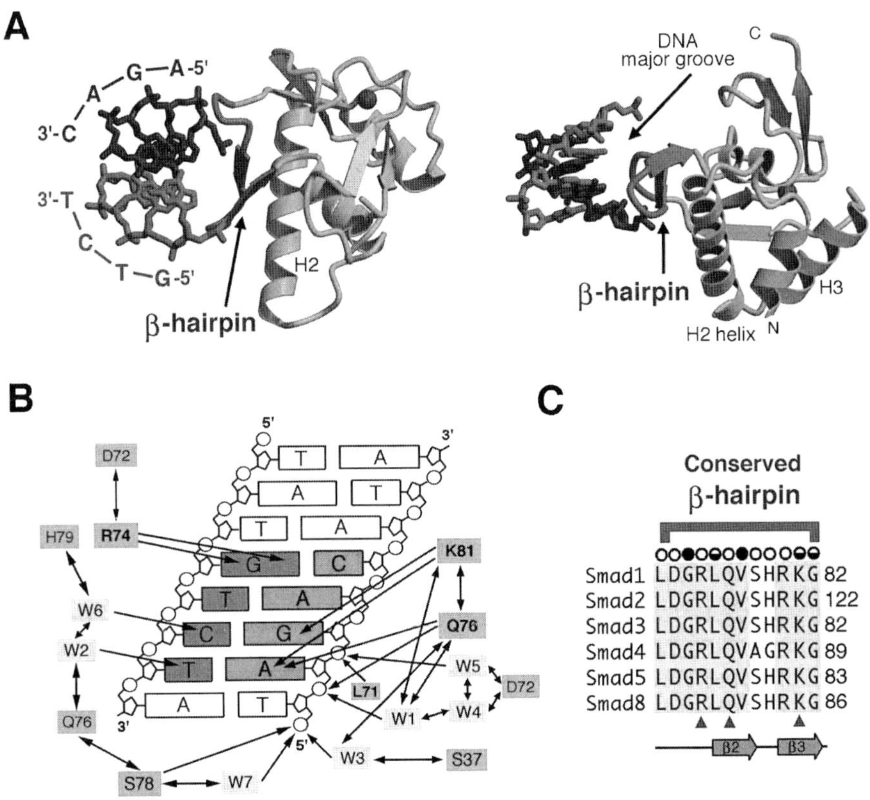

Figure 2. Sequence-specific DNA recognition by the MH1 domain. (A) Structure of the MH1 domain bound to SBE (5'-AGAC-3' or 5'-GTCT-3') shown in two perpendicular views (Chai et al., 2003). The β-hairpin sits asymmetrically in the major groove of DNA. (B) A summary of the MH1-DNA interactions. Arg74, Gln76, and Lys81, all from the β-hairpin, make direct hydrogen bonds to specific DNA bases. These interactions are stabilized by networks of hydrogen bonds involving residues from the MH1 domain, DNA bases and backbone phosphates, and seven well-ordered water molecules. (C) Conservation of the DNA-binding β-hairpin. Sequences of the β-hairpin region are aligned for R-Smads and Smad4. With the exception of two residues at the turn of the β-hairpin in Smad4, all other residues, including those that recognize DNA bases directly (marked by triangles), are identical among R-Smads and Smad4 (A color version of this figure is freely accessible via the website of the book: http://www.springer.com/1-4020-4542-5)

Abbreviations: SBE, Smad binding element

1999; Shi et al., 1998; Zawel et al., 1998), they have also been reported to bind to G/C-rich sequences (Ishida et al., 2000; Kim et al., 1997; Kusanagi et al., 2000; Labbe et al., 1998). This challenge is further confounded by the Lys-rich, positively charged helix H2 of the MH1 domain, which resides next to the DNA-binding β-hairpin in the crystal structure (Fig. 1A; Shi, 2001). Lysine residues in DNA-binding proteins are known to generally prefer G/C-rich sequences because of their favorable stereochemical and geometric parameters for hydrogen bonds (Choo and Klug, 1997). It is possible that the Smad MH1 domain can interact with two distinct sets of DNA sequences using two different sequence motifs, although this has not been shown for any other transcriptional factor.

Caution must be exercised for the identification of Smad-binding DNA sequences. In particular, decreased responsiveness to Smad in a reporter assay through mutation of a promoter element does not necessarily translate into decreased DNA binding by Smad, because endogeneous transcription factors could be involved. Strictly speaking, the optimal Smad-binding DNA elements can only be obtained through systematic *in vitro* selection and further verified through extensive mutagenesis of bindings sites (Zawel et al., 1998). The promoter element of the human $p21^{Cip1}$ gene contains highly G/C-rich DNA sequences and its activation is mediated by the Smad proteins (Moustakas and Kardassis, 1998). In this case, rather than directly binding to these G/C-rich DNA sequences, Smads exert their effects through interaction with the general transcription factor Sp1 (Pardali et al., 2000).

Most DNA-binding proteins recognize binding sequences of more than four base pairs. The length of the DNA-binding site dictates how frequently a binding sequence can be found in the genome. Obviously, the 4-bp SBE by itself is not sufficient to confer responsiveness to the otherwise exquisitely specific TGF-β signals. Fortunately, SBE can be found in the promoter regions of many known TGF-β responsive elements, and it often appears in multiple copies and is often adjacent to binding sites for other transcription factors. In reality, Smad proteins always cooperate with each other and/or with a large number of nuclear cofactors to modulate transcription of target genes. This function is reinforced by the formation of homomeric and heteromeric Smad complexes in response to phosphorylation of R-Smads (see later). The combinatorial approach allows Smad to achieve faithful responses to a wide range of cytokine signals.

2.3 MH2 Domain: A Multi-functional Motif

2.3.1 Structure of the MH2 domain

The structure of the MH2 domain, as exemplified by Smad4 (Qin et al., 1999; Shi et al., 1997), contains a central β-sandwich, with antiparallel β-sheets of five and six strands each, and two addenda at both ends (Fig. 1). One end of the β-sandwich is capped by an addendum – a three-α-helix bundle that extends over the plane of the six-stranded β-sheet, at a roughly perpendicular angle. The other end of the β-sandwich is capped by another addendum – a group of three large loops and an α-helix, collectively referred to as the loop/helix region (Fig. 1).

This structural arrangement immediately suggests that the central β-sandwich of the MH2 domain may serve as a structural scaffold for the two addenda to play important biological functions through protein-protein interactions (Shi et al., 1997). This hypothesis has been proven by a large body of subsequent experimental observations.

The MH2 domain of Smad4 is known to undergo a process of concentration-dependent homo-trimerization. The crystal structure reveals that homo-trimerization is primarily mediated by the two addenda, with the loop/helix region of one protomer packing against the three-helix bundle of the adjacent protomer (Fig. 3A). This interface is extensive; involving 1600 $Å^2$ buried surface area, and includes both hydrophobic contacts and hydrogen bond networks. These interactions are repeated three times, resulting in the homo-trimeric assembly (Fig. 3A). As will be discussed in detail, the exact architecture and interface of homo-trimeric assembly for Smad4 are recapitulated in the homo-trimeric assembly of phosphorylated R-Smads and in the heteromeric assembly between Smad4 and R-Smads. Information gained from this structure lays the foundation for subsequent biochemical, functional, and structural investigations on the MH2 domain of all three types of Smad.

2.3.2 Cancer-derived mutations at the trimer interface

Because Smad-mediated TGF-β signaling generally has a negative impact on cell growth, inactivation of this pathway contributes to tumorigenesis (see Chapters 4, 6, 20, 21, and 22). Tumor-derived mutations have been observed in both TGF-β family receptors and the Smad proteins. For example, Smad4 mutations have been found in nearly half of all pancreatic carcinomas and to a lesser extent in cancers of the colon, breast, ovary, lung, and head and neck (Schutte et al., 1996). Another member of the Smad family, Smad2, which is a mediator of the antiproliferative TGF-β and Activin responses, has also been identified as a tumor suppressor because of its mutation in colorectal and lung cancers (Eppert et al., 1996). Naturally occurring disease mutations in Smads and TGF-β family receptors have significantly facilitated the analysis of the structure and function of these proteins.

The majority of tumor-derived mutations map to the MH2 domain. The residues that are mutated in cancer make important contributions to stabilize the trimeric interface as observed in Smad4 (Fig. 3A; Shi et al., 1997). Of particular note is a network of hydrogen bonds involving three residues, Asp351 and Arg361 from one protomer and Asp537 from the adjacent protomer (Fig. 3B). These three residues are invariant among all Smads; Asp450 in Smad2 (corresponding to Asp537 in Smad4) and Asp351 and Arg361 in Smad4 are targeted for recurrent tumorigenic mutations. This observation suggests that such mutations may have a deleterious effect on the formation of Smad homomeric and heteromeric complexes. Supporting this conclusion, tumor-derived missense mutations appear to inactivate trimeric assembly both *in vitro* and *in vivo*.

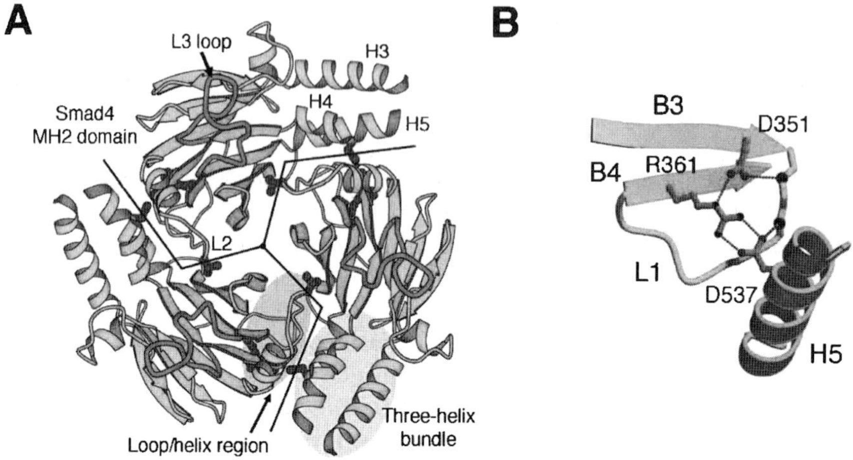

Figure 3. Structure of the Smad4 MH2 domain in a homo-trimeric assembly. (A) Structure of the MH2 domain of Smad4 in a homo-trimeric assembly (Shi et al., 1997). The loop/helix region of one Smad4 protomer packs against the three-helix bundle of the adjacent protomer. The L3 loops and a few secondary structural elements are labeled. Several residues targeted for inactivation in cancer map to the trimeric interface. (B) A close-up view on the key interactions at the trimeric interface (Shi et al., 1997). Shown here is a network of hydrogen bonds involving three residues, Asp351 and Arg361 from one protomer and Asp537 from the adjacent protomer. These three residues are invariant among all Smads; Asp450 in Smad2 (corresponding to Asp537 in Smad4) and Asp351 and Arg361 in Smad4 are targeted for recurrent tumorigenic mutations. These interactions are also preserved in the homo-trimeric interface of R-Smads (Wu et al., 2001b) and the heteromeric interfaces between R-Smad and Smad4 (Chacko et al., 2004; Wu et al., 2001b) (A color version of this figure is freely accessible via the website of the book: http://www.springer.com/1-4020-4542-5)

3. STRUCTURAL INSIGHTS INTO SMAD SIGNALING

The power of structural biology is embedded in the detailed three-dimensional images of macromolecules, which reveal conclusive mechanisms of action and offer readily testable hypotheses. Structural investigation has been applied to several important complexes of the Smad signaling pathway, providing important functional insights.

3.1 Recognition of Smad2 or Smad3 by SARA

The recognition and phosphorylation of R-Smads by the type I receptors are facilitated by auxiliary proteins. Smad2 and Smad3 can be specifically immobilized near the cell surface by the Smad Anchor for Receptor Activation (SARA) through direct interactions (Tsukazaki et al., 1998) (see Chapter 9). The interaction between SARA and Smad2 or Smad3 occurs via a Smad-binding domain (SBD) in SARA and the MH2 domain in Smads (Wu et al., 2000). This interaction is extremely specific since SARA does not recognize either Smad1 or Smad5, which mediates the BMP signaling pathways and shares greater than 80 percent sequence similarity with Smad2/Smad3. SARA contains a phospholipid-binding FYVE domain, which targets the molecule to the membrane of early endosomes (Tsukazaki et al., 1998) (see Chapter 9). These interactions allow more efficient recruitment of Smad2 or Smad3 to the receptors for phosphorylation (Tsukazaki et al., 1998). At steady state, the bulk of SARA and SARA-bound Smad2 are located in early endosomes. Receptor-mediated phosphorylation of SARA-bound Smad2 occurs at the plasma membrane but is more efficient in SARA-rich early endosomes to which the activated receptor complex is internalized via clathrin-coated pits (Di Guglielmo et al., 2003; Hayes et al., 2002; Lu et al., 2002).

The crystal structure of a Smad2 MH2 domain in complex with the SARA SBD reveals a striking arrangement (Fig. 4; Wu et al., 2000). The Smad2 MH2 exhibits a generally similar structure to that of Smad4 except that an amino-terminal β-strand forms an extension to the MH2 scaffold. The 40-residue SARA SBD, comprised of a rigid coil, an α-helix, and a β-strand, assumes an extended conformation traversing a distance of greater than 40 Å. SARA SBD recognizes one end of the β-sandwich in Smad2 with its rigid coil and forms a heterologous anti-parallel β-sheet with its β-strand, which packs against the three-helix bundle in Smad2. The interactions between Smad2 and SARA are predominantly hydrophobic in nature, leading to the burial of approximately 2600 $Å^2$ surface area. The intensity of interaction at the interface is reminiscent of protein interior. Five Smad2 residues that interact with SARA are subtype-conserved in Smad3, but are replaced by other residues

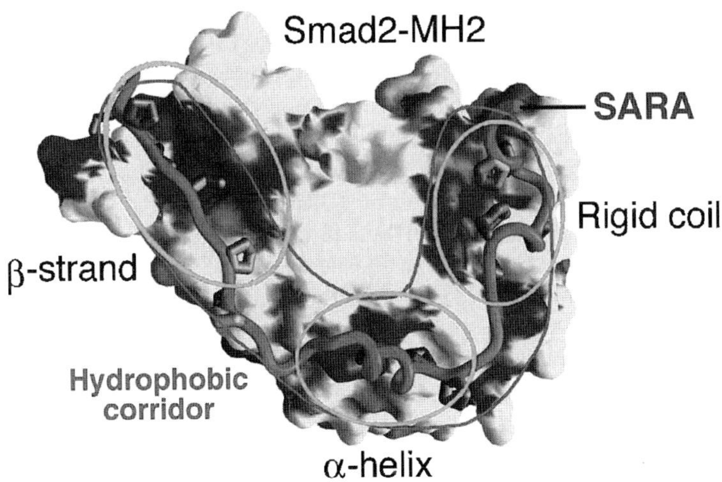

Figure 4. The hydrophobic corridor of the MH2 domain in R-Smads is the binding site for multiple proteins. Shown here is a surface representation of Smad2 MH2 bound to the SBD of SARA (Wu et al., 2000). Note that the extended SARA fragment follows the hydrophobic surface. Binding to Smad2 is anchored by a rigid coil at the amino-terminus and a β-strand at the carboxy terminus of SARA SBD. The rigid coil and the β-strand are connected by an α–helix, which also make interactions to specific residues in the MH2 domain. Most proteins that bind to the hydrophobic corridor contain SIM and/or FM motifs (Randall et al., 2002; Randall et al., 2004) (A color version of this figure is freely accessible via the website of the book: http://www.springer.com/1-4020-4542-5)

Abbreviations: SBD, Smad binding domain; SIM, Smad interaction motif; FM, FoxH1 or Fast motif in Smad1, Smad5, or Smad8, thus explaining the specificity of Smad2-SARA interaction (Wu et al., 2000).

The surface patch in the MH2 domain of Smad2 that is responsible for binding to SARA SBD is lined with hydrophobic amino acids and subsequently has been termed "hydrophobic corridor" (Fig. 4). A large body of evidence demonstrates that the hydrophobic corridor on Smad2 or Smad3 is the binding site for a number of proteins, both in the cytoplasm and in the nucleus, and often in a mutually exclusive manner (ten Dijke and Hill, 2004) (see Chapter 10). In addition to SARA, binding proteins include nucleoporins such as Nup214 and Nup153, and nuclear proteins such as FAST-1 and Mixer. Sequence comparison among these proteins revealed the presence of two short motifs, named Smad interaction motif (SIM) and Fast (or FoxH1) motif (FM), which are responsible for binding to the hydrophobic corridor (Randall et al., 2002; Randall et al., 2004).

3.2 R-Smad Recognition by the Activated Receptor Complex

Only the activated receptors efficiently phosphorylate R-Smads. In contrast to the inability of the unphosphorylated TβRI to interact with its downstream target Smad2, the phosphorylated TβRI protein binds efficiently to Smad2 *in vitro* and exhibits a dramatically enhanced phosphorylation specificity for the carboxy-terminal serine residues of Smad2 (Huse et al., 2001). How does the phosphorylated type I receptor exhibit an enhanced binding affinity for the R-Smad? One important clue came from a structural comparison of the MH2 domains of Smad2 and Smad4, which revealed the presence of a much more positively charged surface patch on Smad2 than that on Smad4, located next to the L3 loop (Wu et al., 2000). This basic surface contains residues that are subtype-conserved in all other R-Smads, suggesting a model in which this region is involved in receptor binding (Fig. 5). In this model, the phosphorylated GS region of TβRI directly interacts with the basic surface region next to the L3 loop of Smad2 or Smad3. Indeed, mutation of one of the invariant residues, His331, in the basic surface region of Smad2 leads to a reduction in its affinity for and phosphorylation by TβRI (Huse et al., 2001).

Although the predicted interaction between the phosphorylated GS region of the type I receptor and the basic patch of an R-Smad enhances binding affinity, it does little to control the signaling specificity. How is then a specific R-Smad chosen by the activated receptors? The answer lies in the L45 loop of the receptor kinase domain, which is located immediately adjacent to the GS region and specifies interactions with the R-Smads (Fig. 5; Chen et al., 1998; Feng and Derynck, 1997). The corresponding specificity-determinant in the R-Smad primarily involves the L3 loop (Chen et al., 1998; Lo et al., 1998). Thus matching sets of Smad L3 loop and receptor L45 loop determine the receptor-Smad choices. Other sequence elements of R-Smads may also play a role in this interaction (Huse et al., 2001). The detailed mechanism for this recognition awaits a structure of an activated type I receptor bound to an R-Smad.

The basic surface region in R-Smad that is responsible for binding to the activated type I receptor coincides with the loop-strand pocket that is required for binding to the pS-X-pS motif in the carboxy-terminus of an R-Smad protein (Fig. 5; Wu et al., 2001b). Thus, mutually exclusive binding to the same surface area between the pS motif in the receptor kinase and the competing pS-X-pS motif in an R-Smad protein likely accounts for the dissociation of the phosphorylated R-Smad from the receptor kinases (Fig. 5). The competing pS-X-pS motif could come from the

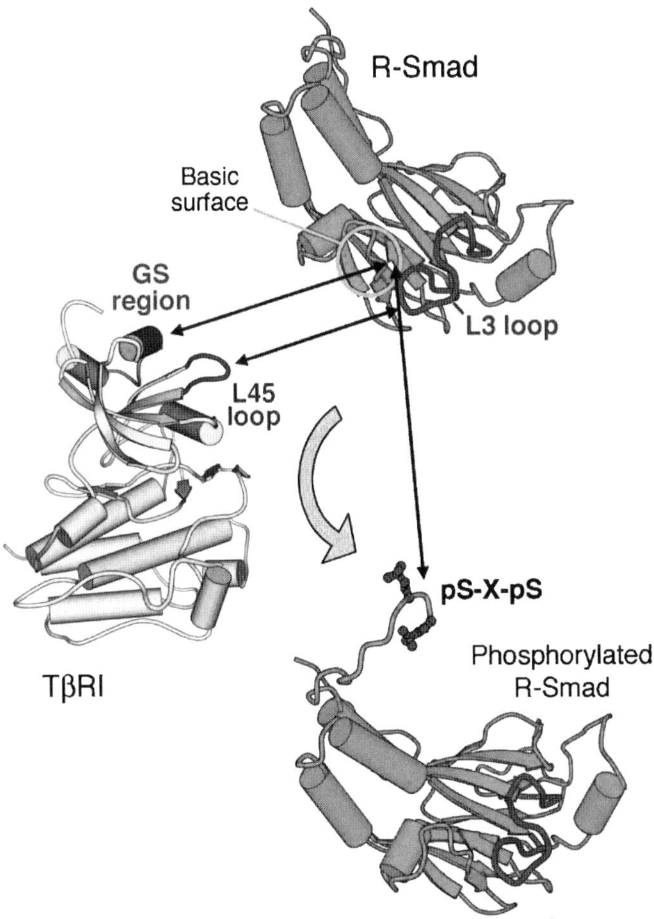

Figure 5. Recognition and phosphorylation of R-Smads by the type I receptor. Binding of Smad2 or Smad3 by the type I receptor (TβRI) has two matching components: the L3 loop and the adjacent basic loop/strand pocket of Smad2/Smad3 are recognized by the L45 loop and the phosphorylated GS region of TβRI (Chen et al., 1998; Feng and Derynck, 1997; Lo et al., 1998; Wu et al., 2000; Wu et al., 2001b). After phosphorylation, the pS-X-pS motif of Smad2/Smad3 likely competes with the phosphorylated GS region of TβRI for binding to the basic loop/strand pocket, thus releasing the phosphorylated Smads from the activated receptors (Wu et al., 2001b) (A color version of this figure is freely accessible via the website of the book: http://www.springer.com/1-4020-4542-5)

same dissociating R-Smad itself or from the adjacent R-Smad, as an activated TGF-β receptor complex contains two type I receptors and likely binds to and phosphorylates two R-Smad molecules simultaneously (Wu et al., 2001b).

3.3 Phosphorylation of R-Smads: Identification of a General Signaling Pattern

3.3.1 Structure of a phosphorylated R-Smad

R-Smads are directly phosphorylated by the activated type I receptors (Kretzschmar et al., 1997; Macias-Silva et al., 1996) (see Chapter 12). In the crystal structures of the MH2 domain from the unphosphorylated R-Smads, the carboxy-terminal 10 amino acids, including the characteristic SXS motif at the extreme carboxy-terminus, are completely flexible and disordered (Qin et al., 2001; Shi, 2001; Wu et al., 2000). Several lines of evidence demonstrate that phosphorylation of an R-Smad takes place in the carboxy-terminal two serine residues within the flexible SXS motif (Abdollah et al., 1997; Souchelnytskyi et al., 1997). Phosphorylation destabilizes R-Smad interaction with SARA, allowing dissociation of R-Smad from the complex (Xu et al., 2000). In addition, R-Smad phosphorylation augments its affinity for Smad4. The association of these two proteins nucleates the assembly of transcriptional regulation complexes.

Recombinant proteins of Smad2 and Smad3 with their carboxy-terminal SXS motif homogeneously phosphorylated were generated using an expressed protein ligation strategy (Wu et al., 2001b). Although the unphosphorylated Smad proteins exhibit a weak tendency for homo-trimerization, the phosphorylated Smad2 or Smad3 forms an extremely stable homo-trimer. Structural analysis of the phosphorylated Smad2-MH2 reveals the molecular basis for the formation of this stable homotrimer (Fig. 6A; Wu et al., 2001b). The C-terminal pS-X-pS motif of one Smad2 molecule is nestled in a positively charged surface pocket of the adjacent molecule through multiple hydrogen bonds (Fig. 6B). In addition to these phosphorylation-specific contacts, there is also a large protein-protein interface between adjacent Smad2 molecules, nearly identical to that observed in the homo-trimeric structure of Smad4 (Shi, 2001).

Importantly, the conformation of all structural elements in the MH2 domain of Smad2, except the amino- and carboxy-termini, remains unchanged before and after phosphorylation (Wu et al., 2000; Wu et al., 2001b). The observed conformational change at the amino-terminus of Smad2-MH2, involving a large movement of the B1′ strand, provides a plausible explanation to the decreased binding affinity for SARA (Wu et al., 2001b), as this β-strand is also required for interaction with SARA (Wu et al., 2000).

3.3.2 pS-X-pS as a signaling motif

The phosphorylated Smad2 structure not only explains the molecular mechanism for the formation of a homo-trimer for R-Smad but also reveals significant insights into the mechanism of R-Smad binding to Smad4 (see Section 3.4). In the structure of the phosphorylated Smad2, the pS-X-pS motif is primarily coordinated by four

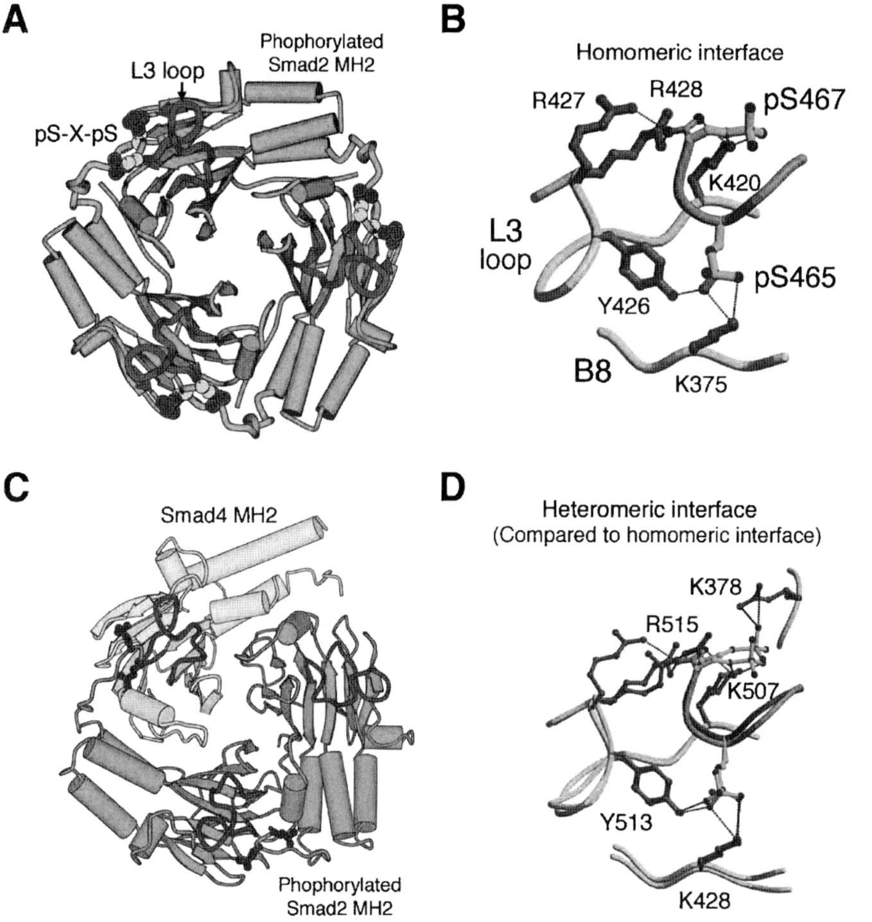

Figure 6. Features of homo- and heteromeric Smad complexes. (A) Structure of the phosphorylated MH2 domain of Smad2 (Wu et al., 2001b). The pS-X-pS motif (in ball-and-stick) drives the formation of a homo-trimer. The L3 loop is labeled. (B) Coordination of the pS-X-pS motif in the homomeric interface (Wu et al., 2001b). Five residues that play important roles in binding to the pS-X-pS motif are shown. Four of the five residues are conserved in Smad4. Hydrogen bonds are represented by dashed lines. (C) Structure of a hetero-trimer between the MH2 domains of Smad4 and the phosphorylated Smad2 MH2 domain (Chacko et al., 2004). The packing interactions are nearly identical to those reported for the phosphorylated Smad2 homotrimer (panel A) (Wu et al., 2001b). (D) Comparison of the coordination of the pS-X-pS motif in the heteromeric interface (Chacko et al., 2004) with that in the homomeric interface. Only Smad4 residues are labeled (A color version of this figure is freely accessible via the website of the book: http://www.springer.com/1-4020-4542-5)

residues in the positively charged loop-strand pocket, Lys375 on the β-strand B8 and Lys420/Tyr426/Arg428 on the L3 loop (Fig. 6B; Wu et al., 2001b). All four residue are invariant not only among all R-Smads but also in Smad4, strongly indicating that a similar surface pocket on Smad4 serves as the binding site for the pS-X-pS motif of the R-Smads in the heteromeric Smad complex (Wu et al., 2001b). This prediction is supported by biochemical and mutational analyses (Chacko et al., 2001; Wu et al., 2001b) and has been recently confirmed (Chacko et al., 2004).

The carboxy-terminal pS-X-pS motif in the R-Smads represents a general signaling motif (Wu et al., 2001b). The GS region of the type I receptor contains a signature sequence TXSXSXS, which becomes pT-X-pS-X-pS-X-pS upon phosphorylation-induced activation. Consequently, the interaction between the type I receptor and R-Smad is predicted to be mediated by the two contiguous pT/pS-X-pS motifs of the phosphorylated GS domain and the basic surface pocket of the MH2 domain (Fig. 5). Therefore, the pS-X-pS interaction with the MH2 basic surface pocket represents the signature interaction of the receptor serine/threonine kinase signaling pathway.

3.4 Hetero-oligomerization of Smads: The Role of Phosphorylation and Specificity

3.4.1 Structure of the heteromeric Smad complex

Structure of a heterotrimeric complex between one molecule of Smad4 MH2 and two molecules of the phosphorylated MH2 domain of Smad2 or Smad3 has been reported (Fig. 6C; Chacko et al., 2004). The packing interactions are nearly identical to those previously reported for the phosphorylated Smad2 homotrimer (Wu et al., 2001b). In addition to confirming an earlier model on the detailed interactions at the Smad heteromeric interface (Wu et al., 2001b), structural analysis also reveals two previously unpredicted findings (Chacko et al., 2004). First, Arg378 of Smad4, but not Lys507 (which corresponds to Lys420 in Smad2), directly binds to the pS-X-pS motif by donating two hydrogen bonds to the terminal pS in Smad2/Smad3 (Fig. 6D). Nonetheless, the coordination of the pS-X-pS motif by Smad4 is highly similar to that at the homomeric interface (Fig. 6D). Second, Asp493 in Smad4, which is mutated in cancer, makes a pair of charge-stabilized hydrogen bonds to two conserved Arg residues (Arg321/Arg279 and Arg329/Arg287) in Smad2/Smad3. Although mutation of Arg378 only exhibited a moderate effect on the formation of a heteromeric Smad complex, mutation of Arg493 to Ala in Smad4 nearly abrogated its ability to disrupt the pre-formed homo-trimer of the phosphorylated Smad3 (Chacko et al., 2004).

3.4.2 Stoichiometry of the heteromeric Smad complex

Using isolated MH2 domains, the stoichiometry of the heteromeric Smad complexes has been extensively investigated, giving rise to either a hetero-trimer or a hetero-dimer model (Chacko et al., 2001; Qin et al., 2001; Shi, 2001; Wu et al., 2001a; Wu et al., 2001b). Investigation of the Smad stoichiometry in cells suggests a complex situation (Inman and Hill, 2002; Jayaraman and Massagué, 2000; Kawabata et al., 1998). It was proposed that, depending on the gene promoter context, both a hetero-trimer and a hetero-dimer are possible (Inman and Hill, 2002). This proposal is supported by the extensive interface at the homo-trimeric assembly, which predicts that one interface is sufficient to maintain stable association.

It is important to note that both hetero-trimer and hetero-dimer employ the same docking architecture, the same interface, and the same detailed interactions, as those observed for the homo-trimeric assembly of Smad4 and the phosphorylated Smad2 (Chacko et al., 2004; Shi et al., 1997; Wu et al., 2001b). These observations dictate that the stoichiometry issue can only be addressed in a biologically relevant manner by investigating the active heteromeric Smad complexes in the nucleus. The conservation of the homo- and heteromeric Smad interface strongly suggests that, in the case of a hetero-trimer, three distinct Smad proteins can be incorporated. For example, Smad4 could form a hetero-trimer with both phosphorylated Smad2 and Smad3 as previously suggested (Feng et al., 2000). Alternatively, more than one type of phosphorylated R-Smads could come together to form a stable hetero-trimer (Wu et al., 2001b). In these cases, formation of a hetero-trimeric Smad complex likely allows integration of more input into the regulation of gene expression.

3.5 Smad Interactions with Ski and SnoN

Among the negative regulators of Smad transcriptional function, c-Ski and SnoN are two highly conserved members of the Ski family of proto-oncoproteins (see Chapter 14). Ski or SnoN antagonize TGF-β signaling through direct interactions with Smad4 and the R-Smads (Liu et al., 2001; Wang et al., 2000). The mechanism

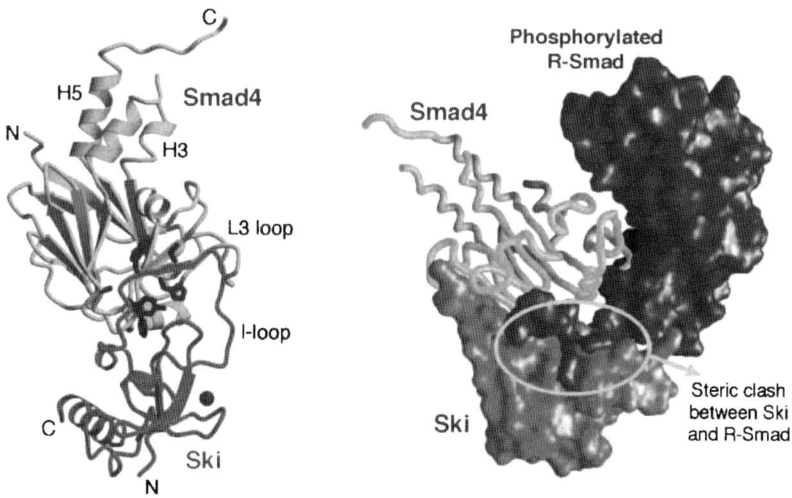

Figure 7. Smad4 recognition by the Ski/SnoN family of proteins (Wu et al., 2002). Structure of the Smad4 MH2 domain bound to the Smad4-binding domain of c-Ski (left panel) was used to model whether Ski can bind to the heteromeric Smad complex, which reveals that c-Ski significantly overlaps with R-Smad (right panel). This structural observation partly explains how Ski can antagonize Smad function – by preventing and disrupting the formation of a functional heteromeric Smad complex (A color version of this figure is freely accessible via the website of the book: http://www.springer.com/1-4020-4542-5)

of Ski-mediated repression of TGF-β signaling was primarily attributed to transcriptional modulation, through recruitment of the nuclear transcriptional co-repressor (N-CoR) and histone deacetylase (HDAC) as well as interference of Smad-mediated binding to the transcriptional co-activator, p300/CBP (Liu et al., 2001).

One additional mechanism was provided by the crystal structure of the Smad4-binding domain of c-Ski in complex with the MH2 domain of Smad4, which reveals a significant overlap between the Ski-binding and the R-Smad-binding surfaces on Smad4 (Fig. 7; Wu et al., 2002). Indeed, Ski competes with R-Smads for interaction with Smad4 and disrupts the formation of a functional complex between Smad4 and R-Smads (Wu et al., 2002). It is important to note that, because Ski contains separate sequence elements for interaction with Smad4 and R-Smads, both Smad4 and R-Smad may remain attached to Ski after disruption of the functional complex (Luo et al., 1999; Stroschein et al., 1999; Wu et al., 2002) but are unable to act as a functional unit to effect transcription.

4. SUMMARY AND FUTURE PROSPECTS

In the last eight years, structural biology of Smad proteins has contributed significantly to the elucidation of the underlying molecular mechanisms of TGF-β signaling from cell membrane to the nucleus. Visualizing the Smad proteins at an atomic resolution in three-dimension has a lasting impact on understanding their functions. The first piece of structural information on Smads comes from the crystal structure of the Smad4 MH2 domain at 2.5 Å resolution (Shi et al., 1997), which reveals a homo-trimeric assembly (Fig. 3A). Information gained from this structure lays the foundation for subsequent biochemical, functional, and structural investigations on the MH2 domain of all three types of Smad. The basic architecture and the vast majority of the detailed interactions observed in the Smad4 MH2 structure are recapitulated in the structures of the phosphorylated Smad2 (Wu et al., 2001b) and the heteromeric Smad complexes (Chacko et al., 2004). The rich surface features (loop/helix region, three-helix bundle, L3 loop, loop/strand pocket, etc) of the MH2 domain allows it to specifically complex with other MH2 domains and with a number of other proteins. These interactions in turn mediate the signals from cell membrane to the nucleus.

Despite major progress in mechanistic understanding of the Smad/TGF-β signaling, a number of biologically significant questions remain to be investigated by structural biology. At present, we do not yet understand, at a molecular and mechanistic level, how the R-Smads are recognized by the activated receptor complex and how this recognition is coupled to efficient phosphorylation of the SXS motif. It is also unclear how R-Smads and I-Smads are recognized by the Smurfs; available evidence suggests that the PY motif in R-Smads is necessary but not sufficient for Smurf-mediated negative regulation of Smads. There is no structural information on I-Smads – does the amino-terminal domain of I-Smads adopt a similar structure as the MH1 domain and how do the I-Smads compete with Smad4 for binding to R-Smads? Importantly, how do these pair-wise interactions play out in the context of cellular environment? Finally, how do Smad proteins cooperate with other transcription factors in the nucleus to recognize specific promoter DNA sequences? The answers to these questions not only rely on rigorous structural biology of the cognate protein-protein and protein-DNA complexes but also require meaningful integration of structural biology with biophysics, biochemistry, and cell biology.

ACKNOWLEDGEMENTS

I would like to thank two former members of my laboratory, Dr. Geng Wu and Dr. Jia-Wei Wu, for their critical contribution to aspects of this chapter. Research on TGF-β/Smad signaling in my laboratory is supported by grants from the National Institutes of Health (R01-CA082171).

REFERENCES

Abdollah, S., Macias-Silva, M., Tsukazaki, T., Hayashi, H., Attisano, L., and Wrana, J.L., 1997, TbRI phosphorylation of Smad2 on Ser465 and Ser467 is required for Smad2-Smad4 complex formation and signaling. *J Biol Chem* 272: 27678-27685.

Attisano, L., and Wrana, J.L., 2002, Signal transduction by the TGF-β superfamily. *Science* 296: 1646-1647.

Chacko, B.M., Qin, B., Correia, J.J., Lam, S.S., de Caestecker, M.P., and Lin, K., 2001, The L3 loop and C-terminal phosphorylation jointly define Smad protein trimerization. *Nat Struct Biol* 8: 248-253.

Chacko, B.M., Qin, B.Y., Tiwari, A., Shi, G., Lam, S., Hayward, L.J., De Caestecker, M., and Lin, K., 2004, Structural basis of heteromeric smad protein assembly in TGF-β signaling. *Mol Cell* 15: 813-823.

Chai, J., Wu, J.-W., Massagué, J., Pavletich, N.P., and Shi, Y., 2003, Features of a Smad3 MH1-DNA Complex: roles of water and zinc in DNA binding at 2.4 Å resolution. *J Biol Chem* 278: 20327-20331.

Chen, Y.-G., Hata, A., Lo, R.S., Wotton, D., Shi, Y., Pavletich, N., and Massasgue, J., 1998, Determinants of specificity in the TGF-β signal transduction. *Genes Develop* 12: 2144-2152.

Choo, Y., and Klug, A., 1997, Physical basis of a protein-DNA recognition code. *Curr Opin Struct Biol* 7: 117-125.

Dennler, S., Itoh, S., Vivien, D., ten Dijke, P., Huet, S., and Gauthier, J.-M., 1998, Direct binding of Smad3 and Smad4 to critical TGFβ-inducible elements in the promoter of human plasminogen activator inhibitor-type 1 gene. *EMBO J* 17: 3091-3100.

Derynck, R., and Zhang, Y.E., 2003, Smad-dependent and Smad-independent pathways in TGF-β family signalling. *Nature* 425: 577-584.

Di Guglielmo, G.M., Le Roy, C., Davidson, A.F., and Wrana, J.L., 2003, Distinct endocytic pathways regulate TGFβ receptor signaling and turnover. *Nature Cell Biol* 5: 410-421.

Eppert, K., Scherer, S.W., Ozcelik, H., Pirone, R., Hoodless, P., Kim, H., Tsui, L.-C., Bapat, B., Gallinger, S., Andrulis, I.L., Thomsen, G.H., Wrana, J.L., and Attisano, L., 1996, MADR2 maps to 18q21 and encodes a TGFβ-regulated MAD-related protein that is functionally mutated in colorectal carcinoma. *Cell* 86: 543-552.

Feng, X.-H., and Derynck, R., 1997, A kinase subdomain of transforming growth factor-β (TGF-β) type I receptor determines the TGF-β intracellular signaling specificity. *EMBO J* 16: 3912-3923.

Feng, X.H., Lin, X., and Derynck, R., 2000, Smad2, Smad3 and Smad4 cooperate with Sp1 to induce p15(Ink4B) transcription in response to TGF-β. *EMBO J* 19: 178-193.

Grishin, N.V., 2001, MH1 domain of Smad is a degraded homing endonuclease. *J Mol Biol* 307: 31-37.

Hata, A., Lagna, G., Massagué, J., and Hemmati-Brivalou, A., 1998, Smad6 inhibits BMP/Smad1 signaling by specifically competing with the Smad4 tumour suppressor. *Genes Dev* 12: 186-197.

Hayes, S., Chawla, A., and Corvera, S., 2002, TGF β receptor internalization into EEA1-enriched early endosomes: role in signaling to Smad2. *J Cell Biol* 158: 1239-1249.

Huse, M., Muir, T.W., Xu, L., Chen, Y.-G., Kuriyan, J., and Massagué, J., 2001, The TGFβ receptor activation process: An inhibitor- to substrate-binding switch. *Mol Cell* 8: 671-682.

Inman, G.J., and Hill, C.S., 2002, Stoichiometry of active Smad-transcription factor complexes on DNA. *J Biol Chem* 277: 51008-51016.

Ishida, W., Hamamoto, T., Kusanagi, K., Yagi, K., Kawabata, M., Takehara, K., Sampath, T.K., Kato, M., and Miyazono, K., 2000, Smad6 is a Smad1/5-induced smad inhibitor. Characterization of bone morphogenetic protein-responsive element in the mouse Smad6 promoter. *J Biol Chem* 275: 6075-6079.

Jayaraman, L., and Massagué, J., 2000, Distinct oligomeric states of SMAD proteins in the TGF-β pathway. *J Biol Chem* 275: 40710-40717.

Johnson, K., Kirkpatrick, H., Comer, A., Hoffmann, F.M., and Laughon, A., 1999, Interaction of Smad complexes with tripartite DNA-binding sites. *J Biol Chem* 274: 20709-20716.

Kawabata, M., Inoue, H., Hanyu, A., Imamura, T., and Miyazono, K., 1998, Smad proteins exist as monomers *in vivo* and undergo homo- and hetero-oligomerization upon activation by serine/threonine kinase receptors. *EMBO J* 17: 4056-4065.

Kim, J., Johnson, K., Chen, H.J., Carroll, S., and Laughon, A., 1997, Drosophila Mad binds to DNA and directly mediates activation of vestigial by Decapentaplegic. *Nature* 388: 304-308.

Kraulis, P.J., 1991, Molscript: a program to produce both detailed and schematic plots of protein structures. *J Appl Crystallogr* 24: 946-950.

Kretzschmar, M., Liu, F., Hata, A., Doody, J., and Massagué, J., 1997, The TGF-β family mediator Smad1 is phosphorylated directly and activated functionally by the BMP receptor kinase. *Genes & Development* 11: 984-995.

Kusanagi, K., Inoue, H., Ishidou, Y., Mishima, H.K., Kawabata, M., and Miyazono, K., 2000, Characterization of a Bone Morphogenetic Protein-responsive Smad-binding element. *Mol Biol Cell* 11: 555-565.

Labbe, E., Silvestri, C., Hoodless, P.A., Wrana, J.L., and Attisano, L., 1998, Smad2 and Smad3 Positively and Negatively Regulate TGFb-Dependent Transcription through the Forkhead DNA-Binding Protein FAST2. *Mol Cell* 2: 109-120.

Liu, X., Sun, Y., Weinberg, R.A., and Lodish, H.F., 2001, Ski/Sno and TGF-β signaling. *Cytokine Growth factor Rev* 12: 1-8.

Lo, R.S., Chen, Y.-G., Shi, Y., Pavletich, N.P., and Massagué, J., 1998, The L3 loop: a structural motif determining specific interactions between SMAD proteins and TGF-β receptors. *EMBO J* 17: 996-1005.

Lu, Z., Murray, J.T., Luo, W., Li, H., Wu, X., Xu, H., Backer, J.M., and Chen, Y.G., 2002, Transforming growth factor β activates Smad2 in the absence of receptor endocytosis. *J Biol Chem* 277: 29363-29368.

Luo, K., Stroschein, S.L., Wang, W., Chen, D., Martens, E., Zhou, S., and Zhou, Q., 1999, The Ski oncoprotein interacts with the Smad proteins to repress TGFβ signaling. *Genes Dev* 13: 2196-2206.

Macias-Silva, M., Abdollah, S., Hoodless, P.A., Pirone, R., Attisano, L., and Wrana, J.L., 1996, MADR2 Is a substrate of the TGFβ receptor and its phosphorylation is required for nuclear accumulation and signalling. *Cell* 87: 1215-1224.

Massagué, J., 1998, TGF-β signal transduction. *Annu Rev Biochem* 67: 753-791.

Moustakas, A., and Kardassis, D., 1998, Regulation of the human p21/WAF1/Cip1 promoter in hepatic cells by functional interactions between Sp1 and Smad family members. *J Biol Chem* 95: 6733-6738.

Moustakas, A., Pardali, K., Gaal, A., and Heldin, C.-H., 2002, Mechanisms of TGF-β signaling in regulation of cell growth and differentiation. *Immunol letters* 82: 85-91.

Nicholls, A., Sharp, K.A., and Honig, B., 1991, Protein folding and association: insights from the interfacial and thermodynamic properties of hydrocarbons. *Proteins: Struct Funct Genet* 11: 281-296.

Pardali, K., Kurisaki, A., Morén, A., ten Dijke, P., Kardassis, D., and Moustakas, A., 2000, Role of Smad proteins and transcription factor Sp1 in p21/WAF-1/Cip-1 regulation by transforming growth factor-β. *J Biol Chem* 275: 29244-29256.

Patterson, G.I., and Padgett, R.W., 2000, TGFβ-related pathways. Roles in *Caenorhabditis elegans* development. *Trends Genet* 16: 27-33.

Qin, B., Lam, S.S.W., and Lin, K., 1999, Crystal structure of a transcriptionally active Smad4 fragment. *Structure* 7: 1493-1503.

Qin, B.Y., Chacko, B.M., Lam, S.S., de Caestecker, M.P., Correia, J.J., and Lin, K., 2001, Structural basis of Smad1 activation by receptor kinase phosphorylation. *Mol Cell* 8: 1303-1312.

Randall, R.A., Germain, S., Inman, G.S., Bates, P.A., and Hill, C.S., 2002, Different Smad2 partners bind a common hydrophobic pocket in Smad2 via a defined proline-rich motif. *EMBO J* 21: 145-156.

Randall, R.A., Howell, M., Page, C.S., Daly, A., Bates, P.A., and Hill, C.S., 2004, Recognition of phosphorylated-Smad2-containing complexes by a novel Smad interaction motif. *Mol Cell Biol* 24: 1106-1121.

Schutte, M., Hruban, R.H., Hedrick, L., Cho, K.R., Nadasdy, G.M., Weinstein, C.L., Bova, G.S., Isaacs, W.B., Cairns, P., Nawroz, H., Sidransky, D., Casero, R.A., Meltzer, P.S., Hahn, S.A., and Kern, S.E., 1996, DPC4 gene in various tumor types. *Cancer Research* 56: 2527-2530.

Shi, Y., 2001, Structural insights on Smad function in TGFβ signaling. *BioEssays* 23: 223-232.

Shi, Y., Hata, A., Lo, R.S., Massagué, J., and Pavletich, N.P., 1997, A structural basis for mutational inactivation of the tumour suppressor Smad4. *Nature* 388: 87-93.

Shi, Y., and Massagué, J., 2003, Mechanisms of TGF-β signaling from cell membrane to the nucleus. *Cell* 113: 685-700.

Shi, Y., Wang, Y.-F., Jayaraman, L., Yang, H., Massagué, J., and Pavletich, N.P., 1998, Crystal structure of a Smad MH1 domain bound to DNA: Insights on DNA binding in TGF-β signaling. *Cell* 94: 585-594.

Souchelnytskyi, S., Tamaki, K., Engström, U., Wernstedt, C., ten Dijke, P., and Heldin, C.-H., 1997, Phosphorylation of Ser465 and Ser467 in the C terminus of Smad2 mediates interaction with Smad4 and is required for transforming growth factor-β signaling. *J Biol Chem* 272: 28107-28115.

Stroschein, S.L., Wang, W., Zhou, S., Zhou, Q., and Luo, K., 1999, Negative feedback regulation of TGF-β signaling by the SnoN oncoprotein. *Science* 286: 771-774.

ten Dijke, P., Goumans, M.-J., Itoh, F., and Itoh, S., 2002, Regulation of cell proliferation by Smad proteins. *J Cell Physiol* 191: 1-16.

ten Dijke, P., and Hill, C.S., 2004, New insights into TGF-β-Smad signalling. *Trends Biochem Sci* 29: 265-273.

Tsukazaki, T., Chiang, T.A., Davison, A.F., Attisano, L., and Wrana, J.L., 1998, SARA, a FYVE domain protein that recruits Smad2 to the TGFβ receptor. *Cell* 95: 779-791.

Wang, W., Mariani, F.V., Harland, R.M., and Luo, K., 2000, Ski represses bone morphogenetic protein signaling in *Xenopus* and mammalian cells. *Proc Natl Acad Sci USA* 97: 14394-14399.

Wu, G., Chen, Y.-G., Ozdamar, B., Gyuricza, C.A., Chong, P.A., Wrana, J.L., Massagué, J., and Shi, Y., 2000, Structural basis of Smad2 recognition by the Smad anchor for receptor activation. *Science* 287: 92-97.

Wu, J.-W., Fairman, R., Penry, J., and Shi, Y., 2001a, Formation of a stable heterodimer between Smad2 and Smad4. *J Biol Chem* 276: 20688-20694.

Wu, J.-W., Hu, M., Chai, J., Seoane, J., Huse, M., Li, C., Rigotti, D.J., Kyin, S., Muir, T.W., Fairman, R., Massagué, J., and Shi, Y., 2001b, Crystal structure of a phosphorylated Smad2: recognition of phosphoserine by the MH2 domain and insights on Smad function in TGF-β signaling. *Mol Cell* 8: 1277-1289.

Wu, J.-W., Krawitz, A.R., Chai, J., Li, W., Zhang, F., Luo, K., and Shi, Y., 2002, Structural mechanism of Smad4 recognition by the nuclear oncoprotein Ski: insights on Ski-mediated repression of TGF-β signaling. *Cell* 111: 357-367.

Xu, L., Chen, Y.-G., and Massagué, J., 2000, Smad2 nuclear import function masked by SARA and unmasked by TGFβ-dependent phosphorylation. *Nat Cell Biol* 2: 559-562.

Yingling, J.M., Datto, M.B., Wong, C., Frederick, J.P., Liberati, N.T., and Wang, X.-F., 1997, Tumor suppressor Smad4 is a transforming growth factor β-inducible DNA binding protein. *Mol Cell Biol* 17: 7019-7028.

Zawel, L., Dai, J.L., Buckhaults, P., Zhou, S., Kinzler, K.W., Vogelstein, B., and Kern, S.E., 1998, Human Smad3 and Smad4 are sequence-specific transcription activators. *Molecular Cell* 1: 611-617.

CHAPTER 12

REGULATION OF SMAD FUNCTION BY PHOSPHORYLATION

IHOR YAKYMOVYCH AND SERHIY SOUCHELNYTSKYI
Lugwig Institute for Cancer Research, Box 595, 751 24 Uppsala, Sweden

Abstract: Phosphorylation is a dynamic and reversible post-translational modification which has a strong impact on structural features of proteins. All Smad proteins have been reported as phosphoproteins. Phosphorylation of receptor-activated Smad proteins by type I receptors initiates TGF-β family signaling. Phosphorylation of Smads by a number of other kinases is a mechanism of cross-talk between Smads and other signaling pathways. Smad phosphorylation initiates a cascade of protein-protein and protein-DNA interactions, and affects localization of Smads. This chapter reviews the role of phosphorylation in Smad signaling

Keywords: cross-talk; MAP kinase; phosphorylation; TGF-β

1. INTRODUCTION

TGF-β superfamily members elicit their cellular response through ligand-induced formation of heteromeric complexes of specific transmembrane type I and II kinase receptors. The activity of this pathway is tightly controlled by serine/threonine phosphorylation, which plays a key role in regulating protein–protein interactions that are critical in the elaboration of signaling responses. The type II receptor is a constitutively active kinase, which upon ligand-mediated heteromeric complex formation phosphorylates particular serine and threonine residues in the type I receptor juxtamembrane region (GS domain), resulting in the activation of the type I receptor (Souchelnytskyi et al., 1996; Wieser et al., 1995; Wrana et al., 1994). The activated type I receptor then transiently associates with and phosphorylates Smad proteins (reviewed in Heldin et al., 1997). Smad phosphorylation by activated receptors triggers intracellular signaling mechanisms, which include interaction of Smads with a number of proteins. Phosphorylation of Smads regulates their intracellular localization and activity. Smads are also phosphorylated by kinases

P. ten Dijke and C.-H. Heldin (eds.), Smad Signal Transduction, 235–252.

other than TGF-β family receptors, thus serving as a convergence point for different signaling pathways. The role of phosphorylation in Smad function is discussed in this chapter.

2. RECEPTOR-ACTIVATED SMADS

2.1 Activation by Type I Receptors

Phosphorylation of the receptor-regulated Smads (R-Smads) by type I receptor kinases is a crucial step in TGF-β family signaling (Abdollah et al., 1997; Macias-Silva et al., 1996; Souchelnytskyi et al., 1997). TGF-β and Activin stimulate phosphorylation of Smad2 and Smad3 (Chen et al., 1996; Eppert et al., 1996; Nakao et al., 1997; Zhang et al., 1996). Smad1, Smad5, and Smad8 are phosphorylated after stimulation with BMPs (Hoodless et al., 1996; Kretzschmar et al., 1997b; Liu et al., 1996; Tamaki et al., 1998). By interacting with the phosphoserines and phosphothreonines of the GS domain of TGF-β receptor type I (TβR-I), a positively charged groove (Arg330, His331, Lys420, Arg427, Arg428, Thr430) next to the L3 loop region of the Smad2 MH2 domain is required for Smad2 docking to activated TβR-I (Huse et al., 2001; Wu et al., 2000). Substrate specificity is determined by sequences in both the receptor and the Smad molecule. Selection of an R-Smad by a receptor is specified by the type I receptor L45 loop (Chen et al., 1998) and the R-Smad L3 loop (Feng and Derynck, 1997; Lo et al., 1998). However, other elements of R-Smads may also contribute to the specificity of the receptor-Smad interaction (Huse et al., 2001; Shi and Massagué, 2003), including the C-terminal sequences which are substrates for receptor kinases (Lo et al., 1998; Yakymovych et al., 2004).

At their C-termini, R-Smads have a characteristic Ser-Ser-X-Ser (SSXS) motif, and the two most C-terminal serine residues are directly phosphorylated by type I receptor kinases (Abdollah et al., 1997; Kretzschmar et al., 1997b; Macias-Silva et al., 1996; Souchelnytskyi et al., 1997). A prominent role of C-terminal phosphorylation for Smad signaling was demonstrated with Smad1 mouse lines that carry mutations at the C-terminal phosphorylatable serine residues. These mutants die in utero and display defects in alantois formation and in primordial germ cells specification like Smad1 null embryos (Aubin et al., 2004).

It has been suggested that phosphorylation of R-Smads by type I receptor proceed through two successive phosphorylation reactions. These reactions are performed in a defined order: first Ser467 in the C-terminus of Smad2 is phosphorylated, and only then can Ser465 be phosphorylated (Souchelnytskyi et al., 1997).

Ser464 is not phosphorylated upon ligand activation, but it is important for efficient phosphorylation of Smad2 (Abdollah et al., 1997). In the phosphorylated Smad2 the side chain of Ser464 accepts a hydrogen bond from the backbone amide of Ser467 while making two additional contacts to its phosphate (Wu et al., 2001). It is possible that phosphorylation of this serine residue may disrupt binding by the phosphorylated C-terminus, likely impeding TGF-β signaling (Wu et al., 2001).

Phosphorylation at the SSXS motif destabilizes R-Smad interaction with the receptors and accessory proteins (*e.g.* SARA), allowing dissociation of activated Smad and formation of homo- or heteromeric Smad complexes. It has been shown that phosphorylation on both serine residues is necessary for dissociation from the receptor. Mutation of either Ser465 or Ser467 to Ala in Smad2 prevents its dissociation from the activated TβR-I due to a stable interaction between the mutant Smad2 and the TGF-β receptor complex (Abdollah et al., 1997; Souchelnytskyi et al., 1997).

There are two models which explain how the recruited R-Smads are released after phosphorylation. One of them predict that, upon activation, the L3 loop of Smad undergoes a conformation change and can serve as a switch for R-Smad dissociation from the receptor (Qin et al., 2001). However, analysis of the crystal structure of the MH2 domain of Smad2 after phosphorylation displayed that the conformation of all structural elements, except the N- and C-termini, remains unchanged (Wu et al., 2000; Wu et al., 2001). It was also suggested, that phosphorylation of the C-terminus increases its affinity for a positively charged surface pocket formed by the L3 loop and the B8 strand in the adjacent monomer (Chacko et al., 2001; Wu et al., 2001). The same pocket is required for binding to the phosphorylated GS domain of TβR-I (Huse et al., 2001; Wu et al., 2001). Thus, competition for binding to this surface between the pSer motif in the receptor kinase and the pSer-containing C-terminus in R-Smad likely accounts for the dissociation of phosphorylated Smad from the receptor kinases (Wu et al., 2001). In fact, as the receptor complex contains two molecules of TβR-I (Luo and Lodish, 1996), which likely binds and phosphorylates two Smad molecules simultaneously, the phosphorylated R-Smad molecules might help each other to dissociate from the receptor complex (Qin et al., 2001; Wu et al., 2001).

Phosphorylation of Smad2 also induces its dissociation from SARA (Tsukazaki et al., 1998). In the Smad2-SARA complex the Smad-binding domain (SBD) of SARA interacts with the B1′-strand of the Smad2 MH2 domain (Wu et al., 2000). Upon phosphorylation of the SSXS motif the N-terminal extension of the MH2 domain of Smad2 moves more than 12 Å away from the position of the B1′ strand, thus destabilizing the SARA-Smad2 interface (Wu et al., 2001). As SARA also interacts with the receptor complex, phosphorylation-driven dissociation of Smad2 from SARA may be one of the key events in releasing activated Smad2 from the receptor complex (Wu et al., 2001). In the experiments with purified proteins it has been shown that Smad2 can dissociate from SARA only when both Ser465 and Ser467 are phosphorylated (Ottesen et al., 2004).

Concurrently with releasing activated R-Smads from the receptor complex, phosphorylation at the SSXS motif induces formation of homo- and heteromeric Smad complexes (Jayaraman and Massagué, 2000; Kawabata et al., 1998; Qin et al., 2001). In contrast to the disordered C-terminal residues in the unphosphorylated Smad2 (Wu et al., 2000), phosphorylation on Ser465 and Ser467 triggers their interactions with four residues, Lys375 on the β strand B8 and Lys420/Tyr426/Arg428 on the L3 loop of an adjacent Smad2, stabilizing the well-defined conformation of this fragment (Wu et al., 2001). Importantly, these four residues are invariant not only among all R-Smads but also for the Co-Smad (Smad4), suggesting critical roles in hetero-oligomerization (Wu et al., 2001).

The phosphorylated C-terminus also makes extensive van der Waals contacts, strengthening the formation of the complexes. Thus, Cys463, which is important for specific phosphorylation of Smad2 by TβR-I (Yakymovych et al., 2004), is completely buried in a hydrophobic pocket formed by five residues (Met327, Thr328, His331, Val419, and Lys420) in the neighbouring Smad2 molecule (Wu et al., 2001).

The available experimental data suggest that C-terminal serine residues are not involved directly in the delivery of Smads into the nucleus (Moustakas et al., 2001; Shi and Massagué, 2003). However, their phosphorylation induces conformational changes which release Smads from cytoplasmic retention proteins or expose signals for association with importin carriers (Moustakas et al., 2001; Reguly and Wrana, 2003).

In the nucleus, Smads interact with a wide array of specific DNA binding proteins to regulate transcriptional responses (Inman and Hill, 2002; Massagué and Wotton, 2000; Moustakas et al., 2001). In most of the cases this interaction is mediated by the MH2 domain of Smads (Kato et al., 2002; Massagué and Wotton, 2000; Shi and Massagué, 2003). It was reported that mutation of the C-terminal serines prevents Smad2 association with the winged-helix transcription factor FAST-I (Liu et al., 1997). However, it is not clear if FAST-I directly interacts with phosphorylated serines, or whether this inhibition is due to disruption of Smad complexes.

Activation of Smad2 and Smad3 by TGF-β signaling induces their ubiquitin-mediated proteolysis (Fukuchi et al., 2001; Lo and Massagué, 1999). In the nucleus, Smad3, through its C-terminal MH2 domain, interacts with a RING finger protein, ROC1, in a ligand-dependent manner, and this results in Smad3 export from the nucleus to the cytoplasm for proteasomal degradation (Fukuchi et al., 2001). However, it was not shown that phosphorylated serines are directly involved in this interaction. Degradation of Smad2 seems not to be dependent on phosphorylation of Smad2, as such, but occurs when Smad2 enters the nucleus (Lo and Massagué, 1999). Nonetheless, it was noted that a receptor-phosphorylated Smad2 was a better substrate for the ubiquitin/proteasome system (Lo and Massagué, 1999).

Thus, multiple protein interactions triggered by receptor-mediated R-Smad phosphorylation control their subcellular localization and functions. These may in turn control the duration of Smad phosphorylation and activation, and thus may give rise to qualitatively different responses resulting from different signaling thresholds (Derynck and Zhang, 2003). The same region of R-Smad is used often for interaction with different components of the TGF-β signaling pathway. Thus, phosphorylation at the C-terminus reduces R-Smads' affinity for the receptor complex and, at the same time, increases their ability to form complexes with other R-Smads and/or Co-Smad; this phosphorylation thus facilitates nuclear entry and regulation of gene expression.

2.2 Phosphorylation of R-Smads by Other Kinases

Smads can be phosphorylated by other kinases than TGF-β receptors, which results in complex phosphorylation patterns of endogenous proteins (Souchelnytskyi et al., 1997; Yakymovych et al., 2001).

The R-Smads are targets for the Erk mitogen-activated protein kinase (MAPK), stimulated by the activation of tyrosine kinase receptors and/or Ras. Activation of Erk kinase by hepatocyte growth factor (HGF) and epidermal growth factor (EGF), and in particular by activated Ras, can induce the phosphorylation of serine residues in S/TP or PXS/TP motifs in R-Smad linker regions (Calonge and Massagué, 1999; Kretzschmar et al., 1997a; Kretzschmar et al., 1999) (Figs. 1 and 2).

In vitro, Erk1, in the presence of activated MEK1, phosphorylates Smad2 on serine and threonine, but not on tyrosine, residues (Funaba et al., 2002). It was proposed that Smad2 can be phosphorylated by the Erk MAPK in the linker region on Thr220, Ser245, Ser250, and Ser255 residues (Funaba et al., 2002; Kretzschmar et al., 1999). In the linker region of Smad3 this kinase can phosphorylate Thr178, Ser203, Ser207, and Ser212 residues (Kretzschmar et al., 1999; Matsuura et al., 2004). Mutation of these residues reduced Smad phosphorylation by recombinant activated Erk2 in *in vitro* kinase assays and in Ras-transfected cells (Kretzschmar et al., 1999). Stimulation of mink lung epithelial cells with EGF has been shown to induce phosphorylation of Ser187, Ser195, Ser206 and Ser214 in the linker region of Smad1 (Kretzschmar et al., 1997a).

Erk phosphorylation stabilizes Smad2 protein, and this effect seems not to depend on Smad phosphorylation at the C-terminal serines (Funaba et al., 2002). The TβR-I phosphorylation sites are not required for Ras-induced phosphorylation (Kretzschmar et al., 1999), however, mutation of C-terminal serine residues prevented phosphorylation of Smad2 upon treatment of cells with HGF (de Caestecker et al., 1998).

It has been reported that Erk1 can phosphorylate also the MH1 domain in Smad2 (Thr8) and, possibly, in Smad1 (Ser11) (Funaba et al., 2002). Phosphorylation of Thr8 in Smad2 may be necessary for its maximal transcriptional activation and is important for interaction with calmodulin (Funaba et al., 2002).

There are conflicting reports about the effect of R-Smad phosphorylation in the linker region by Erk MAPK. Thus, it has been shown that phosphorylation of R-Smads by Erk MAPK inhibits their nuclear localization induced by TGF-β or BMP (Calonge and Massagué, 1999; Kretzschmar et al., 1997a; Kretzschmar et al., 1999). This may explain the loss of TGF-β antiproliferative responses in cells with hyperactivated Ras (Kretzschmar et al., 1999). Similar effect of phosphorylation in the linker region on the intracellular localization of Smads has been shown during embryonal development (Grimm and Gurdon, 2002). Phosphorylation of three serine residues (Ser245, Ser250 and Ser255) of *Xenopus* Smad2 caused the temporal restriction on the ability of Smad2 to accumulate in the nucleus and the loss of mesodermal competence (Grimm and Gurdon, 2002).

Other studies have not found impaired nuclear translocation of Smads in Ras-transformed cells or in cells with activated MAPK signaling (de Caestecker et al., 1998; Engel et al., 1999; Funaba et al., 2002). On the contrary, the MEK1-Erk

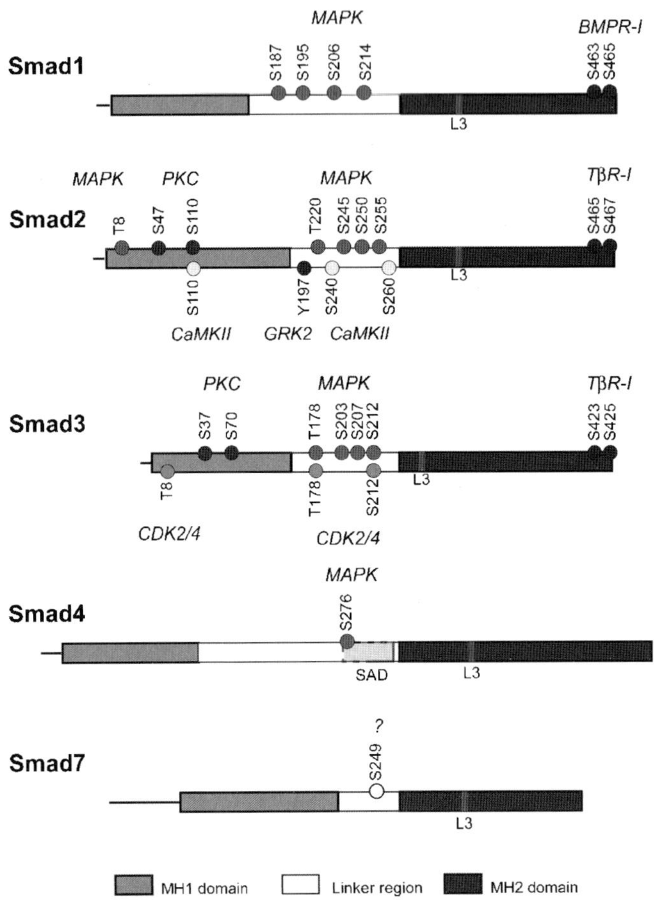

Figure 1. Phosphorylation of Smad proteins. Schematic presentation of phosphorylation sites in Smad proteins. Kinases which phosphorylate Smads at indicated positions, are mentioned (see text for references) (A color version of this figure is freely accessible via the website of the book: http://www.springer.com/1-4020-4542-5)

Abbreviations: BMPR-I, BMP receptor type I; CaMKII, Ca/calmodulin dependent kinase II; CDK2/4, cyclin-dependent kinase 2/4; GRK2, protein G-coupled receptor kinase 2; L3, L3 loop; MAPK, mitogen-activated protein kinase; PKC, protein kinase C; SAD, Smad-activation domain; TβR-I, TGFβ receptor type I

pathway has been shown to increase Smad2/Smad4 complex formation and is required for optimal TGF-β-induced transcriptional activation of R-Smads (Brown et al., 1999; de Caestecker et al., 1998; Funaba et al., 2002; Hayashida et al., 2003). Mutation of the sites phosphorylated by ERK MAPK in the linker region of Smad1 affected cell contacts, actin cytoskeleton, and nuclear β-catenin accumulation, which correlated with retention of Smad1 at the plasma membrane (Aubin et al., 2004). This suggests that sites of Erk MAP kinase-induced phosphorylation may also be

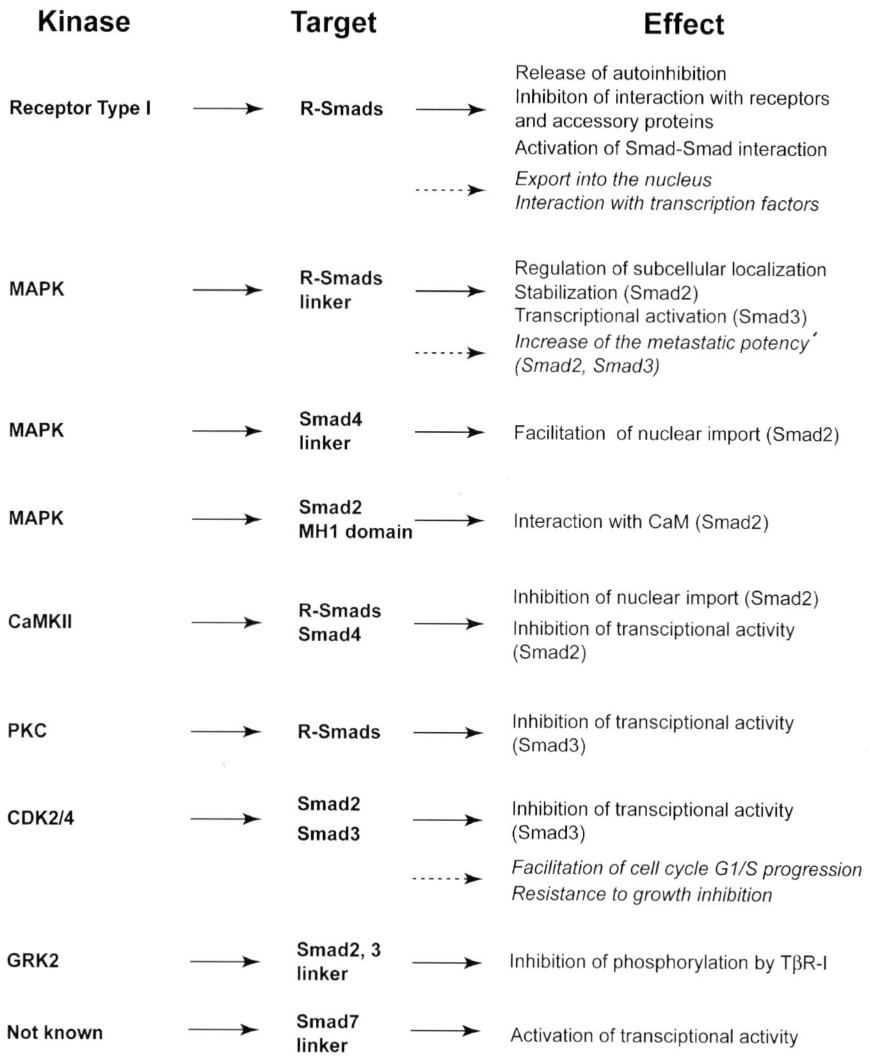

Figure 2. Functional effects of Smad phosphorylation. The impact of Smad phosphorylation of cellular functions (Effect) is described, as the result of phosphorylation of the Smad proteins (Target) by the kinases (Kinase), as indicated. Dashed arrows point to the cellular functions affected by the phosphorylation, though immediate involvement of the phosphorylated residues has not been shown (see text for references)

Abbreviations are as in Figure 1

required for TGF-β-dependent nuclear translocation of the R-Smad/Smad4 complex (de Caestecker et al., 2000; Mori et al., 2004).

It was also shown, that Ser203 and Ser207 residues in Smad3 are targets of the p38 MAP kinase and Rho/ROCK pathways (Kamaraju and Roberts, 2005). Their phosphorylation is required for the full transcriptional activation of Smad3 by

TGF-β, and is necessary for down-regulation of c-Myc protein and up-regulation of p21^{waf1} protein (Kamaraju and Roberts, 2005).

These contrasting findings indicate that activation of MAP kinase pathways may have positive or negative regulatory effects on R-Smads. It should be noted that TGF-β itself induces activation of the MAP kinases (Kamaraju and Roberts, 2005; Mulder, 2000; Nohe et al., 2004; Yamaguchi et al., 1995). The "cellular context", in turn, may affect both the specificity and multiplicity of MAP kinase-dependent phosphorylation events. For example, Erk-dependent activation of Smad2 and Smad3 by TGF-β was detected in human mesangial cells, and it was not seen in the mouse mammary epithelial NMuMG cell line (Hayashida et al., 2003). Interestingly, Smad2 phosphorylated in the linker region was retained in the cytoplasm, whereas linker-phosphorylated Smad3 was located exclusively in the nuclei of Ki-67-immunoreactive adenocarcinomas (Yamagata et al., 2005). Thus, results of phosphorylation in the linker region may be different for different R-Smads.

A high degree of phosphorylation of Smad2 and Smad3 in the linker regions by activated c-Jun NH_2-terminal kinase (JNK) was detected in sporadic colorectal adenocarcinomas (Yamagata et al., 2005). Such phosphorylation was increased in late-stage invasive and metastatic cancers, whereas in normal colorectal epithelial cells Smad2 and Smad3 were phosphorylated at C-terminal regions but not at linker regions (Yamagata et al., 2005). This observation corresponds to the observation, that in rat normal gastric mucosa cells, activation of the JNK pathway by HGF and TGF-β resulted in increased infiltration potency, which was caused partly by induction of plasminogen activator inhibitor type 1 (PAI-1) via Smad3 phosphorylation in the linker region (Mori et al., 2004).

Recently, it was shown that amino acid residues 200–230 in the linker region of Smad3 are required for the transcriptional activity of Smad3 and its interaction with the co-activator pCAF (Prokova et al., 2005). It is possible that the phosphorylation of these Smad3 motifs by MAP kinases may regulate, in addition to the intracellular distribution, the transactivation function of Smad3 by influencing its interactions with co-factors that bind to this region.

Erk-dependent phosphorylation of Smad1 and Smad2 is inhibited by calmodulin (CaM) (Funaba et al., 2002; Scherer and Graff, 2000). This results in reduced stability and hence decreased level of Smad proteins (Funaba et al., 2002; Scherer and Graff, 2000). CaM directly binds to many Smads (Zimmerman et al., 1998) at two distinct regions – the very amino-terminal segment, consisting of amino acid residues 1-49, and the amino acid residues 86-95 of the MH1 domain in Smad1, and the corresponding regions in Smad2 (Scherer and Graff, 2000). Thus, the CaM binding region overlaps with at least one of the sites of Erk-dependent phosphorylation on Smad2 (Thr8) (Funaba et al., 2002). Mutations of Erk1 phosphorylation sites in Smad2 did not affect its interaction with CaM (Funaba et al., 2002). However, others had shown that phosphorylation of Smad1 and Smad2 by Erk2 reduced their potential to associate with CaM (Scherer and Graff, 2000). Thus, CaM binding to Smads and Erk2 phosphorylation of Smads may be reciprocally regulated.

CaM may inhibit TGF-β signaling by engaging the ubiquitously expressed Ca^{2+}/calmodulin-dependent CaM kinase II (CaMKII). Activation of this kinase results in Smad2, Smad3 and Smad4 phosphorylation, inhibits TGF-β-induced nuclear import and transcriptional activity of Smad2, and affects formation of heteromeric complexes (Wicks et al., 2000). CaMKII phosphorylates Smad2 *in vitro* at Ser110 of the MH1 domain and at linker-region residues Ser240 and Ser260 (Wicks et al., 2000). Phosphorylation of Ser240 was observed *in vivo* upon activation of epidermal growth factor or platelet-derived growth factor signaling pathways in 293-HEK cells (Wicks et al., 2000).

Activation of phospholipase Cγ (PLCγ) in Mv1Lu cells by serum treatment induced phosphorylation of Smad3 by protein kinase C (PKC) (Yakymovych et al., 2001). Activation of PKC by phorbol 12-myristate 13-acetate (PMA) resulted also in phosphorylation of endogenous Smad1, Smad2 and Smad3 (Yakymovych et al., 2001). It has been shown that PKC phosphorylates the MH1 domain of Smad2 *in vivo* and *in vitro* at Ser47 and Ser110, and Smad3 at the analogous Ser37 and Ser70 (Yakymovych et al., 2001). In Smad3 these residues are located close to the epitope involved in DNA binding (Shi et al., 1998), and their phosphorylation abrogates Smad3 binding to DNA and its transcriptional activity (Yakymovych et al., 2001). This impairment of Smad3 function by PKC phosphorylation results in selective suppression of growth-inhibitory and pro-apoptotic signals of TGF-β (Yakymovych et al., 2001). How phosphorylation by PKC affects Smad2 functions remains to be elucidated.

TGF-β growth-inhibitory signaling can be selectively downregulated by G1 cyclin-dependent kinases (CDKs). Thr8, Thr178 and Ser212 in Smad3 have been mapped as phosphorylation sites of CDK4 and CDK2 (Matsuura et al., 2004). Smad2, but not Smad4, can also be phosphorylated by immunoprecipitated CDK2 and bacterially expressed and *in vitro* reconstituted cyclin E–CDK2 or cyclin A–CDK2 complexes (Matsuura et al., 2004). Phosphorylation by CDK inhibits Smad3 transcriptional activity, contributing to a decreased level of p15 and an increased level of c-Myc (Matsuura et al., 2004). Thus, under physiological conditions it facilitates cell cycle progression from G1 to S phase. On the other hand, in cancers inactivation of Smads by extensive CDK phosphorylation (Lee and Yang, 2003) may provide an important mechanism for resistance to the TGF-β growth-inhibitory effects.

Recently, the protein G-coupled receptor kinase 2 (GRK2), a kinase involved in the desensitization of G protein-coupled receptors, was identified as a downstream target and regulator of the TGF-β signaling cascade in human hepatocarcinoma cells (Ho et al., 2005). Upon TGF-β stimulation of HepG2 and HuH7 cells, GRK2 associates with Smad2 and Smad3 through their MH1 and MH2 domains and phosphorylates their linker regions. In Smad2 the phosphorylation site was identified as Tyr197 (Ho et al., 2005). To date, this is the only tyrosine residue documented to be phosphorylated in Smads. GRK2 phosphorylation of the GST-Smad2 and GST-Smad3 inhibited their C-terminal phosphorylation by the TβR-I kinase *in vitro* (Ho et al., 2005). This can explain the mechanism by which GRK2 prevents nuclear translocation of the Smad complex, leading to the inhibition of TGF-β-mediated target gene expression, cell growth inhibition and apoptosis.

An activation of the insulin pathway stimulates the interaction of serine/threonine protein kinase Akt/PKB with Smad3 and prevents Smad3 phosphorylation, binding to Smad4 and nuclear translocation. Interestingly, this effect of Akt does not depend on its kinase activity, suggesting that Akt inhibits Smad3 function by physically sequestering it. Phosphorylation of Smad3 decreases the Akt–Smad3 interaction (Conery et al., 2004; Remy et al., 2004).

An activation of the IL-1β signaling leads to interaction of Smad3 with TAK1 kinase (Benus et al., 2005). This resulted in the fast TAK1-dependent phosphorylation of Smad3 at residues other than the C-terminal serines. TAK1 phosphorylation inhibited also Smad3-dependent transcriptional activity, while it did not affect Smad3 nuclear translocation or its direct binding to DNA. Sites of TAK1 phosphorylation in Smad3 protein have not been reported, though a mechanism for IL-1β/TAK1 inhibition of Smad3 pathway has been described (Benus et al., 2005).

The fact that a broad spectrum of kinases, which are activated by different signaling routes, can phosphorylate R-Smads, suggests that Smads represent a point of convergence of various signals. Phosphorylation not only activates Smad proteins but also modulates their activity.

In turn, Smad proteins may regulate activity of other protein kinases. Thus, Smad3 and Smad4 were shown to interact with the regulatory subunits of PKA (Zhang et al., 2004). This interaction was specific, ligand-dependent, and occurred via formation of an activated Smad3/Smad4 complex (Zhang et al., 2004).

Other non-kinase proteins were reported to regulate further R-Smad phosphorylation. Thus, the immunophilin FKBP12 and the integral membrane protein caveolin-1 are able to suppress TGF-β-mediated Smad phosphorylation by interaction with TβR-I (Charng et al., 1996; Chen et al., 1997; Razani et al., 2001). The association of microtubules with Smads may provide a negative regulatory mechanism for the TGF-β signaling pathway, controlling the TGF-β-induced Smad phosphorylation (Dong et al., 2000).

Some studies have indicated that the R-Smads are continuously being dephosphorylated at a low rate by an as yet unidentified phosphatase. This dephosphorylation allows them to dissociate from Smad4 and to be exported to the cytoplasm (Inman et al., 2002; Xu et al., 2002).

3. COMMON-MEDIATOR SMAD4

So far, only Smad4 has been identified as a common-mediator Smad in mammals. This protein lacks the C-terminal SSXS motif found in R-Smads and is not phosphorylated following activation of TGF-β receptors (Heldin et al., 1997; Massagué, 1998; Roelen et al., 2003). However, it has been shown that Smad4 is a phosphoprotein (Lagna et al., 1996; Nakao et al., 1997; Roelen et al., 2003). Smad4 phosphorylation has been reported to be increased after Activin stimulation (Lagna et al., 1996). Phosphoamino acid analysis revealed phosphorylation predominantly on serine residues and on threonine residues to a lesser extent (Roelen et al., 2003).

Examination of the structure of Smad4 revealed three candidate MAPK phosphorylation sites in the linker region (Derynck and Zhang, 2003).

It has been shown that residue Thr276 in the C-terminal part of the linker region of Smad4 can be phosphorylated by Erk2 *in vitro* and in cells (Roelen et al., 2003) (Fig. 1). The phosphorylation of Thr276 can promote nuclear accumulation of Smad4 in response to TGF-β1 and therefore to enhance Smad signaling (Roelen et al., 2003). However, mutation of Thr276 to Ala did not decrease the gross phosphorylation level of Smad4 (Roelen et al., 2003). These results indicate that there are also other residues in Smad4 that are targets for phosphorylation, however, the functional importance of this remains to be determined.

4. INHIBITORY SMADS

Inhibitory SMADs (I-Smads) – Smad6 and Smad7 – also lack a C-terminal SSXS phosphorylation motif, and their N-terminal region has only short segments of similarity to MH1 domain (Derynck and Zhang, 2003; Massagué, 1998). It has been shown that they are phosphoproteins (Imamura et al., 1997; Pulaski et al., 2001). I-Smads bind to the type I receptors, but are not phosphorylated by their kinases (Imamura et al., 1997; Pulaski et al., 2001).

There are at least two phosphorylation sites in Smad7, and one of them has been mapped to Ser249 (Pulaski et al., 2001) (Fig. 1). This residue is located in the C-terminal part of the region corresponding to the linker in other Smads, and its phosphorylation affects transcriptional activity but not stability or intracellular distribution of Smad7 (Pulaski et al., 2001). The kinase(s) that mediate I-Smad phosphorylation and the physiological importance of this phosphorylation remain to be determined.

5. MH2 DOMAIN HAS PROPERTIES OF A PHOSPHOSERINE/PHOSPHOTHREONINE BINDING DOMAIN

Phosphorylation of amino acid residues is known to create recognition sites for protein-protein interaction, in addition to changing of structural properties of phosphorylated molecule. A number of phosphoserine/phosphothreonine-binding domains have been described, *e.g.* FHA, WD40 repeats, 14-3-3, WW, CBP and F-box protein domains for phosphoprotein binding (reviewed in Li et al., 2000; Yaffe and Smerdon, 2004). These domains were found in proteins, which regulate the cell cycle, DNA damage response, transcriptional activation and protein degradation (Li et al., 2000). The first indication of a direct involvement of phosphoryl groups of Smad proteins in a protein-protein interaction came from the observation that the phosphoryl groups at the C-terminal serine residues of Smad2 strongly contributed to the binding of Smad4 to Smad2 C-terminal peptides (Souchelnytskyi et al., 1997). This indicated that the MH2 domain of Smad4 have properties of a phosphoserine-binding domain. Study of the forkhead-associated (FHA) domain unraveled a similarity in the structures of the FHA1 domain of Rad53p and the

MH2 domain of Smad2 (Durocher et al., 2000). Taking into account the extensive similarity between the MH2 domains of R-Smads and Smad4, R-Smads may have the same phosphoserine/phosphothreonine binding ability.

Further structural studies consolidated the importance of the phosphorylation at the C-terminus of Smad1, Smad2 and Smad3 for the complex formation between Smad proteins (Chacko et al., 2001; Qin et al., 2001; Wu et al., 2001). The analysis of surfaces in one Smad molecule which recognize and interact with the phosphorylated C-terminus of another Smad molecule, confirmed an overall similarity with the FHA phosphothreonine binding site. The L3 loop and β8 sheet in the MH2 domain provide residues which contribute to phosphoserine binding (see Chapter 11; Chacko et al., 2001; Qin et al., 2001; Wu et al., 2001). The observed interactions of the MH2 domain with phosphoserine residues, and the preference of the FHA1 domain for phosphothreonine residues indicated that these two domains may represent different types of phosphoserine-binding domains. The mechanism of phosphoserine recognition by the MH2 domain of Smad proteins is also different from the mechanism employed by the phosphoserine-binding domain of 14-3-3 (reviewed in Yaffe and Smerdon, 2004). 14-3-3 binds phosphoserine-containing peptide in a cleft between two helices, while the MH2 domain binds phosphorylated peptide by a surface which is formed by the L3 loop and β8 sheet (Chacko et al., 2001; Qin et al., 2001; Wu et al., 2001). This suggests that the MH2 domain of Smad2 may represent a novel type of phosphoserine-binding domains. The functional importance of the phosphoserine binding properties of the MH2 domains of Smads has to be explored further.

6. CROSS-TALK BETWEEN PHOSPHORYLATION AND OTHER POST-TRANSLATIONAL MODIFICATIONS

Post-translational modifications described for Smad proteins include phosphorylation, ubiquitylation, sumoylation and acetylation. Phosphorylation at serine and threonine residues have been studied most extensively, as this modification triggers Smad-dependent signaling. Treatment of cells with TGF-β and BMP also induced ubiquitylation, sumoylation and acetylation of a number of proteins (see Chapters 13 and 14), suggesting that similar post-translational modifications of Smads may be affected in parallel or in coordination with phosphorylation. As example, ubiquitylation of Smad1, Smad2 and Smad3 is strongly enhanced upon ligand-dependent activation of these Smads (see Chapter 13; Izzi and Attisano, 2004). The exact mechanism of TGF-β-dependent regulation of ubiquitylation has not been reported. However, the presence of WW-domains in E3 ubiquitin ligases of Smads suggests that the ligand-dependent phosphorylation of Smads may create recognition sites for WW-domains of E3 ubiquitin ligases (Bonni et al., 2001). Sumoylation is another post-translational modification which is dependent on activation of Smad proteins (see Chapter 13; Imoto et al., 2003; Lin et al., 2003).

TGF-β initiates interactions of Smad proteins with transcriptional co-activators with acetyltansferase activities, *e.g.* CBP and p300 (see Chapters 14 and 15; Derynck and Zhang, 2003). This may promote acetylation of proteins by Smad-recruited co-activators. TGF-β-induced deacetylation has also been associated with Smad recruitment of histone deacetylases (Bai and Cao, 2002; Simonsson et al., 2005; Wotton et al., 1999). Though Smad phosphorylation initiates Smad interactions with co-activators or deacetylases, there is no evidence that this recruitment is directly mediated by phosphorylated residues in Smad proteins.

The functions of Smad proteins are affected upon activation of tyrosine kinase receptors. Thus rapid phosphorylation of Smad3 and Smad2 upon treatment of cells with nerve growth factor or hepatocyte growth factor was described (de Caestecker et al., 1998; Lutz et al., 2004). The fact that no thorough phosphoamino acid analysis of all generated phosphopeptides was performed leaves open the possibility that Smads can be tyrosine phosphorylated. Notably proteomics-based search has shown that Smad proteins may undergo additional types of post-translational modifications, over those that already have been described. This suggestion is based on the observation that a substantial proportion of Smad tryptic peptides showed masses which suggested modified amino acid residues (Iwahana, Bhaskaran and S.S., unpublished data).

7. FUTURE PERSPECTIVES

Phosphorylation is a potent mechanism to modulate the functional status of a protein. It is fast, reversible, changes the chemical features of proteins and coordinates protein-protein interactions. Phosphorylation of Smad proteins shows all these features. Smads can be phosphorylated within few minutes upon treatment with a ligand. The receptor-dependent phosphorylation of R-Smads at the C-terminal residues changes the structure of Smads to accomodate their complex formation and other protein-protein interactions. This leads to Smad translocations and initiation of signaling processes. The MH2 domain of Smads shows features of a phosphoserine/phosphothreonine binding domain, which indicates an involvement of phosphorylation in Smad-protein interactions.

Smad proteins are phosphorylated also by a variety of other kinases. Identification of the phosphorylated sites is essential for elucidation of Smads' physiological importance as a convergence point for different signaling pathways. Manipulation of Smad phosphorylation by affecting Smad kinases may have a strong impact on TGF-β signaling and may be beneficial for treatment of diseases, *e.g.* cancer and fibrosis (see Chapters 21 and 22).

REFERENCES

Abdollah, S., Macias-Silva, M., Tsukazaki, T., Hayashi, H., Attisano, L., and Wrana, J.L., 1997, TβRI phosphorylation of Smad2 on Ser^{465} and Ser^{467} is required for Smad2-Smad4 complex formation and signaling. *J Biol Chem* 272: 27678-27685.

Aubin, J., Davy, A., and Soriano, P., 2004, In vivo convergence of BMP and MAPK signaling pathways: impact of differential Smad1 phosphorylation on development and homeostasis. *Genes Dev* 18: 1482-1494.

Bai, S., and Cao, X., 2002, A nuclear antagonistic mechanism of inhibitory Smads in transforming growth factor-β signaling. *J Biol Chem* 277: 4176-4182.

Benus, G.F.J.D., Wierenga, A.T.J., de Gorter, D.J.J., Schuringa, J.J., van Bennekum, A.M., Drenth-Diephuis, L., Vellenga, E., and Eggen, B.J.L., 2005, Inhibition of the transforming growth factor β (TGFβ) pathway by interleukin-1β is mediated through TGFβ-activated kinase 1 phosphorylation of SMAD3. *Mol Biol Cell* 16: 3501-3510.

Bonni, S., Wang, H.-R., Causing, C.G., Kavsak, P., Stroschein, S.L., Luo, K., and Wrana, J.L., 2001, TGF-β induces assembly of a Smad2-Smurf2 ubiquitin ligase complex that targets SnoN for degradation. *Nat Cell Biol* 3: 587-595.

Brown, J.D., DiChiara, M.R., Anderson, K.R., Gimbrone, M.A., Jr., and Topper, J.N., 1999, MEKK-1, a component of the stress (stress-activated protein kinase/c-Jun N-terminal kinase) pathway, can selectively activate Smad2-mediated transcriptional activation in endothelial cells. *J Biol Chem* 274: 8797-8805.

Calonge, M.J., and Massagué, J., 1999, Smad4/DPC4 silencing and hyperactive Ras jointly disrupt transforming growth factor-β antiproliferative responses in colon cancer cells. *J Biol Chem* 274: 33637-33643.

Chacko, B.M., Qin, B., Correia, J.J., Lam, S.S., de Caestecker, M.P., and Lin, K., 2001, The L3 loop and C-terminal phosphorylation jointly define Smad protein trimerization. *Nat Struct Biol* 8: 248-253.

Charng, M.-J., Kinnunen, P., Hawker, J., Brand, T., and Schneider, M.D., 1996, FKBP-12 recognition is dispensable for signal generation by type I transforming growth factor-β receptors. *J Biol Chem* 271: 22941-22944.

Chen, Y., Lebrun, J.-J., and Vale, W., 1996, Regulation of transforming growth factor β – and activin-induced transcription by mammalian Mad proteins. *PNAS* 93: 12992-12997.

Chen, Y.G., Hata, A., Lo, R.S., Wotton, D., Shi, Y., Pavletich, N., and Massagué, J., 1998, Determinants of specificity in TGF-β signal transduction. *Genes Dev* 12: 2144-2152.

Chen, Y.G., Liu, F., and Massagué, J., 1997, Mechanism of TGFβ receptor inhibition by FKBP12. *EMBO J* 16: 3866-3876.

Conery, A.R., Cao, Y., Thompson, E.A., Townsend, C.M., Ko, T.C., and Luo, K., 2004, Akt interacts directly with Smad3 to regulate the sensitivity to TGF-β-induced apoptosis. *Nat Cell Biol* 6: 366-372.

de Caestecker, M.P., Parks, W.T., Frank, C.J., Castagnino, P., Bottaro, D.P., Roberts, A.B., and Lechleider, R.J., 1998, Smad2 transduces common signals from receptor serine-threonine and tyrosine kinases. *Genes Dev* 12: 1587-1592.

de Caestecker, M.P., Piek, E., and Roberts, A.B., 2000, Role of transforming growth factor-β signaling in cancer. *J Natl Cancer Inst* 92: 1388-1402.

Derynck, R., and Zhang, Y.E., 2003, Smad-dependent and Smad-independent pathways in TGF-β family signalling. *Nature* 425: 577-584.

Dong, C., Li, Z., Alvarez, J., Rene, Feng, X.-H., and Goldschmidt-Clermont, P.J., 2000, Microtubule binding to Smads may regulate TGFβ activity. *Mol Cell* 5: 27-34.

Durocher, D., Taylor, I.A., Sarbassova, D., Haire, L.F., Westcott, S.L., Jackson, S.P., Smerdon, S.J., and Yaffe, M.B., 2000, The molecular basis of FHA domain:phosphopeptide binding specificity and implications for phospho-dependent signaling mechanisms. *Mol Cell* 6: 1169-1182.

Engel, M.E., McDonnell, M.A., Law, B.K., and Moses, H.L., 1999, Interdependent SMAD and JNK signaling in transforming growth factor-β-mediated transcription. *J Biol Chem* 274: 37413-37420.

Eppert, K., Scherer, S.W., Ozcelik, H., Pirone, R., Hoodless, P., Kim, H., Tsui, L.-C., Bapat, B., Gallinger, S., and Andrulis, I.L., 1996, MADR2 maps to 18q21 and encodes a TGFβ-regulated MAD-related protein that is functionally mutated in colorectal carcinoma. *Cell* 86: 543-552.

Feng, X.-H., and Derynck, R., 1997, A kinase subdomain of transforming growth factor-β (TGF-β) type I receptor determines the TGF-β intracellular signaling specificity. *EMBO J* 16: 3912-3923.

Fukuchi, M., Imamura, T., Chiba, T., Ebisawa, T., Kawabata, M., Tanaka, K., and Miyazono, K., 2001, Ligand-dependent degradation of Smad3 by a ubiquitin ligase complex of ROC1 and associated proteins. *Mol Biol Cell* 12: 1431-1443.

Funaba, M., Zimmerman, C.M., and Mathews, L.S., 2002, Modulation of Smad2-mediated signaling by extracellular signal-regulated kinase. *J Biol Chem* 277: 41361-41368.
Grimm, O.H., and Gurdon, J.B., 2002, Nuclear exclusion of Smad2 is a mechanism leading to loss of competence. *Nat Cell Biol* 4: 519-522.
Hayashida, T., Decaestecker, M., and Schnaper, H.W., 2003, Cross-talk between ERK MAP kinase and Smad signaling pathways enhances TGF-β-dependent responses in human mesangial cells. *FASEB J* 17: 1576-1578.
Heldin, C.-H., Miyazono, K., and ten Dijke, P., 1997, TGF-β signalling from cell membrane to nucleus through SMAD proteins. *Nature* 390: 465-471.
Ho, J., Cocolakis, E., Dumas, V.M., Posner, B.I., Laporte, S.A., and Lebrun, J.J., 2005, The G protein-coupled receptor kinase-2 is a TGFβ-inducible antagonist of TGFβ signal transduction. *EMBO J* 24: 3247-3258.
Hoodless, P.A., Haerry, T., Abdollah, S., Stapleton, M., O'Connor, M.B., Attisano, L., and Wrana, J.L., 1996, MADR1, a MAD-related protein that functions in BMP2 signaling pathways. *Cell* 85: 489-500.
Huse, M., Muir, T.W., Xu, L., Chen, Y.-G., Kuriyan, J., and Massagué, J., 2001, The TGFβ receptor activation process: an inhibitor- to substrate-binding switch. *Mol Cell* 8: 671-682.
Imamura, T., Takase, M., Nishihara, A., Oeda, E., Hanai, J.-I., Kawabata, M., and Miyazono, K., 1997, Smad6 inhibits signalling by the TGF-β superfamily. *Nature* 389: 622-626.
Imoto, S., Sugiyama, K., Muromoto, R., Sato, N., Yamamoto, T., and Matsuda, T., 2003, Regulation of transforming growth factor-β signaling by protein inhibitor of activated STAT, PIASy through Smad3. *J Biol Chem* 278: 34253-34258.
Inman, G.J., and Hill, C.S., 2002, Stoichiometry of active smad-transcription factor complexes on DNA. *J Biol Chem* 277: 51008-51016.
Inman, G.J., Nicolas, F.J., and Hill, C.S., 2002, Nucleocytoplasmic shuttling of Smads 2, 3, and 4 permits sensing of TGF-β receptor activity. *Mol Cell* 10: 283-294.
Izzi, L., and Attisano, L., 2004, Regulation of the TGFβ signalling pathway by ubiquitin-mediated degradation. *Oncogene* 23: 2071-2078.
Jayaraman, L., and Massagué, J., 2000, Distinct oligomeric states of SMAD proteins in the transforming growth factor-β pathway. *J Biol Chem* 275: 40710-40717.
Kamaraju, A.K., and Roberts, A.B., 2005, Role of Rho/ROCK and p38 MAP kinase pathways in transforming growth factor-β-mediated Smad-dependent growth inhibition of human breast carcinoma cells *in vivo*. *J Biol Chem* 280: 1024-1036.
Kato, Y., Habas, R., Katsuyama, Y., Naar, A.M., and He, X., 2002, A component of the ARC/Mediator complex required for TGFβ/Nodal signalling. *Nature* 418: 641-646.
Kawabata, M., Inoue, H., Hanyu, A., Imamura, T., and Miyazono, K., 1998, Smad proteins exist as monomers *in vivo* and undergo homo- and hetero-oligomerization upon activation by serine/threonine kinase receptors. *EMBO J* 17: 4056-4065.
Kretzschmar, M., Doody, J., and Massagué, J., 1997a, Opposing BMP and EGF signalling pathways converge on the TGF-β family mediator Smad1. *Nature* 389: 618-622.
Kretzschmar, M., Doody, J., Timokhina, I., and Massagué, J., 1999, A mechanism of repression of TGFβ/ Smad signaling by oncogenic Ras. *Genes Dev* 13: 804-816.
Kretzschmar, M., Liu, F., Hata, A., Doody, J., and Massagué, J., 1997b, The TGF-β family mediator Smad1 is phosphorylated directly and activated functionally by the BMP receptor kinase. *Genes Dev* 11: 984-995.
Lagna, G., Hata, A., Hemmati-Brivanlou, A., and Massagué, J., 1996, Partnership between DPC4 and SMAD proteins in TGF-β signalling pathways. *Nature* 383: 832-836.
Lee, M.-H., and Yang, H.-Y., 2003, Regulators of G1 cyclin-dependent kinases and cancers. *Cancer Metastasis Rev* 22: 435-449.
Li, J., Lee, G., Van Doren, S., and Walker, J., 2000, The FHA domain mediates phosphoprotein interactions. *J Cell Sci* 113: 4143-4149.
Lin, X., Liang, M., Liang, Y.-Y., Brunicardi, F.C., Melchior, F., and Feng, X.-H., 2003, Activation of transforming growth factor-β Signaling by SUMO-1 modification of tumor suppressor Smad4/DPC4. *J Biol Chem* 278: 18714-18719.

Liu, F., Hata, A., Baker, J.C., Doody, J., Carcamo, J., Harland, R.M., and Massagué, J., 1996, A human Mad protein acting as a BMP-regulated transcriptional activator. *Nature* 381: 620-623.

Liu, F., Pouponnot, C., and Massagué, J., 1997, Dual role of the Smad4/DPC4 tumor suppressor in TGFβ-inducible transcriptional complexes. *Genes Dev* 11: 3157-3167.

Lo, R.S., Chen, Y.G., Shi, Y., Pavletich, N.P., and Massagué, J., 1998, The L3 loop: a structural motif determining specific interactions between SMAD proteins and TGF-β receptors. *EMBO J* 17: 996-1005.

Lo, R.S., and Massagué, J., 1999, Ubiquitin-dependent degradation of TGF-β-activated Smad2. *Nat Cell Biol* 1: 472-478.

Luo, K., and Lodish, H.F., 1996, Signaling by chimeric erythropoietin-TGF-β receptors: homodimerization of the cytoplasmic domain of the type I TGF-β receptor and heterodimerization with the type II receptor are both required for intracellular signal transduction. *EMBO J* 15: 4485-4496.

Lutz, M., Krieglstein, K., Schmitt, S., ten Dijke, P., Sebald, W., Wizenmann, A., and Knaus, P., 2004, Nerve growth factor mediates activation of the Smad pathway in PC12 cells. *Eur J Biochem* 271: 920-931.

Macias-Silva, M., Abdollah, S., Hoodless, P.A., Pirone, R., Attisano, L., and Wrana, J.L., 1996, MADR2 is a substrate of the TGFβ receptor and its phosphorylation is required for nuclear accumulation and signaling. *Cell* 87: 1215-1224.

Massagué, J., 1998, TGF-β signal transduction. *Annu Rev Biochem* 67: 753-791.

Massagué, J., and Wotton, D., 2000, New EMBO member's review: Transcriptional control by the TGF-β/Smad signaling system. *EMBO J* 19: 1745-1754.

Matsuura, I., Denissova, N.G., Wang, G., He, D., Long, J., and Liu, F., 2004, Cyclin-dependent kinases regulate the antiproliferative function of Smads. *Nature* 430: 226-231.

Mori, S., Matsuzaki, K., Yoshida, K., Furukawa, F., Tahashi, Y., Yamagata, H., Sekimoto, G., Seki, T., Matsui, H., Nishizawa, M., Fujisawa, J., and Okazaki, K., 2004, TGF-β and HGF transmit the signals through JNK-dependent Smad2/3 phosphorylation at the linker regions. *Oncogene* 23: 7416-7429.

Moustakas, A., Souchelnytskyi, S., and Heldin, C.-H., 2001, Smad regulation in TGF-β signal transduction. *J Cell Sci* 114: 4359-4369.

Mulder, K.M., 2000, Role of Ras and Mapks in TGFβ signaling. *Cytokine & Growth Factor Reviews* 11: 23-35.

Nakao, A., Imamura, T., Souchelnytskyi, S., Kawabata, M., Ishisaki, A., Oeda, E., Tamaki, K., Hanai, J., Heldin, C.-H., Miyazono, K., and ten Dijke, P., 1997, TGF-β receptor-mediated signalling through Smad2, Smad3 and Smad4. *EMBO J* 16: 5353-5362.

Nohe, A., Keating, E., Knaus, P., and Petersen, N.O., 2004, Signal transduction of bone morphogenetic protein receptors. *Cellular Signalling* 16: 291-299.

Ottesen, J.J., Huse, M., Sekedat, M.D., and Muir, T.W., 2004, Semisynthesis of phosphovariants of Smad2 reveals a substrate preference of the activated TβRI kinase. *Biochemistry* 43: 5698-5706.

Prokova, V., Mavridou, S., Papakosta, P., and Kardassis, D., 2005, Characterization of a novel transcriptionally active domain in the transforming growth factorβ-regulated Smad3 protein. *Nucl Acids Res* 33: 3708-3721.

Pulaski, L., Landström, M., Heldin, C.-H., and Souchelnytskyi, S., 2001, Phosphorylation of Smad7 at Ser-249 does not interfere with its inhibitory role in transforming growth factor-β-dependent signaling but affects Smad7-dependent transcriptional activation. *J Biol Chem* 276: 14344-14349.

Qin, B.Y., Chacko, B.M., Lam, S.S., de Caestecker, M.P., Correia, J.J., and Lin, K., 2001, Structural basis of Smad1 activation by receptor kinase phosphorylation. *Mol Cell* 8: 1303-1312.

Razani, B., Zhang, X.L., Bitzer, M., von Gersdorff, G., Böttinger, E.P., and Lisanti, M.P., 2001, Caveolin-1 regulates transforming growth factor (TGF)-β/SMAD signaling through an interaction with the TGF-β type I receptor. *J Biol Chem* 276: 6727-6738.

Reguly, T., and Wrana, J.L., 2003, In or out? The dynamics of Smad nucleocytoplasmic shuttling. *Trends Cell Biol* 13: 216-220.

Remy, I., Montmarquette, A., and Michnick, S.W., 2004, PKB/Akt modulates TGF-β signalling through a direct interaction with Smad3. *Nat Cell Biol* 6: 358-365.

Roelen, B.A.J., Cohen, O.S., Raychowdhury, M.K., Chadee, D.N., Zhang, Y., Kyriakis, J.M., Alessandrini, A.A., and Lin, H.Y., 2003, Phosphorylation of threonine 276 in Smad4 is involved in transforming growth factor-β-induced nuclear accumulation. *Am J Physiol Cell Physiol* 285: C823-830.

Scherer, A., and Graff, J.M., 2000, Calmodulin differentially modulates Smad1 and Smad2 signaling. *J Biol Chem* 275: 41430-41438.

Shi, Y., and Massagué, J., 2003, Mechanisms of TGF-β signaling from cell membrane to the nucleus. *Cell* 113: 685-700.

Shi, Y., Wang, Y.-F., Jayaraman, L., Yang, H., Massagué, J., and Pavletich, N.P., 1998, Crystal structure of a Smad MH1 domain bound to DNA: insights on DNA binding in TGF-β signaling. *Cell* 94: 585-594.

Simonsson, M., Heldin, C.-H., Ericsson, J., and Grönroos, E., 2005, The balance between acetylation and deacetylation controls Smad7 stability. *J Biol Chem* 280: 21797-21803.

Souchelnytskyi, S., Tamaki, K., Engström, U., Wernstedt, C., ten Dijke, P., and Heldin, C.-H., 1997, Phosphorylation of Ser^{465} and Ser^{467} in the C terminus of Smad2 mediates interaction with Smad4 and is required for transforming growth factor-β signaling. *J Biol Chem* 272: 28107-28115.

Souchelnytskyi, S., ten Dijke, P., Miyazono, K., and Heldin, C.-H., 1996, Phosphorylation of Ser165 in TGF-β type I receptor modulates TGF-β1-induced cellular responses. *EMBO J* 15: 6231-6240.

Tamaki, K., Souchelnytskyi, S., Itoh, S., Nakao, A., Sampath, K., Heldin, C.-H., and ten Dijke, P., 1998, Intracellular signaling of osteogenic protein-1 through Smad5 activation. *J Cell Physiol* 177: 355-363.

Tsukazaki, T., Chiang, T.A., Davison, A.F., Attisano, L., and Wrana, J.L., 1998, SARA, a FYVE domain protein that recruits Smad2 to the TGFβ receptor. *Cell* 95: 779-791.

Wicks, S.J., Lui, S., Abdel-Wahab, N., Mason, R.M., and Chantry, A., 2000, Inactivation of Smad-transforming growth factor β signaling by Ca^{2+}-calmodulin-dependent protein kinase II. *Mol Cell Biol* 20: 8103-8111.

Wieser, R., Wrana, J., and Massagué, J., 1995, GS domain mutations that constitutively activate TβR-I, the downstream signaling component in the TGF-β receptor complex. *EMBO J* 14: 2199-2208.

Wotton, D., Lo, R.S., Lee, S., and Massagué, J., 1999, A Smad transcriptional corepressor. *Cell* 97: 29-39.

Wrana, J.L., Attisano, L., Wieser, R., Ventura, F., and Massagué, J., 1994, Mechanism of activation of the TGF-β receptor. *Nature* 370: 341-347.

Wu, G., Chen, Y.G., Ozdamar, B., Gyuricza, C.A., Chong, P.A., Wrana, J.L., Massagué, J., and Shi, Y., 2000, Structural basis of Smad2 recognition by the Smad anchor for receptor activation. *Science* 287: 92-97.

Wu, J.W., Hu, M., Chai, J., Seoane, J., Huse, M., Li, C., Rigotti, D.J., Kyin, S., Muir, T.W., Fairman, R., Massagué, J., and Shi, Y., 2001, Crystal structure of a phosphorylated Smad2. Recognition of phosphoserine by the MH2 domain and insights on Smad function in TGF-β signaling. *Mol Cell* 8: 1277-1289.

Xu, L., Kang, Y., Col, S., and Massagué, J., 2002, Smad2 nucleocytoplasmic shuttling by nucleoporins CAN/Nup214 and Nup153 feeds TGFβ signaling complexes in the cytoplasm and nucleus. *Mol Cell* 10: 271-282.

Yaffe, M.B., and Smerdon, S.J., 2004, The use of *in vitro* peptide-library screens in the analysis of phosphoserine/threonine-binding domain structure and function. *Annu Rev Biophys Biomol Struct* 33: 225-244.

Yakymovych, I., Heldin, C.-H., and Souchelnytskyi, S., 2004, Smad2 Phosphorylation by Type I Receptor: Contribution of arginine 462 and cysteine 463 in the C terminus of Smad2 for specificity. *J Biol Chem* 279: 35781-35787.

Yakymovych, I., ten Dijke, P., Heldin, C.-H., and Souchelnytskyi, S., 2001, Regulation of Smad signaling by protein kinase C. *FASEB J* 15: 553-555.

Yamagata, H., Matsuzaki, K., Mori, S., Yoshida, K., Tahashi, Y., Furukawa, F., Sekimoto, G., Watanabe, T., Uemura, Y., Sakaida, N., Yoshioka, K., Kamiyama, Y., Seki, T., and Okazaki, K., 2005, Acceleration of Smad2 and Smad3 phosphorylation via c-Jun NH_2-terminal kinase during human colorectal carcinogenesis. *Cancer Res* 65: 157-165.

Yamaguchi, K., Shirakabe, K., Shibuya, H., Irie, K., Oishi, I., Ueno, N., Taniguchi, T., Nishida, E., and Matsumoto, K., 1995, Identification of a member of the MAPKKK family as a potential mediator of TGF-β signal transduction. *Science* 270: 2008-2011.

Zhang, L., Duan, C.J., Binkley, C., Li, G., Uhler, M.D., Logsdon, C.D., and Simeone, D.M., 2004, A transforming growth factor β-induced Smad3/Smad4 complex directly activates protein kinase A. *Mol Cell Biol* 24: 2169-2180.

Zhang, Y., Feng, X., We, R., and Derynck, R., 1996, Receptor-associated Mad homologues synergize as effectors of the TGF-β response. *Nature* 383: 168-172.

Zimmerman, C.M., Kariapper, M.S.T., and Mathews, L.S., 1998, Smad proteins physically interact with calmodulin. *J Biol Chem* 273: 677-680.

CHAPTER 13

REGULATION OF SMAD FUNCTIONS THROUGH UBIQUITINATION AND SUMOYLATION PATHWAYS

XIN-HUA FENG[1,2] AND XIA LIN[2]

[1] *Department of Molecular & Cellular Biology, Baylor College of Medicine, Houston, TX 77030, USA*

[2] *Michael E. DeBakey Department of Surgery, Baylor College of Medicine, Houston, TX 77030, USA*

Abstract: Post-translational modifications by the ubiquitin and ubiquitin-like pathways have recently emerged as common mechanisms to regulate the turnover and functions of many transcription factors. Improper modifications of these regulatory proteins may play an important role in the initiation and/or progression of human diseases. Recent evidence suggests that stability, activity and signaling strength of Smads are fine-tuned via various post-translational modifications including, ubiquitination and sumoylation. Critical functions of the modifications are underscored by their functional consequence in TGF-β antiproliferative and transcriptional responses. Thus the modification of Smads may control diverse developmental processes and the pathogenesis of many diseases including cancer, autoimmune diseases, and fibrotic diseases. In this chapter, we will focus on the functions and underlying mechanisms of ubiquitin and SUMO modifications on Smads

Keywords: TGF-β; ubiquitin; SUMO; modification; E3 ligase; degradation; proteasome

1. INTRODUCTION

The ubiquitin-proteasome system was discovered more than 20 years ago by Ciechanover, Hershko, and Rose, who were recently awarded the Nobel Prize in Chemistry for their pioneering work (Hershko and Ciechanover, 1982). Ubiquitin-mediated proteolysis plays important roles in the turnover of many regulatory proteins that are involved in the regulation of critical cellular processes including DNA repair, cell cycle control, oncogenesis, viral infection, cellular differentiation, and stress response.

P. ten Dijke and C.-H. Heldin (eds.), Smad Signal Transduction, 253–276.

Ubiquitination is an evolutionarily conserved process for covalently attaching ubiquitin to short-lived proteins for targeting their degradation by the 26S proteasome (Ciechanover, 2005; Hershko, 2005). Conjugation of ubiquitin is catalyzed by a series of enzymatic reactions (Fig. 1). The first step is initiated by the activity of the ubiquitin-activating enzyme (Uba1, E1) that requires ATP, forming a high-energy thioester bond between E1 and ubiquitin. Ubiquitin is then transferred to one of distinct ubiquitin-conjugating enzymes (Ubc, E2) also through thioester bond formation. Finally, ubiquitin-protein conjugates are generated via isopeptide bond formation between the acceptor lysine ε-amino group of the substrate protein and the C-terminal exposed glycine carboxyl group of ubiquitin, either by a direct transfer of the ubiquitin onto substrates or through a specific ubiquitin ligase E3 that interacts physically with the substrate. The human genome encodes one E1, tens of E2, and an extensive array of ubiquitin-protein E3 ligases. There are various classes of E3 ligases. The specific E3 ligase recognizes its substrate through a specific sequence or phosphorylated motif. For example, the HECT-class E3 ligases use WW domains to bind to PPXY motifs substrates, whereas the SCF-class E3 ligases often recognize in phosphorylated substrates.

The best-defined role for protein ubiquitination is linked to the degradation of the ubiquitinated proteins by the 26S proteasome. Mechanistically, proteins destined for degradation possess a polyubiquitin chain that is formed on the residue Lys48 of ubiquitin through the action of E3 ligases, and possibly so-called E4 ligases. E4 ligases catalyze the isopeptide bond formation between one ubiquitin and another, *i.e.* "multimerization" of ubiquitin. The polyubiquitin chain serves as a recognition signal for the 26S proteasome to degrade the modified proteins. The proteasome is a large, multimeric protease that catalyzes the final step of the ubiquitin-dependent intracellular protein degradation (Adams, 2003).

Despite the connection of ubiquitination to proteasomal degradation, accumulating evidence also suggests proteasome-independent regulatory functions of ubiquitin. Monoubiquitination does not target modified proteins to the proteasome, but instead it serves as a signal for diverse functions such as cell surface receptor internalization, vesicle sorting, viral budding, transcription, DNA repair, and caspase recruitment in apoptosis (Sigismund et al., 2004; Sun and Chen, 2004). Furthermore, non-classically polyubiquitinated proteins, through a polyubiquitin chain not formed on Lys48 of ubiquitin, are also not destined to the proteasome. Lys6- and Lys63-linked polyubiquitin chains have been shown to mediate DNA repair, protein kinase regulation, and cell signaling (Sun and Chen, 2004).

A number of ubiquitin-related proteins are also present in eukaryotic cells. These proteins, including the small ubiquitin-like modifiers (SUMOs), do not target modified proteins for proteasomal degradation (Hay, 2005; Johnson, 2004). Two independent laboratories discovered the SUMO-1, the first member of the SUMO family that controls the import of proteins into the nucleus through direct modification of RanGAP1, a Ran-specific GTPase-activating protein (Mahajan et al., 1997; Matunis et al., 1996). To date, the SUMO family consists of four members, *i.e.* SUMO-1 to -4 in mammals. SUMOs utilize a conjugation system that is similar to ubiquitination, with Aos1/Uba2 as the E1 activating enzyme and Ubc9 as the E2 conjugating enzyme (Hay, 2005; Johnson, 2004). While ubiquitin E3 ligases are highly selective in substrate recognition, SUMO E3 ligases appear to have a broader specificity. Three different classes of proteins have been identified as SUMO E3 ligases, including nucleoporin RanBP2, polycomb protein Pc2, Protein inhibitors of activated STAT (PIAS) (Hochstrasser, 2001; Jackson, 2001; Kagey et al., 2003; Pichler et al., 2002; Potts and Yu, 2005). Elucidating exactly what SUMO modification does turns out to be rather challenging. In some cases, sumoylation protects proteins from degradation (Desterro et al., 1998; Lin et al., 2003a). Sumoylation also regulates subcellular localization of proteins, *e.g.* CtBP (Lin et al., 2003d), PML (Muller and Dejean, 1999; Zhong et al., 2000) and RanGAP1 (Mahajan et al., 1997; Matunis et al., 1996), and modulates biological activities (Girdwood et al., 2004; Muller et al., 2004; Schmidt and Muller, 2003; Yang et al., 2003).

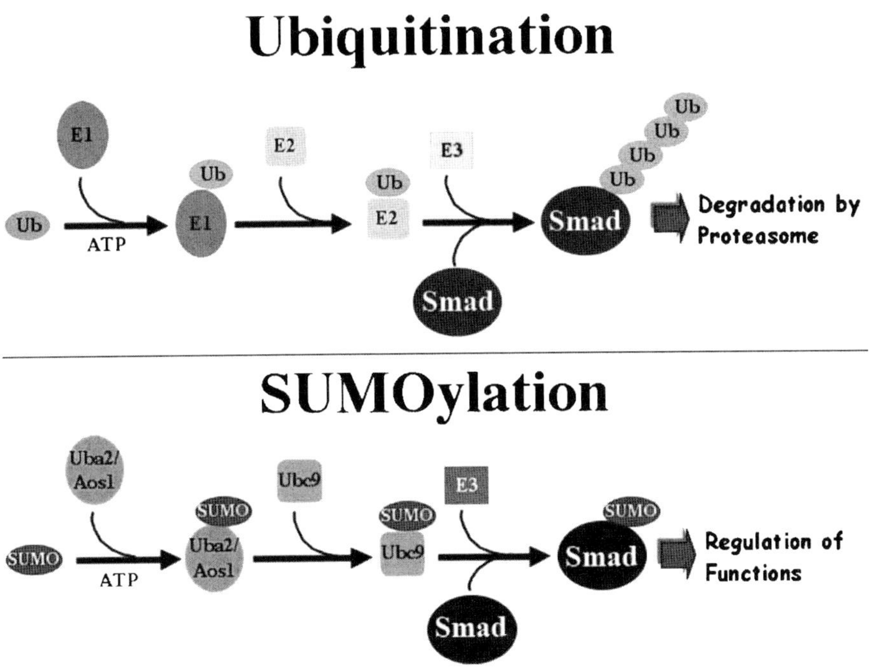

Figure 1. The pathways of ubiquitin and SUMO conjugation. Mature ubiquitin (76 amino acids) and SUMO (97 amino acids) are conjugated to substrate proteins via three classes of enzymes. During ubiquitination, the E3 ligase determines the substrate specificity. Majority of ubiquitin substrates undergo polyubiquitination prior to their proteasome-dependent degradation. Some proteins such as Smad4 are also mono-ubiquitinated. Monoubiquitination and sumoylation do not serve as a means to promote protein degradation. Instead, they regulate the functionality of substrates (A color version of this figure is freely accessible via the website of the book: http://www.springer.com/1-4020-4542-5)

2. TERMINATION OF TGF-β SIGNALS THROUGH UBIQUITINATION AND PROTEASOMAL DEGRADATION OF R-SMADS

Like many key signal molecules, Smads are under stringent control at every level, particularly through post-translational modifications. Besides dephosphorylation of R-Smads that may negatively impact TGF-β signaling, turnover of Smads via proteasomal degradation should also rapidly terminate TGF-β signaling (Fig. 2). Recent evidence supports the notion that Smads are ubiquitinated by various classes of ubiquitin E3 ligases.

2.1 Smurf Family E3 Ligases and Proteasome-dependent Degradation of R-Smads

Smurf (Smad-ubiquitin regulatory factor) proteins belong to the HECT (homologous to the E6AP C terminus) E3 ligase family. Structurally, Smurfs possess an N-terminal C2 domain, 2-3 WW domains, and a C-terminal HECT catalytic domain

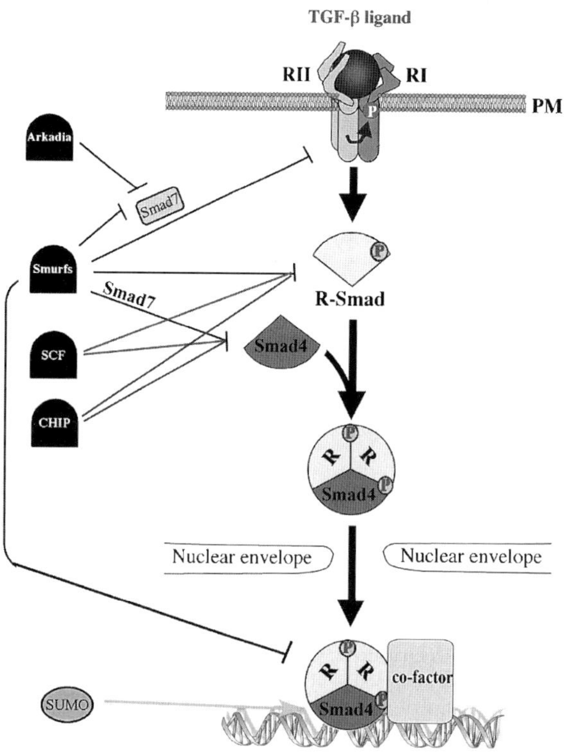

Figure 2. The role of ubiquitination and sumoylation in TGF-β signaling. Major classes of ubiquitin E3 ligases involved in the regulation of TGF-β signaling are shown. Some E3 ligases such as Smurfs and SCF may also function in the nucleus. SUMO modification on Smad4 positively regulates TGF-β signaling (A color version of this figure is freely accessible via the website of the book: http://www.springer.com/1-4020-4542-5)

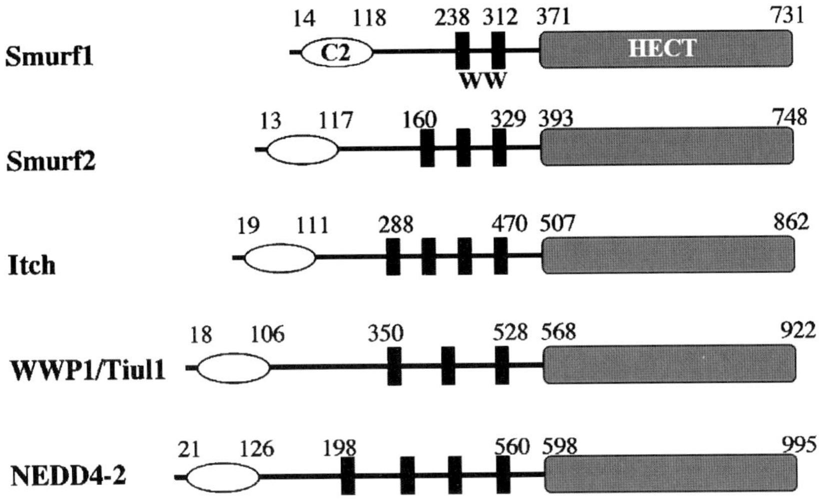

Figure 3. Schematic representation of Smurf-related E3 ligases. Five Smurf-related proteins, their structural domains and the relative positions of the domains are shown

(Fig. 3). Smurf1 and Smurf2 have two and three WW domains, respectively, which are responsible for the interaction with the PPXY motif present in the linker region of R-Smads (except Smad8) and I-Smads. The C2 domain may be responsible for subcellular targeting of Smurf proteins (Suzuki et al., 2002). The HECT domain forms a thioester bond with ubiquitin transferred from E2 enzyme and then catalyzes the ligation of ubiquitin with substrates via an isopeptide bond.

Smurf1 was initially identified in a yeast two-hybrid screen for Smad1-interacting protein (Zhu et al., 1999). Subsequent sequence analysis led to the identification of mammalian Smurf2 and *Drosophila* dSmurf (Liang et al., 2003; Podos et al., 2001). Although Smurfs bind to Smads through the WW domains, there is specificity in Smad binding and subsequent degradation of Smads. Smurf1 only targets BMP-specific Smad1 and Smad5 for degradation (Zhu et al., 1999). While dSmurf is largely similar to Smurf1 in Smad degradation in the BMP/Dpp pathway, Smurf2 appears to have a broader specificity toward Smads in both the TGF-β and BMP pathways. Smurf2 is physically associated with both Smad1/5 and Smad2, hence target them for degradation (Lin et al., 2000; Zhang et al., 2001). Interestingly, Smurf2 does not degrade Smad3, although it binds to Smad2 and Smad3 with similar affinity (Lin et al., 2000; Zhang et al., 2001), indicating that E3 ligase binding is not sufficient for degradation. The Smurf2-Smad3 interaction may be responsible for ubiquitination and degradation of Smad3-associated proteins. Indeed, the Smurf2-Smad3 complex accelerates turnover of other proteins such as SnoN (Bonni et al., 2001).

One striking feature of Smurf proteins is that Smurf binding to the PPXY motif of Smads requires two WW domains in tandem in Smurfs. Since single WW domain suffices for other HECT E3 ligase to bind their substrates (Ingham et al.,

2005), the requirement of two WW domains in Smurf is rather intriguing and likely due to the imperfectness in these WW domains. In addition, the receptor-induced phosphorylation of the C-terminal SXS motif has been shown to be critical for Smurf2-Smad2 association in cells (Lin et al., 2000). Mutation in the SXS motif of Mad abolished its binding to dSmurf, high-lighting the important role of Mad phosphorylation in its degradation (Liang et al., 2003; Podos et al., 2001). The phosphorylation-dependent mechanism is consistent with the disappearance of pMad in dSmurf-overexpressing transgenic flies and also corresponds to the increased pMad level in dSmurf-null mutants (Podos et al., 2001). It will be interesting to determine whether SXS phosphorylation simply activates Smad to expose the PPXY motif or whether one of the WW domains of Smurf can bind directly to the phosphorylated SXS motif of Smads.

Several Smurf-like molecules were found to participate in the regulation of TGF-β signaling. These include WWP1/Tiul1, NEDD4-2, and Itch. Like Smurf2, WWP/Tiul1 interacts with all R-Smads and I-Smads, albeit with different affinity; however, strong binding of WWP1 and NEDD4-2 to Smad2 and Smad3 does not lead to ubiquitin-mediated degradation of Smad2/3 (Komuro et al., 2004; Kuratomi et al., 2005). WWP1/Tiul1 can also interact with the nuclear corepressor TGIF in response to TGF-β stimulation, and interestingly, TGIF allows WWP1/Tiul1 to mediate the degradation of Smad2 (Seo et al., 2004). Noteworthy, the steady-state level of TGIF itself is not affected. An interesting question arises: does TGIF associate with Smurfs or other Smurf-like molecules in degrading R-Smads? Despite the effects on Smads or receptors (discussed later), the utmost effects of both WWP1 and NEDD4-2 are inhibitory to TGF-β signaling. Conversely, Itch positively regulates TGF-β responses by forming a complex with Smad2 and type I receptor (Bai et al., 2004). Itch polyubiquitinates Smad2 through Lys48 chain, but oddly enough, does not promote Smad2 degradation. In human, there are a few more Smurf-like proteins, but no attempts have been reported on whether other WW-HECT E3 ligases such as E6AP, NEDD4, NEDL1, NEDL2, and WWP2 also exert regulatory functions on turnover of Smads and associated proteins.

Thus, rather than any one E3 ligase acting in isolation, the actions of multiple Smurf-related proteins may be necessary to effectively regulate TGF-β responses. In addition, differential subcellular locations of these proteins and differential affinities to R-Smads or associated proteins or receptors, determine the timing of termination of Smad signaling.

2.2 Degradation of R-Smads by SCF and U-box E3 Ligases

The Skp1-Cul1-F-box-protein (SCF) ubiquitin ligases are important players in many mammalian cellular processes. SCF ubiquitin E3 ligases are composed of Skp1, Cul-1, a Roc/Rbx RING finger protein, and a modular F box protein. The F box protein functions as a targeting subunit to recognize its substrate and thus determines the specificity in ubiquitination. Although the F box protein βTrCP1 weakly interacts with Smad3, the RING-containing component Rbx1/Roc1 of the $SCF^{\beta TrCP}$

complex binds to Smad3 and the interaction can target Smad3 for degradation (Fukuchi et al., 2001). Since Rbx1 is a common subunit of many SCF E3 ligases, it raises a specificity issue as to whether other SCF E3 ligases also target Smad3 for degradation. An independent study later demonstrated a strong association of Smad3 with βTrCP2 (Ray et al., 2005). It is unclear, however, how the interactions between Smad3 and Rbx1 or βTrCPs determine the selectivity in the degradation of Smad3 (Fukuchi et al., 2001) or Smad3-associated proteins (Ray et al., 2005).

An alternative pathway to regulate Smad3 turnover is the CHIP (carboxy terminus of Hsc70 interacting protein)-dependent degradation. CHIP belongs to the U-box E3 ligase family, characterized by a tetratricopeptide repeat (TPR) domain that binds to the molecular chaperones Hsc70-Hsp70 and Hsp90, and a U-box domain with its E3 ubiquitin ligase activity (McDonough and Patterson, 2003). CHIP mediates ubiquitination and degradation of Smad3 independently of TGF-β stimulation (Xin et al., 2005). A recent study showed that Smad3 interacts with Hsp70, indicating that Smad3 is an Hsp70 client (Knuesel et al., 2003). This is consistent with the function of CHIP-Hsp interaction that is responsible for ubiquitination and proteasomal degradation of chaperone client proteins.

An unanswered question pertaining to the relationship between CHIP and Smads is the lack of specificity since CHIP also regulates the turnover of Smad1, Smad2 and Smad4. Unlike Smad3, Smad1 and Smad4 directly interact with CHIP (Li et al., 2004), which then may not require the function of Hsp70. Indeed, Smad1 does not interact with Hsp70 (Knuesel et al., 2003). Whether the direct physical contact between Smads and CHIP or between Hsp proteins and Smads determines the time and efficiency of Smad degradation remains unclear.

2.3 Ligand-induced Degradation of R-Smads

Smurf proteins are localized both in the nucleus and cytoplasm. Given the nuclear presence and their preferred interaction with phosphorylated Smads, Smurfs may remove the activated Smads from the nucleus to terminate TGF-β/BMP signaling. Smad3 also undergoes TGF-β-dependent degradation via the $SCF^{\beta TrCP}$ E3 ligase, which binds to and ubiquitinates activated Smad3 in the nucleus, and the ubiquitinated Smad3 is then exported out of the nucleus and degraded in the cytoplasm (Fukuchi et al., 2001).

Another link between ubiquitination and TGF-β signaling termination was revealed in an earlier study by Lo and Massagué, who observed that TGF-β induces proteasomal degradation of phosphorylated Smad2 in the nucleus (Lo and Massagué, 1999). Even the isolated MH2 domain (without the PPXY-containing linker) of activated Smad2 undergoes proteasomal degradation in the nucleus (Lo and Massagué, 1999). Although destruction of an isolated MH2 domain implies that an E3 ligase different from Smurf2 may reside in the nucleus and control degradation of activated Smad2, it is still possible that the exogenously expressed MH2 domain may form a complex with endogenous Smad2 proteins and become degraded through the Smurf pathway.

In BMP signaling, Smad1 is also targeted to proteasomal degradation in response to BMP receptor activation. Aside its unknown mechanism, the proteasomal degradation of Smad1 involves the targeting functions of the ornithine decarboxylase antizyme (Az) and the proteasome β subunit HsN3 (Gruendler et al., 2001). BMP induces Smad1 to form a complex with HsN3 and Az, and subsequently the nuclear translocation of the complex (Lin et al., 2002). These BMP-dependent events occur prior to the incorporation of HsN3 into the mature 20S proteasome. These studies suggest an interesting mechanism for proteasomal degradation of Smad1.

3. DEGRADATION OF I-SMAD, RECEPTORS AND ASSOCIATED PROTEINS

The expression and activities of I-Smads, including Smad6 and Smad7 in vertebrates and presumably Dad in *Drosophila*, constitute a critical point for negative control of TGF-β signaling. Various external signals, including TGF-β ligands themselves, induce expression of the inhibitory Smads and functionally block the responsiveness of cells to TGF-β (Bitzer et al., 2000; Park, 2005; Ulloa et al., 1999). Smad6 and Smad7 inhibit TGF-β family signaling through multiple mechanisms, including competitive interference with activation of R-Smads (Hayashi et al., 1997; Imamura et al., 1997; Nakao et al., 1997), and in the case of Smad6, inhibition of the formation of R-Smad-Smad4 complex (Hata et al., 1998), as well as direct repression of target gene promoters in the nucleus (Bai et al., 2000; Lin et al., 2003c). The last and possibly not the least mechanism for inhibition of TGF-β family signaling by Smad6 and 7 is provided by their ability to recruit Smurf E3 ubiquitin ligases to the type I receptors, leading to proteasomal degradation of the receptors (Ebisawa et al., 2001; Kavsak et al., 2000; Murakami et al., 2003; Suzuki et al., 2002).

3.1 Smad7 Degradation

Both Smad6 and Smad7 bind to Smurf1/2 with higher affinity than R-Smads, and Smurfs appear to target Smad6/7 for degradation. The Smad7-Smurf2 complex is first formed in the nucleus through both the paired WW-PPXY interaction (Kavsak et al., 2000) and a direct contact between the N-terminal domain of Smad7 and the HECT domain of Smurf (Ogunjimi et al., 2005). The Smurf-Smad7 complex is then exported out of the nucleus and subsequently targeted to membranes via the C2 domain of Smurfs (Suzuki et al., 2002; Tajima et al., 2003). Smad7 is only profoundly degraded by Smurf2 in conjunction with TGF-β signaling, and likely through targeting the entire receptor-Smad7 complex. Smurf-mediated ubiquitination of Smad7 is also regulated by acetylation. The transcriptional coactivator p300 interacts with Smad7 and results in the acetylation of Smad7 on Lys64 and Lys70 (Grönroos et al., 2002). This is functionally critical since Smurf and Smad7 mutually regulate their targeting to the cell surface (Suzuki et al., 2002). Obviously, a stable Smad7-Smurf complex is required for Smad7 to block R-Smad activation or for Smurfs to degrade TGF-β receptors. Notably, these two lysines are not present in Smad6, indicating a mechanistic difference in ubiquitination of Smad6 and Smad7.

In addition to Smurfs, Arkadia and Jab1/CSN5 has been shown to target Smad7 for proteasomal degradation. Arkadia, originally identified as a positive regulator of Nodal signaling, physically interacts with Smad7 (Koinuma et al., 2003). Arkadia contains a RING domain that is essential in inducing Smad7 ubiquitination and degradation. Unlike Smurfs, Arkadia does not target TGF-β receptors for destruction, perhaps because the binding of Arkadia to the MH2 domain of Smad7 blocks their recruitment to the receptor. Recently, it was shown that Jab1/CSN5, a component of the COP9 signalosome complex, also interacts with Smad7. Jab1/CSN5 promotes nuclear export and degradation of Smad7, in a similar fashion to that of the Cdk inhibitor p27, thus removing the inhibitory activity of Smad7 on TGF-β signaling (Kim et al., 2004). Since Jab1/CSN5 is not a ubiquitin E3 ligase, it suggests that Jab1/CSN5 serves as an adapter for Smad7 degradation, and consequently enhances TGF-β signaling.

3.2 Degradation of TGF-β Receptors

The consequence of Smurf-Smad7 interaction is far more complex than the degradation of Smad7 itself. Smad7 induces the autoubiquitination and degradation of Smurf2. More importantly, through Smad7 (presumably the more stable acetylated form), Smurf1 and Smurf2 are recruited to the type I receptor on the surface, which become ubiquitinated (Ebisawa et al., 2001; Kavsak et al., 2000). The Smurf-Smad7-receptor complex residing in the lipid raft domains of membranes is then internalized into a caveolin-positive, but EEA1-negative, compartment and subsequently degraded (Di Guglielmo et al., 2003). In addition to the PPXY motif of Smad7 that interacts with the WW domains of Smurfs, the N-terminal domain of Smad7 can directly recruit the E2 enzyme UbcH7 to Smurf2 and enhance the E3 ligase activity of Smurf2 (Ogunjimi et al., 2005). This explains why the N-terminal domain of Smad7 is required for efficient inhibition of the receptor signaling (Hanyu et al., 2001). Like Smad7, Smad6 also binds strongly to Smurfs and may function in a similar mechanism. Smad6/7 also recruit Smurfs to the BMP receptor complexes for their turnover (Murakami et al., 2003). Moreover, other Smurf-like molecules, *e.g.* WWP1/Tiul1 and NEDD4-2, also associate with Smad7 to target the TGF-β receptor for degradation (Komuro et al., 2004; Kuratomi et al., 2005; Seo et al., 2004). It will be interesting to determine whether the N-terminal domain of Smad7 (and Smad6) also regulate the activity of the Smurf-like E3 ligases towards substrate ubiquitination.

TGF-β receptors are destined not only for proteasomal degradation, but their turnover also occurs in the lysosome. Recently, Zhang et al. (2004) elegantly demonstrated that Dapper2, a PDZ-binding protein, suppresses mesoderm induction activities of Nodal signaling by inducing lysosomal degradation of Nodal receptors in zebrafish (Zhang et al., 2004). The mechanism for how Dapper2 promotes TGF-β receptor degradation remains to be elucidated. Does Smurf-Smad7 play a role in assisting Dapper2 to function? Does Dapper2 have ubiquitin E3 ligase activity or act as an adaptor for an E3 ligase? Needless to say, it will be of interest to determine if Dapper2 induces receptor turnover in a ubiquitin-dependent manner.

3.3 Degradation of Smad-interacting Intracellular Proteins

Activation of the TGF-β signaling pathway induces phosphorylation of R-Smads and subsequent assembly of larger signaling complexes. Some associated proteins in the Smad complex have been shown to be targeted for ubiquitin-mediated proteolysis. Besides I-Smads, R-Smads can also recruit Smurfs or other E3 ligases to the Smad-associated proteins.

The transcriptional co-repressor SnoN is closely related to Ski (proto-oncogene *c-ski* product) that binds to various Smads. SnoN functions to maintain the repressed state of TGF-β target genes in the absence of TGF-β ligand. Upon TGF-β stimulation, Smad2/3 induce a rapid degradation of SnoN, allowing activation of TGF-β target genes. One mechanism for SnoN degradation is through the Smurf2 E3 ligase. TGF-β induces the formation of a ternary complex consisting of Smad2/3, SnoN and Smurf2. Smad2/3 serve as adaptors to allow Smurf2 E3 ligase to target SnoN for ubiquitin-mediated degradation. This effect appears to be restricted to Smurf2 since WWP1/Tiul1 and NEDD4-2 do not promote SnoN degradation (Komuro et al., 2004; Kuratomi et al., 2005). Similarly, Smad1 recruits Smurf1 to mediate the ubiquitination and degradation of osteoblast-specific transcription factor Cbfa1/Runx2 (Zhao et al., 2003). Besides its association with Smads, Smurf1 has been shown to interact with atypical protein kinase C zeta (PKCζ), a regulator of cell polarity. Through PKCζ, Smurf1 is recruited to lamellipodia and filopodia, where it promotes the degradation of RhoA, contributing to cell motility and epithelial-mesenchymal transition (Wang et al., 2003).

Other non-Smurf E3 ligases also use Smads as adaptors. SnoN degradation is also regulated through the anaphase-promoting complex (APC) (Stroschein et al., 2001; Wan et al., 2001). The APC is a multisubunit ubiquitin ligase complex that regulates mitotic exit by targeting cyclins and other regulators for degradation in a destruction box-dependent manner (Zachariae and Nasmyth, 1999). Smad3 binds both SnoN and APC, and synergizes with the APC activator CDH1 to promote SnoN ubiquitination and degradation. It is clear that Smad3 itself is not ubiquitinated by APC. Interestingly, a similar complex of Smad3, HEF1 (human enhancer of filamentation 1, a Cas family cytoplasmic docking protein), APC10 and CDH1, controls the degradation of the Smad3-interacting protein HEF1 (Nourry et al., 2004). APC10 may also be the subunit in the APC that binds to Smad3 for SnoN degradation.

A recent report showed that TGF-β promotes $SCF^{\beta TrCP}$-mediated Cdc25A ubiquitination and degradation (Ray et al., 2005). The Cdc25A phosphatase promotes cell cycle progression by dephosphorylating and activating cyclin-dependent kinases (Cdks). In response to TGF-β, Smad3 physically interacts with Cdc25A, and the $SCF^{\beta TrCP}$ ligase. Through multiple contacts, Smad3 is a high-affinity binding partner with the SCF$^{\beta TrCP}$ ligase. Smad3 appears to bind more efficiently to β-TrCP2 than β-TrCP1, the substrate binding subunit of the SCF$^{\beta TrCP}$ ligase (Ray et al., 2005), and meanwhile, Smad3 can bind to Rbx1 of the ligase (Fukuchi et al., 2001). Both overexpression and knock-down experiments showed that the Smad3-SCF$^{\beta TrCP}$ complex induces ubiquitination and degradation of Cdc25A in a phosphorylation-dependent manner (Ray et al., 2005). Although Cdc25A is also targeted by the APC ligase during cell cycle, the Smad3-SCF$^{\beta TrCP}$ complex appears to be solely responsible for TGF-β-induced Cdc25A degradation.

Notably, most Smad3-associated proteins, *e.g*. SnoN and Cdc25A that are targeted to proteasomal degradation by Smad3, are growth-promoting factors that dampen the antiproliferative response of TGF-β. Thus, Smad3-dependent degradation of these proteins provides one mechanism for tumor suppression. An important question is how the specificity in ubiquitination and degradation of these proteins is determined, specifically, why certain and not all associated proteins are targeted? Are R-Smads and association proteins degraded together or in a sequential order? These questions remain unanswered.

3.4 Biological Functions of Smurfs

The supporting evidence that Smurfs serve as ubiquitin E3 ligases for Smads and/or Smad-associated proteins initially emerged from biochemical studies. Overexpression of Smurfs blocks TGF-β-induced responses in cultured cells and *Xenopus* embryos. However, the physiological functions of Smurfs remain poorly understood. A number of attempts have been made to address the functions of Smurfs in various cell types or tissues.

Smurfs are upregulated and correlate with the reduced level of Smads under certain pathophysiological conditions. In esophageal cancers, there is an increase in endogenous Smurf2 expression and concomitantly a decreased level of phosphorylated Smad2 (Fukuchi et al., 2002). Similarly, increased expression of Smurf2 parallels the ubiquitination and degradation of endogenous Smad2 in rat glomeruli with nephritis (Togawa et al., 2003). Upregulation of Smurf1 and Smurf2 may also be responsible for Smad7 degradation in renal fibrosis (Fukasawa et al., 2004). These are consistent with the finding that TGF-β, upregulated in nephritic and fibrotic tissues, induces transcription of Smurf2 (Ohashi et al., 2005).

The role of Smurf1 has also been analyzed through targeted expression of Smurf1 in a variety of tissues. Transgenic analysis further supports the role of Smurf1 in the destruction of Smad1/5. Osteoblast-specific expression of Smurf1 transgene leads to significant reduction in bone mass during postnatal life (Zhao et al., 2004). In contrast, transgenic mice overexpressing Smurf1 in chondrocytes fail to show clear abnormalities (Horiki et al., 2004). However, bigenic expression of both Smad6 and Smurf1 casuses more delayed endochondral ossification than Smad6 expression alone (Horiki et al., 2004). These studies provided evidence that Smurf1 alone or together with I-Smads, plays a specific role in bone and cartilage development. In addition, delivery of Smurf1 in airway epithelial cells via an adenoviral vector inhibits mouse lung branching morphogenesis (Shi et al., 2004). Consistent with its function to degrade Smad1/5, the effect of Smurf1 can be overcome by coexpression of Smad1.

Finally, targeted disruption of the *Smurf1* gene in mice further confirms the negative role of Smurf1 in BMP-dependent bone formation (Yamashita et al., 2005). At birth, the knock-out mice are normal but exhibit age-dependent increase in bone mass that results from enhanced osteoblast activity. Exhaustive analysis did not detect increased levels of any of previously defined Smurf1 substrates, including Smad1/5, Smad7 and Runx2, but instead established MEKK2 as a physiological substrate for Smurf1, highlighting the importance of JNK in TGF-β superfamily signaling. Still, this work does not exclude the E3 ligase function of Smurfs on Smads, particularly under the condition that Smurf2 might compensate for the loss of Smurf1. Indeed, the level of Smurf2 mRNA is increased about 2-fold in Smurf1-null mice, and the compounded Smurf1/2 knock-out leads to early embryonic lethality (Yamashita et al., 2005).

4. UBIQUITINATION AND SUMOYLATION OF SMAD4

Smad4, the common Smad for entire TGF-β family signaling, is frequently altered in cancers and other diseases. Recent evidence demonstrates that Smad4 can be phosphorylated, ubiquitinated and sumoylated (Fig. 4).

4.1 Smad4 Polyubiquitination by Distinct E3 Ligases

Although Smad4 is a fairly stable protein in comparison to R-Smads, its stability is regulated by ubiquitination-mediated proteasomal degradation. Smad4 ubiquitination can be catalyzed by at least three distinct types of E3 ligases, *i.e.* HECT, SCF and single RING proteins (Dupont et al., 2005; Liang et al., 2004a; Morén et al., 2005; Wan et al., 2004).

It is reported that $SCF^{\beta TrCP}$ promotes degradation of Smad4 (Wan et al., 2004). Another SCF E3 ligase, SCF^{Skp2}, seems not to affect wild-type Smad4 stability, but it promotes degradation of cancer-derived Smad4 mutants (Liang et al., 2004a), which will be discussed later. Most recently, a novel RING E3 ligase called Ectodermin (or Ecto) can specifically bind to Smad4 and reduce the level of nuclear Smad4 in the ectoderm, achieving a low level of TGF-β signaling and indirectly promoting mesodermal fate. Depletion of mammalian Ecto mimics a phenotype similar to that caused by overexpression of Smad4, and its overexpression inhibits the transcription of Smad4 target genes and consequently stimulate cell proliferation (Dupont et al., 2005). Like $SCF^{\beta TrCP}$ and SCF^{Skp2}, Ecto exhibits increased expression in tumors, suggesting that destruction of the tumor suppressor Smad4 may be essential for tumorigenesis. Another RING finger E3 ligase called Highwire binds to Medea, the Smad4 equivalent (*i.e.* co-Smad) in fly and perhaps antagonizes BMP-dependent synaptic growth through a ubiquitination-dependent mechanism (McCabe et al.,

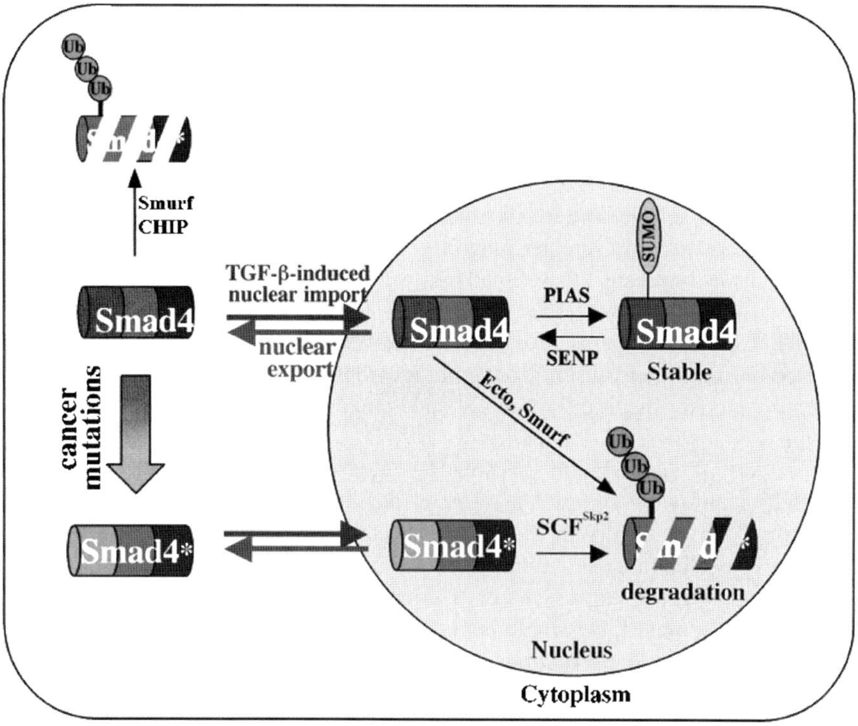

Figure 4. Regulation of Smad4 by ubiquitination and sumoylation. Smad4 undergoes nuclear import/export. Wild-type Smad4 turnover is regulated by Ecto, Smurfs and CHIP. It is also sumoylated in the nucleus, which further enhances its nuclear retention. Cancer-derived mutants of Smad4 (Smad4*), especially those with mutations in the MH1 domain, increase their affinity to SCF^{Skp2} E3 ligase. SCF^{Skp2} induces polyubiquitination and mediates proteasome-dependent rapid proteolysis of these mutants (A color version of this figure is freely accessible via the website of the book: http://www.springer.com/1-4020-4542-5)

2004). It will also be of interest to determine if PAM (protein associated with Myc), the human homolog of Highwire, similarly regulates Smad4 activity in vertebrates.

Although Smurf E3 ligases are initially found to directly ubiquitinate R-Smads and I-Smads, Smurfs appear to mediate polyubiquitination and proteasomal degradation of Smad4 via I-Smads as adaptors (Morén et al., 2005). The interaction of Smad4 with Smad2 or Smad6/7 is critical for Smurf-mediated Smad4 degradation. Finally, Jab1/CSN5 also promotes Smad4 degradation at least in cultured cells (Wan et al., 2002). Since Jab1 also targets Smad7 for degradation, the biological outcome of Jab1 in TGF-β responses may depend on specific cell types or stimuli.

Despite increasing numbers of E3 ligases involved in Smad4 regulation, no attempts have been made to map the ubiquitination acceptor sites. The obvious questions are: do distinct E3 ligases target different lysine residues on Smad4? Do they overlap with sumoylation and monoubiquitination sites? How do these E3 ligases spatially and temporally regulate the turnover of Smad4 during embryonic development and tumorigenesis? Nonetheless, the fact that there are an increasing number of E3 ligases responsible for Smad4 degradation suggests that proper control of Smad4 is essential for cellular functions.

4.2 Regulation of Stability, Subcellular Localization and Transactivation Activity of Smad4 by Sumoylation

SUMO-1 and its E2 Ubc9 were identified as Smad4-interacting proteins in independent yeast two-hybrid screens, respectively (Lee et al., 2003; Lin et al., 2003b). Further analysis indicates that Smad4 is sumoylated on Lys113 and Lys159. A comparison of the two sumoylation sites of human Smad4 with the same region from other organisms revealed that the two sumoylated lysines are highly conserved in Smad4, and absent from other members of Smad family. This suggests that sumoylation is specific for Smad4 and remains highly conserved throughout evolution.

The effects of sumoylation on Smad4 are multifaceted, with an increase in the Smad4 stability in the nucleus (Lee et al., 2003; Lin et al., 2003a). Smad4 sumoylation, which occurs in the nucleus, decreases its nuclear export, thereby enhancing its nuclear accumulation (Lin et al., 2003a). Loss of Smad4 sumoylation, either through ectopic expression of a dominant negative mutant of PIAS1 or depletion of E2 enzyme Ubc9, inhibits TGF-β transcriptional responses (Liang et al., 2004b). In contrast to these findings, Long et al. (2004) proposed that sumoylation inhibits the transactivation of Smad4 because the mutations at Lys113/159 render Smad4 more transcriptionally active (which, however, can be explained by increased stability) (Long et al., 2004a). The controversial conclusions await clarifications with further experimentations in a physiologically relevant environment.

4.3 Complex Roles of PIAS Proteins in TGF-β Signaling

The mammalian PIAS protein family consists of five members: PIAS1, PIAS3, PIASxα, PIASxβ and PIASy. Structurally, PIAS proteins contain a highly conserved SP-RING domain with homology to the yeast Siz proteins that function as SUMO E3 ligases (Schmidt and Muller, 2003). Mammalian protein MMS21/NSE2 and its yeast homologs also contain the SP-RING domain and serves as SUMO E3 ligases (Andrews et al., 2005; Potts and Yu, 2005; Zhao and Blobel, 2005). The SP-RING domain binds to Ubc9 (Kotaja et al., 2002) and thus is indispensable for stimulating sumoylation of Smad4 (Adams, 2003) and other targets (Schmidt and Muller, 2003).

An intriguing aspect of the roles of PIAS proteins in TGF-β signaling is that they not only act as specific E3 ligases in Smad4 sumoylation but also exert sumoylation-independent functions in TGF-β signaling. First, Smad4 sumoylation requires or is strongly induced by the presence of a PIAS protein (Lee et al., 2003; Liang et al., 2004b; Ohshima and Shimotohno, 2003). PIAS1, PIASx and PIASy bind to Smad4 and promote Smad4 sumoylation (Liang et al., 2004b; Long et al., 2004a; Ohshima and Shimotohno, 2003). Consistent with these observations,

PIAS1 mutants lacking functional RING domain (point mutations or deletions), but retaining the Smad4-binding domain (upstream region adjacent to the RING domain), dominant-negatively block Smad4 sumoylation and TGF-β signaling responses (Liang et al., 2004b). Second, PIAS proteins regulate the transcriptional activity of Smad3, either positively or negatively, depending on the recruitment of coactivators or corepressors. Both PIAS3 and PIASy interact with the MH2 domain of Smad3 (Imoto et al., 2003; Long et al., 2003; Long et al., 2004b). PIAS3 recruits the coactivator p300/CBP and activates Smad-dependent TGF-β signaling (Long et al., 2004b). In contrast, PIASy interacts with HDAC1 and represses the transcriptional activity of Smad3 (Long et al., 2004b). Apparently these opposing effects of PIAS3 and PIASy on Smad signaling are not through Smad4 sumoylation. Therefore, PIAS proteins have dual regulatory effects on TGF-β signaling. Examination of TGF-β signaling in cells or animals with a deficiency of PIAS proteins, which are now available (Liu et al., 2004; Roth et al., 2004; Santti et al., 2005), may clarify the physiological role of PIAS proteins or sumoylation in Smad signaling.

4.4 Smad4 Monoubiquitination

Monoubiquitination regulates the signaling functions of the substrates, in a manner similar to sumoylation and non-peptidyl modification such as phosphorylation. Among the components of TGF-β signaling pathways, only Smad4 is reported to be monoubiquitinated at the Lys507 in the MH2 domain (Morén et al., 2003). This lysine, conserved among all members of the Smad family, resides in the L3 loop region that is involved in recognition of the phospho-SXS of R-Smads. This modification appears to play a positive role in Smad oligomerization and subsequent signaling responses, as a substitution of Lys507 with Arg blocks the nuclear accumulation and transcriptional responses of Smad4 (Morén et al., 2003). However, the presumptive E3 ligase(s) that catalyze Smad4 monoubiquitination have not been identified. Furthermore, it will be interesting to explore whether the conserved lysines in other Smads are also monoubiquitinated and whether such modification of R-Smads/I-Smads affects their binding to the receptor.

5. UBIQUITIN/SUMO MODIFICATIONS OF SMADS IN CANCERS

The ubiquitin/proteasome pathway controls the turnover of products of many oncogenes and tumor suppressors, thus playing a major role in cancer development (Burger and Seth, 2004). Removal of Smads should provide a growth advantage to cancer cells. There are two ways to remove Smads via proteasomal degradation to dampen TGF-β tumor suppressing signaling. One is the increased expression or activation of specific ubiquitin-Smad E3 ligases. Recently, Fukuchi and coworkers reported an inverse correlation between expression levels of Smurf2 and phosphorylated Smad2 in esophageal cancers, and more importantly, a correlation between Smurf2 level and esophageal cancers (Fukuchi et al., 2002). The other mechanism underlying rapid Smad degradation, which is further discussed below, is through specific mutations in Smads that accelerate their turnover.

5.1 Accelerated Degradation of Cancer-derived Mutants: Smad2 and Smad4

Consistent with its role in tumor suppression, alterations in the *Smad4* gene frequently occur in human cancers. Whereas many Smad4 mutations or deletions fall into the MH2 domain that render Smad4 inactive in its signaling activity, a number of Smad4 missense mutations, primarily in the MH1 domain found in cancer cause their accelerated proteolysis (Morén et al., 2000; Xu and Attisano, 2000). For example, in pancreatic cancers, the R100T missense mutation of Smad4 (Arg to Thr at codon 100) results in a faster degradation of the mutant than wild-type Smad4 (Xu and Attisano, 2000). In Smad2, the R133C mutant, which carries a missense mutation at residue 133 that is analogous to the position 100 in Smad4, also has a shorter half-life, presumably through a similar destruction pathway (Xu and Attisano, 2000). These mutants are heavily ubiquitinated and rapidly degraded by the proteasome (Lin et al., 2003a; Morén et al., 2003; Xu and Attisano, 2000). In addition, a few other mutations in the Smad4 MH1 domain such as L43S, G65V and P130L are also unstable. It has been previously reported that the mutants are predominantly localized in the cytoplasm even in the presence of TGF-β, and some mutants such as G65V and P130L undergo partial nuclear import induced by TGF-β (Morén et al., 2003). However, MG-132 treatment not only causes increases in the total levels of these mutants in cells but also significantly promotes their nuclear localization, and furthermore, stimulation with leptomycin B, an inhibitor to CRM1-dependent nuclear export, leads to accumulation of these mutants in the nucleus (Liang et al., 2004a). These findings suggest that these MH1 domain mutants still have the ability to undergo nuclear localization, and thus may possess partial, if not complete, activities to mediate TGF-β responses. Therefore, accelerated degradation of these mutants by ubiquitin-mediated proteolysis may be a mechanism to ensure complete removal of any residual Smad4 tumor suppressor activity from cancer cells.

5.2 Altered Phosphorylation, Ubiquitination and Sumoylation of Smad4 mutants

Consistent with the decreased stability of the MH1 domain mutants, the Smad4 cancer mutants display higher affinity for two known ubiquitin E3 ligases, SCF^{Skp2} and $SCF^{\beta TrCP}$ E3 ligases (Liang et al., 2004a). Both SCF^{Skp2} and $SCF^{\beta TrCP}$ are oncogenic and often upregulated in cancer cells. SCF^{Skp2} and $SCF^{\beta TrCP}$ usually bind to their substrates through two non-related domains, *i.e.* the LRR and WD repeats, respectively. The ambiguity in use of more than one SCF E3 ligases by Smad4 mutants may be explained by the observation that Smad4 is capable of interacting with not just the LRR motif, but also the F-box domain of SCF^{Skp2}, which is present in all SCF ligases (Liang et al., 2004a).

Like most substrates of SCF E3 ligases, the MH1 mutants require prior phosphorylation for their proteasomal degradation. Thus, the addition of JNK/p38 kinase inhibitors prevents degradation of the Smad4 mutants to the same extent as proteasomal inhibitors. Mechanistically, missense mutations in the MH1 domain probably introduce conformational changes that induce massive phosphorylation of Smad4 mutants by JNK/p38 kinases, resulting in an increased affinity for the E3 ligase $SCF^{Skp2}/SCF^{\beta TrCP}$ (Liang et al., 2004a). JNK/p38-mediated phosphorylation is apparently low or absent in wild-type Smad4, explaining that Smad4 is a very stable protein in normal cells. Given the analogous position of Arg133 in Smad2 (as Arg100 in Smad4), it is conceivable that the Smad2 R133C mutant may exhibit a phosphorylation-dependent ubiquitination by the same or similar SCF E3 ligases.

Whereas phosphorylation is a prerequisite for ubiquitination of Smad4 mutants, another layer of complexity comes from the fact that the Smad4 MH1 mutants failed to be sumoylated (Liang et al., 2004a). The inverse correlation between sumoylation and ubiquitination could be due to the competition for the same acceptor sites, that is, Lys113 and/or Lys159 might also be ubiquitination sites. Indeed, concurrent mutations of Lys113/159 increase the stability of the R100T mutant (Lin et al., 2003a). However, these studies do not rule out another possibility that loss of sumoylation is attributed to a conformational change induced by JNK/p38-mediated phosphorylation. Or conversely, the presence of SUMO moiety in wild-type Smad4 prevents JNK/p38-mediated phosphorylation and subsequent SCF-mediated ubiquitination. In these latter cases, Lys113/159 are not necessarily targeted by ubiquitin. Thus, further identification of the phosphorylated serine residues and structural analysis of the mutant proteins and/or SUMO/ubiquitin-conjugated Smad4 proteins should gain insights into the relationship among these modifications.

6. UBIQUITIN AND SUMO PROTEASES

Most post-translational covalent modifications are reversible processes, and so are ubiquitin and ubiquitin-like conjugation on proteins. Thus, ubiquitin and SUMO substrates are regulated by not only ubiquitination and sumoylation, but also deubiquitination and desumoylation. There are large repertoires of deubiquitinating enzymes, classified into either the ubiquitin C-terminal hydrolase (UCH) or the ubiquitin-specific processing protease (UBP), which specifically cleave the isopeptide bond at the C-terminus of ubiquitin-protein conjugates. The conjugation and deconjugation of ubiquitin or SUMO are highly sophisticated and fine-tune the degradation and functions of targeted proteins.

Recently, Wicks and colleagues reported that the deubiquitinating enzyme UCH37/UCHL5 physiologically interacts with Smad3 and, more strongly, with Smad7. Smad7 recruits UCH37 to the type I TGF-β receptor to counteract the action of Smurf2. As a consequence, overexpression of UCH37 enhances, and UCH37 siRNA reduces, TGF-β transcriptional responses (Wicks et al., 2005). It is still not mechanistically clear how both UCH37 and Smurf2 use Smad7 to regulate TGF-β receptor turnover and whether UCH37 regulates turnover of Smad3 and other associated partners.

The proteases that process the C-terminal cleavage of SUMOs have also been identified and show no sequence similarity to any known deubiquitinating enzymes (Gong et al., 2000; Li and Hochstrasser, 1999). In human, the SUMO proteases, SENP1 to 7, are responsible for both SUMO processing and protein desumoylation (Melchior et al., 2003). Biochemical evidence suggests that at least SENP1 and SENP2 can remove the SUMO moiety from conjugated Smad4 (unpublished data in the authors' laboratories). Unexpectedly, overexpression of SENP1/SuPr-1 can increase the TGF-β response (Long et al., 2004a), perhaps due to the pleiotropic effects of SENPs. Thus, effects of SENPs in Smad4 sumoylation and functions need further investigation under physiological conditions.

7. OVERVIEW AND PERSPECTIVES

In summary, advances in recent years have illustrated a complex pattern in Smad-mediated transcriptional responses to TGF-β family growth factors. Like many key regulatory proteins, Smads are regulated by target gene product (feedback) and post-translational modifications. Accumulating evidence has clearly demonstrated the important roles of phosphorylation, ubiquitin and ubiquitin-like modifications in regulating the activity and stability of all Smads. These modifications integrate multiple signals to allow a fine-tuning of Smad signaling.

Despite the great progress in ubiquitin and ubiquitin-like modifications of Smads, further investigation is warranted to identify precise modification sites (that relies on mutagenesis, mass spectrometry, and generation of modification-specific antibodies) and to analyze how these modifications modulate Smad functions, especially in a physiological setting (*e.g.* knock-in animal models). Furthermore, much work is needed to explore other types of modifications (*e.g.* methylation, lipidation, nitrosylation, and neddylation) and investigate synergistic or opposing effects of different types of modifications at multiple sites. Since most of these modifications are reversible, identification and characterization of enzymes that counteract modifications (*e.g.* ubiquitin proteases) will provide further insights into the functions of these modifications in TGF-β physiology and disease development.

ACKNOWLEDGEMENT

Research on TGF-β/Smad signaling in our laboratories has been supported by grants from the National Institutes of Health (GM63773, CA108454, DK073932, CA112939) and American Cancer Society (RPG-00-214-01-CCG, RSG-00-214-04-CCG, RSG-02-145-01-CCG). X.-H.F. is a Leukemia & Lymphoma Society Scholar. X.L. is a recipient of the Baylor Breast Center SPORE career development award.

REFERENCES

Adams, J., 2003, The proteasome: structure, function, and role in the cell. *Cancer Treat Rev* 29 Suppl 1: 3-9.

Andrews, E.A., Palecek, J., Sergeant, J., Taylor, E., Lehmann, A.R., and Watts, F.Z., 2005, Nse2, a component of the Smc5-6 complex, is a SUMO ligase required for the response to DNA damage. *Mol Cell Biol* 25: 185-196.

Bai, S., Shi, X., Yang, X., and Cao, X., 2000, Smad6 as a transcriptional corepressor. *J Biol Chem* 275: 8267-8270.

Bai, Y., Yang, C., Hu, K., Elly, C., and Liu, Y.C., 2004, Itch E3 ligase-mediated regulation of TGF-β signaling by modulating smad2 phosphorylation. *Mol Cell* 15: 825-831.

Bitzer, M., von Gersdorff, G., Liang, D., Dominguez-Rosales, A., Beg, A.A., Rojkind, M., and Böttinger, E.P., 2000, A mechanism of suppression of TGF-β/SMAD signaling by NF-kappa B/RelA. *Genes Dev* 14: 187-197.

Bonni, S., Wang, H.R., Causing, C.G., Kavsak, P., Stroschein, S.L., Luo, K., and Wrana, J.L., 2001, TGF-β induces assembly of a Smad2-Smurf2 ubiquitin ligase complex that targets SnoN for degradation. *Nat Cell Biol* 3: 587-595.

Burger, A.M., and Seth, A.K., 2004, The ubiquitin-mediated protein degradation pathway in cancer: therapeutic implications. *Eur J Cancer* 40: 2217-2229.

Ciechanover, A., 2005, Proteolysis: from the lysosome to ubiquitin and the proteasome. *Nat Rev Mol Cell Biol* 6: 79-87.

Desterro, J.M., Rodriguez, M.S., and Hay, R.T., 1998, SUMO-1 modification of IκBα inhibits NF-κB activation. *Mol Cell* 2: 233-239.

Di Guglielmo, G.M., Le Roy, C., Goodfellow, A.F., and Wrana, J.L., 2003, Distinct endocytic pathways regulate TGF-β receptor signalling and turnover. *Nat Cell Biol* 5: 410-421.

Dupont, S., Zacchigna, L., Cordenonsi, M., Soligo, S., Adorno, M., Rugge, M., and Piccolo, S., 2005, Germ-layer specification and control of cell growth by Ectodermin, a Smad4 ubiquitin ligase. *Cell* 121: 87-99.

Ebisawa, T., Fukuchi, M., Murakami, G., Chiba, T., Tanaka, K., Imamura, T., and Miyazono, K., 2001, Smurf1 interacts with transforming growth factor-β type I receptor through Smad7 and induces receptor degradation. *J Biol Chem* 276: 12477-12480.

Fukasawa, H., Yamamoto, T., Togawa, A., Ohashi, N., Fujigaki, Y., Oda, T., Uchida, C., Kitagawa, K., Hattori, T., Suzuki, S., Kitagawa, M., and Hishida, A., 2004, Down-regulation of Smad7 expression by ubiquitin-dependent degradation contributes to renal fibrosis in obstructive nephropathy in mice. *Proc Natl Acad Sci U S A* 101: 8687-8692.

Fukuchi, M., Fukai, Y., Masuda, N., Miyazaki, T., Nakajima, M., Sohda, M., Manda, R., Tsukada, K., Kato, H., and Kuwano, H., 2002, High-level expression of the Smad ubiquitin ligase Smurf2 correlates with poor prognosis in patients with esophageal squamous cell carcinoma. *Cancer Res* 62: 7162-7165.

Fukuchi, M., Imamura, T., Chiba, T., Ebisawa, T., Kawabata, M., Tanaka, K., and Miyazono, K., 2001, Ligand-dependent degradation of Smad3 by a ubiquitin ligase complex of Roc1 and associated proteins. *Mol Biol Cell* 12: 1431-1443.

Girdwood, D.W., Tatham, M.H., and Hay, R.T., 2004, SUMO and transcriptional regulation. *Semin Cell Dev Biol* 15: 201-210.

Gong, L., Millas, S., Maul, G.G., and Yeh, E.T., 2000, Differential regulation of sentrinized proteins by a novel sentrin- specific protease. *J Biol Chem* 275: 3355-3359.

Grönroos, E., Hellman, U., Heldin, C.-H., and Ericsson, J., 2002, Control of Smad7 stability by competition between acetylation and ubiquitination. *Mol Cell* 10: 483-493.

Gruendler, C., Lin, Y., Farley, J., and Wang, T., 2001, Proteasomal degradation of Smad1 induced by bone morphogenetic proteins. *J Biol Chem* 276: 46533-46543.

Hanyu, A., Ishidou, Y., Ebisawa, T., Shimanuki, T., Imamura, T., and Miyazono, K., 2001, The N domain of Smad7 is essential for specific inhibition of transforming growth factor-β signaling. *J Cell Biol* 155: 1017-1027.

Hata, A., Lagna, G., Massagué, J., and Hemmati-Brivanlou, A., 1998, Smad6 inhibits BMP/Smad1 signaling by specifically competing with the Smad4 tumor suppressor. *Genes Dev* 12: 186-197.

Hay, R.T., 2005, SUMO: a history of modification. *Mol Cell* 18: 1-12.

Hayashi, H., Abdollah, S., Qiu, Y., Cai, J., Xu, Y.-Y., Grinnell, B., Richardson, M., Topper, J., Gimbrone, M., Jr., Wrana, J.L., and Falb, D., 1997, The MAD-related protein Smad7 associates with the TGFβ receptor and functions as an antagonist of the TGFβ signaling. *Cell* 89: 1165-1173.

Hershko, A., 2005, The ubiquitin system for protein degradation and some of its roles in the control of the cell division cycle. *Cell Death Differ* 12: 1191-1197.

Hershko, A., and Ciechanover, A., 1982, Mechanisms of intracellular protein breakdown. *Annu Rev Biochem* 51: 335-364.

Hochstrasser, M., 2001, SP-RING for SUMO: new functions bloom for a ubiquitin-like protein. *Cell* 107: 5-8.

Horiki, M., Imamura, T., Okamoto, M., Hayashi, M., Murai, J., Myoui, A., Ochi, T., Miyazono, K., Yoshikawa, H., and Tsumaki, N., 2004, Smad6/Smurf1 overexpression in cartilage delays chondrocyte hypertrophy and causes dwarfism with osteopenia. *J Cell Biol* 165: 433-445.

Imamura, T., Takase, M., Nishihara, A., Oeda, E., Hanai, J., Kawabata, M., and Miyazono, K., 1997, Smad6 inhibits signalling by the TGF-β superfamily. *Nature* 389: 622-626.

Imoto, S., Sugiyama, K., Muromoto, R., Sato, N., Yamamoto, T., and Matsuda, T., 2003, Regulation of transforming growth factor-β signaling by protein inhibitor of activated STAT, PIASy through Smad3. *J Biol Chem* 278: 34253-34258.

Ingham, R.J., Colwill, K., Howard, C., Dettwiler, S., Lim, C.S., Yu, J., Hersi, K., Raaijmakers, J., Gish, G., Mbamalu, G., Taylor, L., Yeung, B., Vassilovski, G., Amin, M., Chen, F., Matskova, L., Winberg, G., Ernberg, I., Linding, R., O'Donnell, P., Starostine, A., Keller, W., Metalnikov, P., Stark, C., and Pawson, T., 2005, WW domains provide a platform for the assembly of multiprotein networks. *Mol Cell Biol* 25: 7092-7106.

Jackson, P.K., 2001, A new RING for SUMO: wrestling transcriptional responses into nuclear bodies with PIAS family E3 SUMO ligases. *Genes Dev* 15: 3053-3058.

Johnson, E.S., 2004, Protein modification by SUMO. *Annu Rev Biochem* 73: 355-382.

Kagey, M.H., Melhuish, T.A., and Wotton, D., 2003, The polycomb protein Pc2 is a SUMO E3. *Cell* 113: 127-137.

Kavsak, P., Rasmussen, R.K., Causing, C.G., Bonni, S., Zhu, H., Thomsen, G.H., and Wrana, J.L., 2000, Smad7 binds to Smurf2 to form an E3 ubiquitin ligase that targets the TGF β receptor for degradation. *Mol Cell* 6: 1365-1375.

Kim, B.C., Lee, H.J., Park, S.H., Lee, S.R., Karpova, T.S., McNally, J.G., Felici, A., Lee, D.K., and Kim, S.J., 2004, Jab1/CSN5, a component of the COP9 signalosome, regulates transforming growth factor β signaling by binding to Smad7 and promoting its degradation. *Mol Cell Biol* 24: 2251-2262.

Knuesel, M., Wan, Y., Xiao, Z., Holinger, E., Lowe, N., Wang, W., and Liu, X., 2003, Identification of novel protein-protein interactions using a versatile Mammalian tandem affinity purification expression system. *Mol Cell Proteomics* 2: 1225-1233.

Koinuma, D., Shinozaki, M., Komuro, A., Goto, K., Saitoh, M., Hanyu, A., Ebina, M., Nukiwa, T., Miyazawa, K., Imamura, T., and Miyazono, K., 2003, Arkadia amplifies TGF-β superfamily signalling through degradation of Smad7. *EMBO J* 22: 6458-6470.

Komuro, A., Imamura, T., Saitoh, M., Yoshida, Y., Yamori, T., Miyazono, K., and Miyazawa, K., 2004, Negative regulation of transforming growth factor-β (TGF-β) signaling by WW domain-containing protein 1 (WWP1). *Oncogene* 23: 6914-6923.

Kotaja, N., Karvonen, U., Janne, O.A., and Palvimo, J.J., 2002, PIAS proteins modulate transcription factors by functioning as SUMO-1 ligases. *Mol Cell Biol* 22: 5222-5234.

Kuratomi, G., Komuro, A., Goto, K., Shinozaki, M., Miyazawa, K., Miyazono, K., and Imamura, T., 2005, NEDD4-2 (neural precursor cell expressed, developmentally down-regulated 4-2) negatively regulates TGF-β (transforming growth factor-β) signalling by inducing ubiquitin-mediated degradation of Smad2 and TGF-β type I receptor. *Biochem J* 386: 461-470.

Lee, P.S., Chang, C., Liu, D., and Derynck, R., 2003, Sumoylation of Smad4, the common Smad mediator of transforming growth factor-β family signaling. *J Biol Chem* 278: 27853-27863.

Li, L., Xin, H., Xu, X., Huang, M., Zhang, X., Chen, Y., Zhang, S., Fu, X.Y., and Chang, Z., 2004, CHIP mediates degradation of Smad proteins and potentially regulates Smad-induced transcription. *Mol Cell Biol* 24: 856-864.

Li, S.J., and Hochstrasser, M., 1999, A new protease required for cell-cycle progression in yeast. *Nature* 398: 246-251.

Liang, M., Liang, Y.Y., Wrighton, K., Ungermannova, D., Wang, X.P., Brunicardi, F.C., Liu, X., Feng, X.H., and Lin, X., 2004a, Ubiquitination and proteolysis of cancer-derived Smad4 mutants by SCFSkp2. *Mol Cell Biol* 24: 7524-7537.

Liang, M., Melchior, F., Feng, X.H., and Lin, X., 2004b, Regulation of Smad4 sumoylation and transforming growth factor-β signaling by protein inhibitor of activated STAT1. *J Biol Chem* 279: 22857-22865.

Liang, Y.Y., Lin, X., Liang, M., Brunicardi, F.C., ten Dijke, P., Chen, Z., Choi, K.W., and Feng, X.H., 2003, dSmurf selectively degrades decapentaplegic-activated MAD, and its overexpression disrupts imaginal disc development. *J Biol Chem* 278: 26307-26310.

Lin, X., Liang, M., and Feng, X.-H., 2000, Smurf2 is a ubiquitin E3 ligase mediating proteasome-dependent degradation of Smad2 in TGF-β signaling. *J Biol Chem* 275: 36818-36822.

Lin, X., Liang, M., Liang, Y.Y., Brunicardi, F.C., and Feng, X.H., 2003a, SUMO-1/Ubc9 promotes nuclear accumulation and metabolic stability of tumor suppressor Smad4. *J Biol Chem* 278: 31043-31048.

Lin, X., Liang, M., Liang, Y.Y., Brunicardi, F.C., Melchior, F., and Feng, X.H., 2003b, Activation of transforming growth factor-β signaling by SUMO-1 modification of tumor suppressor Smad4/DPC4. *J Biol Chem* 278: 18714-18719.

Lin, X., Liang, Y.-Y., Sun, B., Liang, M., Brunicardi, F.C., Shi, Y., Shi, Y., and Feng, X.-H., 2003c, Smad6 recruits transcription corepressor CtBP to repress bone morphogenetic protein-induced transcription. *Mol Cell Biol* 23: 9081-9093.

Lin, X., Sun, B., Liang, M., Liang, Y.Y., Gast, A., Hildebrand, J., Brunicardi, F.C., Melchior, F., and Feng, X.H., 2003d, Opposed regulation of corepressor CtBP by SUMOylation and PDZ binding. *Mol Cell* 11: 1389-1396.

Lin, Y., Martin, J., Gruendler, C., Farley, J., Meng, X., Li, B.Y., Lechleider, R., Huff, C., Kim, R.H., Grasser, W.A., Paralkar, V., and Wang, T., 2002, A novel link between the proteasome pathway and the signal transduction pathway of the bone morphogenetic proteins (BMPs). *BMC Cell Biol* 3: 15.

Liu, B., Mink, S., Wong, K.A., Stein, N., Getman, C., Dempsey, P.W., Wu, H., and Shuai, K., 2004, PIAS1 selectively inhibits interferon-inducible genes and is important in innate immunity. *Nat Immunol* 5: 891-898.

Lo, R.S., and Massagué, J., 1999, Ubiquitin-dependent degradation of TGF-β-activated Smad2. *Nat Cell Biol* 1: 472-478.

Long, J., Matsuura, I., He, D., Wang, G., Shuai, K., and Liu, F., 2003, Repression of Smad transcriptional activity by PIASy, an inhibitor of activated STAT. *Proc Natl Acad Sci U S A* 100: 9791-9796.

Long, J., Wang, G., He, D., and Liu, F., 2004a, Repression of Smad4 transcriptional activity by SUMO modification. *Biochem J* 379: 23-29.

Long, J., Wang, G., Matsuura, I., He, D., and Liu, F., 2004b, Activation of Smad transcriptional activity by protein inhibitor of activated STAT3 (PIAS3). *Proc Natl Acad Sci U S A* 101: 99-104.

Mahajan, R., Delphin, C., Guan, T., Gerace, L., and Melchior, F., 1997, A small ubiquitin-related polypeptide involved in targeting RanGAP1 to nuclear pore complex protein RanBP2. *Cell* 88: 97-107.

Matunis, M.J., Coutavas, E., and Blobel, G., 1996, A novel ubiquitin-like modification modulates the partitioning of the Ran-GTPase-activating protein RanGAP1 between the cytosol and the nuclear pore complex. *J Cell Biol* 135: 1457-1470.

McCabe, B.D., Hom, S., Aberle, H., Fetter, R.D., Marques, G., Haerry, T.E., Wan, H., O'Connor, M.B., Goodman, C.S., and Haghighi, A.P., 2004, Highwire regulates presynaptic BMP signaling essential for synaptic growth. *Neuron* 41: 891-905.

McDonough, H., and Patterson, C., 2003, CHIP: a link between the chaperone and proteasome systems. *Cell Stress Chaperones* 8: 303-308.

Melchior, F., Schergaut, M., and Pichler, A., 2003, SUMO: ligases, isopeptidases and nuclear pores. *Trends Biochem Sci* 28: 612-618.

Morén, A., Hellman, U., Inada, Y., Imamura, T., Heldin, C.-H., and Moustakas, A., 2003, Differential ubiquitination defines the functional status of the tumor suppressor Smad4. *J Biol Chem* 278: 33571-33582.

Morén, A., Imamura, T., Miyazono, K., Heldin, C.-H., and Moustakas, A., 2005, Degradation of the tumor suppressor Smad4 by WW and HECT domain ubiquitin ligases. *J Biol Chem* 280: 22115-22123.

Morén, A., Itoh, S., Moustakas, A., Dijke, P., and Heldin, C.-H., 2000, Functional consequences of tumorigenic missense mutations in the amino- terminal domain of Smad4. *Oncogene* 19: 4396-4404.

Muller, S., and Dejean, A., 1999, Viral immediate-early proteins abrogate the modification by SUMO-1 of PML and Sp100 proteins, correlating with nuclear body disruption. *J Virol* 73: 5137-5143.

Muller, S., Ledl, A., and Schmidt, D., 2004, SUMO: a regulator of gene expression and genome integrity. *Oncogene* 23: 1998-2008.

Murakami, G., Watabe, T., Takaoka, K., Miyazono, K., and Imamura, T., 2003, Cooperative inhibition of bone morphogenetic protein signaling by Smurf1 and inhibitory Smads. *Mol Biol Cell* 14: 2809-2817.

Nakao, A., Afrakhte, M., Morén, A., Nakayama, T., Christian, J., Heuchel, R., Itoh, S., Kawabata, M., Heldin, N.-H., Heldin, C.-H., and ten Dijke, P., 1997, Identification of Smad7, a TGFβ-inducible antagonist of TGF-β signalling. *Nature* 389: 631-635.

Nourry, C., Maksumova, L., Pang, M., Liu, X., and Wang, T., 2004, Direct interaction between Smad3, APC10, CDH1 and HEF1 in proteasomal degradation of HEF1. *BMC Cell Biol* 5: 20.

Ogunjimi, A.A., Briant, D.J., Pece-Barbara, N., Le Roy, C., Di Guglielmo, G.M., Kavsak, P., Rasmussen, R.K., Seet, B.T., Sicheri, F., and Wrana, J.L., 2005, Regulation of Smurf2 ubiquitin ligase activity by anchoring the E2 to the HECT domain. *Mol Cell* 19: 297-308.

Ohashi, N., Yamamoto, T., Uchida, C., Togawa, A., Fukasawa, H., Fujigaki, Y., Suzuki, S., Kitagawa, K., Hattori, T., Oda, T., Hayashi, H., Hishida, A., and Kitagawa, M., 2005, Transcriptional induction of Smurf2 ubiquitin ligase by TGF-β. *FEBS Lett* 579: 2557-2563.

Ohshima, T., and Shimotohno, K., 2003, TGF-b mediated signaling via the p38 MAP kinase pathway activates Smad-dependent transcription through SUMO-1 modification of Smad4. *J Biol Chem* 278: 50833-50842.

Park, S.H., 2005, Fine tuning and cross-talking of TGF-β signal by inhibitory Smads. *J Biochem Mol Biol* 38: 9-16.

Pichler, A., Gast, A., Seeler, J.S., Dejean, A., and Melchior, F., 2002, The nucleoporin RanBP2 has SUMO1 E3 ligase activity. *Cell* 108: 109-120.

Podos, S.D., Hanson, K.K., Wang, Y.C., and Ferguson, E.L., 2001, The DSmurf ubiquitin-protein ligase restricts BMP signaling spatially and temporally during *Drosophila* embryogenesis. *Dev Cell* 1: 567-578.

Potts, P.R., and Yu, H., 2005, Human MMS21/NSE2 is a SUMO ligase required for DNA repair. *Mol Cell Biol* 25: 7021-7032.

Ray, D., Terao, Y., Nimbalkar, D., Chu, L.H., Donzelli, M., Tsutsui, T., Zou, X., Ghosh, A.K., Varga, J., Draetta, G.F., and Kiyokawa, H., 2005, Transforming growth factor β facilitates β-TrCP-mediated degradation of Cdc25A in a Smad3-dependent manner. *Mol Cell Biol* 25: 3338-3347.

Roth, W., Sustmann, C., Kieslinger, M., Gilmozzi, A., Irmer, D., Kremmer, E., Turck, C., and Grosschedl, R., 2004, PIASy-deficient mice display modest defects in IFN and Wnt signaling. *J Immunol* 173: 6189-6199.

Santti, H., Mikkonen, L., Anand, A., Hirvonen-Santti, S., Toppari, J., Panhuysen, M., Vauti, F., Perera, M., Corte, G., Wurst, W., Janne, O.A., and Palvimo, J.J., 2005, Disruption of the murine PIASx gene results in reduced testis weight. *J Mol Endocrinol* 34: 645-654.

Schmidt, D., and Muller, S., 2003, PIAS/SUMO: new partners in transcriptional regulation. *Cell Mol Life Sci* 60: 2561-2574.

Seo, S.R., Lallemand, F., Ferrand, N., Pessah, M., L'Hoste, S., Camonis, J., and Atfi, A., 2004, The novel E3 ubiquitin ligase Tiul1 associates with TGIF to target Smad2 for degradation. *EMBO J* 23: 3780-3792.

Shi, W., Chen, H., Sun, J., Chen, C., Zhao, J., Wang, Y.L., Anderson, K.D., and Warburton, D., 2004, Overexpression of Smurf1 negatively regulates mouse embryonic lung branching morphogenesis by specifically reducing Smad1 and Smad5 proteins. *Am J Physiol Lung Cell Mol Physiol* 286: L293-300.

Sigismund, S., Polo, S., and Di Fiore, P.P., 2004, Signaling through monoubiquitination. *Curr Top Microbiol Immunol* 286: 149-185.

Stroschein, S.L., Bonni, S., Wrana, J.L., and Luo, K., 2001, Smad3 recruits the anaphase-promoting complex for ubiquitination and degradation of SnoN. *Genes Dev* 15: 2822-2836.

Sun, L., and Chen, Z.J., 2004, The novel functions of ubiquitination in signaling. *Curr Opin Cell Biol* 16: 119-126.

Suzuki, C., Murakami, G., Fukuchi, M., Shimanuki, T., Shikauchi, Y., Imamura, T., and Miyazono, K., 2002, Smurf1 regulates the inhibitory activity of Smad7 by targeting Smad7 to the plasma membrane. *J Biol Chem* 277: 39919-39925.

Tajima, Y., Goto, K., Yoshida, M., Shinomiya, K., Sekimoto, T., Yoneda, Y., Miyazono, K., and Imamura, T., 2003, Chromosomal region maintenance 1 (CRM1)-dependent nuclear export of Smad ubiquitin regulatory factor 1 (Smurf1) is essential for negative regulation of transforming growth factor-β signaling by Smad7. *J Biol Chem* 278: 10716-10721.

Togawa, A., Yamamoto, T., Suzuki, H., Fukasawa, H., Ohashi, N., Fujigaki, Y., Kitagawa, K., Hattori, T., Kitagawa, M., and Hishida, A., 2003, Ubiquitin-dependent degradation of Smad2 is increased in the glomeruli of rats with anti-thymocyte serum nephritis. *Am J Pathol* 163: 1645-1652.

Ulloa, L., Doody, J., and Massagué, J., 1999, Inhibition of transforming growth factor-β/SMAD signalling by the interferon-gamma/STAT pathway. *Nature* 397: 710-713.

Wan, M., Cao, X., Wu, Y., Bai, S., Wu, L., Shi, X., and Wang, N., 2002, Jab1 antagonizes TGF-β signaling by inducing Smad4 degradation. *EMBO Rep* 3: 171-176.

Wan, M., Tang, Y., Tytler, E.M., Lu, C., Jin, B., Vickers, S.M., Yang, L., Shi, X., and Cao, X., 2004, Smad4 protein stability is regulated by ubiquitin ligase SCF β-TrCP1. *J Biol Chem* 279: 14484-14487.

Wan, Y., Liu, X., and Kirschner, M.W., 2001, The anaphase-promoting complex mediates TGF-β signaling by targeting SnoN for destruction. *Mol Cell* 8: 1027-1039.

Wang, H.R., Zhang, Y., Ozdamar, B., Ogunjimi, A.A., Alexandrova, E., Thomsen, G.H., and Wrana, J.L., 2003, Regulation of cell polarity and protrusion formation by targeting RhoA for degradation. *Science* 302: 1775-1779.

Wicks, S.J., Haros, K., Maillard, M., Song, L., Cohen, R.E., Dijke, P.T., and Chantry, A., 2005, The deubiquitinating enzyme UCH37 interacts with Smads and regulates TGF-β signalling. *Oncogene*.

Xin, H., Xu, X., Li, L., Ning, H., Rong, Y., Shang, Y., Wang, Y., Fu, X.Y., and Chang, Z., 2005, CHIP controls the sensitivity of transforming growth factor-β signaling by modulating the basal level of Smad3 through ubiquitin-mediated degradation. *J Biol Chem* 280: 20842-20850.

Xu, J., and Attisano, L., 2000, Mutations in the tumor suppressors Smad2 and Smad4 inactivate transforming growth factor β signaling by targeting Smads to the ubiquitin-proteasome pathway. *Proc Natl Acad Sci U S A* 97: 4820-4825.

Yamashita, M., Ying, S.X., Zhang, G.M., Li, C., Cheng, S.Y., Deng, C.X., and Zhang, Y.E., 2005, Ubiquitin ligase Smurf1 controls osteoblast activity and bone homeostasis by targeting MEKK2 for degradation. *Cell* 121: 101-113.

Yang, S.H., Jaffray, E., Senthinathan, B., Hay, R.T., and Sharrocks, A.D., 2003, SUMO and transcriptional repression: dynamic interactions between the MAP kinase and SUMO pathways. *Cell Cycle* 2: 528-530.

Zachariae, W., and Nasmyth, K., 1999, Whose end is destruction: cell division and the anaphase-promoting complex. *Genes Dev* 13: 2039-2058.

Zhang, L., Zhou, H., Su, Y., Sun, Z., Zhang, H., Zhang, Y., Ning, Y., Chen, Y.G., and Meng, A., 2004, Zebrafish Dpr2 inhibits mesoderm induction by promoting degradation of nodal receptors. *Science* 306: 114-117.

Zhang, Y., Chang, C., Gehling, D.J., Hemmati-Brivanlou, A., and Derynck, R., 2001, Regulation of Smad degradation and activity by Smurf2, an E3 ubiquitin ligase. *Proc Natl Acad Sci U S A* 98: 974-979.

Zhao, M., Qiao, M., Harris, S.E., Oyajobi, B.O., Mundy, G.R., and Chen, D., 2004, Smurf1 inhibits osteoblast differentiation and bone formation *in vitro* and *in vivo*. *J Biol Chem* 279: 12854-12859.

Zhao, M., Qiao, M., Oyajobi, B.O., Mundy, G.R., and Chen, D., 2003, E3 ubiquitin ligase Smurf1 mediates core-binding factor alpha1/Runx2 degradation and plays a specific role in osteoblast differentiation. *J Biol Chem* 278: 27939-27944.

Zhao, X., and Blobel, G., 2005, A SUMO ligase is part of a nuclear multiprotein complex that affects DNA repair and chromosomal organization. *Proc Natl Acad Sci U S A* 102: 4777-4782.

Zhong, S., Muller, S., Ronchetti, S., Freemont, P.S., Dejean, A., and Pandolfi, P.P., 2000, Role of SUMO-1-modified PML in nuclear body formation. *Blood* 95: 2748-2752.

Zhu, H., Kavsak, P., Abdollah, S., Wrana, J.L., and Thomsen, G.H., 1999, A SMAD ubiquitin ligase targets the BMP pathway and affects embryonic pattern formation. *Nature* 400: 687-693.

CHAPTER 14

SMAD TRANSCRIPTIONAL CO-ACTIVATORS AND CO-REPRESSORS

KOHEI MIYAZONO[1,2], SHINGO MAEDA[2], AND TAKESHI IMAMURA[2]

[1] *Department of Molecular Pathology, Graduate School of Medicine, University of Tokyo, 7-3-1 Hongo, Bunkyo-ku, Tokyo 113-0033, Japan*

[2] *Department of Biochemistry, The Cancer Institute of the Japanese Foundation for Cancer Research (JFCR), 3-10-6 Ariake, Koto-ku, Tokyo 135-8550, Japan*

Abstract: Upon activation of TGF-β superfamily signaling, R-Smads and Co-Smads translocate into the nucleus where they regulate transcription of target genes. Smads interact with various transcriptional co-activators, including p300, CBP, PCAF, and GCN5, most of which have intrinsic acetyl transferase activities and induce acetylation of histones and other proteins. Interaction of Smads with transcriptional co-activators is regulated by various cellular and viral proteins. Smads also bind to transcriptional co-repressors, including Ski, SnoN, and TGIF, and recruit histone deacetylases, resulting in deacetylation of histones and repression of transcription. Transcriptional regulation may also occur independently of regulation of histone acetylation. Thus, Smad-mediated transcription is induced by the transcriptional co-activators, and its magnitude is fine-tuned by them as well as by co-repressors

Keywords: acetyl transferase; histone deacetylase: p300; CBP; Ski; SnoN; TGIF

1. INTRODUCTION

Members of the transforming growth factor-β (TGF-β) superfamily regulate various cellular responses, including cell proliferation, differentiation, migration, and apoptosis. They bind to type II and I serine/threonine kinase receptors, and transduce intracellular signals through Smad-dependent and -independent signaling pathways. Of the three distinct classes of Smads, receptor-regulated Smads (R-Smads) are directly phosphorylated by type I receptors. R-Smads then form heteromeric complexes with common-partner Smad (Co-Smad), and these complexes translocate into the nucleus, where they regulate transcription of target genes.

P. ten Dijke and C.-H. Heldin (eds.), Smad Signal Transduction, 277–293.

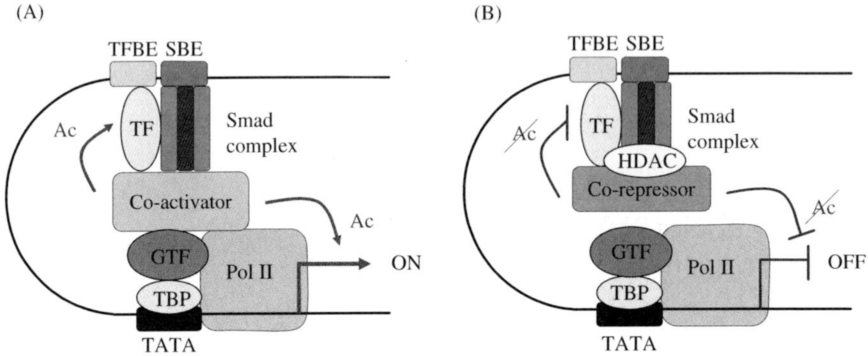

Figure 1. Regulation of Smad-mediated transcription by transcriptional co-activators with HAT activity (A) and transcriptional co-repressors which recruit HDACs to the Smad complexes (B). TF, transcription factor; TFBE, transcription factor binding element; SBE, Smad binding element; TBP, TATA binding protein; GTF, general transcription factors; Pol II, RNA polymerase II; Ac, acetylation of histones and/or TF (A color version of this figure is freely accessible via the website of the book: http://www.springer.com/1-4020-4542-5)

The Smad complexes interact with various transcription factors, including FAST1/FoxH1, TFE3, c-Jun, Sp1, and Runx proteins, and exhibit a wide variety of biological effects (Fig. 1A). These transcription factors directly bind to specific DNA sequences. Smad3 (TGF-β/Activin-specific R-Smad) and Smad4 (Co-Smad) also bind to DNA through their N-terminal MH1 domains, although direct DNA binding of Smads is not essential for transcription of certain target genes.

In addition to binding to various transcription factors, Smads interact with transcriptional co-activators in the nucleus. Some of these transcriptional co-activators have intrinsic histone acetyltransferase (HAT) activity, and facilitate transcription by acetylation of histones and loosening of chromatin condensation (Fig. 1A). In addition, these transcriptional co-activators bring the Smad-transcription factor complexes into close proximity with the basal transcriptional machinery. Other molecules without intrinsic HAT activity may support the interaction between Smad complexes and transcriptional co-activators with HATs, and thereby facilitate transcription of target genes. In contrast, some transcriptional co-repressors interact with Smads, disrupt the interaction between Smads and transcriptional co-activators, and/or recruit histone deacetylases (HDACs) to Smad complexes, resulting in transcriptional repression (Fig. 1B). In this chapter, we discuss the roles of transcriptional co-activators and co-repressors in transcriptional regulation by Smad proteins.

2. TRANSCRIPTIONAL ACTIVATION INDUCED BY SMAD COMPLEXES

During the initiation of transcription and elongation, a number of proteins are recruited to the Smad transcriptional complexes and participate in this process. These include transcriptional co-activators with HAT activity, including p300,

CREB binding protein (CBP), p300/CBP associated factor (PCAF), GCN5, and steroid receptor co-activator (SRC)-1. The activator recruited co-factor (ARC)/Mediator complexes, which may associate with HATs, are also involved in the transcription induced by Smad complexes. These HATs may have distinct substrate specificities. Thus, these transcriptional co-activators are recruited to Smad complexes either sequentially or simultaneously, facilitate acetylation of histones and certain other proteins, and induce efficient transcription of target genes. In addition to transcriptional co-activators with HAT activity, transcription can be induced by certain activators which may stabilize the Smad complexes and facilitate interaction with the basal transcriptional machinery. Transcription can also be induced by Smads through dissociation of transcriptional repressors from cognate DNA binding sites.

2.1 p300 and CBP

2.1.1 *Functions of p300/CBP*

p300 and CBP were the first co-activators to be shown to interact with Smads and have been the most extensively studied. CBP was originally isolated as a protein that binds to the cAMP-responsive element binding protein (CREB), whereas p300 was identified as a protein that interacts with the adenoviral oncoprotein E1A. p300 and CBP are structurally related proteins with 63% amino acid sequence identity, and function as transcriptional co-activators for various nuclear proteins, including Jun, CREB, p53, nuclear hormone receptors, and STAT proteins, through physical interaction with them (Fig. 2). p300 and CBP have HAT domains in their middle regions, and can also interact with other HATs, *e.g.* PCAF and SRC-1.

The intracellular amounts of p300/CBP appear to be limiting and p300 and CBP are each unable to compensate for the insufficiency of the other. Germline mutations of one human *CBP* allele cause the Rubinstein-Taybi syndrome (RTS) characterized by mental retardation, craniofacial abnormalities, and broad thumbs with an increased predisposition to malignancies. *CBP*-null mice are embryonic lethal, and mice lacking one *CBP* allele exhibit certain phenotypes similar to RTS (Tanaka et al., 1997). *p300-/-* mice are also embryonic lethal, and *p300+/-* mice manifest considerable embryonic lethality, supporting the notion that the intracellular pool of p300 is limiting (Yao et al., 1998). p300 and CBP may function as tumor suppressors in certain cancers (reviewed in Iyer et al., 2004). Somatic mutations of the *p300* gene have been found in some cancers, including colon and breast cancers, and translocations of the *CBP* gene has been found in hematological malignancies.

2.1.2 *p300/CBP as Smad transcriptional co-activators*

p300 and CBP physically interact with R-Smads (Smads 1, 2, and 3) in a ligand-dependent fashion and enhance Smad-mediated transcription of target genes (Feng et al., 1998). Smad4 has also been shown to interact with p300/CBP (de Caestecker et al., 2000).

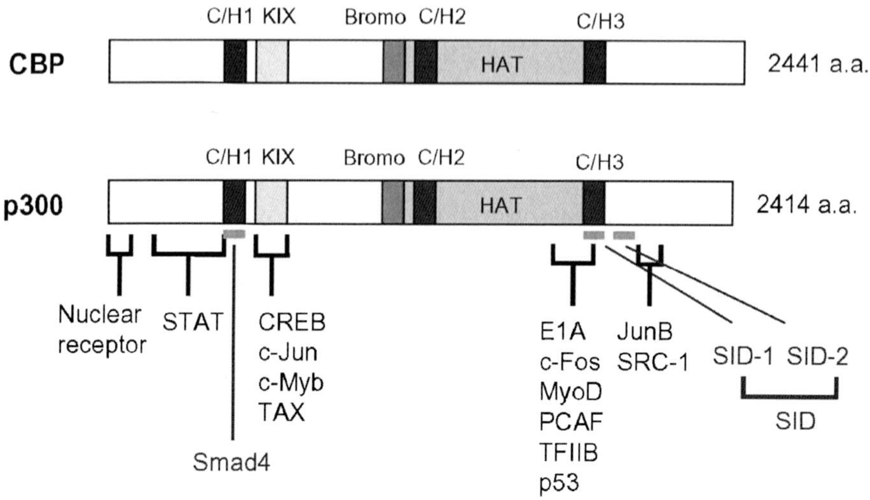

Figure 2. Structures of p300 and CBP. Domains in p300 responsible for interaction withSmads, other transcription factors and viral oncoproteins are shown. SID, Smad-interacting domain; C/H, cysteine/histidine-rich domain; Kix, CREB binding domain; HAT, histone acetyltransferase domain (A color version of this figure is freely accessible via the website of the book: http://www.springer.com/1-4020-4542-5)

R-Smads directly interact with p300/CBP through their C-terminal MH2 domains. The proline-rich linker region of Smad3 has also been shown to interact with p300 (Wang et al., 2005). The domain in p300/CBP responsible for interaction with Smads is located in their C-terminal regions (Fig. 2). Two adjacent regions in p300 are responsible for interaction with Smad3; one is located in the C/H3 domain, conserved between p300 and CBP, and the other in a nonconserved region (Nishihara et al., 1999). In addition, the C/H1 domain interacts with the Smad activation domain (SAD) of Smad4 (de Caestecker et al., 2000). Thus, Smads interact with p300/CBP through multiple domains, though the major binding site in p300/CBP is located in the C-terminal region including C/H3.

2.1.3 Convergence of Smad and other signals on p300/CBP

Signaling cross-talk between Smad and other signaling pathways is positively and negatively regulated by p300/CBP. LIF (leukemia inhibitory factor) and BMP synergistically induce differentiation of neuroepithelial cells to astrocytes (Nakashima et al., 1999). LIF activates gp130 and transduces signals through STAT3. STAT3 activated by LIF and Smad1 activated by BMP translocate into the nucleus. STAT3 interacts with the N-terminal region of p300, whereas Smad1 binds to the C-terminal region of p300. Thus, STAT3 and Smad1 are anchored to distinct regions of p300, and bridged by p300 in the nucleus, resulting in activation of transcription in a coordinated fashion (Nakashima et al., 1999) (Fig. 2). Some other transcription factors have been shown to interact with both p300/CBP and Smads; the interaction with Smads facilitates and stabilizes the interaction with p300/CBP.

In addition to inducing acetylation of histones, p300 acetylates certain non-histone proteins, *e.g.* E2F1 and Runx3 (Fig. 1A). Acetylation occurs on certain lysine residues of E2F1 and Runx3, resulting in protection from protein ubiquitination and stabilization of the proteins. p300/CBP thus serve as scaffold proteins, recruiting Smads and other transcription factors upon TGF-β signaling, and stabilizing the transcriptional complexes for efficient transcriptional activation of target genes.

Since the amounts of intracellular p300/CBP proteins are limited, competition for p300/CBP between Smads and other transcription factors plays important roles in negative cross-talk between these pathways. Expression of E-selectin in endothelial cells is induced by NF-κB activated by inflammatory stimuli, whereas Smad3 activated by TGF-β suppresses the transcription of the *E-selectin* gene. Smad proteins compete with NF-κB for binding to CBP/p300, and TGF-β thus inhibits the E-selectin expression stimulated by NF-κB by sequestering p300/CBP to the activated Smad complexes.

2.1.4 Disruption of the interaction between p300/CBP and Smads

Interaction between Smads and p300/CBP is disrupted by viral oncoproteins and certain intracellular proteins, leading to inhibition of TGF-β signaling. Of the various molecules that disrupt Smad-p300/CBP interaction, the most extensively studied protein is the adenoviral oncoprotein E1A, which directly interacts with R-Smads and the C/H3 domain of p300/CBP, and thereby inhibits TGF-β signaling through prevention of interaction between Smads and p300/CBP (Feng et al., 1998; Nishihara et al., 1999).

In addition to adenoviral E1A, the oncoproteins Tax and K-bZIP derived from the human T-cell leukemia virus type I (HTLV-1) and Kaposi's sarcoma-associated herpes virus (KSHV, also known as human herpesvirus-8), respectively, have been shown to disrupt the interaction between Smads and p300/CBP. HTLV-1-infected T-cells are resistant to the growth-inhibitory effect of TGF-β and to the transcriptional activity induced by TGF-β. Tax regulates transcription of various genes, and has been proposed to be involved in the leukemogenesis of adult T-cell leukemia (ATL). Tax physically interacts with p300/CBP, and competes with Smads for recruitment of p300/CBP into transcriptional complexes, although the binding sites of Tax and Smads in p300 appear to be different (Mori et al., 2001; Lee et al., 2002). In addition, Tax physically interacts with Smads 2, 3, and 4 through their MH2 domains, and inhibits the complex formation between Smad3 and Smad4 (Lee et al., 2002). Similar to these viral oncoproteins, some intracellular proteins, including SNIP1, p53, YB-1, and MdmX, have been reported to disrupt the interaction between Smads and p300/CBP.

2.2 PCAF and GCN5

PCAF and GCN5 (also termed PCAF-B) are transcriptional co-activators containing HAT domains, and belong to the GNAT (GCN5-related N-acetyl transferase) superfamily. PCAF was originally isolated through its ability to associate with p300/CBP. In addition to p300/CBP, PCAF and GCN5 also function as transcriptional co-activators of the TGF-β superfamily proteins (Itoh et al., 2000; Kahata et al., 2004). Human GCN5 has a HAT domain in its middle region of 83% sequence identity with PCAF, and a bromo domain in its C-terminal part that binds to acetylated lysine residues.

Both GCN5 and PCAF bind to TGF-β/Activin-specific R-Smads, Smad2 and Smad3, and enhance the transcriptional activity induced by TGF-β. In addition, GCN5, but not PCAF, interacts with BMP-specific R-Smads, *i.e.* Smad1/5/8, and enhances BMP-induced transcriptional activity (Kahata et al., 2004), suggesting that GCN5 and PCAF may have distinct physiological functions *in vivo*. In fact, although both GCN5 and PCAF are ubiquitously expressed in adult mammalian tissues, they have distinct profiles of expression in many tissues. Both the N-terminal and C-terminal regions of GCN5 interact with Smad3 in TGF-β-signal-dependent fashion, whereas PCAF has been reported to interact with Smad3 only through its N-terminal region (Itoh et al., 2000). The MH2 domains of Smad2/3 are responsible for interaction with the N-terminal region of PCAF.

Although both PCAF and p300/CBP have HAT domains and physically interact with each other, they exhibit different substrate specificities and have distinct effects. PCAF enhances the transcriptional activity of Smads activated by TGF-β, and this effect is further enhanced by p300 and Smad4 (Itoh et al., 2000). Moreover, silencing of the GCN5 gene by RNA interference results in repression of Smad signaling (Kahata et al., 2004), suggesting that PCAF and other transcriptional co-activators may not substitute for GCN5. *PCAF*-null mice show no apparent abnormalities in phenotype, whereas *GCN5*-null mice are embryonic lethal. The phenotype of *GCN5*-null mice is distinct from that of the *p300*-null mice, indicating distinct functional roles of these HATs in mammals (Yamauchi et al., 2000). Thus, PCAF and GCN5 may activate Smad-induced transcription in cooperation with, as well as independently of, p300/CBP.

2.3 Other Transcriptional Co-activators

SRC-1, SRC-2/GRIP-1/TIF-2, and SRC-3/RAC-3/ACTR/AIB-1 are structurally related 160-kDa proteins, and form the SRC family (or p160 family) of co-activators. They bind to nuclear steroid receptors as well as to other transcription factors, *e.g.* AP-1 and NF-κB. Through interaction with SRCs and stabilization of the complexes, Smad3 and Vitamin D receptor (VDR) functionally cooperate in transcription of target genes (Yanagisawa et al., 1999). In addition to binding to nuclear steroid receptors, SRC-1 physically interacts with p300/CBP, but not with Smad3, and augments the transcription of some genes, *e.g. PAI-1*, induced by the Smad-p300/CBP complexes (Dennler et al., 2005).

ARC105 is a component of the ARC complex (also termed the metazoan Mediator complex). The ARC complexes contain 20-25 polypeptides, and interact with various transcription factors and stimulate transcription through linking the transcription complexes to the basal transcription machinery. ARC105, a component of the ARC complex, physically interacts with the Smad2/3-Smad4 complex but not with Smad1, and is specifically involved in Nodal/Activin/TGF-β signaling, but not in BMP signaling, in *Xenopus* embryos (Kato et al., 2002). Thus, through interaction with RNA polymerase II, ARC complexes may regulate Smad signaling through binding to other transcriptional co-activators and co-repressors.

In addition to transcriptional co-activators with intrinsic HAT domains, certain proteins, including SMIF, MSG-1, Swift, and the hepatitis B virus-encoded oncoprotein pX, have been reported to interact with Smads and enhance the transcriptional activity of Smad complexes.

2.4 Transcription Through Dissociation of Transcriptional Repressors by Smads

Certain nuclear proteins, *e.g.* Hoxc-8 and *Drosophila* Brinker, function as transcriptional repressors in the basal state, and the repression due to them may be relieved by Smads. TGF-β superfamily signaling thus induces transcriptional activation through this mechanism. A homeodomain transcription factor, Hoxc-8, binds to the promoters of the *osteopontin* and *osteoprotegerin* genes at specific Hox binding sequences and represses transcription of them. BMP-activated Smad1 interacts with Hoxc-8 and dissociates it from DNA binding elements, leading to transcriptional activation in response to BMP signaling (Wan M. et al., 2001).

3. SMAD-MEDIATED TRANSCRIPTIONAL REPRESSION

Some transcriptional co-repressors interact with Smad proteins, and recruit class I HDACs to Smad complexes directly, or indirectly via N-CoR and/or mSin3A. In addition, these co-repressors may compete with transcriptional co-activators for interaction with Smads, and disrupt the Smad-co-activator complexes. These Smad transcriptional co-repressors inhibit transcription, and suppress the anti-proliferative effects of TGF-β. Thus, the transcriptional activities of Smads are determined by intracellular amounts of transcriptional co-activators, *e.g.* p300 and CBP, and transcriptional co-repressors, *e.g.* Ski/SnoN and TGIF. However, it is currently unclear whether the transcriptional co-repressors abolish the growth-inhibitory effects of TGF-β through recruitment of HDACs to certain gene promoters and repression of gene transcription, or through other mechanisms.

3.1 Ski and SnoN

3.1.1 Functions of Ski and SnoN

Ski was originally identified as the oncogene of the avian Sloan-Kettering retrovirus. v-Ski is truncated in both the N- and C-terminal regions of its cellular homologue, c-Ski (Fig. 3). SnoN (Ski-related novel gene) is structurally closely related to c-Ski, and three alternative variants of SnoN, *i.e.* SnoN2, SnoI, and SnoA, have been isolated. Ski and SnoN are unique proto-oncogene products that induce

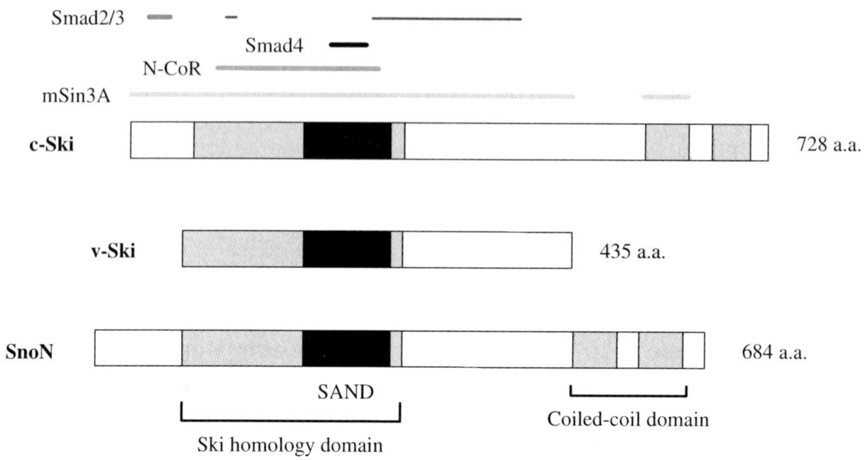

Figure 3. Structures of transcriptional co-repressors, c-Ski, v-Ski, and SnoN. Domains in c-Ski responsible for interaction with Smad2/3, Smad4, N-CoR, and mSin3A are shown. The most N-terminal part is the major binding site to Smad2/3 (bold line on the top), while other parts (shown in thin lines) are weak contact sites to them. Coiled-coil domains are responsible for dimerization of Ski and SnoN. The SAND domain (named after Sp100, AIRE-1, NucP41/75, DEAF-1) includes the I loop which is responsible for interaction with Smad4 (A color version of this figure is freely accessible via the website of the book: http://www.springer.com/1-4020-4542-5)

oncogenic transformation of certain fibroblasts and terminal muscle differentiation of mesenchymal cells. The N-terminal ~270 amino-acid-residue region (termed the Ski homology region) is very similar in Ski and SnoN, and is essential for the transforming and differentiation activities of Ski/SnoN (reviewed in Luo, 2004). The C-terminal region of these molecules is less well conserved, but the coiled-coil domains are important for homo- and hetero-oligomer formation of Ski/SnoN.

Ski and SnoN are highly expressed in some human cancer cells, *e.g.* those of melanoma, breast cancer, and esophageal cancer (reviewed in Medrano, 2003). *Ski*-null mice are embryonic lethal, and exhibit multiple defects, including a cranial neural tube defect and marked reduction in skeletal muscle mass (reviewed in Luo, 2004). *Sno-/-* mice were found to die at an early stage of embryogenesis, although other lines of *Sno-/-* mice were viable and exhibited a defect only in T cell activation. Heterozygous mice deficient in the *Ski* or *SnoN* showed increased chemical carcinogen-induced tumor formation compared to wild-type mice, suggesting that Ski and SnoN play dual roles in tumorigenesis, and act as tumor suppressors in certain types of cells (Shinagawa et al., 2000; 2001).

3.1.2 Multiple mechanisms of Ski/Sno-induced repression of TGF-β superfamily signaling

Ski and SnoN suppress the Smad-mediated transcription induced by TGF-β/Activin and BMPs, and block the growth arrest induced by TGF-β. Ski and SnoN inhibit TGF-β superfamily signaling through multiple mechanisms: 1) c-Ski and SnoN interact with Smads as well as with the nuclear co-repressor (N-CoR)/SMRT, mSin3A, HIPK2, and MeCP2, and form complexes containing HDACs (Akiyoshi et al., 1999). c-Ski and SnoN thereby induce deacetylation of histones and repress transcription of target genes (see Fig. 1B). 2) c-Ski competes with transcriptional co-activators p300/CBP for interaction with Smad complexes, similar to some viral and intracellular proteins, including E1A and SNIP1 (see 2.1.4). 3) c-Ski disrupts R-Smad-Co-Smad complexes (Wu et al., 2002). 4) c-Ski enhances the binding of Smads to DNA and stabilizes the inactive Smad complexes on Smad-binding elements (Suzuki et al., 2004).

SnoN has been reported to exhibit dual effects on TGF-β signaling in cell-type-specific fashion, although its mechanism is unknown. SnoN may enhance TGF-β-mediated responses at low levels but suppress them at high levels (Sarker et al., 2005). In addition to regulating TGF-β superfamily signaling, Ski has effects independent of Smad pathways, and represses transcription induced by Mad, thyroid hormone receptor-β, and Rb protein (Nomura et al., 1999).

3.1.3 Binding of c-Ski/SnoN to Smads

The interaction of c-Ski with Smad2/3 and Smad4 occurs through distinct regions. c-Ski and SnoN interact with Smad2 and Smad3 in their most N-terminal regions (amino acid residues 17-45 in c-Ski) and other minor sites of contact (Fig. 3) (Akiyoshi et al., 1999; Qin et al., 2002; Ueki and Hayman, 2003a). Ski/SnoN interact with the MH2 domain of Smad3, which overlaps with the p300 interaction domain of Smad3 (Mizuide et al., 2003). In contrast, they interact with Smad4 through the interaction loops (I-loops) located in the SAND domain (Fig. 3) (Wu et al., 2002). The I-loop of Ski binds to the L3 loop in the MH2 domain of Smad4, which is also required for interaction with phosphorylated R-Smads. Thus, Ski/SnoN disrupt the complex formation between R-Smads and Co-Smad. However, other studies have shown that the c-Ski mutant W274E, which is unable to interact with Smad4, and the ΔS2/3 mutant, which lacks the most N-terminal part and fails to interact with Smad2/3, do not disrupt Smad3-Smad4 heteromers (Ueki and Hayman, 2003a; Takeda et al., 2004). Moreover, the N-terminal region of c-Ski efficiently interacts with the trimeric form of Smad3 (Qin et al., 2002). These findings suggest that Ski can under certain conditions interact with Smad oligomers without disrupting complex formation.

c-Ski and SnoN inhibit TGF-β signaling, for which binding of c-Ski/SnoN to either R-Smads or Smad4 is sufficient. Through interaction with Smad4, c-Ski represses BMP signaling both *in vitro* and *in vivo*. c-Ski recruits class I HDAC complexes through multiple domains. The N-terminal portion of c-Ski including Leu110 is essential for interaction with N-CoR, but a c-Ski mutant that is unable to interact with N-CoR still interacts with mSin3A and inhibits TGF-β signaling (Ueki and Hayman, 2003b).

Ski binds to the Smad binding element (SBE) sequences, GTCT or AGAC, and this occurs indirectly through binding of Smads to the SBE sequences. c-Ski strongly binds DNA independently of phosphorylation of R-Smads, and represses the basal activity of certain gene promoters, *e.g.* Smad7 (Denissova and Liu, 2004).

3.1.4 Ski-interacting proteins

Ski has been reported to interact with several proteins, including C184M and Skip, which may regulate the functions of Ski. Ski is located in the nucleus of normal cells and cells in early-stage tumors, but is located in the cytoplasm as well as the nucleus in invasive and metastatic malignant melanoma cells (Reed et al., 2001). The Ski-interacting protein C184M, the expression of which is induced by TGF-β stimulation, plays an important role in the cytoplasmic localization of c-Ski. C184M interacts with the coiled-coil motif of c-Ski, inhibits nuclear translocation of R-Smads, and thereby represses TGF-β signaling (Kokura et al., 2003).

3.1.5 Convergence of TGF-β and other signals on Ski/SnoN

SnoN has been shown to actively repress Smad-mediated transcription of the *α-fetoprotein* (AFP) gene (Wilkinson et al., 2005). TGF-β induces the binding of phospho-Smad2 and Smad4 to SBE, and p53 to the p53 regulatory element (p53RE), at the overlapping SBE/p53RE site in the *AFP* gene promoter. SnoN and HDAC are recruited to, and stabilized on, the p53 and phospho-Smad2/Smad4 complex, leading to repression of *AFP* expression. Since Ski cannot substitute for the function of SnoN on transcription of the *AFP* gene, SnoN specifically regulates this process in cooperation with Smad2/4 and p53.

JNK signaling represses TGF-β signaling pathways through the activity of c-Ski. c-Jun interacts with c-Ski, and enhances the interaction between c-Ski and Smad2, leading to suppression of the transcriptional activity induced by TGF-β (Pessah et al., 2002). TGF-β signaling induces dissociation of c-Jun from c-Ski, and the effect of TGF-β on dissociation of c-Jun from c-Ski is antagonized by JNK signaling.

3.1.6 Regulation of expression of Ski/SnoN

TGF-β signaling regulates the expression of Ski and SnoN in a biphasic fashion. c-Ski and SnoN are rapidly degraded upon TGF-β stimulation through a ubiquitin-dependent degradation pathway (Stroschein et al., 1999). SnoN is more rapidly degraded than c-Ski after TGF-β stimulation. Since transcription of SnoN is induced by TGF-β, expression of the SnoN protein is elevated after 2 hours, and represses TGF-β/Smad signaling in a negative-feedback loop (Stroschein et al., 1999).

TGF-β induces degradation of SnoN by two distinct mechanisms, *i.e.* recruitment of anaphase-promoting complex (APC) and of Smurf2 (Stroschein et al., 2001; Wan Y. et al., 2001, Bonni et al., 2001). APC is the E3 ligase responsible for the metaphase/anaphase transition in mitosis. The APC activator CDH1 interacts with the conserved destruction box in SnoN. Activated Smad2/3 interact with SnoN as well as with APC, resulting in induction of the degradation of SnoN. The level of SnoN protein increases in the G2 phase of the cell cycle and decreases as cells enter G1 phase, suggesting that TGF-β signals may be efficiently transmitted only in G1 or S phase through the decrease in SnoN protein (Stroschein et al., 2001; Wan Y. et al., 2001).

Smurf2 is a HECT type E3 ubiquitin ligase, which induces degradation of R-Smads as well as that of the I-Smad and type I receptor complexes, and augments TGF-β superfamily signaling. The WW domain of Smurf2 interacts with the PY motif (PPXY sequence) in the linker region of R-Smads and I-Smads. Although Smurf2 does not directly interact with SnoN, TGF-β induces indirect association of Smurf2 with SnoN via phospho-Smad2, resulting in degradation of SnoN in a TGF-β signal-dependent fashion (Bonni et al., 2001).

3.2 Dach and its Related Proteins

Drosophila Dachshund (DAC) is a nuclear protein that is structurally related to Ski/SnoN. Dach1 and Dach2 are the vertebrate homologues of *Drosophila* DAC. Dach proteins contain a region termed the DS domain, which is approximately 30% identical to the Ski homology region of Ski and SnoN. Dach1 is highly expressed in some breast cancer cell lines, and represses TGF-β signaling by interacting with Smad4, Smad3, mSin3A, and N-CoR (Wu et al., 2003). Since Dach1 interacts with Smad1 as well as with Smad4, it also functions as a BMP antagonist, and controls apical ectodermal ridge (AER) formation and limb development in chick embryos (Kida et al., 2004). The DS domain is responsible for interaction with N-CoR, HDAC, and Smad3, whereas the C-terminal region interacts with Smad4, Smad1, and mSin3A (Wu et al., 2003; Kida et al., 2004). Thus, Dach1 and Ski may function in similar pathways to repress TGF-β signaling, but the Smad-interacting domains appear to differ between Ski/SnoN and Dach proteins.

3.3 TGIF (TG-interacting factor)

TGIF is an atypical three-amino-acid loop extension (TALE) homeodomain protein that was the first transcriptional co-repressor shown to interact with Smads (Wotton et al., 1999). In addition to suppressing Smad-mediated signaling, TGIF binds directly to DNA and represses transcription through competing with other transcription factors, including retinoid receptors. Holoprosencephaly (HPE) is a severe human genetic disorder characterized by structural defects of craniofacial development. HPE is an etiologically heterogenous disease, and mutations of TGIF are responsible for some types of human HPE (Gripp et al., 2000).

TGIF inhibits Smad-dependent transcription by recruitment of mSin3 and HDACs to the Smad2/3 complexes through its C-terminal repression domain and competition with p300 for interaction with Smads (Wotton et al., 1999; Wotton et al., 2001). TGIF2 is structurally similar to TGIF and represses TGF-β signal through interaction with Smads. In addition to associating with mSin3A/HDACs, TGIF may recruit HDACs through interaction with carboxyl-terminus-binding protein (CtBP) via a PLDLS motif located in the N-terminal repression domain. A mutation in the PLDLS motif of TGIF has been identified in HPE. However, TGIF2 does not interact with CtBP, but does recruit HDACs. TGIF also blocks the anti-proliferative activity of TGF-β, and suppression of the increase of p15 has been suggested to be responsible for blockade of TGF-β-mediated growth inhibition (Lo et al., 2001).

TGIF is a short-lived protein, the stability of which is regulated by phosphorylation by Erk MAP kinases, *e.g.* following epidermal growth factor stimulation (Lo et al., 2001). In addition to repressing Smad-dependent transcription and directly binding to DNA, TGIF may also regulate ubiquitin-dependent degradation of Smad2 through interaction with a HECT type E3 ubiquitin ligase Tiul1/WWP1 (Seo et al., 2004).

3.4 Evi-1 and Related Proteins

Evi-1 is a ten Cys2-His2 zinc-finger-containing nuclear protein which is highly expressed in some human myeloid leukemias and myelodysplastic syndromes as a result of chromosomal rearrangements involving 3q26. Evi-1 is a sequence-specific transcriptional repressor, and represses TGF-β signaling through direct interaction with Smad3 through its zinc-finger motif (Kurokawa et al., 1998). Evi-1 represses the anti-proliferative activity of TGF-β. Evi-1 also binds to Smad2 and Smad1, and represses BMP as well as TGF-β signaling (Alliston et al., 2005). Evi-1 does not disrupt the interaction of Smads with CBP, and Evi-1 itself has the ability to interact with co-activators CBP and PCAF. However, it recruits CtBP through its PLDLS-like motifs and decreases acetylation of histones upon ligand stimulation (Palmer et al., 2001; Alliston et al., 2005). An inhibitor of HDACs alleviates Evi-1-mediated repression of TGF-β signaling.

MEL1 (MDS1/Evi-1) is a protein structurally related to Evi-1, and represses TGF-β signaling. MEL1S, an alternatively spliced form of MEL1 lacking the most N-terminal region, is aberrantly expressed in ATL, and confers resistance in ATL cells to TGF-β-induced transcription and growth inhibition (Yoshida et al., 2004).

3.5 Other Transcriptional Repressors

Several proteins, including SIP1 and YY1, have been shown to interact with Smads and repress transcriptional activity of TGF-β superfamily members. SIP1 (Smad-interacting protein-1)/ZEB2 and δEF1/ZEB1 are members of the δEF1/Zfh-1 family of two-handed zinc-finger proteins. SIP1 binds to R-Smads through the MH2 domain in a ligand-dependent manner, and inhibits TGF-β as well as BMP signaling (Verschueren et al., 1999). SIP1 directly binds to DNA and functions as a transcriptional repressor. SIP1 downregulates the expression of *E-cadherin*, and is involved in epithelial-mesenchymal transition. Mutations in *SIP1* cause a form of Hirschsprung's disease with mental retardation. SIP1 and δEF1/ZEB1 display opposite effects in regulation of TGF-β/BMP signaling: δEF1/ZEB1 synergizes with Smad-induced transcription by recruitment of p300 and PCAF, whereas SIP1 represses it through recruitment of CtBP to Smad complexes (Postigo et al.,

2003). δEF1/ZEB1 also has a CtBP binding site, but PCAF bound to δEF1/ZEB1 induces acetylation of several lysine residues in δEF1/ZEB1, dissociating CtBP from δEF1/ZEB1. However, since the interaction between CtBP and SIP1 is dispensable for transcriptional repression of *E-cadherin* (van Grunsven et al., 2003), the functional roles of CtBP in SIP1-mediated transcriptional repression may be context-dependent, and remain to be determined.

Smad6 also serves as a transcriptional co-repressor through recruiting CtBP (Lin et al., 2003). Smad6, but not Smad7, has a PLDLS motif in the linker region and recruits CtBP through this domain. Smad6 binds to DNA possibly through interaction with activated Smad1, and thus inhibits the transcription of the *Id1* gene induced by BMP in the nucleus.

In addition to acetylation, other post-transcriptional modifications of histones, including methylation, phosphorylation, and ubiquitination, lead to alterations of chromatin structure and regulation of gene transcription. Suv39h is a methyltransferase which induces methylation of histone H3 and represses transcription of target genes. Smad1 and Smad5 activated by BMPs, but not Smad2/3 activated by TGF-β, interact with two mammalian homologues of Suv39h, *i.e.* Suv39h1 and Suv39h2, and repress the transcription of muscle creatine kinase (MCK) (Frontelo et al., 2004).

3.6 Ligand-induced Transcriptional Repression by Smad Complexes

Although transcriptional co-repressors are able to suppress TGF-β superfamily signaling, they antagonize the transcription induced by TGF-β superfamily members on most occasions, and exhibit ligand-mediated suppressive effects only under certain occasions, *e.g.* in cooperation between p53, Smad4, and SnoN. TGF-β signaling represses the transcription of some target genes, *e.g. c-Myc* and cyclin A, though the mechanisms of ligand-induced transcriptional repression are still largely unknown.

At the promoter of the *osteocalcin* gene, the Smad3 and Runx2 complex binds to the Runx2-binding DNA sequence, and the class IIa histone deacetylases, HDAC4 and HDAC5, are recruited by activated Smad3 to Runx2, forming a stable complex between Smad3, Runx2, and HDACs. HDAC4/5 directly bind to the MH2 domain of Smad3, but not to other Smads. Smad3-Runx2-HDAC4/5 thus represses transcription of the *osteocalcin* gene, and inhibits osteoblast differentiation (Kang et al., 2005).

Some TGF-β target genes may be repressed independently of the functions of HDACs or methyltransferases. Smad3 activated by TGF-β suppresses the transcription of myogenin through disruption of the complex between MyoD and MEF2 as well as by dissociation of MEF2 from SRC family co-activator SRC-2/GRIP-1, leading to blockade of myogenic differentiation (Liu et al., 2004). For repression of *c-Myc* by TGF-β, Smad3 interacts with E2F4/5 and an Rb family co-repressor p107 and binds to a specific *c-Myc* regulatory sequence, though the involvement of HDACs in this complex has not been determined (Chen et al., 2002).

4. CONCLUSION AND PERSPECTIVES

Smad-mediated transcription is induced by various transcriptional co-activators. Many transcriptional co-activators have been shown to interact with Smads. However, it is currently unclear how each transcriptional co-activator induces transcription of Smad target genes. It is possible that these transcriptional co-activators act in concert with each other or sequentially for transcription of target genes. It is also possible that they act specifically on certain Smad target genes. Although transcription of many genes has been shown to be induced by Smad signaling, the mechanisms of repression of some Smad target genes have not been fully elucidated. Transcriptional co-repressors can reduce the transcription induced by Smads and transcriptional co-activator complexes, though the mechanisms of direct repression of genes by Smads are not fully understood. Since repression of *c-Myc* and *Id* genes may play important roles in the growth inhibition induced by TGF-β, the mechanisms of Smad-mediated transcriptional repression are an important subject for future research. It is also unclear how Smads choose transcriptional co-activators or co-repressors as their partners. Since Ski and SnoN have high affinities for Smads, the intracellular amounts of them may be the major determinant of choice of Smad partners. However, it is also possible that certain post-transcriptional modifications of Smad proteins alter their affinities for transcriptional co-activators and co-repressors. Since alterations of the Smad-mediated signaling system are closely linked to the pathogeneses of various clinical diseases, it is important to determine the mechanisms of interaction of Smads with various transcriptional co-activators and co-repressors.

ACKNOWLEDGEMENTS

We deeply apologize to those colleagues in the field whose manuscripts could not be cited due to space limitations. Research on TGF-β superfamily signaling in our laboratories is supported by grants from the Ministry of Education, Culture, Sports, Science, and Technology of Japan, and the Ludwig Institute for Cancer Research.

REFERENCES

Akiyoshi, S., Inoue, H., Hanai, J., Kusanagi, K., Nemoto, N., Miyazono, K., and Kawabata, M., 1999, c-Ski acts as a transcriptional co-repressor in transforming growth factor-β signaling through interaction with Smads. *J Biol Chem* 274: 35269-35277.

Alliston, T., Ko, T.C., Cao, Y., Liang, Y.Y., Feng, X.H., Chang, C., and Derynck, R., 2005, Repression of bone morphogenetic protein and activin-inducible transcription by Evi-1. *J Biol Chem* 280: 24227-24237.

Bonni, S., Wang, H.R., Causing, C.G., Kavsak, P., Stroschein, S.L., Luo, K., and Wrana, J.L., 2001, TGF-β induces assembly of a Smad2-Smurf2 ubiquitin ligase complex that targets SnoN for degradation. *Nat Cell Biol* 3: 587-595.

Chen, C.R., Kang, Y., Siegel, P.M., and Massagué, J., 2002, E2F4/5 and p107 as Smad cofactors linking the TGFβ receptor to *c-myc* repression. *Cell* 110: 19-32.

de Caestecker, M.P., Yahata, T., Wang, D., Parks, W.T., Huang, S., Hill, C.S., Shioda, T., Roberts, A.B., and Lechleider, R.J., 2000, The Smad4 activation domain (SAD) is a proline-rich, p300-dependent transcriptional activation domain. *J Biol Chem* 275: 2115-2122.

Denissova, N.G., and Liu, F., 2004, Repression of endogenous Smad7 by Ski. *J Biol Chem* 279: 28143-28148.

Dennler, S., Pendaries, V., Tacheau, C., Costas, M.A., Mauviel, A., and Verrecchia, F., 2005, The steroid receptor co-activator-1 (SRC-1) potentiates TGF-β/Smad signaling: role of p300/CBP. *Oncogene* 24: 1936-1945.

DiChiara, M.R., Kiely, J.M., Gimbrone, M.A. Jr., Lee, M.E., Perrella, M.A., and Topper, J.N., 2000, Inhibition of *E-selectin* gene expression by transforming growth factor β in endothelial cells involves coactivator integration of Smad and nuclear factor κB-mediated signals. *J Exp Med* 192: 695-704.

Feng, X.H., Zhang, Y., Wu, R.Y., and Derynck, R., 1998, The tumor suppressor Smad4/DPC4 and transcriptional adaptor CBP/p300 are coactivators for Smad3 in TGF-β-induced transcriptional activation. *Genes Dev* 12: 2153-2163.

Frontelo, P., Leader, J.E., Yoo, N., Potocki, A.C., Crawford, M., Kulik, M., and Lechleider, R.J., 2004, Suv39h histone methyltransferases interact with Smads and cooperate in BMP-induced repression. *Oncogene* 23: 5242-5251.

Gripp, K.W., Wotton, D., Edwards, M.C., Roessler, E., Ades, L., Meinecke, P., Richieri-Costa, A., Zackai, E.H., Massagué, J., Muenke, M., and Elledge, S.J., 2000, Mutations in TGIF cause holoprosencephaly and link NODAL signalling to human neural axis determination. *Nat Genet* 25: 205-208.

Itoh, S., Ericsson, J., Nishikawa, J., Heldin, C.-H., and ten Dijke, P., 2000, The transcriptional co-activator P/CAF potentiates TGF-β/Smad signaling. *Nucleic Acids Res* 28: 4291-4298.

Iyer, N.G., Ozdag, H., and Caldas, C., 2004, p300/CBP and cancer. *Oncogene* 23: 4225-4231.

Kahata, K., Hayashi, M., Asaka, M., Hellman, U., Kitagawa, H., Yanagisawa, J., Kato, S., Imamura, T., and Miyazono, K., 2004, Regulation of transforming growth factor-β and bone morphogenetic protein signalling by transcriptional coactivator GCN5. *Genes Cells* 9: 143-151.

Kang, J.S., Alliston, T., Delston, R., and Derynck, R., 2005, Repression of Runx2 function by TGF-β through recruitment of class II histone deacetylases by Smad3. *EMBO J* 24: 2543-2555.

Kato, Y., Habas, R., Katsuyama, Y., Naar, A.M., and He, X., 2002, A component of the ARC/Mediator complex required for TGF-β/Nodal signalling. *Nature* 418: 641-646.

Kida, Y., Maeda, Y., Shiraishi, T., Suzuki, T., and Ogura, T., 2004, Chick Dach1 interacts with the Smad complex and Sin3a to control AER formation and limb development along the proximodistal axis. *Development* 131: 4179-4187.

Kokura, K., Kim, H., Shinagawa, T., Khan, M.M., Nomura, T., and Ishii, S., 2003, The Ski-binding protein C184M negatively regulates tumor growth factor-β signaling by sequestering the Smad proteins in the cytoplasm. *J Biol Chem* 278: 20133-20139.

Kurokawa, M., Mitani, K., Irie, K., Matsuyama, T., Takahashi, T., Chiba, S., Yazaki, Y., Matsumoto, K., and Hirai, H., 1998, The oncoprotein Evi-1 represses TGF-β signalling by inhibiting Smad3. *Nature* 394: 92-96.

Lee, D.K., Kim, B.C., Brady, J.N., Jeang, K.T., and Kim, S.J., 2002, Human T-cell lymphotropic virus type 1 tax inhibits transforming growth factor-β signaling by blocking the association of Smad proteins with Smad-binding element. *J Biol Chem* 277: 33766-33775.

Lin, X., Liang, Y.Y., Sun, B., Liang, M., Shi, Y., Brunicardi, F.C., Shi, Y., and Feng, X.H., 2003, Smad6 recruits transcription corepressor CtBP to repress bone morphogenetic protein-induced transcription. *Mol Cell Biol* 23: 9081-9093.

Liu, D., Kang, J.S., and Derynck, R., 2004, TGF-β-activated Smad3 represses MEF2-dependent transcription in myogenic differentiation. *EMBO J* 23: 1557-1566.

Lo, R.S., Wotton, D., and Massagué, J., 2001, Epidermal growth factor signaling via Ras controls the Smad transcriptional co-repressor TGIF. *EMBO J* 20: 128-136.

Luo, K., 2004, Ski and SnoN: negative regulators of TGF-β signaling. *Curr Opin Genet Dev* 14: 65-70.

Medrano, E.E., 2003, Repression of TGF-β signaling by the oncogenic protein SKI in human melanomas: consequences for proliferation, survival, and metastasis. *Oncogene* 22: 3123-3129.

Mizuide, M., Hara, T., Furuya, T., Takeda, M., Kusanagi, K., Inada, Y., Mori, M., Imamura, T., Miyazawa, K., and Miyazono, K., 2003, Two short segments of Smad3 are important for specific interaction of Smad3 with c-Ski and SnoN. *J Biol Chem* 278: 531-536.

Mori, N., Morishita, M., Tsukazaki, T., Giam, C.Z., Kumatori, A., Tanaka, Y., and Yamamoto, N., 2001, Human T-cell leukemia virus type I oncoprotein Tax represses Smad-dependent transforming growth factor β signaling through interaction with CREB-binding protein/p300. *Blood* 97: 2137-2144.

Nakashima, K., Yanagisawa, M., Arakawa, H., Kimura, N., Hisatsune, T., Kawabata, M., Miyazono, K., and Taga, T., 1999, Synergistic signaling in fetal brain by STAT3-Smad1 complex bridged by p300. *Science* 284: 479-482.

Nishihara, A., Hanai, J., Imamura, T., Miyazono, K., and Kawabata, M., 1999, E1A inhibits transforming growth factor-β signaling through binding to Smad proteins. *J Biol Chem* 274: 28716-28723.

Nomura, T., Khan, M.M., Kaul, S.C., Dong, H.D., Wadhwa, R., Colmenares, C., Kohno, I., and Ishii, S., 1999, Ski is a component of the histone deacetylase complex required for transcriptional repression by Mad and thyroid hormone receptor. *Genes Dev* 13: 412-423.

Palmer, S., Brouillet, J.P., Kilbey, A., Fulton, R., Walker, M., Crossley, M., and Bartholomew, C., 2001, Evi-1 transforming and repressor activities are mediated by CtBP co-repressor proteins. *J Biol Chem* 276: 25834-25840.

Pessah, M., Marais, J., Prunier, C., Ferrand, N., Lallemand, F., Mauviel, A., and Atfi, A., 2002, c-Jun associates with the oncoprotein Ski and suppresses Smad2 transcriptional activity. *J Biol Chem* 277: 29094-29100.

Postigo, A.A., Depp, J.L., Taylor, J.J., and Kroll K.L., 2003, Regulation of Smad signaling through a differential recruitment of coactivators and corepressors by ZEB proteins. *EMBO J* 22: 2453-2462.

Qin, B.Y., Lam, S.S., Correia, J.J., and Lin, K., 2002, Smad3 allostery links TGF-β receptor kinase activation to transcriptional control. *Genes Dev* 16: 1950-1963.

Reed, J.A., Bales, E., Xu, W., Okan, N.A., Bandyopadhyay, D., and Medrano, E.E., 2001, Cytoplasmic localization of the oncogenic protein Ski in human cutaneous melanomas *in vivo*: functional implications for transforming growth factor β signaling. *Cancer Res* 61: 8074-8078.

Sarker, K.P., Wilson, S.M., and Bonni, S., 2005, SnoN is a cell type-specific mediator of transforming growth factor-β responses. *J Biol Chem* 280: 13037-13046.

Seo, S.R., Lallemand, F., Ferrand, N., Pessah, M., L'Hoste, S., Camonis, J., and Atfi A., 2004, The novel E3 ubiquitin ligase Tiul1 associates with TGIF to target Smad2 for degradation. *EMBO J* 23: 3780-3792.

Shinagawa, T., Dong, H.D., Xu, M., Maekawa, T., and Ishii, S., 2000, The sno gene, which encodes a component of the histone deacetylase complex, acts as a tumor suppressor in mice. *EMBO J* 19: 2280-2291.

Shinagawa, T., Nomura, T., Colmenares, C., Ohira, M., Nakagawara, A., and Ishii, S., 2001, Increased susceptibility to tumorigenesis of ski-deficient heterozygous mice. *Oncogene* 20: 8100-8108.

Stroschein, S.L., Bonni, S., Wrana, J.L., and Luo, K., 2001, Smad3 recruits the anaphase-promoting complex for ubiquitination and degradation of SnoN. *Genes Dev* 15: 2822-2836.

Stroschein, S.L., Wang, W., Zhou, S., Zhou, Q., and Luo, K., 1999, Negative feedback regulation of TGF-β signaling by the SnoN oncoprotein. *Science* 286: 771-774.

Suzuki, H., Yagi, K., Kondo, M., Kato, M., Miyazono, K., and Miyazawa, K., 2004, c-Ski inhibits the TGF-β signaling pathway through stabilization of inactive Smad complexes on Smad-binding elements. *Oncogene* 23: 5068-5076.

Takeda, M., Mizuide, M., Oka, M., Watabe, T., Inoue, H., Suzuki, H., Fujita, T., Imamura, T., Miyazono, K., and Miyazawa, K., 2004, Interaction with Smad4 is indispensable for suppression of BMP signaling by c-Ski. *Mol Biol Cell* 15: 963-972.

Tanaka, Y., Naruse, I., Maekawa, T., Masuya, H., Shiroishi, T., and Ishii, S., 1997, Abnormal skeletal patterning in embryos lacking a single Cbp allele: a partial similarity with Rubinstein-Taybi syndrome. *Proc Natl Acad Sci U S A* 94: 10215-10220.

Ueki, N., and Hayman, M.J., 2003a, Direct interaction of Ski with either Smad3 or Smad4 is necessary and sufficient for Ski-mediated repression of transforming growth factor-β signaling. *J Biol Chem* 278: 32489-32492.

Ueki, N., and Hayman, M.J., 2003b, Signal-dependent N-CoR requirement for repression by the Ski oncoprotein. *J Biol Chem* 278: 24858-24864.

van Grunsven, L.A., Michiels, C., Van de Putte, T., Nelles, L., Wuytens, G., Verschueren, K., and Huylebroeck, D., 2003, Interaction between Smad-interacting protein-1 and the corepressor C-terminal binding protein is dispensable for transcriptional repression of *E-cadherin*. *J Biol Chem* 278: 26135-26145.

Verschueren, K., Remacle, J.E., Collart, C., Kraft, H., Baker, B.S., Tylzanowski, P., Nelles, L., Wuytens, G., Su, M.T., Bodmer, R., Smith, J.C., and Huylebroeck, D., 1999, SIP1, a novel zinc finger/homeodomain repressor, interacts with Smad proteins and binds to 5'-CACCT sequences in candidate target genes. *J Biol Chem* 274: 20489-20498.

Wan, M., Shi, X., Feng, X., and Cao, X., 2001, Transcriptional mechanisms of bone morphogenetic protein-induced osteoprotegrin gene expression. *J Biol Chem* 276: 10119-10125.

Wan, Y., Liu, X., and Kirschner, M.W., 2001, The anaphase-promoting complex mediates TGF-β signaling by targeting SnoN for destruction. *Mol Cell* 8: 1027-1039.

Wang, G., Long, J., Matsuura, I., He, D., and Liu, F., 2005, The Smad3 linker region contains a transcriptional activation domain. *Biochem J* 386: 29-34.

Wilkinson, D.S., Ogden, S.K., Stratton, S.A., Piechan, J.L., Nguyen, T.T., Smulian, G.A., and Barton, M.C., 2005, A direct intersection between p53 and transforming growth factor β pathways targets chromatin modification and transcription repression of the α-fetoprotein gene. *Mol Cell Biol* 25: 1200-1212.

Wotton, D., Knoepfler, P.S., Laherty, C.D., Eisenman, R.N., and Massagué, J., 2001, The Smad transcriptional corepressor TGIF recruits mSin3. *Cell Growth Differ* 12: 457-463.

Wotton, D., Lo, R.S., Lee, S., and Massagué, J., 1999, A Smad transcriptional corepressor. *Cell* 97: 29-39.

Wu, J.W., Krawitz, A.R., Chai, J., Li, W., Zhang, F., Luo, K., and Shi, Y., 2002, Structural mechanism of Smad4 recognition by the nuclear oncoprotein Ski: insights on Ski-mediated repression of TGF-β signaling. *Cell* 111: 357-367.

Wu, K., Yang, Y., Wang, C., Davoli, M.A., D'Amico, M., Li, A., Cveklova, K., Kozmik, Z., Lisanti, M.P., Russell, R.G., Cvekl, A., and Pestell, R.G., 2003, DACH1 inhibits transforming growth factor-β signaling through binding Smad4. *J Biol Chem* 278: 51673-51684.

Yamauchi, T., Yamauchi, J., Kuwata, T., Tamura, T., Yamashita, T., Bae, N., Westphal, H., Ozato, K., and Nakatani, Y., 2000, Distinct but overlapping roles of histone acetylase PCAF and of the closely related PCAF-B/GCN5 in mouse embryogenesis. *Proc Natl Acad Sci U S A* 97: 11303-11306.

Yanagisawa, J., Yanagi, Y., Masuhiro, Y., Suzawa, M., Watanabe, M., Kashiwagi, K., Toriyabe, T., Kawabata, M., Miyazono, K., and Kato, S., 1999, Convergence of transforming growth factor-β and vitamin D signaling pathways on SMAD transcriptional coactivators. *Science* 283: 1317-1321.

Yao, T.P., Oh, S.P., Fuchs, M., Zhou, N.D., Ch'ng, L.E., Newsome, D., Bronson, R.T., Li, E., Livingston, D.M., and Eckner, R., 1998, Gene dosage-dependent embryonic development and proliferation defects in mice lacking the transcriptional integrator p300. *Cell* 93: 361-372.

Yoshida, M., Nosaka, K., Yasunaga, J., Nishikata, I., Morishita, K., and Matsuoka, M., 2004, Aberrant expression of the MEL1S gene identified in association with hypomethylation in adult T-cell leukemia cells. *Blood* 103: 2753-2760.

CHAPTER 15

INTEGRATION OF SIGNALING PATHWAYS VIA SMAD PROTEINS

ETIENNE LABBÉ[1] AND LILIANA ATTISANO[1,2]
[1] *Department of Medical Biophysics*
[2] *Department of Biochemistry, University of Toronto, Toronto, Ontario, Canada*

Abstract: TGF-β ligands are involved in most aspects of embryonic development and post-natal homeostasis and numerous examples of crosstalk with other signaling pathways have been described. The Smad proteins are the most extensively characterized signal transducers downstream of the TGF-β superfamily of secreted growth factors. In most cases, Smad proteins have been shown to mediate this crosstalk via the formation of DNA-bound transcriptional complexes with transducers of other pathways, thereby modifying promoter selectivity and transcriptional output. We will examine the ability of Smads to integrate the signaling output of TGF-β ligands with other signaling pathways through direct interactions with other signal transducers

Keywords: BMP; crosstalk; growth factors; TGF-β; transcription

1. INTRODUCTION

Members of the transforming growth factor β (TGF-β) family, including the Activins, bone morphogenetic proteins (BMPs), growth and differentiation factors (GDFs), and TGF-βs are pleiotropic secreted growth factors involved in all stages of embryonic development, homeostasis in adult tissues, and a variety of pathologies. As such, these multifunctional proteins exhibit exquisite cell and context-specific roles, seemingly paradoxical with the apparent simplicity of their intracellular signal transduction pathway (Feng and Derynck, 2005; Massagué, 2000). Accumulating evidence indicates that there are numerous cell-specific intracellular mediators of TGF-β signals, such as RhoGTPases and MAPKs, contributing to the multifunctionality of TGF-β superfamily ligands. However, the most extensively characterized pathway downstream of the TGF-β ligands in all cell types examined involves Smad proteins. Smads are activated by receptor-mediated phosphorylation followed

P. ten Dijke and C.-H. Heldin (eds.), Smad Signal Transduction, 295–316.

by relocalization of Smad complexes to the nucleus where transcription of target genes is modulated through direct DNA contact. Promoter selectivity is influenced by the partnering of Smads with other sequence-specific transcription factors, a number of which mediate transcriptional output from other pathways. The ability of Smads to bind to numerous other protein partners, their ability to associate with the basal transcriptional machinery through CBP/p300 and histone deacetylases (HDACs) (see Chapter 14), and their low DNA-binding site selectivity are key factors through which Smads integrate TGF-β signaling with other modulators of gene transcription.

The rapidly expanding repertoire of Smad-interacting proteins is beginning to explain the pervasive influence of TGF-β ligands. The high degree of promiscuity with which Smads associate with other proteins suggests that the TGF-β pathway can modify or be modified by numerous cellular events in response to a wide variety of co-existing signaling molecules. Smads are known to directly interact with a large number of other proteins including transcription factors such as FoxH1, Runx, OAZ, Hoxc8, Milk/Mixer, and GATA proteins, to mention a few (Feng and Derynck, 2005). Although these interacting partners are important modulators of TGF-β superfamily signaling, they are not considered components of signal transduction pathways *per se*, and will not be discussed here. Rather, this review will focus on the role of Smads as integrators of TGF-β signaling within the cellular signaling environment. More specifically, we will examine direct protein interactions involving Smads and other signaling molecules and the current state of knowledge regarding the specific effects associated with these interactions.

2. TGF-β AND BMP CROSSTALK WITH THE WNT PATHWAY

Wnts are secreted glycoprotein growth factors involved in embryonic growth and patterning, as well as post-natal homeostasis in a variety of tissues (Chang et al., 2002; Nusse, 2005). Wnt ligands have been shown to signal via three intracellular pathways, the canonical Wnt/β-catenin pathway, the Wnt/calcium pathway and the Planar cell polarity (PCP) pathway, with the latter two also referred to as non-canonical Wnt pathways. Non-signaling cells normally target cytoplasmic β-catenin for ubiquitin-mediated proteasomal degradation via the F-box protein β-TrCP in response to phosphorylation by Casein kinase 1α (CKIα) and Glycogen synthase kinase 3β (GSK3β). Wnt stimulation inhibits GSK3β activity via the action of Dishevelled (Dvl) proteins. β-catenin then disengages from the destruction complex which also contains the Adenomatous polyposis coli (APC) and Axin proteins, accumulates in the cytoplasm and then nucleus where it activates gene transcription in concert with high-mobility-group LEF/TCF proteins and other co-factors (Fig. 1; Nusse, 2005). Comparatively little is known about non-canonical Wnt signaling at the biochemical level (Veeman et al., 2003) and there is no evidence of molecular crosstalk with Smads to date.

TGF-β and Wnt ligands both regulate numerous common developmental processes ranging from patterning of imaginal discs in *Drosophila* to tissue specification and organogenesis in vertebrate embryos (Nusse, 2005). Several studies have established that these pathways coordinate developmental events by regulating the expression of target genes such as *vestigial* in *Drosophila* wing development, *cerberus, chordin, crescent, goosecoid, noggin, siamois, Xtwn, Xnr6*, in the *Xenopus* Spemann organizer, and *Id2* and *Msx1* in mammalian cells (Kimelman and Griffin, 2000; Klein and Arias, 1999; Willert et al., 2002; Xanthos et al., 2002).

2.1 Smads and LEF/TCF

Insights into the molecular mechanism for the cooperation of the Wnt and TGF-β pathways first came with the demonstration that Smad4 (Nishita et al., 2000) or R-Smads (Smad2, 3) and Smad4 (Labbé et al., 2000) physically associate with members of the LEF/TCF family of HMG-box proteins to activate transcription of the *Xenopus* homeobox gene *Xtwn* (Fig. 1). Synergistic gene induction required the presence of closely spaced DNA binding elements for LEF/TCF and Smad proteins. No interaction between Smads and β-catenin was detected in the absence of LEF/TCF proteins, suggesting that integration of the two pathways occurs in the nucleus following ligand-induced nuclear translocation of β-catenin and R-Smad/Smad4 complexes.

A similar mechanism of cooperation between Wnt and TGF-β signaling was shown for the activation of the gastrin gene in mouse gastric adenocarcinoma cells, although the presence of functional LEF/TCF binding sites was not absolutely required for cooperative transcriptional activation when β-catenin, Smad3 and Smad4 were over-expressed (Lei et al., 2004). Gastrin is a mitogenic growth factor for gastrointestinal epithelial cells and its synergistic induction by the Wnt and TGF-β pathways suggests that this cooperation may be an important contributor to cancers of the gastrointestinal tract.

The *c-Myc* gene is also a common target of both pathways, though Wnt usually induces, while TGF-β inhibits, *c-Myc* expression (Fernandez-Pol et al., 1987; He et al., 1998). Interestingly, binding of distinct members of the LEF/TCF family to the *c-Myc* promoter differentially modulates its inhibition by TGF-β (Sasaki et al., 2003). More specifically, when TCF4 was bound to the *c-Myc* promoter, TGF-β signaling caused dissociation of β-catenin from TCF4 and resulted in inhibition of *c-Myc* expression, whereas TGF-β-mediated inhibition did not occur when LEF1 was bound to *c-Myc* promoter. The role of Smads in triggering repression only in the presence of TCF4 is unknown and appears paradoxical with other studies showing Smad cooperation with either TCF4 or LEF1 in the activation of *c-Myc* and other genes (see below). Smad binding to TCF4, and not to LEF1, may specifically recruit transcriptional co-repressors that cause dissociation of β-catenin. Nevertheless, these results may suggest an interesting mechanism by which mitogenic Wnt signaling may override growth inhibition by TGF-β in the presence of LEF1.

Cooperation between β-catenin signaling and Smads is not exclusive to the Smad2/3 subfamily of R-Smads (Hu et al., 2003; Hu and Rosenblum, 2005; Hussein et al., 2003). Smad1 was recently shown to form a transcriptionally active complex

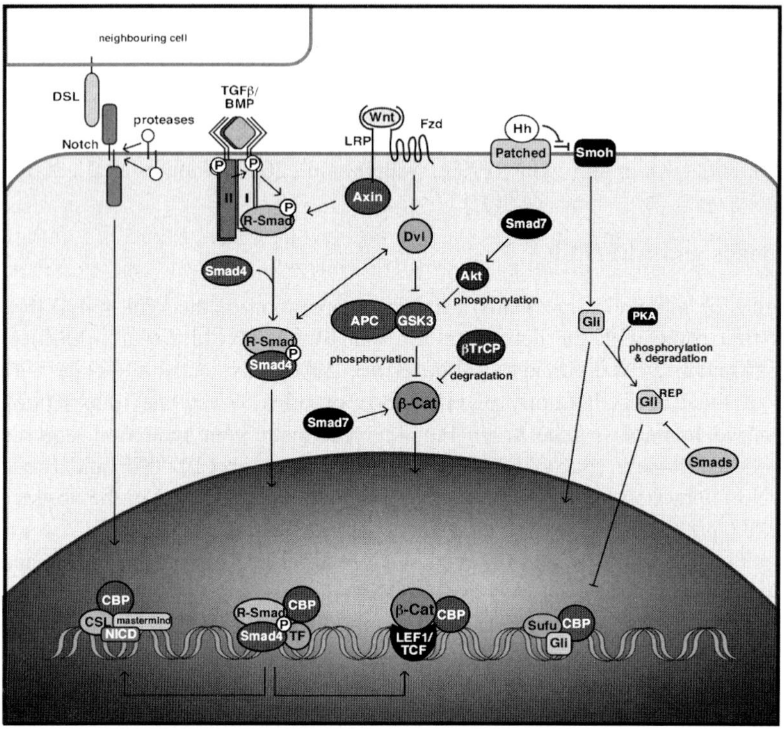

Figure 1. Smad-mediated crosstalk events between TGF-β/BMP signaling and the Wnt, Notch, and Hedgehog pathways. R-Smad/Smad4 form transcriptional complexes with NICD and LEF/TCF proteins to modulate Notch and Wnt-responsive genes, respectively. Axin interacts with and enhances activation of Smad3 by TGF-β, while Smad7 enhances stabilization of β-catenin downstream of TGF-β via GSK3β phosphorylation by Akt. Smad7 also directly interacts with β-catenin through the MH1 domain. Dvl proteins interact with Smads with unknown consequences on signaling output. Smad1-4 also interact with the truncated repressor form of Gli3 (Gli^{REP}), preventing its nuclear translocation

Abbreviations: Akt, alpha serine/threonine-protein kinase/murine thymoma oncogene; APC, adenomatous polyposis coli; β-Cat, β-catenin; βTrCP, β-transducin repeat containing protein; CSL, c-promoter binding factor/suppressor of hairless/lag-1; DSL, delta/serrate/lag-3; BMP, bone morphogenetic protein; CBP, CREB-binding protein; Dvl, dishevelled; Fzd, frizzled; Gli, glioma-associated oncogene; GSK3, glycogen synthase kinase-3; Hh, hedgehog; LEF/TCF, lymphoid enhancer binding factor/t-cell-specific transcription factor; LRP, low-density lipoprotein receptor-related protein; NICD, notch intracellular domain; P, phosphorylated amino acid; PKA, protein kinase-A; R-Smad, receptor-regulated Smad; Smoh, smoothened; Sufu, suppressor of fused; TF, transcription factor; TGF-β, transforming growth factor-β; Wnt, wingless/int-1

with β-catenin and TCF4 leading to activation of the *c-Myc* gene in the $TgAlk3^{QD}$ mouse model of renal medullary cystic dysplasia (Hu et al., 2003). These transgenic mice express a constitutively active BMP type I receptor in the ureteric bud lineage of the kidney and show a loss of branching morphogenesis and formation of medullary cysts during embryogenesis (Hu et al., 2003; Hu and Rosenblum, 2005). In dysplastic renal tissue from these mice, aberrant induction of *c-Myc* leads to a loss of epithelial differentiation markers and an increase in proliferation. *In vitro*, BMP2-mediated up-regulation of a *c-Myc* reporter in collecting duct cells was dependent on the presence of Smad1, β-catenin, and TCF4, as well as functional Smad and TCF-binding elements. Interestingly, mutation of the SBE resulted in an increase in basal reporter activity in unstimulated cells, suggesting that BMP2 treatment also relieves SBE-mediated transcriptional repression (Hu et al., 2003).

Although Smad4 has been firmly established as a component of the TGF-β superfamily pathway, some studies suggest that Smad4 might also function within the Wnt pathway, independently of TGF-β. For example, Smad4 can potentiate transcriptional activation by β-catenin and LEF1 on the *Xtwn* promoter in the absence of a TGF-β signal (Nishita et al., 2000). Moreover, endogenous Smad4 protein was seen to translocate to the nucleus in dissociated *Xenopus* animal cap explants from Xwnt8-injected embryos (Nishita et al., 2000). The mechanism for nuclear translocation of Smad4 in response to Xwnt8 is still unknown, but suggests an additional interaction between Smad4 and another Wnt-regulated protein. A requirement for Smad4 has also been demonstrated for Wnt-dependent activation of the *Msx2* gene in mouse ES cells (Hussein et al., 2003). Specifically, Wnt-dependent induction was lost in Smad4-deficient cells or when Smad4 binding sites were mutated. BMP2 also induces *Msx2* expression in mouse ES cells via Smad1 and Smad4, in a manner requiring adjacent Smad and LEF/TCF sites. Furthermore, Wnt and BMP can synergistically activate *Msx2*, possibly through enhanced recruitment of CBP (Hussein et al., 2003).

A number of studies have also shown that TGF-β ligands can activate LEF/TCF target genes in the absence of β-catenin. More specifically, deletion of the β-catenin binding domain in LEF1 did not abrogate the ability of Smads and this LEF1 mutant to synergistically activate *Xtwn* in response to TGF-β, or Msx2 in response to BMP stimulation (Labbé et al., 2000; Hussein et al., 2003). TGF-β3 is essential for palatogenesis by activating an epithelial-to-mesenchymal transition of medial edge epithelial cells and a recent study showed that while LEF1 and Smad4 were required for TGF-β-induced palatal EMT, β-catenin was not (Nawshad et al., 2004). Thus, these studies demonstrate that TGF-β can activate LEF1 target genes independently of Wnt and β-catenin.

From these studies, it is clear that the TGF-β and Wnt pathways can cooperate to activate target gene promoters containing adjacent Smad and LEF/TCF binding sites. However, a few reports have suggested that simultaneous DNA-binding by both Smads and LEF/TCF may not be required for the cooperative induction of all target promoters or in all cell types (Lei et al., 2004; Warner et al., 2005a). Furthermore, there is evidence that in some contexts Smad4 can act as a component of the Wnt pathway, while LEF/TCF proteins may act within the TGF-β pathway. Taken together, this data suggests that LEF/TCF and Smad4 proteins may be shared DNA-binding transcription factors of the TGF-β and Wnt signaling pathways. In addition, it appears that these proteins can serve to integrate both signals in cases of co-stimulation or to bias promoter selection in cases where only one pathway is activated. The identification of additional proteins capable of interacting with

Smads, β-catenin, and LEF/TCF proteins will certainly provide a clearer picture of the transcriptional cooperation between these two pathways.

2.2 Smads and β-catenin

A recent study by Jian and co-workers reports that TGF-β induces rapid Smad3-dependent nuclear accumulation of β-catenin (Jian et al., 2006). Interestingly, the authors show that this phenomenon appears to be restricted to bone marrow-derived mesenchymal stem cells (MSCs) and that β-catenin translocation occurs in the absence of the typical increase in protein stability observed following stimulation by Wnt. Using a dominant-negative C-terminal truncation of LEF1 that sequesters β-catenin in the cytoplasm, the authors showed that β-catenin function is required for the ability of TGF-β to stimulate proliferation and to inhibit osteogenic differentiation (Jian et al., 2006). Although this study shows that β-catenin can exist in a complex with Smad3 in MSCs, this interaction may be mediated by another protein. It is intriguing to note that the authors also described a novel interaction between Smad3 and GSK3β, and that this interaction disappears following treatment with TGFβ (Jian et al., 2006). Taken together, these results may suggest that Smad3 is an integral component of the β-catenin destruction complex, at least in MSCs.

The inhibitory Smads, Smad6 and Smad7, are noted for their antagonistic effects on TGF-β superfamily signaling via their ability to block R-Smad/receptor interaction (Feng and Derynck, 2005; Massagué, 2000; Chapter 20) and induce proteasomal-mediated degradation of the receptors (Feng and Derynck, 2005). However, a recent study showed that Smad7 promotes the formation of β-catenin/LEF/TCF complexes and that this interaction is required for the TGF-β-mediated induction of *c-Myc* associated with apoptosis in human PC-3U prostate cancer cells (Fig. 1; Edlund et al., 2005). Targeted knock-down of Smad7 expression using siRNA abrogated TGF-β-dependent β-catenin accumulation and redistribution to the perinuclear and nuclear regions, formation of β-catenin/LEF/TCF complexes, and induction of apoptosis. Smad7 and β-catenin were shown to physically interact and this was mediated primarily by the amino-terminal portion of Smad7, a region that diverges in sequence from the MH1 domain of R- and Co-Smads. In addition to this direct interaction, the authors showed that Smad7 was required for TGF-β-dependent stabilization of β–catenin. Smad7 was previously shown to mediate activation of Akt by the TGF-β-TAK1-MKK3 pathway (Edlund et al., 2003). Together with activation of P13K, this was shown to lead to inhibitory phosphorylation of GSK3β on serine-9 (Edlund et al., 2005). These results suggest that Smad7 may therefore potentiate cooperative transcriptional activation by β-catenin and R-Smads by facilitating β-catenin stabilization downstream of TGF-β and independently of Wnt ligands. The impact of Smad7 on β-catenin in the context of co-stimulation with TGF-β and Wnt, or in cells harbouring APC or β-catenin mutations, remains unknown. In addition, a potential transcriptional role for Smad7 within the DNA-bound β-catenin/LEF/TCF complex will require further investigation.

2.3 Smads and Wnt Cytoplasmic Components

In addition to cooperation within DNA-binding complexes, Smads have also been shown to associate with various cytoplasmic components of the Wnt pathway. As a component of the β-catenin destruction complex, Axin negatively impacts Wnt signaling through interactions with proteins such as APC, GSK3β and Dvl (Nusse, 2005). However, in the context of TGF-β signaling, Axin association with Smad3 enhances TGF-β-dependent transcription (Furuhashi et al., 2001). Smad3/Axin complexes are present in the cytoplasm in the absence of ligand and Axin appears to function as an adaptor that facilitates Smad3 phosphorylation and subsequent transcriptional output (Fig. 1). Smad3 binds Axin in a region distinct from that for APC, GSK3β, β-catenin and Dvl, however, it is not clear whether Smad3 coexists in a complex with these proteins. CKIε, another Axin binding protein, is a positive regulator of Wnt signaling and has recently been shown to interact with Smads and modulate TGF-β signaling (Waddell et al., 2004). The molecular mechanism of CKIε action, in either TGF-β or Wnt signaling, remains elusive, but it is intriguing to speculate that CKIε-mediated phosphorylation of common targets may contribute to pathway interplay.

A direct interaction between Smad3 and Dvl1 (Fig. 1), a key component of canonical and non-canonical Wnt pathways, was recently uncovered in a yeast two-hybrid screen (Warner et al., 2003b). *In vitro* analysis also demonstrated binding of Dvl1 to Smad2, Smad4, and Smad7, and the interaction of overexpressed Dvl1 with Smad3 was confirmed in mammalian cells (Warner et al., 2005b). Characterization of the functional consequence of this novel link between the two pathways is bound to reveal another interesting level of crosstalk.

The examples described above suggest extensive intermingling of the Wnt and TGF-β pathways at multiple levels. Intriguing questions remain concerning the physiological roles of Wnt and TGF-β crosstalk. For example, to what extent can β-catenin transduce TGF-β signals independently of Wnt and of the R-Smad/Smad4 complex? Given the role of TGF-β and Wnt pathways in cancer, are genes targeted synergistically by Wnt and TGF-β involved in tumor formation and progression?

3. TGF-β AND BMP CROSSTALK WITH THE NOTCH PATHWAY

The Notch pathway is activated in a juxtacrine fashion, requiring close proximity between the ligand-producing cell and the receiving cell (Kadesch, 2004; Weng and Aster, 2004). Membrane-bound ligands of the Delta and Jagged/Serrate families bind to single-pass transmembrane Notch receptors on neighbouring cells triggering proteolytic cleavage of Notch mediated by TACE (TNF-α converting enzyme) and γ-secretase, releasing the Notch intracellular domain (NICD). NICD translocates to the nucleus where it interacts with the transcription factor CSL (C-promoter binding factor 1/Suppressor of Hairless/LAG-1). This binding converts CSL from a repressor to an activator of transcription by displacing a CSL co-repressor and is followed by recruitment of co-activators such as Mastermind, p/CAF and p300

(Fig. 1). Target genes of the NICD-CSL-co-activator complex include members of the Hairy/Enhancer of Split (*HES*) and the *HRT/HERP/Hey* families of transcriptional repressors (Kadesch, 2004; Weng and Aster, 2004).

The TGF-β and Notch signaling pathways are involved in controlling a number of developmental processes including the differentiation of myocytes, neurons and endothelial cells. While it was generally believed that these pathways functioned independently, emerging evidence suggests that they are intricately intermingled. TGF-β is noted for its ability to induce epithelial-to-mesenchymal transition (EMT) and a recent report has demonstrated that Smad3 and Jagged1/Notch are required for certain aspects of TGF-β-induced EMT (Zavadil et al., 2004). In this system, TGF-β was shown to induce the expression of the transcriptional repressor, Hey1, in a biphasic pattern. While the first phase was dependent only on Smad3, delayed expression of *Hey1* also required Jagged1/Notch signaling. Chemical inactivation of Notch signaling prevented TGF-β-induced EMT, demonstrating a functional integration of the two pathways.

Further insight into the molecular nature of the cooperative interaction of these two pathways was gained with the demonstration of a direct interaction between Smads and NICD (Fig. 1). TGF-β-dependent activation of *Hes-1* in adult neural stem cells and C2C12 myoblasts required Notch signaling and it was shown that NICD directly recruited Smad3 to CSL-bound promoters (Blokzijl et al., 2003). Analysis of a CSL-containing artificial promoter revealed that while Smad4 was essential for promoter activation, the DNA-binding activity of Smad3 was dispensable (Blokzijl et al., 2003). Furthermore, overexpression of NICD enhanced the activity of the artificial Smad reporter CAGA-luc suggesting the possibility that NICD and Smads may synergistically activate transcription even on promoters lacking CSL DNA-binding sites. Interactions between the BMP and Notch pathways have also been reported. BMP-dependent inhibition of myogenesis and activation of *Hes-1* and *Hey-1* in C2C12 cells required Notch signaling and a physical interaction between the BMP-regulated R-Smad, Smad1, and NICD was observed (Dahlqvist et al., 2003). Moreover, in mouse neuroepithelial cells, enhancement of Notch-induced *Hes-5* expression by BMP2 required CSL DNA-binding and complex formation between Smad1 and NICD (Takizawa et al., 2003). For both Smad1 and Smad3, association with NICD appears to be stabilized in the presence of the transcriptional co-activators p/CAF and p300 (Itoh et al., 2004; Takizawa et al., 2003).

BMP and Notch signaling have also both been implicated in vasculogenesis in endothelial cells (EC) and a recent report has revealed a complex interaction of both synergy and antagonism between the two pathways in the control of EC migration (Itoh et al., 2004). Previous work has shown that BMP stimulates EC migration via up-regulation of *Id1* (Valdimarsdottir et al., 2002). The authors examined the effect of Notch on this process and demonstrated that BMP and Notch cooperate to induce *Herp2* expression via a physical interaction of Smad1 and NICD. Furthermore, it was shown that the increase in *Herp2* expression blocked Id1-induced EC migration. This model in which BMP-induced EC migration is inhibited in the presence of Notch suggests the intriguing possibility that a complex cooperation of BMP and

Notch pathways might also contribute to other developmental processes requiring modulation of cell migration.

4. BMP CROSSTALK WITH THE HEDGEHOG PATHWAY

The Hedgehog (Hh) family of growth factors was originally identified in *Drosophila* and is now known to be critical for cell growth and differentiation and numerous developmental processes in vertebrates (Ingham and McMahon, 2001). Mutations affecting Hh signaling give rise to several human developmental abnormalities as well as malignancies in the adult (Mullor et al., 2002). The main intracellular transducers of Hh signals are the Gli proteins (Cohen, 2003). In the absence of a signal, Glis are phosphorylated by PKA and processed by the proteasome to generate transcriptionally repressive forms lacking the C-terminus (at least for Gli2 and Gli3), or retained in the cytoplasm (for Gli1). Activation of the pathway results in the inhibition of Gli phosphorylation, which then leads to translocation to the nucleus and activation of gene transcription (Fig. 1).

Common patterns of expression of BMPs and Hhs, as well as demonstrated genetic interactions in mouse models have suggested that these two pathways may directly cooperate to control certain cellular processes (Bastida et al., 2004; Dunn et al., 1997). Consistent with this possibility, Smads 1, 2, 3 and 4 were found to specifically interact with a C-terminally deleted form of Gli3 that mimicks $Gli3^{REP}$, the endogenous processed repressor form (Liu et al., 1998). These interactions only occurred in the absence of the appropriate BMP or TGF-β ligands, suggesting that release of $Gli3^{REP}$ from Smads upon TGF-β/BMP signaling may potentiate inhibition of the Hh pathway, whereas lack of stimulation may allow for Smad-mediated segregation of $Gli3^{REP}$ in the cytoplasm, facilitating Hh signaling (Liu et al., 1998). One caveat of this model is that over-expression of full-length or truncated Gli3 had no effect on signaling output from TGF-β or BMP-specific reporters.

An additional level of crosstalk may exist between Glis and Smads via the Ski family of transcriptional co-repressors. Ski was shown to interact with Gli3 as well as with Smad proteins (Dai et al., 2002; Luo, 2004). Gli3-mediated transcriptional repression was impaired in Ski-null fibroblasts, suggesting a requirement for Ski function (Dai et al., 2002). Interestingly, recent evidence suggests that the related protein SnoN acts as a necessary transcriptional co-activator for TGF-β/Smad signaling in a cell-specific manner (Sarker et al., 2005). Whether Ski and SnoN modulate the interplay of Gli and Smad proteins at the transcriptional level is an interesting subject for further investigation.

5. TGF-β CROSSTALK WITH JAK/STAT SIGNALING

STAT (Signal transducer and activator of transcription) proteins are mediators of signaling by cytokines which play key roles in cell growth, differentiation, and homeostasis (Williams, 2000). Classically, the pathway downstream of these cytokines involves ligand-induced dimerization of receptors and activation of associated Janus Kinases (JAK). JAKs are tyrosine kinases that, upon activation, undergo autophosphorylation, then phosphorylate the receptors to reveal SH2-binding docking sites for STAT proteins. Tyrosine phosphorylation of STATs by JAKs induces their release from the receptors and promotes homo- or heterodimerization and nuclear translocation where they activate transcription via direct DNA-binding and co-activator partnering (Fig. 2).

Even though STAT and Smad signaling exert opposite effects on many biological processes (Iwamoto et al., 2002), no evidence exists of a direct negative modulation of Smad function by STAT signaling. Rather, inhibition of TGF-β signaling by JAK/STAT in response to Interferon-γ or Interleukin-6 (IL-6) involves indirect downregulation of TGF-β-receptor function via transcriptional up-regulation of the I-Smad, Smad7 (Ulloa et al., 1999; Jenkins et al., 2005).

A more direct mode of interaction between Smad and STAT signaling has also been reported. STAT activation downstream of IL-6 and the related cytokine, Leukemia inhibiting factor (LIF), were both shown to synergize with BMP2 in astrocyte differentiation from neural precursors (Nakashima et al., 1999; Yanagisawa et al., 2001). The molecular mechanism of cooperation involves formation of a transcriptional complex between Smad1 and STAT3 bridged by the co-activator p300 (Nakashima et al., 1999).

Intriguing reports have recently revealed that Smads and STATs interact with common proteins, such as the Protein-inhibitor of activated STATs (PIAS) and the Runt-domain-containing Runx proteins (Kim et al., 2003; Long et al., 2004). It will be interesting to examine whether these interactions represent additional levels of crosstalk between the two pathways.

6. TGF-β AND BMP CROSSTALK WITH NUCLEAR HORMONE RECEPTOR SIGNALING

The nuclear hormone receptor superfamily consists of structurally-related ligand-activated transcription factors numbering approximately 30 members according to a search of the Interpro database (Mulder et al., 2005). These receptors directly interact with cognate DNA sites and transcriptional co-factors following ligand stimulation (Fig. 2; Aranda and Pascual, 2001). They are involved in a wide variety of biological processes important to embryonic development, tissue growth and homeostasis, immune responses, and more (Aranda and Pascual, 2001).

A number of nuclear receptors interact with Smads and do so in either a cooperative or inhibitory fashion. For example, overexpression of BMP4 by pituitary prolactinoma cells stimulates a physical interaction of Smad1/Smad4 with the estrogen receptor (ER) and transcriptional activation of *c-Myc* leading to increased cell proliferation (Paez-Pereda et al., 2003). Conversely, 17-β-estradiol (E2) inhibits BMP2-induced gene expression through a similar interaction of ER and Smads in breast cancer and mesangial cells (Yamamoto et al., 2002). While E2 also inhibits TGF-β-mediated activation of 3TP-luc in transfected HEK293T cells, TGF-β/Smad3 and E2 synergistically stimulate the estrogen-responsive Vit-Luc reporter (Matsuda et al., 2001). Smad4 was also found to physically associate with ERα, inhibiting ER-mediated gene activation in breast cancer cells (Wu et al., 2003). Thus, the transcriptional endpoint of Smad-ER association is modified by cell-specific factors and appears to be promoter-specific.

Smad3 and Smad4 also interact with the androgen receptor (AR) leading to either positive or negative regulation of TGF-β or AR-stimulated gene transcription (Hayes et al., 2001; Chipuk et al., 2002; Kang et al., 2002). TGF-β signaling in PC-3 human prostate cells can inhibit AR-mediated activation of the MMTV and *prostate specific antigen* promoters via a direct interaction of Smad3 and AR (Hayes et al., 2001). Androgen signaling in human prostate adenocarcinoma LNCaP cells can inhibit TGF-β/Smad-mediated expression of c-Fos, Egr-1 and TGF-β-specific reporter genes via a direct interaction between AR and Smad3 that inhibits Smad3 binding to DNA (Chipuk et al., 2002). In the same LNCaP cells, Smad3 was shown to enhance AR-mediated transactivation, while co-transfection with Smad4 mitigated this Smad3-dependent effect (Kang et al., 2002). Differential expression of the Smad3-binding AR-associated protein 55 (Ara55/Hic-5) provides a possible explanation for cell-type specific effects (Wang et al., 2005).

Positive and negative interactions between TGF-β/Smad and glucocorticoid receptor (GR) signaling also appear to involve direct association of Smads and GR. In murine M12 B-cell lymphomas, TGF-β enhanced the induction of a minimal MMTV-luciferase reporter by dexamethasone. This effect was dependent on the interaction of Smad3/4 with GA-binding proteins (GABP) α and β, as well as on direct binding of Smad3 to the GR (Aurrekoetxea-Hernandez and Buetti, 2004). In contrast, TGF-β antagonized dexamethasone/GR-induced MMTV reporter activity in L929 mouse fibrosarcoma cells in a manner dependent on AP1 binding to the MMTV promoter (Periyasamy and Sanchez, 2002). Moreover, in transfected Hep3B cells, dexamethasone treatment inhibited TGF-β-induced PAI-1 reporter activity through a direct physical interaction between GR and Smad3 (Song et al., 1999).

Extensive evidence exists for regulation of TGF-β signaling by retinoic acid (RA) (Roberts and Sporn, 1992). However, only recently has an intracellular mechanism for crosstalk between retinoid and TGF-β signaling been described. Specifically, direct physical association of retinoic acid receptor (RAR) and Smad3/4 was shown to potentiate TGF-β signaling in WI-26 human lung fibroblasts (Pendaries et al., 2003). Interestingly, this interaction and transcriptional cooperation was inhibited by RAR agonists, but enhanced by RAR antagonists. Active RAR signaling may therefore exert a dominant effect on TGF-β signaling leading to dampening of transcriptional output. BMP ligands also appear to interact with RA signaling during osteogenic and chondrogenic differentiation (Drissi et al., 2003; Skillington et al., 2002). In the latter case, RA/BMP synergy appears to be indirectly mediated by the interaction of RA-induced Runx2 with BMP-stimulated Smad1 (Drissi et al., 2003).

The lipid-sensing peroxisome-proliferator-activator receptor proteins (PPARα, δ, and γ) control nutrient metabolism, energy homeostasis, adipocyte differentiation, and modulate inflammation (Berger et al., 2005). Synthetic agonists of PPARs have been the subject of intense research as they have demonstrated therapeutic value in the treatment of atherosclerosis, inflammation, diabetes, lipid metabolism imbalance, and other diseases (Berger et al., 2005). A number of reports have documented crosstalk between TGF-β signaling and PPARs in the context of TGF-β's pro-fibrotic activity. For example, PPARγ stimulation in vascular smooth muscle cells (VSMCs) inhibits the Smad3-dependent induction of connective tissue growth factor (CTGF) expression, a mediator of fibroblast growth and extracellular matrix production (Fu et al., 2001). Moreover, Smad3 but not Smad4 was shown to physically interact with ligand-stimulated PPARγ, suggesting a direct molecular mechanism for this inhibition (Fu et al., 2001). In contrast, association of PPARα with Smad4 was shown to inhibit Smad-Sp1 complexes leading to abrogation of TGF-β-dependent induction of β5 integrin in a rat VSMC system (Kintscher et al., 2002).

The first member of the nuclear receptor superfamily shown to interact with Smads was the vitamin D receptor (VDR; Subramaniam et al., 2001; Yanagisawa et al., 1999). Smad3 physically associated with VDR and potentiated transcription from the vitamin D response elements (VDRE) of *osteopontin* and *osteocalcin* genes in response to TGF-β and vitamin D. Stimulation with $1,25(OH)_2D_3$ enhanced the association of Smad3 with VDR, heterodimerization of VDR with RXR, and recruitment of the SRC-1/TIF2 co-activator. Interestingly, the Smad3/VDR interaction required an intact Smad MH1 domain, in contrast to many other Smad interactions mediated by the MH2 domain.

Although HNF4 remains an orphan nuclear receptor, its interaction with Smad3/4 has been shown to regulate the activation of the *apolipoprotein C-III* promoter in hepatocytes in response to TGF-β (Chou et al., 2003; Kardassis et al., 2000). Interestingly, this transcriptional cooperation did not require DNA binding by Smads, as deletion of the β-hairpin required for DNA contacts did not abrogate Smad-dependent enhancement (Chou et al., 2003).

Thus, the convergence of Smads and members of the nuclear receptor family on target genes provides for a multifaceted mechanism to direct precise patterns of gene expression in diverse physiological and pathological contexts.

7. TGF-β CROSSTALK WITH NF-κB SIGNALING

NF-κB is a transcriptional complex consisting of homo- or heterodimers of p50, p52, p65/RelA, c-Rel, or RelB, although the major form is a dimer of p50 and p65 (Ghosh et al., 1998). These dimeric transcription factors are involved in transducing signals in response to pro-inflammatory and immune-stimulatory cytokines including TNFα (tumor necrosis factor alpha) and interleukins, and stress stimuli such as bacterial toxins, ultraviolet light, reactive oxygen species, double-stranded RNA, and viral transactivating proteins (Ghosh et al., 1998). NF-κB is sequestered

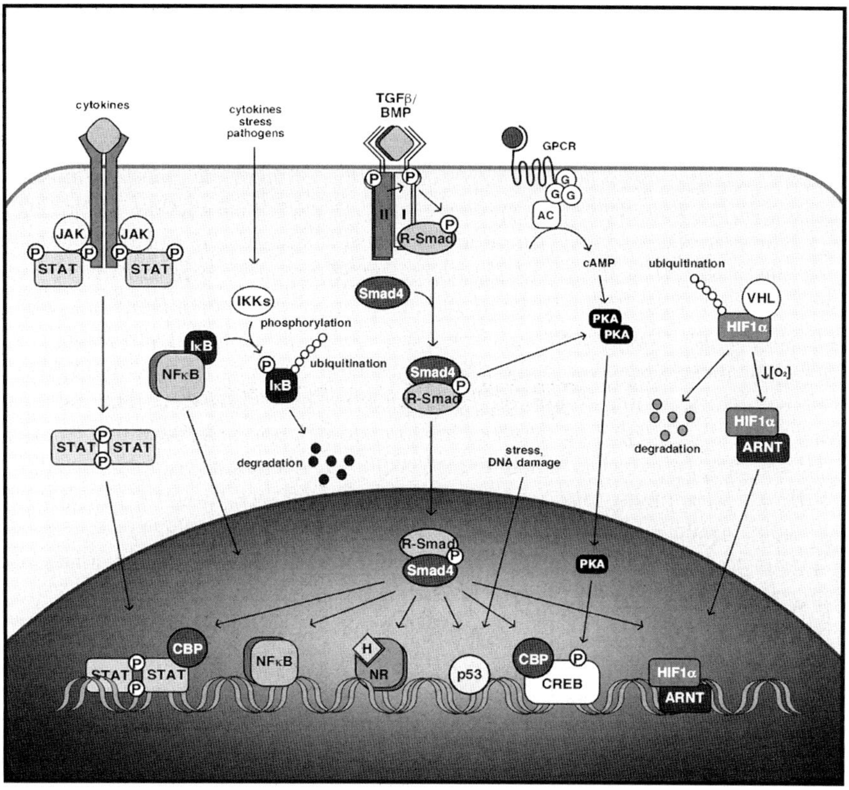

Figure 2. Smad-mediated crosstalk events between TGF-β/Smad signaling and the JAK/STAT, NF-κB, Nuclear Hormone, HIF1, and p53 pathways. Smad1 can interact and synergize with STAT3 protein via the co-activator, CBP. Activated Smads may also form a DNA-transcription complex with p52/NFκB to synergistically activate JunB transcription. Multiple positive and negative interactions between Smads and members of the nuclear hormone receptor (NR) family have been reported that affect TGF-β or NR target genes expression. Smads also synergize with p53 and HIF1 transcriptional complexes through physical interactions. Direct interaction of activated Smads with the regulatory subunit of PKA releases and activates the catalytic subunit. Further cooperation between Smad and PKA signaling occurs in the nucleus via CBP

Abbreviations: AC, adenylate cyclase; ARNT, aryl receptor nuclear translocator; BMP, bone morphogenetic protein; cAMP, cyclic adenosine monophosphate; CBP, CREB-binding protein; CREB, cAMP response element binding protein; G, GTP-binding protein; GPCR, G-protein-coupled receptor; H, hormone; HIF1α, hypoxia-inducible factor-1-alpha; IκB, inhibitor of kappa-B; IKK, IκB kinase; JAK, janus kinase; NFκB, nuclear factor kappa-B; NR, nuclear hormone receptor; PKA, protein kinase A; R-Smad, receptor-regulated Smad; STAT, signal transducer and activator of transcription; TGF-β, transforming growth factor-β; VHL, von Hippel-Lindau protein

in the cytosol by IκB (Inhibitor of κB) in the absence of stimulatory stimuli. Activation of NF-κB involves phosphorylation of IκB by IκB kinases (IKK), ubiquitin-dependent degradation of IκB followed by release of NF-κB and translocation to the nucleus where it modulates target gene expression (Fig. 2).

While NF-κB is a key mediator of pro-inflammatory and pathogenic signals, TGF-β is noted for its anti-inflammatory and immunosuppressive activity. Thus, it was no surprise to find an interaction between these two pathways in controlling cellular responses (Letterio and Roberts, 1998). TNFα, IL-1β or bacterial lipopolysaccharides can inhibit TGF-β-mediated signaling, and in fibroblasts this was shown to occur via an NF-κB-dependent up-regulation of the *Smad7* promoter which in turn prevents TGF-β receptor signaling (Bitzer et al., 2000). TNFα and TGF-β can have both additive and antagonistic effects on the expression of collagens. In the case of the *COL1A2* promoter, TNFα-mediated activation of JNK, and not NF-κB, was shown to be required for TNFα to block TGF-β's transcriptional activity (Verrecchia and Mauviel, 2004; Verrecchia et al., 2000). In contrast with these studies, a functional cooperation between NF-κB and Smads on a 3'-regulatory region of the *junB* promoter has been described. The authors reported a physical interaction between Smad3 and p52 and suggested that Smads and NF-κB may form a DNA-bound transcriptional activation complex at adjacent SBE and κB binding sites on the *junB* promoter (Lopez-Rovira et al., 2000). Although the presence of a DNA-bound Smad/NF-κB was not shown, a cooperation between Smads and NF-κB in pathogen-induced MUC2 expression in a variety of human epithelial cells has also been observed (Jono et al., 2002). The basis for these contrasting results is unclear but likely represent cell and promoter-specific differences.

8. TGF-β INTERACTION WITH HYPOXIA AND p53 PATHWAYS

8.1 Smads and HIF-1

Cells have evolved sophisticated mechanisms to allow for sensing and adapting to perturbations in oxygen concentration. Hypoxia-inducible factor-1 (HIF-1) is the primary sensor of cellular hypoxia and is a critical effector of cellular oxygen homeostasis (Ruas and Poellinger, 2005). Under normal oxygen conditions, the oxygen-sensing subunit HIF-1α is hydroxylated and targeted for ubiquitin-mediated proteasomal degradation. Hypoxic conditions inhibit HIF-1α degradation allowing it to associate with HIF-1β (also known as ARNT), the DNA-binding subunit of the HIF-1 dimer (Fig. 2; Ruas and Poellinger, 2005). Although HIF-1α is ubiquitously expressed and responds to hypoxia in all cells, interaction with tissue-specific transcriptional partners selectively activates target genes under hypoxia.

TGF-β ligands play important roles in vasculogenesis as well as in the response of endothelial cells to stress such as inflammation and hypoxia. Under hypoxic conditions, TGF-β2 production is induced in an autocrine loop dependent on activation of latent TGF-β2, stimulation of Smad signaling and cooperation of Smad3 with HIF-1α (Akman et al., 2001; Zhang et al., 2003). In fact, a direct physical interaction between Smad3 and HIF-1α mediating cooperative induction of vascular endothelial growth factor and erythropoietin has been reported (Sanchez-Elsner et al., 2001; Sanchez-Elsner et al., 2004). Moreover, hypoxia was shown to stimulate TGF-β production and to cooperate with TGF-β/Smad3 signaling to inhibit adipogenic differentiation in human and mouse marrow stromal cells (Zhou et al., 2005). Thus, these results suggest that HIF-1-mediated responses to hypoxia may be extensively modulated by tissue-specific interactions with the TGF-β/Smad pathway. TGF-β may act to prolong hypoxic transcriptional responses via a positive feedback autocrine loop.

8.2 Smads and p53

p53, the protein product of the most commonly mutated gene in human malignancies, is an important sensor of cellular stress and DNA damage (Harris and Levine, 2005). In response to various types of stress or damage, p53 activity is modulated by an impressive array of post-translational modifications such as phosphorylation, methylation, acetylation, ubiquitination, and sumoylation. Each type of damage appears to result in specific p53 modifications, in a code-like fashion, that trigger appropriate response pathways leading to DNA repair, metabolic changes, or apoptosis (Harris and Levine, 2005). Additional evidence suggests that p53 is also involved in embryonic development in the frog and probably in mammals, although redundancy with p63 and p73 in the latter may mask early p53 functions (Cordenonsi et al., 2003).

Two recent studies reported a developmental interplay of p53 and TGF-β/Activin in *Xenopus* embryogenesis, including a specific cooperation and physical interaction of Smads with p53 (Fig. 2; Cordenonsi et al., 2003; Takebayashi-Suzuki et al., 2003). Injection of mouse p53 in *Xenopus* blastomeres was found to induce a gene expression signature characteristic of TGF-β/Activin signaling (Cordenonsi et al., 2003). It was demonstrated that p53 is a DNA-binding Smad transcriptional partner that physically interacts with TGF-β and BMP-regulated Smads to specifically enhance the activation of promoters that contained both Smad and p53-binding elements. This cooperation appears critical for both dorsal and ventral mesoderm formation during *Xenopus* development (Cordenonsi et al., 2003; Takebayashi-Suzuki et al., 2003). The interaction also appears to be relevant to mammalian cells where p53 silencing with siRNA significantly reduced TGF-β-induced growth inhibition, as well $asp21^{WAF1}$ and *MMP2* expression (Cordenonsi et al., 2003). A role for p53 and TGF-β/Smad in negatively regulating transcription in hepatocytes has also been described (Wilkinson et al., 2005). p53 could recruit the transcriptional co-repressor SnoN and histone deacetylases via the Smad-binding element of the tumor marker α–fetoprotein (AFP) promoter, causing AFP repression in response to TGF-β treatment.

These new connections between two major tumor-suppressor pathways during embryonic development and postnatal homeostasis, is quite unexpected. It may suggest a novel mechanism for insensitivity to TGF-β growth inhibition in p53-mutated tumors. This aspect of TGF-β crosstalk is likely to show exciting developments in the near future.

9. TGF-β CROSSTALK WITH PKA SIGNALING

Cyclic AMP (cAMP) is an intracellular secondary messenger molecule produced following activation of adenylate cyclase (AC) by G-protein coupled receptors (reviewed in Taskén and Aandahl, 2004). Binding of cAMP to the protein kinase A (PKA) regulatory subunit (PKA-R) releases the PKA catalytic subunit (PKA-C) allowing it to phosphorylate a plethora of substrates in both the cytoplasm and nucleus and thereby modulate numerous physiological processes (Fig. 2). Thus, the potential for interaction between TGF-β and PKA is quite large. For instance, PKA activation of Erk leads to Smad1 phosphorylation and cytoplasmic retention which causes inhibition of BMP2 signaling in neural crest cells (Ji and Andrisani, 2005).

As many PKA substrates are components of other signaling pathways, they will not be considered further. However, an important PKA substrate is the cAMP-response element binding (CREB) protein whose phosphorylation promotes recruitment of the co-activators CBP (CREB-binding protein) and p300. A critical role for CBP/p300 as a co-activator for Smads has been amply demonstrated (Chapter 14). Thus, PKA-dependent phosphorylation of CREB could represent a point of positive transcriptional cooperation between TGF-β and PKA, a suggestion that has been supported by a number of studies (Ionescu et al., 2004; Warner et al., 2003a). For instance, TGF-β and Forskolin-induced cAMP signaling in primary mouse embryonic palate mesenchymal cells induces the formation of an SBE-bound complex that contains phosphorylated Smad2, phosphorylated CREB, CBP, and the co-repressors Ski and SnoN (Warner et al., 2003a). In contrast, the interaction can be inhibitory as seen in HaCaT keratinocytes where elevated cAMP/PKA signaling blocks Smad3/4-driven TGF-β signaling by preventing the recruitment CBP/p300 to the DNA-bound Smad complex (Schiller et al., 2003).

A more direct means of pathway crosstalk was recently observed with the demonstration that a TGF-β-activated Smad3/Smad4 complex, but not a Smad2/Smad4 complex, directly associated with PKA-R, thus liberating the PKA-C for downstream signaling in the absence of elevated cellular cAMP (Fig. 2; Zhang et al., 2004). This activation of PKA activity was found to be required for the TGF-β-mediated activation of CREB, as well as for the induction of $p21^{CIP1}$ and TGF-β-induced growth inhibition in Mv1Lu mink lung epithelial cells (Zhang et al., 2004). Following release of the catalytic subunit, it is not known whether Smad/PKA-R complexes persist and regulate transcription together with or independently of PKA-C.

10. SMAD CROSSTALK VIA NON-RECEPTOR KINASES

A number of other intracellular kinases including MAPK, JNK, PKC, CaMKII, GRK2, and CKIε are known to phosphorylate Smads at sites distinct from those targeted by the TGF-β/BMP receptors. These modifications alter Smad function by modulating receptor-dependent activation, nuclear translocation, or transcriptional activity. In addition, downstream DNA-binding targets of these pathways such as jun and fos are Smad partners that cooperate to modulate expression patterns in several contexts. A detailed analysis on the regulation of Smads by phosphorylation is provided in Chapter 12, while the interplay between Smads and MAPK pathways is discussed in Chapter 16.

11. A LOOK FORWARD: THE SMAD SIGNALING INTERACTOME

In recent years, we have witnessed an explosion in TGF-β and Smad-related literature. In this chapter, we have reviewed the elucidation of numerous molecular crosstalk events between Smads and other signaling pathways. The recent advent of high-throughput screening methods now allows for a rapid and dynamic analysis of protein-protein interaction networks. Such screens centered on the TGF-β pathway using luciferase-tagging (Barrios-Rodiles et al., 2005), peptide mass-fingerprinting (Kanamoto et al., 2002), and yeast two-hybrid/RNAi analyses (Colland et al., 2004; Tewari et al., 2004) have revealed a plethora of novel Smad interactors, identifying new potential points of crosstalk with other signaling pathways. Following this phase of proteomic data generation, the next challenge will be to validate each of these new interactions and, more importantly, to place them within the physiological context of a functioning transduction network within a cellular environment. Proteome-wide analyses of splice variant functions and post-translational modifications are also highly relevant to the study of signal crosstalk and represent key areas of research for the near future.

ACKNOWLEDGEMENTS

Research in the lab was supported by grants from the Canadian Institute for Health Research and the National Cancer Institute of Canada with funds from the Cancer Research Society. We thank S. Perusini and C. Silvestri for useful comments on this manuscript.

REFERENCES

Akman, H.O., Zhang, H., Siddiqui, M.A., Solomon, W., Smith, E.L., and Batuman, O.A., 2001, Response to hypoxia involves transforming growth factor-β2 and Smad proteins in human endothelial cells. *Blood* 98: 3324-3331.

Aranda, A., and Pascual, A., 2001, Nuclear hormone receptors and gene expression. *Physiol Rev* 81: 1269-1304.

Aurrekoetxea-Hernandez, K., and Buetti, E., 2004, Transforming growth factor β enhances the glucocorticoid response of the mouse mammary tumor virus promoter through Smad and GA-binding proteins. *J Virol* 78: 2201-2211.

Barrios-Rodiles, M., Brown, K.R., Ozdamar, B., Bose, R., Liu, Z., Donovan, R.S., Shinjo, F., Liu, Y., Dembowy, J., Taylor, I.W., Luga, V., Przulj, N., Robinson, M., Suzuki, H., Hayashizaki, Y., Jurisica, I., and Wrana, J.L., 2005, High-throughput mapping of a dynamic signaling network in mammalian cells. *Science* 307: 1621-1625.

Bastida, M.F., Delgado, M.D., Wang, B., Fallon, J.F., Fernandez-Teran, M., and Ros, M.A., 2004, Levels of Gli3 repressor correlate with Bmp4 expression and apoptosis during limb development. *Dev Dyn* 231: 148-160.

Berger, J.P., Akiyama, T.E., and Meinke, P.T., 2005, PPARs: therapeutic targets for metabolic disease. *Trends Pharmacol Sci* 26: 244-251.

Bitzer, M., von Gersdorff, G., Liang, D., Dominguez-Rosales, A., Beg, A.A., Rojkind, M., and Böttinger, E.P., 2000, A mechanism of suppression of TGF-β/SMAD signaling by NF-κB/RelA. *Genes Dev* 14: 187-197.

Blokzijl, A., Dahlqvist, C., Reissmann, E., Falk, A., Moliner, A., Lendahl, U., and Ibanez, C.F., 2003, Cross-talk between the Notch and TGF-β signaling pathways mediated by interaction of the Notch intracellular domain with Smad3. *J Cell Biol* 163: 723-728.

Chang, H., Brown, C.W., and Matzuk, M.M., 2002, Genetic analysis of the mammalian transforming growth factor-β superfamily. *Endocr Rev* 23: 787-823.

Chipuk, J.E., Cornelius, S.C., Pultz, N.J., Jorgensen, J.S., Bonham, M.J., Kim, S.J., and Danielpour, D., 2002, The androgen receptor represses transforming growth factor-β signaling through interaction with Smad3. *J Biol Chem* 277: 1240-1248.

Chou, W.C., Prokova, V., Shiraishi, K., Valcourt, U., Moustakas, A., Hadzopoulou-Cladaras, M., Zannis, V.I., and Kardassis, D., 2003, Mechanism of a transcriptional cross talk between transforming growth factor-β-regulated Smad3 and Smad4 proteins and orphan nuclear receptor hepatocyte nuclear factor-4. *Mol Biol Cell* 14: 1279-1294.

Cohen, M.M., Jr., 2003, The hedgehog signaling network. *Am J Med Genet A* 123: 5-28.

Colland, F., Jacq, X., Trouplin, V., Mougin, C., Groizeleau, C., Hamburger, A., Meil, A., Wojcik, J., Legrain, P., and Gauthier, J.M., 2004, Functional proteomics mapping of a human signaling pathway. *Genome Res* 14: 1324-1332.

Cordenonsi, M., Dupont, S., Maretto, S., Insinga, A., Imbriano, C., and Piccolo, S., 2003, Links between tumor suppressors: p53 is required for TGF-β gene responses by cooperating with Smads. *Cell* 113: 301-314.

Dahlqvist, C., Blokzijl, A., Chapman, G., Falk, A., Dannaeus, K., Ibanez, C.F., and Lendahl, U., 2003, Functional Notch signaling is required for BMP4-induced inhibition of myogenic differentiation. *Development* 130: 6089-6099.

Dai, P., Shinagawa, T., Nomura, T., Harada, J., Kaul, S.C., Wadhwa, R., Khan, M.M., Akimaru, H., Sasaki, H., Colmenares, C., and Ishii, S., 2002, Ski is involved in transcriptional regulation by the repressor and full-length forms of Gli3. *Genes Dev* 16: 2843-2848.

Drissi, M.H., Li, X., Sheu, T.J., Zuscik, M.J., Schwarz, E.M., Puzas, J.E., Rosier, R.N., and O'Keefe, R.J., 2003, Runx2/Cbfa1 stimulation by retinoic acid is potentiated by BMP2 signaling through interaction with Smad1 on the collagen X promoter in chondrocytes. *J Cell Biochem* 90: 1287-1298.

Dunn, N.R., Winnier, G.E., Hargett, L.K., Schrick, J.J., Fogo, A.B., and Hogan, B.L., 1997, Haploinsufficient phenotypes in Bmp4 heterozygous null mice and modification by mutations in Gli3 and Alx4. *Dev Biol* 188: 235-247.

Edlund, S., Bu, S., Schuster, N., Aspenström, P., Heuchel, R., Heldin, N.-E., ten Dijke, P., Heldin, C.-H., and Landström, M., 2003, Transforming growth factor-β1 (TGF-β)-induced apoptosis of prostate cancer cells involves Smad7-dependent activation of p38 by TGF-β-activated kinase 1 and mitogen-activated protein kinase kinase 3. *Mol Biol Cell* 14: 529-544.

Edlund, S., Lee, S.Y., Grimsby, S., Zhang, S., Aspenström, P., Heldin, C.-H., and Landström, M., 2005, Interaction between Smad7 and β-catenin: importance for transforming growth factor β-induced apoptosis. *Mol Cell Biol* 25: 1475-1488.

Feng, X.H., and Derynck, R., 2005, Specificity and versatility in TGF-β signaling through Smads. *Annu Rev Cell Dev Biol* 21: 659-693.

Fernandez-Pol, J.A., Talkad, V.D., Klos, D.J., and Hamilton, P.D., 1987, Suppression of the EGF-dependent induction of *c-Myc* proto-oncogene expression by transforming growth factor β in a human breast carcinoma cell line. *Biochem Biophys Res Commun* 144: 1197-1205.

Fu, M., Zhang, J., Zhu, X., Myles, D.E., Willson, T.M., Liu, X., and Chen, Y.E., 2001, Peroxisome proliferator-activated receptor gamma inhibits transforming growth factor β-induced connective tissue growth factor expression in human aortic smooth muscle cells by interfering with Smad3. *J Biol Chem* 276: 45888-45894.

Furuhashi, M., Yagi, K., Yamamoto, H., Furukawa, Y., Shimada, S., Nakamura, Y., Kikuchi, A., Miyazono, K., and Kato, M., 2001, Axin facilitates Smad3 activation in the transforming growth factor β signaling pathway. *Mol Cell Biol* 21: 5132-5141.

Ghosh, S., May, M.J., and Kopp, E.B., 1998, NF-kappa B and Rel proteins: evolutionarily conserved mediators of immune responses. *Annu Rev Immunol* 16: 225-260.

Harris, S.L., and Levine, A.J., 2005, The p53 pathway: positive and negative feedback loops. *Oncogene* 24: 2899-2908.

Hayes, S.A., Zarnegar, M., Sharma, M., Yang, F., Peehl, D.M., ten Dijke, P., and Sun, Z., 2001, SMAD3 represses androgen receptor-mediated transcription. *Cancer Res* 61: 2112-2118.

He, T.C., Sparks, A.B., Rago, C., Hermeking, H., Zawel, L., da Costa, L.T., Morin, P.J., Vogelstein, B., and Kinzler, K.W., 1998, Identification of *c-Myc* as a target of the APC pathway. *Science* 281: 1509-1512.

Hu, M.C., Piscione, T.D., and Rosenblum, N.D., 2003, Elevated SMAD1/β-catenin molecular complexes and renal medullary cystic dysplasia in ALK3 transgenic mice. *Development* 130: 2753-2766.

Hu, M.C., and Rosenblum, N.D., 2005, Smad1, β-catenin and Tcf4 associate in a molecular complex with the Myc promoter in dysplastic renal tissue and cooperate to control Myc transcription. *Development* 132: 215-225.

Hussein, S.M., Duff, E.K., and Sirard, C., 2003, Smad4 and β-catenin co-activators functionally interact with lymphoid-enhancing factor to regulate graded expression of Msx2. *J Biol Chem* 278: 48805-48814.

Ingham, P.W., and McMahon, A.P., 2001, Hedgehog signaling in animal development: paradigms and principles. *Genes Dev* 15: 3059-3087.

Ionescu, A.M., Drissi, H., Schwarz, E.M., Kato, M., Puzas, J E., McCance, D.J., Rosier, R.N., Zuscik, M.J., and O'Keefe, R.J., 2004, CREB Cooperates with BMP-stimulated Smad signaling to enhance transcription of the Smad6 promoter. *J Cell Physiol* 198: 428-440.

Itoh, F., Itoh, S., Goumans, M.-J., Valdimarsdottir, G., Iso, T., Dotto, G.P., Hamamori, Y., Kedes, L., Kato, M., and ten Dijke P., 2004, Synergy and antagonism between Notch and BMP receptor signaling pathways in endothelial cells. *EMBO J* 23: 541-551.

Iwamoto, T., Oshima, K., Seng, T., Feng, X., Oo, M.L., Hamaguchi, M., and Matsuda, S., 2002, STAT and SMAD signaling in cancer. *Histol Histopathol* 17: 887-895.

Jenkins, B.J., Grail, D., Nheu, T., Najdovska, M., Wang, B., Waring, P., Inglese, M., McLoughlin, R.M., Jones, S.A., Topley, N., Baumann, H., Judd, L.M., Giraud, A.S., Boussioutas, A., Zhu, H.J., and Ernst, M., 2005, Hyperactivation of Stat3 in gp130 mutant mice promotes gastric hyperproliferation and desensitizes TGF-β signaling. *Nat Med* 11: 845-852.

Ji, M., and Andrisani, O.M., 2005, High-level activation of cyclic AMP signaling attenuates bone morphogenetic protein 2-induced sympathoadrenal lineage development and promotes melanogenesis in neural crest cultures. *Mol Cell Biol* 25: 5134-5145.

Jian, H., Shen, X., Liu, I., Semenov, M., He, X., and Wang, X.F. 2006, Smad3-dependent nuclear translocation of beta-catenin is required for TGF-beta1-induced proliferation of bone marrow-derived adult human mesenchymal stem cells. *Genes Dev* 20: 666-674.

Jono, H., Shuto, T., Xu, H., Kai, H., Lim, D.J., Gum, J.R., Jr., Kim, Y.S., Yamaoka, S., Feng, X.-H., and Li, J.D., 2002, Transforming growth factor-β-Smad signaling pathway cooperates with NF-kappa B to mediate nontypeable Haemophilus influenzae-induced MUC2 mucin transcription. *J Biol Chem* 277: 45547-45557.

Kadesch, T., 2004, Notch signaling: the demise of elegant simplicity. *Curr Opin Genet Dev* 14: 506-512.

Kanamoto, T., Hellman, U., Heldin, C.-H., and Souchelnytskyi, S., 2002, Functional proteomics of transforming growth factor-β1-stimulated Mv1Lu epithelial cells: Rad51 as a target of TGFβ1-dependent regulation of DNA repair. *EMBO J* 21: 1219-1230.

Kang, H.Y., Huang, K.E., Chang, S.Y., Ma, W.L., Lin, W.J., and Chang, C., 2002, Differential modulation of androgen receptor-mediated transactivation by Smad3 and tumor suppressor Smad4. *J Biol Chem* 277: 43749-43756.

Kardassis, D., Pardali, K., and Zannis, V.I., 2000, SMAD proteins transactivate the human ApoCIII promoter by interacting physically and functionally with hepatocyte nuclear factor 4. *J Biol Chem* 275: 41405-41414.

Kim, S., Koga, T., Isobe, M., Kern, B.E., Yokochi, T., Chin, Y.E., Karsenty, G., Taniguchi, T., and Takayanagi, H., 2003, Stat1 functions as a cytoplasmic attenuator of Runx2 in the transcriptional program of osteoblast differentiation. *Genes Dev* 17: 1979-1991.

Kimelman, D., and Griffin, K.J., 2000, Vertebrate mesendoderm induction and patterning. *Curr Opin Genet Dev* 10: 350-356.

Kintscher, U., Lyon, C., Wakino, S., Bruemmer, D., Feng, X., Goetze, S., Graf, K., Moustakas, A., Staels, B., Fleck, E., Hsueh, W.A., and Law, R.E., 2002, PPARα inhibits TGF-β-induced β_5-integrin transcription in vascular smooth muscle cells by interacting with Smad4. *Circ Res* 91: e35-44.

Klein, T., and Arias, A.M., 1999, The vestigial gene product provides a molecular context for the interpretation of signals during the development of the wing in *Drosophila. Development* 126: 913-925.

Labbé, E., Letamendia, A., and Attisano, L., 2000, Association of Smads with lymphoid enhancer binding factor 1/T cell-specific factor mediates cooperative signaling by the transforming growth factor-β and wnt pathways. *Proc Natl Acad Sci U S A* 97: 8358-8363.

Lei, S., Dubeykovskiy, A., Chakladar, A., Wojtukiewicz, L., and Wang, T.C., 2004, The murine gastrin promoter is synergistically activated by transforming growth factor-β/Smad and Wnt signaling pathways. *J Biol Chem* 279: 42492-42502.

Letterio, J.J., and Roberts, A.B., 1998, Regulation of immune responses by TGF-β. *Annu Rev Immunol* 16: 137-161.

Liu, F., Massagué, J., and Ruiz i Altaba, A., 1998, Carboxy-terminally truncated Gli3 proteins associate with Smads. *Nat Genet* 20: 325-326.

Long, J., Wang, G., Matsuura, I., He, D., and Liu, F., 2004, Activation of Smad transcriptional activity by protein inhibitor of activated STAT3 (PIAS3). *Proc Natl Acad Sci U S A* 101: 99-104.

Lopez-Rovira, T., Chalaux, E., Rosa, J.L., Bartrons, R., and Ventura, F., 2000, Interaction and functional cooperation of NF-κB with Smads. Transcriptional regulation of the junB promoter. *J Biol Chem* 275: 28937-28946.

Luo, K., 2004, Ski and SnoN: negative regulators of TGF-β signaling. *Curr Opin Genet Dev* 14: 65-70.

Massagué, J., 2000, How cells read TGF-β signals. Nat Rev *Mol Cell Biol* 1: 169-178.

Matsuda, T., Yamamoto, T., Muraguchi, A., and Saatcioglu, F., 2001, Cross-talk between transforming growth factor-β and estrogen receptor signaling through Smad3. *J Biol Chem* 276: 42908-42914.

Mulder, N.J., Apweiler, R., Attwood, T.K., Bairoch, A., Bateman, A., Binns, D., Bradley, P., Bork, P., Bucher, P., Cerutti, L., Copley, R., Courcelle, E., Das, U., Durbin, R., Fleischmann, W., Gough, J., Haft, D., Harte, N., Hulo, N., Kahn, D., Kanapin, A., Krestyaninova, M., Lonsdale, D., Lopez, R., Letunic, I., Madera, M., Maslen, J., McDowall, J., Mitchell, A., Nikolskaya, A.N., Orchard, S., Pagni, M., Ponting, C.P., Quevillon, E., Selengut, J., Sigrist, C.J., Silventoinen, V., Studholme, D.J., Vaughan, R., and Wu, C.H., 2005, InterPro, progress and status in 2005. *Nucleic Acids Res* 33: D201-205.

Mullor, J.L., Sanchez, P., and Altaba, A.R., 2002, Pathways and consequences: Hedgehog signaling in human disease. *Trends Cell Biol* 12: 562-569.

Nakashima, K., Yanagisawa, M., Arakawa, H., Kimura, N., Hisatsune, T., Kawabata, M., Miyazono, K., and Taga, T., 1999, Synergistic signaling in fetal brain by STAT3-Smad1 complex bridged by p300. *Science* 284: 479-482.

Nawshad, A., LaGamba, D., and Hay, E. D, 2004, Transforming growth factor β (TGFβ) signaling in palatal growth, apoptosis and epithelial mesenchymal transformation (EMT). *Arch Oral Biol* 49: 675-689.

Nishita, M., Hashimoto, M.K., Ogata, S., Laurent, M.N., Ueno, N., Shibuya, H., and Cho, K.W., 2000, Interaction between Wnt and TGF-β signaling pathways during formation of Spemann's organizer. *Nature* 403: 781-785.

Nusse, R., 2005, Wnt signaling in disease and in development. *Cell Res* 15: 28-32.

Paez-Pereda, M., Giacomini, D., Refojo, D., Nagashima, A.C., Hopfner, U., Grubler, Y., Chervin, A., Goldberg, V., Goya, R., Hentges, S.T., Low, M.J., Holsboer, F., Stalla, G.K., and Arzt, E.,

2003, Involvement of bone morphogenetic protein 4 (BMP-4) in pituitary prolactinoma pathogenesis through a Smad/estrogen receptor crosstalk. *Proc Natl Acad Sci U S A* 100: 1034-1039.

Pendaries, V., Verrecchia, F., Michel, S., and Mauviel, A., 2003, Retinoic acid receptors interfere with the TGF-β/Smad signaling pathway in a ligand-specific manner. *Oncogene* 22: 8212-8220.

Periyasamy, S., and Sanchez, E.R., 2002, Antagonism of glucocorticoid receptor transactivity and cell growth inhibition by transforming growth factor-β through AP-1-mediated transcriptional repression. *Int J Biochem Cell Biol* 34: 1571-1585.

Roberts, A.B., and Sporn, M.B., 1992, Mechanistic interrelationships between two superfamilies: the steroid/retinoid receptors and transforming growth factor-β. *Cancer Surv* 14: 205-220.

Ruas, J.L., and Poellinger, L., 2005, Hypoxia-dependent activation of HIF into a transcriptional regulator. *Semin Cell Dev Biol* 16: 514-522.

Sanchez-Elsner, T., Botella, L.M., Velasco, B., Corbi, A., Attisano, L., and Bernabeu, C., 2001, Synergistic cooperation between hypoxia and transforming growth factor-β pathways on human vascular endothelial growth factor gene expression. *J Biol Chem* 276: 38527-38535.

Sanchez-Elsner, T., Ramirez, J.R., Sanz-Rodriguez, F., Varela, E., Bernabeu, C., and Botella, L. M, 2004, A cross-talk between hypoxia and TGF-β orchestrates erythropoietin gene regulation through SP1 and Smads. *J Mol Biol* 336: 9-24.

Sarker, K.P., Wilson, S.M., and Bonni, S., 2005, SnoN is a cell type-specific mediator of transforming growth factor-β responses. *J Biol Chem* 280: 13037-13046.

Sasaki, T., Suzuki, H., Yagi, K., Furuhashi, M., Yao, R., Susa, S., Noda, T., Arai, Y., Miyazono, K., and Kato, M., 2003, Lymphoid enhancer factor 1 makes cells resistant to transforming growth factor β-induced repression of *c-Myc*. *Cancer Res* 63: 801-806.

Schiller, M., Verrecchia, F., and Mauviel, A., 2003, Cyclic adenosine 3′, 5′-monophosphate-elevating agents inhibit transforming growth factor-β-induced SMAD3/4-dependent transcription via a protein kinase A-dependent mechanism. *Oncogene* 22: 8881-8890.

Skillington, J., Choy, L., and Derynck, R., 2002, Bone morphogenetic protein and retinoic acid signaling cooperate to induce osteoblast differentiation of preadipocytes. *J Cell Biol* 159: 135-146.

Song, C.Z., Tian, X., and Gelehrter, T.D., 1999, Glucocorticoid receptor inhibits transforming growth factor-β signaling by directly targeting the transcriptional activation function of Smad3. *Proc Natl Acad Sci U S A* 96: 11776-11781.

Subramaniam, N., Leong, G.M., Cock, T.A., Flanagan, J.L., Fong, C., Eisman, J.A., and Kouzmenko, A.P., 2001, Cross-talk between 1,25-dihydroxyvitamin D3 and transforming growth factor-β signaling requires binding of VDR and Smad3 proteins to their cognate DNA recognition elements. *J Biol Chem* 276: 15741-15746.

Takebayashi-Suzuki, K., Funami, J., Tokumori, D., Saito, A., Watabe, T., Miyazono, K., Kanda, A., and Suzuki, A., 2003, Interplay between the tumor suppressor p53 and TGF β signaling shapes embryonic body axes in *Xenopus*. *Development* 130: 3929-3939.

Takizawa, T., Ochiai, W., Nakashima, K., and Taga, T., 2003, Enhanced gene activation by Notch and BMP signaling cross-talk. *Nucleic Acids Res* 31: 5723-5731.

Taskén, K., and Aandahl, E.M., 2004, Localized effects of cAMP mediated by distinct routes of protein kinase A. *Physiol Rev* 84: 137-167.

Tewari, M., Hu, P.J., Ahn, J.S., Ayivi-Guedehoussou, N., Vidalain, P.O., Li, S., Milstein, S., Armstrong, C.M., Boxem, M., Butler, M.D., Busiguina, S., Rual, J.F., Ibarrola, N., Chaklos, S.T., Bertin, N., Vaglio, P., Edgley, M.L., King, K.V., Albert, P.S., Vandenhaute, J., Pandey, A., Riddle, D.L., Ruvkun, G., and Vidal, M., 2004, Systematic interactome mapping and genetic perturbation analysis of a *C. elegans* TGF-β signaling network. *Mol Cell* 13: 469-482.

Ulloa, L., Doody, J., and Massagué, J., 1999, Inhibition of transforming growth factor-β/SMAD signaling by the interferon-gamma/STAT pathway. *Nature* 397: 710-713.

Valdimarsdottir, G., Goumans, M.-J., Rosendahl, A., Brugman, M., Itoh, S., Lebrin, F., Sideras, P., and ten Dijke, P., 2002, Stimulation of Id1 expression by bone morphogenetic protein is sufficient and necessary for bone morphogenetic protein-induced activation of endothelial cells. *Circulation* 106: 2263-2270.

Veeman, M.T., Axelrod, J.D., and Moon, R.T., 2003, A second canon. Functions and mechanisms of β-catenin-independent Wnt signaling. *Dev Cell* 5: 367-377.

Verrecchia, F., and Mauviel, A., 2004, TGF-β and TNF-α: antagonistic cytokines controlling type I collagen gene expression. *Cell Signal* 16: 873-880.

Verrecchia, F., Pessah, M., Atfi, A., and Mauviel, A., 2000, Tumor necrosis factor-α inhibits transforming growth factor-β/Smad signaling in human dermal fibroblasts via AP-1 activation. *J Biol Chem* 275: 30226-30231.

Waddell, D.S., Liberati, N.T., Guo, X., Frederick, J.P., and Wang, X.F., 2004, Casein kinase Iepsilon plays a functional role in the transforming growth factor-β signaling pathway. *J Biol Chem* 279: 29236-29246.

Wang, H., Song, K., Sponseller, T.L., and Danielpour, D., 2005, Novel function of androgen receptor-associated protein 55/Hic-5 as a negative regulator of Smad3 signaling. *J Biol Chem* 280: 5154-5162.

Warner, D.R., Greene, R.M., and Pisano, M.M., 2005a, Cross-talk between the TGFβ and Wnt signaling pathways in murine embryonic maxillary mesenchymal cells. *FEBS Lett* 579: 3539-3546.

Warner, D.R., Greene, R.M., and Pisano, M.M., 2005b, Interaction between Smad 3 and Dishevelled in murine embryonic craniofacial mesenchymal cells. *Orthod Craniofac Res* 8: 123-130.

Warner, D.R., Pisano, M.M., and Greene, R.M., 2003a, Nuclear convergence of the TGFβ and cAMP signal transduction pathways in murine embryonic palate mesenchymal cells. *Cell Signal* 15: 235-242.

Warner, D.R., Roberts, E.A., Greene, R.M., and Pisano, M.M., 2003b, Identification of novel Smad binding proteins. *Biochem Biophys Res Commun* 312: 1185-1190.

Weng, A.P., and Aster, J.C., 2004, Multiple niches for Notch in cancer: context is everything. *Curr Opin Genet Dev* 14: 48-54.

Wilkinson, D.S., Ogden, S.K., Stratton, S.A., Piechan, J.L., Nguyen, T.T., Smulian, G.A., and Barton, M.C., 2005, A direct intersection between p53 and transforming growth factor β pathways targets chromatin modification and transcription repression of the alpha-fetoprotein gene. *Mol Cell Biol* 25: 1200-1212.

Willert, J., Epping, M., Pollack, J.R., Brown, P.O., and Nusse, R., 2002, A transcriptional response to Wnt protein in human embryonic carcinoma cells. *BMC Dev Biol* 2: 8.

Williams, J.G., 2000, STAT signaling in cell proliferation and in development. *Curr Opin Genet Dev* 10: 503-507.

Wu, L., Wu, Y., Gathings, B., Wan, M., Li, X., Grizzle, W., Liu, Z., Lu, C., Mao, Z., and Cao, X., 2003, Smad4 as a transcription corepressor for estrogen receptor alpha. *J Biol Chem* 278: 15192-15200.

Xanthos, J.B., Kofron, M., Tao, Q., Schaible, K., Wylie, C., and Heasman, J., 2002, The roles of three signaling pathways in the formation and function of the Spemann Organizer. *Development* 129: 4027-4043.

Yamamoto, T., Saatcioglu, F., and Matsuda, T., 2002, Cross-talk between bone morphogenic proteins and estrogen receptor signaling. *Endocrinology* 143: 2635-2642.

Yanagisawa, J., Yanagi, Y., Masuhiro, Y., Suzawa, M., Watanabe, M., Kashiwagi, K., Toriyabe, T., Kawabata, M., Miyazono, K., and Kato, S., 1999, Convergence of transforming growth factor-β and vitamin D signaling pathways on SMAD transcriptional coactivators. *Science* 283: 1317-1321.

Yanagisawa, M., Takizawa, T., Ochiai, W., Uemura, A., Nakashima, K., and Taga, T., 2001, Fate alteration of neuroepithelial cells from neurogenesis to astrocytogenesis by bone morphogenetic proteins. *Neurosci Res* 41: 391-396.

Zavadil, J., Cermak, L., Soto-Nieves, N., and Böttinger, E.P., 2004, Integration of TGF-β/Smad and Jagged1/Notch signaling in epithelial-to-mesenchymal transition. *EMBO J* 23: 1155-1165.

Zhang, H., Akman, H.O., Smith, E.L., Zhao, J., Murphy-Ullrich, J.E., and Batuman, O.A., 2003, Cellular response to hypoxia involves signaling via Smad proteins. *Blood* 101: 2253-2260.

Zhang, L., Duan, C.J., Binkley, C., Li, G., Uhler, M.D., Logsdon, C.D., and Simeone, D.M., 2004, A transforming growth factor β-induced Smad3/Smad4 complex directly activates protein kinase A. *Mol Cell Biol* 24: 2169-2180.

Zhou, S., Lechpammer, S., Greenberger, J.S., and Glowacki, J., 2005, Hypoxia inhibition of adipocytogenesis in human bone marrow stromal cells requires transforming growth factor-β/Smad3 signaling. *J Biol Chem* 280: 22688-22696.

CHAPTER 16

INTERPLAYS BETWEEN THE SMAD AND MAP KINASE SIGNALING PATHWAYS

DELPHINE JAVELAUD AND ALAIN MAUVIEL

INSERM, U697, Université Paris 7-Denis Diderot, Faculté de médecine Hôpital Saint-Louis, 1 avenue Claude Vellefaux, 75010 Paris, France

Abstract: Canonical transcriptional responses to transforming growth factor-β (TGF-β) superfamily members occur via rapid nuclear translocation of cytoplasmic proteins of the Smad family, which are activated by ligand-activated membrane-bound heteromeric serine-threonine kinase receptor complexes. Smad-driven gene expression is strongly dependent upon interactions of with other intracellular signaling mechanisms, initiated or not by TGF-β itself, that may potentiate, synergize, or antagonize, the TGF-β/Smad pathway. Among pathways identified to modulate Smad responses are mitogen-activated protein kinases (MAPKs), a large family of kinases involved in the transmission of diverse extracellular signals from the plasma membrane to the cell nucleus. In this chapter, we describe how MAPKs modify the outcome of Smad activation by TGF-β, and how crosstalk mechanisms between the Smad and MAPK pathways take place and affect cellular behavior and TGF-β target gene expression

Keywords: TGF-β; Smad; MAP kinases; gene regulation; signal transduction; carcinogenesis; embryonic development; tumor suppressor; oncogene; wound repair

1. INTRODUCTION

The Smad pathway may not be viewed as a unique mean for TGF-βs to regulate cellular functions, as other signaling pathways including the mitogen-activated protein kinase (MAPK), the NF-κB or PI3 kinase/Akt pathways can either be induced by TGF-β, or can modulate the outcome of TGF-β-induced Smad signaling (Javelaud et al., 2005; Derynck and Zhang, 2003; Lutz and Knaus, 2002; Massagué and Chen, 2000). Indeed, broad evidence exists for a tight integration of Smad signaling within a complex network of crosstalks with other signaling pathways that largely contribute to modify the initial Smad signal and allow the pleiotropic activities of TGF-β. Also, there may be instances whereby Smad signaling may even be dispensable for some TGF-β responses, as exemplified by Smad-independent

P. ten Dijke and C.-H. Heldin (eds.), Smad Signal Transduction, 317–334.

activation of the cyclin kinase inhibitors p15 and p21 in HaCaT keratinocytes (Hu et al., 1999), or transcriptional activation of the fibronectin promoter via MAPK-dependent mechanisms (Hocevar et al., 1999).

Interestingly, it appears now clearly that Smad proteins are not only the primary substrates for the TGF-β receptor kinases, but may also be phosphorylated by MAPKs, in response to either TGF-β itself or to various cytokines. Such R-Smad phosphorylation by MAPKs may serve to regulate either Smad transcriptional activity or capacity to translocate into the cell nucleus (Derynck and Zhang, 2003; Lutz and Knaus, 2002; Massagué and Chen, 2000). Smad proteins are also capable of physically interacting with transcription factors, themselves substrates of MAPKs, adding more complexity to the intricate relationship between MAPKs and the Smad pathway. This chapter will summarize some of the recent literature regarding MAPK and Smad interactions, and their involvement in physiopathology.

2. MAP KINASES

MAP kinases are a large group of serine-threonine kinases that allow numerous extracellular signals reaching the cell surface to rapidly activate nuclear transcription factors (Whitmarsh and Davis, 1998). Their activation could be the result of various extracellular stimuli, including cytokines, ultraviolet irradiations, cell-cell or cell-matrix contacts, to cite a few. They mainly consist in three subfamilies: the extracellular signal-regulated kinases (Erk1 and Erk2), the stress activated protein (SAP) kinases known as c-Jun N-terminal Kinase (JNK1, JNK2 and JNK3), and the p38 MAPKs (α, β, γ and δ) (Chang and Karin, 2001). Erk5, described as a mediator of Src activation (Zhou et al., 1995), is another member of the MAP kinase superfamily but thus far, unlike the first three groups of MAPKs, it has not been shown to be activated by TGF-β or to interfere with Smad signaling.

Activation of MAP kinases is the result of the induction of sequential activation by phosphorylation of different upstream kinases (Fig. 1): a MAP kinase kinase kinase (MAPKKK) that phosphorylates and activates a MAP kinase kinase (MAPKK), which in turn phosphorylates the TXY motif in the catalytic domain of the MAP kinase (Chang and Karin, 2001). Of note, MAPKKs are dual specificity enzyme that can phosphorylate serine/threonine and tyrosine residues of their MAP kinase substrates.

Erks are phosphorylated by MEK1 and MEK2, themselves substrates of the Raf-1 MAPKKK, the latter being activated by transmembrane-bound small G-protein Ras, for example, following induction by mitogenic stimuli such as epidermal growth factor (EGF) upon binding and activation of their receptors. JNK family members are the substrates of MKK4 (also known as SEK1) and MKK7, themselves targets of several MAPKKKs, including, but not restricted to, apoptosis signal-regulating kinase-1 (ASK-1), mixed lineage kinase (MLK) and TGF-β-activated kinase-1 (TAK1) (Ip and Davis, 1998; Davis, 2000). These MAPKKK can also activate MKK3 and MKK6, which in turn phosphorylate p38 MAPKs. It has been shown that MKK4, in some cell types, can also activate p38 (Chang and Karin, 2001).

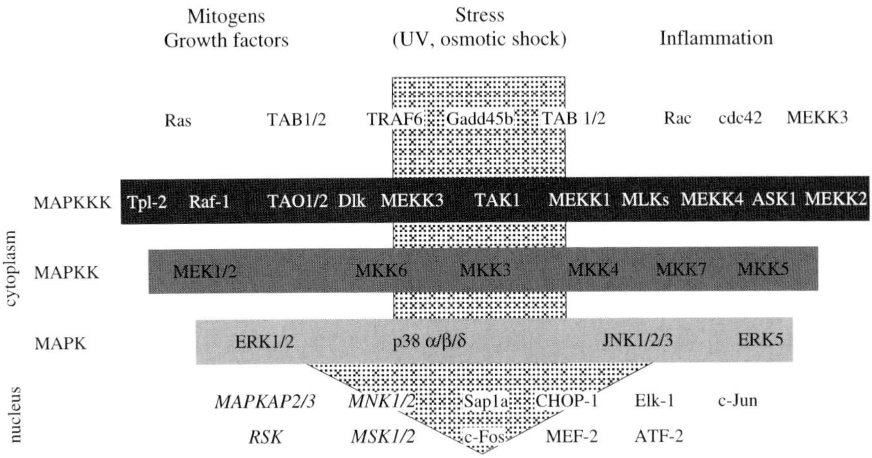

Figure 1. MAP kinases network. Transduction of the signal from the plasma membrane to the nucleus occurs through the sequential activation of MAPKKK, MAPKK, MAPK and their direct nuclear targets (*kinases* and transcription factors). For further details, see the text

Also contributing to the p38 and JNK activation upstream of MAPKKKs are low molecular weight GTP-binding proteins in the Rho family such as Rac1 and Cdc42. Rac1 can bind to MEKK1 or MLK1 while Cdc42 can only bind to MLK1 and both result in activation of p38 via MAPKKs (Lopez-Ilasaca, 1998). p21-activated kinase (PAKs), targets of Rho family proteins can participate in the activation of JNK and p38 MAPKs (Bagrodia et al., 1995; Zhang et al., 1995).

Erk-mediated pathways are mostly involved in proliferation and differentiation and generally considered as anti-apoptotic. On the other hand, JNK and p38 signaling pathways are activated by various stress stimuli, many of which induce apoptosis, but in some cellular context they have been implicated in proliferation, migration and differentiation as well (Eferl and Wagner, 2003; Javelaud et al., 2003).

MAP kinase activation leads to the downstream phosphorylation/activation of nuclear kinases or, most commonly, transcription factors (Treisman, 1996). Among them is activating protein-1 (AP-1), a family of pleiotropic transcription factors comprised of homo- and heterodimers of Fos, Jun and ATF family members, involved in the control of cell proliferation, death and survival, as well as tumorigenesis (Eferl and Wagner, 2003; Shaulian and Karin, 2002).

Erk1/2 phosphorylate TCF/Elk1, SAP1a and SAP1b, and activate CREB and c-Fos through the MSK-1 (mitogen- and stress-activated protein Kinase) and RSK (ribosomal S6 kinase family) proteins, respectively. Thus, MAP kinases are able to modulate gene expression by phosphorylating transcription factors directly and by activating other protein kinases, which then phosphorylate proteins involved in gene expression regulation (Pearson et al., 2001). For instance, Erk and p38 MAPKs can activate MNK1 and MNK2 (MAP kinase-interacting protein), which in turn phosphorylate the eukaryotic translation initiation factor 4F (eIF4E); activated RSK proteins by Erk regulated gene expression via association and phosphorylation of transcription factor/ coactivator, such as c-Fos, estrogen receptor, NF-κB, CREB and CBP (Frodin and Gammeltoft, 1999). Transcription factors activated by p38 MAPK are CHOP, ATF-2, CREB and MEF2C (Zarubin and Han, 2005). JNK is the only MAPK able to phosphorylate c-Jun, the main component of AP-1 complexes, and also has ATF-2 and Elk-1 as substrates (Chang and Karin, 2001).

3. ACTIVATION OF MAPKS BY TGF-β

TGF-β is able to activate all Erk, JNK and p38 MAP kinases (Fig. 2; reviewed in Derynck and Zhang, 2003; Wakefield and Roberts, 2002). Not only is MAPK activation by TGF-β cell-type specific, but the activation of a given MAPK combination by TGF-β is also cell-type dependent. For example, in mink lung epithelial cells, TGF-β induced activation of JNK mediates Smad3 phosphorylation, which is required for the transcriptional activation of Smad3-dependent responses (Engel et al., 1999). On the other hand, in rat articular chondrocytes, TGF-β induces a rapid activation of Erk1/2, but not that of either p38 or JNK MAPKs (Yonekura et al., 1999). In C2C12 cells, TGF-β induces a specific activation of MKK6/p38 pathway without any effect on JNK pathway (Hanafusa et al., 1999). Activation of MAPK by TGF-β seems to be a conserved phenomenon since it has been reported that the TGF-β-like secreted ligand, Decapentaplegic, activates the p38 homologue in *Drosophila* wing morphogenesis (Adachi-Yamada et al., 1999).

Activation of MAPK by TGF-β has been described to occur either with rapid kinetics similar to those observed downstream of cytokines receptors, but also with slow kinetics, suggesting a Smad-dependent transcription responses. The rapid activation (5-15 min) of MAPK phosphorylation strongly suggest on the one hand independence of Smad-driven transcription, and on the other hand a direct activation of MAPKKKs from the TGF-β receptors. Initial evidence for Smad-independent activation of MAPK by TGF-β was obtained in Smad-4 deficient cells, or in cells overexpressing dominant-negative Smads, where activation of the JNK/MAPK pathway was still possible in response to TGF-β despite the deficient Smad cascade (Itoh et al., 2003). Also, it has been shown that mutated TGF-β type I receptors that lack the ability to activate Smads but retain kinase activity, still activate p38 MAPK signaling in response to TGF-β. The ability of the TGF-β type I receptor to activate the p38 most likely requires the kinase activity of the receptor (Yu et al., 2002). The mechanisms of Erk, JNK and p38 MAPK activation by TGF-β and their biological consequences are not fully characterized. Erk activation by TGF-β in epithelial cells may implicate Ras signaling (Yue and Mulder, 2000a), while JNK and p38 signaling could be activated by multiple MAPKKKs in response to the TGF-β.

The first known MAPKKK to be activated by TGF-β family members is TGF-β-activated kinase-1 (TAK1), originally identified as a MAPKKK positively regulating JNK and p38 kinase pathways and activated by TAB1 (for TGF-β-activated kinase binding protein-1) downstream of TGF-β/ BMP receptors (Yamaguchi et al.,

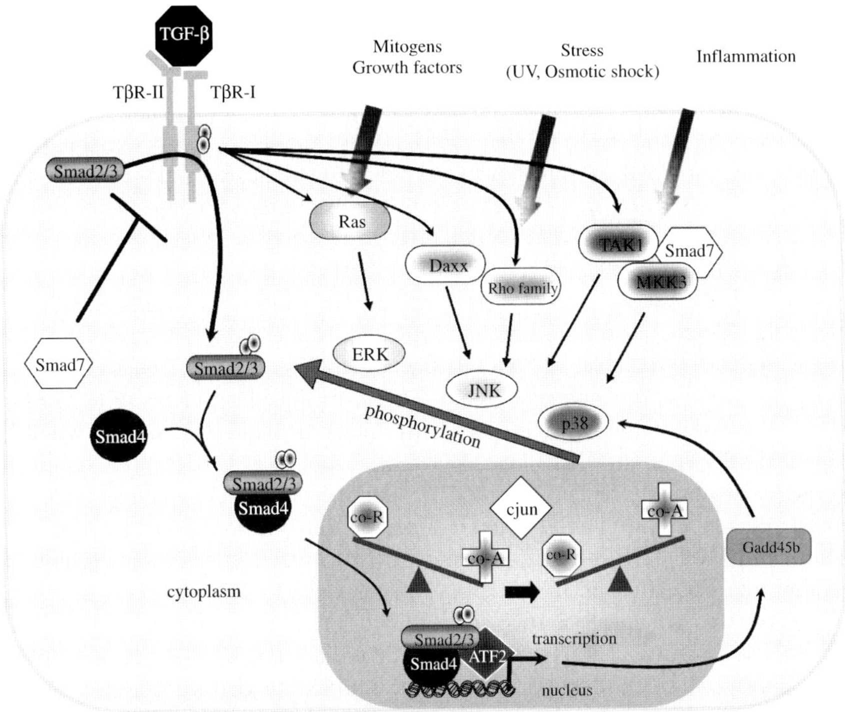

Figure 2. Activation of Smad and MAP kinases by TGF-β and their interactions. Upon TGF-β ligation to TβRII, the latter phosphorylates TβRI, which in turn phosphorylates Smad2/3. Activated R-Smads bind Smad4 and translocate to the nucleus to act as transcription factors, controlled by a balance between transcriptional co-activators (co-A) or co-repressors (co-R). The three principal MAP kinases can be activated by TGF-β by either Smad-dependent or Smad-independent ways (refer to the text for further details). Activated MAP kinases by TGF-β or others stimuli such as growth factors or pro-inflammatory cytokines can regulate the Smad activation by a direct phosphorylation or through their downstream effectors. For example, activated Jun and ATF-2 modulate the Smad transcriptional activity through direct physical contacts or by altering the balance between transcriptional co-activators or co-repressors

1995). TAB-1 is able to associate with the inhibitory Smads, Smad6 and Smad7, a phenomenon that may lead to the inhibition of TAK-1-dependent p38 activation (Yanagisawa et al., 2001). Therefore Smad6 has been reported to inhibit BMP/TAK1-induced phosphorylation of p38 in MH60 cells (Kimura et al., 2000). Alternatively, it has also been suggested that Smad7 could act as a scaffolding protein to provide structural support for MKK3/p38 activation by TAK1 (Edlund et al., 2003). Moreover, overexpression of Smad7 in MvLu1 and MDCK cells has been shown to cause activation of JNK pathway (Mazars et al., 2001). Of note, because TAK1 also activates NF-κB, TGF-β/BMP receptors, due to their ability to activate TAK1, may under certain circumstances, also induce NF-κB signaling (Arsura et al., 2003).

XIAP (X-linked inhibitor of apoptosis), was then identified as a bridging molecule between TGF-β/BMP receptors and TAK1/TAB1, serving as a cofactor for TAK1-dependent signaling (Yamaguchi et al., 1999). Uncertainty remains, however, as (a) some of the cooperative activities of XIAP and TGF-β are not mediated by TAK1 dependent signals, and (b) XIAP may also be dispensable for TGF-β signaling. For example, activation of TGF-β responsive genes by XIAP has been shown to depend on Smad4, while the anti-apoptotic effects of XIAP are Smad4-independent (Birkey et al., 2001). Furthermore, XIAP-deficient mice respond to TGF-β (Harlin et al., 2001). This could be explained if a direct link between TAK1 (in association with the cofactor TAB1) and the receptors is established by another upstream kinase, such as the hematopoietic progenitor kinase-1 (HPK-1), which allows JNK activation by bridging the TGF-β receptors to TAK1 independently from XIAP (Zhou et al., 1999).

MEKK1 may also function upstream of TGF-β-mediated activation of MAPKKs (Brown et al., 1999); thus, MEKK1 and TAK1 could activate JNK through MKK4, and p38 MAPK through MKK3 and MKK6, in response to TGF-β.

A yeast two-hybrid screen designed to identify proteins interacting with the cytoplasmic tail of the TGF-β type II receptor, has allowed isolating Daxx, a protein involved in TGF-β-induced apoptosis in B-cell lymphomas, and capable of mediating the activation of JNK pathway in response to TGF-β (Perlman et al., 2001). Daxx was previously known as an adaptor protein for the Fas receptor, that mediates Fas activation of JNK and programmed cell death (Yang et al., 1997). Daxx thus appears to also function as an adaptor protein linking the TGF-β receptor complex to the apoptotic machinery and the JNK pathway.

Depending on the cell line, TGF-β can rapidly activate the Rho-like GTPases, RhoA, RhoB, Rac and Cdc42 (Mucsi et al., 1996; Atfi et al., 1997; Edlund et al., 2002; Kamaraju and Roberts, 2005). Ras activation in response to TGF-β may also lead to the activation of Rho-like GTPases. Rac and Cdc42 regulate JNK and p38 MAPK pathway activation, presumably by directly interacting with MAPKKKs upstream of JNK and p38 MAPK.

Recently, is has been reported that TGF-β can rapidly activate PAK2, in a Smad-independent manner in fibroblasts but not in epithelial cells (Wilkes et al., 2003). In the past, it has been reported that PAK1 and PAK2 can modulate JNK and p38 MAPK activation (Bagrodia et al., 1995; Zhang et al., 1995), suggesting a supplementary mechanism involved in the TGF-β-MAPK activation.

Different mechanisms have been described for the rapid activation of p38 by TGF-β in numerous cell types. On the other hand, delayed and sustained p38 activation observed in pancreatic carcinoma cells, hepatocytes or osteoblasts requires Smad signaling. Smad activation results in the induction of the expression of Gadd45b, an upstream activator of MKK4, which thus promotes the delayed activation of p38 (Yue and Mulder, 2000b). Likewise, JNK activation may occur rapidly or in a delayed manner (Engel et al., 1999), and among the potential candidates that could mediate this delayed activation of JNK in response to TGF-β could be the inhibitory Smad, Smad7, a target of the TGF-β/BMP pathway whose overexpression has been shown to induce persistent JNK activation in HepG2 cells (Mazars et al., 2001).

4. MODULATION OF SMADS ACTIVITY BY MAP KINASES

MAP kinases can modify signaling by phosphorylation-dependent modification of ligand- dependent R-Smad nuclear translocation. Thus, Ras signaling has been proposed to inhibit TGF-β signaling via the Erk pathway. Specifically, Erk has been shown to phosphorylate the linker region of Smad1, Smad2 and Smad3, which results in blocking the nuclear translocation of these TGF-β/BMP-activated Smads (Kretzschmar et al., 1997, 1999). It was proposed that such mechanism might explain why some cells with hyperactive Ras pathway do not respond anymore to TGF-β. However, others reports have demonstrated efficient nuclear translocation of Smad in Ras-transformed cells or in cells with activated MAP kinase signaling. Thus, Smad2 phosphorylation by Erk1 has been shown to increased Smad2 half-life, complex formation with Smad4 and nuclear translocation, all contribute to an enhancement of the transcriptional activity of Smad2 (de Caestecker et al., 1998). Furthermore, it has been reported that Erk phosphorylates Smad4, leading to enhanced TGF-β-induced nuclear accumulation and, as a consequence, increase of transcription activity of R-Smad/Smad4 complexes (Roelen et al., 2003).

Thus, while such cooperativity between the Ras/MAP kinase pathway and TGF-β signaling has been observed during tumor cell differentiation and behaviour, Ras may also interfere with cell cycle arrest and apoptotic responses to TGF-β, indicative of the complexity of Smad-MAPK outcome which largely depend on the cellular context.

Erks are not the only MAPKs capable of phosphorylating Smads, since it has been reported a JNK-dependent phosphorylation of Smad2 and Smad3 in response to HGF, PDGF and TGF-β itself, with no loss of R-Smad capacity to translocate into the nucleus or to transactivate target genes. On the contrary, these phosphorylations positively regulate the R-Smad/Smad4 complex formation and the Smad transcriptional response (Engel et al., 1999; Mori et al., 2004; Yoshida et al., 2005). Recently, it was shown in the human breast cancer cell line MCF10CA1h that the Rho/ROCK and p38 pathways cooperate to allow TGF-β-induced growth arrest (Kamaraju and Roberts, 2005). This effect is achieved by phosphorylation of the R-Smad linker-region by both kinases, resulting in an increased transactivation potential of R-Smads, ultimately leading to cell cycle withdrawal.

Different sites of phosphorylation on Smad by MAPKs have been identified, and most of them are located in the linker region (Fig. 3). Four Erk kinase sites in the linker region of Smad1 have been characterized (Ser187, Ser195, Ser206, Ser214). Smad2 has only one Erk site (Thr220) in this region and Smad3 has two (Thr178, Ser212) (Kretzschmar et al., 1999).

However, the Smad2 linker region has three Ser-Pro sequence (Ser245, Ser250, Ser255) and Smad3 has two (Ser203, Ser207). These sequences can serve as phosphorylation sites for proline-directed protein kinase such as MAPK. For Smad4, only one Erk phosphorylation consensus site in Thr276 has been isolated (Roelen et al., 2003). Moreover Funaba et al., have observed a phosphorylation of Smad2 on Thr8 by Erks (Funaba et al., 2002). Of note, interactions with other proteins can modulate R-Smad activity, for example Calmodulin is able to bind Smad2 on

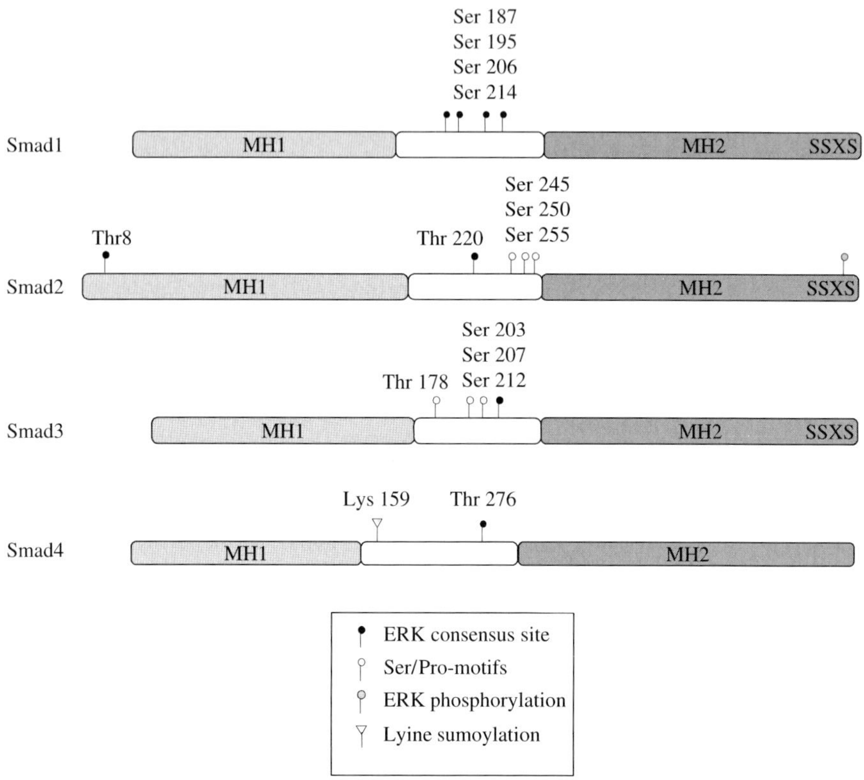

Figure 3. Smad post-translational modifications induced by MAPK. MAPKs can modulate Smad activity with several posttranslational modification such as phosphorylation or sumoylation. Sites of direct MAPK phosphorylation and SUMO-1 conjugation have been identified on R-Smads and Co-Smad. For further details, see the text

two different sites in the MH1 domain, leading to the decreased of Smad2 protein level or an inactivating phosphorylation of Smad2 by Ca/calmodulin dependent kinase II (Funaba et al., 2002; Wicks et al., 2000). The latter introduces a new level of crosstalk, since it has been reported that calmodulin binding to Smad inhibits subsequent Erk2 dependent phosphorylation of Smad and vice versa (Funaba et al., 2002; Scherer and Graff, 2000). Furthermore, it has been shown that Smad2 may be phosphorylated by Erk in response to HGF or EGF in its C-terminal SSXS motif, specific sequence usually targeted by TβRI (de Caestecker et al., 1998).

MAPKs may also indirectly affect Smad signaling by controlling Smad7 expression. Initial observations indicated that TGF-β-induced Smad7 expression depends on cooperative interactions between AP-1, Sp1 and Smad proteins (Brodin et al., 2000). More recently, both JNK and TAK1/p38 pathways were shown to regulate Smad7 expression, in a cell-type-specific manner (Dowdy et al., 2003; Uchida et al.,

2001). Also, ultraviolet irradiation, a well-known activator of MAPKs, induces Smad7 gene transcription in a c-Jun-dependent way (Quan et al., 2005).

p38 MAPK activation by TGF-β is also involved in the induction of PIAS protein family expression and protein stabilization. These proteins function as E3 ligase for SUMO-1 conjugation and allow the sumoylation of Smad4 on Lys159 in the linker region (Ohshima and Shimotohno, 2003). This supplementary post-translational modification facilitates Smad-dependent transcription activation.

Together, these results indicate that activation of MAPK pathways may have positive or negative regulatory effects on R-Smads, depending on the nature of MAPK activation.

5. NUCLEAR INTERACTIONS BETWEEN SMAD COMPLEXES AND MAPK-ACTIVATED TRANSCRIPTION FACTORS/KINASES

Activated R-Smad proteins have been shown to participate in a number of heterogeneous transcription complexes bound to DNA, as they exhibit a broad capacity to interact with numerous transcription factors, such as Sp1. Downstream components of MAP kinase signaling pathways, and especially transcription factors of the AP-1 family, may also interact with R-Smad/Smad4 complex in the nucleus, providing an additional level of crosstalk between these pathways (Zhang et al., 1998). c-Jun and JunB, both downstream substrates of JNK, are components of the AP-1 complex that are transcriptionally regulated by the TGF-β/Smad pathway (Mauviel et al., 1996; Jonk et al., 1998; Wong et al., 1999), and contribute to an autocrine regulatory loop of Smad activity (Verrecchia et al., 2001a). Interestingly, transcriptional cooperation depends on the structure of the target promoters, as Smad and AP-1 cooperate to activate AP-1-dependent promoters, while they tend to antagonize each other with regard to Smad-specific transcription dependent on Smad-binding site (Verrecchia et al., 2001b). While Fos/Jun-Smad3/4 physical interactions may participate in a hetero-tetrameric complex bound to AP-1 elements or their adjacent nucleotides on DNA (Zhang et al., 1998), data from our own laboratory indicate that both c-Jun and JunB are capable of interrupting Smad3-mediated transcription, as Jun/Smad3 complexes may form off-DNA, preventing Smad3 binding to cognate DNA sequences (Fig. 4) (Verrecchia et al., 2000). We also showed that JNK activity promotes such off-DNA association of Jun proteins with Smad3 (Verrecchia et al., 2003). Accordingly, in *jnk1*$^{-/-}$, *jnk2*$^{-/-}$ (*jnk*$^{-/-}$) fibroblasts, TNF-α had no effect on TGF-β driven, Smad-dependent, gene transactivation unless *jnk1* was introduced exogenously. Aside from preventing Smad3 from binding to its cognate DNA-binding sites, the JNK pathway may also regulate Smad2/3-dependent transcription via alternate mechanisms, for example by facilitating c-Jun association with Itranscriptional co-partners, such as the co-activators p300/CBP, or the co-repressors c-Ski and TGIF: c-Jun association with p300/CBP has been shown to interrupt Smad3-driven transcription by squelching of p300/CBP away from Smad complexes; also c-Jun may physically associate with c-Ski and

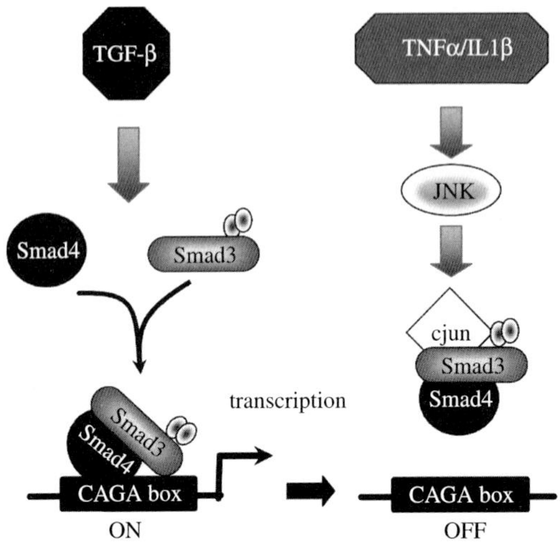

Figure 4. Inhibition of Smad-driven transcription by JNK/c-Jun pathway. Activated c-Jun by JNK in response to the proinflammatory cytokines such as TNF-α is able to interrupt Smad3-mediated transcription: Jun/Smad3 complexes prevent Smad3 binding to cognate DNA sequence

TGIF and allow the latter to exert their repressor activity by interfering with the assembly of Smad2/p300 complexes (Pessah et al., 2001, 2002). Recently, it has been shown that MSK1, a downstream target of p38, regulates the association of Smad3 and the co-activator p300 in response to TGF-β signaling (Abecassis et al., 2004).

It is interesting to note that the JNK/Jun axis is instrumental to the HTLV-1 Tax oncoprotein in repressing TGF-β signaling, a mechanism that may contribute to leukemogenesis (Arnulf et al., 2002). Also, the JNK pathway may contribute to regulate autocrine TGF-β1 expression, as *jnk*-deficient fibroblasts constitutively express TGF-β1, expression that can be repressed by complementation of the cells by JNK (Ventura et al., 2004).

ATF-2, a downstream substrate of both JNK and p38 MAPKs participates in certain AP-1 complexes. Its expression is induced by TGF-β and transcriptionally regulated by both Smad- and TAK1-dependent mechanisms (Hanafusa et al., 1999; Sano et al., 1999). Furthermore, it is also possible that ATF-2 participates in transcription complexes in association with Smad proteins (Sano et al., 1999).

Although these are just a few examples of the intricacy of transcriptional control by AP-1 and Smad complexes, they suggest a very complex level of integration of the signaling pathways resulting in activation of AP-1 by MAPKs with Smad-driven signals originating from TGF-β receptors, leading to either amplification or negative feedback loops controlling TGF-β effects. Naturally, the outcome of these interactions is further diversified by the presence a nature of distinct regulatory sequences within target gene regulatory sequences.

6. INTEGRATION OF SMAD AND MAPK PATHWAYS: ROLES IN PHYSIOPATHOLOGY

BMP and other TGF-β family members are critical regulators of many processes in vertebrate development, including the specification of ectodermic fate, cell proliferation in the limb bud, and patterning within the somite, among many others. Crosstalk between intracellular signaling pathways could be a key mechanism for the modulation of specific developmental responses. In *Xenopus*, MAPK-mediated phosphorylation blocks the ability of Smad2 to mediate ectoderm differentiation into mesoderm by Activin-like factors during early embryonic development. This inhibition correlates with phosphorylation of the linker and cytoplasmic retention of Smad2 (Grimm and Gurdon, 2002). Similarly, neural induction in *Xenopus* embryos also involves crosstalk between MAPK and TGF-β signaling (Pera et al., 2003), and in particular the inhibition of Smad1 in its linker region by Erk phosphorylation (Kuroda et al., 2005). Moreover, Aubin *et al.* have generated Smad1 mutant mice carrying mutations that prevent phosphorylation of either the C-terminal motifs required for BMP downstream transcriptional activation (Smad1^{C} mutation) or of the MAPK motifs in the linker region (Smad1^{L} mutation) (Aubin et al., 2004). Smad1$^{C/C}$ mutants recapitulate many Smad1$^{-/-}$ phenotypes (death in utero and defects in allantois formation and in primordial germ cells specification), while Smad1$^{L/L}$ mutants survive but exhibit defects in gastric epithelial homeostasis that correlated with changes in cell contacts, actin cytoskeleton remodelling, and nuclear β-catenin accumulation. This study suggests that MAPK-dependent Smad1 phosphorylation may not only serve to inhibit BMP signaling but may serve other important cellular functions as well.

TGF-β is a crucial regulator of ECM deposition, as it control both expression of components of ECM network, such as fibrillar collagens and fibronectin, and the expression of protease inhibitors, including PAI-1 or TIMPs. These combined anabolic and anti-catabolic effects of TGF-β make it a key growth factor in the development of tissue fibrosis (Verrecchia et al., 2002). The relevance of the crosstalk between MAPK and TGF-β pathways in the context of tissues fibrosis is still not fully understood, but *in vitro* studies suggest that TGF-β-activated p38 phosphorylation may be implicated in COL1A1 gene expression in dermal fibroblasts (Sato et al., 2002). Conversely, blocade of JNK activity in fibroblasts not only alters their migratory potential in response to TGF-β (Javelaud et al., 2003), but also prevents TNF-α- and 5-fluoro-uracyl-driven inhibition of TGF-β-induced transactivation of the fibrillar collagen genes COL1A1, COL1A2 and COL3A1 and other Smad-dependent gene targets (Verrecchia et al., 2000; Verrecchia et al., 2002; Verrecchia et al., 2003; Wendling et al., 2003). Moreover, after liver injury, TGF-β and PDGF regulate the activation of hepatic stellate cells and tissues remodelling via JNK-phosphorylation linker region of Smad2/3 (Yoshida et al., 2005).

The MAPK pathway controls the growth and survival of a broad spectrum of human tumors. Activating mutation in Ras and Raf result in activation of Erk pathway and are present in a large percentage of solid tumors. Because the TGF-β plays a dual role during tumorigenesis, there is a particularly complicated and intimate interrelationship between TGF-β and Ras/MAPK pathways in tumor development.

Responses that are directly proportional to the level of Smad activity in the nucleus may be attenuated by the opposing effects of Ras signaling, as is the case with the antiproliferative response to TGF-β in epithelial cells (Oft et al., 1996; Calonge and Massagué, 1999). Ras-Smad antagonism may for example occur at the level of Smad nuclear accumulation (Kretzschmar et al., 1999), but other mechanisms such as opposite regulation of cyclin-dependent kinases (Cdks) during G1 phase of the cell cycle may also contribute to attenuate TGF-β tumor suppressive activities (Hannon and Beach, 1994; Reynisdottir et al., 1995).

Nevertheless, aberrant activation of MAPK pathways may play an important role in diverting the TGF-β response towards a pro-oncogenic outcome, and TGF-β and activated Ras may cooperate to promote invasive, metastatic disease. For example, in the presence of oncogenic Ha-Ras or Ki-Ras, the growth-inhibitory response of human prostate and colon cancer cells to TGF-β is converted to a Smad-independent mitogenic response. In kidney epithelial cells, activation of Raf confers protection against TGF-β-induced apoptosis while enhancing its pro-invasive effects (Lehmann et al., 2000). Induction of epithelial-to-mesenchymal transition (EMT), which marks the acquisition of an aggressive phenotype in epithelial cancers, has been shown to require cooperation between Ras/MAPK and TGF-β/Smad cascades (Derynck et al., 2001). Induction of EMT in breast tumor cells is dependent on the presence of both activated Ras and a functional TGF-β autocrine loop, that is enhanced by Ras (Lehmann et al., 2000; Xie et al., 2004). Gene array data obtained from human keratinocytes induced by TGF-β to undergo EMT has provided the first insights into Erk-dependent gene targets with roles in cell-matrix interaction and cell motility (Xie et al., 2003). Also, TβRI that selectively disable Smad binding and activation but not signaling through the MAPK pathways, it has been shown that both Smad and MAPK signaling are required for EMT (Itoh et al., 2003; Yu et al., 2002).

In numerous tumors, acquisition of MMP activity is associated with increased migration and invasiveness of cancer cells. Studies have shown an involvement of p38 MAPK activity in TGF-β induced several MMP (Matrix Metalloproteinase) biosynthesis in stromal fibroblasts, breast epithelial cells or in transformed keratinocytes (Ravanti et al., 1999; Johansson et al., 2000; Kim et al., 2004).

Recent studies suggest that TGF-β plays a specific role in directing metastatic cells to particular organ sites such as bone, which is a common site of metastatic foci of breast and prostate cancer, a phenomenon that may require cooperation of Smad and MAPKs. For example, TGF-β/Smad and p38 MAPK signaling pathways cooperate to induce the production of PTHrP (parathyroid hormone-related protein) by breast tumor cells, PTHrP, in turn, induces RANK ligand expression by osteoblasts, leading to osteoclast activation and subsequent osteolytic bone metastasis (Yin et al., 1999; Kakonen et al., 2002). MAP kinase activation can also contribute to the TGF-β metastasis promotion by stimulating migration of tumor cells (Dumont et al., 2003). Furthermore, it has been shown that expression of a mutant TGF-β type I receptor unable to bind R-Smad but maintaining kinase activity, enhance tumorigenesis but suppresses metastasis of MCF10A-derived cell lines (Tian et al., 2004), clearly emphasizing the bifunctional role of TGF-β in carcinogenesis.

TGF-β-regulated apoptosis is cell type and context dependent, indeed TGF-β provides signals for both cell survival and death (Siegel and Massagué, 2003; Sanchez-Capelo, 2005). Activation of p38 and JNK MAP kinases participates in the TGF-β induced apoptosis in numerous cell types, such as in prostate cancer cells, the murine myeloid cell line M1 and the human hepatoma cell line Hep3B (Edlund et al., 2003). Delayed TGF-β-induced p38 activation, rather than the rapid Smad-independent p38 stimulation, participates in the induction of apoptosis by TGF-β in AML12 murine hepatocytes (Yoo et al., 2003). Nevertheless, TGF-β can also rescue several cell type from serum withdrawal-induced apoptosis, and activation of c-Jun contribute to this rescue (Sanchez-Capelo, 2005).

7. CONCLUSIONS AND PERSPECTIVES

TGF-β superfamily members exert a wide range of functions including control of proliferation, migration, terminal differentiation and cell death, processes that are also under the tight influence of sources of MAPK signaling. Understanding of the integration of different crosstalks between MAP kinases and TGF-β/Smad pathways by the cell is of prime importance for the complete comprehension of the pleiotropic TGF-β-induced responses. Thus, this advances are the more important as TGF-β has a dual role in tumorigenesis (tumor suppressor vs. tumor promoting activities). The fine line drawn between mechanisms involving TGF-β signaling that are either deleterious or beneficial in the context of tumor progression and the complexity of the interactions with other signaling cascades make it extremely difficult to identify the proper context in which inhibition of TGF-β signaling will be really advantageous to the patient, that is, restore TGF-β tumor suppressive functions.

REFERENCES

Abecassis, L., Rogier, E., Vazquez, A., Atfi, A., and Bourgeade, M.F., 2004, Evidence for a role of MSK1 in transforming growth factor-β-mediated responses through p38α and Smad signaling pathways. *J Biol Chem* 279: 30474-30479.

Adachi-Yamada, T., Nakamura, M., Irie, K., Tomoyasu, Y., Sano, Y., Mori, E., Goto, S., Ueno, N., Nishida, Y., and Matsumoto, K. 1999, p38 mitogen-activated protein kinase can be involved in transforming growth factor β superfamily signal transduction in *Drosophila* wing morphogenesis. *Mol Cell Biol* 19: 2322-2329.

Arnulf, B., Villemain, A., Nicot, C., Mordelet, E., Charneau, P., Kersual, J., Zermati, Y., Mauviel, A., Bazarbachi, A., and Hermine, O., 2002, Human T-cell lymphotropic virus oncoprotein Tax represses TGF-β1 signaling in human T cells via c-Jun activation: a potential mechanism of HTLV-I leukemogenesis. *Blood* 100: 4129-4138.

Arsura, M., Panta, G.R., Bilyeu, J.D., Cavin, L.G., Sovak, M.A., Oliver, A.A., Factor, V., Heuchel, R., Mercurio, F., Thorgeirsson, S.S., and Sonenshein, G.E., 2003, Transient activation of NF-kappaB through a TAK1/IKK kinase pathway by TGF-β1 inhibits AP-1/SMAD signaling and apoptosis: implications in liver tumor formation. *Oncogene* 22: 412-425.

Atfi, A., Djelloul, S., Chastre, E., Davis, R., and Gespach, C., 1997, Evidence for a role of Rho-like GTPases and stress-activated protein kinase/c-Jun N-terminal kinase (SAPK/JNK) in transforming growth factor β-mediated signaling. *J Biol Chem* 272: 1429-1432.

Aubin, J., Davy, A., and Soriano, P., 2004, In vivo convergence of BMP and MAPK signaling pathways: impact of differential Smad1 phosphorylation on development and homeostasis. *Genes Dev* 18: 1482-1494.

Bagrodia, S., Derijard, B., Davis, R.J., and Cerione, R.A., 1995, Cdc42 and PAK-mediated signaling leads to Jun kinase and p38 mitogen-activated protein kinase activation. *J Biol Chem* 270: 27995-27998.

Birkey Reffey, S., Wurthner, J.U., Parks, W.T., Roberts, A.B., and Duckett, C.S., 2001, X-linked inhibitor of apoptosis protein functions as a cofactor in transforming growth factor-β signaling. *J Biol Chem* 276: 26542-26549.

Brodin, G., Åhgren, A., ten Dijke, P., Heldin, C.-H., and Heuchel, R., 2000, Efficient TGF-β induction of the Smad7 gene requires cooperation between AP-1, Sp1, and Smad proteins on the mouse Smad7 promoter. *J Biol Chem* 275: 29023-29030.

Brown, J.D., DiChiara, M.R., Anderson, K.R., Gimbrone, M.A., Jr., and Topper, J.N., 1999, MEKK-1, a component of the stress (stress-activated protein kinase/c-Jun N-terminal kinase) pathway, can selectively activate Smad2-mediated transcriptional activation in endothelial cells. *J Biol Chem* 274: 8797-8805.

Calonge, M.J., and Massagué, J., 1999, Smad4/DPC4 silencing and hyperactive Ras jointly disrupt transforming growth factor-β antiproliferative responses in colon cancer cells. *J Biol Chem* 274: 33637-33643.

Chang, L., and Karin, M., 2001, Mammalian MAP kinase signalling cascades. *Nature* 410: 37-40.

Davis, R.J., 2000, Signal transduction by the JNK group of MAP kinases. *Cell* 103: 239-252.

de Caestecker, M.P., Parks, W.T., Frank, C.J., Castagnino, P., Bottaro, D.P., Roberts, A.B., and Lechleider, R.J., 1998, Smad2 transduces common signals from receptor serine-threonine and tyrosine kinases. *Genes Dev* 12: 1587-1592.

Derynck, R., Akhurst, R.J., and Balmain, A., 2001, TGF-β signaling in tumor suppression and cancer progression. *Nat Genet* 29: 117-129.

Derynck, R., and Zhang, Y.E., 2003, Smad-dependent and Smad-independent pathways in TGF-β family signalling. *Nature* 425: 577-584.

Dowdy, S.C., Mariani, A., and Janknecht, R., 2003, HER2/Neu- and TAK1-mediated up-regulation of the transforming growth factor β inhibitor Smad7 via the ETS protein ER81. *J Biol Chem* 278: 44377-44384.

Dumont, N., Bakin, A.V., and Arteaga, C.L., 2003, Autocrine transforming growth factor-β signaling mediates Smad-independent motility in human cancer cells. *J Biol Chem* 278: 3275-3285.

Edlund, S., Bu, S., Schuster, N., Aspenström, P., Heuchel, R., Heldin, N.E., ten Dijke, P., Heldin, C.-H., and Lanström, M., 2003, Transforming growth factor-β1 (TGF-β)-induced apoptosis of prostate cancer cells involves Smad7-dependent activation of p38 by TGF-β-activated kinase 1 and mitogen-activated protein kinase kinase 3. *Mol Biol Cell* 14: 529-544.

Edlund, S., Landström, M., Heldin, C.-H., and Aspenström, P., 2002, Transforming growth factor-β-induced mobilization of actin cytoskeleton requires signaling by small GTPases Cdc42 and RhoA. *Mol Biol Cell* 13: 902-914.

Eferl, R., and Wagner, E.F., 2003, AP-1: a double-edged sword in tumorigenesis. *Nat Rev Cancer* 3: 859-868.

Engel, M.E., McDonnell, M.A., Law, B.K., and Moses, H.L., 1999, Interdependent SMAD and JNK signaling in transforming growth factor-β-mediated transcription. *J Biol Chem* 274: 37413-37420.

Frodin, M., and Gammeltoft, S., 1999; Role and regulation of 90 kDa ribosomal S6 kinase (RSK) in signal transduction. *Mol Cell Endocrinol* 151: 65-77.

Funaba, M., Zimmerman, C.M., and Mathews, L.S., 2002, Modulation of Smad2-mediated signaling by extracellular signal-regulated kinase. *J Biol Chem* 277: 41361-41368.

Grimm, O.H., and Gurdon, J.B., 2002, Nuclear exclusion of Smad2 is a mechanism leading to loss of competence. *Nat Cell Biol* 4: 519-522.

Hanafusa, H., Ninomiya-Tsuji, J., Masuyama, N., Nishita, M., Fujisawa, J., Shibuya, H., Matsumoto, K., and Nishida, E., 1999, Involvement of the p38 mitogen-activated protein kinase pathway in transforming growth factor-β-induced gene expression. *J Biol Chem* 274: 27161-27167.

Hannon, G.J., and Beach, D., 1994, p15INK4B is a potential effector of TGF-β-induced cell cycle arrest. *Nature* 371: 257-261.

Harlin, H., Reffey, S.B., Duckett, C.S., Lindsten, T., and Thompson, C.B., 2001, Characterization of XIAP-deficient mice. *Mol Cell Biol* 21: 3604-3608.

Hocevar, B.A., Brown, T.L., and Howe, P.H., 1999, TGF-β induces fibronectin synthesis through a c-Jun N-terminal kinase-dependent, Smad4-independent pathway. *EMBO J* 18: 1345-1356.

Hu, P.P., Shen X., Huang D., Liu Y., Counter C., and Wang X.-F., 1999, The MEK pathway is required for stimulation of p21(WAF1/CIP1) by transforming growth factor-β. *J Biol Chem* 274: 35381-35387.

Ip, Y.T., and Davis, R.J., 1998, Signal transduction by the c-Jun N-terminal kinase (JNK) – from inflammation to development. *Curr Opin Cell Biol* 10: 205-219.

Itoh, S., Thorikay, M., Kowanetz, M., Moustakas, A., itoh, F., Heldin, C.-H., and ten Dijke, P., 2003, Elucidation of Smad requirement in transforming growth factor-β type I receptor-induced responses. *J Biol Chem* 278: 3751-3761.

Javelaud, D., and Mauviel, A., 2005, Crosstalk mechanisms between the mitogen-activated protein kinase pathways and Smad signaling downstream of TGF-β: implications for carcinogenesis. *Oncogene* 24: 5742-5750.

Javelaud, D., Laboureau, J., Gabison, E., Verrecchia, F., and Mauviel, A., 2003, Disruption of basal JNK activity differentially affects key fibroblast functions important for wound healing. *J Biol Chem* 278: 24624-24628.

Johansson, N., Ala-aho, R., Uitto, V., Grenman, R., Fusenig, N.,E., Lopez-Otin, C., and Kähäri, V.-M.., 2000, Expression of collagenase-3 (MMP-13) and collagenase-1 (MMP-1) by transformed keratinocytes is dependent on the activity of p38 mitogen-activated protein kinase. *J Cell Sci* 113 Pt 2: 227-235.

Jonk, L.J., Itoh, S., Heldin, C.-H., ten Dijke, P., and Kruijer, W., 1998, Identification and functional characterization of a Smad binding element (SBE) in the JunB promoter that acts as a transforming growth factor-β, activin, and bone morphogenetic protein-inducible enhancer. *J Biol Chem* 273: 21145-1152.

Kakonen, S.M., Selander, K.S., Chirgwin, J.M., Yin, J.J., Burns, S., Rankin, W.A., Grubbs, B.G., Dallas, M., Cui, Y., and Guise, T.A., 2002, Transforming growth factor-β stimulates parathyroid hormone-related protein and osteolytic metastases via Smad and mitogen-activated protein kinase signaling pathways. *J Biol Chem* 277: 24571-24578.

Kamaraju, A.K., and Roberts, A.B., 2005, Role of Rho/ROCK and p38 MAP kinase pathways in transforming growth factor-β-mediated Smad-dependent growth inhibition of human breast carcinoma cells *in vivo*. *J Biol Chem* 280: 1024-1036.

Kim, E.S., Kim, M.S., and Moon, A., 2004, TGF-β-induced upregulation of MMP-2 and MMP-9 depends on p38 MAPK, but not Erk signaling in MCF10A human breast epithelial cells. *Int J Oncol* 25: 1375-1382.

Kimura, N., Matsuo, R., Shibuya, H., Nakashima, K., and Taga T., 2000, BMP2-induced apoptosis is mediated by activation of the TAK1-p38 kinase pathway that is negatively regulated by Smad6. *J Biol Chem* 275: 17647-17652.

Kretzschmar, M., Doody, J., and Massagué, J., 1997, Opposing BMP and EGF signalling pathways converge on the TGF-β family mediator Smad1. *Nature* 389: 618-622.

Kretzschmar, M., Doody, J., Timokhina, I., and Massagué, J., 1999, A mechanism of repression of TGFβ/ Smad signaling by oncogenic Ras. *Genes Dev* 13: 804-816.

Kuroda, H., Fuentealba, L., Ikeda, A., Reversade, B., and De Robertis, E.M., 2005, Default neural induction: neuralization of dissociated *Xenopus* cells is mediated by Ras/MAPK activation. *Genes Dev* 19: 1022-1027.

Lehmann, K., Janda, E., Pierreux, C.E., Rytomaa, M., Schulze, A., McMahon, M., Hill, C.S., Beug, H., and Downward, J., 2000, Raf induces TGFβ production while blocking its apoptotic but not invasive responses: a mechanism leading to increased malignancy in epithelial cells. *Genes Dev* 14: 2610-2622.

Lopez-Ilasaca, M., 1998, Signaling from G-protein-coupled receptors to mitogen-activated protein (MAP)-kinase cascades. *Biochem Pharmacol* 56: 269-277.

Lutz, M., and Knaus, P., 2002, Integration of the TGF-β pathway into the cellular signalling network. *Cell Signal* 14: 977-988.

Massagué, J., and Chen, Y.G., 2000, Controlling TGF-β signaling. *Genes Dev* 14: 627-644.

Mauviel, A., Chung, K.Y., Agarwal, A., Tamai, K., and Uitto, J., 1996, Cell-specific induction of distinct oncogenes of the Jun family is responsible for differential regulation of collagenase gene expression by transforming growth factor-β in fibroblasts and keratinocytes. *J Biol Chem* 271: 10917-10923.

Mazars, A., Lallemand, F., Prunier, C., Marais, J., Ferrand, N., Pessah, M., Cherqui, G., and Atfi, A., 2001, Evidence for a role of the JNK cascade in Smad7-mediated apoptosis. *J Biol Chem* 276: 36797-36803.

Mori, S., Matsuzaki, K., Yoshida, K., urukawa, F., Tahashi, Y., Yamagata, H., Sekimoto, G., Seki, T., Matsui, H., Nishizawa, M., Fujizawa, J., and Okazaki, K., 2004, TGF-β and HGF transmit the signals through JNK-dependent Smad2/3 phosphorylation at the linker regions. *Oncogene* 23: 7416-7429.

Mucsi, I., Skorecki, K.L., and Goldberg, H.J., 1996, Extracellular signal-regulated kinase and the small GTP-binding protein, Rac, contribute to the effects of transforming growth factor-β1 on gene expression. *J Biol Chem* 271: 16567-16572.

Oft, M., Peli, J., Rudaz, C., Schwarz, H., Beug, H., and Reichmann, E., 1996, TGF-β1 and Ha-Ras collaborate in modulating the phenotypic plasticity and invasiveness of epithelial tumor cells. *Genes Dev* 10: 2462-2477.

Ohshima, T., and Shimotohno, K., 2003, Transforming growth factor-β-mediated signaling via the p38 MAP kinase pathway activates Smad-dependent transcription through SUMO-1 modification of Smad4. *J Biol Chem* 278: 50833-50842.

Pearson, G., Robinson, F., Beers Gibson, T., Xu, B.E., Karandikar, M., Berman, K., and Cobb, M.H., 2001, Mitogen-activated protein (MAP) kinase pathways: regulation and physiological functions. *Endocr Rev* 22: 153-183.

Pera, E.M., Ikeda, A., Eivers, E., and De Robertis, E.M., 2003, Integration of IGF, FGF, and anti-BMP signals via Smad1 phosphorylation in neural induction. *Genes Dev* 17: 3023-3028.

Perlman, R., Schiemann, W.P., Brooks, M.W., Lodish, H.F., and Weinberg, R.A., 2001, TGF-β-induced apoptosis is mediated by the adapter protein Daxx that facilitates JNK activation. *Nat Cell Biol* 3: 708-714.

Pessah, M., Marais, J., Prunier, C., Ferrand, N., Lallemand, F., Mauviel, A., and Atfi, A., 2002, c-Jun associates with the oncoprotein Ski and suppresses Smad2 transcriptional activity. *J Biol Chem* 277: 29094-29100.

Pessah, M., Prunier, C., Marais, J., Ferrand, N., Mazars, A., Lallemand, F., Gauthier, J.-M., and Atfi, A., 2001, c-Jun interacts with the corepressor TG-interacting factor (TGIF) to suppress Smad2 transcriptional activity. *Proc Natl Acad Sci U S A* 98: 6198-6203.

Quan, T., He, T., Voorhees, J.J., and Fisher, G.J., 2005, Ultraviolet irradiation induces Smad7 via induction of transcription factor AP-1 in human skin fibroblasts. *J Biol Chem* 280: 8079-8085.

Ravanti, L., Hakkinen, L., Larjava, H., Saarialho, U., Foschi, M., Han, J., and Kähäri, V.-M., 1999, Transforming growth factor-β induces collagenase-3 expression by human gingival fibroblasts via p38 mitogen-activated protein kinase. *J Biol Chem* 274: 37292-37300.

Reynisdottir, I., Polyak, K., Iavarone, A., and Massagué, J., 1995, Kip/Cip and Ink4 Cdk inhibitors cooperate to induce cell cycle arrest in response to TGF-β. *Genes Dev* 9: 1831-1845.

Roelen, B.A., Cohen, O.S., Raychowdhury, M.K., Chadee, D.N., Zhang, Y., Kyriadis, J.M., Alessandrini, A.A., and Lin, H.Y., 2003, Phosphorylation of threonine 276 in Smad4 is involved in transforming growth factor-β-induced nuclear accumulation. *Am J Physiol Cell Physiol* 285: C823-C830.

Sanchez-Capelo, A., 2005, Dual role for TGF-β1 in apoptosis. *Cytokine Growth Factor Rev* 16: 15-34.

Sano, Y., Harada, J., Tashiro, S., Gotoh-Mandeville, R., Maekawa, T., and Ishii, S., 1999, ATF-2 is a common nuclear target of Smad and TAK1 pathways in transforming growth factor-β signaling. *J Biol Chem* 274: 8949-8957.

Sato, M., Shegogue, D., Gore, E.A., Smith, E.A., McDermott, P.J., and Trojanowska, M., 2002, Role of p38 MAPK in transforming growth factor β stimulation of collagen production by scleroderma and healthy dermal fibroblasts. *J Invest Dermatol* 118: 704-711.

Scherer, A., and Graff, J.M., 2000, Calmodulin differentially modulates Smad1 and Smad2 signaling. *J Biol Chem* 275: 41430-41438.

Shaulian, E., and Karin, M., 2002, AP-1 as a regulator of cell life and death. *Nat Cell Biol* 4: E131-E136.

Siegel, P.M., and Massagué, J., 2003, Cytostatic and apoptotic actions of TGF-β in homeostasis and cancer. *Nat Rev Cancer* 3: 807-821.

Tian, F., Byfield, S.D., Parks, W.T., et al., 2004, Smad-binding defective mutant of transforming growth factor β type I receptor enhances tumorigenesis but suppresses metastasis of breast cancer cell lines. *Cancer Res* 64: 4523-4530.

Treisman, R., 1996, Regulation of transcription by MAP kinase cascades. *Curr Opin Cell Biol* 8: 205-215.

Uchida, K., Suzuki, H., Ohashi, T., Nitta, K., Yumura, W., and Nihei, H., 2001, Involvement of MAP kinase cascades in Smad7 transcriptional regulation. *Biochem Biophys Res Commun* 289: 376-381.

Ventura, J.J., Kennedy, N.J., Flavell, R.A., and Davis, R.J., 2004, JNK regulates autocrine expression of TGF-β1. *Mol Cell* 15: 269-278.

Verrecchia, F., and Mauviel, A., 2002, Transforming growth factor-β signaling through the Smad pathway: role in extracellular matrix gene expression and regulation. *J Invest Dermatol* 118: 211-215.

Verrecchia, F., Wagner, E.F., and Mauviel, A., 2002, Distinct involvement of the Jun-N-terminal kinase and NF-κB pathways in the repression of the human COL1A2 gene by TNF-α. *EMBO Rep* 3: 1069-1074.

Verrecchia, F., Pessah, M., Atfi, A., and Mauviel, A., 2000, Tumor necrosis factor-α inhibits transforming growth factor-β/Smad signaling in human dermal fibroblasts via AP-1 activation. *J Biol Chem* 275: 30226-30231.

Verrecchia, F., Tacheau, C., Schorpp-Kistner, M., Angel, P., and Mauviel, A., 2001a, Induction of the AP-1 members c-Jun and JunB by TGF-β/Smad suppresses early Smad-driven gene activation. *Oncogene* 20: 2205-2211.

Verrecchia, F., Tacheau, C., Wagner, E.F., and Mauviel, A., 2003, A central role for the JNK pathway in mediating the antagonistic activity of pro-inflammatory cytokines against transforming growth factor-β-driven SMAD3/4-specific gene expression. *J Biol Chem* 278: 1585-1593.

Verrecchia, F., Vindevoghel, L., Lechleider, R.J., Uitto, J., Roberts, A.B., and Mauviel, A., 2001b, Smad3/AP-1 interactions control transcriptional responses to TGF-β in a promoter-specific manner. *Oncogene* 20: 3332-3340.

Wakefield, L.M., and Roberts, A.B., 2002, TGF-β signaling: positive and negative effects on tumorigenesis. *Curr Opin Genet Dev* 12: 22-29.

Wendling, J., Marchand, A., Mauviel, A., and Verrecchia, F., 2003, 5-fluorouracil blocks transforming growth factor-β-induced α2 type I collagen gene (COL1A2) expression in human fibroblasts via c-Jun NH2-terminal kinase/activator protein-1 activation. *Mol Pharmacol* 64: 707-713.

Whitmarsh, A.J., and Davis, R.J., 1998; Structural organization of MAP-kinase signaling modules by scaffold proteins in yeast and mammals. *Trends Biochem Sci* 23: 481-485.

Wicks, S.J., Lui, S., Abdel-Wahab, N., Mason, R.M., and Chantry, A., 2000, Inactivation of smad-transforming growth factor β signaling by Ca(2+)-calmodulin-dependent protein kinase II. *Mol Cell Biol* 20: 8103-8111.

Wilkes, M.C., Murphy, S.J., Garamszegi, N., and Leof, E.B., 2003, Cell-type-specific activation of PAK2 by transforming growth factor β independent of Smad2 and Smad3. *Mol Cell Biol* 23: 8878-8889.

Wong, C., Rougier-Chapman, E.M., Frederick, J.P., Datto, M.B., Liberati, N.T., Li, J.M., and Wang, X.F., 1999, Smad3-Smad4 and AP-1 complexes synergize in transcriptional activation of the c-Jun promoter by transforming growth factor β. *Mol Cell Biol* 19: 1821-1830.

Xie, L., Law, B.K., Aakre, M.E., Edgerton, M., Shyr, Y., Bhowmick, N.A., and Moses, H.L., 2003, Transforming growth factor β–regulated gene expression in a mouse mammary gland epithelial cell line. *Breast Cancer Res* 5: R187-R198.

Xie, L., Law, B.K., Chytil, A.M., Brown, K.A., Aakre, M.E., and Moses, H.L., 2004, Activation of the Erk pathway is required for TGF-β1-induced EMT *in vitro*. *Neoplasia* 6: 603-610.

Yamaguchi, K., Nagai, S., Ninomiya-Tsuji, J., Nishita, M., Tamai, K., Irie, K., Ueno, N., Nishida, E., Shibuya, H., and Matsumoto, K., 1999, XIAP, a cellular member of the inhibitor of apoptosis protein family, links the receptors to TAB1-TAK1 in the BMP signaling pathway. *EMBO J* 18: 179-187.

Yamaguchi, K., Shirakabe, K., Shibuya, H., Irie, K., Oishi, I., Ueno, N., Taniguchi, T., Nishida, E., and Matsumoto, K., 1995, Identification of a member of the MAPKKK family as a potential mediator of TGF-β signal transduction. *Science* 270: 2008-2011.

Yanagisawa, M., Nakashima, K., Takeda, K., Ochiai, W., Takizawa, T., Ueno, M., Takizawa, M., Shibuya, H., and Taga, T., 2001, Inhibition of BMP2-induced, TAK1 kinase-mediated neurite outgrowth by Smad6 and Smad7. *Genes Cells* 6: 1091-1099.

Yang, X., Khosravi-Far, R., Chang, H.Y., and Baltimore, D., 1997, Daxx, a novel Fas-binding protein that activates JNK and apoptosis. *Cell* 89: 1067-1076.

Yin, J.J., Selander, K., Chirgwin, J.M., Dallas, M., Grubbs, B.G., Wieser, R., Massague, J., Mundy, G.R., and Guise, T.A., 1999, TGF-β signaling blockade inhibits PTHrP secretion by breast cancer cells and bone metastases development. *J Clin Invest* 103: 197-206.

Yonekura, A., Osaki, M., Hirota, Y., Tsukazaki, T., Miyazaki, Y., Matsumoto, T., Ohtsuru, A., Namba, H., Shindo, H., and Yamashita, S., 1999, Transforming growth factor-β stimulates articular chondrocyte cell growth through p44/42 MAP kinase (Erk) activation. *Endocr J* 46: 545-553.

Yoo, J., Ghiassi, M., Jirmanova, L., Balliet, A.G., Hoffman, B., Fornace, A.J.Jr, Liebermann, D.A., Bottinger, E.P., and Roberts, A.B., 2003, Transforming growth factor-β-induced apoptosis is mediated by Smad-dependent expression of GADD45b through p38 activation. *J Biol Chem* 278: 43001-43007.

Yoshida, K., Matsuzaki, K., Mori, S., Tahashi, Y., Yamagata, H., Furukawa, F., Seki, T., Nishizawa, M., Fujisawa, J., and Okazaki, K., 2005, Transforming growth factor-β and platelet-derived growth factor signal via c-Jun N-terminal kinase-dependent Smad2/3 phosphorylation in rat hepatic stellate cells after acute liver injury. *Am J Pathol* 166: 1029-1039.

Yu, L., Hebert, M.C., and Zhang, Y.E., 2002; TGF-β receptor-activated p38 MAP kinase mediates Smad-independent TGF-β responses. *EMBO J* 21: 3749-3759.

Yue, J., and Mulder, K.M., 2000a, Requirement of Ras/MAPK pathway activation by transforming growth factor β for transforming growth factor β1 production in a Smad-dependent pathway. *J Biol Chem* 275: 35656.

Yue, J., and Mulder, K.M., 2000b, Activation of the mitogen-activated protein kinase pathway by transforming growth factor-β. *Methods Mol Biol* 142: 125-131.

Zarubin, T., and Han, J., 2005, Activation and signaling of the p38 MAP kinase pathway. *Cell Res* 15: 11-18.

Zhang, S., Han, J., Sells, M.A., Chernoff, J., Knaus, U.G., Ulevitch, R.J., and Bokoch, G.M., 1995, Rho family GTPases regulate p38 mitogen-activated protein kinase through the downstream mediator Pak1. *J Biol Chem* 270: 23934-23936.

Zhang, Y., Feng, X.-H., and Derynck, R., 1998, Smad3 and Smad4 cooperate with c-Jun/c-Fos to mediate TGF-β-induced transcription. *Nature* 394: 909-913.

Zhou, G., Bao, Z.Q., and Dixon, J.E., 1995, Components of a new human protein kinase signal transduction pathway. *J Biol Chem* 270: 12665-12669.

Zhou, G., Lee, S.C., Yao, Z., and Tan, T.H., 1999, Hematopoietic progenitor kinase 1 is a component of transforming growth factor β-induced c-Jun N-terminal kinase signaling cascade. *J Biol Chem* 274: 13133-13138.

CHAPTER 17

GENE EXPRESSION SIGNATURES OF TGF-β/SMAD-INDUCED RESPONSES

ERWIN P. BÖTTINGER AND WENJUN JU

Mount Sinai School of Medicine, One Gustave L. Levy Pl., New York City, New York, USA

Abstract: Microarray technology has enabled large-scale discovery of transcriptional targets of important signaling pathways, including signal transducers of the TGF-β/Smad family. Thus, numerous studies present gene expression profiles obtained by microarray approaches from a range of normal and malignant cell types exposed to TGF-β1, BMP7, BMP2, and other TGF-β family proteins. Here we evaluate newly synthesized TGF-β/Smad gene expression signatures identified by systematic cross-referencing and functional annotation of published lists of TGF-β/Smad target genes. This work provides a valuable compendium for context-dependent analysis of transcriptional profiles of TGF-β/Smad signaling

Keywords: development; gene regulation; genomics; Microarray; mRNA; pathogenesis; TGF-β; transcription; signal transduction

1. INTRODUCTION

Microarray technology enables the parallel interrogation of mRNA species present in a biological sample providing a genome-wide snapshot of the transcriptome and its modulation by developmental cues and extracellular signals in cells or tissues (Brown and Botstein, 1999; Lockhart and Winzeler, 2000). An immediate major impact of microarray studies of TGF-β family signaling in many different experimental models is the discovery of hundreds of new downstream target genes which led researchers to generate new hypotheses about their individual functional roles as effectors of diverse biological activities of TGF-βs. Collectively, these target gene discoveries are being explored to unravel further poorly understood biological activities of TGF-βs and to pinpoint critical target mediators for novel functional roles.

In addition to single gene discovery, advanced computational analysis of gene expression patterns across numerous physiological states has the potential to reveal

P. ten Dijke and C.-H. Heldin (eds.), Smad Signal Transduction, 335–360.

distinctive molecular signatures and data environments to develop innovative, complex mathematical models to characterize states of TGF-β signaling at a systems biology level. Together, gene discovery and systems level analysis of TGF-β/Smad pathways, enabled by microarray technology and advanced computation, will be valuable in clinical settings in development of molecular diagnostics for diseases associated with aberrant TGF-β signaling, and of molecular markers to guide clinical trials of compounds targeting TGF-β pathways.

In-depth reviews of the role and regulation of Smad proteins in TGF-β family member signaling are available in specialized chapters provided in this volume. The goal for the present chapter is to review and synthesize the massive amount of largely unvalidated gene expression information presented in reports of microarray-based analysis of TGF-β/Smad pathways. We will describe gene expression signatures that characterize discrete cellular responses induced by TGF-β/Smad pathway signaling in different cell types in the absence or presence of experimental interference with receptor or Smad function. Moreover, we will discuss how TGF-β/Smad gene expression signatures may become useful tools in clinical settings as resource for molecular diagnostics and pharmacogenomics.

2. DATABASES AND METHODS

2.1 Public Gene Expression Data Repositories and TGF-β Gene Expression Signatures

Since their introduction nearly a decade ago (Lockhart et al., 1996; Schena et al., 1995), microarrays have been increasingly used to characterize gene expression controlled by TGF-β/Smad pathways. For example, recent PubMed searches for keywords "microarray" AND "transforming growth factor beta" produced 165 hits, published between 2000 and 2005. Similarly, searches for keywords "microarray" AND "bone morphogenetic protein" or "Smad" produced 56 or 35 hits, respectively. A fraction of the microarray datasets from these studies are available for viewing or download in public microarray data repositories. The Gene Expression Omnibus (GEO) repository is supported by the National Center for Biotechnology Information (NCBI) and contains 73,679 individual measurements of gene expression (GEO Profiles) related to "transforming growth factor beta" across 2,446 complete experiments (GEO Dataseries). Additional "transforming growth factor beta" datasets are deposited at ArrayExpress and CIBEX (Center for Information Biology Gene Expression Database), repositories supported by the European Bioinformatics Institute (EBI) and the DNA Data Bank of Japan (DDBJ), respectively.

Most reports of TGF-β/Smad gene expression signatures are based on defined cell culture systems where recombinant TGF-β family members were applied to cells of various origins in time series analyses. The most commonly used experimental systems involve either nonmalignant or malignant epithelial cells treated with TGF-β1. Fibroblasts of various origins and the murine mesenchymal progenitor cell line

C2C12 represent the most common *in vitro* mesenchymal cell systems that were analyzed using microarray technology. A Supporting Table 1 is available online that summarizes the technical and experimental design features of microarray studies that report significant lists of TGF-β responsive genes. We retrieved the gene expression data reported in those studies where available, to a custom-designed local database for comparative data analysis to identify gene expression signatures controlled by TGF-β/Smad pathways.

2.2 Impact of Study Designs and Analytical Methods

To achieve a uniform, up-to-date nomenclature, we standardized the disparate gene identifiers, gene symbols, and gene names provided in the original reports by referencing all genes against the Unigene database of the National Center for Biotechnology Information (NCBI). This allowed us to cross-reference systematically all gene expression profiles across all available studies. Next, we generated a data matrix to determine the overlap between gene lists in individual gene profiling studies by pairwise comparisons. Three distinct data matrices were generated: a) epithelial cells treated with TGF-β1 or BMP7 (Supporting Table 2); b) fibroblasts treated with TGF-β1 (Supporting Table 3); and c) C2C12 murine mesenchymal progenitor cells treated with either BMP2, Activin, or TGF-β1 (Supporting Table 4).

The data matrices are informative as a measure of consistency across studies. For example, out of ten gene profiling studies using TGF-β1 in epithelial cells, two studies revealed overlapping genes with only one or two of the other studies, respectively (Jazag et al., 2005; Ijichi et al., 2004), while all other studies revealed varying degrees of overlap (Zavadil et al., 2001; Valcourt et al., 2005; Xie et al., 2003; Chen et al., 2001; Kang et al., 2003; Levy and Hill, 2005; Wu et al., 2003; Deacu et al., 2004). These findings indicate that the studies by Jazag et al. and Ijichi et al. may be confounded by experimental design parameters, which caused us not to consider these studies for comparative analysis.

In addition, experimental designs of the analyzed studies vary considerably. In general, validity and reliability of gene expression profiling studies improve with an increasing number of biological replicates of all experimental conditions, and with application of specialized statistical analysis of gene expression data across experimental conditions. Nine studies used a minimum of three biological replicates of all experimental conditions together with statistical data analysis to identify genes with differential expression profiles depending on TGF-β exposure (Supporting Table 1). Another eight studies applied two biological replicates, with or without additional technical replication, and statistical data analysis (Supporting Table 1). Ten studies report differential gene expression patterns that were determined based on arbitrarily defined expression thresholds (fold-difference) between experimental conditions without or with biological replication (Supporting Table 1).

3. TGF-β/SMAD GENE EXPRESSION SIGNATURES IN CELLS OF EPITHELIAL ORIGINS

3.1 TGF-β Responsiveness in Nontransformed and Malignant Cells

Human HaCaT keratinocytes and murine NMuMG mammary epithelial cells were originally derived from normal epidermis or mammary gland, respectively. TGF-βs are potent inducers of epithelial-mesenchymal transition (EMT) (see Chapter 7), apoptosis (see Chapter 6), and cell cycle arrest (see Chapter 4) in these nontransformed epithelial cell lines. In addition, both lines have been widely-used to investigate molecular mechanisms of Smad-dependent and Smad-independent signaling and transcriptional regulation by TGF-β family members (Derynck and Zhang, 2003). Thus, HaCaT cells (Zavadil et al., 2001; Levy and Hill, 2005; Kang et al., 2003), or NMuMG cells (Valcourt et al., 2005; Xie et al., 2003) were used by several groups to determine TGF-β-dependent gene expression responses. HMEC and MCF10A are well-characterized TGF-β-responsive cells derived from normal human breast (Chen et al., 2001; Kang et al., 2003; Kloeker et al., 2004). Importantly, the genes encoding TGF-β receptors and Smad proteins are typically intact in nontransformed epithelial cells.

In contrast, cell lines derived from malignant tumors may vary considerably in their responsiveness to TGF-β family members and are frequently characterized by a selective loss of growth inhibitory responses (Massagué, 2004). For example, epithelial MDA-MB-231 breast cancer cells have a hyperactive Ras pathway and are poorly growth inhibited by TGF-β, but are stimulated to form bone metastasis in athymic mice. MDA-MB-468 breast cancer cells are Smad4-deficient, similar to pancreatic cancers, including BxPC3, which are frequently characterized by chromosomal deletions at the *SMAD4* gene locus (Kowanetz et al., 2004; Ijichi et al., 2004). HCT-15 is a colon cancer cell line that is deficient for Activin receptor type II (Deacu et al., 2004).

3.2 Epithelial Gene Expression Signatures

A total of 1,458 unique, named TGF-β-responsive target genes were reported in eight quality-verified studies (Supporting Table 2). 157 TGF-β target genes were present in at least 2 studies and were considered as epithelial gene expression signature of TGF-β (Table 1). To provide a measure of robustness, each target gene was assigned a score where 1.0 (highest) indicates that the target gene was listed in 8 out of 8 cross-referenced studies, while a score of 0.25 (lowest) indicates that the gene was present in 2 out of 8 cross-referenced studies respectively (Table 1, column 1).

Biological annotation of the 157 epithelial signature genes, using Gene Ontology tables, demonstrates a highly significant enrichment with known and novel TGF-β target genes that function in cell cycle control, in actin cytoskeletal organization and biosynthesis, cell adhesion, and apoptosis (Table 1). Thus, the epithelial TGF-β gene expression signature derived from microarray approaches extends considerably the scope of molecular mediators that may underlie characteristic TGF-β

Table 1. TGF-β/Smad gene expression signatures in cells of epithelial origins (The online version of this supplementary table is freely accessible via the website of the book: http://www.springer.com/1-4020-4542-5)

Score	Gene	Factor	Cell Type	Biological Process	Source Ref
0,75	JUN	T1	E	transcription	V, X, C, K, L, D
0,63	**ID3**	**T1, B7**	**E, F, C2C12**	**transcription**	Z, V, X, K, L
0,63	ATF3	T1	E	transcription	Z, K, L, D, W
0,63	**COL4A1**	**T1**	**E, F**	**cell-matrix adhesion**	**Z, X, C, K, L**
0,63	EDN1	T1	E	cell-cell signaling	Z, C, L, W, D
0,63	SOX4	T1	E	transcription	Z, V, C, K, L
0,50	ETS2	T1	E	transcription	Z, C, K, L
0,50	FOS	T1, B7	E	transcription	Z, V, C, W
0,50	**ID1**	**T1, B7**	**E, F**	**transcription**	X, C, K, W
0,50	ID2	T1, B7	E	transcription	Z, V, K, L
0,50	**JUNB**	**T1, B7**	**E, F**	**transcription**	C, K, L, D
0,50	**MYC**	**T1, B7**	**E, F**	**cell cycle arrest**	**Z, V, C, K**
0,38	**BHLHB2**	**T1**	**E, F**	**transcription**	**C, K, L**
0,38	BMP1	T1	E	cell differentiation	Z, V, X
0,38	BPGM	T1	E	carbohydrate metabolism	V, X, C
0,38	C1S	T1	E	complement activation	V, X, D
0,38	CCND2	T1	E	regulation of cell cycle	Z, V, X
0,38	CDKN1A	T1, B7	E	cell cycle arrest	Z, K, L
0,38	CDKN2B	T1	E	cell cycle arrest	Z, K, L
0,38	CITED2	T1	E	transcription	C, K, L
0,38	**CTGF**	**T1**	**E, F, C2C12**	**cell-cell signaling**	**Z, C, K**
0,38	CXCL1	T1	E	chemokine activity	Z, V, X
0,38	DLX2	T1	E	transcription	C, K, L
0,38	**FHL2**	**T1**	**E, F**	**transcription**	**Z, V, L**
0,38	IL11	T1	E	cell-cell signaling	C, K, W
0,38	ITGA2	T1, B7	E	cell-matrix adhesion	Z, C, W
0,38	ITGB5	T1	E	cell-matrix adhesion	Z, V, X
0,38	JAG1	T1	E	Notch signaling pathway	Z, C, K
0,38	LDLR	T1	E	O-linked glycosylation	Z, K, L
0,38	P2RY2	T1	E	cell ion homeostasis	Z, K, W
0,38	PDGFA	T1	E	cell-cell signaling	V, X, C
0,38	**PPAP2B**	**T1**	**E, C2C12**	**lipid metabolism**	**Z, V, D**
0,38	SEMA3C	T1	E	development	Z, V, L
0,38	**SERPINE1**	**T1**	**E, F**	**blood coagulation**	Z, C, K
0,38	SGK	T1	E	apoptosis	V, K, D
0,38	**SMAD7**	**T1**	**E, F, C2C12**	**signal transduction**	Z, C, K
0,38	**TGIF**	**T1, B2**	**E, C2C12**	**transcription**	**Z, V, X**
0,38	**TIMP3**	**T1, B2, B7**	**E, C2C12**	**apoptosis**	**Z, V, L**
0,38	**TXNIP**	**T1**	**E, F**	**oxidative stress**	**V, C, K**
0,38	UGCG	T1	E	glucosylceramide biosynthesis	Z, K, W
0,38	VEGF	T1	E	cell-cell signaling	Z, C, K
0,25	Ablim1	T1	E	cytoskeleton organization and biogenesis	Z, X
0,25	**Actn1**	**T1**	**E, F**	**cytoskeleton organization and biogenesis**	**X, L**

(Continued)

Table 1. (Continued)

Score	Gene	Factor	Cell Type	Biological Process	Source Ref
0,25	Actr1a	T1	E	cytoskeleton organization and biogenesis	X, D
0,25	**ADAM19**	**T1**	**E, C2C12**	**metalloendopeptidase activity**	**K, L**
0,25	ARHGEF18	T1	E	signal transduction	K, L
0,25	ARPC4	T1	E	cytoskeleton organization and biogenesis	X, D
0,25	Atp1a1	T1	E	hydrogen ion homeostasis	X, D
0,25	ATP2A2	T1	E	calcium ion transport	Z, D
0,25	BCL2A1	T1	E	anti-apoptosis	Z, W
0,25	BIRC3	T1	E	anti-apoptosis	C, K
0,25	Bmp4	T1	E	cell-cell signaling	V, K
0,25	BMP7	T1	E	cell-cell signaling	Z, W
0,25	BMPR2	T1	E	signal transduction	Z, K
0,25	Cfh	T1	E	complement activation, alternative pathway	Z, V
0,25	CLU	T1	E	apoptosis	V.L
0,25	CMKOR1	T1	E	signal transduction	Z, V
0,25	COPZ2	T1	E	clathrin vesicle coat	Z, W
0,25	Crot	T1	E	fatty acid metabolism	V, X
0,25	CSE1L	T1	E	apoptosis	Z, W
0,25	Csf1	T1	E	cell-cell signaling	X, W
0,25	CYP1B1	T1	E	electron transport	Z, K
0,25	DHRS3	T1	E	fatty acid metabolism	Z, X
0,25	DLC1	T1	E	cytoskeleton organization and biogenesis	Z, L
0,25	DUSP5	T1	E	MAP kinase phosphatase activity	Z, W
0,25	E2F3	T1	E	regulation of cell cycle	Z, W
0,25	EEF1A1	T1	E	translation	Z, X
0,25	EIF2S1	T1	E	translation	Z, D
0,25	ERBB3	T1	E	epidermal growth factor receptor activity	Z, W
0,25	Ets1	T1	E	negative regulation of cell proliferation	V, X
0,25	FGFR2	T1	E	cell growth	Z, V
0,25	FOSB	T1	E	development	Z, D
0,25	FOSL1	T1	E	cellular defense response	Z, V
0,25	FSTL3	T1	E	extracellular space	K, L
0,25	FZD1	T1	E	signal transduction	Z, W
0,25	**GADD45B**	T1	**E, F**	**apoptosis**	**K, L**
0,25	GATA3	T1	E	defense response	Z, L
0,25	Gclc	T1	E	glutathione synthesis	V, X
0,25	Gcnt2	T1	E	O-linked glycosylation	V, X
0,25	GPC1	T1	E	extracellular space	V, D
0,25	GPX1	T1	E	response to oxidative stress	Z, V

(Continued)

Table 1. (Continued)

Score	Gene	Factor	Cell Type	Biological Process	Source Ref
0,25	GSN	T1	E	cytoskeleton organization and biogenesis	Z, V
0,25	Gsta4	T1	E	glutathione transferase activity	V, X
0,25	GSTM2	T1	E	glutathione transferase activity	Z, V
0,25	HES1	T1	E	transcription	Z, L
0,25	HGD	T1	E	oxidoreductase activity	Z, W
0,25	HS3ST1	T1	E	sulfotransferase activity	V, K
0,25	HSPA8	T1	E	protein folding	Z, X
0,25	IER2	T1	E	transcription	Z, D
0,25	IL15	T1	E	cell-cell signaling	Z, C
0,25	Iqgap1	T1	E	signal transduction	V, X
0,25	ITGB1	T1	E	cell-matrix adhesion	Z, X
0,25	Jak2	T1	E	signal transduction	V, X
0,25	KAL1	T1, B7	E	cell adhesion	Z, L
0,25	KRT15	T1	E	epidermis development	Z, W
0,25	LAMC2	T1	E	cell-matrix adhesion	Z, C
0,25	LCN2	T1	E	transport	Z, V
0,25	LTBP2	T1	E	extracellular signaling	Z, L
0,25	MAP2K3	T1	E	MAP kinase kinase activity	X, D
0,25	MAP2K5	T1	E	MAP kinase kinase activity	Z, D
0,25	MAP3K4	T1	E	MAP kinase kinase kinase activity	Z, K
0,25	MBTPS1	T1	E	lipid metabolism	Z, W
0,25	MEF2C	T1	E	transcription	Z, W
0,25	MMP1	T1	E	collagen catabolism	Z, W
0,25	MN1	T1	E	cell cycle	C, K
0,25	MPHOSPH1	T1	E	cell cycle arrest	Z, D
0,25	MSN	T1	E	cell motility	Z, V
0,25	MUC1	T1	E	cytoskeleton organization and biogenesis	Z, V
0,25	NEDD9	T1	E	cytoskeleton organization and biogenesis	K, L
0,25	NET1	T1	E	signal transduction	Z, L
0,25	NFE2L1	T1	E	heme biosynthesis	Z, X
0,25	NR2F1	T1	E	transcription	D, V
0,25	ODC1	T1	E	polyamine biosynthesis	Z, X
0,25	OGT	T1	E	O-linked glycosylation	Z, W
0,25	PAPPA	T1	E	cell-cell signaling	Z, L
0,25	PDHA1	T1	E	acetyl-CoA metabolism	Z, X
0,25	PEA15	T1	E	anti-apoptosis	Z, W, D
0,25	PER2	T1	E	transcription	Z, W
0,25	PIM1	T1	E	development	Z, K
0,25	PLAU	T1, B7	E	blood coagulation	C, K
0,25	PLCB4	T1	E	signal transduction	Z, D
0,25	Plcl2	T1	E	signal transduction	V, X

(Continued)

Table 1. (Continued)

Score	Gene	Factor	Cell Type	Biological Process	Source Ref
0,25	POLR2B	T1	E	transcription	Z, X
0,25	PRSS8	T1	E	extracellular space	D, W
0,25	PSMD13	T1	E	proteasome regulation	X, D
0,25	PTGS2	T1	E	cyclooxygenase pathway	Z, L
0,25	PTPRB	T1	E	protein amino acid dephosphorylation	Z, L
0,25	PTPRS	T1	E	cell adhesion	Z, X
0,25	PTPRU	T1	E	cell adhesion	Z, D
0,25	Ralb	T1	E	intracellular protein transport	V, X
0,25	REST	T1	E	transcription	Z, W
0,25	RRS1	T1	E	ribosome biogenesis	Z, K
0,25	RUNX1	T1	E	transcription	Z, L
0,25	SDC1	T1	E	cytoskeleton organization and biogenesis	Z, V
0,25	SERPINB2	T1	E	anti-apoptosis	Z, C
0,25	SFPQ	T1	E	RNA splicing	Z, W
0,25	SFRS3	T1	E	RNA splicing	Z, W
0,25	SKIL	T1	E	transcription	K, L
0,25	SLC29A1	T1	E	nucleoside transport	Z, X
0,25	SLC2A1	T1	E	glucose transport	Z, X
0,25	Slc9a1	T1	E	ion transport	V, X
0,25	SNAI2	T1	E	transcription	Z, L
0,25	SPRY2	T1	E	signal transduction	Z, K
0,25	**TAGLN**	**T1**	**E, F**	**cytoskeleton organization and biogenesis**	Z, L
0,25	TBX3	T1	E	transcription	K, L
0,25	Tgfb1i4	T1	E	transcription	V, X
0,25	TGM1	T1, B7	E	peptide cross-linking	C, L
0,25	**TIEG**	**T1**	**E, C2C12**	**transcription**	L, W
0,25	TLE3	T1	E	transcription	Z, D
0,25	**TNFRSF1B**	T1	**E, F**	**apoptosis**	**Z, W**
0,25	TP73L	T1	E	apoptosis	Z, L
0,25	ULK1	T1	E	signal transduction	Z, X
0,25	VCP	T1	E	ER-associated protein catabolism	Z, X
0,25	VIL2	T1	E	cellular morphogenesis	V, D
0,25	VMP1	T1	E	–	Z, L
0,25	WNT5B	T1	E	development	Z, K
0,25	ZYX	T1	E	cell adhesion	Z, V

V = Valcourt, et al., Mol.Biol.Cell, 2005; X = Xie, et al., Breast Cancer Res., 2003; C = Chen, et al., Proc.Natl.Acad.Sci.U.S.A, 2001;
K = Kang, et al., Mol.Cell, 2003; Levy, et al., Mol.Cell Biol., 2005; Deacu, et al., Cancer Res., 2004; Z = Zavadil, et al., Proc.Natl.Acad.Sci.U.S.A, 2001.
bold face indicates target genes in more than one cell type; T1 indicates TGF-β1; B7 indicates BMP7; E indicates epithelial cell; F indicates fibroblasts; C2C12 indicates mesenchymal progenitor cell.

responses in epithelial cells, including cell cycle arrest, apoptosis, or EMT. In addition, the epithelial expression signature comprises gene products that point to less well-characterized functional roles of TGF-β pathways, including i) oxidative response (glutamate-cysteine ligase, catalytic subunit (GCLC), glutathione S-transferase A4 (GSTA4), glutathione S-transferase, mu 2 (GSTM2), homogentisate 1, 2-dioxygenase (HGD), thioredoxin interacting protein (TXNIP)); ii) fatty acid and lipid metabolism (carnitine O-octanoyltransferase (CROT), dehydrogenase/reductase (SDR family) member 3 (DHRS3), phosphatidic acid phosphatase type 2B (PPAP2B), membrane-bound transcription factor peptidase, site 1 (MBTPS1)), and iii) complement activation (Complement component (C1s), complement component factor h (CFH)).

Similarly, functional annotations of the epithelial gene signature reveal an extraordinary enrichment of transcriptional regulators and signal transducers, providing strong support for a central role of TGF-β/Smad pathways in the specification of epithelial cell homeostasis and stress responses. Thus, the epithelial signature is characterized by broad control of AP1 family transcription factors, kinases and phosphatases involved in MAPK pathways, other members within the TGF-β/Smad family (BMP4, BMP7), and multiple secreted signaling proteins (interleukin 11 (IL11), interleukin 15 (IL15), colony-stimulating factor 1 (CSF1), endothelin 1 (EDN1), platelet-derived growth factor-A (PDGFA), and vascular endothelial growth factor (VEGF)). Highly robust signature genes, *i.e.* identified in at least 4 out of 8 studies, encode well-known TGF-β regulated proteins such as Jun oncogene (JUN), FBJ osteosarcoma viral (v-fos) oncogene homolog (FOS), Jun-B proto-oncogene (JUNB), c-Myc proto-oncogene (MYC), MAD homolog 7 (SMAD7), connective tissue growth factor (CTGF), TIMP metallopeptidase inhibitor 3 (TIMP3), procollagen, type IV, alpha 1 (COL4A1), and Inhibitor of DNA binding proteins (ID1, ID2, ID3), but also a large number of new TGF-β target genes, including basic helix-loop-helix domain containing, class B2 (BHLHB2), 2,3-bisphosphoglycerate mutase (BPGM), Distal-less homeobox 2 (DLX2), four and a half LIM domains 2 (FHL2), low density lipoprotein receptor (LDLR), sema domain, immunoglobulin domain, 3C (SEMA3C), serum/glucocorticoid regulated kinase (SGK), TXNIP, and UDP-glucose ceramide glucosyltransferase (UGCG).

3.3 Pathway-dependency of Epithelial Gene Signature

Several studies attempted to delineate the functional requirement of Smad4 for gene expression signatures controlled by TGF-β1 (Kowanetz et al., 2004; Levy and Hill, 2005), using either knock-down of Smad4 by RNA interference (Levy and Hill, 2005), or Smad4-deficient cells with rescue of Smad4 expression (Kowanetz et al., 2004). Levy et al. used tetracycline-inducible expression of Smad4 siRNA to knock-down Smad4 expression in HaCaT cells. Smad4 knock-down caused a loss of growth inhibitory and migratory responses, while the EMT response remained intact. Interestingly, of 114 target genes responsive to 1 h or 6 h of TGF-β1 treatment, the majority (65) was scored as Smad4-independent, and the authors draw associations between Smad4-independent gene and phenotypic responses.

In contrast, Kowanetz et al. report that all but 12 of the 66 TGF-β1 and 114 BMP7 target genes identified in Smad4-deficient MDA-MB-468 breast cancer cells were dependent on ectopic expression of Smad4 protein, indicating that the vast majority of TGF-β family target genes is Smad4-dependent. Significant methodological differences between these two studies may explain the discrepancy in their results and conclusions. For example, it is possible that Levy et al. considerably overestimate the class of Smad4-independent genes in their study, because the classification is based on narrowly-defined threshold criteria without statistical verification (Levy and Hill, 2005). Additional studies of TGF-β responses in epithelial cells with Smad4 alterations will be needed to resolve this issue. Other studies describe Smad4-dependent and Smad4-independent gene expression profiles, but lack statistical analysis and biological replication (Ijichi et al., 2004; Jazag et al., 2005; Wu et al., 2003).

Strong evidence exists for cross-talk and cooperation between TGF-β and Erk MAPK (see Chapter 16). Erk is rapidly activated by TGF-β in the context of growth arrest (Hartsough and Mulder, 1995) and EMT *in vitro* (Ellenrieder et al., 2001; Zavadil et al., 2001; Xie et al., 2004). Pharmacological inhibition of MEK upstream of Erk1/2 blocked key morphological features of EMT such as disassembly of E-cadherin mediated adherens junctions in all these models. A transcriptome screen of HaCaT keratinocytes stimulated to undergo EMT with TGF-β in the absence or presence of inhibitor of MEK/Erk identified ~80 EMT-related targets of Erk (Zavadil et al., 2001). This subset is enriched for genes with defined roles in cell-matrix interactions, cell motility, and endocytosis, suggesting that Erk controls cell motility and disruption of adherens junctions. Interestingly, Erk1/2 function is required for induction of the Notch ligand Jagged1 by TGF-β1, which was required for activation of Notch1 receptor signaling and EMT in HaCaT cells (Zavadil et al., 2004).

Kowanetz et al. identified 65 TGF-β1 and 114 BMP7 target genes in MDA-MB-468 cells with ectopic expression of Smad4 (Kowanetz et al., 2004). 25 genes were coregulated by TGF-β1 and BMP7 (Table 1). In particular, inhibitors of differentiation ID2 and ID3 showed longterm repression by TGF-β1, but sustained induction by BMP7. The opposing regulation of ID2 and ID3 was Smad4-dependent. Interestingly, forced expression of ID2 and ID3 abrogated growth inhibition and EMT induced by TGF-β, while knock-down of endogenous ID2 and ID3 sensitizes epithelial cells to growth inhibition and transdifferentiation induced by BMP. Thus, ID2 and ID3 may provide molecular sensors at the intersection of opposing Smad pathways controlled by TGF-β and BMP7 signaling (Kowanetz et al., 2004).

4. TGF-β/SMAD GENE EXPRESSION SIGNATURES IN FIBROBLASTS

4.1 Overview

TGF-β/Smad signaling in fibroblasts has been tightly associated with activation of profibrotic responses (Schnaper and Kopp, 2003). Dysregulated TGF-β1 signaling is widely considered as a critical stimulus for the progression of chronic disease characterized by tissue fibrosis (see Chapter 22) (Border and Noble, 1994). Five studies report gene expression responses to TGF-β1 treatment in mouse or human fibroblasts of various origins and genotype (Yang et al., 2003; Karlsson et al., 2005; Renzoni et al., 2004; Untergasser et al., 2005; Chambers et al., 2003). Two studies used at least three biological replicates and statistical testing to identify TGF-β-responsive genes (Yang et al., 2003; Renzoni et al., 2004). Yang et al. (2003) used a systematic microarray design to identify TGF-β1 gene responses in mouse embryonic fibroblasts derived from wildtype, Smad2 knock-out and Smad3 knock-out embryos, respectively. Similarly, Karlsson et al. (2005) used mouse embryonic fibroblasts with genetic deletion of TGF-β type I receptor. Renzoni et al. (2004) compared TGF-β1 gene responses in fibroblasts derived from normal and fibrotic human lungs, respectively. In contrast, Untergasser et al. (2005) characterized gene responses associated with TGF-β1-induced myofibroblastic differentiation of stromal fibroblasts derived from prostate. Finally, Chambers et al. (2003) performed a time series analysis of TGF-β gene responses in human embryonic lung fibroblasts. Experimental design and microarray features of all studies are summarized in Supporting Table 1.

4.2 Fibroblast Gene Expression Signatures

A total of 758 unique, named TGF-β1-responsive genes were identified in five studies of gene expression in fibroblasts (Supporting Table 3). The fibroblast gene expression signature consists of 56 TGF-β1-responsive genes that were identified in 2 or more studies (Table 2). *CTGF* and *BHLHB2* were the most robust signature genes, present in five or four studies, respectively, followed by serine (or cysteine) peptidase inhibitor, clade E, member 1 (*SERPINE1*), *MYC*, cysteine-rich angiogenic inducer 61 (*CYR61*), cartilage oligomeric matrix protein (*COMP*), and tropomyosin 1 (*TPM1*), which were present in three studies (Table 2). Interestingly, *SERPINE1, MYC, CTGF*, and *BHLHB2* are also among the most robust signature genes in epithelial cells (see Table 3). While CTGF, MYC, and SERPINE1 are well-characterized TGF-β target proteins involved in cell cycle and extracellular matrix turnover, very little is known about the putative roles of BHLHB2, CYR61, COMP, and TPM1 in TGF-β responses in either epithelial cells or fibroblasts. BHLHB2, also known as STRA-13 or DEC-1, is a widely-expressed transcription factor that promotes chondrogenic differentiation (Shen et al., 2002). In addition, BHLHB2 may mediate anti-apoptotic activity by blocking mitochondrial apoptosis pathways (Li et al., 2002), and may play an important role in the adaptation of cells to hypoxia (Miyazaki et al., 2002). CYR61 is a member of the CCN (Cyr61, Ctgf, Nov) family of signaling proteins and is closely related to CTGF (Perbal, 2004). Acting as an extracellular, matrix-associated signaling molecule, CYR61 promotes the adhesion of endothelial cells through interaction with integrin and augments growth factor-induced DNA synthesis in the same cell type. COMP is a noncollagenous extracellular matrix (ECM) protein (Hecht et al., 2005). Mutations can cause the osteochondrodysplasias pseudochondroplasia (PSACH) and multiple

Table 2. TGF-β/Smad gene expression signatures in fibroblasts (The online version of this supplementary table is freely accessible via the website of the book: http://www.springer.com/1-4020-4542-5)

Score	Gene	Factor	Cell Type	Biological Process	Source Ref
1	**Ctgf**	**T1**	**F, E, C2C12**	**cell-cell signaling**	**Y, R, U, K, C**
0,8	**Bhlhb2**	**T1**	**F, E**	**transcription**	**Y, R, K, C**
0,6	Comp	T1	F	cell-matrix adhesion	R, U, C
0,6	Cyr61	T1	F	cell-cell signaling	Y, R, C
0,6	**Myc**	**T1**	**F, E**	**cell cycle arrest**	**Y, K, C**
0,6	**SERPINE1**	**T1**	**F, E**	**blood coagulation**	**R, U, C**
0,6	Tpm1	T1	F	regulation of muscle contraction	K, R, C
0,4	ACTG2	T1	F	cytoskeletal organization and biosynthesis	C, U
0,4	**ACTN1**	**T1**	**F, E**	**cytoskeletal organization and biosynthesis**	**C, K**
0,4	ANXA2	T1	F	skeletal development	C, K
0,4	Bag2	T1	F	apoptosis	Y, K
0,4	Bcat1	T1	F	cell cycle	Y, K
0,4	Cdkn2c	T1	F	cell cycle arrest	Y, K
0,4	**COL4A1**	**T1**	**F, E**	**cell-matrix adhesion**	**C, R**
0,4	Cry1	T1	F	DNA repair	Y, R
0,4	**CSRP2**	**T1**	**F, E**	**cell growth**	**C, Y**
0,4	Ddx21	T1	F	mRNA processing	R, K
0,4	DOC1	T1	F	biological process unknown	C, R
0,4	ELN	T1	F	cell prolifereation	C, R
0,4	F11r	T1	F	cell motility	Y, K
0,4	**Fhl2**	**T1**	**F, E**	**transcription**	Y, U
0,4	Flnb	T1	F	cytoskeletal organization and biosynthesis	Y, K
0,4	**Gadd45b**	**T1**	**F, E**	**apoptosis**	Y, R
0,4	Gas1	T1	F	cell cycle arrest	R, K
0,4	Glul	T1	F	glutamine biosynthesis	Y, K
0,4	Hbp1	T1	F	transcription	Y, K
0,4	**ID1**	**T1**	**F, E**	**transcription**	**C, K**
0,4	**ID3**	**T1**	**F, E, C2C12**	**transcription**	**C, K**
0,4	IER3	T1	F	anti-apoptosis	C, R
0,4	IGFBP3	T1	F	cell-cell signaling	C, U
0,4	**JUNB**	**T1**	**F, E**	**transcription**	**C, R**
0,4	Lama5	T1	F	cell-matrix adhesion	Y, K
0,4	Lrrfip1	T1	F	transcription	Y, R
0,4	Maged1	T1	F	transcription	Y, K
0,4	MAPK6	T1	F	signal transduction	C, Y
0,4	NME1	T1	F	GTP biosynthesis	C, K
0,4	Nolc1	T1	F	cell cycle	Y, K
0,4	Pdgfra	T1	F	signal transduction	Y, K
0,4	Plk2	T1	F	signal transduction	R, K
0,4	PLOD2	T1	F	protein modification	C, R
0,4	Pprc1	T1	F	transcription	Y, K
0,4	Ptprf	T1	F	cell adhesion	Y, K

(Continued)

Table 2. (Continued)

Score	Gene	Factor	Cell Type	Biological Process	Source Ref
0,4	Sesn1	T1	F	cell cycle arrest	Y, K
0,4	Shrm	T1	F	cytoskeletal organization and biosynthesis	Y, K
0,4	Slc7a1	T1	F	amino acid metabolism	Y, R
0,4	**SMAD7**	**T1**	**F, E, C2C12**	**signal transduction**	**C, R**
0,4	Smurf2	T1	F	transcription	Y, R
0,4	Syne2	T1	F	cytoskeletal organization and biosynthesis	Y, K
0,4	**TAGLN**	**T1**	**F, E**	**cytoskeletal organization and biosynthesis**	**C, U**
0,4	Tgfbr3	T1	F	signal transduction	Y, R
0,4	TIMP1	T1	F	cell-matrix adhesion	K, U
0,4	**Tnfrsf1b**	**T1**	**F, E**	**apoptosis**	Y, R
0,4	**Txnip**	**T1**	**F, E**	**oxidative stress**	Y, R
0,4	Zfp36l1	T1	F	mRNA catabolism	Y, K
0,4	Zfp395	T1	F	transcription	Y, K

Y = Yang, et al., Proc.Natl.Acad.Sci.U.S.A, 2003; R = Renzoni, et al., Respir.Res., 2004; U = Untergasser, et al., Mech.Ageing Dev., 2005; K = Karisson, et al., Physiol Genomics, 2005; C = Chambers, et al., Am.J.Pathol., 2003.
bold face indicates target genes in more than one cell type; T1 indicates TGF-β1; E indicates epithelial cell; F indicates fibroblasts; C2C12 indicates mesenchymal progenitor cells.

epiphyseal dysplasia (MED). TPM1 is a member of the tropomyosin family of highly conserved, widely distributed actin-binding proteins involved in the contractile system of striated and smooth muscles and the cytoskeleton of non-muscle cells (Perry, 2001). Mutations in this gene are associated with type 3 familial hypertrophic cardiomyopathy. Interestingly, epigenetic suppression of TPM1 may alter TGF-β tumor suppressor function and contribute to metastatic properties of tumor cells (Varga et al., 2005).

Approximately one quarter of fibroblast expression signature genes are also included in the epithelial TGF-β1 expression signature (*CTGF, BHLHB2, ID3, ID1, MYC, JUNB, SERPINE1, SMAD7*, growth arrest and DNA-damage-inducible, beta (*GADD45B*), *FHL2*, transgelin (*TAGLN*), tumor necrosis factor receptor superfamily, member 1b (*TNFRSF1B*), and *TXNIP*). Products of these genes may be good candidates for ubiquitous TGF-β1 expression signatures with primary functional roles in cell cycle arrest (MYC, ID1, ID3), apoptosis (GADD45B, TNFRSF1B), cytoskeletal organization and biogenesis (FHL2, TAGLN), cell-cell signaling (CTGF, SERPINE1), and stress responses (BHLHB2, TXNIP). In general, some of the major functional roles of fibroblast expression signature genes appear similar to epithelial signature genes, including transcription, signal transduction, cell cycle arrest and cytoskeletal organization and biosynthesis. However, the TGF-β responsive fibroblast target genes in those categories are surprisingly distinct from their functionally-related classes of epithelial target genes. For example, AP1 transcription factors (*JUN, FOS, JUNB, FOSB,* fos-like antigen 1 (*FOSL1*)) are characteristic of the epithelial expression signature, but only *JUNB* appears to be a target of the TGF-β response in fibroblasts. Similarly, numerous MAPK pathway-related genes and genes encoding secreted factors involved in cell-cell signaling are characteristic of the epithelial expression signature, but are rare in the fibroblast expression signature. It is also surprising that the fibroblast expression signature is relatively devoid of well-characterized TGF-β target genes involved in cell-matrix adhesion and extracellular matrix synthesis (with the exception of *COL4A1*), as demonstrated previously by matrix-focused macroarray-based gene expression profiling in dermal fibroblasts (Verrecchia et al., 2001).

4.3 Distinct Roles of Smad2, Smad3 and Erk1/2 in Control of TGF-β Gene Expression Responses in Fibroblasts

The relative ease of establishing fibroblast cultures from mouse embryos enabled Yang et al. (2003) and Karlsson et al. (2005) to derive fibroblasts from Smad2 or Smad3 knock-out embryos, and TGF-β type I receptor knock-out embryos, respectively. Karlsson et al. (2005) identified 290 unique, named TGF-β1 target genes in wildtype mouse embryonic fibroblasts (MEFs), but none in TβR1 knock-out MEFs, indicating that all TGF-β1 responsive target genes require TβR1 for activation. Yang et al. (2003) investigated expression profiles of genes controlled by TGF-β in fibroblasts with genetic ablations of Smad2 or Smad3, and chemical ablation of Erk MAPK. Of 150 TGF-β-responsive genes in wildtype MEFs treated up to 4 hours, only 9 were also regulated in Smad3 knock-out MEFs. In contrast, 161 genes and 238 genes were TGF-β-responsive in Smad2 knock-out MEFs, or MEFs pretreated with U0126 MEK/Erk inhibitor. These results suggest that Smad3 is the essential mediator of TGF-β signaling and directly activates genes encoding regulators of transcription and signal transducers through Smad3/Smad4 DNA-binding motif repeats. In contrast, Smad2 and Erk may predominantly function as transmodulators of regulation of both immediate-early and intermediate genes by TGF-β/Smad3. Yang et al. (2003) proposed a hierarchical model of gene regulation in which TGF-β causes direct activation by Smad3 of cascades of regulators of transcription and signaling that are transmodulated by Smad2 and/or Erk.

5. BMP/SMAD GENE EXPRESSION SIGNATURES IN MURINE MESENCHYMAL PRECURSOR CELLS C2C12

5.1 Overview

TGF-β superfamily members have distinct roles in controlling differentiation of mouse mesenchymal precursor cells into osteoblastic cells (see Chapters 5 and 14). Regulation and transcriptional targets of BMP receptor signaling have been extensively studied in mesenchymal precursors such as C2C12 cells (Miyazono et al., 2005). For example, BMP/Smad pathways cooperate through direct interaction with the transcription factor CBFA1/RUNX2 to regulate key effectors of osteoblast

Table 3. BMP2/Smad gene expression signatures in C2C12 mesenchymal progenitor cells (The online version of this supplementary table is freely accessible via the website of the book: http://www.springer.com/1-4020-4542-5)

Score	Gene	Biological Process	Factor	Cell Type	Source Ref
0.80	Cd97	signal transduction	B2, 3, 6, 9, 12	C2C12	K, D(b), V, P
0.60	Chrnb1	ion transport	B2, 3, 6, 9, 12	C2C12	K, V, P
0.60	Col6a1	cell-matrix adhesion	B2, 3, 6, 9, 12	C2C12	D(b), V, P
0.60	Dpysl3	neurogenesis	B2, 3, 6, 9, 12	C2C12	D(b), V, P
0.60	F3	blood coagulation	B2, 3, 6, 9, 12	C2C12	D(b), V, P
0.60	Gsto1	metabolism	T1, B2	C2C12	D(a), D(b), V
0.60	Krt1-19	cytoskeleton organization and biogenesis	B2	C2C12	K, D(b), V
0.60	Lfng	development	B2, 3, 6, 9, 12	C2C12	D(b), V, P
0.60	Lox	oxidoreductase activity	B2, 3, 6, 9, 12	C2C12	K, V, P
0.60	Lxn	metalloendopeptidase inhibitor activity	B2, 3, 6, 9, 12	C2C12	D(b), V, P
0.60	**Ppap2b**	**development**	**B2**	**C2C12, E**	**D(b), V, P**
0.60	Prrx2	development	B2, 3, 6, 9, 12	C2C12	D(b), V, P
0.60	Tfrc	endocytosis	B2	C2C12	K, D(b), V
0.40	Adam19	heart development	B2, 3, 6, 9, 12	C2C12	K, P
0.40	Antxr1	biological process unknown	B2	C2C12	K, V
0.40	Bteb1	transcription	B2	C2C12	D(a), V
0.40	Car3	metabolism	B2	C2C12	D(b), V
0.40	Ccnd1	cell cycle	B2	C2C12	D(b), V
0.40	Cd44	cell adhesion	B2	C2C12	K, V
0.40	Chrng	ion transport	B2, 3, 6, 9, 12	C2C12	K, P
0.40	Cmya1	cell-cell signaling	B2, 3, 6, 9, 12	C2C12	K, P
0.40	Col18a1	cell-matrix adhesion	B2	C2C12	D(b), V
0.40	**Csrp2**	cell differentiation	**B2, 3, 6, 9, 12**	**C2C12, E, F**	**V, P**
0.40	**Ctgf**	**cell-cell signaling**	**B2, 3, 6, 9, 12**	**C2C12, E, F**	**D(a), P**
0.40	Dnpep	proteolysis and peptidolysis	B2	C2C12	D(b), V
0.40	Ecm1	transport	B2	C2C12	K, D(b)
0.40	Emilin2	cell adhesion	B2	C2C12	D(b), V
0.40	Enah	cytoskeleton organization and biogenesis	B2	C2C12	K, V
0.40	Gatm	creatine biosynthesis	B2	C2C12	K, V
0.40	Grem2	cell-cell signaling	B2, 3, 6, 9, 12	C2C12	V, P
0.40	Gse1	biological process unknown	B2	C2C12	D(a), V
0.40	**Idb3**	**transcription**	**B2, 3, 6, 9, 12**	**C2C12, E, F**	**V, P**
0.40	Igf1	anti-apoptosis	B2, 3, 6, 9, 12	C2C12	K, P
0.40	Itga7	cell-matrix adhesion	B2, 3, 6, 9, 12	C2C12	K, P
0.40	Lamb3	cell-matrix adhesion	B2, 3, 6, 9, 12	C2C12	V, P
0.40	Ly6a	external side of plasma membrane	B2	C2C12	D(b), V
0.40	Mcm4	DNA replication	B2, 3, 6, 9, 12	C2C12	K, P

(Continued)

Table 3. (Continued)

Score	Gene	Biological Process	Factor	Cell Type	Source Ref
0.40	Msln	extracellular space	B2, 3, 6, 9, 12	C2C12	V, P
0.40	Mthfd2	metabolism	B2	C2C12	D(b), V
0.40	Musk	development	B2	C2C12	D(b), V
0.40	Myl4	cytoskeleton organization and biogenesis	B2, 3, 6, 9, 12	C2C12	K, P
0.40	Nsg1	dopamine receptor signaling pathway	B2	C2C12	D(b), V
0.40	Ogn	extracellular space	B2	C2C12	D(b), V
0.40	Omd	cell-matrix adhesion	B2	C2C12	D(b), V
0.40	Ostf1	transcription	B2	C2C12	K, V
0.40	Pdlim3	NOT muscle development	B2, 3, 6, 9, 12	C2C12	V, P
0.40	Pkia	signal transduction	B2, 3, 6, 9, 12	C2C12	V, P
0.40	Postn	cell-matrix adhesion	B2	C2C12	K, V
0.40	Ppap2a	metabolism	B2	C2C12	D(b), V
0.40	Pparg	cell fate commitment	B2, 3, 6, 9, 12	C2C12	K, P
0.40	Psat1	metabolism	B2, 3, 6, 9, 12	C2C12	V, P
0.40	Rgs2	signal transduction	B2, 3, 6, 9, 12	C2C12	V, P
0.40	Rhou	cell cycle	B2, 3, 6, 9, 12	C2C12	K, P
0.40	Rrm2	DNA replication	B2, 3, 6, 9, 12	C2C12	V, P
0.40	Serpinf1	cell-cell signaling	B2, 3, 6, 9, 12	C2C12	V, P
0.40	Sertad4	biological process unknown	T1, B2	C2C12	D(a), D(b)
0.40	Slc20a1	transport	B2, 3, 6, 9, 12	C2C12	D(a), V
0.40	Smad6	signal transduction	B2, 3, 6, 9, 12	C2C12	D(b), P
0.40	**Smad7**	**signal transduction**	**B2, 3, 6, 9, 12**	**C2C12, E, F**	**D(a), P**
0.40	Snapap	exocytosis	B2	C2C12	D(b), V
0.40	Sox11	transcription	B2	C2C12	K, D(a)
0.40	Tcn2	ion transport	B2	C2C12	K, V
0.40	Tgfbi	cell adhesion	B2	C2C12	K, V
0.40	**Tgif**	**transcription**	**B2**	**C2C12, E**	**K, D(a)**
0.40	Tgm2	signal transduction	B2	C2C12	K, V
0.40	**Tieg1**	**bone mineralization**	**B2, 3, 6, 9, 12**	**C2C12, E**	**D(b), P**
0.40	**Timp3**	**basement membrane**	**B2**	**C2C12, E**	**D(a), V**
0.40	Ugp2	metabolism	B2	C2C12	K, V

K = Korchynskyi, et al., J.Bone Miner.Res., 2003; D(a) = de Jong, et al., J.Bone Miner.Res., 2002; D(b) = de Jong, et al., Biochem.Biophys.Res.Commun., 2004; V = Vaes, et al., J.Bone Miner.Res., 2002; Peng, et al., J.Cell Biochem., 2003.
bold face indicates target genes in more than one cell type; T1 indicates TGF-β1; B2 indicates BMP2; E indicates epithelial cell; F indicates fibroblasts; C2C12 indicates mesenchymal progenitor cell.

differentiation, and mice lacking Runx2 show complete loss of bone formation (Miyazono et al., 2005). Microarray-based investigations of BMP reponses in C2C12 led to the identification of new genes which may play important roles in osteoblast differentiation and skeletal development (de Jong et al., 2002; de Jong et al., 2004; Vaes et al., 2002; Korchynskyi et al., 2003; Peng et al., 2003).

5.2 BMP Responsive Gene Expression Signatures Associated with Osteogenic Differentiation of Mesenchymal Progenitors

Combined 582 unique, named BMP responsive genes were identified and reported in five studies where C2C12 mesenchymal precursor cells are induced into osteoblastic differentiation by osteogenic BMPs, mostly BMP2 (Supporting Table 4). Pairwise cross-referencing of BMP responsive genes revealed 68 genes that were reported in at least two different studies (Table 3). The highest scoring genes were identified in four (*Cd97*) or three out of five studies and represent relatively new BMP target genes, including CD97, cholinergic receptor, nicotinic, beta polypeptide 1 (CHRNB1), procollagen, type VI, alpha 1 (*Col6a1*), dihydropyrimidinase-like 3 (*Dpysl3*), coagulation factor III (*F3*), glutathione S-transferase omega 1 (*Gsto1*), keratin complex 1, acidic gene 19 (*Krt1-19*), latexin (*Lxn*), lunatic fringe gene homolog (*Lfng*), lysyl oxidase (*Lox*), *Ppap2b*, paired related homeobox 2 (Prrx2) and transferrin receptor 1 (Tfrc) (Table 3). Only few of the BMP target genes in C2C12 cells are also regulated by TGF-β1 in epithelial cells (*Id3*), cysteine and glycine-rich protein 2 (*Csrp2*), *Ppap2b*, TGF-β-induced factor (*Tgif*), TGF-β inducible early growth response (*Tieg*), and *Timp3*).

Overall, however, the biological and functional classifications of BMP responsive genes in mesenchymal precursor cells are fundamentally different from those induced by TGF-β1 or BMP7 in epithelial cells and fibroblasts. Thus, in contrast with the characteristic epithelial and fibroblastic TGF-β1 expression signatures, genes with functional roles in transcription, signal transduction, cell cycle, apoptosis, cell-cell signaling (growth factors/cytokines), and oxidative response are rare or absent in the BMP expression signature in C2C12 cells (Table 3). For example, there are no AP1 or Ets transcription factors or MAPK pathway related genes. In contrast, there is significant enrichment for target genes with roles in cytoskeleton organization and biosynthesis (*Krt1-19*, enabled homolog (*Enah*), myosin light polypeptide 4 (*Myl4*), thymosin-β4, X-linked (*Tmsb4x*)), and cell adhesion (*Cd44, Col6a1, Col18a1,* elastin microfibril interfacer 2 (*Emilin2*), integrin-α7 (*Itga7*), laminin-β3 (*Lamb3*), periostin (*Postn*) and transforming growth factor, beta induced (*Tgfbi*)), or development and cell differentiation (A disintegrin and metallopeptidase domain 19 (*Adam19*), *Dpysl3, Lfng*, muscle skeletal receptor tyrosine kinase (*Musk*), PDZ and LIM domain 3 (*Pdlim3*), *Ppap2b* and *Prrx2*).

In addition, several BMP responsive genes are known to be involved in osteogenesis and/or skeletal development, including osteoglycin (*Ogn*), osteomodulin (*Omd*), *Postn*, osteoclast stimulating factor 1 (*Ostf1*), *Pdlim3* and Tieg1. Osteoglycin and osteomodulin encode small proteoglycans which contain tandem leucine-rich repeats. Osteoglycin induces ectopic bone formation in conjunction with TGF-β (Ge et al., 2004). TGF-β1 may induce osteomodulin (also known as osteoadherin), contributing to mineralization of odontoblasts (Lucchini et al., 2002). Periostin is also known as osteoblast-specific factor 2 or OSF2 and is a direct transcriptional target of the key osteoblastic transcription factor Runx2 (Stock et al., 2004). Finally, several of the BMP-responsive targets in C2C12 cells have also been identified as BMP2 target genes in another microarray screen of osteoblast differentiation reported by Balint et al. (2003), including *Ogn, Omd, Col6a1*, and *Timp3*.

Kahai et al. (2004) analyzed gene expression in a different osteoblast differentiation model *in vitro* using clonal MC3T3-E1 cells treated with TGF-β1. In contrast with the presented shortterm C2C12 models, this study uses longterm exposure (6 days) to very high concentrations of TGF-β1. Inspite of the significantly different experimental designs, several genes of the BMP2 expression signature in C2C12 were also considered responsive to longterm TGF-β1 treatment, including *Tgfbi, Dpysl3*, solute carrier family 20 member 1 (*Slc20a1*), *Ppap2b*, and lymphocyte antigen 6 complex locus A (*Ly6a*). In addition, TGF-β1 also stimulated genes of the epithelial and fibroblastic TGF-β1 expression signature, including *Bhlhb2, Txnip*, hairy and enhancer of split 1 (*Hes1*) and *Bmp1*.

6. TGF-β/SMAD FAMILY GENE EXPRESSION SIGNATURES IN SPECIALIZED PHYSIOLOGICAL CONTEXTS

TGF-β signaling is typically mediated through ALK5, a type I receptor of the ALK family that activates Smad2 and Smad3 (Miyazono et al., 2000). However, another type I receptor expressed in endothelial cells, ALK1, binds TGF-β ligands and activates BMP pathway Smads, including Smad1, 5 and 8 (Oh et al., 2000). Ota et al. (2002) reported gene expression profiles induced in human umbilical endothelial cells by adenovirus-mediated expression of constitutively active ALK1 or ALK5. These receptors mediated differential regulation of target genesets. For example, ALK1 specifically regulated expression of *SMAD6, SMAD7, ID1, ID2*, endoglin (*ENG*), signal transducer and activator of transcription 1 (*STAT1*), interleukin 1 receptor-like 1 (*IL1RL1*), Smad1, chemokine (C-X-C motif) receptor 4 (*CXCR4*), ephrin-A1 (*EFNA1*), and plakoglobin (*JUP*), whereas ALK5 controlled expression of phosphatidylinositol glycan, class F (*PIGF*), smooth muscle protein 22-alpha (*SM22a*), connexin 37 (*GJA4*), *TGFBI*, latent transforming growth factor-β binding protein 1 (*LTBP1*), claudin 5 (*CLDN5*), and integrin-β5 (*ITGB5*) (Ota et al., 2002).

Matsuyama et al. (2003) used MG63 human osteosarcoma cells as a model for TGF-β1 induced cell proliferation and analyzed gene expression by microarray to identify downstream mediators. Several mitogenic growth factors, including platelet-derived growth factor-A (PDGFA) were induced by TGF-β1. In addition, MYC expression was upregulated in MG63 cells, in contrast with TGF-β1-stimulated downregulation of MYC commonly observed in cells with growth inhibitory responses.

Luo et al. (2005) compared gene expression profiles of leiomyoma and matched myometrial smooth muscle cells in response to TGF-β1 and identified 310 genes that were differentially regulated, including *IL11, TGFBI, TIEG*, CBP/p300-interacting transactivator, with Glu/Asp-rich carboxy-terminal domain 2 (*CITED2*), growth arrest specific 1 (*GAS1*), and runt-related transcription factor 2 (*RUNX2*). The majority of TGF-β-responsive genes in leiomyoma was dependent on TGF-β type II receptor.

7. APPLICATION OF TGF-β/SMAD EXPRESSION GENE EXPRESSION SIGNATURES *IN VIVO*

7.1 TGF-β/Smad Markers in EMT Profiles of Medial Edge Epithelium in Palatogenesis

TGF-βs are widely considered important regulators of epithelial-mesenchymal transitions during development, tissue remodeling after injury, and invasive/metatastic progression of tumors (Zavadil and Böttinger, 2005). While EMT-associated TGF-β target gene responses have been well-characterized using *in vitro* cell culture models (Zavadil et al., 2001; Zavadil et al., 2004; Kowanetz et al., 2004; Valcourt et al., 2005; Levy and Hill, 2005; Xie et al., 2003), little is known concerning EMT gene expression signatures *in vivo*. For example, TGF-β2 and TGF-β3 have been strongly implicated in EMT during palatogenesis and endocardiac cushion formation, respectively (Proetzel et al., 1995; Camenisch et al., 2002). LaGamba et al. (2005) isolated medial edge epithelium from palatal shelf during critical stages of phenotypic EMT in mouse and determined the associated gene expression profiles by microarray analysis. During the first 12 hour period, functional expression profiles for cellular adhesion, desmosome formation, focal adhesion kinase complex, and AP1 transcription factor FOS are characteristic. Between 12 and 24 h, the loss of adherent morphology is observed and the mesenchymal phenotype begins to manifest (LaGamba et al., 2005). The phenotype conversion is associated with cell cycle arrest and anti-apoptotic profiles, as well as extracellular matrix and cytoskeletal remodeling. Interestingly, a considerable number of the key genes observed in developmental EMT *in vivo* are also represented on the *in vitro* epithelial and fibroblast gene expression signatures, including cyclin D2 (*CCND2*), cyclin-dependent kinase inhibitor 1A (*CDKN1A*), cyclin-dependent kinase inhibitor 2B (*CDKN2B*), *BIRC, FOS, ITGB1, COL6A1* (epithelial TGF-β signature) and actin-γ2 (*ACTG2*) and filamin beta (*FLNB*) (mesenchymal TGF-β1 signature). However, discrepancies between expression profiles of EMT during palatogenesis and TGF-β-induced EMT *in vitro* were noted. For example, *in vitro* EMT profiles indicating previously recognized regulation of RHO, NOTCH, and NFkB dependent gene expression profiles were not observed (LaGamba et al., 2005). Similarly, TGF-β-inducible repressors of E-cadherin that are typically described in EMT *in vitro*, such as SNAIL, SLUG, and TWIST, were not upregulated during palatogenesis (LaGamba et al., 2005).

7.2 TGF-β/Smad Gene Expression Signatures in Tumor Invasion and Metastasis

A growing body of evidence suggests that TGF-β/Smad signaling exerts oncogenic properties promoting tumor invasiveness and metastasis formation associated with EMT of malignant cells of epithelial origins (Cui et al., 1996; Derynck et al., 2001;

Wakefield and Roberts, 2002; Valcourt et al., 2005). Massagué and coworkers reported a series of elegant studies using microarray approaches to identify and validate genes that differentially mediate breast cancer metastasis to bone or lung *in vivo*, respectively (Kang et al., 2003; Minn et al., 2005b; Minn et al., 2005a; Kang et al., 2005). By *in vivo* selection and gene expression profiling of subpopulations of MDA-MB-231 breast cancer cells for osteolytic metastasis formation in bone, they identified a bone metastasis signature which is highly enriched for TGF-β expression signature genes *CTGF*, deleted in liver cancer 1 (*DLC1*), *IL11*, and cytochrome P450-B1 (*CYP1B1*) (Kang et al., 2003). In a separate report, a lung metastasis signature of 54 genes was identified that includes TGF-β expression signature genes *ID1*, prostaglandin endoperoxide synthase 2 (*PTGS2*), chemokine (C-X-C motif) ligand 1 (*CXCL1*), and neural precursor cell expressed, developmentally down-regulated 9 (*NEDD9*) (Minn et al., 2005a). Matrix metallopeptidase 1 (*MMP1*) and SRY (sex determining region Y)-box 4 (*SOX4*) are included in both, osteolytic bone metastasis and lung metastasis gene signatures of human breast cancer cells (Minn et al., 2005a). Minn et al. (2005b) further discriminate a subset (Subset A) consisting of four genes that confer both breast tumorigenicity and basal lung metastagenicity, including *ID1, CXCL1, PTGF2*, and *MMP1*. Interestingly, all four genes are also represented in the epithelial TGF-β/SMAD gene expression signature in Table 1. Moreover, induction of the osteolytic bone metastasis expression marker IL11 requires cooperation of SMAD and AP1 transcription factors for strong expression (Kang et al., 2005). JUNB was consistently identified as member of the AP1 complex. In contrast, there was no overlap of the TGF-β/SMAD gene expression signatures described here and a set of 70 genes that was previously reported to indicate poor prognosis of primary breast cancer (van 't Veer et al., 2002). Thus, the *in vitro* TGF-β/Smad gene expression signatures identified here are considerably enriched for highly-specific marker gene expression signatures in human breast cancer cells that mediate lung metastasis or osteolytic bone metastasis, respectively, consistent with the proposed oncogenic role of TGF-β pathways in tumor metastasis formation.

Similarly, comparison of gene expression profiling of human noninvasive and invasive lung adenocarcinoma demonstrated an enrichment of TGF-β/Smad target genes among genes that are differentially-expressed according to tumor invasiveness (Borczuk et al., 2005). These include *DLC1*, mucin 1 (*MUC1*), laminin gamma 2 (*LAMC2*), *TNFRSF1B*, protease serine 8 (*PRSS8*), CSE1 chromosome segregation 1 (*CSE1L*), V-ets erythroblastosis virus E26 oncogene homolog 2 (*ETS2*), nonmetastatic cells 1 (*NME1*), and period homolog 2 (*PER2*).

Thus, the examples presented in section 8 demonstrate that TGF-β gene expression signatures that were identified by cross-referencing of microarray reports of TGF-β-dependent gene regulation *in vitro*, can be useful to identify known TGF-β target genes among gene expression profiles derived from developmental or disease model systems *in vivo*. Of note, numerous additional gene expression profiling studies using *in vivo* models or human disease samples have been reported. Their detailed discussion would be beyond the scope of this chapter.

7.3 Use of TGF-β/Smad Gene Expression Signatures as Molecular Markers to Support Evaluation of Therapeutics Targeting TGF-β/Smad Pathways

Large molecule TGF-β antagonists and small molecule inhibitors of the TGF-β receptor type 1 kinase are under development for primary indications in fibrotic diseases and metastatic cancers (Yingling et al., 2004). Systemic, long-term inhibition of TGF-β has long been considered as problematic because of the anticipation of adverse affects due to loss of immunosuppressive and tumor suppressor activities of TGF-β. However, recent reports demonstrate that long-term inhibition of TGF-β in rodent models by transgenic overexpression of soluble inhibitors (soluble TGF-β receptor type 2 Fc chimera (SR2F)) or use of pan-neutralizing antibody 1D11 is well-tolerated (Yang et al., 2002; Ruzek et al., 2003). As inhibitors of TGF-β signaling are advancing from preclinical studies to phase I and phase II trials in humans, it will be essential to validate informative molecular markers and biomarkers as quantitative, molecular endpoints to evaluate efficacy and safety of therapeutic compounds. Cross-referencing emerging candidate disease markers, such as lung and bone metastasis signatures (Kang et al., 2003; Minn et al., 2005a), with TGF-β/Smad gene expression signatures will likely identify promising molecular markers to assess TGF-β/Smad pathway activities *in vivo* quantitatively and qualitatively.

8. IN SILICO ADVANCES FOR TGF-β/SMAD GENE EXPRESSION SIGNATURES AND SIGNALING NETWORKS BY ADVANCED GENE EXPRESSION DATA MINING

Recent advances in informatics tools and statistical methods enable increasingly reliable and informative gene expression data mining strategies, in which a handful of known pathway members can be used as 'baits' to identify relevant gene expression data sets and assign new members to a pathway of interest. Hu et al. applied such a strategy to elucidate an interconnected TGF-β signaling network in human kidney (Hu et al., 2005). Five genes were identified by literature review, that represent well-established TGF-β signaling responses in kidney, including *JUNB, SERPINE1, GADD45B, CYR61,* and *CTGF*. All five genes were among the TGF-β target genes frequently identified by microarray analysis in epithelial cells and/or fibroblasts (see Table 1 and 2). Additional TGF-β signaling genes were identified by searching a proprietary human kidney gene expression database for coregulation using a non-parametric statistic-based gene scoring system (NP score) (Hu et al., 2005). Remarkably, the identified top-ranking new candidate members of the TGF-β signaling network were highly-significantly enriched with TGF-β target genes that were also identified in several studies by experimental *in vitro* approaches, including *JUN*, activating transcription factor 3 (*ATF3*), *FOS, COL4A1, TIMP3,* serum/glucocorticoid regulated kinase (*SGK*), *GADD45B, TAGLN, TPM1,* actinin alpha1 (*ACTN1, TIMP1*, and glutamate-ammonia ligase (*GLUL*) (see Table 1 and 2).

TGF-β signaling subnetworks connect the identified genes based on their reported biological relationships such as protein binding and gene regulation. This analysis indicates novel signaling interactions between TGF-β/Smad signaling and complement pathways, prostaglandin D2 biosynthesis, and interleukin 6 (IL6) signaling pathway, in addition to known interactions, such as integrin signaling pathways and small leucine-rich proteoglycans (SLRP) molecules (Hu et al., 2005).

9. SUMMARY AND FUTURE PERSPECTIVES

Enabling cross-referencing of lists of experimentally-defined genes of interest with the TGF-β gene expression signatures for epithelial cells (Table 1) and fibroblasts (Table 2), as well as BMP gene expression signatures in mesenchymal progenitors (Table 3), investigators now have access to a curated, quality-controlled reference system for TGF-β/Smad family gene expression data. Thus, tables presented in this chapter provide several important features: i) a measure of robustness of TGF-β target genes by indicating the fraction of studies that identified a particular target gene (SCORE); ii) standardized, updated database identifiers (UNIGENE ID) and nomenclature (NAME, SYMBOL); iii) TGF-β family member known to regulate a particular gene (FACTOR); iv) cell type(s) in which regulation was observed (CELL TYPE); v) primary biological processes or functions of the respective gene products (BIOLOGICAL PROCESS); and vi) the source references where regulation of a particular gene has been observed (SOURCE REF). Supporting Tables are available online to provide this information in addition to the information contained in Tables 1, 2, and 3.

However, the reader also needs to understand that the gene expression signatures described in our chapter can not be considered definitive or comprehensive for the following reasons: i) gene expression profiling is not comprehensive because of the considerable variation of microarray platforms and total number of probesets used in various studies; and ii) most TGF-β target gene expression profiles have not been validated experimentally to date. Nevertheless, new datasets of expression patterns characterizing TGF-β/Smad-responsive genes by using up-to-date, genome-wide microarray platforms are published at an accelerated pace and are expected to allow ever more comprehensive and increasingly reliable and robust prediction of TGF-β/Smad gene expression signatures. Together with advances in informatics, proteomics, and genomics, a rich and reliable data environment of TGF-β/Smad gene expression signatures will be essential for understanding TGF-β signaling networks and biological responses at a systems biology level.

REFERENCES

Balint, E., Lapointe, D., Drissi, H., van der, M.C., Young, D.W., Van Wijnen, A.J., Stein, J.L., Stein, G.S., and Lian, J.B., 2003. Phenotype discovery by gene expression profiling: mapping of biological processes linked to BMP-2-mediated osteoblast differentiation. *J Cell Biochem* 89: 401-426.

Borczuk, A.C., Kim, H.K., Yegen, H.A., Friedman, R.A., and Powell, C.A., 2005. Lung adenocarcinoma global profiling identifies type II transforming growth factor-β receptor as a repressor of invasiveness. *Am J Respir Crit Care Med* 172: 729-737.

Border, W.A., and Noble, N.A., 1994. Transforming growth factor β in tissue fibrosis. *N Engl J Med* 331: 1286-1292.

Brown, P.O., and Botstein, D., 1999. Exploring the new world of the genome with DNA microarrays. *Nat Genet* 21: 33-37.

Camenisch, T.D., Molin, D.G., Person, A., Runyan, R.B., Gittenberger-de Groot, A.C., McDonald, J.A., and Klewer, S.E., 2002. Temporal and distinct TGFβ ligand requirements during mouse and avian endocardial cushion morphogenesis. *Dev Biol* 248: 170-181.

Chambers, R.C., Leoni, P., Kaminski, N., Laurent, G.J., and Heller, R.A., 2003. Global expression profiling of fibroblast responses to transforming growth factor-β1 reveals the induction of inhibitor of differentiation-1 and provides evidence of smooth muscle cell phenotypic switching. *Am J Pathol* 162: 533-546.

Chen, C.R., Kang, Y., and Massagué, J., 2001. Defective repression of c-myc in breast cancer cells: A loss at the core of the transforming growth factor β growth arrest program. *Proc Natl Acad Sci U S A* 98: 992-999.

Cui, W., Fowlis, D.J., Bryson, S., Duffie, E., Ireland, H., Balmain, A., and Akhurst, R.J., 1996. TGFβ1 inhibits the formation of benign skin tumors, but enhances progression to invasive spindle carcinomas in transgenic mice. *Cell* 86: 531-542.

de Jong, D.S., Steegenga, W.T., Hendriks, J.M., van Zoelen, E.J., Olijve, W., and Dechering, K.J., 2004. Regulation of Notch signaling genes during BMP2-induced differentiation of osteoblast precursor cells. *Biochem Biophys Res Commun* 320: 100-107.

de Jong, D.S., van Zoelen, E.J., Bauerschmidt, S., Olijve, W., and Steegenga, W.T., 2002. Microarray analysis of bone morphogenetic protein, transforming growth factor β, and activin early response genes during osteoblastic cell differentiation. *J Bone Miner Res* 17: 2119-2129.

Deacu, E., Mori, Y., Sato, F., Yin, J., Olaru, A., Sterian, A., Xu, Y., Wang, S., Schulmann, K., Berki, A., Kan, T., Abraham, J.M., and Meltzer, S.J., 2004. Activin type II receptor restoration in ACVR2-deficient colon cancer cells induces transforming growth factor-β response pathway genes. *Cancer Res* 64:, 7690-7696.

Derynck, R., Akhurst, R.J., and Balmain, A., 2001. TGF-β signaling in tumor suppression and cancer progression. *Nat Genet* 29: 117-129.

Derynck, R. and Zhang, Y.E., 2003. Smad-dependent and Smad-independent pathways in TGF-β family signalling. *Nature* 425: 577-584.

Ellenrieder, V., Hendler, S.F., Boeck, W., Seufferlein, T., Menke, A., Ruhland, C., Adler, G., and Gress, T.M., 2001. Transforming growth factor β1 treatment leads to an epithelial-mesenchymal transdifferentiation of pancreatic cancer cells requiring extracellular signal-regulated kinase 2 activation. *Cancer Res* 61: 4222-4228.

Ge, G., Seo, N.S., Liang, X., Hopkins, D.R., Hook, M., and Greenspan, D.S., 2004. Bone morphogenetic protein-1/tolloid-related metalloproteinases process osteoglycin and enhance its ability to regulate collagen fibrillogenesis. *J Biol Chem* 279: 41626-41633.

Hartsough, M.T. and Mulder, K.M., 1995. Transforming growth factor β activation of p44mapk in proliferating cultures of epithelial cells. *J Biol Chem* 270: 7117-7124.

Hayashi, H., Abdollah, S., Qiu, Y., Cai, J., Xu, Y.Y., Grinnell, B.W., Richardson, M.A., Topper, J.N., Gimbrone, M.A., Jr., Wrana, J.L., and Falb, D., 1997. The MAD-related protein Smad7 associates with the TGFβ receptor and functions as an antagonist of TGFβ signaling. *Cell* 89: 1165-1173.

Hecht, J.T., Hayes, E., Haynes, R., and Cole, W.G., 2005. COMP mutations, chondrocyte function and cartilage matrix. *Matrix Biol* 23: 525-533.

Hu, G., Jain, K., and Hurle, M., 2005. Revealing transforming growth factor-β signaling transduction in human kidney by gene expression data mining. *OMICS* 9: 266-280.

Ijichi, H., Otsuka, M., Tateishi, K., Ikenoue, T., Kawakami, T., Kanai, F., Arakawa, Y., Seki, N., Shimizu, K., Miyazono, K., Kawabe, T., and Omata, M., 2004. Smad4-independent regulation of p21/WAF1 by transforming growth factor-β. *Oncogene* 23: 1043-1051.

Jazag, A., Ijichi, H., Kanai, F., Imamura, T., Guleng, B., Ohta, M., Imamura, J., Tanaka, Y., Tateishi, K., Ikenoue, T., Kawakami, T., Arakawa, Y., Miyagishi, M., Taira, K., Kawabe, T., and Omata, M., 2005. Smad4 silencing in pancreatic cancer cell lines using stable RNA interference and gene expression profiles induced by transforming growth factor-β. *Oncogene* 24: 662-671.

Kahai, S., Vary, C.P., Gao, Y., and Seth, A., 2004. Collagen, type V, alpha1 (COL5A1) is regulated by TGF-β in osteoblasts. *Matrix Biol* 23: 445-455.

Kang, Y., He, W., Tulley, S., Gupta, G.P., Serganova, I., Chen, C.R., Manova-Todorova, K., Blasberg, R., Gerald, W.L., and Massagué, J., 2005. Breast cancer bone metastasis mediated by the Smad tumor suppressor pathway *Proc Natl Acad Sci U S A* 102: 13909-13914.

Kang, Y., Siegel, P.M., Shu, W., Drobnjak, M., Kakonen, S.M., Cordon-Cardo, C., Guise, T.A., and Massagué, J., 2003. A multigenic program mediating breast cancer metastasis to bone. *Cancer Cell* 3: 537-549.

Karlsson, G., Liu, Y., Larsson, J., Goumans, M.J., Lee, J.S., Thorgeirsson, S.S., Ringner, M., and Karlsson, S., 2005. Gene expression profiling demonstrates that TGF-β1 signals exclusively through receptor complexes involving Alk5 and identifies targets of TGF-β signaling. *Physiol Genomics* 21: 396-403.

Kloeker, S., Major, M.B., Calderwood, D.A., Ginsberg, M.H., Jones, D.A., and Beckerle, M.C., 2004. The Kindler syndrome protein is regulated by transforming growth factor-β and involved in integrin-mediated adhesion. *J Biol Chem* 279: 6824-6833.

Korchynskyi, O., Dechering, K.J., Sijbers, A.M., Olijve, W., and Ten Dijke, P., 2003. Gene array analysis of bone morphogenetic protein type I receptor-induced osteoblast differentiation. *J Bone Miner Res* 18: 1177-1185.

Kowanetz, M., Valcourt, U., Bergström, R., Heldin, C.-H., and Moustakas, A., 2004. Id2 and Id3 define the potency of cell proliferation and differentiation responses to transforming growth factor β and bone morphogenetic protein. *Mol Cell Biol* 24: 4241-4254.

LaGamba, D., Nawshad, A., and Hay, E.D., 2005. Microarray analysis of gene expression during epithelial-mesenchymal transformation. *Dev Dyn* 234: 132-142.

Levy, L. and Hill, C.S., 2005. Smad4 dependency defines two classes of transforming growth factor β (TGF-β) target genes and distinguishes TGF-{β}-induced epithelial-mesenchymal transition from its antiproliferative and migratory responses. *Mol Cell Biol* 25: 8108-8125.

Li, Y., Zhang, H., Xie, M., Hu, M., Ge, S., Yang, D., Wan, Y., and Yan, B., 2002. Abundant expression of Dec1/stra13/sharp2 in colon carcinoma: its antagonizing role in serum deprivation-induced apoptosis and selective inhibition of procaspase activation. *Biochem. J* 367: 413-422.

Lockhart, D.J., Dong, H., Byrne, M.C., Follettie, M.T., Gallo, M.V., Chee, M.S., Mittmann, M., Wang, C., Kobayashi, M., Horton, H., and Brown, E.L., 1996. Expression monitoring by hybridization to high-density oligonucleotide arrays [see comments]. *Nat Biotechnol* 14: 1675-1680.

Lockhart, D.J. and Winzeler, E.A., 2000. Genomics, gene expression and DNA arrays. *Nature* 405: 827-836.

Lucchini, M., Romeas, A., Couble, M.L., Bleicher, F., Magloire, H., and Farges, J.C., 2002. TGF β1 signaling and stimulation of osteoadherin in human odontoblasts in vitro. *Connect Tissue Res* 43: 345-353.

Luo, X., Ding, L., Xu, J., and Chegini, N., 2005. Gene expression profiling of leiomyoma and myometrial smooth muscle cells in response to transforming growth factor-β. *Endocrinology* 146: 1097-1118.

Massagué, J., 2004. G1 cell-cycle control and cancer. *Nature* 432: 298-306.

Matsuyama, S., Iwadate, M., Kondo, M., Saitoh, M., Hanyu, A., Shimizu, K., Aburatani, H., Mishima, H.K., Imamura, T., Miyazono, K., and Miyazawa, K., 2003. SB-431542 and Gleevec inhibit transforming growth factor-β-induced proliferation of human osteosarcoma cells. *Cancer Res* 63: 7791-7798.

Minn, A.J., Gupta, G.P., Siegel, P.M., Bos, P.D., Shu, W., Giri, D.D., Viale, A., Olshen, A.B., Gerald, W.L., and Massagué, J., 2005a. Genes that mediate breast cancer metastasis to lung. *Nature* 436: 518-524.

Minn, A.J., Kang, Y., Serganova, I., Gupta, G.P., Giri, D.D., Doubrovin, M., Ponomarev, V., Gerald, W.L., Blasberg, R., and Massagué, J., 2005b. Distinct organ-specific metastatic potential of individual breast cancer cells and primary tumors. *J Clin Invest* 115: 44-55.

Miyazaki, K., Kawamoto, T., Tanimoto, K., Nishiyama, M., Honda, H., and Kato, Y., 2002. Identification of functional hypoxia response elements in the promoter region of the DEC1 and DEC2 genes. *J Biol Chem* 277: 47014-47021.

Miyazono, K., Maeda, S., and Imamura, T., 2005. BMP receptor signaling: transcriptional targets, regulation of signals, and signaling cross-talk. *Cytokine Growth Factor Rev* 16: 251-263.
Miyazono, K., ten Dijke, P., and Heldin, C.-H., 2000. TGF-β signaling by Smad proteins. *Adv Immunol* 75: 115-157.
Nakao, A., Afrakhte, M., Morén, A., Nakayama, T., Christian, J.L., Heuchel, R., Kawabata, M., Heldin, N.-E., Heldin, C.-H., and ten Dijke, P., 1997. Identification of Smad7, a TGFβ-inducible antagonist of TGF-β signalling [see comments]. *Nature* 389: 631-635.
Oh, S.P., Seki, T., Goss, K.A., Imamura, T., Yi, Y., Donahoe, P.K., Li, L., Miyazono, K., ten Dijke, P., Kim, S., and Li, E., 2000. Activin receptor-like kinase 1 modulates transforming growth factor-β1 signaling in the regulation of angiogenesis. *Proc Natl Acad Sci U S A* 97: 2626-2631.
Ota, T., Fujii, M., Sugizaki, T., Ishii, M., Miyazawa, K., Aburatani, H., and Miyazono, K., 2002. Targets of transcriptional regulation by two distinct type I receptors for transforming growth factor-β in human umbilical vein endothelial cells. *J Cell Physiol* 193: 299-318.
Peng, Y., Kang, Q., Cheng, H., Li, X., Sun, M.H., Jiang, W., Luu, H.H., Park, J.Y., Haydon, R.C., and He, T.C., 2003. Transcriptional characterization of bone morphogenetic proteins (BMPs)-mediated osteogenic signaling. *J Cell Biochem* 90: 1149-1165.
Perbal, B., 2004. CCN proteins: multifunctional signalling regulators. *Lancet* 363: 62-64.
Perry, S.V., 2001. Vertebrate tropomyosin: distribution, properties and function. *J Muscle Res Cell Motil* 22: 5-49.
Proetzel, G., Pawlowski, S.A., Wiles, M.V., Yin, M., Boivin, G.P., Howles, P.N., Ding, J., Ferguson, M.W.J., and Doetschman, T., 1995. Transforming growth factor-b3 is required for secondary palate fusion. *Nature Genet* 11: 409-414.
Renzoni, E.A., Abraham, D.J., Howat, S., Shi-Wen, X., Sestini, P., Bou-Gharios, G., Wells, A.U., Veeraraghavan, S., Nicholson, A.G., Denton, C.P., Leask, A., Pearson, J.D., Black, C.M., Welsh, K.I., and du Bois, R.M., 2004. Gene expression profiling reveals novel TGFβ targets in adult lung fibroblasts. *Respir Res* 5: 24.
Ruzek, M.C., Hawes, M., Pratt, B., McPherson, J., Ledbetter, S., Richards, S.M., and Garman, R.D., 2003. Minimal effects on immune parameters following chronic anti-TGF-β monoclonal antibody administration to normal mice. Immunopharmacol. *Immunotoxicol* 25: 235-257.
Schena, M., Shalon, D., Davis, R.W., and Brown, P.O., 1995. Quantitative monitoring of gene expression patterns with a complementary DNA microarray. *Science* 270: 467-470.
Schnaper, H.W. and Kopp, J.B., 2003. Renal fibrosis. *Front Biosci* 8: e68-e86.
Shen, M., Yoshida, E., Yan, W., Kawamoto, T., Suardita, K., Koyano, Y., Fujimoto, K., Noshiro, M., and Kato, Y., 2002. Basic helix-loop-helix protein DEC1 promotes chondrocyte differentiation at the early and terminal stages. *J Biol Chem* 277: 50112-50120.
Stock, M., Schafer, H., Fliegauf, M., and Otto, F., 2004. Identification of novel genes of the bone-specific transcription factor Runx2. *J Bone Miner Res* 19: 959-972.
Untergasser, G., Gander, R., Lilg, C., Lepperdinger, G., Plas, E., and Berger, P., 2005. Profiling molecular targets of TGF-β1 in prostate fibroblast-to-myofibroblast transdifferentiation. *Mech Ageing Dev* 126: 59-69.
Vaes, B.L., Dechering, K.J., Feijen, A., Hendriks, J.M., Lefevre, C., Mummery, C.L., Olijve, W., van Zoelen, E.J., and Steegenga, W.T., 2002. Comprehensive microarray analysis of bone morphogenetic protein 2-induced osteoblast differentiation resulting in the identification of novel markers for bone development. *J Bone Miner Res* 17: 2106-2118.
Valcourt, U., Kowanetz, M., Niimi, H., Heldin, C.-H., and Moustakas, A., 2005. TGF-β and the Smad signaling pathway support transcriptomic reprogramming during epithelial-mesenchymal cell transition. *Mol Biol Cell* 16: 1987-2002.
Varga, A.E., Stourman, N.V., Zheng, Q., Safina, A.F., Quan, L., Li, X., Sossey-Alaoui, K., and Bakin, A.V., 2005. Silencing of the Tropomyosin-1 gene by DNA methylation alters tumor suppressor function of TGF-β. *Oncogene* 24: 5043-5052.
van 't Veer, L.J., Dai, H., van de Vijver, M.J., He, Y.D., Hart, A.A., Mao, M., Peterse, H.L., van der, K.K., Marton, M.J., Witteveen, A.T., Schreiber, G.J., Kerkhoven, R.M., Roberts, C., Linsley, P.S.,

Bernards, R., and Friend, S.H., 2002. Gene expression profiling predicts clinical outcome of breast cancer. *Nature* 415: 530-536.

Verrecchia, F., Chu, M.L., and Mauviel, A., 2001. Identification of novel TGF-β/Smad gene targets in dermal fibroblasts using a combined cDNA microarray/promoter transactivation approach. *J Biol Chem* 276: 17058-17062.

Wakefield, L.M. and Roberts, A.B., 2002. TGF-β signaling: positive and negative effects on tumorigenesis. *Curr Opin Genet Dev* 12: 22-29.

Wu, K., Yang, Y., Wang, C., Davoli, M.A., D'Amico, M., Li, A., Cveklova, K., Kozmik, Z., Lisanti, M.P., Russell, R.G., Cvekl, A., and Pestell, R.G., 2003. DACH1 inhibits transforming growth factor-β signaling through binding Smad4. *J Biol Chem* 278: 51673-51684.

Xie, L., Law, B.K., Aakre, M.E., Edgerton, M., Shyr, Y., Bhowmick, N.A., and Moses, H.L., 2003. Transforming growth factor β-regulated gene expression in a mouse mammary gland epithelial cell line. *Breast Cancer Res* 5: R187-R198.

Xie, L., Law, B.K., Chytil, A.M., Brown, K.A., Aakre, M.E., and Moses, H.L., 2004. Activation of the Erk pathway is required for TGF-β1-induced EMT *in vitro. Neoplasia* 6: 603-610.

Yang, Y.A., Dukhanina, O., Tang, B., Mamura, M., Letterio, J.J., MacGregor, J., Patel, S.C., Khozin, S., Liu, Z.Y., Green, J., Anver, M.R., Merlino, G., and Wakefield, L.M., 2002. Lifetime exposure to a soluble TGF-β antagonist protects mice against metastasis without adverse side effects. *J Clin Invest* 109: 1607-1615.

Yang, Y.C., Piek, E., Zavadil, J., Liang, D., Xie, D., Heyer, J., Pavlidis, P., Kucherlapati, R., Roberts, A.B., and Böttinger, E.P., 2003. Hierarchical model of gene regulation by transforming growth factor β *Proc Natl Acad Sci U S A* 100: 10269-10274.

Yingling, J.M., Blanchard, K.L., and Sawyer, J.S., 2004. Development of TGF-β signalling inhibitors for cancer therapy. *Nat Rev Drug Discov* 3: 1011-1022.

Zavadil, J., Bitzer, M., Liang, D., Yang, Y.C., Massimi, A., Kneitz, S., Piek, E., and Böttinger, E.P., 2001. Genetic programs of epithelial cell plasticity directed by transforming growth factor-β. *Proc Natl Acad Sci U S A* 98: 6686-6691.

Zavadil, J. and Böttinger, E.P., 2005. TGF-β and epithelial-to-mesenchymal transitions. *Oncogene* 24: 5764-5774.

Zavadil, J., Cermak, L., Soto-Nieves, N., and Böttinger, E.P., 2004. Integration of TGF-β/Smad and Jagged1/Notch signalling in epithelial-to-mesenchymal transition. *EMBO J* 23: 1155-1165.

CHAPTER 18

SYSTEMS BIOLOGY APPROACHES TO TGF-β/SMAD SIGNALING

MUNEESH TEWARI[1] AND ARVIND RAO[2]

[1] *Fred Hutchinson Cancer Research Center, Seattle WA, USA*

[2] *University of Michigan, Ann Arbor MI, USA*

Abstract: TGF-β/Smad signaling and its biological responses are clearly complex and not fully understood. The developing field of systems biology might be highly relevant to ultimately reconciling and understanding the many diverse observations pertaining to TGF-β and Smads. Although this is still a nascent field, new tools and approaches have produced recent advances, particularly in mapping of signaling networks and in computational modeling of signaling cascades. Some of these have already been applied to study Smad signaling. We will review these studies, as well as discuss some emerging approaches in systems biology that may find future application to the TGF-β/Smad signaling field

Keywords: TGF-β; yeast two-hybrid; protein interaction mapping; interactome; siRNA; reverse genetics; network; modeling; differential equations; high-throughput; morpholino

1. INTRODUCTION

The last decade has seen dramatic progress in our knowledge of the molecules that transduce TGF-β signals. Although there have been many key advances, including most prominently the delineation of the Smad pathway, many aspects of TGF-β signaling still remain obscure. A common theme emerging from studies in diverse organisms and systems is that outputs of TGF-β/Smad signaling are highly context-dependent. Variation in both the environment and genetic background of the organism can lead to very divergent responses. Although some of this context-dependence can now be explained by accumulating knowledge on this topic (Massagué, 2000), much of it remains mysterious.

Although TGF-β/Smad signaling may early on have been thought of as a "linear" pathway, it is becoming clear that this representation falls short in capturing reality, and that a more accurate model is that of a "signaling network" (Massagué, 2000).

P. ten Dijke and C.-H. Heldin (eds.), Smad Signal Transduction, 361–378.

A network is defined quite simply as a system of components (called nodes) that interact with each other (these interactions are represented as links, or "edges" between the nodes) (Barabasi and Oltvai, 2004). In recent years, there has been an explosion of basic research on networks, with the demonstration that the network mode of organization is ubiquitous in biological systems of all sorts. Furthermore, theoretical analyses have suggested that organizing biological components in the form of networks can lead to special system-level properties such as robustness to perturbation, context-dependence and the ability to fine-tune responses, all of which are relevant to TGF-β/Smad signaling.

With this increasing appreciation of and interest in biological networks, TGF-β/Smad signaling has been subjected to ambitious studies that seek to delineate increasingly comprehensive "maps" of molecules and interactions that may be relevant to TGF-β/Smad signaling. Experimental studies have focused on both transcriptional networks (which focus on connections between transcription factors and their target genes), and on protein-interaction networks with the large-scale discovery of novel proteins that interact with known TGF-β/Smad signaling components. Transcriptional regulatory networks will not be addressed in detail here, since transcriptional responses to TGF-β are covered in Chapter 17. We will, however, review large-scale efforts to identify protein interaction networks relevant to TGF-β/Smad signaling, and discuss recent advances in functional analysis of networks, made possible by new tools such as RNA interference. We will also discuss how these tools might be applied at large-scale in genetic screens to identify additional components of Smad regulatory networks.

All of the approaches mentioned above will ultimately lead to increasingly complete "wiring diagrams" of TGF-β/Smad networks. Although they are useful tools, wiring diagrams by themselves do not provide understanding of system behavior. The next frontier, therefore, is learning how to extract biological meaning from these network maps, and ultimately make useful predictions about system behavior. This area of investigation is still in its infancy and the directions are not yet clear. Computational models, however, are likely to be an important part of this. We will discuss a recent study that sought to apply computational modeling to Smad signaling, as well as mention some other generic modeling approaches that might find application in the future.

2. FUNCTIONAL GENOMIC APPROACHES FOR MAPPING SMAD SIGNALING NETWORKS

2.1 Protein Interaction Maps and their Functional Analysis

Although large-scale approaches for developing network maps are a recent development, it must be noted that the cumulative work over many years of hundreds of laboratories has already defined an archetypal TGF-β/Smad signaling network with many dozens (at least) of components and interactions. Some of this information has in fact been compiled in review articles (Massagué, 2000) as well as in on-line databases.

More systematic attempts at delineating the TGF-β/Smad network, however, have been made possible by the development of methods that permit the large-scale identification of protein-protein interactions. These methods include the yeast two-hybrid system as well as co-purification/mass spectroscopy-based approaches. The general approach is to begin with a known protein component of a process under study, and to use it to "fish" for novel interacting proteins. One approach to this is screening proteins of interest against libraries of expressed proteins as in the yeast two-hybrid system (Walhout et al., 2000; Walhout and Vidal, 2001). Another is to seek biochemical co-purification and identification of interacting proteins as in the tandem affinity purification (TAP-tag)/mass spectrometry approach (Puig et al., 2001), as well as in the more recently developed LUMIER assay (Barrios-Rodiles et al., 2005). The yeast two-hybrid and LUMIER approaches have been applied to mapping large TGF-β/Smad networks and will be discussed in detail here.

Two recent studies used large-scale yeast two-hybrid screens to identify novel proteins that interact with known TGF-β/Smad network components. The yeast two-hybrid (Y2H) system is a method that uses a genetic selection to discover protein interactions (Fields and Sternglanz, 1994). Though the assay takes place in yeast, the interacting proteins, which are expressed from exogenously introduced plasmids, can be derived from any species. An open reading frame (ORF) corresponding to a protein of interest, referred to as the "bait", is expressed from a plasmid as an in-frame fusion to the *DNA-binding domain* of a yeast transcriptional activator protein such as GAL4. This hybrid protein (DB-bait), when expressed in yeast, is capable of binding target sites in promoters, but on its own cannot cause transcriptional activation of GAL4-responsive genes because the activation domain of the GAL4 transcription factor is not present. The other "hybrid" component of the system is typically a library of cDNAs cloned as fusions with the *activation domain* of the GAL4 protein. When this library of "prey" hybrid proteins (AD-preys) is transformed into a population of yeast cells expressing the DB-bait fusion protein, in cases where the prey protein interacts with the bait protein, the activation domain of GAL4 is brought into physical proximity of the DB-bait protein, and the complex is then able to activate transcription at GAL-4-responsive promoter elements. This is convenient, as the yeast strains used are engineered to have GAL4 target sites placed upstream of genes such as HIS3 which can be used for growth selection (*i.e.* only yeast in which the HIS3 gene becomes transcriptionally activated are able to grow on medium lacking histidine), as well as placed upstream of chromogenic reporter genes such as *lacZ*. This system has been adapted for high-throughpout assays in which hundreds of baits can be screened against libraries of many thousands of AD-prey proteins to discover protein interactions at a large-scale, ultimately generating protein interaction network maps from this data (Walhout et al., 2000; Walhout and Vidal, 2001).

The first study to use this technique in the context of TGF-β/Smad signaling investigated the proteins encoded by the *daf-7* pathway of *C. elegans* (Tewari et al., 2004). As discussed in Chapter 2, the *daf-7* gene encodes a homolog of TGF-β that signals via type I and type II receptors (DAF-1 and DAF-4 in the worm, respectively), as well as 2 regulatory Smad homologs (DAF-8 and DAF-14), a Smad4 homolog (DAF-3) and a nuclear hormone receptor (DAF-12). This set of proteins, identified as a result of genetic screens, regulates a post-embryonic developmental process in *C. elegans* known as dauer formation (see Chapter 2). The authors utilized each of these known components as baits in high-stringency, sequential two-hybrid screens that ultimately defined a network of 59 interacting proteins all linked by one or two connections to the known DAF-7/TGF-β signaling components. In a few cases, interactions were found that were known from other systems, including interaction with a newly defined *C. elegans* homolog of mammalian Sno/Ski proteins. However, the vast majority were proteins without a prior known role in TGF-β/Smad signaling in any species. Thus, although a putative interaction network was defined based on protein-protein interactions, further investigation was necessary to define the roles of novel proteins found.

For such functional analysis, the authors used RNA interference (RNAi), which relies on the delivery of double-stranded RNA to the cells of the animal to effect specific degradation of the cognate endogenous mRNA (Hannon, 2002). After perturbing nearly all the interactors in the protein interaction map and measuring the propensity toward dauer formation as a readout for DAF-7/TGF-β pathway activity, three genes were found to give strong dauer formation phenotypes following RNAi. These included an ADP-ribosylation factor as well as two other proteins without characterized roles in TGF-β signaling.

Interpretation of the lack of RNAi phenotype with the majority of protein-interaction components was complicated, however, by the fact that RNAi is frequently not fully penetrant, and also by the existence of genetic buffering mechanisms in cells, which would suggest that the role of many relevant components might not be revealed by single gene inactivations. In order to overcome the latter issue, and also to begin to order the various components in the network on the basis of genetic interactions (*i.e.* epistasis and/or synthetic enhancement), the authors next performed a systematic double genetic perturbation experiment in which nearly all the novel components of the protein interaction network were individually subjected to RNAi in the context of known mutations of each of the known components of TGF-β signaling. Taken together, this delineated a clear role for 9 novel components of the network in TGF-β/Smad signaling, and identified 13 genetic interactions between network components. By combining the protein interaction and genetic interaction data, an integrated though preliminary wiring diagram of a TGF-β/Smad signaling network was generated. In addition, many of the novel components identified have clear human homologs, suggesting that this diagram may be a useful scaffold for the study of mammalian TGF-β/Smad signaling.

A second study took a similar approach to define a mammalian Smad signaling network, in this case combining yeast two-hybrid-based discovery of protein interactors with functional analysis of a small subset of the novel interactors (Colland et al., 2004). The investigators used 11 human Smad pathway-related baits (Smad2, 3,

4, 5, 7, 8, SARA, Smurf2, SNIP1 and SnoN) to screen a human placenta cDNA library for interacting clones. From the novel interacting proteins identified in these screens, 12 were chosen as baits and used to again screen the human cDNA library for interacting proteins. At completion of all the two-hybrid screens, 755 protein-protein interactions had been identified, defining a network containing 591 proteins. Of these, at least 27 had been previously shown to be involved in Smad signaling, providing additional validation for the general approach.

The authors decided to study in detail 14 of the newly-defined interactors (chosen on the basis of high-confidence scores in the yeast two-hybrid assay and attractiveness of functional domains or novelty). They used a variety of assays, the most notable of which was an analysis of the effect of RNAi directed against the novel interactors on TGF-β or BMP signaling, either using responsive promoter-luciferase assays in HepG2 cells or by assaying TGF-β or BMP-responsive endogenous target genes in human cells. Seven of the 14 genes selected were found to modulate TGF-β or BMP responses. One of these (LAPTm5) appears to be a lysosomal protein, and based on this and other data the authors suggested there may be a link between Smad signaling and endocytosis.

Ultimately, this study provided an additional demonstration of the utility of the yeast two-hybrid-based protein interaction network mapping approach, and showed that such maps can be useful for identifying novel Smad signaling modulators. It is also of note, that since only a small subset of the novel interactors identified in the study were analyzed functionally using RNAi, it is likely that many additional novel TGF-β/Smad signaling modifiers may remain to be identified from the protein interaction network delineated in the study.

The third, and most recent, study takes a conceptually similar but experimentally distinct approach. Barrios-Rodiles et al. developed a high-throughput approach called luminescence-based mammalian interactome mapping (LUMIER) in which protein pairs are assayed for interaction in human 293T cells (Barrios-Rodiles et al., 2005). The ORF encoding the bait protein is cloned into a mammalian expression plasmid as a fusion to the *Renilla* luciferase coding sequence. The gene encoding the prey protein is cloned into an expression plasmid as a fusion to an epitope tag (Flag). Both plasmids are transfected into cells, following which immunoprecipitation is performed using an antibody directed against the FLAG tag. Co-precipitation of the bait protein is then conveniently detected by measurement of luciferase activity in the immunoprecipitates. After showing proof-of-principle at a small scale, the authors automated the procedure to allow it to be performed at high-throughput in 96-well plates.

The baits chosen in this study corresponded to 10 core components of Smad signaling (Smad 1, 2, 3, 4, 7, Smurf1, Smurf2, TβRI, ALK-2, BMPR1B), as well as various point mutants of some of these molecules. For preys, they assembled 518 cDNAs of proteins selected largely on the basis of the presence of interesting functional domains or annotations that could be relevant to intracellular signaling. Pairwise combinations of each bait and prey (11,914 interaction tests in total) were examined for interaction using automated LUMIER. A further, powerful feature added by the authors was that interactions were examined in the presence or absence of exogenously added TGF-β, allowing them to determine the TGF-β-dependence of interactions. Indeed, in a positive control experiment, Smad2 was shown to be associated with Smad4 only following TGF-β treatment, presumably as a result of Smad2 phosphorylation. And finally, the semi-quantitative nature of the LUMIER assay allowed annotation of results with respect to confidence level.

After carrying out systematic screens, a network of 947 interactions was delineated. The network was highly interconnected, allowing a clustering analysis of network components based on their profile of interaction partners, much like gene expression profiles can be clustered. To carry this out, baits with a similar profile of interaction with preys were first arranged such that baits that tended to interact with a similar set of preys were grouped closer to each other (clustering by interaction partners). Next, a similar clustering was done in a second dimension for the preys, this time using a self-organizing map (SOM) based algorithm. As a result of this analysis, some notable clusters (or "subnetworks") became apparent. One of these was a subnetwork centered on PAK1, a protein that is a member of a family of proteins involved in regulating cytoskeletal dynamics and motility, as well as cell proliferation and survival. In fact, another family member, PAK2, had already been implicated functionally in TGF-β signaling but no physical association with TGF-β pathway proteins had been identified.

Further cluster analysis revealed strong links between TGF-β/Smad pathway components and a subnetwork of proteins centered around Occludin (OCLN) and PAR6. This subnetwork was rich in proteins involved in regulation of tight junctional complexes and cell polarity. In fact, detailed analysis of OCLN demonstrated that this protein associates with TGF-β receptors in a ligand-dependent manner, and that this interaction is required for localization of the receptor to tight junctions. This immediately suggested a role for OCLN in dissolution of tight junctions, which is an early event during epithelial-mesenchymal transition (EMT) which can be induced by TGF-β in certain cell types (see Chapter 7 for further discussion on TGF-β and EMT). And in fact, expression of an interaction-defective mutant allele of OCLN blocked dissolution of tight junctions during EMT, suggesting that targeting of the receptor complex by OCLN is a critical step in TGF-β-induced EMT.

In a companion paper, the authors went on to study PAR6 in the context of TGF-β-induced EMT (Ozdamar et al., 2005). PAR6, a known regulator of epithelial cell polarity and tight junction assembly, was shown to be a phosphorylation substrate of the TGF-β type II receptor (TβRII). This phosphorylation was also shown to be required for TGF-β-induced EMT. Furthermore, phosphorylation of Par6 was shown to lead to its association with the E3 ubiquitin ligase Smurf1 and relocalization of Smurf1 to tight junctions. This was particularly important for EMT, since the Smurf1 target RhoA is important for regulating cytoskeletal changes (protrusive activity in particular) of cells undergoing EMT. The investigators further demonstrated that Smurf1-mediated targeting of RhoA for degradation was critical for dissolution of tight junctions and the process of EMT.

The LUMIER approach is clearly an important advance in protein interaction mapping. It has the advantage of providing a potentially more native environment for detecting interactions, since it is carried out in human cells (prior interaction mapping approaches in human cells had been reported, but not at high-throughput). It also permitted the detection of interactions that only occur following post-translational modifications induced dynamically by TGF-β ligand binding, which is a feature that yeast two-hybrid screens lack. The approach is currently limited, however, by requiring that prey ORFs be cloned individually into epitope-tagged expression vectors for interaction assay. Unlike the case with yeast two-hybrid screens, high complexity cDNA libraries cannot be used, limiting the universe of preys that can be tested for interaction to those which have been cloned one-at-a-time into the FLAG-tagged vector. However, with ongoing projects that aim to clone all the ORFs in the human genome, it is likely that this limitation will diminish rapidly with time. It is also notable that the 518 preys in the Barrios-Rodiles study represent only a small fraction of the human ORFeome, suggesting that there are many more interactions yet to be discovered. It is likely, therefore, that LUMIER-based protein-interaction mapping, combined with functional analysis using RNAi and other methods, will continue to be a powerful and important method for mapping TGF-β/Smad protein interaction networks in the future.

2.2 Large-scale "Reverse Genetic" Screening to Identify Components of the Smad Signaling Network

The protein interaction mapping approaches described above identify proteins that are linked to the core Smad pathway by virtue of protein interactions, and then functionally analyze them using perturbation approaches, typically RNA interference or in some cases expression of dominant negative versions of the novel proteins. Although such approaches have the advantage of not only identifying new components but also demonstrating their physical links to known members of the pathway, they do miss components that are not linked closely by protein interactions or are linked by protein interactions that are not detectable easily in the assay systems used. Recently, new tools have been developed for large-scale analysis using loss-of-function methods that can be applied generically across the genome (Brummelkamp and Bernards, 2003). These "reverse genetic" screens, made possible by the elucidation of complete genome sequences and the advent of tools such as RNA interference, offer a powerful approach for genome-wide analysis.

"Forward genetic screening" typically involves random mutation of the genomes of a population of organisms (typically simpler model organisms such as yeast or flies), followed by identification of individual organisms with a heritable alteration in phenotype. This ultimately requires laborious positional cloning to identify the altered gene. Because of the difficulty in positional cloning, this has been largely restricted to easily tractable model genetic organisms such as yeast or *C. elegans*.

"Reverse genetic" approaches, in contrast, involve perturbing a gene or genes of interest in a more directed fashion and then examining the effect on phenotype. A classic example of this is the creation of targeted gene knock-outs by homologous recombination in mice. Although very powerful, this approach is laborious as well, and is not easily taken in cultured cell lines or at high-throughput. Recently, this situation has been dramatically changed with the generation of resources and approaches for large-scale gene knock-down in cultured mammalian cells (and potentially animals) using RNA interference. Although these approaches have not yet been applied at genome-scale to TGF-β/Smad signaling, they have great potential and it is only a matter of time before they will yield fruit for this field as well. We will discuss these screens now.

RNA interference (RNAi) was a phenomenon first demonstrated clearly in *C. elegans*, when Fire et al. injected double-stranded RNA (dsRNA) into the animals and showed specific degradation of the cognate mRNA (Fire et al., 1998). The early availability of the *C. elegans* genome sequence and good-quality gene predictions permitted the rapid generation of resources for genome-wide RNAi, allowing one-by-one knock-down of gene activities (Kamath et al., 2003). In addition, high-throughput methods for delivery of dsRNA were developed, including soaking of animals in double-stranded RNA (Maeda et al., 2001) and feeding of worms with bacteria expressing double-stranded RNA (Timmons and Fire, 1998; Kamath and Ahringer, 2003). Several screens rapidly followed, leading to the genome-scale identification of genes involved in many processes, including growth and development (Kamath et al., 2003), fat metabolism (Ashrafi et al., 2003), and genetic instability (Pothof et al., 2003). Almost concurrently, genome-wide screens were being carried out using cultured *Drosophila* cell lines, which are highly proficient at taking up double-stranded RNA. The early screens in *Drosophila* successfully identified genes involved in many processes, including cell shape (Kiger et al., 2003), cell viability (Boutros et al., 2004), and Wnt signaling (DasGupta et al., 2005). Although such screens have not yet been carried out at genome-scale to study TGF-β/Smad signaling, it is clear that both *C. elegans* and *Drosophila* are model organisms for which the RNAi reagents are available to do so quite rapidly.

The early attempts at genome-scale screens in model organisms set the precedent for more recent mammalian RNAi screens. Initially, it was not clear whether gene-specific RNAi would even be possible in human or mouse cells, since mammalian cells typically have a strong interferon-based response to exogenous dsRNA, which leads to nonspecific degradation of cellular RNA. However, Elbashir et al. demonstrated in 2001 that short (*i.e.* 21 nucleotide) dsRNAs (known as short interfering RNAs, or siRNAs) could mediate RNAi in mammalian cells without activating the interferon response (Elbashir et al., 2001). This paved the way for large-scale RNAi analysis, and since then several large-scale RNAi screens have been carried out in mammalian cells.

One of the earliest of these screens was a report by Aza-Blanc et al. in which the authors used one-at-a-time transfection of siRNAs directed against a set of 510 genes, enriched for kinases and other so-called "signaling" molecules (Aza-Blanc et al., 2003). In this study, the interest was in identifying genes that might modify the apoptosis-inducing effects of the extracellular ligand TRAIL. They used a simple, dye-based assay for viability and ultimately identified several enhancers and suppressors of TRAIL-induced apoptosis, many of which were novel.

The approach of using transfected siRNA is a simple and powerful one, because the reagents can be chemically synthesized at will and be directed against any sequence. If the optimal sequence is chosen, knock-down can be profound because a substantial amount of siRNA can be delivered. But a major limitation of this method is that knock-down is transient, lasting typically 3-7 days depending on the particular target, making phenotypes that become apparent on a longer time-scale problematic, because as cells proliferate the siRNA activity is diluted out. Furthermore, screening using this approach is limited to those cell lines that are transfectable at sufficiently high efficiency.

An alternative approach seeks to overcome these limitations be achieving stable, long-term expression of DNA cassettes that integrate into the genome and direct the expression of self-complementary molecules that fold to form hairpin dsRNA structures (called short hairpin RNAs, or shRNAs) (Paddison et al., 2002). These hairpin structures are processed in cells to produce siRNAs that can direct RNAi against specific target genes. The cassettes are delivered into cells by transfection, or by infection using retroviral or lentiviral vectors, and stable integration is selected for using antibiotic selection (*i.e.* puromycin).

Several groups have begun to generate genome-wide collections of such constructs, and several screens have recently been reported. The general principle is to transfect or infect the shRNA constructs either singly or in pools, select stable integrants as colonies, and then assay these in some way (reviewed in Brummelkamp and Bernards, 2003). The assays may be promoter:luciferase reporters, morphological assays, or even growth selection such as the ability to grow in soft agar. The range of assays is almost limitless, and it is easy to see how cellular phenotypes related to TGF-β/Smad signaling could be adapted to such a screen. The identity of the shRNA construct is either known in the case of one-shRNA-at-a-time experiments, or in the case where pools are used one can go back to the pool and test subpools to quickly narrow down the active shRNA. In addition, in some cases it is also possible to separate the cells showing the positive phenotype and amplify the integrated shRNA constructs and determine their identities by sequencing. A further variation on this idea embeds unique identifying sequences in the shRNA constructs ("barcodes") that are delivered into cells. Thus, following a phenotypic selection of transfected/infected cells, rather than sequencing many clones to determine identities of the selected shRNAs, the barcodes can instead be PCR-amplified from the cells and hybridized to a microarray that has the uniquely identifying barcode sequences pre-printed as oligonucleotides (Brummelkamp and Bernards, 2003). Early studies using this approach have been promising (Berns et al., 2004; Paddison et al., 2004; Westbrook et al., 2005).

In summary, there are several collections of reagents now available for large-scale RNAi screens in cultured mammalian cells. Though conceptually similar, each approach and reagent set has its advantages and limitations. Thus it is important to choose the correct approach depending on the specific question and application. Although it is not clear yet which (if any) technology is going to emerge as broadly the best approach, it seems quite clear that RNAi-based reverse genetic screens will be very useful in future studies aiming to globally define regulators of mammalian Smad signaling.

In addition to RNAi, another emerging technology for gene knock-down is the use of morpholino antisense oligonucleotides. These reagents, which have been shown to have potent knock-down activity in a variety of organisms, have found particularly widespread application in zebrafish, where they can be injected into developing embryos. Morpholino oligos are modified DNA oligonucleotides that are complementary to the region flanking the translational start site of endogenous mRNAs (Corey and Abrams, 2001). They appear to function by binding to the cognate mRNA and blocking translation, rather than promoting mRNA degradation, as is the case with siRNAs. A recent review of over 200 papers in which these reagents have used indicates that in 74% of cases tested, the oligos phenocopied what would be expected from a genetic mutation (Iversen and Newbry, 2005). Morpholinos in fact may have some advantages over siRNAs, in that the knock-down effect appears to be sustained for a longer period of time since morpholino oligos appear to be quite stable inside cells. Yet like siRNAs, they have the advantage of being able to be synthesized at will against any known sequence.

With the zebrafish genome sequencing project well underway, genome-wide screens are clearly on the horizon. In fact a recent study used morpholino oligos to perturb the activity of 61 genes, selected as zebrafish orthologs of genes enriched in human hematopoietic stem cells. Knock-down of 23% of the genes tested resulted in hematopoietic defects in zebrafish embryos (Eckfeldt et al., 2005). Zebrafish is a particularly attractive organism for reverse genetic analysis because much more of the tissue organization and phenotypic complexity of humans is conserved, as compared with worms and flies. At the same time, the shorter gestation period, ease of cultivation and large brood size makes it more tractable to work with than mice (Detrich et al., 1999; Shin and Fishman, 2002). And what is perhaps most attractive is that complex phenotypes that exist only in a whole animal can be observed, which is not the case with the cell culture-based screens described earlier. Some of the major barriers to genome-wide screens with morpholino oligos are cost and labor involved with delivering (*i.e.* injecting) the oligos into cells. However, it is likely that automation and improvements in production methods and scale may in the future allow genome-scale studies using this approach. Given the complex developmental phenotypes seen with perturbation of TGF-β/Smad signaling in diverse organisms, reverse genetic analysis in zebrafish is likely to be an attractive model for the future system-level study of Smad signaling.

3. FROM NETWORK WIRING DIAGRAMS TO COMPUTATIONAL MODELS OF TGF-β/SMAD SIGNALING

Although having good network wiring diagrams of TGF-β/Smad signaling is an important goal, there is still a large gap between such diagrams and deep understanding of the system-level behavior of this signaling network. Arguably, the most difficult challenge facing systems biology is learning how to move from wiring diagrams to an understanding of the dynamic and global properties of the system that permits accurate predictions about system behavior. Although this area of research is still in its infancy, it appears likely that computational models of network behavior will play a central role.

Computational modeling in systems biology is an evolving discipline (see review by Kitano, 2002). The basic principle, however, is to use detailed knowledge of a biological process (*i.e.* whatever is known about a pathway or network) to generate formal models that are computationally accessible. Based on these models, software can be written to run "simulations", in which various parameters of the model (*i.e.* concentrations of components or strengths of their interactions) are varied, and the resulting effect on other parameters or on the output of the system is calculated. These simulations can be compared to existing experimental data and/or be used to guide the choice of further experiments, which in turn leads to further refinement of the model. Sometimes, the simulations yield surprising and non-canonical predictions about system behavior. And when such unexpected predictions are tested experimentally and prove to be true, the real beauty and power of computational modeling are realized.

There are many approaches for building models, and some of these will be touched upon briefly later. Now, however, we will discuss in more detail one specific modeling approach that has already been applied to analyze Smad pathway signaling. This method, which uses Ordinary Differential Equations (ODEs) to build a model of Smad signal transduction (Kyoda et al., 2000), represents an excellent example of systems modeling in biology.

3.1 Computational Modeling of Smad Signaling using an Ordinary Differential Equations-based Simulator

During fruitfly development, a gradient of expression of the BMP family member Dpp is established along the anterior-posterior axis of the wing disc (see Chapter 3). It is also known that ligand (Dpp) binding to type I and II receptors leads to phosphorylation of Mad (an R-Smad), which, together with Med (a co-Smad) acts to modulate expression of target genes. One of these target genes is Dad, which is an inhibitory Smad (I-Smad) that functions to inhibit signaling by stimulating degradation of the activated receptor (among other mechanisms; see Chapter 19 for further details), providing a negative feedback loop.

Using this cumulative knowledge of *Drosophila* Smad signaling components and interactions/reactions, Kyoda et al. built a computational model of the canonical

Smad signaling pathway (Kyoda et al., 2000) using an Ordinary Differential Equations approach. Simply stated, ODEs are mathematical expressions that can relate the change in concentration (usually over time) of various quantities in a system. As an example, an ODE could be written that describes the relationship between the rate of change in concentration of a ligand-receptor complex, and the diffusion rate of the ligand and concentrations of free ligand and receptor. Additional parameters (*i.e.* kinetic binding constants for a ligand-receptor interaction) that are specific to the particular process (and often experimentally determined) are usually required to more accurately describe the temporal and spatial variation between the different quantities.

The authors wrote a series of ODEs to represent the major events in *Drosophila* Smad signaling, including:

- *Diffusion of ligand (Dpp) along a gradient*: an ODE was written to relate the rate of change in concentration of Dpp ligand over time to its concentration at its physical location, as described by (x,y) coordinates. More intuitively, closest to where the concentration of Dpp is high, the rate of change in Dpp concentration will be faster, whereas in areas of low Dpp concentration the rate of change due to diffusion will be slower.
- *Ligand-receptor binding*: an ODE was written that relates the rate at which ligand-receptor complexes are formed to the ligand concentration at the adjacent cell, the number of surrounding cells, and constants specific to the ligand-receptor pair that describe rates of binding and dissociation.
- *Physical interactions and biochemical reactions*: ODEs were formulated to relate the rate of change of concentration of a protein-protein complex (*i.e.* phospho-Mad forming a complex with the co-Smad Med) or of a reaction product (*i.e.* phospho-Mad formed as a result of a reaction between Mad and ATP, catalyzed by the receptor Ser/Thr kinase activity), to concentrations of the interactors/reactants. Kinetic constants specific to each reaction or interaction were also included.
- *Changes in target gene expression*: ODEs were written to relate the rate of change in concentration of a target gene product (*i.e.* the I-Smad Dad) to the concentration of the activator (*i.e.* transcription factor).

The set of equations was used to write a computer program known as BioDrive that simulates the changes in concentration over time (and over space, in the case of Dpp diffusion), which vary depending on various parameters that can be set (*i.e.* what the initial concentrations are, what the rate constants are, etc.).

They then used the simulator to determine whether the model that had been defined could reproduce one of the known properties of the system – the specific spatial and temporal patterns of transcriptional target gene expression that are seen in the developing *Drosophila* wing disc. The initial parameters were set such that Dpp was produced in the anterior-posterior boundary region of the wing disc, and allowed to diffuse to the anterior and posterior compartments. The simulation was run, and the spatial and temporal changes in concentration of Dad, as well as of the phospho-Mad-Med complex, were graphed. It was satisfying to observe that the model was able to recapitulate the experimentally observed spatial and temporal patterns of activity of these molecules.

The authors then used the simulations to study the consequences of the apparent negative feedback loop that exists in this signaling cascade. From theoretical (and experimental) studies in other systems, it is known that having a negative feedback loop can allow a system to vary the dynamics of its response (*i.e.* rate of production of an activated transcription factor complex) in a way that is determined by the strength of the stimulus (*i.e.* the concentration of a ligand). The authors tested this possibility in the framework of their ODE-based model of Smad signaling. Using BioDrive, they varied the rate of formation of the Dpp-receptor complex (which is determined by the local concentration of Dpp ligand at a given point in the gradient) and simulated how this would affect signaling. In effect, they asked whether the rate of Dpp-receptor complex formation affects the dynamics of response (*i.e.* production of the active phospho-Mad-Med complex), as might be expected in a system with a negative feedback loop. The simulations showed that when ligand binding is slow (*i.e.* where the morphogen concentration is low), the concentration of active phospho-Mad-Med oscillates over time before stabilizing, whereas when Dpp ligand binding is fast (*i.e.* where the morphogen is at high concentration), it quickly reaches steady-state. Thus, the system has evolved in a way that allows positional information (local concentration of Dpp, which is determined by the position in the gradient) to modulate the dynamic pattern of response within a cell.

From a conceptual standpoint, the work of Kyoda et al. has implications for both the modeling and biology communities. First, using the simulator they have constructed, one can potentially discover unexpected, non-canonical behavior of a signaling pathway. Although BioDrive was applied to Smad signaling in this study, it has been written as a generic program that can be applied to many signaling pathways. An additional important feature (not explored in this study), is that the simulator can examine phenomenon at different "scales", meaning that it can be just as easily used to model interactions and events that occur between cells, as those that occur between molecules. This paves the way for potential analysis of Smad signaling in higher order processes that depend upon cell-cell interactions, such as embryonic development or tumorigenesis.

More recently, another study of the role of Dpp in *Drosophila* development has added to understanding of this pathway. Shimmi et al. studied patterning of the dorsal surface of the *Drosophila* blastoderm embryo, a process in which both Dpp and another BMP-family ligand known as Scw are known to be required (Shimmi et al., 2005). Heterozygous knock-out *scw* mutants, however, were known to undergo normal patterning, suggesting that fluctuations in dosage of *scw* are remarkably buffered. The authors first made the surprising discovery that Dpp not only signals as a homodimer, but can also form a heterodimer with Scw which is both more spatially restricted and significantly more potent in stimulating SMAD signaling than Dpp homodimers. They went on to use a mathematical modeling approach to explore the implications of heterodimer formation with respect to the experimentally observed robustness to alteration in Scw gene dosage. By formulating a set of kinetic equations that included all possible homo- and hetero-dimerization reactions involving Dpp and Scw, they were able to run simulations that analyzed the sensitivity of heterodimer output to changes in the concentration of monomer. Their analysis showed that if Scw is present in slight excess with respect to Dpp, reduction in the level of Scw has little effect on the output of the Dpp/Scw heterodimer, even across wide ranges of variation of parameters such as reaction rates, transport coefficients and production rates. Thus, the Dpp/Scw heterodimer model was able to provide a plausible explanation for the robustness of blastomere patterning to reduction in Scw gene dosage.

3.2 Some Other Modeling Approaches with Potential Application to TGF-β/Smad Signaling

Although the ODE-based simulator built by Kyoda et al. and the work of Shimmi et al. are outstanding examples of how mathematical modeling can be applied to Smad signaling, these approaches do have their limitations. For instance, for these methods to work well, it is important that quantitative details of several parameters (*i.e.* relating to rates of association/dissociation, phosphorylation/dephosphorylation, etc.) be fairly well characterized experimentally. Thus, these are potentially useful approaches in cases where a pathway has been investigated in great detail by conventional biochemical methods.

Differential equations are also not the only approach to computational modeling in biology. In this section we will discuss briefly two very distinct kinds of modeling that could find applications in studying various aspects of TGF-β/Smad signaling. This area is still in its infancy, however, and only through further work will the potential utility of these and other modeling approaches become known.

One approach, which, interestingly, has been used successfully in the social and ecological sciences, is called *agent-based modeling*. Here, each component in the system is considered to be an "agent", and "rules of interaction" between agents can be defined based on the best available experimental knowledge (in this way it is similar to the ODE-based study (Kyoda et al., 2000)). Typically, modeling software (*i.e.* Swarm, Repast, Netlogo) allows the agents to be represented as symbols on 2-dimensional grids. Rules of interaction can be defined between agents both with respect to time as well as to spatial location on the grid. Starting parameters (*i.e.* how many agents, their locations, behaviors that depend on time and scale, etc.) are set and then the system is allowed to evolve in accordance with the pre-defined rules. This approach has commonly been used to model interactions between individuals in populations (as an example, to examine population dynamics in predator-prey systems), but could find application to Smad signaling if molecules were, for example, represented as agents. In contrast to ODEs, agent-based models are more qualitative, and therefore do not require as much detailed characterization of parameters such as kinetic rate constants or binding affinities. The advantage of this technique is that it allows the investigation of "synthetic" behaviors or cellular-states which might be important to simulate but experimentally difficult to approach.

Another approach is *probabilistic modeling*. This method requires a large number of experimental observations, and therefore may be particularly suited for interpretation of results from high-throughput experiments. The primary idea here is to collect a large number of observations (*i.e.* from gene expression profiling microarrays, etc.) with varying experimental perturbations. Then, one of several probabilistic models (*i.e.* Bayesian networks (Shmulevich et al., 2002)) can be used to assign a joint probability score to each perturbation-response combination. When these combinations are tabulated, network graph structure algorithms can allow inference of both direct and indirect influences operative between the genes/proteins participating in the process. In addition, other factors that pertain to cellular context (*i.e.* stage of the cell cycle, state of differentiation) can be incorporated. This approach may be particularly relevant for understanding the well-known context-dependent responses of TGF-β.

There are some general principles that apply to the formulation of computational models. First, it is important to have a good idea of the likely relevant parameters in a system (*i.e.* diffusion, ligand binding, etc.), and to also have an idea of what the real-life range of variation is for these parameters, in order to help bound the computational problem. It is notable that this kind of information comes from biological knowledge and insight; there is no substitute for having a deep biological knowledge of the process under study. Once a model is developed based on best available existing knowledge, it is used to run simulations, and results from simulations may then guide the choice of biological experiments. Thus, this kind of investigation relies on its own feedback loop in which experiment guides modeling and vice-versa.

It is also important to understand some of the current challenges in modeling. Prominent among these is the difficulty in integrating very diverse data sources (expression, proteomic, localization, literature, etc.) from experiments of varying throughput, cell context, biological relevance and reliability. To make approaches like the ones listed above computationally viable, we need high quality data as well as some established gold standards to validate the modeling methods. Smad signaling, in this regard, may be a good system for work in this area because of the intensive experimental investigation that it has been subjected to. It is not surprising, then, that Kyoda et al. chose this signaling pathway for their first application of the BioDrive simulation program. Another challenge is in moving from modeling of relatively simple "pathways" of a small number of components and feedback loops, to producing reliable models of large, complex networks of the kind being revealed by the network mapping studies discussed earlier in this chapter. The path in this regard is not at all clear, but given the burgeoning growth in the field of systems biology, progress in this direction can be expected.

4. CONCLUSION AND PERSPECTIVES

In many ways, system-level analysis of TGF-β/Smad signaling has come a long way in the last few years. The field has progressed from a state in which the molecules and interactions that populate the Smad signaling network were discovered one-at-a-time by laborious approaches, to a situation in which new high-throughput tools allow these experiments to be done in rapid order. This has led, and will continue to lead, to increasingly comprehensive network maps comprised of molecules and interactions that modulate Smad signaling. The short- and mid-term yield of system-level approaches to Smad signaling will likely come from these maps, by their utility in facilitating a variety of deeper, smaller-scale investigations related to this pathway in a variety of contexts. Much like having a completed genome sequence for an organism provides the infrastructure to make cloning and studying individual genes easier, having these maps is likely to aid investigators whose work focuses on particular proteins or "small neighborhoods" within the larger network.

A much more formidable challenge is learning how to make global sense of these networks such that reliable predictions about overall system behavior can be made based upon knowledge of the individual components and interactions. Current approaches are limited to predicting behavior of smaller, "local" parts of the larger network. This problem is of substantial interest to the systems biology community, however, and will continue to be one of the areas of long-term investigation in the systems biology field.

ACKNOWLEDGEMENTS

We thank Brad Helbing for his expert assistance in preparation of this manuscript. M.T. is supported by NIH award K08-AG21613 from the National Institute on Aging.

REFERENCES

Ashrafi, K., Chang, F.Y., Watts, J.L., Fraser, A.G., Kamath, R.S., Ahringer, J., and Ruvkun, G., 2003, Genome-wide RNAi analysis of *Caenorhabditis elegans* fat regulatory genes. *Nature* 421: 268-272.

Aza-Blanc, P., Cooper, C.L., Wagner, K., Batalov, S., Deveraux, Q.L., and Cooke, M.P., 2003, Identification of modulators of TRAIL-induced apoptosis via RNAi-based phenotypic screening. *Mol Cell* 12: 627-637.

Barabasi, A.L. and Oltvai, Z.N., 2004, Network biology: understanding the cell's functional organization. *Nat Rev Genet* 5: 101-113.

Barrios-Rodiles, M., Brown, K.R., Ozdamar, B., Bose, R., Liu, Z., Donovan, R.S., Shinjo, F., Liu, Y., Dembowy, J., Taylor, I.W., Luga, V., Przulj, N., Robinson, M., Suzuki, H., Hayashizaki, Y., Jurisica, I., and Wrana, J.L., 2005, High-throughput mapping of a dynamic signaling network in mammalian cells. *Science* 307: 1621-1625.

Berns, K., Hijmans, E.M., Mullenders, J., Brummelkamp, T.R., Velds, A., Heimerikx, M., Kerkhoven, R.M., Madiredjo, M., Nijkamp, W., Weigelt, B., Agami, R., Ge, W., Cavet, G., Linsley, P.S., Beijersbergen, R.L., and Bernards, R., 2004, A large-scale RNAi screen in human cells identifies new components of the p53 pathway. *Nature* 428: 431-437.

Boutros, M., Kiger, A.A., Armknecht, S., Kerr, K., Hild, M., Koch, B., Haas, S.A., Consortium, H.F., Paro, R., and Perrimon, N., 2004, Genome-wide RNAi analysis of growth and viability in *Drosophila* cells. *Science* 303: 832-835.

Brummelkamp, T.R. and Bernards, R., 2003, New tools for functional mammalian cancer genetics. Nat. Rev. Cancer 3: 781-789.

Colland, F., Jacq, X., Trouplin, V., Mougin, C., Groizeleau, C., Hamburger, A., Meil, A., Wojcik, J., Legrain, P., and Gauthier, J.M., 2004, Functional proteomics mapping of a human signaling pathway. *Genome Res* 14: 1324-1332.

Corey, D.R. and Abrams, J.M., 2001, Morpholino antisense oligonucleotides: tools for investigating vertebrate development. *Genome Biol* 2: 1015.

DasGupta, R., Kaykas, A., Moon, R.T., and Perrimon, N., 2005, Functional genomic analysis of the Wnt-wingless signaling pathway. *Science* 308: 826-833.

Detrich, H.W., 3rd, Westerfield, M., and Zon, L.I., 1999, Overview of the Zebrafish system. *Methods Cell Biol* 59: 3-10.

Eckfeldt, C.E., Mendenhall, E.M., Flynn, C.M., Wang, T.F., Pickart, M.A., Grindle, S.M., Ekker, S.C., and Verfaillie, C.M., 2005, Functional analysis of human hematopoietic stem cell gene expression using zebrafish. *PLoS Biol* 3: e254.

Elbashir, S.M., Harborth, J., Lendeckel, W., Yalcin, A., Weber, K., and Tuschl, T., 2001, Duplexes of 21-nucleotide RNAs mediate RNA interference in cultured mammalian cells. *Nature* 411: 494-498.

Fields, S. and Sternglanz, R., 1994, The two-hybrid system: an assay for protein-protein interactions. *Trends Genet* 10: 286-292.

Fire, A., Xu, S., Montgomery, M.K., Kostas, S.A., Driver, S.E., and Mello, C.C., 1998, Potent and specific genetic interference by double-stranded RNA in *Caenorhabditis elegans*. *Nature* 391: 806-811.

Hannon, G.J., 2002, RNA interference. *Nature* 418: 244-251.

Iversen, P.L. and Newbry, S., 2005, Manipulation of zebrafish embryogenesis by phosphorodiamidate morpholino oligomers indicates minimal non-specific teratogenesis. *Curr Opin Mol Ther* 7: 104-108.

Kamath, R.S. and Ahringer, J., 2003, Genome-wide RNAi screening in *Caenorhabditis elegans*. *Methods* 30: 313-321.

Kamath, R.S., Fraser, A.G., Dong, Y., Poulin, G., Durbin, R., Gotta, M., Kanapin, A., Le Bot, N., Moreno, S., Sohrmann, M., Welchman, D.P., Zipperlen, P., and Ahringer, J., 2003, Systematic functional analysis of the *Caenorhabditis elegans* genome using RNAi. *Nature* 421: 231-237.

Kiger, A.A., Baum, B., Jones, S., Jones, M.R., Coulson, A., Echeverri, C., and Perrimon, N., 2003, A functional genomic analysis of cell morphology using RNA interference. *J Biol Chem* 2: 27.

Kitano, H., 2002., Computational systems biology. *Nature* 420: 206-210.

Kyoda, K.M., Muraki, M., and Kitano, H., 2000, Construction of a generalized simulator for multi-cellular organisms and its application to SMAD signal transduction. *Pac. Symp. Biocomput* 5: 314-325.

Maeda, I., Kohara, Y., Yamamoto, M., and Sugimoto, A., 2001, Large-scale analysis of gene function in *Caenorhabditis elegans* by high-throughput RNAi. *Curr Biol* 11: 171-176.

Massagué, J., 2000, How cells read TGF-β signals. *Nat Rev Mol Cell Biol* 1: 169-178.

Ozdamar, B., Bose, R., Barrios-Rodiles, M., Wang, H.R., Zhang, Y., and Wrana, J.L., 2005, Regulation of the polarity protein Par6 by TGFβ receptors controls epithelial cell plasticity. *Science* 307: 1603-1609.

Paddison, P.J., Caudy, A.A., Bernstein, E., Hannon, G.J., and Conklin, D.S., 2002, Short hairpin RNAs (shRNAs) induce sequence-specific silencing in mammalian cells. *Genes Dev* 16: 948-958.

Paddison, P.J., Silva, J.M., Conklin, D.S., Schlabach, M., Li, M., Aruleba, S., Balija, V., O'Shaughnessy, A., Gnoj, L., Scobie, K., Chang, K., Westbrook, T., Cleary, M., Sachidanandam, R., McCombie, W.R., Elledge, S.J., and Hannon, G.J., 2004, A resource for large-scale RNA-interference-based screens in mammals. *Nature* 428: 427-431.

Pothof, J., van Haaften, G., Thijssen, K., Kamath, R.S., Fraser, A.G., Ahringer, J., Plasterk, R.H., and Tijsterman, M., 2003, Identification of genes that protect the *C. elegans* genome against mutations by genome-wide RNAi. *Genes Dev* 17: 443-448.

Puig, O., Caspary, F., Rigaut, G., Rutz, B., Bouveret, E., Bragado-Nilsson, E., Wilm, M., and Seraphin, B., 2001, The tandem affinity purification (TAP) method: a general procedure of protein complex purification. *Methods* 24: 218-229.

Rual, J.F., Hirozane-Kishikawa, T., Hao, T., Bertin, N., Li, S., Dricot, A., Li, N., Rosenberg, J., Lamesch, P., Vidalain, P.O., Clingingsmith, T.R., Hartley, J.L., Esposito, D., Cheo, D., Moore, T., Simmons, B., Sequerra, R., Bosak, S., Doucette-Stamm, L., Le Peuch, C., Vandenhaute, J., Cusick, M.E., Albala, J.S., Hill, D.E., and Vidal, M., 2004, Human ORFeome version 1.1: a platform for reverse proteomics. *Genome Res* 14: 2128-2135.

Shimmi, O., Umulis, D., Othmer, H., and O'Connor, M.B., 2005, Facilitated transport of a Dpp/Scw heterodimer by Sog/Tsg leads to robust patterning of the *Drosophila* blastoderm embryo. *Cell* 120: 873-886.

Shin, J.T. and Fishman, M.C., 2002, From Zebrafish to human: modular medical models. *Annu Rev Genomics Hum Genet* 3: 311-340.

Shmulevich, I., Dougherty, E.R., Kim, S., and Zhang, W., 2002, Probabilistic boolean networks: a rule-based uncertainty model for gene regulatory networks. *Bioinformatics* 18: 261-274.

Tewari, M., Hu, P.J., Ahn, J.S., Ayivi-Guedehoussou, N., Vidalain, P.O., Li, S., Milstein, S., Armstrong, C.M., Boxem, M., Butler, M.D., Busiguina, S., Rual, J.F., Ibarrola, N., Chaklos, S.T., Bertin, N., Vaglio, P., Edgley, M.L., King, K.V., Albert, P.S., Vandenhaute, J., Pandey, A., Riddle, D.L., Ruvkun, G., and Vidal, M., 2004, Systematic interactome mapping and genetic perturbation analysis of a *C. elegans* TGF-β signaling network. *Mol. Cell* 13: 469-482.

Timmons, L. and Fire, A., 1998, Specific interference by ingested dsRNA. *Nature* 395: 854.

Walhout, A.J., Boulton, S.J., and Vidal, M., 2000, Yeast two-hybrid systems and protein interaction mapping projects for yeast and worm. *Yeast* 17: 88-94.

Walhout, A.J. and Vidal, M., 2001, High-throughput yeast two-hybrid assays for large-scale protein interaction mapping. *Methods* 24: 297-306.

Westbrook, T.F., Martin, E.S., Schlabach, M.R., Leng, Y., Liang, A.C., Feng, B., Zhao, J.J., Roberts, T.M., Mandel, G., Hannon, G.J., Depinho, R.A., Chin, L., and Elledge, S.J., 2005, A genetic screen for candidate tumor suppressors identifies REST. *Cell* 121: 837-848.

CHAPTER 19

INHIBITORY SMADS: MECHANISMS OF ACTION AND ROLES IN HUMAN DISEASES

ATSUHITO NAKAO[1,2]
[1] *Department of Immunology, Faculty of Medicine, University of Yamanashi, Yamanashi 409-3898, Japan*
[2] *Atopy (Allergy) Research Center, Juntendo University School of Medicine, Tokyo 113-8421, Japan*

Abstract: Inhibitory Smads are one of the major inhibitory regulators of intracellular signaling mediated by TGF-β superfamily proteins. Inhibitory Smads repress TGF-β superfamily signaling by several different mechanisms at the levels of cellular membrane, cytoplasm, and nucleus. Importantly, expression of I-Smads is induced by TGF-β superfamily proteins themselves and other signaling pathways, indicating a key role for I-Smads in negative feedback or cross-talk control of TGF-β superfamily signaling. Probably, I-Smads tightly control intensity and duration of TGF-β superfamily signaling and maintain homeostasis. Indeed, there is accumulating evidence that aberrant I-Smads expression and activity contributes to human diseases induced by TGF-β superfamily proteins. Better understanding of the mechanisms underlying I-Smads regulation may thus provide novel therapeutic targets for diseases associated with TGF-β superfamily proteins

Keywords: cancer; fibrosis; inflammatory bowel disease; Smad6; Smad7; TGF-β

1. INTRODUCTION

Intracellular signal transduction pathways initiated by cytokines have been studied extensively. As a result, negative feedback by pathway-specific inhibitory molecules has been identified as an important regulatory mechanism to control cytokine activity. For example, signaling mediated by nuclear factor κB (NF-κB) or signal transducer and activator of transcription (STAT) pathways induce expression of inhibitor of κBα (IκBα) or suppressor of cytokine signaling (SOCS) family proteins, respectively, which in turn act to inhibit signaling that led to their activation (Yoshimura et al., 2003). This inhibitory mechanism is perhaps essential to regulate the intensity and duration of intracellular signaling responses and maintain homeostasis.

P. ten Dijke and C.-H. Heldin (eds.), Smad Signal Transduction, 379–395.

Intracellular signaling mediated by TGF-β superfamily is not an exception of this rule. Following the identification of receptor-restricted Smads (R-Smads) and common-mediator Smads (Co-Smads), a unique subset of Smad proteins, that inhibits signaling by TGF-β superfamily (inhibitory Smads: I-Smads), was identified in late 90's (Heldin et al., 1997). In vertebrates Smad6 (Topper et al., 1997; Imamura et al., 1997; Nakayama et al., 1998a), and Smad7 (Topper et al., 1997; Hayashi et al., 1997; Nakao et al., 1997; Nakayama et al., 1998b) and in *Drosophila* daughters against decapentaplegic (Dad) (Tsuneizumi et al., 1997) function as I-Smads.

Expressions of I-Smads are induced by members of the TGF-β superfamily and turn off their signaling via several different mechanisms (Miyazono, 2002). Importantly, I-Smads have been shown to be induced by other signaling pathways such as NF-κB or STATs, indicating a key role for I-Smads in cross-talk of TGF-β superfamily signaling with other signaling pathways (Miyazono, 2002; Nakao et al., 2002). Indeed, although many negative regulators of TGF-β superfamily signaling have been identified so far, the control of TGF-β superfamily signaling by regulating I-Smad expression appears to be particularly important because, 1) potent induction of I-Smads in response to TGF-β family members is frequently observed in many different cell types, 2) knock-down of endogenous I-Smads increases cellular responses to TGF-β superfamily ligands, supporting their role as a physiological negative regulator in the signaling, 3) aberrant expression of I-Smads has been shown to be involved in pathology of some human diseases associated with TGF-β.

This chapter aims to discuss the current state of knowledge concerning structure, function, regulation, and physiological/pathological roles of I-Smads, important regulators of TGF-β superfamily signaling.

2. STRUCTURE OF I-SMADS

The signal transducing R- and Co-Smads have structurally highly conserved N- and C-terminal regions, termed MH1 and MH2 domains, respectively, which are linked by linker regions (Heldin et al., 1997). The MH2 domain plays important roles in receptor recognition, interaction with transcription factors, and homo- and hetero-oligomerization among R-Smads and Co-Smad (Heldin et al., 1997). The MH1 domain exhibits sequence-specific DNA binding activity and negatively regulates the functions of the MH2 domain through physical interaction (Heldin et al., 1997). This physical interaction between MH1 and MH2 domains is released upon receptor activation.

I-Smads (Smad6 and Smad7) have conserved MH2 domains, but their N-terminal domains are highly divergent from the MH1 domains and linker regions of R-Smads and Co-Smads (Heldin et al., 1997). Moreover, amino acid sequences of the N-terminal domains are only partially conserved between Smad6 and 7 (36.7%). These unique structural differences between R-Smads/Co-Smads and I-Smads, and also between Smad6 and Smad7 have been implicated in their functional differences, which are discussed below.

3. MECHANISMS OF I-SMADS ACTION

3.1 Antagonistic Functions

I-Smads inhibit signaling by members of TGF-β superfamily in several different ways (Fig. 1). Originally, I-Smads were reported to repress the signaling by competing with R-Smads for their binding to activated type I receptors and preventing phosphorylation of R-Smads (Imamura et al., 1997; Hayashi et al., 1997; Nakao et al., 1997). For example, Smad7 stably and strongly binds to the activated type I receptors for BMP and TGF-β/Activin and inhibits phosphorylation of Smad1/5 or Smad2/3, thereby repressing BMP and TGF-β/Activin signaling. In contrast, Smad6 potently binds only to the activated type I receptors for BMP, and only weakly to TGF-β or

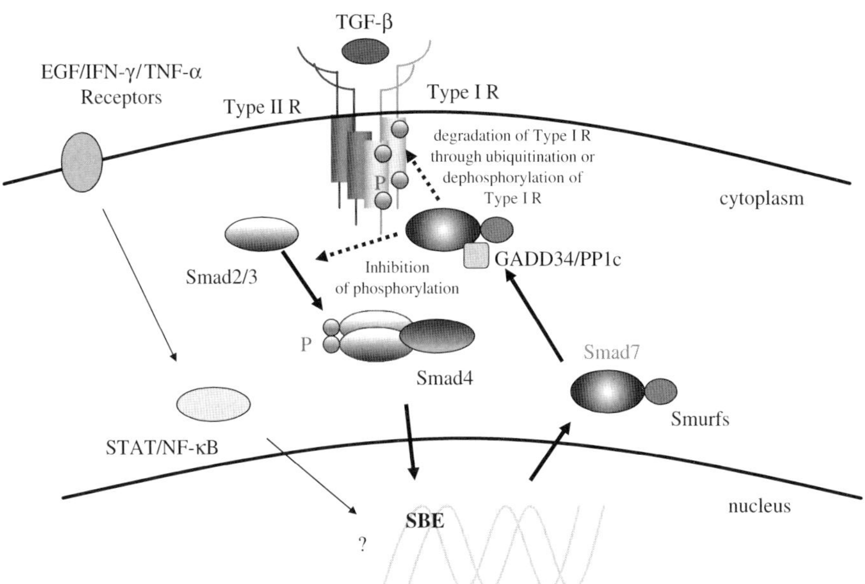

Figure 1. Smad7 is an inhibitory regulator in feedback or cross-control of TGF-β signal transduction. At the cell surface, the ligand (TGF-β) binds a complex of transmembrane receptor serine/threonine kinases (type I and II) and induces transphosphorylation of the GS domain in the type I receptor by the type II receptor kinases. The activated type I receptors phosphorylate R-Smads (Smad2/3), which then form a complex with Co-Smad (Smad4). Activated Smad complexes translocate into the nucleus, where they regulate transcription of target genes including the Smad7 gene acting on Smad binding element (SBE) in the promoter regions. Smad7 is then exported from the nucleus with the help of Smurfs and specifically associates with the activated type I receptors and inhibits TGF-β signaling in different ways, such as 1) interfering with phosphorylation of Smad2/3, 2) inducing degradation of the receptors through ubiquitin/proteosomal pathways with Smurfs, and 3) dephosphorylating the activated type I receptors by recruitment of specific phosphatases (GADD34/PPlc). Other signaling pathways, such as EGF/IFN-γ/TNF-α pathways, are reported to induce Smad7 in certain cell types. There are also other mechanisms of inhibition by I-Smads (please see the text) (A color version of this figure is freely accessible via the website of the book: http://www.springer.com/1-4020-4542-5)

Activin receptors, which may explain the functional differences between Smad6 and Smad7, *i.e.* Smad7 can inhibit both TGF-β and BMP signaling whereas Smad6 preferentially inhibits BMP signaling (Hanyu et al., 2001). Smad6 can also inhibit BMP signaling by competing with Smad4 for the binding to Smad1, thereby preventing formation of a Smad1/Smad4 complex (Hata et al., 1998).

Subsequently, ubiquitin-mediated proteolysis of the TGF-β family receptor, has also been shown to be involved in the inhibitory mechanisms of I-Smads (Kavsak et al., 2000; Ebisawa et al., 2001). I-Smads have been found to strongly interact with Smad ubiquitination regulatory factor 1 (Smurf1) and Smurf2, that are HECT type E3 ubiquitin ligases, containing an N-terminal C2 domain, followed by WW domains and a C-terminal HECT domain (Kavsak et al., 2000; Ebisawa et al., 2001). The HECT domain mediates the E3 ligase activity of Smurfs and the WW domains physically interact with the PY motifs found in I-Smads. Interaction of Smurfs with Smad7 leads to its nuclear export (Itoh et al., 1998) and recruitment of the complex to the activated TGF-β receptors, where it causes degradation of the receptors and Smad7 via ubiquitin/proteasomal pathways, thereby terminating signaling (see Chapter 13; Kavsak et al., 2000; Ebisawa et al., 2001).

In addition, the recruitment of phosphatases to the activated type I receptors by I-Smads is also an effective mechanism for the inhibition of signaling. Smad7 interacts with growth arrest and DNA damage protein, (GADD34), a regulatory subunit of the protein phosphatase 1 (PP1) holoenzyme, which subsequently recruits catalytic subunit of PP1 (PP1c) to dephosphorylate TGF-β type I receptor, thereby inactivating the receptor (Shi et al., 2004).

Furthermore, I-Smads have been suggested to serve as transcriptional repressors, silencing the transcription of target genes in the nucleus. Smad6 represses BMP-induced Id1 transcription through recruiting transcriptional corepressor C-terminal binding protein (CtBP) (Lin et al., 2003). Consistent with these findings, Smad6 has a consensus CtBP-binding motif, PLDLS, in the linker region of Smad6 and mutation in the motif abolishes the Smad6 binding to CtBP and subsequently its transcription repressor activity (Lin et al., 2003). Moreover, Smad6 is reported to interact with the homeodomain transcriptional factor HoxC-8 as a heterodimer when binding to DNA and the Smad6-Hoxc-8 complex inhibits interaction of Smad1 and Smad1/Hoxc-8-induced transcription activity, thereby inhibiting BMP signaling in the nucleus (Bai et al., 2000). Smad6 is also shown to interact with class I histon deacetylases (HDACs) and represses gene transcription by modifying chromatin conformation (Bai et al., 2002).

Thus, through multiple modes of inhibition I-Smads may tightly antagonize the signaling activity mediated by TGF-β superfamily ligands.

3.2 Agonistic Functions

I-Smads may also have agonistic mediator functions downstream of TGF-β superfamily receptors. TGF-β induces apoptosis in several cell types including the human prostate cancer cell line PC-3U cells. Interestingly, TGF-β-induced apoptosis in PC-3U cells is prevented by inhibition of Smad7 expression by anti-sense mRNA, indicating that Smad7 is essential for the induction of apoptosis (Landström et al., 2000). Consistent with this finding, ectopic expression of Smad7 in PC-3U cells induces apoptosis in PC-3U cells. Subsequent studies have shown that TGF-β-induced apoptosis in PC-3U cells depends on activation of the TAK1/MKK3/p38 kinase pathway and Smad7 may be required for their activation as a scaffolding protein (Edlund et al., 2003). In the other hand, in Madin-Darby canine kidney (MDCK) cells, Smad7-induced apoptosis depends on JNK activation (Mazars et al., 2001). Smad7-induced apoptosis is also observed in podocytes, which is associated with suppression of transcriptional activity of the cell survival factor NF-κB and may be involved in podocyte depletion in progressive glomerulosclerosis (Schiffer et al., 2001). Induction of apoptosis through inhibition of NF-κB by Smad7 is reported in MDCK cells as well. (Lallemand et al., 2001). Taken together, Smad7 may potentiate TGF-β-induced apoptosis through distinct mechanisms in certain cell types.

By using the PC-3U cells, Smad7 is also shown to be required for the TGF-β-induced activation of small GTPases Cdc42 and RhoA and rearrangement of cytoskeleton (Edlund et al., 2004). Reduction of Smad7 expression by anti-sense mRNA abolishes the TGF-β-induced activation of Cdc42 and the concomitant reorganization of the actin filament system (Edlund et al., 2004). Collectively, these findings suggest novel agonistic functions of I-Smads in TGF-β superfamily signaling.

3.3 Relevance of the Structure

Isolated MH2 domains of mammalian I-Smads can bind to activated type I receptors for TGF-β superfamily ligands and inhibit their signaling, indicating that the MH2 domains of I-Smads are sufficient for interaction with activated type I receptors and for inhibitory functions (Hanyu et al., 2001; Souchelnytskyi et al., 1998). Indeed, the MH2 domain of Smad6 (Smad6C) or Smad7 (Smad7C) bind to the activated type I receptors for BMP and inhibits BMP signaling as potently as full-length Smad6 or Smad7. Similarly, the MH2 domain of Smad7 (Smad7C) binds to the activated type I receptors for TGF-β/Activin and inhibits TGF-β/Activin signaling (although less efficiently than full-length Smad7; see below). In addition, Smad6C is also required for binding to Smad1, although the oligomeric complex did not enter the nucleus nor serve as a transcriptionally active complex (Hanyu et al., 2001). Thus, the MH2 domain of I-Smads is a major effector domain, serving as a docking site for serine/threonine kinase receptors and as a region responsible for oligomer formation with other Smads similar to the MH2 domains of R-Smads.

On the other hand, the roles of the N-terminal domains of I-Smads remain largely unclear, but they may confer specificity for inhibition of signaling. Reporter assays using TGF-β or BMP responsive reporter constructs reveal that the isolated MH2 domains of Smad7 (Smad7C) are less potent than the full-length Smad7 in inhibiting TGF-β signaling, but not BMP signaling, as described above. In addition, a chimeric molecule containing the N-terminal domain of Smad7 and MH2 domain of Smad6 (Smad7/6), but not one containing the N-terminal domain of Smad6 and MH2 domain of Smad7 (Smad6/7), is as potent as full-length Smad7 in the inhibition of TGF-β signaling. Thus, the N-terminal domain of Smad7 is important for the efficient inhibition of TGF-β signaling by Smad7 (Hanyu et al., 2001). Furthermore, cross-linking assays reveal that TGF-β type II/type I receptor complexes is efficiently co-immunoprecipitated by full length Smad7 and Smad7/6 chimera, but not by full-length Smad6, Smad6/7 chimera, and Smad7C, suggesting that the N-terminal domain of Smad7 determines the affinity of Smad7 for activated TGF-β type I receptors (Hanyu et al., 2001). Consistent with these results, phosphorylation of Smad2 by TGF-β is also efficiently inhibited by full length Smad7 and Smad7/6 chimera, but not by full-length Smad6, Smad6/7 chimera, and Smad7C. Collectively, the N-domain of Smad7 may play a role in the specific inhibition of TGF-β signaling by Smad7 through its strong affinity to TGF-β receptors (Hanyu et al., 2001). However, a recent report shows that the MH2 domains of I-Smads may also take part in the specific inhibition of TGF-β superfamily signaling. Mutations of specific lysine residues within the basic surface groove (Lys312 and Lys316) of the Smad7 MH2 domain abolished inhibition of TGF-β, but not BMP signaling (Mochizuki et al., 2004).

4. REGULATION OF I-SMADS EXPRESSION AND ACTIVITY

4.1 Transcriptional Regulation

Transcriptional regulation of I-Smads has been extensively studied and expression of I-Smads mRNA has been found to be upregulated by various stimuli (Fig. 1; Nakao et al., 2002). Originally, the mRNAs of Smad6 and Smad7 were reported to be rapidly and directly induced by TGF-β, Activin and BMPs in cells, which express functional receptors for these ligands (Nakao et al., 1997; Afrakhte et al., 1998; Takase et al., 1998). Induction of Smad7 mRNAs by TGF-β is mediated by R-Smad/Co-Smad complex, acting on the 8-base pair (GTCTAGAC) palindromic Smad-binding element (SBE) in the promoter region of Smad7 (Nagarajan et al., 1999; Denissova et al., 2000; Brodin et al., 2000). Efficient induction of Smad7 by TGF-β appears to require cooperation of Smad, Sp1, and AP-1 transcriptional factors (Brodin et al., 2000).

The Smad6 promoter contains a 28-base pair GC-rich sequence including four overlapping copies of the GCCGnCGC-like motif, which is similar to a binding site for *Drosophila* Mad and Maedea (Ishida et al., 2000). BMP-activated Smad1/5 and Smad4 can bind to the GC-rich motifs that are important for BMP-induced activation of the Smad6 promoter (Ishida et al., 2000). Multiple BMP responsive elements (BREs) have been identified in the Smad7 gene (Benchabane et al., 2003). Two BREs (BRE1 and 2) that reside in the promoter region contain several Smad1/4 sites, and a the third BRE (I-BRE) that is located in the first intron, contains GATA factor binding sites. The GATA and Smad1-dependent enhancers in the Smad7 gene are important for the differential interpretation of BMP concentrations (Benchabane et al., 2003).

Importantly, Smad7 has been shown to be induced by other signaling pathways, including epidermal growth factor (EGF) (Afrakhte et al., 1998), interferon-γ (IFN-γ)/STAT1 (Ulloa et al., 1999), tumor necrosis factor-α (TNF-α)/NF-κB (Bitzer et al., 2000). In addition, mechanical stimulations, such as vascular laminar shear stress, also rapidly induce Smad6 and Smad7 expression (Topper et al., 1997). Furthermore, ultraviolet irradiation induces Smad7 via activation of AP-1 in human skin fibroblasts (Quan et al., 2005). These findings suggest that I-Smads also serve to mediate cross-talk control between TGF-β signaling and other signaling pathways. Similar to mammalian I-Smads, in *Drosophila* activation of Mothers against decapentaplegic (Mad) by decapentaplegic (Dpp) induces the expression of Dad leading to negative feed-back regulation of Dpp signaling (Tsuneizumi et al., 1997).

Although there are many reports regarding upregulation of I-Smad mRNAs, there is little information on down-regulation of I-Smad mRNAs. The Ski protein has been proposed to serve as a corepressor for Smad4 to maintain a TGF-β-responsive promoter at a repressed, basal level. Indeed, one of the target genes of Ski repression is Smad7 (Denissova and Liu, 2004). Ski together with Smad4 binds to endogenous Smad7 promoter in a SBE-dependent manner and RNAi-mediated knock-down of Ski increases the endogenous level of Smad7 mRNA. Thus, Ski acts as a corepressor for Smad4, which can inhibit transcription of Smad7 at the basal state. Thus, the normal physiological function of Ski may be to repress Smad7 expression and the low level of Smad7 maintained by the repressive action of Ski may greatly facilitate, at least in initial stage, propagation of TGF-β signaling (Fig. 2).

4.2 Posttranslational Regulation

Expression of I-Smads is also tightly controlled by post-translational protein modifications. As described earlier, both Smurf1 and Smurf2 form a stable complex with Smad7 and target the Smurf-Smad7 complex to the plasma membrane, where Smad7 directly binds to the activated type I TGF-β receptor and inhibits phosphorylation of R-Smads (Hata et al., 1998; Kavsak et al., 2000). Smad7 itself undergoes ubiquitination and degradation in this process. Thus, Smurfs negatively regulate protein levels of Smad7 by promoting degradation.

Recent work has shown that the level of Smad7 is also regulated by another E3-ubiquitin ligase, Arkadia (Koinuma et al., 2003). Arkadia was originally identified as a protein that enhances signaling activity of Nodal and induces mammalian nodes during early embryogenesis. Arkadia physically interacts with Smad7, and induces its poly-ubiquitination and degradation. In contrast to Smurf1, which interacts with TGF-β receptor complexes through Smad7 and mediates TGF-β receptor degradation, Arkadia fails to associate with TGF-β receptors, suggesting that Arkadia only targets Smad7. Importantly, silencing of the Arkadia gene results in repression of transcriptional activities induced by TGF-β and BMP, and accumulation of the Smad7 protein. Arkadia may therefore play a role as an amplifier of TGF-β superfamily signaling.

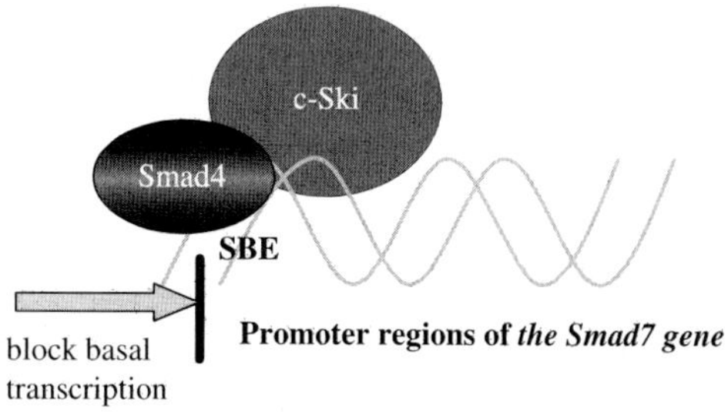

Figure 2. c-Ski is a corepressor of Smad4 that inhibits Smad7 gene in the basal state. c-Ski is recruited by Smad4 to the Smad7 promoter and inhibits the basal promoter activity on Smad binding element (SBE) dependent manner. It is possible that the basal level of Smad7 expression is tightly maintained at very low level by c-Ski and the repression of the Smad7 gene may be useful to greatly facilitate propagation of the TGF-β signal at least in the initial stage (A color version of this figure is freely accessible via the website of the book: http://www.springer.com/1-4020-4542-5)

Most recently, an E3-ubiquitin ligase Atrophin 1-interacting protein 4 (AIP4) is also shown to specifically target Smad7 for ubiquitin-dependent degradation, but without affecting turnover of the activated TGF-β type I receptor (Lallemand et al., 2005). Interestingly, despite the ability to degrade Smad7, AIP4 can inhibit TGF-β signaling, presumably by enhancing the association of Smad7 with the activated TGF-β type I receptor. This may suggest a new mechanism by which E3 ubiquitin ligase function to turn off TGF-β signaling.

Interestingly, the degradation of Smad7 protein by E3-ubiquitin ligase may be regulated by competition between ubiquitination and acetylation of overlapping lysine residues in Smad7 (Fig. 3; Grönroos et al., 2002). In the nucleus, Smad7 is associated with the nuclear acetyl transferase p300, which acetylates lysine residues 64 and 70 in the N-terminus of Smad7. The acetylation of Smad7 protects it against ubiquitination and degradation mediated by the ubiquitin ligase Smurf1 as

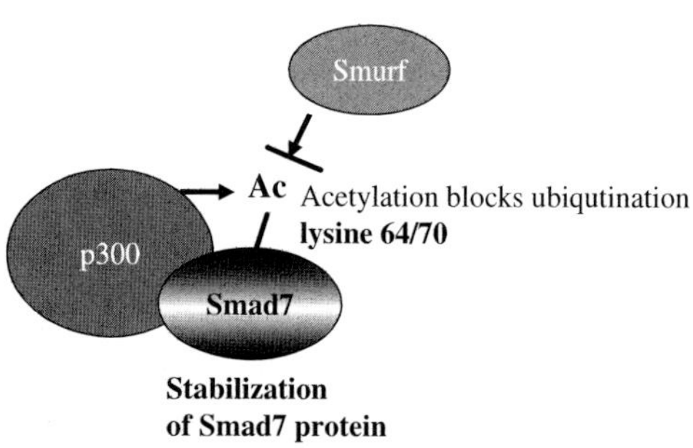

Figure 3. The balance between acetylation and ubiquitination controls the stability of Smad7. In the nucleus, Smad7 interacts with the nuclear acetyl transferase p300, which acetylates lysine residues 64 and 70 in the N-terminus of Smad7. The acetylation of Smad7 protects it against Smurf-mediated ubiquitination (and subsequent degradation) targeting the same lysine residues (A color version of this figure is freely accessible via the website of the book: http://www.springer.com/1-4020-4542-5)

acetylation of the same lysine residues in Smad7 prevent ubiquitination. Smad7 is deacetylatated as the protein leaves the nucleus in response to TGF-β stimulation and dissociates from p300.

4.3 I-Smads Regulators

Accumulating evidence suggests that I-Smad activity can also be affected independently of their protein levels. A cytoplasmic protein, previously termed Associated molecule with the SH3 domain of STAM (AMSH), inhibits the binding of Smad6 to activated type I receptors or activated R-Smads and thereby, potentiates BMP signaling (Itoh et al., 2001). A WD40 repeat protein, Serine-threonine kinase receptor-associated protein (STRAP), synergizes specifically with Smad7 in the inhibition of TGF-β signaling by stabilizing the association between Smad7 and the activated receptor (Datta and Moses, 2000). Yes-associated protein (YAP65) also interacts with Smad7 and augments the association of Smad7 to activated TGF-β type I receptor, thereby enhancing the inhibitory activity of Smad7 (Ferrigno et al., 2002). In contrast, Tid1, a regulator of apoptosis, can bind to Smad7 and blocks BMP-dependent regulatory activity of Smad7 in developing *Xenopus* embryos.

5. *IN VIVO* FUNCTIONS OF I-SMADS

The *in vivo* physiological functions of I-Smads remain largely unknown. However, gene targeting experiments in mice have begun to reveal specific developmental and physiological functions of I-Smads. Functonal inactivation of Smad6 through insertion of a LacZ reporter in the Smad6 gene locus demonstrated that Smad6 expression is largely restricted to the heart and blood vessels, and that Smad6 null mice have multiple cardiovascular abnormalities including hyperplasia of the cardiac valves and outflow tract septation defects, indicating a function for Smad6 in the regulation of endocardial cushion transformation (Galvin et al., 2000). Smad6 viable null mice also show the development of aortic ossification and elevated blood pressure. Thus, endogenous Smad6 appears to be crucial for the normal development and homeostasis of cardiovascular system.

Mice null for Smad7 gene has not been reported yet. However, transgenic mice expressing Smad7 in a tissue-specific way have been reported. Transgenic mice expressing Smad7 selectively in the epidermis or other stratified epithelia under the control of a keratin K5 promoter (K5.Smad7 Tg mice) exhibit pathological changes in multiple tissues and die within 10 days after birth (He et al., 2002). These mice are born with open eyelids and corneal defects, significantly delayed and aberrant hair follicle morphogenesis, and, importantly, hyperproliferation in the epidermis and other stratified epithelia. Thus, Smad7 may play an important role in the development and maintenance of homeostasis of multiple epithelial tissues.

6. INVOLVEMENT OF I-SMADS IN PATHOLOGIES

6.1 Fibrosis

Tissue fibrosis has been firmly associated with overactivity of TGF-β signaling activity in animal models or human diseases, although numerous additional molecules modulate this pathway or have a direct effect on fibrosis. Excessive activity of TGF-β induces pathological fibrosis of various tissues including lung, kidney, skin and liver, in part, through its stimulatory effects on extracellular matrix production such as collagens by fibroblasts (Leask and Abraham, 2004). Thus, antagonists of the TGF-β activity may have therapeutic potential for tissue fibrosis. Indeed, ectopic overexpression of Smad7 by adenoviral genetransfer, or other methods in target tissues can effectively block the tissue fibrosis in rodents such as lung fibrosis induced by bleomycin (Nakao et al., 1999), renal fibrosis after unilateral urethral obstruction (UUO) (Lan et al., 2003). or liver fibrosis induced by chemical reagents (Dooley et al., 2003). Most recently, vascular Smad7 overexpression is shown to attenuate remodeling and contribution of adventitial fibroblasts to neointima formation after ballon angioplasty (Mallawaarachchi et al., 2005). Thus, direct administration of Smad7 or reagents that can induce expression of Smad7 may be useful for treatment of fibrotic diseases mediated by over-expression of TGF-β.

Not only excessive activity of TGF-β, but also a disturbed negative regulation of TGF-β signaling by I-Smads may also play a role in the pathogenesis of tissue fibrosis (Fig. 4). It is recently reported that a decrease in the level of Smad7 protein, but not mRNA, results in enhanced TGF-β signaling and increases progressive tubulointerstitial fibrosis in UUO kidney model (Fukasawa et al., 2004). The reduction of Smad7 protein is suggested to result from enhanced ubiquitination by Smurfs and subsequent Smad7 degradation. This may be a common event in tissue fibrosis because reduction of Smad7 protein levels is also reported in the skin of scleroderma (Dong et al., 2002), which is a chronic systemic disease that leads to fibrosis of affected organs, and in asthmatic epithelium undergoing remodeling (Nakao et al., 2002). In scleroderma fibroblasts, impaired Smad7-Smurf-mediated inhibitory effect on TGF-β signaling is also reported (Asano et al., 2004). However, it remains unclear how Smad7-Smurf system is impaired in such disorders.

6.2 Inflammatory Bowel Disease (IBD)

It is well known that the local production of proinflammatory cytokines, such as TNF-α, IL-1β, and IFN-γ, is increased in inflammatory bowel disease (IBD) tissue and studies in experimental models of IBD indicate that intestinal mucosal inflammation depends at least in part on a balance between proinflammatory cytokines and TGF-β (Strober et al., 1997). Paradoxically, however, TGF-β and other anti-inflammatory cytokines (*e.g.* IL-10 or IL-1 receptor antagonist) are increased in IBD tissue, and yet mucosal inflammation proceeds uncontrolled. Smad7 may be a key to explain this paradoxical finding (Fig. 4; Monteleone et al., 2001). In

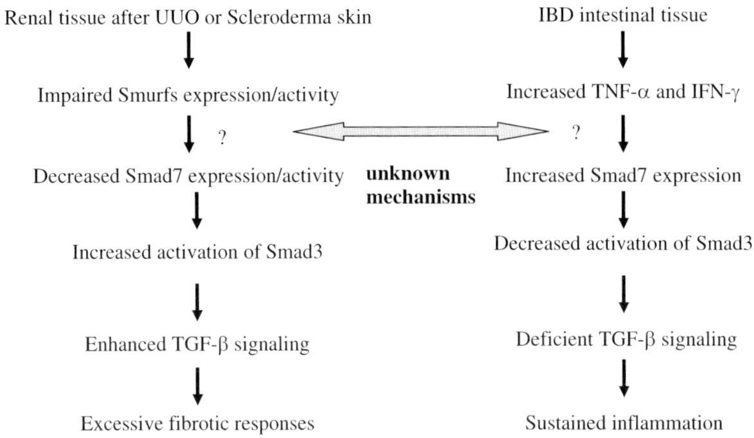

Figure 4. Effect of aberrant expression/activity of Smad7 in renal fibrotic lesions after induction of unilateral urethral obstruction (UUO), skin lesions of scleroderma, and the intestinal lesions of inflammatory bowel diseases (IBD). TGF-β functions as a pro-fibrotic and anti-inflammatory agent. Aberrant Smad7 (or Smurf) expression/activity occurs at the target sites of UUO, scleroderma, and IBD by as yet undefined mechanisms. The decreased or increased Smad7 expression/activity eventually results in augmentation or suppression of TGF-β signaling in the target sites, thereby contributing to enhanced fibrotic responses or sustained inflammation

IBD mucosa and purified mucosal T cells from the IBD lesions, overexpression of Smad7 is observed (Monteleone et al., 2001). Consistent with this finding, both whole tissue and isolated cells exhibit defective TGF-β/Smad signaling pathway, as measured by phospho-Smad3 immunoreactivity. Importantly, specific antisense oligonucleotides for Smad7 reduce Smad7 protein expression in cells isolated from patients with IBD, permitting the cells to respond to exogenous TGF-β. Thus, the overexpression of Smad7 in IBD tissues or mucosal T cells compromises the ability of TGF-β to downregulate inflammatory/immune responses in this condition and Smad7 blockade of TGF-β signaling may help maintain the chronic inflammatory process in IBD. It remains to be determined why Smad7 is overexpressed in IBD mucosa. Similar observations are also reported in whole gastric biopsy specimens from *Helicobacter pylori*-infected patients (Monteleone et al., 2004). In the specimens, there is defective TGF-β-associated Smad3 phosphorylation, which is associated with high expression of Smad7.

6.3 Cancer

Resistance to TGF-β-induced anti-mitogenic action is a frequently found characteristic of cancer. Therefore, TGF-β and its signaling intermediates have been considered as tumor suppressor genes. Indeed, inactivating mutations in the TGF-β type I and type II receptors, as well as several Smad proteins (Smad2 and Smad4) have been reported in several human cancers although the potential implications of the mutations in carcinogenesis appear to be complicated and remain elusive (Roberts and Wakefield, 2003).

Since I-Smads antagonize TGF-β signaling, it is possible that constitutive activation of I-Smads by a somatic mutation may impair TGF-β signaling, resulting in TGF-β resistance and is involved in carinogenesis. However, so far, genomic mutations of Smad6 and Smad7 have not been detected in colon, breast, lung, liver, and pancreatic cancers (Kleeff et al., 1999; Javelaud et al., 2005).

Nonetheless, it is reported that Smad7 is more frequently amplified (10%) than Smad2, Smad4 and DCC genes (4-7%) in pancreatic cancer (Boulay et al., 2001). In addition, increased Smad7 mRNA levels were reported in human pancreatic cancer in comparison with the normal pancreas (Kleeff et al., 1999). Furthermore, several experimental studies suggest the possibilities that TGF-β resistance induced by I-Smads contributes to carcinogenesis. For example, ectopic expression of Smad7 in human pancreatic or colon cancer cell lines abolishes their growth inhibitory or apoptotic response to TGF-β and, in the case of pancreatic cancer cell lines, they display enhanced anchorage-independent growth and accelerated growth in nude mice (Kleeff et al., 1999; Halder et al., 2005). Overexpression of Smad7 and oncogenic ras in primary mouse keratinocytes accelerates progression to squamous cell carcinoma whereas overexpresion of oncogenic ras alone forms only benign papilloma *in vivo* (Liu et al., 2003). Furthermore, hyperactivation of Stat3 in gp130 mutant (Y757F) mice promotes gastric adenomas associated with TGF-β resistance induced by increased Smad7 in gastric epithelial cells (Jenkins et al., 2005). Conversely, however, it is reported that Smad7-expressing melanoma cell lines exhibit a dramatically reduced capacity to form colonies under anchorage-independent culture conditions, and, when injected subcutaneously into nude mice, are largely delayed in their ability to form tumors (Javelaud et al., 2005). Thus, roles of I-Smads in carcinogenesis are not fully understood.

7. FUTURE PERSPECTIVES

Extensive progress has provided insight into the mechanisms of action, regulation, and physiological/pathological roles of I-Smads. The accumulating evidence tells us that the expression level of I-Smads is a major determinant for transcriptional responses mediated by TGF-β superfamily, which would be essential for balanced activity of TGF-β superfamily ligands through "feedback" or "cross-control" mechanisms. Many issues on I-Smads, however, remain to be clarified, including the following examples.

1) There are many ways of inhibition of signaling by I-Smads. Then, what is the major inhibitory mechanism of I-Smads? Alternatively, are the inhibitory mechanisms chosen depending on cell types or certain cellular contexts?
2) Inhibitory or nuclear functions of I-Smads should be closely linked to cytoplasmic, cell membrane, or nuclear localization of I-Smads and, therefore, it is important to reveal molecular basis by which the subcellular localization of I-Smads is controlled.

3) What are the *in vivo* functions of I-Smads? Generation of tissue- or cell type-specific conditional knock-out mice of Smad6 or Smad7 genes is eagerly awaited.
4) How do perturbations of I-Smads expression/activity occur in or contribute to certain TGF-β-associated human diseases? In particular, it is interesting to know the molecular basis for Smad7 down-regulation in fibrotic responses or Smad7 up-regulation in inflammatory conditions such as IBD. The former study may include a search for single nucleotide polymorphisms (SNPs) in the promoter region of Smad7, and the latter study may provide a molecular basis how chronic inflammation persists in certain cases. In addition, roles of I-Smads in carcinogenesis remain to be determined.

We hope that these basic studies on the regulation or function of I-Smads will be translated into the development of new approaches for the treatment of TGF-β superfamily-associated diseases, which are refractory to current protocols.

ACKNOWLEDGEMENTS

I appreciate all my colleagues at Department of Immunology, Faculty of Medicine, University of Yamanashi, and the Atopy Research Center, Juntendo University, School of Medicine, Japan, for their support and assistance, and apologize to the many researchers whose work could not be cited because of space limitations, or was only cited indirectly by referring to reviews, or more recent papers.

REFERENCES

Afrakhte, M., Morén, A., Jossan, S., Itoh, S., Sampath, K., Westermark, B., Heldin, C.-H., Heldin, N.-E., and ten Dijke, P., 1998, Induction of inhibitory Smad6 and Smad7 mRNA by TGF-β family members. *Biochem Biophys Res Commun* 249: 505-511.

Asano, Y., Ihn, H., Yamane, K., Kubo, M., and Tamaki, K., 2004, Impaired Smad7-Smurf-mediated negative regulation of TGF-β signaling in scleroderma fibroblasts. *J Clin Invest* 113: 253-264.

Bai, S., and Cao, X., 2002, A nuclear antagonistic mechanism of inhibitory Smads in transforming growth factor-β signaling. *J Biol Chem* 277: 4176-4182.

Bai, S., Shi, X., Yang, X., and Cao, X., 2000, Smad6 as a transcriptional corepressor. *J Biol Chem* 275: 8267-8270.

Benchabane H, Wrana JL. 2003, GATA- and Smad1-dependent enhancers in the Smad7 gene differentially interpret bone morphogenetic protein concentrations. *Mol Cell Biol* 23: 664661.

Bitzer, M., von Gersdorff, G., Liang, D., Dominguez-Rosales, A., Beg, A.A., Rojkind, M., and Böttinger, E.P., 2000, A mechanism of suppression of TGF-β/SMAD signaling by NF-κ B/RelA. *Genes Dev* 14: 187-197.

Boulay, J.L., Mild, G., Reuter, J., Lagrange, M., Terracciano, L., Lowy, A., Laffer, U., Orth, B., Metzger, U., Stamm, B., Martinoli, S., Herrmann, R., and Rochlitz, C., 2001, Combined copy status of 18q21 genes in colorectal cancer shows frequent retention of SMAD7. *Genes Chromosomes Cancer* 31: 240-247.

Brodin, G., Åhgren, A., ten Dijke, P., Heldin, C.-H., and Heuchel, R., 2000, Efficient TGF-β induction of the Smad7 gene requires cooperation between AP-1, Sp1, and Smad proteins on the mouse Smad7 promoter. *J Biol Chem* 275: 29023-29030.

Datta, P.K., and Moses, H.L., 2000, STRAP and Smad7 synergize in the inhibition of transforming growth factor β signaling. *Mol Cell Biol* 20: 3157-3167.

Denissova, N.G., and Liu, F., 2004, Repression of endogenous Smad7 by Ski. *J Biol Chem* 279: 28143-28148.

Denissova, N.G., Pouponnot, C., Long, J., He, D., and Liu, F., 2000, Transforming growth factor β-inducible independent binding of SMAD to the Smad7 promoter. *Proc Natl Acad Sci U S A* 97: 6397-6402.

Dong, C., Zhu, S., Wang, T., Yoon, W., Li, Z., Alvarez, R.J., ten Dijke, P., White, B., Wigley, F.M., and Goldschmidt-Clermont, P.J., 2002, Deficient Smad7 expression: a putative molecular defect in scleroderma. *Proc Natl Acad Sci U S A* 99: 3908-3913.

Dooley, S., Hamzavi, J., Breitkopf, K., Wiercinska, E., Said, H.M., Lorenzen, J., ten Dijke, P., and Gressner, A.M., 2003, Smad7 prevents activation of hepatic stellate cells and liver fibrosis in rats. *Gastroenterology* 125: 178-191.

Ebisawa, T., Fukuchi, M., Murakami, G., Chiba, T., Tanaka, K., Imamura, T., and Miyazono, K., 2001, Smurf1 interacts with transforming growth factor-β type I receptor through Smad7 and induces receptor degradation. *J Biol Chem* 276: 12477-12480.

Edlund, S., Bu, S., Schuster, N., Aspenström, P., Heuchel, R., Heldin, N.-E., ten Dijke, P., Heldin, C.-H., and Landström, M., 2003, Transforming growth factor-β1 (TGF-β)-induced apoptosis of prostate cancer cells involves Smad7-dependent activation of p38 by TGF-β-activated kinase 1 and mitogen-activated protein kinase kinase 3. *Mol Biol Cell* 14: 529-544.

Edlund, S., Landström, M., Heldin, C.-H., and Aspenström, P., 2004, Smad7 is required for TGF-β-induced activation of the small GTPase Cdc42. *J Cell Sci* 117: 1835-1847.

Ferrigno, O., Lallemand, F., Verrecchia, F., L'Hoste, S., Camonis, J., Atfi, A., Mauviel, A., 2002, Yes-associated protein (YAP65) interacts with Smad7 and potentiates its inhibitory activity against TGF-β/Smad signaling. *Oncogene* 21: 4879-4884.

Fukasawa, H., Yamamoto, T., Togawa, A., Ohashi, N., Fujigaki, Y., Oda, T., Uchida, C., Kitagawa, K., Hattori, T., Suzuki, S., Kitagawa, M., and Hishida, A., 2004, Down-regulation of Smad7 expression by ubiquitin-dependent degradation contributes to renal fibrosis in obstructive nephropathy in mice. *Proc Natl Acad Sci U S A* 101: 8687-8692.

Galvin, K.M., Donovan, M.J., Lynch, C.A., Meyer, R.I., Paul, R.J., Lorenz, J.N., Fairchild-Huntress, V., Dixon, K.L., Dunmore, J.H., Gimbrone, M.A., Jr., Falb, D., and Huszar, D., 2000, A role for smad6 in development and homeostasis of the cardiovascular system. *Nat Genet* 24: 171-174.

Grönroos, E., Hellman, U., Heldin, C.-H., and Ericsson, J., 2002, Control of Smad7 stability by competition between acetylation and ubiquitination. *Mol Cell* 10: 483-493.

Halder SK, Beauchamp RD, Datta PK. 2005, Smad7 induces tumorigenicity by blocking TGF-β-induced growth inhibition and apoptosis. *Exp Cell Res* 307: 231-246.

Hanyu, A., Ishidou, Y., Ebisawa, T., Shimanuki, T., Imamura, T., and Miyazono, K., 2001, The N domain of Smad7 is essential for specific inhibition of transforming growth factor-β signaling. *J Cell Biol* 155: 1017-1027.

Hata, A., Lagna, G., Massagué, J., and Hemmati-Brivanlou, A., 1998, Smad6 inhibits BMP/Smad1 signaling by specifically competing with the Smad4 tumor suppressor. *Genes Dev* 12: 186-197.

Hayashi, H., Abdollah, S., Qiu, Y., Cai, J., Xu, Y.Y., Grinnell, B.W., Richardson, M.A., Topper, J.N., Gimbrone, M.A.J., Wrana, J.L., and Falb, D., 1997, The MAD-related protein Smad7 associates with the TGFβ receptor and functions as an antagonist of TGFβ signaling. *Cell* 89: 1165-1173.

He, W., Li, A.G., Wang, D., Han, S., Zheng, B., Goumans, M.-J., ten Dijke, P., and Wang, X.J., 2002, Overexpression of Smad7 results in severe pathological alterations in multiple epithelial tissues. *EMBO J* 21: 2580-2590.

Heldin, C.-H., Miyazono, K., and ten Dijke, P., 1997, TGF-β signalling from cell membrane to nucleus through SMAD proteins. *Nature* 390: 465-471.

Imamura, T., Takase, M., Nishihara, A., Oeda, E., Hanai, J., Kawabata, M., and Miyazono, K., 1997, Smad6 inhibits signalling by the TGF-β superfamily. *Nature* 389: 622-626.

Ishida W, Hamamoto T, Kusanagi K, Yagi K, Kawabata M, Takehara K, Sampath TK, Kato M, Miyazono K., 2000, Smad6 is a Smad1/5-induced Smad inhibitor. Characterization of bone morphogenetic protein-responsive element in the mouse Smad6 promoter. *J Biol Chem* 275: 6075-6079.

Itoh, F., Asao, H., Sugamura, K., Heldin, C.-H., ten Dijke, P., and Itoh, S., 2001, Promoting bone morphogenetic protein signaling through negative regulation of inhibitory Smads. *EMBO J* 20: 4132-4142.

Itoh, S., Landström, M., Hermansson, A., Itoh, F., Heldin, C.-H., Heldin, N.-E., and ten Dijke, P., 1998, Transforming growth factor β1 induces nuclear export of inhibitory Smad7. *J Biol Chem* 273: 29195-29201.

Javelaud, D., Delmas, V., Moller, M., Sextius, P., Andre, J., Menashi, S., Larue, L., and Mauviel, A., 2005, Stable overexpression of Smad7 in human melanoma cells inhibits their tumorigenicity in vitro and in vivo. *Oncogene* 24: 7624-7629.

Jenkins BJ, Grail D, Nheu T, Najdovska M, Wang B, Waring P, Inglese M, McLoughlin RM, Jones SA, Topley N, Baumann H, Judd LM, Giraud AS, Boussioutas A, Zhu HJ, Ernst M., 2005, Hyperactivation of Stat3 in gp130 mutant mice promotes gastric hyperproliferation and desensitizes TGF-β signaling. *Nat Med* 11: 845-852.

Jonson, T., Gorunova, L., Dawiskiba, S., Andrén-Sandberg, Å., Stenman, G., ten Dijke, P., Johansson, B., and Höglund, M., 1999, Molecular analyses of the 15q and 18q *SMAD* genes in pancreatic cancer. *Genes, Chrom and Cancer* 24: 62-71.

Kawate, S., Ohwada, S., Hamada, K., Koyama, T., Takenoshita, S., Morishita, Y., and Hagiwara, K., 2001, Mutational analysis of the Smad6 and Smad7 genes in hepatocellular carcinoma. *Int J Mol Med* 8: 49-52.

Kavsak, P., Rasmussen, R.K., Causing, C.G., Bonni, S., Zhu, H., Thomsen, G.H., and Wrana, J.L., 2000, Smad7 binds to Smurf2 to form an E3 ubiquitin ligase that targets the TGFβ receptor for degradation. *Mol Cell* 6: 1365-1375.

Kleeff, J., Ishiwata, T., Maruyama, H., Friess, H., Truong, P., Büchler, M.W., Falb, D., and Korc, M., 1999, The TGF-β signaling inhibitor Smad7 enhances tumorigenicity in pancreatic cancer. *Oncogene* 18: 5363-5372.

Koinuma, D., Shinozaki, M., Komuro, A., Goto, K., Saitoh, M., Hanyu, A., Ebina, M., Nukiwa, T., Miyazawa, K., Imamura, T., and Miyazono, K., 2003, Arkadia amplifies TGF-β superfamily signalling through degradation of Smad7. *EMBO J* 22: 6458-6470.

Lallemand F, Mazars A, Prunier C, Bertrand F, Kornprost M, Gallea S, Roman-Roman S, Cherqui G, Atfi A. 2001, Smad7 inhibits the survival nuclear factor κB and potentiates apoptosis in epithelial cells. *Oncogene* 20: 879-884.

Lallemand F, Seo SR, Ferrand N, Pessah M, L'Hoste S, Rawadi G, Roman-Roman S, Camonis J, Atfi A., 2005, AIP4 restricts transforming growth factor-β signaling through a ubiquitination-independent mechanism. *J Biol Chem* 280: 27645-27653.

Lan, H.Y., Mu, W., Tomita, N., Huang, X.R., Li, J.H., Zhu, H.J., Morishita, R., and Johnson, R.J., 2003, Inhibition of renal fibrosis by gene transfer of inducible Smad7 using ultrasound-microbubble system in rat UUO model. *J Am Soc Nephrol* 14: 1535-1548.

Landström, M., Heldin, N.-E., Bu, S., Hermansson, A., Itoh, S., ten Dijke, P., and Heldin, C.-H., 2000, Smad7 mediates apoptosis induced by transforming growth factor β in prostatic carcinoma cells. *Curr Biol* 10: 535-538.

Leask, A., and Abraham, D.J., 2004, TGF-β signaling and the fibrotic response. *FASEB J* 18: 816-827.

Lin, X., Liang, Y.Y., Sun, B., Liang, M., Shi, Y., Brunicardi, F.C., and Feng, X.H., 2003, Smad6 recruits transcription corepressor CtBP to repress bone morphogenetic protein-induced transcription. *Mol Cell Biol* 23: 9081-9093.

Liu X, Lee J, Cooley M, Bhogte E, Hartley S, Glick A., 2003, Smad7 but not Smad6 cooperates with oncogenic ras to cause malignant conversion in a mouse model for squamous cell carcinoma. *Cancer Res* 63: 7760-7768.

Mallawaarachchi CM, Weissberg PL, Siow RC., 2005, Smad7 gene transfer attenuates adventitial cell migration and vascular remodeling after balloon injury. *Arterioscler Thromb Vasc Biol* 25: 1383-1387.

Mazars A, Lallemand F, Prunier C, Marais J, Ferrand N, Pessah M, Cherqui G, Atfi A. 2001, Evidence for a role of the JNK cascade in Smad7-mediated apoptosis. *J Biol Chem*. 276: 36797-36803.

Miyazono, K., 2002, A new partner for inhibitory Smads. *Cytokine Growth Factor Rev* 13: 7-9.

Mochizuki, T., Miyazaki, H., Hara, T., Furuya, T., Imamura, T., Watabe, T., and Miyazono, K., 2004, Roles for the MH2 domain of Smad7 in the specific inhibition of transforming growth factor-β superfamily signaling. *J Biol Chem* 279: 31568-31574.

Monteleone, G., Del Vecchio Blanco, G., Palmieri, G., Vavassori, P., Monteleone, I., Colantoni, A., Battista, S., Spagnoli, L.G., Romano, M., Borrelli, M., MacDonald, T.T., and Pallone, F., 2004, Induction and regulation of Smad7 in the gastric mucosa of patients with Helicobacter pylori infection. *Gastroenterology* 126: 674-682.

Monteleone, G., Kumberova, A., Croft, N.M., McKenzie, C., Steer, H.W., and MacDonald, T.T., 2001, Blocking Smad7 restores TGF-β1 signaling in chronic inflammatory bowel disease. *J Clin Invest* 108: 601-609.

Nagarajan, R.P., Zhang, J., Li, W., and Chen, Y., 1999, Regulation of Smad7 promoter by direct association with Smad3 and Smad4. *J Biol Chem* 274: 33412-33418.

Nakao, A., Afrakhte, M., Morén, A., Nakayama, T., Christian, J.L., Heuchel, R., Itoh, S., Kawabata, M., Heldin, N.-E., Heldin, C.-H., and ten Dijke, P., 1997, Identification of Smad7, a TGFβ-inducible antagonist of TGF-β signalling. *Nature* 389: 631-635.

Nakao, A., Fujii, M., Matsumura, R., Kumano, K., Saito, Y., Miyazono, K., and Iwamoto, I., 1999, Transient gene transfer and expression of Smad7 prevents bleomycin-induced lung fibrosis in mice. *J Clin Invest* 104: 5-11.

Nakao, A., Okumura, K., and Ogawa, H., 2002a, Smad7: a new key player in TGF-β-associated disease. *Trends Mol Med* 8: 361-363.

Nakao, A., Sagara, H., Setoguchi, Y., Okada, T., Okumura, K., Ogawa, H., and Fukuda, T., 2002b, Expression of Smad7 in bronchial epithelial cells is inversely correlated to basement membrane thickness and airway hyperresponsiveness in patients with asthma. *J Allergy Clin Immunol* 110: 873-878.

Nakayama, T., Gardner, H., Berg, L.K., and Christian, J.L., 1998a, Smad6 functions as an intracellular antagonist of some TGF-β family members during *Xenopus* embryogenesis. *Genes Cells* 3: 387-394.

Nakayama, T., Snyder, M.A., Grewal, S.S., Tsuneizumi, K., Tabata, T., and Christian, J.L., 1998b, *Xenopus* Smad8 acts downstream of BMP-4 to modulate its activity during vertebrate embryonic patterning. *Development* 125: 857-867.

Quan T, He T, Voorhees JJ, Fisher GJ., 2005, Ultraviolet irradiation induces Smad7 via induction of transcription factor AP-1 in human skin fibroblasts. *J Biol Chem* 280: 8079-8085.

Riggins, R.G., Kinzler, K.W., Vogelstein, B., and Thiagalingam, S., 1997, Frequency of *Smad* gene mutations in human cancers. *Cancer Res* 57: 2578-2580.

Roberts, A.B., and Wakefield, L.M., 2003, The two faces of transforming growth factor β in carcinogenesis. *Proc Natl Acad Sci U S A* 100: 8621-8623.

Schiffer, M., Bitzer, M., Roberts, I.S., Kopp, J.B., ten Dijke, P., Mundel, P., and Böttinger, E.P., 2001, Apoptosis in podocytes induced by TGF-β and Smad7. *J Clin Invest* 108: 807-816.

Shi, W., Sun, C., He, B., Xiong, W., Shi, X., Yao, D., and Cao, X., 2004, GADD34-PP1c recruited by Smad7 dephosphorylates TGFβ type I receptor. *J Cell Biol* 164: 291-300.

Souchelnytskyi, S., Nakayama, T., Nakao, A., Morén, A., Heldin, C.-H., Christian, J.L., and ten Dijke, P., 1998, Physical and functional interaction of murine and *Xenopus* Smad7 with bone morphogenetic protein receptors and transforming growth factor-β receptors. *J Biol Chem* 273: 25364-25370.

Strober, W., Kelsall, B., Fuss, I., Marth, T., Ludviksson, B., Ehrhardt, R., and Neurath, M., 1997, Reciprocal IFN-γ and TGF-β responses regulate the occurrence of mucosal inflammation. *Immunol Today* 18: 61-64.

Takase, M., Imamura, T., Sampath, T.K., Takeda, K., Ichijo, H., Miyazono, K., and Kawabata, M., 1998, Induction of Smad6 mRNA by bone morphogenetic proteins. *Biochem Biophys Res Commun* 244: 26-29.

Topper, J.N., Cai, J., Qiu, Y., Anderson, K.R., Xu, Y.-Y., Deeds, J.D., Feeley, R., Gimeno, C.J., Woolf, E.A., Tayber, O., Mays, G.G., Sampson, B.A., Schoen, F.J., Gimbrone Jr., M.A., and Falb, D., 1997, Vascular *MAD*s: Two novel *MAD*-related genes selectively inducible by flow in human vascular endothelium. *Proc Natl Acad Sci U S A* 94: 9314-9319.

Torregroza, I., and Evans, T., 2006, Tid1 is a Smad-binding protein that can modulate Smad7 activity in developing embryos. *Biochem J* 393: 311-320.

Tsuneizumi, K., Nakayama, T., Kamoshida, Y., Kornberg, T.B., Christian, J.L., and Tabata, T., 1997, *Daughters against dpp* modulates *dpp* organizing activity in *Drosophila* wing development. *Nature* 389: 627-631.

Ulloa, L., Doody, J., and Massagué, J., 1999, Inhibition of transforming growth factor-β/SMAD signalling by the interferon-gamma/STAT pathway. *Nature* 397: 710-713.

Yoshimura, A., Mori, H., Ohishi, M., Aki, D., and Hanada, T., 2003, Negative regulation of cytokine signaling influences inflammation. *Curr Opin Immunol* 15: 704-708.

CHAPTER 20

ALTERATIONS IN SMAD SIGNALING IN CARCINOGENESIS

SEONG-JIN KIM AND JOHN J. LETTERIO

The Laboratory of Cell Regulation and Carcinogenesis, The Center for Cancer Research, The National Cancer Institute, The National Institutes of Health, Bethesda, MD 20892, USA

Abstract: The importance of TGF-β signaling in the suppression of tumorigenesis is supported by the presence of frequent mutations in genes encoding both TGF-β receptors and intermediates in this signaling pathway in cancer. In epithelial cancers sporadic mutations have been found in the genes encoding both the receptor-activated Smad2 (*MADH2*) and the common intermediate, Smad4 (*MADH4*). Germline mutations in *MADH4* and in *BMPR1* (BMP type 1 receptor gene) are the most common mutations in the familial cancer syndrome, Familial Juvenile Polyposis, FJP. More recent studies have revealed epigenetic mechanisms that also play important roles in subverting the function of this pathway. In this chapter, we discuss some of these mechanisms, and provide insight into novel ways in which Smad signaling contributes to the maintenance of tissue homeostasis and ultimately to the suppression of cancer

Keywords: cancer; leukemogenesis; signaling; tumor suppressor

1. INTRODUCTION

The inactivation of tumor suppressor genes and the activation of proto-oncogenes in affected cells are considered to be core events that provide for selective growth advantage and clonal expansion during the multi-step process of carcinogenesis. Classical tumor suppressor genes are by definition inactivated by intragenic mutations or deletions in one allele (typically a germline mutation), and failure to express the second allele is sufficient to initiate the process of carcinogenesis. The second allele is most often disrupted through the loss of a chromosomal region containing the other allele, called a loss of heterozygosity (LOH) (Croce, 1991). *MADH4* fits this definition of a classical tumor suppressor, as germline mutations are found in familial cohorts with FJP, and LOH has been described in intestinal tumors that develop in these patients (Howe et al., 1998; Woodford-Richens et al.,

P. ten Dijke and C.-H. Heldin (eds.), Smad Signal Transduction, 397–413.

2000). However, recent studies have demonstrated that Smad4 can also suppress transformation indirectly, by exerting effects on stromal cells and immune cells within the environment in which the transformed cell arises (Kim, B.G, et al., submitted). The concept that TGF-β signaling modulates cancer progression by exerting control of the microenvironment is well established, yet we are only now beginning to appreciate the contribution of Smad signaling to this mechanism of tumor suppression.

Inactivating mutations in *SMAD* genes are found in a number of human cancers, with the highest frequency observed in pancreatic and colon carcinomas (de Caestecker et al., 2000; Elliott and Blobe, 2005). While these mutations have been typically found in epithelial cancers, disruption of Smad signaling has also been reported in hematopoietic malignancies For example, a missense mutation in the MH1 domain (P102L) of the *MADH4* gene, and a single frame shift mutation resulting in termination in the MH2 domain (Delta 483-552) have been reported in acute myelogeneous leukemia (Imai et al., 2001). Both of the mutated Smad4 proteins lack transcriptional activities. The Delta (483-552) mutant blocks nuclear translocation of wild-type Smad4 and thus disrupts TGF-β signaling. Here too, we are only beginning to understand unique specificities of Smad proteins and the mechanisms leading to their inactivation during the process of carcinogenesis. It is of interest that the receptor-activated Smad3 appears to be a unique pathway target in leukemogenesis, and that epigenetic mechanisms are principally involved in the inactivation of Smad3 function. The data that are emerging from studies in mouse and human cancer suggest that a balance between Smad and non-Smad pathways downstream of the TGF-β receptor may be a critical determinant of transformation. Moreover, the context in which Smad signaling is altered may be a key determinant of cell transformation. Knowledge of the critical oncogenic events that cooperate with loss of Smad function will be important to our understanding of Smad function in the pathogenesis of cancer.

2. GENETIC AND EPIGENETIC MECHANISMS LEADING TO SMAD INACTIVATION IN HUMAN CANCER

2.1 Mutational Inactivation of Smad Genes in Cancers of the Gastrointestinal Tract

SMAD4/DPC4 was originally isolated as a tumor suppressor gene on chromosome 18q21 that is deleted or mutated in nearly half of all human pancreatic carcinomas (Hahn et al., 1996), and much less frequently in other cancers (Barrett et al., 1996; Kawate et al., 1999; Kim et al., 1996; Kong et al., 1997; Nagatake et al., 1996; Schutte et al., 1996). Allelic loss of chromosome 18q21.1 is a relatively common event in a wide range of malignancies (Schutte et al., 1996) and is associated with greater mortality and an increased risk of metastatic spread (Jen et al., 1994; Kern et al., 1989). However, biallelic deletion or inactivation of Smad4 is largely restricted to tumors of the pancreas and gastrointestinal tract (Maurice et al., 2001;

Thiagalingam et al., 1996), with a much lower frequency in lung, breast, ovarian, and squamous cell carcinomas (Schutte, 1999; Takei et al., 1998). This discrepancy could reflect the importance of Smad4 haploid insufficiency in the initiation of carcinogenesis. In addition to the relative infrequency of LOH at this locus, mutational inactivation of Smad4 is often a late event in tumorigenesis associated with pancreatic and gastrointestinal epithelia, which may suggest that there are additional, as yet unidentified, tumor suppressor loci within chromosome 18q21.

The *MADH2* gene encodes Smad2, another member of the TGF-β signaling cascade and potential tumor suppressor also located on chromosome 18q21.1 (Riggins et al., 1996). *MADH2* is mutated in a small percentage of cancers of the colon, head and neck, and lung (Eppert et al., 1996). Unlike the case for Smad4, LOH occurs rarely at the *MADH2* locus, and is described infrequently only in colorectal and lung tumors (Uchida et al., 1996). Furthermore, genomic alterations in the *MADH2* locus have not been found in other types of carcinomas, and they are not a feature of either leukemias or lymphomas (Bevan et al., 1999; Ikezoe et al., 1998b; Latil et al., 1999; Maesawa et al., 1997; Shitara et al., 1999; Wieser et al., 1998). The gene encoding Smad7 is also located on chromosome 18q21.1 (Roijer et al., 1998). As an inhibitor of TGF-β signaling, Smad7 might be viewed as a candidate oncogene, yet activating mutations or amplifications of the Smad7 gene in colorectal or pancreatic carcinomas have not been identified (Jonson et al., 1999). However, increased expression of Smad7 has been reported in pancreatic and gastric cancers (Kleeff et al., 1999), suggesting that aberrant expression of this protein may play a role in subverting TGF-β signaling in cancer cells (see below). Recent studies have bolstered this concept that Smad7 may promote tumorigenesis. For example, a human colon adenocarcinoma engineered to express Smad7 shows anchorage-independent growth and enhanced tumorigenicity in athymic nude mice (Halder et al., 2005). In this TGF-β-sensitive, well-differentiated, and non-tumorigenic cell line, Smad7 inhibits TGF-β-induced, Smad-dependent transcriptional responses without affecting TGF-β-induced activation of p38 MAPK and Erk. The overexpression of Smad7 clearly blocks TGF-β-induced growth inhibition and inhibits the apoptosis of these cells, suggesting a mechanism by which Smad7 could act to enhance tumorigenicity.

The genes encoding additional intermediates in the TGF-β signaling cascade, Smad3 and Smad6, are located on chromosome 15q21–22 (Riggins et al., 1996), and this is another site of frequent allelic loss in breast, colorectal, lung and pancreatic tumors (Park et al., 2000). However, inactivating mutations in *MADH3* and *MADH6* have not been observed in any of a large number of tumors, including those from gastrointestinal, breast, lung, ovarian, and pancreatic cancers (Arai et al., 1998; Ikezoe et al., 1998a; Wang et al., 2000; Wick et al., 1996). These observations made in human cancers are at odds with murine models in which metastatic colorectal tumors have been reported in *SMAD3*$^{-/-}$ mice (Zhu et al., 1998). However, the fact that two distinct *SMAD3*$^{-/-}$ mouse models are not associated with a colon cancer phenotype suggests that this discrepancy could result from the expression of a hypomorphic *SMAD3* allele that may have oncogenic activity in those mice developing colorectal tumors (Datto et al., 1999; Yang et al., 1999). Alternatively, undefined epigenetic factors may have contributed to the variable rates of malignancy. Taken together, these data indicate that Smad3 has a complex role in the regulation of epithelial cell transformation. More recent studies that have directly implicated a tumor suppressor role for Smad3 in hematopoietic malignancies will be discussed below.

2.2 Epigenetic Mechanisms Leading to Reduced Expression of Smads in Cancer

Mutation in genes that inhibit the formation of tumors has long been known to be one of the main driving forces in the development of cancer. However, in recent years it has become increasingly obvious that genetic abnormalities are by no means the only mechanism by which the expression of tumor suppressor proteins can become altered during tumorigenesis. Growing evidence now indicates that epigenetic factors play a major role in carcinogenesis and indeed may be as significant as the more widely studied genetic abnormalities. Tumor cells often become resistant to TGF-β-induced growth inhibition and apoptosis due to functional inactivation of TGF-β receptors and Smads.

Gene silencing has been described in association with Smad4 heterozygosity in pancreatic cancer (Villanueva et al., 1998), but this is a relatively rare event, and the underlying mechanisms have not been determined. In human esophageal squamous cell carcinoma (SCC), variable expression levels of Smad4 have been reported (Fukuchi et al., 2002), and the levels of Smad4 expression were reported to relate inversely to the depth of tumor invasion. Interestingly, the level of expression of Smad4 mRNA was relatively normal, while the expression of Smad4 protein was markedly decreased. This observation can be interpreted to suggest that a decrease in the stability of Smad4 protein, or alternatively, an impaired translation of the Smad4 message, may underlie a loss of Smad4 expression. A reduced expression of Smad4 has also been observed both in the human papillomavirus type 16 (HPV16)-positive head and neck squamous cell carcinomas (HNSCC) tumors and in HPV16-negative tumors, suggesting that the loss of Smad4 expression may be involved in HPV16-induced carcinogenesis of HNSCC (Baez et al., 2005). This data is consistent with recent studies in murine models in which targeted deletion of the Smad4 gene leads to induction of squamous cell carcinomas in mice (Chapter 8).

The compartmental localization and cytoplasmic-nuclear shuttling of Smad proteins may also play an important role in mediating tumor suppressor activity. In prostate cancers, the expression of nuclear Smad8 and nuclear Smad4 is decreased during the progression to prostatic malignancy, with loss of BMP2 and nuclear Smad4 expression correlating with increasing Gleason score (Horvath et al., 2004). However, it is not clear whether this is a primary or secondary phenomenon in prostate carcinogenesis. Decreased levels of Smad2 and Smad4 mRNAs were reported in the uterine tumors infiltrated into the myometrial wall compared with noninfiltrating endometrial cancers (ECs). Significantly higher Smad4 protein levels in the cytoplasmic fraction of ECs were found when tumor grade and depth of myometrial invasion were considered. These data add yet another twist by implicating mislocalization of Smad4 as a mechanism to subvert the function of this pathway. Moreover, the deficiency of nuclear Smad4 found with increased infiltration of the myometrial wall by type I endometrial carcinomas suggests that this accumulation of cytoplasmic Smad4 is a correlate of an invasive phenotype (Piestrzeniewicz-Ulanska et al., 2003; Piestrzeniewicz-Ulanska et al., 2004). An increased turnover of Smad4 and Smad4 adapter complexes has also been implicated in blocking the tumor suppressor function of this pathway. For example, PRAJA, which functions as an E3 ligase, interacts with ELF (a β-Spectrin, Smad4 adaptor protein) in a TGF-β-dependent manner in gastric cancer cell lines (Mishra et al., 2005). PRAJA is increased five-fold in human gastric cancers, and inactivates ELF.

In contrast to the known tumor suppressors Smads 2 and 4, the role of Smad3 as a tumor suppressor has been harder to demonstrate, as mutations in the *MADH3* gene have thus far not been discovered in human tumors. However, the expression of both Smad2 and Smad3 is often decreased in epithelial components of both human skin and rat prostatic carcinomas (Lange et al., 1999), and Smad3 expression is suppressed in a subset of human gastric cancers and human gastric cancer cell lines. Interestingly, it has been suggested that a suppression of Smad3 expression could underlie some of the putative oncogenic properties of TGF-β. To provide just one example, recent studies in mice have demonstrated the loss of Smad3 expression protects against TPA-induced skin carcinogenesis, with decreased susceptibility observed in both *SMAD3*$^{-/-}$ mice and *SMAD*$^{+/-}$ mice (Li et al., 2004). The observation suggests that Smad3-dependent induction of TGF-β responsive genes is required for tumor promotion. In this instance, it may involve a unique role for Smad3 in mediating the effects of TGF-β on the local inflammatory response in the tumor microenvironment.

A role for TGF-β signaling in the link between inflammation and cancer has been further suggested by studies focused on the role of IL-6 in gastric cancer. In mice, expression of a constitutively active gp130 subunit of the IL-6 receptor promotes adenomatous gastric hyperplasia and abrogates the epithelial cell response to TGF-β by inducing the expression of Smad7, an inhibitor of TGF-β-induced Smad signaling. The significance of this observation is supported by studies of gastric biopsy specimens taken from patients with or without *Helicobactor pylori* infection, showing the induction of Smad7 expression by Western blotting associated with *H. pylori* infection (Monteleone et al., 2004). Clearly, in this context, Smad3 functions as a mediator of the tumor suppressor activities of TGF-β. Indeed, the restoration of TGF-β signaling by transfecting Smad3 in Smad3-deficient gastric cancer cells has revealed this tumor suppressive activity (Han et al., 2004). Importantly, restoration of TGF-β signaling induced the tumor suppressor E-cadherin, which correlated perfectly with the TGF-β responsiveness of the gastric cancer cell lines examined in this study. The data suggest that intact TGF-β signaling is important for the expression of E-cadherin, and that this may play a role in controlling the maintenance of mucosal epithelial homeostasis.

The relevance of inhibitory Smad7 has been suggested by other studies. Smad7 is highly expressed in a subset of human primary gastric cancer tissues, and Smad7 expression has been associated with poor outcomes in gastric carcinomas (Kim et al., 2004). Increased expression of Smad6 and Smad7 has also been described in human endometrial, gastric, and pancreatic carcinomas, as well as rat prostatic carcinomas (Brodin et al., 1999; Kim et al., 2004; Korchynskyi et al., 1999). Smad7 expression is significantly upregulated in endometrial adenocarcinomas compared to benign endometrium (Dowdy et al., 2005). In contrast, another study reported that increased levels of Smad7 protein were not observed in endometrial cancers, which may be due to the utilization of immunohistochemical staining that may be less sensitive compared to quantitative RT–PCR. Smad7 expression has also been evaluated in hepatocellular carcinoma, including one study in which immunohistochemistry was performed in low-grade dysplastic nodules (DNs), high-grade DNs, early hepatocellular carcinomas (eHCCs), and in 41 advanced HCCs. Smad7 expression was significantly higher in advanced HCCs, and not found in DNs and eHCCs (Park et al., 2004). The mechanisms underlying these changes in Smad7 expression are unknown.

2.3 Functional Inactivation of Smad3 by Smad3-Interacting Oncoproteins

As noted above, evidence to support the relevance of Smad3 as a mediator of the tumor suppressor activities of TGF-β has been somewhat elusive. The absence of an overt tumor phenotype in the *SMAD3*$^{-/-}$ mouse models, combined with the apparent absence of inactivating mutations of deletions of the *MADH3* gene in human cancers, led to speculation that Smad3 may be more relevant in mediating tumor-promoting activities of this pathway. However, there is now a growing body of evidence indicating that the function of Smad3 is disrupted in a very distinct way, by a variety of oncoproteins that have the capacity to physically associate with Smad3, and to effectively disrupt Smad3-dependent TGF-β-induced gene transcription. This mechanism seems to be particularly common in hematopoietic cancers, suggesting a unique and essential role for Smad3-dependent TGF-β regulation in suppression of leukemogeneisis. The paradigm that is evolving implicates that the loss of Smad3 function cooperates with the known oncogenic activity of oncoproteins (both viral and novel oncoproteins arising from chromosomal translocations), to induce transformation (Fig. 1).

For example, chromosomal rearrangements resulting in the fusion products AML/Evi-1 (Kurokawa et al., 1998) or AML/MDS/Evi-1 (Sood et al., 1999) are frequent in myeloid leukemia and myelodysplasia. Both fusion proteins interact with and functionally inactivate Smad3, interfering with growth inhibition of myeloid cells by TGF-β and suggesting that Evi-1 may interfere with TGF-β signaling mediated by Smad3 in these tumors (Kurokawa et al., 1998). Evi-1 has also been shown to repress Smad-induced transcription by recruiting C-terminal binding protein (CtBP) as a co-repressor through a mechanism that requires histone deacetylase (HDAc).

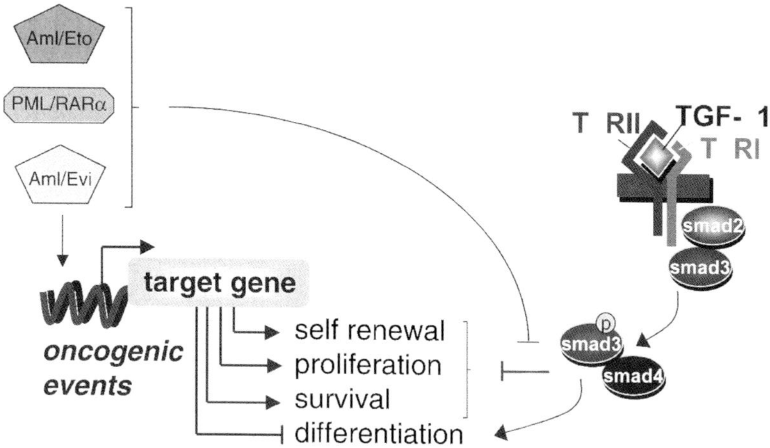

Figure 1. Smad3 is a target of oncoproteins. In the context of leukemogenesis, a variety of established oncoproteins have now been characterized for their ability to physically interact with Smad3 and disrupt TGF-β signaling. Examples include novel oncoproteins AML/Evi-1, AML/ETO, PML/RARα, and the viral oncoprotein, Notch ICN, and Tax. In theory, the loss of Smad3-dependent tumor suppression may cooperate with oncogenic proterties of these oncoproteins to induce leukemia (A color version of this figure is freely accessible via the website of the book: http://www.springer.com/1-4020-4542-5)

These studies not only implicate Evi-1 as a member of co-repressor complexes but also suggest that aberrant recruitment of co-repressors of TGF-β signaling is one of the mechanisms for Evi-1-induced leukemogenesis (Izutsu et al., 2001).

In the acute promyelocitic subtype of myeloid leukemia (APL), a fusion between the promyelocytic leukaemia (PML) tumor suppressor and the Retinoid Acid Receptor-α also gives rise to a novel oncoprotein. This fusion protein has been reported to exert unique transcriptional activity, but has also been shown to act competitively to inhibit the normal transcriptional activity of the RAR-α. A recent study demonstrates that cytoplasmic Pml is an essential modulator of TGF-β signaling (Lin et al., 2004). Through a series of experiments, cytoplasmic Pml was shown to physically interact with Smad2/3 and SARA (Smad anchor for receptor activation). More importantly, in the absence of normal Pml function or in the presence of the PML-RAR-α oncoprotein of APL, Smad2/3 fails to associate with SARA. The implication is that the PML-RAR-α oncoprotein not only can antagonize PML function but also disrupt TGF-β signaling.

Smad3 has also been implicated as a putative target of viral oncoproteins. For example, infection of cells with human T-cell lymphotropic virus-1 (HTLV-1) has been shown to impair TGF-β signaling. The HTLV-I oncoprotein Tax is implicated in the various clinical manifestations associated with infection by HTLV-1, including an aggressive and fatal T-cell malignancy, and HTLV-1-infected T-cell lines are uniformly resistant to TGF-β growth inhibitory activity. Recently, we have shown that Tax suppresses TGF-β signaling through direct interaction with Smad proteins, indicating that the inhibition of TGF-β signaling by Tax may lead to the HTLV-1-associated leukemogenesis. Tax directly interacts with Smad3 via the Smad MH2 domain, inhibiting formation of the Smad3-Smad4 complex leading to a loss of DNA binding (Lee et al., 2002a). It is important to highlight that this effect of viral oncoproteins is not restricted to lymphoid malignancies, as the human papillomavirus (HPV) oncoprotein E7 also can impair Smad function. E7 is implicated in the etiology of cervical cancer associated with infection by HPV, and we have also demonstrated that E7 interacts constitutively with Smads 2, 3, and 4, and blocks Smad3 binding to its target sequence on DNA. These results indicate that suppression of Smad-mediated signaling by E7 may similarly contribute to HPV-associated carcinogenesis (Lee et al., 2002b).

The list of oncoproteins that can affect Smad signaling is ever increasing. The oncoprotein c-Ski is also frequently involved in chromosomal translocations associated with non-Hodgkin's lymphoma (Chaganti et al., 1986) and pre-B acute lymphoblastic leukemia (Kees et al., 1990), potentially suppressing the activity of Smad3 by leading to increased Smad3 turnover in these hematopoietic malignancies (Luo, 2004; Luo et al., 1999). An interesting newly identified class of candidates is the Notch receptors. The Notch proteins are transmembrane receptors, and dysregulated Notch signaling is now known to induce neoplastic transformation. Mutations in the Notch receptor that enhance activation and promote stability have been found in a high percentage of pediatric patients with Acute Lymphoblastic Leukemia of the T cell lineage. Recently, two studies have suggested that disruption of Smad3 function may again be an important mechanism by which excess Notch signaling promotes tumorigenesis (Masuda et al., 2005). Similar to other oncoproteins, intracellular Notch inhibits TGF-β-mediated transcriptional responses, an effect that appears to be linked to sequestration of p300, as an excess of this transcriptional co-activator can restore the TGF-β response. The Notch receptors and their ligands play important roles in regulating cellular differentiation, proliferation and survival in many systems, suggesting this effect on the TGF-β response could underlie TGF-β resistance in epithelial cancers as well. Indeed, MCF-7 human breast cancer cells that consitutively express the Notch4 intracellular domain are resistant to the growth-inhibitory effects of TGF-β. Inhibitors of gamma-secretase impair Notch4 processing and were shown to restore the ability of MCF-7 cells to respond to the growth inhibitory effect by TGF-β (Sun et al., 2005).

In summary, the aberrant expression of Smad3 inhibitors clearly contributes to the functional inactivation of Smad3, and deregulated expression of Smad3 inhibitors may be particularly important for the functional inactivation of Smad3 in hematopoietic malignancies.

3. PRECLINICAL MODELS DEFINE TUMOR SUPPRESSOR ACTIVITY OF SMAD SIGNALING INTERMEDIATES

TGF-β plays an important role both in the early suppression of malignancy and at a later stage in tumor progression. The role of the Smad signaling pathway in the tumor suppressor and pro-oncogenic activities of TGF-β has recently explored in several experimental systems. In a mouse multi-stage skin carcinogenesis model, overexpression of Smad3, but not of Smad2 or Smad4, induced senescence and growth arrest of v-ras^{Ha} transduced *SMAD3*$^{-/-}$ keratinocytes, whereas Smad3 depletion reduced senescence in a gene dose-dependent manner. This suggests that Smad3 can suppress premalignant progression, but that endogenous Smad2 cannot compensate for the loss of Smad3 (Vijayachandra et al., 2003). However, constitutive activation of Smad2/3 drives tumor progression: In squamous carcinoma cells, expression of the activated form of either Smad2 or Smad3 induced overt epithelial-mesenchymal transition (EMT) when the level of H-*ras* was elevated (Oft et al., 2002). The coordinate upregulation of both Ras and TGF-β can be seen as a mechanism that allows the tumor cell to adapt the cell-fate change and invasive properties required for tumor progression, with the increase in TGF-β expression leading to the constitutive activation of Smad2/3.

The *SMAD3*$^{-/-}$ mouse model has also been useful to uncover dose-dependent effects of Smad3 in the suppression of leukemogenesis. In mice, the loss of one allele for Smad3 impairs the inhibitory effect of TGF-β on the proliferation of normal T cells and works in tandem with the homozygous inactivation of the cdk inhibitor, $p27^{Kip1}$, to promote T-cell leukemogenesis (Fig. 2). It is interesting that this cooperativity is observed between a loss of Smad3 and an inhibitor of the retinoblastoma (Rb) pathway. It is likely that the loss of $p27^{Kip1}$, which alone is insufficient to induce leukemia, primarily affects proliferative activity of T cell progenitors. Our speculation is that Smad3 is critical for mediating apoptotic effects of TGF-β, and that loss of Smad3 provides a pro-survival advantage in T cell progenitors. Another

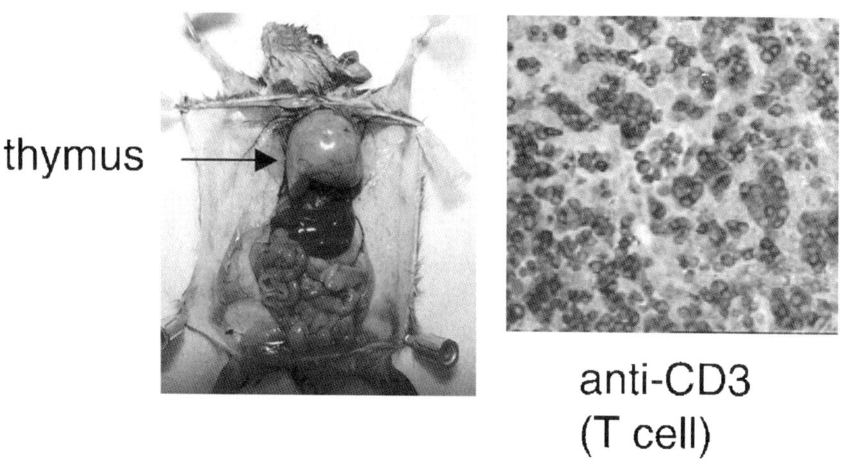

Figure 2. Smad3 is a suppressor of T cell leukemogenesis. In mice, loss of a single *SMAD3* allele cooperates with the loss of the Cdk inhibitor $p27^{Kip1}$ to promote spontaneous T cell leukemia. We also find that loss of Smad3 potentiates retroviral-induced leukemogenesis (manuscript in preparation). Image on the left depicts a large thymic tumor which developed in a *SMAD3*$^{+/-}$ mouse on a background in which both alleles of the Cdk inhibitor $p27^{Kip1}$ are disrupted; the right panel demonstrates positive staining for the T cell receptor CD3 (A color version of this figure is freely accessible via the website of the book: http://www.springer.com/1-4020-4542-5)

interesting coincidence, perhaps, is that both Smad3 and $p27^{Kip1}$ are putative targets of the Notch oncoprotein. Deregulation of Notch leads to suppression of $p27^{Kip1}$ expression and to functional interference with the TGF-β receptor-activated Smad3. Taken together, the data truly highlight a central role for Smad3 in maintenance of T cell homeostasis and in the suppression of T cell transformation. The significance of these observations has been supported by our observation that there is a specific loss of Smad3 expression in human T cell ALL (Wolfraim et al., 2004). Although the mechanism for loss of Smad3 protein expression has not been defined, we did not observe any specific mutations in the *MADH3* gene in any of the patients evaluated, and there were no significant reductions in mRNA for Smad3, implicating a post-transcriptional mechanism leading to loss of Smad3 protein.

The tumor suppressor activity of the endogenous Smad2/3 pathway has also been investigated using a MCF10A human breast cell line system with defined oncogenic/metastatic potential. Interference with endogenous Smad2/3 signaling enhances the malignancy of xenografted tumors of premalignant and well-differentiated tumor cells, but strongly suppresses lung metastases of more aggressive carcinoma cells. Overexpression of Smad3 in the same cells has the opposite effect. These studies show that signaling through the Smad2/3 pathway can mediate tumor suppressor and oncogenic effects of TGF-β, depending on the stage of progression of the cells and other cooperating contextual changes (Tian et al., 2004). In another study, functional imaging was utilized to demonstrate Smad signaling in bone metastasis of xenografts of human MDA-MB-231 breast carcinomas. This cell line, which was derived from the pleural effusion of a breast cancer patient with metastatic disease (Bevan et al., 1999), was used to derive sublines with distinct organ-specific metastatic behavior (Kang et al., 2005). In a subline (SCP2), which is highly metastatic to bone, the expression of a retroviral reporter vector (*cis*-TGF-β1-Smads-HSV1-tk-GFP) is activated specifically in bone metastases. In a series of genetic depletion experiments, the authors of this work were able to demonstrate that Smad4 contributes to the formation of osteolytic bone metastases. A mechanism provided in this report suggests that Smad4 is required for optimal induction of *IL-11*, a gene implicated in bone metastasis in this mouse model system. This study has important implications for other malignant diseases in which expansion of the malignant population within the bone and bone marrow are common. This list includes multiple myeloma, a tumor that has been characterized for production of active TGF-β and for resistance to the inhibitory effects of this cytokine on both proliferation and survival.

Although Smad7 was initially depicted in biochemical studies as simply an inhibitor of TGF-β signaling, dysegulated expression of Smad7 may also play a role in tumorigenesis, as noted above. In preclinical models, overexpression of Smad7 in the context of an activated *ras* oncogene can cause conversion of epithelial cells from a benign to a malignant tumor phenotype in a mouse model for squamous cell carcinoma, suggesting that Smad7 may have an important role in the pathogenesis of epithelial cancers (Liu et al., 2003). In another study, transfection of pancreatic cancer cells with Smad7 leads to enhanced anchorage-independent growth and accelerated tumoigenesis in nude mice (Kleeff et al., 1999). In another model, transgenic mice over-expressing Smad7 under the control of K5 promoter exhibited pathological changes in multiple tissues and died within 10 days after birth. This study provided evidence that Smad7 is a potent *in vivo* inhibitor for signal transduction of the TGF-β superfamily, and can regulate the development and maintain the homeostasis of multiple epithelial tissues (He et al., 2002).

Although Smad signaling clearly acts in a cell autonomous fashion to suppress malignant transformation, recent data highlight the importance of Smad-dependent TGF-β signaling in the stromal cells in the environment in which tumors arise. As discussed above, the development of carcinoma, the most common form of human cancer, is classically viewed as the consequence of somatic mutations that accumulate in epithelial cells. TGF-β signaling plays an important role in this process, with mutations in the TGF-β type II receptor frequently found in intestinal epithelial cells in patients with colon cancer (Markowitz et al., 1995). However, the initiation and progression of carcinomas is also influenced by the tumor microenvironment, which includes extracellular matrix, blood vasculature, inflammatory cells and fibroblasts (Bissell and Labarge, 2005; Bhowmick et al., 2004b). The role of TGF-β in maintaining this epithelial 'landscape' is supported by the observation that a disruption in TGF-β signaling in fibroblasts modulates the oncogenic potential of adjacent epithelia via a mechanism that involves the production of soluble factors that provide mitogenic signals in an abnormal paracrine signaling loop (Bhowmick et al., 2004a).

In our own recent investigation, we have evaluated the role of Smad4-dependent signaling in lymphocytes and have uncovered an important link between Smad4 function in T cells and the process of epithelial carcinogenesis. We focused on Smad4-dependent signaling in T lymphocytes as this lineage plays a major role in maintaining immune homeostasis within the intestinal environment. Utilizing a conditional gene targeting strategy to specifically disrupt expression of Smad4 within the T lineage in mice, we were able to ask whether loss of Smad4-dependent signaling in T cells would be sufficient to induce the formation of hamartomatous lesions and cancer within the gastrointestinal mucosa. This strategy provides the advantage that one can clearly assess the contribution of a disruption in *SMAD4* gene expression within one lineage, while Smad4 expression remains intact in the other. Our results support the concept that cancer, as an outcome, reflects the loss of the normal communication between the cellular constituents of a given organ (Bissell and Labarge, 2005), as Smad4-deficient T cells act to induce the malignant transformation of their epithelial neighbors (manuscript submitted). Although the mechanisms that mediate this epithelial transformation remain unknown, we hypothesize that the loss of Smad4-dependent signaling in T cells results in the elaboration of lymphokines and cytokines that promote epithelial hyperplasia, potentially through an indirect mechanism that involves alterations of stromal-eplithelial interactions.

Finally, it is important to emphasize that we cannot assume that Smads simply function as mediators of TGF-β signaling in this context. Clearly, Smad4 play important roles in both BMP and Activin signaling, and it may be these cytokines that are critical in the suppression of certain forms of cancer. For example, the presence of both MADH4 and BMPR1 mutations in patients with FJP implies that BMP signaling is critically important in the suppression of the hamaromatous polyposis that is seen in this familial cancer syndrome (He et al., 2004; Howe et al., 2004). Activin, another member of the TGF-β superfamily, has also been shown to inhibit the proliferation of breast cancer cells. Activin functions by interacting with type I and type II Activin receptors to induce phosphorylation of Smads. Activin has been shown to inhibit cellular proliferation of T47D breast cancer cells, an effect that could be blocked by an adenoviral dominant-negative Smad3. Similar to TGF-β, Activin acts to maintain expression of $p21^{Cip1}$ and $p27^{Kip1}$ cyclin-dependent kinase inhibitors involved in cell cycle control, and reduced phosphorylation of the retinoblastoma (Rb) protein. This similarity between the Activin and TGF-β signaling system has also been observed in *in vitro* models of colon cancer. Frameshift mutations of ACVR2 have been described in colon cancer and may contribute to MSI-H colon tumorigenesis via disruption of alternate TGF-β effector pathways. Restoration of ACVR2 in such colon cancer cells leads to the expression of genes implicated in the control of cell growth and tumorigenesis, including the activator protein (AP)-1 complex genes JUND, JUN, and FOSB, and others, suggesting that Smad-mediated Activin signaling may serve as an alternative activator of TGF-β effectors.

Similar to TGF-β, Activin expression my suppress tumorigenesis through a Smad-dependent effect on the tumor microenvironment. Activin A is an inhibitor of angiogenesis, and is down-regulated by the N-MYC oncogene, whose expression predicts poor outcome in neuroblastoma. It has been shown that restoration of Activin A expression in neuroblastoma cells inhibited the growth of xenografted neuroblastoma tumors (Panopoulou et al., 2005). The effect was associated with a reduced vascularity, and correlated with the ability of Activin A to inhibit angiogenic responses in cultured endothelial cells. The suppression of proteolytic activity, migration, and proliferation of endothelial was demonstrated to be Smad-dependent.

4. SUMMARY AND PERSPECTIVES

The role of the Smad signaling pathway in tumor suppression is well established, but many significant questions remain, regarding the context in which these intermediates act as tumor suppressors. We have tried to highlight the important concept that mutational inactivation of the genes encoding these intermediates may only be relevant in a small subset of human cancers, and that epigenetic mechanisms that lead to silencing or suppression of Smad signaling play an important role. In this regard, the physical interaction of oncoproteins with Smad3 appears to be an important step in hematopoietic cell transformation (this includes both viral and oncoproteins arising from novel fusion genes resulting from chromosomal translocations). It is interesting that Smad2 has not emerged as a target of these oncoproteins, suggesting a unique role for Smad3 in this context.

In the future, it will be important to address the specificities of these two receptor-activated intermediates in the TGF-β pathway. It is also important to recognize that other members of the TGF-β superfamily, particularly Activin and BMP, also signal via the Smad family, and that it may ultimately be the function of Smads in these pathways that arbitrate the transformation of cells of the epithelial and hematopoietic lineages. The interplay between Smad pathways and MAP Kinase pathways may be important in determining the threshold for transformation, and will continue to be an area of intense investigation. This information will be essential if therapeutics developed to augment Smad function and enhance Smad expression are going to find application in the prevention and treatment of cancer. Finally, while the Smad genes can be viewed as classical tumor suppressors, recent studies are pointing to important roles of Smad-dependent signaling the microenvironment in which a malignant cell arises, and suggest that altered Smad signaling in these supporting cells may be a critical step in the initiation and progression of a variety of cancers.

REFERENCES

Arai, T., Akiyama, Y., Okabe, S., Ando, M., Endo, M., and Yuasa, Y., 1998, Genomic structure of the human Smad3 gene and its infrequent alterations in colorectal cancers. *Cancer Lett* 122: 157-163.

Baez, A., Cantor, A., Fonseca, S., Marcos-Martinez, M., Mathews, L.A., Muro-Cacho, C.A., and Munoz-Antonia, T., 2005, Differences in Smad4 expression in human papillomavirus type 16-positive and human papillomavirus type 16-negative head and neck squamous cell carcinoma. *Clin Cancer Res* 11: 3191-3197.

Barrett, M.T., Schutte, M., Kern, S.E., and Reid, B.J., 1996, Allelic loss and mutational analysis of the DPC4 gene in esophageal adenocarcinoma. *Cancer Res* 56: 4351-4353.

Bevan, S., Woodford-Richens, K., Rozen, P., Eng, C., Young, J., Dunlop, M., Neale, K., Phillips, R., Markie, D., Rodriguez-Bigas, M., Leggett, B., Sheridan, E., Hodgson, S., Iwama, T., Eccles, D., Bodmer, W., Houlston, R., and Tomlinson, I., 1999, Screening SMAD1, SMAD2, SMAD3, and SMAD5 for germline mutations in juvenile polyposis syndrome. *Gut* 45: 406-408.

Bhowmick, N.A., Chytil, A., Plieth, D., Gorska, A.E., Dumont, N., Shappell, S., Washington, M.K., Neilson, E.G., and Moses, H.L., 2004a, TGF-β signaling in fibroblasts modulates the oncogenic potential of adjacent epithelia. *Science* 303: 848-851.

Bhowmick, N.A., Neilson, E.G., and Moses, H.L., 2004b, Stromal fibroblasts in cancer initiation and progression. *Nature* 432: 332-337.

Bissell, M.J., and Labarge, M.A., 2005, Context, tissue plasticity, and cancer: are tumor stem cells also regulated by the microenvironment? *Cancer Cell* 7: 17-23.

Brodin, G., ten Dijke, P., Funa, K., Heldin, C.-H., and Landström, M., 1999, Increased smad expression and activation are associated with apoptosis in normal and malignant prostate after castration. *Cancer Res* 59: 2731-2738.

Chaganti, R.S., Balazs, I., Jhanwar, S.C., Murty, V.V., Koduru, P.R., Grzeschik, K.H., and Stavnezer, E., 1986, The cellular homologue of the transforming gene of SKV avian retrovirus maps to human chromosome region 1q22-q24. *Cytogenet Cell Genet* 43: 181-186.

Croce, C.M., 1991, Genetic approaches to the study of the molecular basis of human cancer. *Cancer Res* 51: 5015-5018.

Datto, M.B., Frederick, J.P., Pan, L., Borton, A.J., Zhuang, Y., and Wang, X.F., 1999, Targeted disruption of Smad3 reveals an essential role in transforming growth factor β-mediated signal transduction. *Mol Cell Biol* 19: 2495-2504.

de Caestecker, M.P., Piek, E., and Roberts, A.B., 2000, Role of transforming growth factor-β signaling in cancer. *J Natl Cancer Inst* 92: 1388-1402.

Dowdy, S.C., Mariani, A., Reinholz, M.M., Keeney, G.L., Spelsberg, T.C., Podratz, K.C., and Janknecht, R., 2005, Overexpression of the TGF-β antagonist Smad7 in endometrial cancer. *Gynecol Oncol* 96: 368-373.

Elliott, R.L., and Blobe, G.C., 2005, Role of transforming growth factor B in human cancer. *J Clin Oncol* 23: 2078-2093.

Eppert, K., Scherer, S.W., Ozcelik, H., Pirone, R., Hoodless, P., Kim, H., Tsui, L.C., Bapat, B., Gallinger, S., Andrulis, I.L., Thomsen, G.H., Wrana, J.L., and Attisano, L., 1996, MADR2 maps to 18q21 and encodes a TGFβ-regulated MAD-related protein that is functionally mutated in colorectal carcinoma. *Cell* 86: 543-552.

Fukuchi, M., Masuda, N., Miyazaki, T., Nakajima, M., Osawa, H., Kato, H., and Kuwano, H., 2002, Decreased Smad4 expression in the transforming growth factor-β signaling pathway during progression of esophageal squamous cell carcinoma. *Cancer* 95: 737-743.

Hahn, S.A., Schutte, M., Hoque, A.T., Moskaluk, C.A., da Costa, L.T., Rozenblum, E., Weinstein, C.L., Fischer, A., Yeo, C.J., Hruban, R.H., and Kern, S.E., 1996, DPC4, a candidate tumor suppressor gene at human chromosome 18q21.1. *Science* 271: 350-353.

Halder, S.K., Beauchamp, R.D., and Datta, P.K., 2005, Smad7 induces tumorigenicity by blocking TGF-β-induced growth inhibition and apoptosis. *Exp Cell Res* 307: 231-246.

Han, S.U., Kim, H.T., Seong do, H., Kim, Y.S., Park, Y.S., Bang, Y.J., Yang, H.K., and Kim, S.J., 2004, Loss of the Smad3 expression increases susceptibility to tumorigenicity in human gastric cancer. *Oncogene* 23: 1333-1341.

He, W., Li, A.G., Wang, D., Han, S., Zheng, B., Goumans, M.J., Ten Dijke, P., and Wang, X.J., 2002, Overexpression of Smad7 results in severe pathological alterations in multiple epithelial tissues. *EMBO J* 21: 2580-2590.

He, X.C., Zhang, J., Tong, W.G., Tawfik, O., Ross, J., Scoville, D.H., Tian, Q., Zeng, X., He, X., Wiedemann, L.M., Mishina, Y., and Li, L., 2004, BMP signaling inhibits intestinal stem cell self-renewal through suppression of Wnt-β-catenin signaling. *Nat Genet* 36: 1117-1121.

Horvath, L.G., Henshall, S.M., Kench, J.G., Turner, J.J., Golovsky, D., Brenner, P.C., O'Neill, G.F., Kooner, R., Stricker, P.D., Grygiel, J.J., and Sutherland, R.L., 2004, Loss of BMP2, Smad8, and Smad4 expression in prostate cancer progression. *Prostate* 59: 234-242.

Howe, J.R., Roth, S., Ringold, J.C., Summers, R.W., Jarvinen, H.J., Sistonen, P., Tomlinson, I.P., Houlston, R.S., Bevan, S., Mitros, F.A., Stone, E.M., and Aaltonen, L.A., 1998, Mutations in the SMAD4/DPC4 gene in juvenile polyposis. *Science* 280: 1086-1088.

Howe, J.R., Sayed, M.G., Ahmed, A.F., Ringold, J., Larsen-Haidle, J., Merg, A., Mitros, F.A., Vaccaro, C.A., Petersen, G.M., Giardiello, F.M., Tinley, S.T., Aaltonen, L.A., and Lynch, H.T., 2004, The prevalence of MADH4 and BMPR1A mutations in juvenile polyposis and absence of BMPR2, BMPR1B, and ACVR1 mutations. *J Med Genet* 41: 484-491.

Ikezoe, T., Takeuchi, S., Arai, T., Yuasa, Y., and Taguchi, H., 1998a, Analysis of the Smad3 gene in hematological malignancies. *Int J Hematol* 68: 457-458.

Ikezoe, T., Takeuchi, S., Kamioka, M., Daibata, M., Kubonishi, I., Taguchi, H., and Miyoshi, I., 1998b, Analysis of the Smad2 gene in hematological malignancies. *Leukemia* 12: 94-95.

Imai, Y., Kurokawa, M., Izutsu, K., Hangaishi, A., Maki, K., Ogawa, S., Chiba, S., Mitani, K., and Hirai, H., 2001, Mutations of the Smad4 gene in acute myelogeneous leukemia and their functional implications in leukemogenesis. *Oncogene* 20: 88-96.

Izutsu, K., Kurokawa, M., Imai, Y., Maki, K., Mitani, K., and Hirai, H., 2001, The corepressor CtBP interacts with Evi-1 to repress transforming growth factor β signaling. *Blood* 97: 2815-2822.

Jen, J., Kim, H., Piantadosi, S., Liu, Z.F., Levitt, R.C., Sistonen, P., Kinzler, K.W., Vogelstein, B., and Hamilton, S.R., 1994, Allelic loss of chromosome 18q and prognosis in colorectal cancer. *N Engl J Med* 331: 213-221.

Jonson, T., Gorunova, L., Dawiskiba, S., Andren-Sandberg, A., Stenman, G., ten Dijke, P., Johansson, B., and Hoglund, M., 1999, Molecular analyses of the 15q and 18q SMAD genes in pancreatic cancer. *Genes Chromosomes Cancer* 24: 62-71.

Kang, Y., He, W., Tulley, S., Gupta, G.P., Serganova, I., Chen, C.R., Manova-Todorova, K., Blasberg, R., Gerald, W.L., and Massagué, J., 2005, Breast cancer bone metastasis mediated by the Smad tumor suppressor pathway. *Proc Natl Acad Sci U S A* 102: 13909-13914.

Kawate, S., Takenoshita, S., Ohwada, S., Mogi, A., Fukusato, T., Makita, F., Kuwano, H., and Morishita, Y., 1999, Mutation analysis of transforming growth factor β type II receptor, Smad2, and Smad4 in hepatocellular carcinoma. *Int J Oncol* 14: 127-131.

Kees, U.R., Lukeis, R., Ford, J., Willoughby, M.L., and Garson, O.M., 1990, Establishment and characterization of a childhood pre-B acute lymphoblastic leukemia cell line, PER-278, with chromosome translocations t(1;19) and t(1;9). *Cancer Genet Cytogenet* 46: 201-208.

Kern, S.E., Fearon, E.R., Tersmette, K.W., Enterline, J.P., Leppert, M., Nakamura, Y., White, R., Vogelstein, B., and Hamilton, S.R., 1989, Clinical and pathological associations with allelic loss in colorectal carcinoma [corrected]. *Jama* 261: 3099-3103.

Kim, S.K., Fan, Y., Papadimitrakopoulou, V., Clayman, G., Hittelman, W.N., Hong, W.K., Lotan, R., and Mao, L., 1996, DPC4, a candidate tumor suppressor gene, is altered infrequently in head and neck squamous cell carcinoma. *Cancer Res* 56: 2519-2521.

Kim, Y.H., Lee, H.S., Lee, H.J., Hur, K., Kim, W.H., Bang, Y.J., Kim, S.J., Lee, K.U., Choe, K.J., and Yang, H.K., 2004, Prognostic significance of the expression of Smad4 and Smad7 in human gastric carcinomas. *Ann Oncol* 15: 574-580.

Kleeff, J., Ishiwata, T., Maruyama, H., Friess, H., Truong, P., Buchler, M.W., Falb, D., and Korc, M., 1999, The TGF-β signaling inhibitor Smad7 enhances tumorigenicity in pancreatic cancer. *Oncogene* 18: 5363-5372.

Kong, X.T., Choi, S.H., Inoue, A., Xu, F., Chen, T., Takita, J., Yokota, J., Bessho, F., Yanagisawa, M., Hanada, R., Yamamoto, K., and Hayashi, Y., 1997, Expression and mutational analysis of the DCC, DPC4, and MADR2/JV18-1 genes in neuroblastoma. *Cancer Res* 57: 3772-3778.

Korchynskyi, O., Landström, M., Stoika, R., Funa, K., Heldin, C.-H., ten Dijke, P., and Souchelnytskyi, S., 1999, Expression of Smad proteins in human colorectal cancer. *Int J Cancer* 82: 197-202.

Kurokawa, M., Mitani, K., Imai, Y., Ogawa, S., Yazaki, Y., and Hirai, H., 1998, The t(3;21) fusion product, AML1/Evi-1, interacts with Smad3 and blocks transforming growth factor-β-mediated growth inhibition of myeloid cells. *Blood* 92: 4003-4012.

Lange, D., Persson, U., Wollina, U., ten Dijke, P., Castelli, E., Heldin, C.-H., and Funa, K., 1999, Expression of TGF-β related Smad proteins in human epithelial skin tumors. *Int J Oncol* 14: 1049-1056.

Latil, A., Pesche, S., Valeri, A., Fournier, G., Cussenot, O., and Lidereau, R., 1999, Expression and mutational analysis of the MADR2/Smad2 gene in human prostate cancer. *Prostate* 40: 225-231.

Lee, D.K., Kim, B.C., Brady, J.N., Jeang, K.T., and Kim, S.J., 2002a, Human T-cell lymphotropic virus type 1 tax inhibits transforming growth factor-β signaling by blocking the association of Smad proteins with Smad-binding element. *J Biol Chem* 277: 33766-33775.

Lee, D.K., Kim, B.C., Kim, I.Y., Cho, E.A., Satterwhite, D.J., and Kim, S.J., 2002b, The human papilloma virus E7 oncoprotein inhibits transforming growth factor-β signaling by blocking binding of the Smad complex to its target sequence. *J Biol Chem* 277: 38557-38564.

Li, A.G., Lu, S.L., Zhang, M.X., Deng, C., and Wang, X.J., 2004, Smad3 knockout mice exhibit a resistance to skin chemical carcinogenesis. *Cancer Res* 64: 7836-7845.

Lin, H.K., Bergmann, S., and Pandolfi, P.P., 2004, Cytoplasmic PML function in TGF-β signalling. *Nature* 431: 205-211.

Liu, X., Lee, J., Cooley, M., Bhogte, E., Hartley, S., and Glick, A., 2003, Smad7 but not Smad6 cooperates with oncogenic ras to cause malignant conversion in a mouse model for squamous cell carcinoma. *Cancer Res* 63: 7760-7768.

Luo, K., 2004, Ski and SnoN: negative regulators of TGF-β signaling. *Curr Opin Genet Dev* 14: 65-70.

Luo, K., Stroschein, S.L., Wang, W., Chen, D., Martens, E., Zhou, S., and Zhou, Q., 1999, The Ski oncoprotein interacts with the Smad proteins to repress TGFβ signaling. *Genes Dev* 13: 2196-2206.

Maesawa, C., Tamura, G., Nishizuka, S., Iwaya, T., Ogasawara, S., Ishida, K., Sakata, K., Sato, N., Ikeda, K., Kimura, Y., Saito, K., and Satodate, R., 1997, MAD-related genes on 18q21.1, Smad2 and Smad4, are altered infrequently in esophageal squamous cell carcinoma. *Jpn J Cancer Res* 88: 340-343.

Markowitz, S., Wang, J., Myeroff, L., Parsons, R., Sun, L., Lutterbaugh, J., Fan, R.S., Zborowska, E., Kinzler, K.W., Vogelstein, B., and et al., 1995, Inactivation of the type II TGF-β receptor in colon cancer cells with microsatellite instability. *Science* 268: 1336-1338.

Masuda, S., Kumano, K., Shimizu, K., Imai, Y., Kurokawa, M., Ogawa, S., Miyagishi, M., Taira, K., Hirai, H., and Chiba, S., 2005, Notch1 oncoprotein antagonizes TGF-β/Smad-mediated cell growth suppression via sequestration of coactivator p300. *Cancer Sci* 96: 274-282.

Maurice, D., Pierreux, C.E., Howell, M., Wilentz, R.E., Owen, M.J., and Hill, C.S., 2001, Loss of Smad4 function in pancreatic tumors: C-terminal truncation leads to decreased stability. *J Biol Chem* 276: 43175-43181.

Mishra, L., Katuri, V., and Evans, S., 2005, The role of PRAJA and ELF in TGF-β signaling and gastric cancer. *Cancer Biol Ther* 4: 694-699.

Monteleone, G., Del Vecchio Blanco, G., Palmieri, G., Vavassori, P., Monteleone, I., Colantoni, A., Battista, S., Spagnoli, L.G., Romano, M., Borrelli, M., MacDonald, T.T., and Pallone, F., 2004, Induction and regulation of Smad7 in the gastric mucosa of patients with Helicobacter pylori infection. *Gastroenterology* 126: 674-682.

Nagatake, M., Takagi, Y., Osada, H., Uchida, K., Mitsudomi, T., Saji, S., Shimokata, K., and Takahashi, T., 1996, Somatic *in vivo* alterations of the DPC4 gene at 18q21 in human lung cancers. *Cancer Res* 56: 2718-2720.

Oft, M., Akhurst, R.J., and Balmain, A., 2002, Metastasis is driven by sequential elevation of H-ras and Smad2 levels. *Nat Cell Biol* 4: 487-494.

Panopoulou, E., Murphy, C., Rasmussen, H., Bagli, E., Rofstad, E.K., and Fotsis, T., 2005, Activin A suppresses neuroblastoma xenograft tumor growth via antimitotic and antiangiogenic mechanisms. *Cancer Res* 65: 1877-1886.

Park, W.S., Park, J.Y., Oh, R.R., Yoo, N.J., Lee, S.H., Shin, M.S., Lee, H.K., Han, S., Yoon, S.K., Kim, S.Y., Choi, C., Kim, P.J., Oh, S.T., and Lee, J.Y., 2000, A distinct tumor suppressor gene locus on chromosome 15q21.1 in sporadic form of colorectal cancer. *Cancer Res* 60: 70-73.

Park, Y.N., Chae, K.J., Oh, B.K., Choi, J., Choi, K.S., and Park, C., 2004, Expression of Smad7 in hepatocellular carcinoma and dysplastic nodules: resistance mechanism to transforming growth factor-β. *Hepatogastroenterology* 51: 396-400.

Piestrzeniewicz-Ulanska, D., Brys, M., Semczuk, A., Jakowicki, J.A., and Krajewska, W.M., 2003, Expression and intracellular localization of Smad proteins in human endometrial cancer. *Oncol Rep* 10: 1539-1544.

Piestrzeniewicz-Ulanska, D., Brys, M., Semczuk, A., Rechberger, T., Jakowicki, J.A., and Krajewska, W.M., 2004, TGF-β signaling is disrupted in endometrioid-type endometrial carcinomas. *Gynecol Oncol* 95: 173-180.

Riggins, G.J., Thiagalingam, S., Rozenblum, E., Weinstein, C.L., Kern, S.E., Hamilton, S.R., Willson, J.K., Markowitz, S.D., Kinzler, K.W., and Vogelstein, B., 1996, Mad-related genes in the human. *Nat Genet* 13: 347-349.

Roijer, E., Morén, A., ten Dijke, P., and Stenman, G., 1998, Assignment1 of the Smad7 gene (MADH7) to human chromosome 18q21.1 by fluorescence in situ hybridization. *Cytogenet Cell Genet* 81: 189-190.

Schutte, M., 1999, DPC4/SMAD4 gene alterations in human cancer, and their functional implications. *Ann Oncol* 10 Suppl 4: 56-59.

Schutte, M., Hruban, R.H., Hedrick, L., Cho, K.R., Nadasdy, G.M., Weinstein, C.L., Bova, G.S., Isaacs, W.B., Cairns, P., Nawroz, H., Sidransky, D., Casero, R.A., Jr., Meltzer, P.S., Hahn, S.A., and Kern, S.E., 1996, DPC4 gene in various tumor types. *Cancer Res* 56: 2527-2530.

Shitara, Y., Yokozaki, H., Yasui, W., Takenoshita, S., Kuwano, H., Nagamachi, Y., and Tahara, E., 1999, No mutations of the Smad2 gene in human sporadic gastric carcinomas. *Jpn J Clin Oncol* 29: 3-7.

Sood, R., Talwar-Trikha, A., Chakrabarti, S.R., and Nucifora, G., 1999, MDS1/EVI1 enhances TGF-β1 signaling and strengthens its growth-inhibitory effect but the leukemia-associated fusion protein

AML1/MDS1/EVI1, product of the t(3;21), abrogates growth-inhibition in response to TGF-β1. *Leukemia* 13: 348-357.

Sun, Y., Lowther, W., Kato, K., Bianco, C., Kenney, N., Strizzi, L., Raafat, D., Hirota, M., Khan, N.I., Bargo, S., Jones, B., Salomon, D., and Callahan, R., 2005, Notch4 intracellular domain binding to Smad3 and inhibition of the TGF-β signaling. *Oncogene* 24: 5365-5374.

Takei, K., Kohno, T., Hamada, K., Takita, J., Noguchi, M., Matsuno, Y., Hirohashi, S., Uezato, H., and Yokota, J., 1998, A novel tumor suppressor locus on chromosome 18q involved in the development of human lung cancer. *Cancer Res* 58: 3700-3705.

Thiagalingam, S., Lengauer, C., Leach, F.S., Schutte, M., Hahn, S.A., Overhauser, J., Willson, J.K., Markowitz, S., Hamilton, S.R., Kern, S.E., Kinzler, K.W., and Vogelstein, B., 1996, Evaluation of candidate tumour suppressor genes on chromosome 18 in colorectal cancers. *Nat Genet* 13: 343-346.

Tian, F., Byfield, S.D., Parks, W.T., Stuelten, C.H., Nemani, D., Zhang, Y.E., and Roberts, A.B., 2004, Smad-binding defective mutant of transforming growth factor β type I receptor enhances tumorigenesis but suppresses metastasis of breast cancer cell lines. *Cancer Res* 64: 4523-4530.

Uchida, K., Nagatake, M., Osada, H., Yatabe, Y., Kondo, M., Mitsudomi, T., Masuda, A., and Takahashi, T., 1996, Somatic *in vivo* alterations of the JV18-1 gene at 18q21 in human lung cancers. *Cancer Res* 56: 5583-5585.

Wang, D., Kanuma, T., Mizunuma, H., Takama, F., Ibuki, Y., Wake, N., Mogi, A., Shitara, Y., and Takenoshita, S., 2000, Analysis of specific gene mutations in the transforming growth factor-β signal transduction pathway in human ovarian cancer. *Cancer Res* 60: 4507-4512.

Wick, W., Petersen, I., Schmutzler, R.K., Wolfarth, B., Lenartz, D., Bierhoff, E., Hummerich, J., Muller, D.J., Stangl, A.P., Schramm, J., Wiestler, O.D., and von Deimling, A., 1996, Evidence for a novel tumor suppressor gene on chromosome 15 associated with progression to a metastatic stage in breast cancer. *Oncogene* 12: 973-978.

Wieser, R., Gruber, B., Rieder, H., and Fonatsch, C., 1998, Mutational analysis of the tumor suppressor Smad2 in acute lymphoid and myeloid leukemia. *Leukemia* 12: 1114-1118.

Vijayachandra, K., Lee, J., and Glick, A.B., 2003, Smad3 regulates senescence and malignant conversion in a mouse multistage skin carcinogenesis model. *Cancer Res* 63: 3447-3452.

Villanueva, A., Garcia, C., Paules, A.B., Vicente, M., Megias, M., Reyes, G., de Villalonga, P., Agell, N., Lluis, F., Bachs, O., and Capella, G., 1998, Disruption of the antiproliferative TGF-β signaling pathways in human pancreatic cancer cells. *Oncogene* 17: 1969-1978.

Wolfraim, L.A., Fernandez, T.M., Mamura, M., Fuller, W.L., Kumar, R., Cole, D.E., Byfield, S., Felici, A., Flanders, K.C., Walz, T.M., Roberts, A.B., Aplan, P.D., Balis, F.M., and Letterio, J.J., 2004, Loss of Smad3 in acute T-cell lymphoblastic leukemia. *N Engl J Med* 351: 552-559.

Woodford-Richens, K., Williamson, J., Bevan, S., Young, J., Leggett, B., Frayling, I., Thway, Y., Hodgson, S., Kim, J.C., Iwama, T., Novelli, M., Sheer, D., Poulsom, R., Wright, N., Houlston, R., and Tomlinson, I., 2000, Allelic loss at SMAD4 in polyps from juvenile polyposis patients and use of fluorescence in situ hybridization to demonstrate clonal origin of the epithelium. *Cancer Res* 60: 2477-2482.

Yang, X., Letterio, J.J., Lechleider, R.J., Chen, L., Hayman, R., Gu, H., Roberts, A.B., and Deng, C., 1999, Targeted disruption of SMAD3 results in impaired mucosal immunity and diminished T cell responsiveness to TGF-β. *EMBO J* 18: 1280-1291.

Zhu, Y., Richardson, J.A., Parada, L.F., and Graff, J.M., 1998, Smad3 mutant mice develop metastatic colorectal cancer. *Cell* 94: 703-714.

CHAPTER 21

TGF-β RECEPTOR KINASE INHIBITORS FOR THE TREATMENT OF CANCER

MICHAEL LAHN, BRANDI BERRY, SUSANNE KLOEKER, AND JONATHAN M. YINGLING

Oncology Division, Lilly Research Laboratories, Lilly Corporate Center, Indianapolis, Indiana 46285, USA

Abstract: TGF-β signaling in cancer can be modulated by a variety of different pharmacological agents, including small molecule inhibitors of the TGF-β type I receptor kinase. Such inhibitors are currently being developed by a number of pharmaceutical companies and may soon be clinically investigated. While such inhibitors progress through early clinical development, selecting patients, who most likely will benefit from this novel compound class, will be critical for determining the efficacy of the TβRI kinase inhibitors. In the present article, we review clinical studies examining the association of TGF-β/Smad signaling and clinical outcome. This review may help in the identification of the appropriate patient population to investigate the activity of future TGF-β/Smad inhibitors

Keywords: breast cancer; colorectal cancer; kinase inhibitors; non small cell lung cancer; patient selection; prostate cancer; TGF-β

1. TARGETING THE TGF-β/SMAD SIGNALING PATHWAY FOR NOVEL ANTI-CANCER DRUGS

1.1 Pharmacological Interventions Alter TGF-β-mediated Signaling in Cancer

Based on knowledge of the TGF-β signaling pathway, three pharmacological interventions have been designed to block TGF-β-mediated signaling. The first pharmacologic intervention consists of blocking the generation of TGF-β and its isoforms using either DNA- or RNA-based inhibition. For instance, antisense oligonucleotides have been developed and are currently being investigated clinically to block TGF-β1 and TGF-β2 mRNA production. A second approach is directed towards blocking the TGF-β isoforms. This pharmacological intervention uses large

P. ten Dijke and C.-H. Heldin (eds.), Smad Signal Transduction, 415–442.

molecules, such as monoclonal antibodies. Some of the monoclonal antibodies are currently in advanced clinical investigation, including the monoclonal antibodies lerdelimumab and metelimumab. However, none of these antibodies are being examined in cancer patients, and only the monoclonal antibody GC-1008 directed against all three TGF-β isoforms is being considered for clinical development in oncology (Genzyme, Press Release). The third approach is focusing on the development of small molecule inhibitors which are designed to block the signaling downstream of the TGF-β receptor. Companies such as Glaxo-Smith-Kline (GSK), Scios, Pfizer, Biogen and Lilly have active programs to develop small molecule kinase inhibitors targeting the phosphorylation of the TGF-β receptor type I (TGF-βRI) kinase. In the following, we will focus on this particular approach as it represents the most proximal inhibition to the SMAD proteins.

1.2 Small Molecule TGF-β Inhibitors as Potential Cancer Treatment

Although many individual molecular scaffolds have been developed as small-molecule TGF-βRI kinase inhibitors, most of these have recognizable features that are congruent with a single known pharmacophore. For example, the representative small-molecule ATP-competitive compounds LY550410, LY580276 and SB-505124 each contain a different set of heteroaryl rings appended from a central heterocyclic scaffold, yet also have the key functionality necessary for potent binding to the kinase-domain active site.

Generally, the most essential functionality consists of a crucial 'warhead' group that contains an adequate hydrogen-bond acceptor. Core ring systems such as imidazopyridine and pyrazolopyridine, developed by Biogen Idec and Lilly Research Laboratories, respectively, feature pyrimidinyl and quinolinyl substituents as warheads. Representatives of the dihydropyrrolopyrazole series have shown excellent activity in both enzyme inhibition and cell-based assays. LY580276 inhibits TGF-βRI with an IC50 of 175 nM, and potent activity in a TGF-β-dependent luciferase (p3TP Lux) reporter assay (IC50 = 96 nM). This compound also proved to be highly selective in a panel of 40 kinases, including p38 MAP kinase (IC50 > 10μM). GSK reported on a series of 1,5-naphthyridine aminothiazole and pyrazole derivatives demonstrating an IC50 of 4 nM in a TGF-βRI autophosphorylation assay, an IC50 of 18 nM in a TGF-β-dependent cellular assay and was also found to be highly selective against p38 MAP kinase (IC50 > 16 μM). Compound SB-505124 inhibits the purified kinase domain of TGF-βRI (IC50 value = 47 nM). Further evidence for activity of these small molecules TGF-βRI kinase inhibitor compounds is their ability to modulate TGF-β-induced EMT, which leads to tumor-cell invasion and metastasis. Two compounds, LY580276 and SD-093 (Scios/Johnson & Johnson) both can inhibit EMT in NMuMG (normal murine mammary epithelial) cells at concentration of 2 μM or 1 μM, respectively. Furthermore, SD-093 inhibits the basal migratory and invasive phenotype that is generated by autocrine TGF-β signaling in Smad4-deficient BxPC-3 pancreatic cancer cells, without affecting the morphological characteristics of these cells. Another inhibitor, SD-208, was recently shown to inhibit the growth of intracranial SMA-560 gliomas in syngeneic mice, and also showed *in vivo* kinase inhibition in the spleen and brain. Efficacy analysis in the SMA-560 model showed a survival advantage in the SD-208-treated animals. Interestingly, antitumor efficacy in this model did not correlate with changes in angiogenesis, proliferation or apoptosis, but instead correlated with immune-cell infiltration into responding tumors. This result highlights the therapeutic potential of combating TGF-β-mediated immunosuppression to generate a significant immunological response.

1.3 Translating Small Molecule TGF-β Inhibitors into Clinical Investigation

While progress in medicinal chemistry has led to a wide range of various chemical scaffolds for potential TGF-βRI kinase inhibitors, deciding which of these molecules to advance into clinical investigation remains challenging. To overcome this challenge pharmacokinetic modeling is increasingly being used to select promising compounds for clinical development. For example, predictive pharmacokinetics was successfully applied to develop cytotoxics for clinical investigation. A similar approach seems to have been used for the development of SU11248, a novel kinase inhibitor inhibiting neo-angiogenesis. Likewise, novel TGF-β inhibitors should be developed using pharmacokinetic prediction models as well as biomarkers that allow rapid determination whether the TGF-βRI kinase inhibitors achieve biologically effective doses at the predicted pharmacokinetic exposure. For the TGF-βRI kinase inhibitors the biomarker of choice should measure the phosphorylation of Smad in either tumor or surrogate tissue. This biomarker-based strategy in the clinical development will provide the necessary information on pharmacokinetic (PK) and pharmacodynamic (PD) profile of a TGF-βRI kinase inhibitor, and thus allow a better determination of future Phase II doses. The integration of non-clinical PK and PD to predict clinical anti-tumor effect in patients will be even more important if TGF-βRI kinase inhibitors are being investigated in patients where the TGF-β signaling pathway is critical for tumor growth. Recognizing the appropriate patient population has not only relevance to determine efficacy of a novel TGF-βRI kinase inhibitor, but for understanding its safety profile.

2. TGF-β/SMAD SIGNALING STUDIES IN CANCER PATIENTS

2.1 TGF-β/Smad-Inhibitors in Cancer Clinical Trials Require Selected Patient Populations

In recent years, selective TGF-β/Smad inhibitors have been identified via target-based drug discovery. These target-based TGF-β/Smad inhibitors require a different clinical development strategy than the cytotoxic anti-cancer agents. TGF-β/Smad inhibitors should be investigated in patients with enriched expression of the respective target or where the target plays a critical role in tumor growth. As exemplified by the failure of gefitinib in Phase III trials to show survival benefit in unselected patients with NSCLC, enriching for the appropriate patient population is likely to increase the successful clinical development of target-based inhibitors. Thus, identifying the appropriate patient population will be critical for clinical trials with TGF-β/Smad inhibitors. Based on the hypothesis that TGF-β/Smad expression is associated with clinical outcome, we reviewed clinical studies investigating the impact of TGF-β/Smad expression in cancer patients. We focused our analysis on four tumor types: colorectal, breast, prostate and lung cancer. We also used these studies to determine important selection criteria for cancer patients that might benefit from TGF-β/Smad inhibitors. As result of this review it is clear that the TGF-β/Smad signaling is associated with favorable clinical outcome in early tumors, while in advanced/metastatic cancers the same pathway is associated with tumor growth (Table 1).

2.2 Target Expression of TGF-β/Smad in Colorectal Cancer

In colorectal cancer, TGF-β signaling modulates tumor growth and studies on the expression of each component of the TGF-β signaling pathway have been conducted in malignant tissue and the surrounding microenvironment. In addition, serum or plasma levels of TGF-β isoforms have also been measured in patients with colorectal cancer. The following review is centered on those studies that have correlated the expression of TGF-β/Smad signaling with clinical outcome. These correlations may indicate in which patients the TGF-β-mediated signaling is active, and patients with such a profile may benefit from treatment with a TGF-β/Smad inhibitor.

2.2.1 Tissue expression of TGF-β/Smad signaling pathway in colorectal cancer

One of the earliest studies examined the TGF-β/Smad signaling pathway in patients by using immunohistochemistry (IHC) to detect protein expression of TGF-β1. High TGF-β1 protein expression was found in 59% (20/34) of TGF-β positive tumor specimens, and high expression correlated with an unfavorable outcome. TGF-β2 and TGF-β3 protein expression was not associated with poor outcome in this study suggesting a singular role of TGF-β1 in driving tumor growth. This observation was confirmed by another study comparing the expression of TGF-β1, TGF-β2 and TGF-β3 in tumor tissue from 39 colorectal cancer patients. The three isoforms were measured by IHC and reverse transcription polymerase chain reaction (RT-PCR) in tumor tissue and compared to serum TGF-β1, TGF-β2 and TGF-β3 levels. This comparative study found that both TGF-β1 and TGF-β2 expression in tumor tissue was correlated with higher tumor stage, but only increased TGF-β1 expression was associated with poor outcome. Increased TGF-β2 expression was associated with advanced clinical stages, suggesting a not yet identified

Table 1. TGF-β/SMAD Expression and Its Association with Clinical Outcome Cancer

Tissue Expression or Plasma/Serum Levels		Clinical Outcome		References
Colorectal Cancer		**Favorable**	**Unfavorable**	
Increased	**TGF-β1**	N/A	Advanced, metastatic ca.	(Xiong et al. 2002)
	TGF-β1*	N/A	Advanced, metastatic ca.	(Xiong et al. 2002, Tsushima et al. 2001, Shim et al. 1999)
Decr./Loss	**TGF-βRII**	N/A	Advanced, metastatic ca.	(Xiong et al. 2002)
	SMAD 2	N/A	Advanced, metastatic ca.	(Xiong et al. 2002)
	SMAD 4	N/A	Localized and advanced ca.	(Boulay 2002)
	SMAD 7	Localized and adv. ca.	N/A	(Boulay 2002)
Breast Cancer		**Favorable**	**Unfavorable**	
Increased	**TGF-β1**	Primary ca.	N/A	(Murray et al. 1993)
	TGF-β2	Invasive ca.	N/A	(Auvinen et al. 1995)
	TGF-β3	N/A	Invasive, metastatic ca.	(Ghellal et al. 2000)
Decr./Loss	**TGF-βRII**	N/A	Invasive, metastatic ca.	(Barlow et al. 2003)
	SMAD 2	N/A	Invasive, metastatic ca.	(Xie et al. 2002)
	SMAD 4	N/A	Invasive, metastatic ca.	(Xie et al. 2002)
Prostate Cancer		**Favorable**	**Unfavorable**	
Increased	**TGF-β1**	N/A	Advanced, metastatic ca.	
	TGF-β1*	N/A	Predictor of metastatic disease	(Shariat et al. 2004, Kattan et al. 2003)
	TGF-β2*	Localized ca.	N/A	(Perry et al. 1997)
Decr./Loss	**TGF-βRII**	N/A	Advanced, metastatic ca.	
	SMAD 4	N/A	Advanced, metastatic ca.	(Horvath et al. 2004)
Non Small Cell Lung Ca.		**Favorable**	**Unfavorable**	
Increased	**TGF-β1**	N/A	Early stage and adv. ca.	(Takanami et al. 1997, Hasegawa et al. 2001)
	TGF-β1*	N/A	Advanced ca.	(Barthelemy-Brichant et al. 2002, Kong et al. 1999)
Increased	**TGF-βRI**	N/A	Early and advanced ca.	(Takanami et al. 1997)
	TGF-βRII	N/A	Early and advanced ca.	(Takanami et al. 1997)

N/A, Not available; *, Serum or plasma levels; ca., cancer; decr., decreased role during tumor progression. Sera from all 39 patients also showed significant elevation of TGF-β1 and TGF-β2 indicating that both isoforms were associated with colorectal cancer growth. In addition to these two smaller cohort studies, a large patient cohort study from the United Kingdom (UK) compared TGF-β1 with TGF-β3 expression in tissue of 111 patients with colorectal cancer. This study found that higher TGF-β1 expression was associated with tumor cell growth, while expression of TGF-β3 was not. IHC staining revealed TGF-β1 in tumor cells, while TGF-β3 staining was only detected in the mircoenvironment. In contrast to the previous studies, no correlation was found between TGF-β1 expression and clinical outcome. Because the latter study was larger than the first two studies, one might conclude that there is no correlation between tumor TGF-β1 expression and clinical outcome. However, a number of factors influencing the correlation were not assessed in this study, including metastatic disease or aggressive phenotype. If the authors had correlated such subtypes they might have found a correlation between TGF-β1 expression and clinical outcome.

While the previous tissue expression studies were mainly performed on samples from Caucasians, some studies were conducted in Asian patients with colorectal cancer. For example, a large cohort study in 98 Chinese patients with colorectal cancer found that tissue TGF-β1 was expressed in 38% (37/98) and TGF-βRII in 47% (46/98) of patients. In addition to the isoform this large study also determined the TGF-βRII expression in tumor tissue and found that it was significantly lower in patients with advanced disease, including T3/T4 disease, Duke's stage C/D. The authors determined that the TGF-βRII expression was inversely related to the expression of TGF-β1 in tumor tissue. A smaller comparative study was also performed in Japanese patients measuring mRNA of TGF-β1, TGF-βRI and TGF-βRII in 22 patients with colorectal cancer and in 11 patients with colorectal adenomas. The TGF-β1 mRNA was strongly expressed in tumor cells and moderately in adjacent tissue. Both TGF-βRs were absent in all examined colorectal cancer specimens, but were present in normal colonic mucosa and adenomas. In contrast to malignant cells, fibroblasts and endothelial cells had strong expression of TGF-βRI, TGF-βRII and TGF-β1, which suggested that only the microenvironment had a functional TGF-β signaling pathway. Taken together, these studies imply that in tumor tissue the down-regulation of TGF-βRs may reflect the loss of the tumor tissue to control the tumor growth. This would indicate that the tumor suppressive role of TGF-β is lost in advanced/metastatic tumors. At the same time this loss occurs, the adjacent tumor microenvironment will still have a functional TGF-β/Smad signaling pathway leading to tumor neo-angiogenesis, immune suppression and other tumorigenic mechanisms.

In addition to TGF-β and its receptors, assessment of Smad proteins can also be used for determining the effect of TGF-β signaling on tumor growth. Deletion or structural alteration of chromosome 18q21, which is associated with a poor prognosis in colon cancer contains most of the Smad genes. Because of this observation, the loss of Smads in colorectal cancer is hypothesized as a cause of aggressiveness in specific subsets of patients. In the following we will examine the correlation between deficiency in specific Smad proteins and clinical outcome.

Genomic DNA for Smad2 and Smad7 have been examined in tumor biopsies from 264 colorectal patients and correlated with clinical outcome. All patients examined had previously been enrolled in a randomized trial conducted by the Swiss Association for Clinical Cancer Research (SAKK) investigating the long-term benefit of a single course adjuvant intraportal chemotherapy for the treatment of colorectal cancer (SAKK study 40/81). Deletion of the Smad2 gene was present in 64% (126/197) and Smad7 gene was deleted in 43% (77/178) of the examined cases. In contrast to Smad2, Smad7 gene deletion or amplification was associated with a favorable or unfavorable clinical outcome, respectively. Whether the prognostic impact of Smad7 reflects its function of inhibiting TGF-β signaling in an un-opposed manner could not be determined in this study. Because Smad2 amplification did not confer a beneficial prognosis in this group of patients, the authors concluded that the tumor suppressor activity of the TGF-β signaling pathway was inactive. In a subsequent study Smad4 expression was assessed using samples from patients enrolled in SAKK 40/81. The authors examined Smad4 expression to test the hypothesis that it was a more reliable read-out for determining TGF-β signaling than Smad2 expression. In the tumor tissue of 202 examined patients, normal *Smad4* diploidy was associated with a 3-fold higher benefit for 5-fluorouracil (5-FU)-based adjuvant chemotherapy. Thus, *Smad4* or the presence of the 18q21 locus appears to be associated with improved survival and indicates that a functioning TGF-β pathway may aid 5-FU-based therapy. This observation is supported by another study which reported that the loss of Smad4 expression is associated with a higher risk of liver metastasis. Taken together, these findings support the hypothesis that TGF-β signaling is important for suppressing tumor metastasis in early stage colorectal cancer.

2.2.2 Serum/plasma measurements of TGF-β isoforms in colorectal cancer

Serum measurements of TGF-β also have been used to determine the relevance of the TGF-β signaling pathway in colorectal cancer patients. The underlying rationale for measuring circulating TGF-β levels is based on the hypothesis that tumor tissue releases TGF-β to the circulation. Hence, determining TGF-β levels could provide a biomarker of TGF-β-dependent tumor growth in patients with cancer. To confirm this hypothesis a study was conducted in a cohort of Korean patients and total serum TGF-β1 levels were determined prior and after surgery. In this cohort of 121 Korean colorectal patients with varying stages of colorectal cancer, serum was obtained preoperatively in all patients and collected postoperatively in a subgroup of 50 patients. Compared to a cohort of healthy volunteers elevated mean serum levels of TGF-β1 were present in 57% (69/121) prior to surgery and those levels were significantly reduced after tumor resection. Serum TGF-β1 levels were also correlated with depth of tumor invasion (patients with Duke's stage D had the highest TGF-β1 levels), lymph node metastasis, distant metastasis and elevated carcinoembryonic antigen (CEA). A similar study in Chinese patients also found that 60% (30/50) had increased serum TGF-β1 levels, which also correlated with stage, depth of tumor invasion (T3/T4 disease), and metastasis. In contrast to studies assessing serum TGF-β1 levels, others have used plasma to determine TGF-β1 levels in colorectal cancer patients. Using plasma rather than serum minimizes the presence of platelet-derived TGF-β1, and thus may be more accurate in describing tumor-associated TGF-β1 levels in circulation. Plasma TGF-β1 levels were found to correlate with disease progression and patients with Duke's stage B and C had the highest levels. Using a cut-off of 7.5 ng/mL, plasma TGF-β1 levels were elevated in 66% (77/117) of patients and levels were predictive of tumor progression. These findings were similar to those based on serum TGF-β assessments.

But confounding factors for measuring TGF-β levels are not limited to the difference of serum or plasma collection of the TGF-β isoforms. There are at least two other confounding factors that need to be considered: first, TGF-β exists in an active and in a latent form; second, TGF-β can be bound to various plasma/serum proteins. Both factors may have an impact in measuring change after patients have been treated with a TGF-β/Smad inhibitor. Because of its technical complexity, the active TGF-β isoform assessment has not been extensively investigated. However, one study did compare the plasma levels of total TGF-β1 with those of active TGF-β1 in 45 Japanese patients with colorectal cancer. While the activation rate was not different between colorectal cancer patients and healthy volunteers, plasma levels of active TGF-β1 were significantly elevated in patients compared to healthy volunteers. Elevated active plasma TGF-β1 levels were correlated with clinicopathologic progression. In contrast to total plasma TGF-β1 levels, active plasma TGF-β1 levels were elevated in cancer patients compared to healthy volunteers even in early stage tumors, such as Duke's stage A or B. These studies on the difference of active and total TGF-β1 levels in patients demonstrates that total TGF-β1 levels may not describe the overall activity of circulating TGF-β1 levels. Total TGF-β1 level measurements can therefore either represent a false-positive or false-negative result and must be interpreted with caution as response biomarkers in patients treated with a TGF-β/Smad inhibitor.

In summary, reviewing studies with approximately 100 or more patients imply that TGF-β/Smad signaling is associated with poor clinical outcome in patients with metastatic disease (Table 1). Patients with advanced, metastatic colorectal cancer and poor prognosis tend to have increased tissue expression of TGF-β1 and elevated levels of serum or plasma TGF-β1. The loss of TGF-βRII, Smad2 or Smad4 is also associated with unfavorable outcome in patients with advanced, metastatic condition. By contrast, Smad7 loss is associated with favorable outcome in patients with localized, advanced colorectal cancer consistent with tumor suppressor activity of TGF-β/Smad signaling. Measurements of total TGF-β1 in serum or plasma of patients with colorectal cancer should be used with caution as response biomarkers, because there are a number of confounding factors that can lead to false-positive or false-negative results and in consequence not accurately detect changes following treatment with a TGF-β/Smad inhibitor.

2.3 Target Expression of TGF-β/Smad in Breast Cancer

TGF-β signaling has been recognized as a mediator of tumor progression in breast cancer. In the following we will review the correlation between TGF-β expression and clinical outcome in various stages of breast cancer.

2.3.1 Tissue expression of TGF-β/Smad signaling pathway in breast cancer

Tissue expression studies on the TGF-β/Smad pathway in breast cancer patients were first based on analyzing non-invasive versus invasive breast cancer specimen as defined by standard histopathologic diagnoses. In a British breast cancer study TGF-β1 protein expression was examined in 27 ductal carcinomas in situ and 54 invasive carcinomas by IHC. High intensity staining was present in 11% of in situ tumors, and 20% of invasive carcinomas. A significant correlation between prominent reactivity and positive node status was reported. This finding, along with the differences in reactivity between in situ and invasive carcinomas suggests that TGF-β1 is associated with tumor invasion and metastasis. A similar US study also found that TGF-β1, TGF-β2 and TGF-β3 were expressed in breast cancer, but only TGF-β1 was associated with tumor progression. No correlation was observed between hormone receptor status and TGF-β protein expression. In a third study, the expression of TGF-β1 was compared between primary and metastatic breast cancer. In 68% (19/28) of primary tumors and in 92% (11/12) of metastases demonstrated intracellular TGF-β1 staining was detected. The authors found a preferential expression of secreted TGF-β1 at the advancing tumor edges and in lymph node metastases. This staining pattern is consistent with the role of TGF-β1 in advancing metastatic spread in breast cancer. While these three studies with relatively small cohorts of patients were suggesting a correlation of TGF-β1 with clinical stage, only a larger study conducted in a cohort of 153 patients with invasive breast cancer in Britain established a correlation between TGF-β expression and clinical outcome. In this study, TGF-β1 and TGF-β3 protein expression was determined by using IHC. TGF-β1 protein expression was expressed strongly in 16% (25/153) and moderately in 64% (98/153), while TGF-β3 was expressed strongly in 14% (21/153) and moderately in 68% (104/153) of the cases. Co-expression of both isoforms was present in 72% (111/153). Importantly, the expression of TGF-β3 and not TGF-β1 was inversely correlated with overall survival. Patients with nodal metastasis and TGF-β3 expression had the worst prognosis. Neither isoform was correlated with expression of estrogen (ER) or progesterone (PR) hormone-receptor status.

In contrast to the previous studies, others have found that TGF-β1 expression is related to favorable and not to unfavorable clinical outcome. To date, it is not understood why in some cases TGF-β expression is associated with poor clinical outcome, while in other cases it is associated with favorable outcome. The first study describing a favorable outcome associated with high expression of TGF-β1 was conducted in 167 patients with primary breast cancer. In this large cohort patient study mRNA expression of TGF-β1 was inversely correlated with node status, but not with hormone receptor status, tumor size or menopausal status. Patients with high expression levels of TGF-β1 had a longer disease-free interval with a significantly longer probability of survival at 80 months, although the overall relapse-free survival was not increased. A second study examined TGF-β1 and TGF-β2 protein expression by IHC in a series of 273 breast cancer biopsies. TGF-β1 expression was found in 59% (160/273) and TGF-β2 in 40% (110/273) of the tumor specimens. Concomitant expression of TGF-β1 and TGF-β2 was detected in 33% (89/273), while TGF-β1 alone was expressed in 26% (71/273) and TGF-β2 alone was expressed in 7% (19/273) of cases. The expression of both TGF-β1 and TGF-β2 was associated with a favorable prognostic factor. The expression of TGF-β2 without the expression of TGF-β1 was significantly related to favorable disease outcome.

There are several factors which might explain the discrepancies between high TGF-β expression and clinical outcome. The most important difference between these studies is the tumor stages examined in each of the studies. Clear information was missing so that the definition of advanced or local disease was not transparent. In future studies it will be important to use standard pathological staging and grading information as defined by the TNM classification system. Also, a comprehensive analysis on previous therapies will be important to understand whether previous anti-tumor treatments can confound the expression of TGF-β in cancer patients.

Previous hormone therapy has been associated with altering the natural expression of TGF-β in tumor tissue. For example, treatment of three months with tamoxifen causes a consistent induction of extracellular protein expression of TGF-β1 as detected by IHC. The induced TGF-β1 is localized between and around stromal fibroblasts and appears to be derived from these cells. The increased stromal staining of TGF-β1 occurred in ER-negative as well as ER-positive tumors. TGF-β1, TGF-β2, and TGF-β3 expression levels were lower in epithelial cells and were not altered after tamoxifen treatment. A similar comparative study was performed in breast cancer patients with ER-positive status alone. In tissue from 37 ER-positive breast cancer patients before and during treatment with tamoxifen the majority of tumors showed no change in TGF-β expression. However, TGF-β2 expression did increase and was associated with tumor response in tumors responding to treatment. This study suggests that response to tamoxifen therapy may be associated with an increase in expression of specific TGF-β isoforms in a subset of tumors. Another study assessing the impact of tamoxifen-induced TGF-β expression in breast cancer cells was based on tissue from 11 patients who had received tamoxifen therapy for 3 to 6 months prior to surgery. These patients with increased tumor size and lack of response to tamoxifen expressed unexpectedly high levels of TGF-β1 mRNA. Although the sample size for all three studies is small, these observations suggest that tamoxifen treatment is associated with higher tumor TGF-β1 expression. The relevance of this increased expression remains to be determined and future studies should focus on establishing whether tamoxifen-resistance is influenced by non-functional TGF-β signaling pathway.

In addition to the expression of the cytokine, presence of TGF-βRs is another read-out of the TGF-β/Smad signaling pathway. Some studies have correlated TGF-βR expression with clinical outcome. For instance, the loss of TGF-βRII protein expression was correlated with various grades of breast cancer by IHC. Neoplastic cells showed reduced expression of TGF-βRII in comparison to the normal breast tissue and benign lesions. There was a significant inverse correlation between loss of TGF-βRII expression and tumor grade within both ductal carcinoma in situ (DCIS) and invasive mammary carcinomas (IMC). Another inverse correlation was observed between TGF-βRII expression and mitotic count or clinical stage. Furthermore, TGF-βRII expression appears to have an important role in the transformation of breast epithelial hyperplastic lesions lacking atypia (EHLA), which is associated with a small increase in subsequent invasive breast cancer. When compared to normal breast tissue, expression of TGF-βRII varied between 25% to 75% in specimen obtained from 54 patients with EHLA. Patients with lower TGF-βRII expression had a greater rate of developing invasive breast cancer than those with higher TGF-βRII expression. This differential expression of TGF-βRII was further investigated in 72 patients with breast cancer by comparing TGF-βRII expression in the epithelium and stroma. The percentage of cells expressing TGF-βRII in stroma was higher in patients that had a positive lymph node status and/or negative estrogen and progesterone receptor expression. This observation suggested that high TGF-βRII expression in the stroma was especially associated with poor outcome. As observed for colorectal cancer, loss of TGF-β signaling pathway in tumor cells and/or increased TGF-β signaling in the microenvironment is linked to tumor progression. The tumor promoting effect of TGF-β signaling in the stroma, seems to be especially present in patients with metastatic or aggressive forms of breast cancer.

Examining expression of proteins downstream of the TGF-βRs is another approach to determine the relevance of the TGF-β/Smad signaling pathway in cancer patients. For example, a tissue microarray study investigated the expression of the Smad proteins in breast cancer patients in the United States. Among the 456 examined cases, 92% expressed Smad2, pSmad2 and Smad4, suggesting that the TGF-β signaling pathway was active in the majority of the patients. However, no pSmad2 was detected in 6% (30/456) of the cases, and no Smad4 was observed in 2% (9/456). Loss of Smad4 was inversely correlated with presence of axillary lymph node metastasis and loss of pSmad2 expression was associated with shorter overall survival in stage II breast cancer patients. BRCA1 and BRCA2 genotype, HER2/neu or hormone-receptor status were not correlated with Smad expression. This large study implied that the loss of the active TGF-β/Smad signaling was associated with unfavorable clinical outcome in patients with stage II breast cancer. Smad3 expression also was investigated in breast cancer tissue using IHC. Samples of 59 breast cancer patients were assessed for Smad3 expression. Increased Smad3 expression was correlated with pathologic grade in the peritumoral tissue, but was decreased in the tumor cells. Concomitant expression of Smad2 and Smad4 mirrored the expression pattern of Smad3. Interestingly, ER-positive/PR-positive tumors showed a preponderance of intense nuclear Smad3 staining compared to hormone receptor-negative tumor tissue. Whether the nuclear Smad3 staining reflects the contribution of the TGF-β/Smad signaling in hormonally regulated tumor growth remains to be determined. Lastly, IHC studies have looked at the levels of inhibitory Smads, such as Smad7 in tumor tissue. In a study conducted in the United States, the expression of Smad7 mRNA was compared with the Smad2 mRNA expression in 67 breast cancer patients. Compared to normal breast tissue, Smad7 expression levels were increased in primary tumor tissue and Smad2 levels were reduced. Interestingly, Smad2 mRNA levels were higher in 0 to +2 HER2/neu compared to +3 HER2/neu tumors. All these studies on Smad expression seem to indicate that Smad expression is required to suppress tumor growth in early tumor stages. Its loss in these tumor stages is associated with poor clinical outcome.

2.3.2 *Serum/plasma measurements of TGF-β isoforms in breast cancer*

Circulating TGF-β levels have been investigated as possible markers of disease progression in breast cancer. For example, in 60 Chinese patients with invasive breast cancer serum levels of TGF-β1 were measured prior to breast surgery. Patients with early stage disease (stage I and II) had significantly lower serum levels of TGF-β1 than patients with late stage disease (stage III and IV). Based on a multivariate analysis, advanced TNM staging was associated with higher TGF-β1 levels, which suggests that serum TGF-β1 is linked with transformation and progression in breast cancer. Other TGF-β isoforms have also been studied in breast cancer patients, but were based on plasma rather than serum assessment. In a UK study, plasma samples from 80 early-stage breast cancer patients were examined for TGF-β1, TGF-β3 and CD105/TGF-β3 levels. While plasma TGF-β1 levels did not correlate with positive lymph node status, both plasma TGF-β3 and CD105/TGF-β3 levels were significantly elevated in patients with axillary lymph node metastasis. No correlation with survival was reported at the 5-year follow-up. This study shows that TGF-β1 and TGF-β3 may exert distinct functions and additional studies are needed to further understand the contribution of each TGF-β isoform for tumor growth.

Changes in circulating TGF-β levels have been observed after surgery supporting the hypothesis that circulating TGF-β is tumor derived. Plasma TGF-β1 levels were determined in 26 breast cancer patients prior to and after 2 weeks following surgery and compared to levels in healthy volunteers. In 81% of these patients, TGF-β1 levels were elevated prior to surgery. After surgery the TGF-β1 levels were significantly reduced with the exception of patients with positive lymph nodes or unresectable tumor. This lack of reduction after surgery differs from a previous study where all patients showed a marked reduction in TGF-β1 levels. Perhaps two factors may account for this difference: first, the patients in this study had a more invasive cancer than in the previous study suggesting that surgery in this patient population did not reduce tumor burden sufficiently to cause a marked reduction in TGF-β1 levels; second, measurements of this study were performed at later time points than in the previous study indicating that TGF-β1 levels may need to be assessed no later than 1 week after surgery to obtain an accurate assessment of tumor-derived TGF-β1 levels.

Other interventions associated with reduction of plasma TGF-β1 levels are radiotherapy and chemotherapy. TGF-β1 levels were investigated in breast cancer patients undergoing autologous bone marrow transplantation after induction-radiotherapy followed by high-dose chemotherapy with carmustine, cyclophosphamide, and cisplatin. The risk of developing subsequent hepatic veno-occlusive disease or pulmonary fibrosis was increased in patients with elevated plasma TGF-β1 levels after radiotherapy and prior to chemotherapy. Patients who did not have elevated plasma TGF-β1 levels were less likely to suffer from post-treatment related fibrosis after completion of high-dose chemotherapy. The effect of radiotherapy-induced elevation of TGF-β1 levels has been observed before and is associated with significant morbidity and may explain the increased risk of veno-occlusive disease or pulmonary fibrosis. In pre-menopausal stage II breast cancer patients with axillary lymph node metastases a reduction of plasma TGF-β1 levels was observed after administration of adjuvant chemotherapy containing cyclophosphamide, methotrexate and 5-FU (CMF) regimen. Whereas the elevation of TGF-β1 levels after radiotherapy has been recognized as a pathomechanism in fibrosis, the importance of reduced TGF-β1 levels following chemotherapy is not known.

In summary, TGF-β/Smad signaling is associated with poor clinical outcome in invasive, metastatic breast cancer patients based on large studies with approximately 100 or more patients (Table 1). Patients with invasive, metastatic disease also have an unfavorable prognosis when TGF-β3 is overexpressed in tissue. Furthermore, loss of TGF-βRII or Smad2 or Smad4 results in poor clinical outcome in patients with invasive, metastatic breast cancer. By contrast, in primary breast cancer the increased TGF-β/Smad signaling is associated with a more favorable clinical outcome consistent with its role as a tumor suppressor. Currently, large studies correlating serum/plasma TGF-β levels to outcome are not published in breast cancer.

2.4 Expression of TGF-β/Smad in Prostate Cancer

TGF-β signaling pathway has been recognized as a possible biochemical marker for tumor progression in prostate cancer. Prostate cancer is the only cancer type that displays this strong correlation between plasma TGF-β levels and prediction of recurrence.

2.4.1 Tissue expression of TGF-β/Smad signaling pathway in prostate cancer

One of the earlier IHC studies investigating the presence of TGF-β1 in prostate cancer found that extracellular TGF-β1 was present in tumor and in benign hyperplastic tissue of the prostate. Subsequent studies detected TGF-β1 expression only in primary prostate cancers with a pronounced intracellular accumulation within epithelial cells. The expression of TGF-β1 in prostate cancer increases with advanced stages of prostate cancer. Examining specimens of 59 prostate cancer patients, TGF-β1 immunoreactivity was positively correlated with serum PSA levels and tumor stage. Compared to patients with normal TGF-β1 expression, patients with increased TGF-β1 immunoreactivity were found to have reduced survival.

Differential expression of TGF-β isoforms have also been investigated. Based on IHC, expression of TGF-β1 and TGF-β3 in prostate cancer tissue was observed, with TGF-β1 being mainly present in stromal cells. TGF-β3 was not present in either tumor or stromal cells of the tumor. This finding suggests that TGF-β1 and TGF-β3 are independently regulated in prostate cancer. A similar comparative IHC study was performed on tissue from 25 patients with clinically localized prostate cancer. Compared to normal prostate tissue, malignant prostate epithelial cells had higher TGF-β1 and TGF-β2 immunostaining than the surrounding stromal cells. Although TGF-β3 staining intensity was similar for both malignant and normal prostate epithelial cells, the pattern of staining switched from uniform apical to diffuse protein staining in malignant prostate glands.

TGF-βR expression was investigated in prostate cancer and compared to the expression of the ligands. TGF-βRI, TGF-βRII and TGF-β1 immunoreactivity were examined in 73 cases of prostate cancer and correlated with clinical outcome. Loss of TGF-βRII expression in combination with TGF-β1 overexpression (12/73; 16%) was correlated with a significantly reduced survival compared to patients with normal immunoreactivity (29/73; 40%). Patients with high TGF-β1 expression in tissue had shorter median survival than patients with normal TGF-β1 immunoreactivity. Furthermore, increased TGF-β1 staining was associated with tumor grade, high vascular counts and metastasis. A similar comparative study was performed in 60 Italian prostate cancer patients. TGF-β1, TGF-β2, TGF-β3, TGF-βRI and TGF-βRII were expressed more strongly in prostate carcinoma than in intraepithelial neoplasia (PIN) or benign prostate hyperplasia (BPH). In comparison to the ligands, both types of the TGF-β receptors were expressed at a lower level. This finding supports the hypothesis, that in advanced disease TGF-βR are underexpressed in relationship to the expression of TGF-β ligands. Another study in patients with localized prostate cancer showed that in addition to the loss of TGF-βRII, advanced forms of prostate cancer seem to lose expression of Smad4. Both the loss of Smad4 and TGF-βRII were associated with a decrease of markers of apoptosis, such as p53 and the cell cycle regulator protein $p27^{Kip1}$. A correlation with increased PSA level and the loss of TGF-βRII was also observed. This impact of Smad4 loss on disease progression was confirmed by an Australian group finding that increased Gleason scores were associated with decreased nuclear Smad4 expression in 74 patients with localized prostate cancer.

2.4.2 *Serum/plasma measurements of TGF-β isoforms in prostate cancer*

In addition to tissue expression, circulating TGF-β isoforms were assessed in prostate cancer patients. Serum TGF-β1 levels showed no significant elevation when compared to either healthy volunteers or patients with BPH. This suggests that serum levels of TGF-β1 may not be a sensitive measure of detecting differences in circulating TGF-β1. The following studies used plasma TGF-β1 levels instead of serum TGF-β1 levels. Elevated plasma TGF-β1 levels were found to be an important marker of progression in prostate cancer and elevated plasma TGF-β1 levels in some patients were correlated with rising PSA. In contrast to primary prostate cancer, patients with relapsed prostate cancer had elevated TGF-β1 levels in the absence of detectable PSA levels. In a study of 55 Japanese prostate cancer patients, plasma TGF-β1 levels were elevated only in patients with metastatic disease. In contrast to previous studies, this study did not find a correlation between plasma TGF-β1 level and PSA or tumor grades, suggesting that this correlation cannot be consistently found in all prostate cancer patients. In another study plasma TGF-β1 was compared to plasma TGF-β2 levels. In this study the overall prostate cancer patient population did not show significantly elevated levels of plasma TGF-β2 compared to those with BPH or healthy volunteers. However, a subgroup of patients with early stage prostate cancer, such as stage T2a and Gleason score 3 or less, were found to have elevated TGF-β2 levels. Other studies also compared plasma and urine TGF-β1 levels detecting high urinary TGF-β1 levels in prostate cancer patients. While this study shows a differential secretion of TGF-β1 and TGF-β2 in plasma and urine, the lack of association with prostate cancer progression in the overall patient population may be due to the fact that the study population did not include patients with aggressive prostate cancer.

The correlation of TGF-β1 with other circulating cytokines, such as inflammation markers, was also investigated to identify patients at high risk after prostatectomy. For example, in a study of 69 patients, serum IL-6 and plasma TGF-β1 levels positively correlated with progression of metastatic disease and PSA levels. The authors also observed that platelet activation was reduced when vacuum was not applied during blood draws, resulting in reproducible plasma TGF-β1 levels.

Based on these observations, three large studies have confirmed that plasma TGF-β1 levels can be used as predictor of biochemical progression after surgery. The first study examined plasma TGF-β1 levels in 120 men prior to undergoing radical prostatectomy. The second study examined plasma TGF-β1 levels in 302 prostate cancer patients with clinically localized disease. The third and largest study examined plasma TGF-β1 levels in 814 patients with clinically localized prostate cancer. In all three studies, the blood collection was standardized using Cell Preparation Tubes (CPT) to reduce platelet activation during blood draws. These three studies established a significant correlation between elevated levels of plasma TGF-β1 with metastatic disease, IL-6, sIL-6R and PSA. The conclusion from these studies was that plasma TGF-β1 levels as part of a preoperative nomogram can predict biochemical progression in patients with clinically localized prostate cancer.

In summary, based on studies with approximately 100 or more patients TGF-β/Smad signaling is associated with poor clinical outcome in patients with advanced, metastatic prostate cancer (Table 1). Patients with increased tissue expression of TGF-β1 and elevated levels of serum or plasma TGF-β1 are associated with advanced or aggressive tumor growth. Loss of TGF-βRII or Smad4 also results in unfavorable outcome in patients with advanced, metastatic condition. In localized tumors, however, increased serum or plasma TGF-β2 levels are associated with favorable outcome consistent with the role of TGF-β/Smad signaling as a tumor suppressor.

2.5 Target Expression of TGF-β/Smad in Lung Cancer

The role of TGF-β signaling in lung cancer is not as well characterized as in colorectal, breast and prostate cancer. However, there is a growing interest in characterizing the role of TGF-β/Smad signaling in lung cancer.

2.5.1 Tissue expression of TGF-β/Smad signaling pathway in non-small cell lung cancer (NSCLC)

The expression of TGF-β1, TGF-β2, and TGF-β3 was examined in tissue from 120 patients with pulmonary adenocarcinoma by IHC. TGF-β was present in most samples and the expression of TGF-β1 correlated inversely with survival. This study was extended to compare the concomitant expression of TGF-β1, TGF-βRI and TGF-βRII in the same set of patients. Patients with TGF-β1, TGF-βRI, and TGF-βRII expression had a worse outcome than patients with negative immunoreactivity. Furthermore, TGF-β1 expression was correlated with microvessel density in 53 NSCLC tumor specimen. As in the previous studies, TGF-β1 expression was correlated with disease progression as defined by stage and lymph node metastasis. These studies suggest that TGF-β signaling is active in both the tumor and microenvironment. In contrast to TGF-β signaling in the previously reviewed tumor histologies, TGF-β signaling in NSCLC appears to directly promote tumor growth, perhaps by activating additional signaling proteins downstream of the activation cascade, such as MAPK.

2.5.2 Serum/plasma measurements of TGF-β isoforms in NSCLC

Circulating TGF-β levels were also measured in lung cancer patients. Plasma TGF-β1 levels were assessed in 37 patients with lung cancer compared to healthy volunteers or patients with pulmonary disease and a reduced pulmonary function. In this study, TGF-β1 levels in blood were compared based on collection techniques. Blood was either collected in a tube containing a specific inhibitor of platelet degranulation or EDTA alone. Inhibiting platelet degranulation resulted in lower TGF-β1 levels than in measurements using EDTA tubes alone. This approach of using a specific mixture to inhibit platelet degranulation provided a more accurate measure of tumor-associated TGF-β1 levels than the previous plasma assessments. Based on this special pre-analytic approach healthy volunteers and patients with pulmonary disease had little or no TGF-β1 elevation, and only lung cancer patients had significantly higher levels of plasma TGF-β1.

Circulating TGF-β1 levels also have been observed to change after treatment in patients with NSCLC. Patients who had elevated TGF-β1 levels after radiotherapy tended to have a less favorable outcome than patients with reduced plasma TGF-β1 levels. Furthermore, prior to starting radiotherapy in 59 NSCLC patients, elevated plasma TGF-β1 levels were correlated with unfavorable outcome. There was no association between elevated TGF-β1 levels and a particular histologic subtype in NSCLC. Thus, these two studies suggest that elevated TGF-β1 levels predict a less favorable outcome in NSCLC independent of the therapy.

Chemotherapy also has been shown to reduce plasma TGF-β1 levels in patients with NSCLC. In this study, 15 patients with metastatic NSCLC were followed for survival after treatment with docetaxel and enoxaparin. Among the 15 patients, 13 showed a tumor response or stable disease and 12/13 had a reduction of plasma TGF-β1 levels. Because patients received low-weight heparins in this study, it is also possible that administration of heparin rather than chemotherapy caused the reduction in TGF-β1 levels. Malignancy-associated hypercoagulation is common, and reducing TGF-β1 levels may reflect the activity of heparin on the state of coagulation in cancer patients. Thus, reducing TGF-β1 levels after treatment with heparin could in part explain the beneficial effect of heparin in cancer patients. As this remains speculative, additional studies will be needed to better assess the potential synergy of chemotherapy and heparin therapy in lowering TGF-β1 levels in NSCLC patients.

In summary, patients with NSCLC seem to have an unfavorable clinical outcome associated with increased tissue expression or serum/plasma secretion of TGF-β1 (Table 4). Furthermore, increased expression of TGF-βRI or TGF-βRII results in an unfavorable outcome. In contrast to colorectal, breast and prostate cancer the TGF-β/Smad signaling was not found to have a tumor suppressor role in the reviewed studies in NSCLC. This may be attributed to the following factors: first, the size of the studies conducted in NSCLC patients are smaller than in the previously reviewed cancer types and may not uncover the tumor suppressor activity of TGF-β/Smad signaling; second, patients with NSCLC generally present with more advanced disease limiting the ability to assess the role of TGF-β/Smad signaling in early NSCLC. Thus, larger studies in patients with NSCLC are needed to better characterize the function of TGF-β/Smad signaling in lung cancer and its association with concomitant treatments, such as chemotherapy and low-molecular weight heparins.

3. GERMLINE TGF-β/SMAD EXPRESSION AND THEIR ROLE IN TUMOR PROGRESSION

Genetic polymorphisms have been implicated in drug response and disease progression. Assuming a relationship between a genetic predisposition to increase TGF-β levels and cancer, TGF-β/Smad inhibitors are likely to be beneficial in reducing cancer risk or slowing tumor progression in such genetically predisposed patients.

3.1 Germline TGF-β/Smad Expression in Breast Cancer

Elevated TGF-β levels or an enhanced TGF-β signaling pathway has been associated with specific genetic polymorphisms. A study conducted in 170 pairs of female twins demonstrated that the C-509T polymorphism in the TGF-β1 gene promotor was associated with significantly elevated plasma TGF-β1 levels. Another study demonstrating that genetic TGF-β polymorphisms can play an important role in causing elevated TGF-β1 levels was based on a large cohort-study in breast cancer patients. Using samples of patients with invasive breast cancer obtained in the United Kingdom, Germany and Finland, elevated levels of TGF-β1 were linked to tumor progression. This case-control series of 3987 patients and 3867 controls found, that the C-509T and the T+29C (Leu10Pro) polymorphisms in the TGF-β1 gene promotor were significantly associated with increased incidence of invasive breast cancer in a recessive manner. As a result of their study, the authors estimated the percentage of the T+29C polymorphism homozygosity in breast cancer to be 13%. Based on this estimate, they concluded that patients with either T+29C or C-509T polymorphism had a 1.2-1.3-fold higher risk for developing breast cancer than patients without TGF-β polymorphisms. Furthermore, the authors concluded that 3% of all breast cancer cases may be attributed to T+29C polymorphism homozygosity. Another study assessing the impact of the T+29C polymorphism on developing breast cancer came to an opposite conclusion. Comparing the two studies, the former study was much larger than the latter. Because the estimated risk is small, large studies are needed to more accurately assess the relative risk of T+29C polymorphism in the development of breast cancer. More importantly, Dunning et al. investigated the T+29C polymorphism in young patients with exclusively invasive breast cancer. Another study that found no association of TGF-β1 and TGF-βR gene polymorphisms in breast cancer was conducted in 659 unselected, familial breast cancer patients from Poland, Sweden, Germany and Finland. Based on the earlier twin study, the authors hypothesized that TGF-β1 and TGF-βR gene polymorphisms should be more commonly detected in familial breast cancer patients. However, there was no significant increase of frequency of the examined TGF-β gene polymorphisms and no association with tumor progression. It is possible that in familial breast cancer there are TGF-β-independent risk factors, which could account for the differences seen in those studies. Additional studies are needed to understand whether there are TGF-β-dependent and TGF-β-independent risks in familial breast cancer.

In addition to the aggressiveness of a tumor, the menopausal status of a patient may influence the cancer risk associated with the different TGF-β gene polymorphisms. For example, in premenopausal Japanese women the C/C genotype of the T+29C polymorphism was associated with a significant risk reduction of breast cancer compared with the T/T genotype. Such an association was not observed for postmenopausal women. Compared to 172 patients without cancer, the 232 examined breast cancer patients had a genotype frequency of 28.9% for T/T (23.7% control), 46.1% for T/C (49.2% control), and 25.0% for C/C (27.1% control). The risk reduction associated with the C/C genotype of the T+29C polymorphism was not seen in the large Multiethnic Cohort (MEC) Study investigating 1123 cases of invasive breast cancer and 2314 controls. This large prospective study was conducted in Hawaii and Los Angeles and included Japanese, Caucasian, African American, Latino, and Native Hawaiian women who were predominantly postmenopausal at baseline. After adjustment for breast cancer risk factors, the analyses stratified by race/ethnicity, stage and age. The lack of association between the C/C genotype of the T + 29C polymorphisms and breast cancer risk was also observed in another study examining 500 invasive breast cancer patients. These three large studies contradict the study associating the T + 29C polymorphism with a higher breast cancer risk. One factor that might explain this difference is the fact that patients with T + 29C polymorphism and increased risk for breast cancer have primarily aggressive phenotypes. Because the associated risk for the T + 29C polymorphism was detected in patients with aggressive forms of breast cancer, it is possible that T + 29C polymorphism plays a particular role in patients with a pre-existing or aggressive cancer.

Supporting the hypothesis that certain polymorphisms are more relevant to patients with pre-existing or aggressive breast cancer is provided by the data of the Shanghai Breast Cancer Study. In this study the T + 29C and C-509T polymorphism were investigated for their impact on survival of patients with breast cancer. Following 1111 patients with primary breast cancer, women with a variant allele of the T + 29C (frequency of 50% for T/C; and 26.9% for C/C) or C-509T polymorphism (frequency of 50% for C/T and 26.4% for T/T) had a lower disease-free survival rate than those with both wild-type alleles: 77.5% (TC/CC) vs. 85.2% (TT) for the T + 29C polymorphism; 77.6% (CT/TT) vs. 85.1% (CC) of the C-509T polymorphism. This study suggested that variant alleles of T + 29C or C-509T polymorphism are important factors in progression of breast cancer. Therefore, women with T + 29C or C-509T polymorphism and who have an aggressive breast cancer appear to be at a higher risk of metastatic spread and poor outcome.

3.2 Germline TGF-β/Smad Expression in Prostate Cancer

The impact of TGF-β1 gene polymorphism was also investigated in prostate cancer. In a nested case-control study of 492 men with prostate cancer, patients with homozygous C-509T polymorphism had a 2.4-fold increased risk of developing advanced stage prostate cancer. Changes affecting the T + 29C polymorphism were not associated with an increased risk. One caveat of this study is the number of patients with the homozygous C-509T polymorphism. Only 13% (20/157) of the examined patients were homozygous for C-509T polymorphism. Another study specifically investigated the T + 29C polymorphism in 351 prostate cancer patients, 221 benign prostate hyperplasia patients and 303 male controls in Japan. There were significant differences in the C/C versus T/C+T/T genotype distribution between prostate cancer patients and male controls. Males with the T/C or T/T genotype had a 1.6-fold increased risk of prostate cancer compared to patients with the C/C genotype. Taken together, studies in prostate cancer patients show a small risk of patients with C-509T or T + 29C polymorphism. However, larger studies are needed to better assess the impact of either of these polymorphisms on clinical outcome in prostate cancer.

3.3 Germline Expression of TGF-βR I in Cancer

One of the largest polymorphism assessment studies for a TGF-β signaling component was performed for TGF-βRI polymorphism. Using studies of various tumor types and integrating seven case-control studies in a large meta-analysis, a total of 2438 cases and 1846 controls were assessed for the frequency of the *TGFBR1*6A*. Overall, the *TGFBR1*6A* carriers had a 26% increased risk of developing cancer, such as breast cancer, hematological malignancies and ovarian cancer. Carriers of the *TGFBR1*6A* also showed an increased risk for colorectal cancer if they were from the United States but not from Southern Europe. *TGFBR1*6A* was found to play an important role in hereditary colorectal cancer, where its highest frequency was found in patients with no mismatch repair gene mutation and no microsatellite instability. Similar to the TGF-β1 gene polymorphism, the changes in the *TGFBR1*6A* seem to indicate that such a polymorphism may lead to a less favorable outcome in patients with an established cancer. Prospective studies are needed to confirm the findings on the role of *TGFBR1*6A* polymorphism in the progression of cancer.

3.4 Other Genetic Polymorphisms of TGF-β/Smad Expression in Cancer

In contrast to the polymorphisms of TGF-βRI, germline mutations of the TGF-βRII gene have not been extensively investigated. However, one of the most common observation is that TGF-βRII gene is deleted in tumor tissue of patients with colorectal cancer. This deletion occurs particularly in patients with microsatellite instability. Polymorphism studies for germline mutations have not been studied except in small patient cohorts. For example, in patients with hereditary nonpolyposis colorectal cancer (HNPCC), where TGF-βRII plays an important role in tumorigenesis, no correlation was observed between TGF-βRII gene polymorphism and clinical outcome.

4. PERSPECTIVES

TGF-β/Smad signaling functions as a tumor promoter in patients with invasive cancer (Fig. 1). Thus, initial clinical experience with TGF-β/Smad inhibitors will be in patients with invasive or aggressive cancers. To realize the full potential of TGF-β inhibitors as cancer therapeutics patient selection strategies will be critical. In certain cancers, plasma TGF-β1 levels can predict for tumor progression. Future clinical experience will determine if elevated TGF-β levels can be used to predict the benefit of a TGF-β/Smad inhibitor and thus serve as a patient selection criteria. In addition, the genetic predisposition of certain TGF-β polymorphisms for tumor progression may provide an independent approach to identify cancer patients with advanced disease who could benefit from a TGF-β/Smad inhibitor. Finally, clinical strategies are available to monitor phospho-SMAD in patients to demonstrate TGF-β

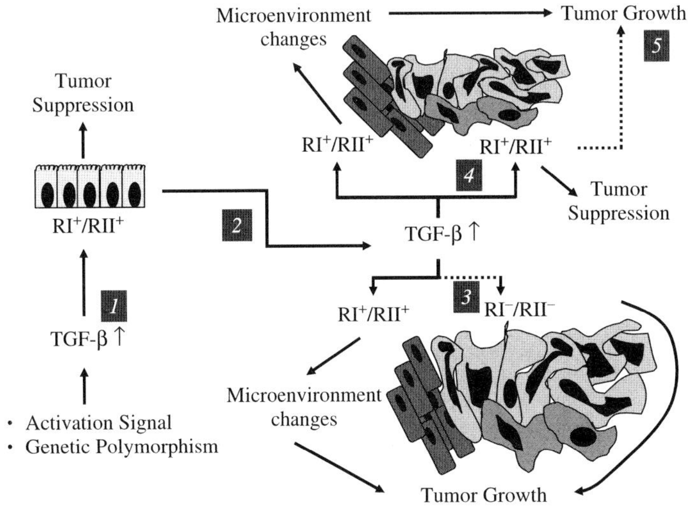

Figure 1. The Role of TGF-β as a Tumor Suppressor and/or Promotor in Cancer Patients. Elevated TGF-β secretion or expression is generally induced by various activation signals and can be endogenously induced in subjects with a genetic polymorphism predisposing them to higher TGF-β production (1). The result of this condition is generally tumor suppression. As tumors become invasive (2), elevated TGF-β levels can influence tumor and its microenvironment. Elevated TGF-β secretion or expression can cause tumor growth in two ways: first, with loss of TGF-βR on tumor cells the tumor suppressive activity is reduced and tumor growth occurs unopposed (3). Second, tumor cells may still express TGF-βRs, but the suppressive effect of TGF-β is overpowered by the TGF-β-mediated growth of the microenvironment (4). In some instances, however, the tumor growth may be promoted by both active signaling of TGF-β in tumor cells and microenvironment cells (dotted line; 5)

pathway modulation in response to therapy. This clinical strategy is important given the pleiotropic nature of TGF-β signaling and the likelihood of adverse events if TGF-β signaling were systemically abrogated. As the first TGF-β pathway small molecule and antibody therapeutics are now entering oncology clinical trials, these important questions are on the verge of being answered in patients who desperately need novel and effective therapies to treat their disease.

REFERENCES

Adler, H.L., McCurdy, M.A., Kattan, M.W., Timme, T.L., Scardino, P.T., and Thompson, T.C., 1999, Elevated levels of circulating interleukin-6 and transforming growth factor-β1 in patients with metastatic prostatic carcinoma. *J Urol* 161: 182-187.

Akiyama, Y., Iwanaga, R., Ishikawa, T., Sakamoto, K., Nishi, N., Nihei, Z., Iwama, T., Saitoh, K., and Yuasa, Y., 1996, Mutations of the transforming growth factor-β type II receptor gene are strongly related to sporadic proximal colon carcinomas with microsatellite instability. *Cancer* 78: 2478-2484.

Anscher, M.S., Marks, L.B., Shafman, T.D., Clough, R., Huang, H., Tisch, A., Munley, M., Herndon, J.E. 2nd, Garst, J., Crawford, J., and Jirtle, R.L., 2001, Using plasma transforming growth factor β-1 during radiotherapy to select patients for dose escalation. *J Clin Oncol* 19: 3758-3765.

Anscher, M.S., Peters, W.P., Reisenbichler, H., Petros, W.P., and Jirtle, R.L., 1993, Transforming growth factor β as a predictor of liver and lung fibrosis after autologous bone marrow transplantation for advanced breast cancer. *N Engl J Med* 328: 1592-1598.

Arai, T., Akiyama, Y., Okabe, S., Ando, M., Endo, M., and Yuasa, Y., 1998, Genomic structure of the human Smad3 gene and its infrequent alterations in colorectal cancers. *Cancer Lett* 122: 157-163.

Auvinen, P., Lipponen, P., Johansson, R., and Syrjanen, K., 1995, Prognostic significance of TGF-β 1 and TGF-β 2 expressions in female breast cancer. *Anticancer Res* 15: 2627-2631.

Barlow, J., Yandell, D., Weaver, D., Casey, T., and Plaut, K., 2003, Higher stromal expression of transforming growth factor-β type II receptors is associated with poorer prognosis breast tumors. *Breast Cancer Res Treat* 79: 149-159.

Barthelemy-Brichant, N., David, J.L., Bosquee, L., Bury, T., Seidel, L., Albert, A., Bartsch, P., Baugnet-Mahieu, L., and Deneufbourg, J.M., 2002, Increased TGFβ1 plasma level in patients with lung cancer: potential mechanisms. *Eur J Clin Invest* 32: 193-198.

Beight, D.W. e. a., 2004, Preparation of pyrazolopyridines as TGF-β signal transduction inhibitors useful against cancer and other diseases in E. L. a. Company, editor.

Bellone, G., Carbone, A., Tibaudi, D., Mauri, F., Ferrero, I., Smirne, C., Suman, F., Rivetti, C., Migliaretti, G., Camandona, M., Palestro, G., Emanuelli, G., and Rodeck, U., 2001, Differential expression of transforming growth factors-β1, -β2 and -β3 in human colon carcinoma. *Eur J Cancer* 37: 224-233.

Benson, J.R., 2004, Role of transforming growth factor β in breast carcinogenesis. *Lancet Oncol* 5: 229-239.

Bian, Y., Caldes, T., Wijnen, J., Franken, P,.Vasen, H., Kaklamani, V., Nafa, K., Peterlongo, P., Ellis, N., Baron, J.A., Burn, J., Moeslein, G., Morrison, P.J., Chen, Y., Ahsan, H., Watson, P., Lynch, H.T., de la Chapelle, A., Fodde, R., and Pasche, B., 2005, TGFBR1*6A may contribute to hereditary colorectal cancer. *J Clin Oncol* 23: 3074-3078.

Boulay, J.L., Mild, G., Lowy, A., Reuter, J., Lagrange, M., Terracciano, L., Laffer, U., Herrmann, R., and Rochlitz, C., 2002, Smad4 is a predictive marker for 5-fluorouracil-based chemotherapy in patients with colorectal cancer. *Br J Cancer* 87: 630-634.

Boulay, J.L., Mild, G., Lowy, A., Reuter, J., Lagrange, M., Terracciano, L., Laffer, U., Herrmann, R., and Rochlitz, C., 2003, Smad7 is a prognostic marker in patients with colorectal cancer. *Int J Cancer* 104: 446-449.

Butta, A., MacLennan, K., Flanders, K.C., Sacks, N.P., Smith, I., McKinna, A., Dowsett, M., Wakefield, L.M., Sporn, M.B., Baum, M., and et al., 1992, Induction of transforming growth factor β 1 in human breast cancer *in vivo* following tamoxifen treatment. *Cancer Res* 52: 4261-4264.

Caine, G.J., Stonelake, P. S., Lip, G.Y., and Kehoe, S.T., 2002, The hypercoagulable state of malignancy: pathogenesis and current debate. *Neoplasia* 4: 465-473.

Callahan, J.F., Burgess, J.L., Fornwald, J.A., Gaster, L.M., Harling, J.D., Harrington, F.P., Heer, J., Kwon, C., Lehr, R., Mathur, A., Olson, B.A., Weinstock, J., and Laping, N.J., 2002, Identification of novel inhibitors of the transforming growth factor β1 (TGF-β1) type 1 receptor (ALK5). *J Med Chem* 45: 999-1001.

Canto, E.I., Shariat, S.F., and Slawin, K.M., 2003, Biochemical staging of prostate cancer. *Urol Clin North Am* 30: 263-277.

Cardillo, M.R., Petrangeli, E., Perracchio, L., Salvatori, L., Ravenna, L., and Di Silverio, F., 2000, Transforming growth factor-β expression in prostate neoplasia. *Anal Quant Cytol Histol* 22: 1-10.

Cordeiro, M.F., 2003, Technology evaluation: lerdelimumab, Cambridge Antibody Technology. *Curr Opin Mol Ther* 5: 199-203.

DaCosta Byfield, S., Major, C., Laping, N.J., and Roberts,.A.B., 2004, SB-505124 is a selective inhibitor of transforming growth factor-β type I receptors ALK4, ALK5, and ALK7. *Mol Pharmacol* 65: 744-752.

Dalal, B.I., Keown, P.A., and Greenberg, A.H., 1993, Immunocytochemical localization of secreted transforming growth factor-β1 to the advancing edges of primary tumors and to lymph node metastases of human mammary carcinoma. *Am J Pathol* 143: 381-389.

Djonov, V., Ball, R.K., Graf, S., Mottaz, A.E., Arnold, A.M., Flanders, K., Studer, U.E., and Merz, V.W., 1997, Transforming growth factor-β 3 is expressed in nondividing basal epithelial cells in normal human prostate and benign prostatic hyperplasia, and is no longer detectable in prostate carcinoma. *Prostate* 31: 103-109.

Dunning, A.M., Ellis, P.D., McBride, S., Kirschenlohr, H.L., Healey, C.S., Kemp, P.R., Luben, R.N., Chang-Claude, J., Mannermaa, A., Kataja, V., Pharoah, P.D., Easton, D.F., Ponder, B.A., and Metcalfe, J.C., 2003, A transforming growth factor β1 signal peptide variant increases secretion *in vitro* and is associated with increased incidence of invasive breast cancer. *Cancer Res* 63: 2610-2615.

Eastham, J.A., Truong, L.D., Rogers, E., Kattan, M., Flanders, K.C., Scardino, P.T., and Thompson, T.C., 1995, Transforming growth factor-β 1: comparative immunohistochemical localization in human primary and metastatic prostate cancer. *Lab Invest* 73: 628-635.

Eppert, K., Scherer, S.W., Ozcelik, H., Pirone, R., Hoodless, P., Kim, H., Tsui, L.C., Bapat, B., Gallinger, S., Andrulis, I.L., Thomsen, G.H., Wrana, J.L., and Attisano, L., 1996, MADR2 maps to 18q21 and encodes a TGFβ-regulated MAD-related protein that is functionally mutated in colorectal carcinoma. *Cell* 86: 543-552.

Ewart-Toland, A., Chan, J.M., Yuan, J., Balmain, A., and Ma, J., 2004, A gain of function *TGFB1* polymorphism may be associated with late stage prostate cancer. *Cancer Epidemiol Biomarkers Prev* 13: 759-764.

Fonsatti, E., Del Vecchio, L., Altomonte, M., Sigalotti, L., Nicotra, M.R., Coral, S., Natali, P.G., and Maio, M., 2001, Endoglin: An accessory component of the TGF-β-binding receptor-complex with diagnostic, prognostic, and bioimmunotherapeutic potential in human malignancies. *J Cell Physiol* 188: 1-7.

Fox, E., Curt, G.A., and Balis, F.M., 2002, Clinical trial design for target-based therapy. *Oncologist* 7: 401-409.

Friedman, E., Gold, L.I., Klimstra, D., Zeng, Z.S., Winawer, S., and Cohen, A., 1995, High levels of transforming growth factor β 1 correlate with disease progression in human colon cancer. *Cancer Epidemiol Biomarkers Prev* 4: 549-554.

Ge, R., Rajeev, V., Subramanian, G., Reiss, K.A., Liu, D., Higgins, L., Joly, A., Dugar, S., Chakravarty, J., Henson, M., McEnroe, G., Schreiner, G., and Reiss, M., 2004, Selective inhibitors of type I receptor kinase block cellular transforming growth factor-β signaling. *Biochem Pharmacol* 68: 41-50.

Gellibert, F., Woolven, J., Fouchet, M.H., Mathews, N., Goodland, H., Lovegrove, V., Laroze, A., Nguyen, V.L., Sautet, S., Wang, R., Janson, C., Smith, W., Krysa, G., Boullay, V., De Gouville, A.C., Huet, S., and Hartley, D., 2004, Identification of 1,5-naphthyridine derivatives as a novel series of potent and selective TGF-β type I receptor inhibitors. *J Med Chem* 47: 4494-4506.

Ghellal, A., Li, C., Hayes, M., Byrne, G., Bundred, N., and Kumar, S., 2000, Prognostic significance of TGF β 1 and TGF β 3 in human breast carcinoma. *Anticancer Res* 20: 4413-4418.

Gobbi, H., Arteaga, C.L., Jensen, R.A., Simpson, J.F., Dupont, W.D., Olson, S.J., Schuyler, P.A., Plummer, Jr., W.D., and Page, D.L., 2000, Loss of expression of transforming growth factor β type II receptor correlates with high tumour grade in human breast in-situ and invasive carcinomas. *Histopathology* 36: 168-177.

Gobbi, H., Dupont, W.D., Simpson, J.F., Plummer, Jr., W.D., Schuyler, P.A., Olson, S.J., Arteaga, C.L., and Page, D.L., 1999, Transforming growth factor-β and breast cancer risk in women with mammary epithelial hyperplasia. *J Natl Cancer Inst* 91: 2096-2101.

Gorsch, S.M., Memoli, V.A., Stukel, T.A., Gold, L.I., and Arrick, B.A., 1992,. Immunohistochemical staining for transforming growth factor β 1 associates with disease progression in human breast cancer. *Cancer Res* 52: 6949-6952.

Grainger, D.J., Heathcote, K., Chiano, M., Snieder, H., Kemp, P.R., Metcalfe, J.C., Carter, N.D., and Spector, T. D., 1999, Genetic control of the circulating concentration of transforming growth factor type β1. *Hum Mol Genet* 8: 93-97.

Hahn, S.A., Bartsch, D., Schroers, A., Galehdari, H., Becker, M., Ramaswamy, A., Schwarte-Waldhoff, I., Maschek, H., and Schmiegel, W., 1998, Mutations of the DPC4/Smad4 gene in biliary tract carcinoma. *Cancer Res* 58: 1124-1126.

Hasegawa, Y., Takanashi, S., Kanehira, Y., Tsushima, T., Imai, T., and Okumura, K., 2001, Transforming growth factor-β1 level correlates with angiogenesis, tumor progression, and prognosis in patients with nonsmall cell lung carcinoma. *Cancer* 91: 964-971.

Hishida, A., Iwata, H., Hamajima, N., Matsuo, K., Mizutani, M., Iwase, T., Miura, S., Emi, N., Hirose, K., and Tajima, K., 2003, Transforming growth factor B1 T29C polymorphism and breast cancer risk in Japanese women. *Breast Cancer* 10: 63-69.

Horvath, L.G., Henshall, S.M., Kench, J.G., Turner, J.J., Golovsky, D., Brenner, P.C., O'Neill, G.F., Kooner, R., Stricker, P.D., Grygiel, J.J., and Sutherland, R.L., 2004, Loss of BMP2, Smad8, and Smad4 expression in prostate cancer progression. *Prostate* 59: 234-242.

Ivanovic, V., Melman, A., Davis-Joseph, B., Valcic, M., and Geliebter, J., 1995, Elevated plasma levels of TGF-β 1 in patients with invasive prostate cancer. *Nat Med* 1: 282-284.

Jeruss, J.S., Sturgis, C.D., Rademaker, A.W., and Woodruff, T.K., 2003, Down-regulation of activin, activin receptors, and Smads in high-grade breast cancer. *Cancer Res* 63: 3783-3790.

Jin, Q., Hemminki, K., Grzybowska, E., Klaes, R., Soderberg, M., Zientek, H., Rogozinska-Szczepka, J., Utracka-Hutka, B., Pamula, J., Pekala, W., and Forsti, A., 2004, Polymorphisms and haplotype structures in genes for transforming growth factor β1 and its receptors in familial and unselected breast cancers. *Int J Cancer* 112: 94-99.

Kajdaniuk, D., Marek, B., Swietochowska, E., Ostrowska, Z., Glogowska-Szelag, J., Kos-Kudla, B., Ciesielska-Kopacz, N., and Wielkoszyski, T., 2000, Plasma transforming growth factor β1 in breast cancer patients treated with CMF chemotherapy. *J Clin Pharm Ther* 25: 291-294.

Kakehi, Y., Oka, H., Mitsumori, K., Itoh, N., Ogawa, O., and Yoshida, O., 1996, Elevation of serum transforming growth factor-β1 Level in patients with metastatic prostate cancer. *Urologic Oncology* 2: 131-135.

Kaklamani, V.G., Hou, N., Bian, Y., Reich, J., Offit, K., Michel, L.S., Rubinstein, W.S., Rademaker, A., and Pasche, B., 2003, TGFBR1*6A and cancer risk: a meta-analysis of seven case-control studies. *J Clin Oncol* 21: 3236-3243.

Kattan, M.W., Shariat, S.F., Andrews, B., Zhu, K., Canto, E., Matsumoto, K., Muramoto, M., Scardino, P.T., Ohori, M., Wheeler, T.M., and Slawin, K. M., 2003, The addition of interleukin-6 soluble receptor and transforming growth factor β1 improves a preoperative nomogram for predicting biochemical progression in patients with clinically localized prostate cancer. *J Clin Oncol* 21: 3573-3579.

Kong, F., Jirtle, R.L., Huang, D.H., Clough, R.W., and Anscher, M.S., 1999, Plasma transforming growth factor-β1 level before radiotherapy correlates with long term outcome of patients with lung carcinoma. *Cancer* 86: 1712-1719.

Kong, F.M., Anscher, M.S., Murase, T., Abbott, B.D., Iglehart, J.D., and Jirtle, R.L., 1995, Elevated plasma transforming growth factor-β 1 levels in breast cancer patients decrease after surgical removal of the tumor. *Ann Surg* 222: 155-162.

Kong, F.M., Washington, M.K., Jirtle, R.L., and Anscher, M.S., 1996, Plasma transforming growth factor-β 1 reflects disease status in patients with lung cancer after radiotherapy: a possible tumor marker. *Lung Cancer* 16: 47-59.

Kretzschmar, M., and Massagué, J., 1998, Smads: mediators and regulators of TGF-β signaling. *Curr Opin Genet Dev* 8: 103-111.

Krippl, P., Langsenlehner, U., Renner, W., Yazdani-Biuki, B., Wolf, G., Wascher, T.C., Paulweber, B., Bahadori, B., and Samonigg, H., 2003, The L10P polymorphism of the transforming growth factor-β1 gene is not associated with breast cancer risk. *Cancer Lett* 201: 181-184.

Laffer, U., 1995, Long-term results of single course of adjuvant intraportal chemotherapy for colorectal cancer. Swiss Group for Clinical Cancer Research (SAKK). *Lancet* 345: 349-353.

Lahn, M., Kloeker, S., and Berry, B.S., 2005, TGF-β inhibitors for the treatment of cancer. *Expert Opin Investig Drugs* 14: 629-643.

Le Marchand, L., Haiman, C.A., van den Berg, D., Wilkens, L.R., Kolonel, L.N., and Henderson, B.E., 2004, T29C polymorphism in the transforming growth factor β1 gene and postmenopausal breast cancer risk: the Multiethnic Cohort Study. *Cancer Epidemiol Biomarkers Prev* 13: 412-415.

Lee, A.Y., Levine, M.N., Baker, R.I., Bowden, C., Kakkar, A.K., Prins, M., Rickles, F.R., Julian, J.A., Haley, S., Kovacs, M.J., and Gent, M., 2003, Low-molecular-weight heparin versus a coumarin for the prevention of recurrent venous thromboembolism in patients with cancer. *N Engl J Med* 349: 146-153.

Lee, W.C., 2004, Preparation of (pyridinyl)(pyrimidinyl) imidazo[1,2-a]pyridines as TGFβ receptor type I antagonists for treatment of fibrotic disorders and tumors. in B. IDEC, editor.

Li, C., Gardy, R., Seon, B.K., Duff, S.E., Abdalla, S., Renehan, A., O'Dwyer, S.T., Haboubi, N., and Kumar, S., 2003, Both high intratumoral microvessel density determined using CD105 antibody and elevated plasma levels of CD105 in colorectal cancer patients correlate with poor prognosis. *Br J Cancer* 88: 1424-1431.

Li, C., Wang, J., Wilson, P.B., Kumar, P., Levine, E., Hunter, R.D., and Kumar, S., 1998, Role of transforming growth factor β3 in lymphatic metastasis in breast cancer. *Int J Cancer* 79: 455-459.

Li, Z., Habuchi, T., Tsuchiya, N., Mitsumori, K., Wang, L., Ohyama, C., Sato, K., Kamoto, T., Ogawa, O., and Kato, T., 2004, Increased risk of prostate cancer and benign prostatic hyperplasia associated with transforming growth factor-β 1 gene polymorphism at codon10. *Carcinogenesis* 25: 237-240.

Losi, L., Luppi, G., and Benhattar, J., 2004, Assessment of K-ras, Smad4 and p53 gene alterations in colorectal metastases and their role in the metastatic process. *Oncol Rep* 12: 1221-1225.

MacCallum, J., Keen, J.C., Bartlett, J.M., Thompson, A.M., Dixon, J.M., and Miller, W.R., 1996, Changes in expression of transforming growth factor β mRNA isoforms in patients undergoing tamoxifen therapy. *Br J Cancer* 74: 474-478.

Matsushita, M., Matsuzaki, K., Date, M., Watanabe, T., Shibano, K., Nakagawa, T., Yanagitani, S., Amoh, Y., Takemoto, H., Ogata, N., Yamamoto, C., Kubota, Y., Seki, T., Inokuchi, H., Nishizawa, M., Takada, H., Sawamura, T., Okamura, A., and Inoue, K., 1999, Down-regulation of TGF-β receptors in human colorectal cancer: implications for cancer development. *Br J Cancer* 80: 194-205.

Mead, A.L., Wong, T.T., Cordeiro, M.F., Anderson, I.K., and Khaw, P.T., 2003, Evaluation of anti-TGF-β2 antibody as a new postoperative anti-scarring agent in glaucoma surgery. *Invest Ophthalmol Vis Sci* 44: 3394-3401.

Murray, P.A., Barrett-Lee, P., Travers, M., Luqmani, Y., Powles, T., and Coombes, R.C., 1993, The prognostic significance of transforming growth factors in human breast cancer. *Br J Cancer* 67: 1408-1412.

Nakao, A., Afrakhte, M., Morén, A., Nakayama, T., Christian, J.L., Heuchel, R., Itoh, S., Kawabata, M., Heldin, N.-E., Heldin, C.-H., and ten Dijke, P., 1997, Identification of Smad7, a TGFβ-inducible antagonist of TGF-β signalling. *Nature* 389: 631-635.

Narai, S., Watanabe, M., Hasegawa, H., Nishibori, H., Endo, T., Kubota, T., and Kitajima, M. 2002, Significance of transforming growth factor β1 as a new tumor marker for colorectal cancer. *Int J Cancer* 97: 508-511.

O'Farrell, A.M., Foran, J.M., Fiedler, W., Serve, H., Paquette, R.L., Cooper, M.A., Yuen, H.A., Louie, S.G., Kim, H., Nicholas, S., Heinrich, M.C., Berdel, W.E., Bello, C., Jacobs, M., Scigalla, P., Manning, W.C., Kelsey, S., and Cherrington, J.M., 2003, An innovative phase I clinical study demonstrates inhibition of FLT3 phosphorylation by SU11248 in acute myeloid leukemia patients. *Clin Cancer Res* 9: 5465-5476.

Ogunbiyi, O.A., Goodfellow, P.J., Gagliardi, G., Swanson, P.E., Birnbaum, E.H., Fleshman, J.W., Kodner, I.J., and Moley, J.F., 1997, Prognostic value of chromosome 1p allelic loss in colon cancer. *Gastroenterology* 113: 761-766.

Ogunbiyi, O.A., Goodfellow, P.J., Herfarth, K., Gagliardi, G., Swanson, P.E., Birnbaum, E.H., Read, T.E., Fleshman, J.W., Kodner, I.J., and Moley, J.F., 1998, Confirmation that chromosome 18q allelic loss in colon cancer is a prognostic indicator. *J Clin Oncol* 16: 427-433.

Park, B.J., Park, J.I., Byun, D.S., Park, J.H., and Chi, S.G., 2000, Mitogenic conversion of transforming growth factor-β1 effect by oncogenic Ha-Ras-induced activation of the mitogen-activated protein kinase signaling pathway in human prostate cancer. *Cancer Res* 60: 3031-3038.

Peng, S.B., Yan, L., Xia, X., Watkins, S.A., Brooks, H.B., Beight, D., Herron, D.K., Jones, M.L., Lampe, J.W., McMillen, W.T., Mort, N., Sawyer, J.S., and Yingling, J.M., 2005, Kinetic characterization of novel pyrazole TGF-β receptor I kinase inhibitors and their blockade of the epithelial-mesenchymal transition. *Biochemistry* 44: 2293-2304.

Perry, K.T., Anthony, C.T., Case, T., and Steiner, M.S.,. 1997a, Transforming growth factor β as a clinical biomarker for prostate cancer. *Urology* 49: 151-155.

Perry, K.T., Anthony, C.T., and Steiner, M.S., 1997b, Immunohistochemical localization of TGF β 1, TGF β 2, and TGF β 3 in normal and malignant human prostate. *Prostate* 33: 133-140.

Raftery, L.A., Twombly, V., Wharton, K., and Gelbart, W.M., 1995, Genetic screens to identify elements of the decapentaplegic signaling pathway in *Drosophila. Genetics* 139: 241-254.

Reinholz, M.M., An, M.W., Johnsen, S.A., Subramaniam, M., Suman, V.J., Ingle, J.N., Roche, P.C., and Spelsberg, T.C., 2004, Differential gene expression of TGF β inducible early gene (TIEG), Smad7, Smad2 and Bard1 in normal and malignant breast tissue. *Breast Cancer Res Treat* 86: 75-88.

Robert, F., Busby, E., Marques, M.B., Reynolds, R.E., and Carey, D.E., 2003, Phase II study of docetaxel plus enoxaparin in chemotherapy-naive patients with metastatic non-small cell lung cancer: preliminary results. *Lung Cancer* 42: 237-245.

Savage, C., Das, P., Finelli, A.L., Townsend, S.R., Sun, C.Y., Baird, S.E., and Padgett, R.W., 1996, *Caenorhabditis elegans* genes sma-2, sma-3, and sma-4 define a conserved family of transforming growth factor β pathway components. *Proc Natl Acad Sci U S A* 93: 790-794.

Sawyer, J.S., Beight, D.W., Britt, K.S., Anderson, B.D., Campbell, R.M., Goodson, Jr., T., Herron, D.K., Li, H.Y., McMillen, W.T., Mort, N., Parsons, S., Smith, E.C., Wagner, J.R., Yan, L., Zhang, F., and Yingling, J.M., 2004, Synthesis and activity of new aryl- and heteroaryl-substituted 5,6-dihydro-4H-pyrrolo[1,2-b]pyrazole inhibitors of the transforming growth factor-β type I receptor kinase domain. *Bioorg Med Chem Lett* 14: 3581-3584.

Schlingensiepen, K.-H., Bischof, A., Egger, T., Hafner, M., Herrmuth, H., Jachimczak, P., Kielmanowicz, M., Niewel, M., Zavadova, E., and Stauder, G., 2004, The TGF-β1 antisense oligonucleotide AP 11014 for the treatment of non-small cell lung, colorectal and prostate cancer: preclinical studies. Abstract No. 3132. *Am Soc Clin Oncol Ann Meet.*

Schlingensiepen, R., Goldbrunner, M., Szyrach, M.N., Stauder, G., Jachimczak, P., Bogdahn, U., Schulmeyer, F., Hau, P., and Schlingensiepen, K.H., 2005, Intracerebral and intrathecal infusion of the TGF-β2-specific antisense phosphorothioate oligonucleotide AP 12009 in rabbits and primates: Toxicology and Safety. *Oligonucleotides* 15: 94-104.

Shariat, S.F., Kattan, M.W., Traxel, E., Andrews, B., Zhu, K., Wheeler, T.M., and Slawin, K.M., 2004, Association of pre- and postoperative plasma levels of transforming growth factor β(1) and interleukin 6 and its soluble receptor with prostate cancer progression. *Clin Cancer Res* 10: 1992-1999.

Shariat, S.F., Shalev, M., Menesses-Diaz, A., Kim, I.Y., Kattan, M.W., Wheeler, T.M., and Slawin, K.M., 2001, Preoperative plasma levels of transforming growth factor β(1) (TGF-β(1)) strongly predict progression in patients undergoing radical prostatectomy. *J Clin Oncol* 19: 2856-2864.

Sheen-Chen, S.M., Chen, H.S., Sheen, C.W., Eng, H.L., and Chen, W.J., 2001, Serum levels of transforming growth factor β1 in patients with breast cancer. *Arch Surg* 136: 937-940.

Shim, K.S., Kim, K.H., Han, W.S., and Park, E.B., 1999, Elevated serum levels of transforming growth factor-β1 in patients with colorectal carcinoma: its association with tumor progression and its significant decrease after curative surgical resection. *Cancer* 85: 554-561.

Shin, K.H., Park, Y.J., and Park, J.G., 2000, Mutational analysis of the transforming growth factor β receptor type II gene in hereditary nonpolyposis colorectal cancer and early-onset colorectal cancer patients. *Clin Cancer Res* 6: 536-540.

Shu, X.O., Gao, Y.T., Cai, Q., Pierce, L., Cai, H., Ruan, Z.X., Yang, G., Jin, F., and Zheng, W., 2004, Genetic polymorphisms in the TGF-β 1 gene and breast cancer survival: a report from the Shanghai Breast Cancer Study. *Cancer Res* 64: 836-839.

Siegel, P.M., and Massagué, J.,. 2003, Cytostatic and apoptotic actions of TGF-β in homeostasis and cancer. *Nat Rev Cancer* 3: 807-821.

Simeoni, M., Magni, P., Cammia, C., De Nicolao, G., Croci, V., Pesenti, E., Germani, M., Poggesi, I., and Rocchetti, M., 2004, Predictive pharmacokinetic-pharmacodynamic modeling of tumor growth kinetics in xenograft models after administration of anticancer agents. *Cancer Res* 64: 1094-1101.

Stravodimos, K., Constantinides, C., Manousakas, T., Pavlaki, C., Pantazopoulos, D., Giannopoulos, A., and Dimopoulos, C., 2000, Immunohistochemical expression of transforming growth factor β 1 and nm-23 H1 antioncogene in prostate cancer: divergent correlation with clinicopathological parameters. *Anticancer Res* 20: 3823-3828.

Subramanian, G., Schwarz, R.E., Higgins, L., McEnroe, G., Chakravarty, S., Dugar, S., and Reiss, M., 2004, Targeting endogenous transforming growth factor β receptor signaling in Smad4-deficient human pancreatic carcinoma cells inhibits their invasive phenotype1. *Cancer Res* 64: 5200-5211.

Takanami, I., Tanaka, F., Hashizume, T., Kikuchi, K., Yamamoto, Y., Yamamoto, T., and Kodaira, S., 1997a, Transforming growth factor-β isoforms expressions in pulmonary adenocarcinomas as prognostic markers: an immunohistological study of one hundred and twenty patients. *Oncology* 54: 122-128.

Takanami, I., Tanaka, F., Hashizume, T., and Kodaira, S., 1997b, Roles of the transforming growth factor β 1 and its type I and II receptors in the development of a pulmonary adenocarcinoma: results of an immunohistochemical study. *J Surg Oncol* 64: 262-267.

Tamura, K., and Fukuoka, M., 2005, Gefitinib in non-small cell lung cancer. *Expert Opin Pharmacother* 6: 985-993.

Thompson, A.M., Kerr, D.J., and Steel, C.M., 1991, Transforming growth factor β 1 is implicated in the failure of tamoxifen therapy in human breast cancer. *Br J Cancer* 63: 609-614.

Truong, L.D., Kadmon, D., McCune, B.K., Flanders, K.C., Scardino, P.T., and Thompson, T.C., 1993, Association of transforming growth factor-β 1 with prostate cancer: an immunohistochemical study. *Hum Pathol* 24: 4-9.

Tsushima, H., Ito, N., Tamura, S., Matsuda, Y., Inada, M., Yabuuchi, I., Imai, Y., Nagashima, R., Misawa, H., Takeda, H., Matsuzawa, Y., and Kawata, S., 2001, Circulating transforming growth factor β 1 as a predictor of liver metastasis after resection in colorectal cancer. *Clin Cancer Res* 7: 1258-1262.

Uhl, M., Aulwurm, S., Wischhusen, J., Weiler, M., Ma, J.Y., Almirez, R., Mangadu, R., Liu, Y.W., Platten, M., Herrlinger, U., Murphy, A., Wong, D.H., Wick, W., Higgins, L.S., and Weller, M., 2004, SD-208, a novel transforming growth factor β receptor I kinase inhibitor, inhibits growth and invasiveness and enhances immunogenicity of murine and human glioma cells *in vitro* and *in vivo*. *Cancer Res* 64: 7954-7961.

van der Poel, H.G., 2004, Smart drugs in prostate cancer. *Eur Urol* 45: 1-17.

Vogelstein, B., Fearon, E.R., Hamilton, S.R., Kern, S.E., Preisinger, A.C., Leppert, M., Nakamura, Y., White, R., Smits, A.M., and Bos, J.L., 1988, Genetic alterations during colorectal-tumor development. *N Engl J Med* 319: 525-532.

Walker, R.A., and Dearing, S.J., 1992, Transforming growth factor β 1 in ductal carcinoma in situ and invasive carcinomas of the breast. *Eur J Cancer* 28: 641-644.

Weinshilboum, R., 2003, Inheritance and drug response. *N Engl J Med* 348: 529-537.

Weinshilboum, R., and Wang, L., 2004, Pharmacogenomics: bench to bedside. *Nat Rev Drug Discov* 3: 739-748.

Wikstrom, P., Stattin, P., Franck-Lissbrant, I., Damber, J.E., and Bergh, A., 1998, Transforming growth factor β1 is associated with angiogenesis, metastasis, and poor clinical outcome in prostate cancer. *Prostate* 37: 19-29.

Wolff, J.M., Fandel, T., Borchers, H., Brehmer, Jr., B., and Jakse, G., 1998, Transforming growth factor-β1 serum concentration in patients with prostatic cancer and benign prostatic hyperplasia. *Br J Urol* 81: 403-405.

Wolff, J.M., Fandel, T.H., Borchers, H., and Jakse, G., 1999, Serum concentrations of transforming growth factor-β 1 in patients with benign and malignant prostatic diseases. *Anticancer Res* 19: 2657-2659.

Xie, W., Mertens, J.C., Reiss, D.J., Rimm, D.L., Camp, R.L., Haffty, B.G., and Reiss, M., 2002, Alterations of Smad signaling in human breast carcinoma are associated with poor outcome: a tissue microarray study. *Cancer Res* 62: 497-505.

Xiong, B., Gong, L.L., Zhang, F., Hu, M.B., and Yuan, H.Y., 2002a, TGF β1 expression and angiogenesis in colorectal cancer tissue. *World J Gastroenterol* 8: 496-498.

Xiong, B., Yuan, H.Y., Hu, M.B., Zhang, F., Wei, Z.Z., Gong, L.L., and Yang, G.L., 2002b, Transforming growth factor-β1 in invasion and metastasis in colorectal cancer. *World J Gastroenterol* 8: 674-678.

Yingling, J.M., Blanchard, K.L., and Sawyer, J.S., 2004, Development of TGF-β signalling inhibitors for cancer therapy. *Nat Rev Drug Discov* 3: 1011-1022.

Zeng, L., Rowland, R.G., Lele, S.M., and Kyprianou, N., 2004, Apoptosis incidence and protein expression of p53, TGF-β receptor II, p27Kip1, and Smad4 in benign, premalignant, and malignant human prostate. *Hum Pathol* 35: 290-297.

Ziv, E., Cauley, J., Morin, P.A., Saiz, R., and Browner, W.S., 2001, Association between the T29--> C polymorphism in the transforming growth factor β1 gene and breast cancer among elderly white women: The Study of Osteoporotic Fractures. *Jama* 285: 2859-2863.

CHAPTER 22

TGF-β RECEPTOR KINASE INHIBITORS FOR TREATMENT OF FIBROSIS

NICHOLAS J. LAPING[1] AND STÉPHANE HUET[2]

[1] *GlaxoSmithKline Pharmaceuticals, UW2521, 709 Swedeland Road, PO Box 1539, King of Prussia, PA 19406-0939, USA*

[2] *GlaxoSmithKline Research Center, 25 avenue du Québec, 91951 Les Ulis, France*

Abstract: Because TGF-β is central to the progression of fibrosis, selective inhibition of this signaling pathway could provide a novel treatment for many fibrotic diseases. Small molecule inhibitors of the kinase activity of the TGF-β type I receptor (ALK5) have been developed by several companies and institutions. These inhibitors prevent the phosphorylation of the Smad proteins and many if not most of the sequelae following TGF-β release and activation. Thus, it has been demonstrated that these inhibitors prevent the extra-cellular matrix accumulation and organ destruction associated with fibrosis. The effectiveness of these inhibitors in several animal models suggests broad application in many fibrotic diseases. However, due to the pleiotropic activity of TGF-β there may be unwanted side effects of TGF-β type I receptor kinase inhibition, especially when considering chronic therapy

Keywords: ALK5; collagen; fibrosis; kinase inhibitor

1. INTRODUCTION

Recent advances in TGF-β inhibitors have identified novel small molecule inhibitors of ALK5, the TGF-β type I receptor kinase. The role of TGF-β as a potent profibrotic cytokine has been demonstrated in a number of animal models (Border and Noble, 1994; Kopp et al., 1996) and over-expression of TGF-β in transgenic mice results in liver and kidney fibrosis (Sanderson et al., 1995). TGF-β being central to the progression of fibrosis in human, selective inhibition of the TGF-β signaling pathway could provide a novel treatment for many fibrotic diseases. Using different approaches such as molecular modelling, cellular or enzyme based assays, several

P. ten Dijke and C.-H. Heldin (eds.), Smad Signal Transduction, 443–459.

pharmaceutical companies have identified ALK5 inhibitor compounds showing structural similarities and comparable *in vitro* potency. A limited number of ALK5 inhibitors have been evaluated *in vivo*, in models of fibrosis. While it is beyond the scope of this review to fully detail the role of TGF-β in fibrosis, this chapter will examine the basic rationale for TGF-β inhibition in fibrotic chronic organ remodelling diseases and describe the impact of ALK5 inhibition in different models of lung, liver and kidney fibrosis. Because the ALK5 kinase shares structural homologies with other TGF-β type I receptors, the pros and cons of ALK5 selectivity of inhibitors versus other type I receptor kinases will be discussed in prospective of their use to treat human chronic fibrotic diseases.

2. RATIONALE FOR TGF-β AS A KEY PLAYER IN FIBROSIS

2.1 Evidence that Ligand, Receptor, and Smads are Expressed in Fibrotic Tissues

It is well documented that TGF-β, its receptors, and the Smad signaling proteins are present in most tissues and cell types and are regarded to be central to the progression of fibrosis. Before a causal involvement of the TGF-β signaling pathway in the development of fibrosis had been established, it became clear that elevated levels of TGF-β were associated with pathologies in animal models that manifest aspects of fibrosis. For example, TGF-β1 levels are elevated in a rat model of liver fibrosis induced by CCl_4 as well as renal fibrosis induced by puromycin amino nucleoside or anti-bodies for glomerular basement membrane (Sun et al., 1990; Coimbra et al., 1991; Jones et al., 1991).

In human, elevated TGF-β1 levels can be measured in the serum and urine of patients with hepatitis (Bayer et al., 1998) diabetic nephropathy as well as other fibrotic diseases (Border and Noble, 1994; Broekelmann et al., 1991; Shah et al., 1999). In liver biopsies from patients with chronic liver diseases of various etiologies, TGF-β expression, receptors and Smad proteins are increased and correlate with inflammation and fibrosis (Paradis et al., 1996; Kanzler et al., 2001; Calabrese et al., 2003). A genetic polymorphism in codon 10 and 25 of the TGF-β1 gene has also been linked with faster fibrosis progression in patients with hepatitis C virus (HCV)-induced liver disease (Gewaltig et al., 2002). Similarly, renal fibrosis associated with diabetic nephropathy appears to be linked to the polymorphism found in codon 10 of the TGF-β1 gene (Patel et al., 2005). These polymorphisms are associated with higher levels of TGF-β1 which is consistent with a causal role of TGF-β in fibrosis. The genetic evidence and expression data describe a strong correlation between TGF-β1 levels and fibrosis. While it has yet to be shown that inhibiting TGF-β signaling will stop the progression of fibrosis in patients, the evidence in animal models is mounting that inhibiting TGF-β is an excellent approach to treat fibrotic disease.

2.2 Manipulation of the Signaling Pathway Influences Fibrosis Development

There are many examples in the literature of anti-fibrotic effects observed after manipulation of the TGF-β signaling pathway. In models of kidney or lung fibrosis, reducing TGF-β levels with ACE inhibitors or neutralizing TGF-β activity with blocking antibodies, soluble TGF-β type II receptors or natural TGF-β binding proteins such as decorin or the TGF-β latency-associated peptide have been associated with beneficial anti-fibrotic effects (Böttinger et al., 1996; Peters et al., 1998; Wang et al., 1999; Ziyadeh et al., 2000; Kolb et al., 2001). Intravenous injection of an adenovirus expressing a dominant-negative TGF-β type II receptor showed anti-fibrotic effects in a rat model of dimethylnitrosamine (DMN)-induced liver fibrosis thereby validating this model as TGF-β-dependent (Nakamura et al., 2000). In a model of bile duct ligation-induced liver fibrosis, adenoviral expression of TGF-β antisense, was proved effective in preventing liver fibrosis (Arias et al., 2003).

Blocking of downstream ALK5 signaling by over-expressing Smad7, a Smad that specifically inhibits Smad2 and Smad3 phosphorylation, has been shown to prevent kidney tubulo-interstitial fibrosis (see Chapter 19; Lan et al., 2003). In a liver model, mice with altered TGF-β signaling due to Smad3 heterozygous or homozygous knock-out show a reduced increase in liver matrix gene expression following acute administration of CCl_4, a liver damaging agent (Schnabl et al., 2001).

2.3 Inhibition of the TGF-β Pathway: A Newly Explored Therapeutic Approach for Fibrotic Diseases

As described in 1.1 and 1.2 there are now evidences that the TGF-β pathway is upregulated in chronic fibrotic diseases in human and observations made in different models of fibrosis strongly support the idea that pharmacological interventions aimed at blocking TGF-β downstream signaling may represent novel therapeutic opportunities. In the late nineties, research programmes looking specifically for TGF-β receptor kinase inhibitors were initiated by several pharmaceutical companies leading to the identification of ALK5 inhibitors which activities in models of liver, lung and kidney fibrosis are described in the section below.

3. KINASE CHARACTERISTICS OF THE TYPE I RECEPTOR AND SMALL MOLECULE INHIBITORS

3.1 Screening Strategies to Inhibit the Kinase that Phosphorylates Smad2/3

With the discovery that the type I receptors phosphorylate the signaling Smads, an opportunity was created to develop small molecule inhibitors of the TGF-β signaling pathway. While TGF-β can signal through ALK5 and ALK1, it was clear that with respect to fibrosis, ALK5 and its primary substrates, Smad2 and Smad3, are the main proteins through which TGF-β influences the development of fibrotic disease. Therefore, as seen by the development strategies of several pharmaceutical companies and institutions, the kinase activity of ALK5 was targeted to identify small molecules that inhibit this enzymatic activity (*e.g.* Callahan et al., 2002; Yakymovych et al., 2002; Sawyer et al., 2003).

Different approaches which converged to the identification of structurally related ALK5 inhibitors have been described in the patent and scientific literature; these include the use of TGF-β responsive cellular assays, kinase activity or binding assays to screen compound libraries, as well as molecular modeling approaches starting from p38 mitogen activated protein kinase (MAPK) inhibitors.

It must be noted that there are other kinases that phosphorylate Smads and thereby modulate the TGF-β signaling pathway, which include members of the Erk MAPK pathway and Rho kinase (see Chapter 12). Furthermore, there is a link with the p38 MAPK pathway (see Chapter 16). However, to specifically and robustly prevent TGF-β signaling in fibrosis, the most direct target is ALK5 phosphorylation of Smad3 (Flanders, 2004).

3.1.1 Cell based assays

Activity of ALK5 inhibitors can be evaluated in a number of TGF-β responsive cellular assays. One well described biological effect of TGF-β is its ability to induce transcription of the *plasminogen activator inhibitor-1 (PAI-1)* promoter in mesenchymal and hepatocyte cells in particular, leading to an increase in PAI-1 protein secretion (Lund et al., 1987; Westerhausen et al., 1991). Alternatively, the p3TP-Lux, an artificial promoter construct containing regions of both the collagenase and *PAI-1* promoters and showing maximal responsiveness to TGF-β (Wrana et al., 1992), can be used to study anti-TGF-β activity of ALK5 inhibitors. Addition of ng concentration of TGF-β1 to cells stably or transiently transfected with a construct containing the wild-type or Smad binding regions of the *PAI-1* promoter upstream of a luciferase reporter gene induces a 10- to 1000-fold increase in luciferase activity (Dennler et al., 1998). This approach has been used by GlaxoSmithKline to screen a compound library leading to the identification of an aminothiazole derivative as a potent inhibitor ($IC_{50} = 429\,nM$) of TGF-β-induced *PAI-1* promoter transcription in the hepatocyte cell line HepG2 (Gellibert et al., 2004). A medicinal chemistry programme then led to the identification of an orally active ALK5 inhibitor from a different chemical series, GW6604 (2-phenyl-4-(3-pyridin-2-yl-1*H*-pyrazol-4-yl)pyridine), which inhibits ALK5 kinase activity in an autophosphorylation assay ($IC_{50} = 140\,nM$). A similar affinity (107 nM) of GW6604 to purified recombinant ALK5 was also measured in a binding assay. In cellular assays, the inhibitory potency (IC50 value) of GW6604 with respect to TGF-β, Activin and BMP signaling was $0.5\,\mu M$, $2.5\,\mu M$ and $> 10\,\mu M$ (IC50 values) respectively. This pyrazole derivative has been described in different models of liver fibrosis (DeGouville et al., 2005, and section 3.1 in this chapter).

Cellular assays based on the reversion of TGF-β induced cell cycle arrest in Mv1Lu cells, or TGF-β proliferation of NIH3T3 cells, have been used by Lilly scientists to demonstrate *in vitro* anti-TGF-β activities of substituted pyrazole series (Sawyer et al., 2003 and 2004; Peng et al., 2005). In the first description of the *in vitro* characterization of an ALK5 inhibitor, the authors showed that TGF-β-induced nuclear translocation of Smad3, and expression of collagen or fibronectin mRNA in the renal epithelial carcinoma cell line A498 was blocked by SB-431542 (Laping et al., 2002). Effects of ALK5 inhibitors on intracellular phosphorylation of Smad2 was used to demonstrate cellular activity of Lilly compounds (Peng et al., 2005).

In conclusion, several end points based on TGF-β-induced intracellular Smad translocation and phosphorylation, gene expression, matrix production, cell cycle regulation are available to evaluate the potency of TGF-β inhibitors and have facilitated the discovery of ALK5 kinase inhibitors.

3.1.2 Isolated kinase assay

Another method to identify small molecule inhibitor of ALK5 is to utilize purified kinase domain of ALK5 and provide either Smad2 or Smad3 as a substrate. It had been shown that the soluble cytoplasmic domain of the type I receptor could be expressed and purified from baculovirus–infected insect cells while the glutathione-S-transferase (GST) — Smad fusion proteins could be purified from transfected *E coli* (Zhang et al., 1996). Furthermore, it was shown that this purified type I receptor could phosphorylate the purified Smad protein.

Thus, an approach to identify small molecule inhibitors of ALK5 kinase activity is to test compounds for their ability to inhibit ALK5 kinase domain phosphorylation of GST-Smad3 (Callahan et al., 2002). The kinase domain of ALK5 without the GS region was cloned and expressed as a GST fusion protein. By expressing the protein without the GS domain, which has been shown to regulate the kinase activity, a weak but constitutively active kinase is obtained that is able to phosphorylate GST-Smad3 *in vitro* (Laping et al., 2002). The screening results identified many inhibitors of p38 MAPK, indicating that the ATP binding site of ALK5 and p38 MAPK bind similar pharmacophores. This was further illustrated by a report that SB-203580 inhibited the autophosphorylation of ALK5 (Eyers et al., 1998). Based on analysis of structure activity relationships, SB-431542, (4-(5-Benzol[1,3]dioxol-5-yl-4-pyridin-2-yl-1H-imidazol-2-yl)-benzamide), was designed to be a relatively selective ALK5 inhibitor with little activity against p38 MAPK (compound 14 of Callahan et al., 2002). However, it retained some activity against ALK4. SB-431542 was identified as a potent ALK5 inhibitor with an IC_{50} value of 90-100 nM (Laping et al., 2002; Inman et al., 2002).

Further development of that series resulted in more potent inhibitors such as SB-502124 (DaCosta-Byfield et al., 2004) and SB-525334, 6-[2-tert-butyl-5-(6-methyl-pyridin-2-yl)-1H-imidazol-4-yl]-quinoxaline (Grygielko et al., 2005). SB-525334 inhibited ALK5 phosphorylation of Smad3 with an IC_{50} value of 14.3 nM. ALK4 phosphorylation of Smad3 was inhibited by SB-525334 with an IC_{50} value of 58.5 nM and ALK2 phosphorylation of Smad1 exhibited an IC_{50} value greater than 10 μM. Furthermore, the phosphorylation of activating transcription factor-2 (ATF-2) by p38α MAPK was inhibited by SB-525334 with an IC_{50} value of 1.5 μM demonstrating that the inhibitor is more than 200-fold more selective for ALK5 than p38α MAPK.

There is evidence at least *in vitro* that the type I receptors can autophosphorylate (Bassing et al., 1994; Wieser et al., 1995). Thus another screening method was to look for small molecules that prevent the autophosphorylation of ALK5 using isolated kinase domain and the T204D mutation (Sawyer et al., 2003; 2004). This strategy led to the identification of pyrazole inhibitors as discussed in the previous chapter. Currently there are no publications using these compounds in fibrosis models.

3.2 Crystallography

The crystal structure of the kinase domain of ALK5 has been solved with both ATP and small molecule inhibitors (Huse et al., 1999; Sawyer et al., 2003; Gellibert et al., 2004). The crystal structures reveal that the 2-pyridyl group, a selectivity requirement for ALK5 inhibitors, fits into a unique pocket of ALK5 that is surrounded by residues Lys232 and Ser280. However, key interactions with the pyridyl nitrogen are made with Tyr249, Glu345 and Asp351 (Gellibert et al., 2004). This is consistent with the structure activity findings that this nitrogen mediates the selectivity of all potent ALK5 inhibitors described to date over p38 MAP kinase. This nitrogen is a feature of GW6604, SB-431542, SB-505124 and SB-525334. GW6604 and SB-525334 are shown in Fig. 1.

3.3 Pharmacokinetics of SB-525334 and GW6604

Although GW6604 has low (13%) oral bioavailability in the rat and a half-life of around 25 minutes, likely due to liver clearance, it was still very effective in a model of liver fibrosis (DeGouville et al., 2005, and section 3.1.1 for *in vitro* properties). SB-525334 is also orally bioavailable (87%) and has a plasma half-life

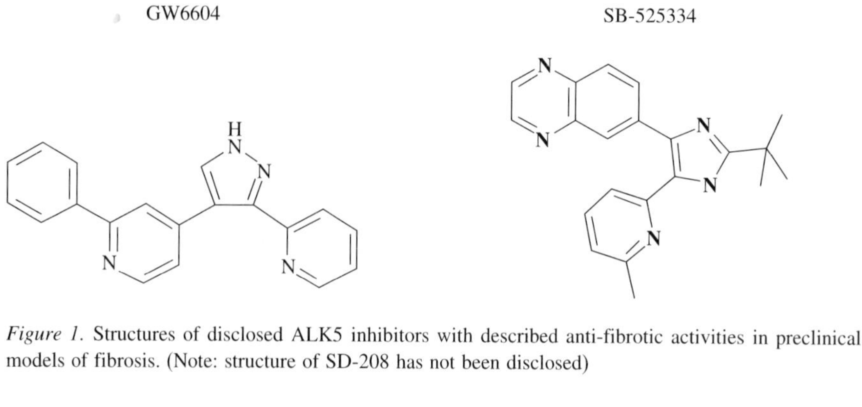

Figure 1. Structures of disclosed ALK5 inhibitors with described anti-fibrotic activities in preclinical models of fibrosis. (Note: structure of SD-208 has not been disclosed)

in the rat of 115 minutes. After a single oral dose of 10 mg/kg the plasma levels of SB-525334 drop from the peak of 4 μM at one hour after dosing to 30 nM at 24 hours after the oral dose in the rat (Grygielko et al., 2005). Therefore, ALK5 kinase activity is expected to be significantly inhibited in most tissues of rats for 24 hours after single oral administration of SB-525334.

4. EFFICACY OF ALK5 INHIBITORS IN ANIMAL MODELS OF FIBROSIS

4.1 Liver Fibrosis

Liver fibrosis models can be induced in rodents by chronic administration of liver damaging agents like dimethylnitrosamine, CCl_4, acetaminophen, or bile duct ligation causing cholestasis (Wu and Norton, 1996). Although these models do not mimic human liver disease such as non-alcoholic steatohepatitis, alcohol abuse, or chronic hepatitis B or C virus infection, which are the major causes of liver cirrhosis, they share some common features of human liver disease. Indeed, following liver damage, hepatic stellate cells (HSC) differentiate into myofibroblast-like cells, which respond to TGF-β signal triggering their proliferation and matrix gene expression and collagen deposit (Li and Fridman, 1999). Thus, measurement of matrix gene expression in these models can be used to evaluate anti-TGF-β activity of ALK5 inhibitors. Recently, DeGouville et al. (2004) described the use of an acute model of DMN-induced liver damage to demonstrate anti-fibrotic activities of an ALK5 inhibitor. DMN was given for three consecutive days to rats, causing an increase in liver collagen IA1 mRNA expression by about 10-fold at day 8. The ALK5 inhibitor, GW6604 was given on days 6, 7 and 8

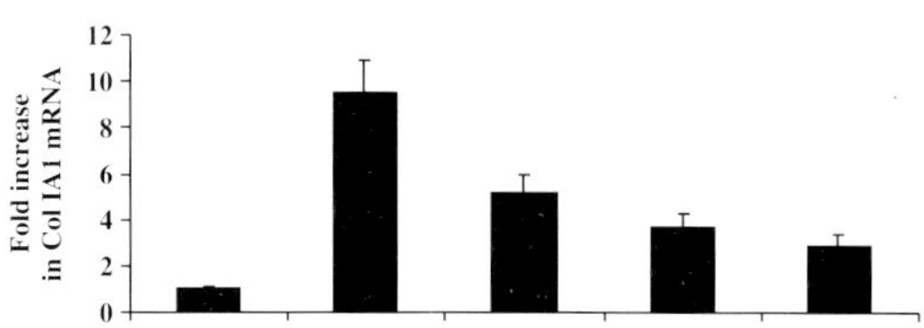

DMN	–	+	+	+	+
GW6604 (mg/kg)	–	–	25	50	80

Figure 2. Effect of GW6604 on liver collagen Iα1 mRNA expression in the acute DMN model. Animals received a daily i.p. injection of DMN (12.5 mg.kg-1) for three consecutive days and treatment with GW6604 or vehicle given orally twice a day was initiated 72 hours after the last injection of DMN. Animals were treated with GW6604 or vehicle for 2 days and collagen Iα1 mRNA levels were measured by quantitative RT-PCR and normalised to ribosomal 18S RNA levels as described in Methods. Difference in collagen Iα1 mRNA between DMN-GW6604 and DMN-vehicle was statistically significant ($P < 0.05$ for the 25 mg.kg-1 dose and $P < 0.01$ for higher doses)

to DMN-pretreated rats causing a dose-dependent inhibition of collagen Iα1 over-expression (Fig. 2).

In a chronic model of DMN-induced fibrosis where DMN was administered for six weeks and GW6604 dosed for the last 3 weeks (80 mg/kg per os, given twice a day), mortality was prevented and DMN-induced elevations of mRNA encoding for collagen Iα1, Iα2, III, TIMP-1 and TGF-β were reduced by 50–75%. Inhibition of the matrix gene expression was accompanied by reduced matrix protein deposition and reduction in liver function deterioration, as assessed by bilirubin and liver enzyme levels (Table 1 and Fig. 3). Immunohistological analysis of liver sections showed a 10-fold increase in PCNA staining in animals given DMN and the ALK5 inhibitor, whereas no increase in PCNA stained cells was observed in the control group treated with GW6604 only, suggesting that, following liver injury in rodents, ALK5 signaling may influence and control parenchymal cell proliferation. Although this was not explored in this study, the beneficial effect of ALK5 inhibition on liver function and fibrosis progression could be, in part, linked to decreased parenchymal cell apoptosis as suggested by the observation that TGF-β has a pro-apoptotic effect on hepatocytes (Oberhammer et al., 1992).

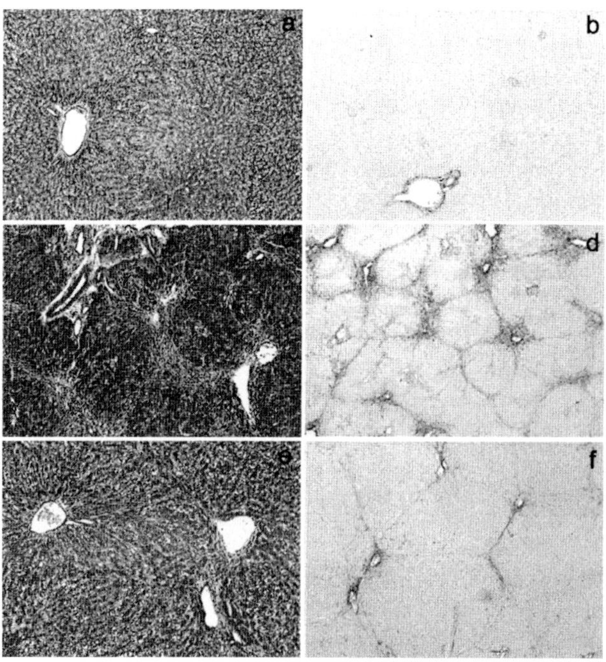

Figure 3. Effect of GW6604 on liver histology in a chronic model of DMN-induced fibrosis. Hematoxilin/eosin staining (a, c, e) and sirius red staining (b, d, f) of liver sections from control (a, b), DMN animals treated with vehicle (c, d), DMN animals treated with GW6604 (e, f). Animals received DMN for 6 consecutive weeks and were given vehicle or GW6604 (80 mg/kg, given twice a day) or the last 3 weeks of DMN administration (A color version of this figure is freely accessible via the website of the book: http://www.springer.com/1-4020-4542-5)

Table 1. Effect of GW6604 in the chronic DMN-induced model of liver fibrosis. mRNA encoding for coll Iα1, coll Iα2, coll III, TIMP-1 and TGF-β are quantified relative to 18S which is not modified by any of the treatments. Statistically (t-test) significant difference between DMN + vehicle and Saline + vehicle groups ($p < 0.01$ for *); Statistically significant difference between DMN + GW6604 and DMN + vehicle groups ($p < 0.01$ for #) and ($p < 0.001$ for ##)

	Saline + Vehicle	Saline + GW6604	DMN + Vehicle	DMN + GW6604
Mortality/number of rat at day1	0/7	0/7	7/16	0/11
Liver weight (g)	15.1 ± 0.7	18.9 ± 1.8	9 ± 1.4*	17.3 ± 0.5#
18S mRNA	1.0 ± 0.01	1.01 ± 0.003	1.03 ± 0.003	1.01 ± 0.006
Coll IA1 mRNA	1 ± 0.3	0.5 ± 0.08	10.9 ± 2.03*	3.7 ± 1.04#
Coll IA2 mRNA	1 ± 0.25	0.5 ± 0.09	6.4 ± 0.82*	2.7 ± 0.72#
Coll III mRNA	1 ± 0.13	0.7 ± 0.09	3.0 ± 0.6*	2.0 ± 0.45
TIMP-1 mRNA	1 ± 0.34	0.7 ± 0.16	14.0 ± 2.72*	4.2 ± 1.08#
TGF-β_1 mRNA	1 ± 0.16	1.0 ± 0.35	6.8 ± 0.81*	2.7 ± 0.37#
ALAT(IU/L)	43 ± 2	47 ± 4	257 ± 36*	82 ± 8#
Bilirubin(mg/L)	1 ± 0.1	2 ± 0.1	16 ± 4*	7 ± 1#
Collagen area (pixel/field)	1571 ± 72	1568 ± 80	34809 ± 2978**	11229 ± 1072##

4.2 Renal Fibrosis

TGF-β is a central factor in the etiology of several aspects of renal disease. As such, TGF-β activation causes the accumulation of extra-cellular matrix components leading to renal fibrosis (Ketteler et al., 1994; Sharma and Ziyadeh, 1994). This was most directly demonstrated in transgenic mice and transfected rats made to over-express TGF-β (Isaka et al., 1993; Kopp et al., 1996). Therefore, pharmacological intervention in models of renal injury and disease should result in reduction of extra-cellular matrix markers and fibrosis.

4.2.1 Minimal change nephropathy

Puromycin aminonucleoside (PAN) induces glomerulosclerosis and tubulointerstitial inflammation that is similar to human minimal change nephropathy. PAN-induced renal injury exhibits a multifactorial etiology with early podocyte depletion leading to marked proteinuria that is accompanied by interstitial influx of mononuclear cells (Eddy and Michael, 1988). There is a decrease in proteinuria and an increase in extracellular matrix (ECM) mRNA in the recovery phase of PAN-induced disease, which is followed by the formation of glomerulosclerosis. It has been shown that TGF-β1 mRNA is one of the markers that increase in the kidneys of PAN-treated rats (Jones et al., 1991). If TGF-β plays a causal role in the proteinuria and ECM elevation that is seen in this model, then inhibiting this pathway should show a reduction in these markers. The ALK5 inhibitor, SB-525334, reduced dose-dependently the PAN-induced proteinuria. At 10 mg/kg there was a significant decrease in proteinuria compared with the PAN only group, although the levels were still above untreated control rats (Grygielko et al., 2005).

Similarly, the PAN-induced procollagen α1(I) mRNA in the kidney was elevated 4-fold compared to controls and SB-525334 administration dose-dependently decreased procollagen α1(I) mRNA. A similar dose-dependent decrease was seen with procollagen III mRNA, and *PAI-1* mRNA. These results illustrate that TGF-β and specifically ALK5 activation not only are causative in extra-cellular matrix production in this model, but also that ALK5 phosphorylation of Smad proteins plays a role in progression of proteinuria and may in fact regulate podocyte function.

4.2.2 Diabetic nephropathy

There is much evidence that TGF-β plays a significant role in the disease process of diabetic nephropathy (Chen et al., 2003). Also, intervention with TGF-β1 neutralizing antibodies has shown reduction of several of key disease markers in a mouse model of diabetic nephropathy (Ziyadeh et al., 2000). Therefore, inhibition of ALK5 should also reduce at the very least extra-cellular matrix accumulation in the kidney in models of diabetic nephropathy. As shown in Fig. 4, SB-525334 significantly reduced procollagen α1(I), mRNA in the kidney of diabetic Zucker rats. This was additive with angiotensin converting enzyme inhibitor suggesting that there would be an additional benefit of treating diabetic nephropathy patients that are already on an ACE-inhibitor also with an ALK5 inhibitor (E. Grygielko, P. Thronton, and N.J.L., unpublished results).

4.3 Lung Fibrosis

Inhibition of ALK5 phosphorylation of Smad2 and Smad3 promises also to be effective for lung fibrosis. As a model of idiopathic lung fibrosis, fibrosis of the lung can be induced by instillation of adenovirus expressing active TGF-β1. The high expression of TGF-β1 in the lungs for 7-10 days then results in the manifestation of lung fibrosis as measured by hydroxyproline concentrations and histological evaluation. The ALK5 kinase inhibitor SD-208 made by Scios Inc. was given for either 7 days at the same time of the TGF-β expressing virus or from 7 to 11 days after the viral dosing at 50 and 100 mg/kg/day (Bonniaud et al., 2005).

While the compound structure has not been revealed, it is somewhat selective for ALK5 but has sub-micromolar activities against p38α MAPK and epidermal growth factor receptor (Bonniaud et al., 2005). The selectivity against ALK4 and ALK7 has not been reported. This compound was able to reduce the TGF-β1-adenovirus-induced increases in collagen and other extra-cellular matrix markers both when given before instillation of the adenovirus or 7 days thereafter (Bonniaud et al., 2005). If this dose was selective for ALK5 then this suggests that ALK5 kinase activity mediates the extra-cellular matrix changes induced by TGF-β including the histological manifestation of lung fibrosis.

Similar therapeutic effects were seen with the ALK5 inhibitor SB-525334 in a different model of lung fibrosis (Underwood et al., 2003). Lung fibrosis can

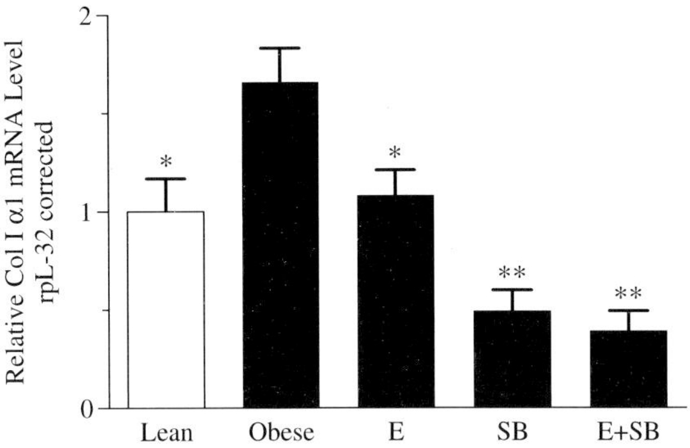

Figure 4. Effect of SB-525334 on renal collagen I α1 mRNA of diabetic Zucker rats. Six month old obese Zucker rats were treated for 5 weeks with Enalapril at 50 mg/L in the drinking water or 200 mg/L SB-525334 in the drinking water or in combination. Collagen Iα1 mRNA was measured from total RNA isolated from whole kidney and compared with ribosomal protein L32 mRNA by real time quantitative PCR. E- ACE inhibitor Enalapril, SB-ALK5 inhibitor. $^{*}p < 0.05$, $^{**}p < 0.01$ (E. Grygielko, P. Thornton, and N.J.L., unpublished results)

be induced transiently in Lewis rats by lung instillation of bleomycin. In this model, lung weight increases maximally by 7 days after bleomycin and then begins to decline while the collagen and structural manifestations of fibrosis continue to develop up to 14–28 days after bleomycin instillation. SB-525334 treatment prevented the bleomycin-induced collagen production when given at the initiation of lung injury, (Fig. 5).

Together these results demonstrate that ALK5 inhibition is efficacious in reducing lung fibrosis and maintaining the functional architecture. While this treatment will certainly stop fibrosis from progressing, it is not clear if the fibrosis can be reversed in patients and some degree of organ function reclaimed.

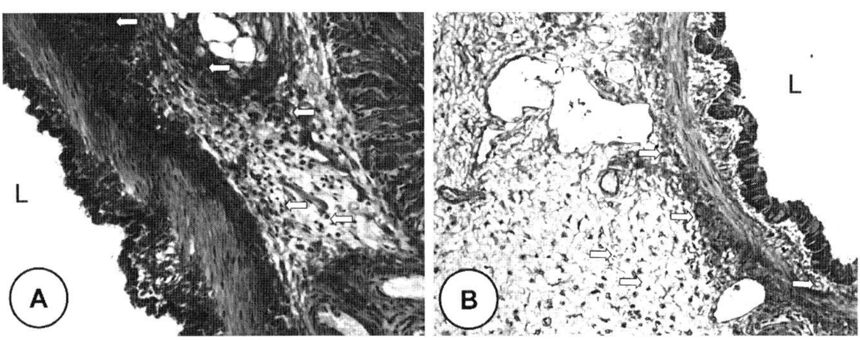

Figure 5. SB-525334 reduces collagen in the lungs of bleomycin treated rats. A, Lung tissue 14 days after bleomycin. B, Lung tissue from bleomycin instilled rats also given SB-525334 at 10 mg/kg/day over the same time period (Underwood et al., 2003) (A color version of this figure is freely accessible via the website of the book: http://www.springer.com/1-4020-4542-5)

5. PERSPECTIVES AND ISSUES FOR LONG TERM THERAPY

Chronic diseases leading to lung, kidney or liver fibrosis can take decades to progress toward organ failure. Halting or reversing fibrosis progression with an ALK5 inhibitor to ultimately preserve organ function will presumably require chronic treatment, thus increasing the risk of unwanted side effects due to a pleiotropic dysregulation of TGF-β signaling.

5.1 Immune Response Dysregulation in HCV-induced Fibrosis

In chronic liver disease caused by HCV infection, the antigen-specific activated CD4(+) T-cell and cytotoxic responses to the HCV antigens are not as strong as the response observed in acutely infected patients who display a more vigorous T-cell response (Cerny and Chisari, 1999; Ulsenheimer et al., 2003; Neumann-Haefelin et al., 2005). Thus, viral persistence during HCV infection may be linked to a weak antiviral immune response to the viral antigens. TGF-β being involved in the mechanism of T cell mediated immunosuppression (von Boehmer, 2005, for review) it is tempting to speculate that an ALK5 inhibitor might, in addition to reducing fibrosis progression, also restore immune response to HCV infected cells.

However, due to the pleiotropic effects of TGF-β on different cell types involved in the progression of liver disease, it is currently difficult to predict the overall impact of *in vivo* inhibition of the TGF-β signaling pathway in patients chronically infected with HCV. Indeed, TGF-β inhibition may influence hepatocyte proliferation and apoptosis, hepatic stellate cell proliferation and extra-cellular matrix production, in addition to T lymphocyte response to virus infected cells. There is also evidence that HCV proteins may directly inhibit TGF-β signaling by a physical interaction between the MH1 region of the Smad3 and the core and NS3 proteins of HCV (Cheng et al., 2004). Whether, in this setting, ALK5 inhibition will cause a positive boost in T cell response to HCV antigens leading to clearance of infected cells, or alternatively, exacerbate liver inflammation to the extend that this will be detrimental to liver function is currently unknown. To address this point, a possibility would be to evaluate ALK5 inhibitors in HCV infected primates which can be used to study the extrahepatic and intrahepatic T cell responses to HCV (Thimme et al., 2002). In conclusion, the progression of ALK5 inhibitors for the treatment of HCV disease will require careful examination of a potential risk associated with a dysregulation of intrahepatic T cell response.

5.2 Tumor Suppression

While the previous chapter describes the utility of inhibiting TGF-β signaling as a therapy to treat cancer and tumor growth, we must bear in mind that for some cell and tumor types, TGF-β plays a tumor suppressive role. Thus, in a chronic treatment paradigm, as would be needed for the treatment of fibrosis, prolonged TGF-β inhibition might confer an increased risk in neoplasms especially of epithelial origin. This poses a serious hurdle for using ALK5 kinase inhibitors in chronic therapy because a kinase inhibitor can completely prevent phosphorylation of a substrate, while other modes of inhibition (such as antibodies, soluble receptors, binding proteins) would allow the basal (unstimulated) kinase activity of the receptor to maintain a low tone of Smad phosphorylation. We speculate that this low tone could maintain a minimum physiological requirement rendering interventions at the ligand with fewer side effects than complete inhibition of the kinase activity. Therefore, careful dose selection or intermittent dosing may be required for such kinase inhibitors.

5.3 Cardiovascular Effects of TGF-β Inhibition

TGF-β has a number of important roles in angiogenesis and cardiovascular development. On the one hand this portends well for treating tumors (see Chapter 21), but might also pose a liability for chronic therapy. During fetal cardiovascular development, TGF-β plays an important role including the epithelial-to-mesenchymal transition and valve development (Potts et al., 1992; Brown et al., 1996).

TGF-β's effects on vascular endothelial cells is complex and is affected by the ratios of its receptors ALK1 and ALK5 (Lebrin et al., 2005). Because ALK5 kinase activity is required for the biological activity of ALK1 (Goumans et al., 2003), the ALK5 inhibitors will probably adversely affect endothelial function, whether the compounds are selective for ALK5 over ALK1 or not. Because many fibrotic disorders are likely to be treated chronically for several years, cardiovascular safety of ALK5 inhibitors must be monitored carefully.

5.4 Consequences of ALK4 and ALK7 Inhibition

Based on the receptor sequence homology it is perhaps not surprising that the few ALK5 inhibitors described in detail so far, have significant activity against the related type I receptors ALK4 and ALK7 in potency that mirror the respective degree of sequence similarity (Inman et al., 2002; DaCosta Byfield et al., 2004). As far what consequences this activity poses in an adult treated for fibrosis or cancer, the likely interference is with Activin signaling (Ten Dijke et al., 1993; Tsuchida et al., 2004). While at therapeutic doses there might be little effect on ALK4 or ALK7, at higher concentrations it is likely that in females there might be inhibition of follicular development. This may be borne out in toxicology dose range studies and would need to be monitored in the clinic. However, inhibition of follicle development in the aged population, which is the main target population for fibrotic diseases, is likely to be a minor drug development issue.

6. SUMMARY AND PERSPECTIVES

It is evident that TGF-β plays a major role in the progression of fibrosis, which has been described in many reviews. With the arrival of specific inhibitors of the TGF-β signaling pathway it is becoming clear that ALK5 phosphorylation of Smad3 and Smad2 mediates many of the actions of TGF-β. Therefore, it appears that ALK5 is well positioned as a target for intervention of TGF-β-mediated diseases and is supported by several studies using ALK5 inhibitors in models of fibrosis. We are reminded, however, that TGF-β is a pleiotropic factor with actions that also modulate cell growth, angiogenesis, and immune functions of both pro- and anti-inflammatory nature. It is therefore a challenge to develop ALK5 inhibitors as therapeutics not only because kinase selectivity is needed, but also because selective actions against the disease relevant fibrotic actions versus the immune-modulatory and cardiovascular roles of TGF-β must be found. It is still an open question if potent kinase inhibitors can serve this purpose or if subtle modulation of the TGF-β pathway is needed by using mechanisms that either target Smad3 directly or prevent active TGF-β1 from acting on the TGF-β type I receptor ALK5.

REFERENCES

Arias, M., Sauer-Lehnen, S., Treptau, J., Janoschek, N., Theuerkauf, I., Buettner, R., Gressner, A.M., and Weiskirchen, R., 2003, Adenoviral expression of a transforming growth factor-β1 antisense mRNA is effective in preventing liver fibrosis in bile-duct ligated rats. *BMC* Gastro*enterol* 3: 29.

Bassing, C.H., Yingling, J.M., Howe, D.J., Wang, T., He, W.W., Gustafson, M.L., Shah, P., Donahoe, P.K., and Wang, X.F., 1994, A transforming growth factor β type I receptor that signals to activate gene expression. *Science* 263: 87-89.

Bayer, E.M., Herr, W., Kanzler, S., Waldmann, C.; Meyer Zum Buschenfelde, K.H., Diesne, H.P., and Lohse, A.W., 1998, Transforming growth factor β1 in autoimmune hepatitis: correlation of liver tissue expression and serum levels with disease activity. *J Hepatol* 28: 803-811.

Bonniaud, P., Margetts, P.J., Kolb, M., Schroeder, J.A., Kapoun, A.M., Damm, D., Murphy, A., Chakravarty, S., Dugar, S., Higgins, L., Protter, A.A., and Gauldie, J., 2005, Progressive transforming growth factor β1-induced lung fibrosis is blocked by an orally active ALK5 kinase inhibitor. *Am J Respir Crit Care Med* 171: 889-898.

Border, W.A., and Noble, N.A., 1994, Transforming growth factor β in tissue fibrosis. *N Eng J Med* 331: 1286-1292.

Böttinger, E.P., Factor, V.M., Tsang, M.L., Weatherbee, J.A., Kopp, J.B., Qian, S.W., Wakefield, L.M., Roberts, A.B., Thorgeirsson, S.S., and Sporn, M.B., 1996, The recombinant proregion of transforming growth factor β1 (latency-associated peptide) inhibits active transforming growth factor β1 in transgenic mice. *Proc Natl Acad Sci USA* 93: 5877-5882.

Broekelmann, T.J., Limper, A.H., Colby, T.V., and McDonald, J.A., 1991, Transforming growth factor β1 is present at sites of extra-cellular matrix gene expression in human pulmonary fibrosis. *Proc Natl Acad Sci U S A* 88: 6642-6646.

Brown, C.B., Boyer, A.S., Runyan, R.B., and Barnett, J.V., 1996, Antibodies to the type II TGFβ receptor block cell activation and migration during atrioventricular cushion transformation in the heart. *Developmental Biology* 174: 248-257.

Chen, S., Jim, B., and Ziyadeh, F.N., 2003, Diabetic nephropathy and transforming growth factor-β: transforming our view of glomerulosclerosis and fibrosis build-up. *Seminars in Nephrology* 23: 532-543.

Calabrese, F, Valente, M, Giacometti, C, Pettenazzo, E, Benvegnu, L, Alberti, A, Gatta, A, and Pontisso, P. (2003) Parenchymal transforming growth factor β-1: its type II receptor and Smad signaling pathway correlate with inflammation and fibrosis in chronic liver disease of viral etiology. *J Gastroenterol Hepatol* 18: 1302-8.

Callahan, J.F., Burgess, J.L., Fornwald, J.A., Gaster, L.M., Harling, J.D., Harrington, F.P., Heer, J., Kwon, C., Lehr, R., Mathur, A., Olson, B.A., Weinstock, J., and Laping, N.J., 2002, Identification of novel inhibitors of the transforming growth factor β1 (TGF-β1) type 1 receptor (ALK5). *J Med Chem* 45: 999-1001.

Cerny, A., and Chisari, F.V., 1999, Pathogenesis of chronic hepatitis C: immunological features of hepatic injury and viral persistence. *Hepatology* 30: 595-601.

Cheng, P.L., Chang, M.H., Chao, C.H., and Lee, Y.H., 2004, Hepatitis C viral proteins interact with Smad3 and differentially regulate TGF-β/Smad3-mediated transcriptional activation. *Oncogene* 23: 7821-7838.

Coimbra, T., Wiggins, R., Noh, J.W., Merritt, S., and Phan, S.H., 1991, Transforming growth factor-β production in anti-glomerular basement membrane disease in the rabbit. *Am J Pathol* 138: 223-234.

DaCosta Byfield, S., Major, C., Laping, N.J., and Roberts, A.B., 2004, SB-505124 is a selective inhibitor of transforming growth factor-β type I receptors ALK4, ALK5, and ALK7. *Mol Pharmacol* 65: 744-752.

Dennler, S., Itoh, S., Vivien, D., ten Dijke, P., Huet, S., and Gauthier, J.M.(1998) Direct binding of Smad3 and Smad4 to critical TGF-β inducible elements in the promoter of human plasminogen activator inhibitor-type1 gene. *EMBO J* 17 3091-3100.

De Gouville, A.C., Boullay, V., Krysa, G., Pilot, J., Brusq, J.M., Loriolle, F., Gauthier, J.M., Papworth, S.A., Laroze, A., Gellibert, F., and Huet, S., 2005, Inhibition of TGF-β signaling by an ALK5 inhibitor protects rats from dimethylnitrosamine-induced liver fibrosis. *Br J Pharmacol* 145: 166-177.

Eddy, A.A., and Michael, A.F., 1988, Acute tubulointerstitial nephritis associated with aminonucleoside nephrosis. *Kidney International* 33: 14-23.

Eyers, P.A., Craxton, M., Morrice, N., Cohen, P., and Goedert, M., 1998, Conversion of SB 203580-insensitive MAP kinase family members to drug-sensitive forms by a single amino-acid substitution. *Chem Biol* 5: 321-328.

Flanders, K.C., 2004, Smad3 as a mediator of the fibrotic response. *Int J Exp Pathol* 85: 47-64.

Gellibert, F., Woolven, J., Fouchet, M.H., Mathews, N., Goodland, H., Lovegrove, V., Laroze, A., Nguyen, V.L., Sautet, S., Wang, R., Janson, C., Smith, W., Krysa, G., Boullay, V., De Gouville, A.C., Huet, S., and Hartley, D., 2004, Identification of 1,5-naphthyridine derivatives as a novel series of potent and selective TGF-β type I receptor inhibitors. *J Med Chem* 47: 4494-4506.

Gewaltig, J., Mangasser-Stephan, K., Gartung, C., Biesterfeld, S., and Gressner, A.M., 2002, Association of polymorphisms of the transforming growth factor-β1 gene with the rate of progression of HCV-induced liver fibrosis. *Clin Chim Acta* 316: 83-94.

Grygielko, E.T., Martin, W.M., Tweed, C., Thornton, P., Harling, J., Brooks, D.P., and Laping, N.J., 2005, Inhibition of gene markers of fibrosis with a novel inhibitor of transforming growth factor-β type I receptor kinase in puromycin-induced nephritis. *J Pharmacol Exp Ther* 313: 943-951.

Goumans, M.J., Valdimarsdottir, G., Itoh, S., Lebrin, F., Larsson, J., Mummery, C., Karlsson, S., and ten Dijke, P., 2003, Activin receptor-like kinase (ALK)1 is an antagonistic mediator of lateral TGFβ/ALK5 signaling. *Mol Cell* 12: 817-828.

Huse, M., Chen, Y.G., Massagué, J., and Kuriyan, J., 1999, Crystal structure of the cytoplasmic domain of the type I TGF β receptor in complex with FKBP12. *Cell* 96: 425-436.

Inman, G.J., Nicolas, F.J., Callahan, J.F., Harling, J.D., Gaster, L.M., Reith, A.D., Laping, N.J., and Hill, C.S., 2002, SB-431542 is a potent and specific inhibitor of transforming growth factor-β superfamily type I activin receptor-like kinase (ALK) receptors ALK4, ALK5, and ALK7. *Mol Pharmacol* 62: 65-74.

Isaka, Y., Fujiwara, Y., Ueda, N., Kaneda, Y., Kamada, T., and Imai, E., 1993, Glomerulosclerosis induced by *in vivo* transfection of transforming growth factor-β or platelet-derived growth factor gene into the rat kidney. *J Clin Invest* 92: 2597-2601.

Jones, C.L., Buch, S., Post, M., McCulloch, L., Liu, E., and Eddy, A.A., 1991, Pathogenesis of interstitial fibrosis in chronic purine aminonucleoside nephrosis. *Kidney International* 40: 1020-1031.

Kanzler, S., Baumann, M., Schirmacher, P., Dries, V., Bayer, E., Gerken, G., Dienes, HP., and Lohse, A.W., 2001, Prediction of progressive liver fibrosis in hepatitis C infection by serum and tissue levels of transforming growth factor-β. *J Viral Hepat* 8: 430-437.

Ketteler, M., Noble, N.A., and Border, W.A., 1994, Increased expression of transforming growth factor-β in renal disease. *Curr Opin Nephrol Hypertens* 3: 446-452.

Kolb, M., Margetts, P.J., Galt, T., Sime, P.J., Xing, Z., Schmidt, M., and Gauldie, J., 2001, Transient gene expression of decorin in the lung reduces the fibrotic response to bleomycin. Am. *Am J Respir Crit Care Med* 163: 770-777.

Kopp, J.B., Factor, V., Mozes, M., Nagy, P., Sanderson, N., Böttinger, E.P., Klotman, P., and Thorgeirsson, S., 1996, Transgenic mice with increased plasma levels of TGF-β1 develop progressive renal disease. *Lab Invest* 74: 991-1003.

Lan, H.Y., Mu, W., Tomita, N., Huang, X.R., Li, J.H., Zhu, H.J., Morishita, R., and Johnson, R.J., 2003, Inhibition of renal fibrosis by gene transfer of inducible Smad7 using ultrasound-microbubble system in rat UUO model. Journal of the *J Am Soc Nephrol* 14: 1535-1548.

Laping, N., Grygielko, E., Mathur, A., Butter, S., Bomberger, J., Tweed, C., Martin, W., Fornwald, J., Lehr, R., Harling, J., Gaster, L., Callahan, J.F., and Olson, B.A.(2002) Inhibition of transforming growth factor (TGF)-β1-induced extra-cellular matrix with a novel inhibitor of the TGF-β type 1 receptor kinase activity: SB-431542. *Mol Pharmacol* 62: 58-64.

Lebrin, F., Deckers, M., Bertolino, P., and ten Dijke, P., 2005, TGF-β receptor function in the endothelium. *Cardiovasc Res* 65: 599-608

Li, D., and Friedman, S.L., 1999, Liver fibrogenesis and the role of hepatic stellate cells: New insights and prospects for therapy. Journal of Gastroenterology and *Hepatology* 14: 618-633.

Lund, L.R., Riccio, A., Andreasen, P.A., Nielsen, L.S., Kristensen, P., Laiho, M., Saksela, O., Blasi, F., and Dano, K., 1987 Transforming growth factor-β is a strong and fast acting positive regulator of the level of type-1 plasminogen activator inhibitor mRNA in WI-38 human lung fibroblasts. *EMBO J* 6: 1281-1286.

Nakamura, T., Sakata, R., Ueno, T., Sata, M., and Ueno, H., 2000, Inhibition of transforming growth factor β prevents progression of liver fibrosis and enhances hepatocyte regeneration in dimethylnitrosamine–treated rats. *Hepatology* 32: 247-255.

Neumann-Haefelin, C., Blum, H.E., Chisari, F.V., and Thimme, R., 2005, T cell response in hepatitis C virus infection. *J Clin Virol* 32: 75-85.

Oberhammer, F.A., Pavelka, M., Sharma, S., Tiefenbacher, R., Purchio, A.F., Bursch, W., and Schulte-Hermann, R., 1992, Induction of apoptosis in cultured hepatocytes and in regressing liver by transforming growth factor β1. *Proc Natl Acad Sci U S A* 89: 43-66.

Paradis, V., Mathurin, P., Laurent, A., Charlotte, F., Vidaud, M., Poynard, T., Hoang, C., Opolon, P., and Bedossa, P., 1996, Histological features predictive of liver fibrosis in chronic hepatitis C infection. *J Clin Pathol* 49: 998-1004.

Patel, A., Scott, W.R., Lympany, P.A., Rippin, J.D., Gill, G.V., Barnett, A.H., Bain, S.C., Warren, and 3/UK GoKind Study Group, 2005, The TGF-β1 gene codon 10 polymorphism contributes to the genetic predisposition to nephropathy in Type 1 diabetes. *Diabet Med* 22: 69-73.

Peters, H., Border, W.A., and Noble, N.A., 1998, Targeting TGF-β overexpression in renal disease: maximizing the antifibrotic action of angiotensin II blockade. *Kidney Int* 54: 1570-1580.

Peng, S.B., Yan, L., Xia, X., Watkins, S.A., Brooks, H.B., Beight, D., Herron, D.K., Jones, M.L., Lampe, J.W., McMillen, W.T., Mort, N., Sawyer, J.S., and Yingling, J.M., 2005, Kinetic characterization of novel pyrazole TGF-β receptor I kinase inhibitors and their blockade of the epithelial-mesenchymal transition. *Biochem* 44: 2293-2304.

Potts, J.D., Vincent, E.B., Runyan, R.B., and Weeks, D.L., 1992, Sense and antisense TGFβ 3 mRNA levels correlate with cardiac valve induction. *Dev Dyn* 193: 340-345

Sanderson, N., Factor, V., Nagy, P., Kopp, J., Kondaiah, P., Wakefield, L., Roberts, A., Sporn, M., and Thorgeirsson, S., 1995, Hepatic expression of mature transforming growth factor β1 in transgenic mice results in multiple tissue lesions. *Proc Natl Acad Sci U S A* 92: 2572-2576.

Sawyer, J.S., Anderson, B.D., Beight, D.W., Campbell, R.M., Jones, M.L., Herron, D.K., Lampe, J.W., McCowan, J.R., McMillen, W.T., Mort, N., Parsons, S., Smith, E.C.R., Vieth, M., Weir, L.C., Yan, L., Zhang, F., and Yingling, J.M., 2003, Synthesis and activity of new aryl- and heteroaryl-substituted pyrazole inhibitors of the transforming growth factor-β type I Receptor Kinase Domain. *J Med Chem* 46: 3953-3956.

Sawyer, J.S., Beight, D.W., Britt, K.S., Anderson, B.D., Campbell, R.M., Goodson, T. Jr, Herron, D.K., Li, H.Y., McMillen, W.T., Mort, N., Parsons, S., Smith, E.C., Wagner, J.R., Yan, L., Zhang, F., and Yingling, J.M., 2004, Synthesis and activity of new aryl- and heteroaryl-substituted 5,6-dihydro-4H-pyrrolo[1,2-b]pyrazole inhibitors of the transforming growth factor-β type I receptor kinase domain. *Bioorg Med Chem Lett* 14: 3581-3584.

Shah, M., Revis, D., Herrick, S., Baillie, R., Thorgeirson, S., Ferguson, M., and Roberts, A., 1999, Role of elevated plasma TGF-β1 levels in wound healing. *Am J Pathol* 154: 1115-1124.

Schnabl, B., Kweon, Y.O., Frederick, J.P., Wang, X.F., Rippe, R.A., and Brenner, D.A., 2001, The role of Smad3 in mediating mouse hepatic stellate cell activation. *Hepatology* 34: 89-100.

Sharma, K., and Ziyadeh, F.N., 1994, The emerging role of transforming growth factor-β in kidney diseases. *Am J Physiol* 266: F829-842.

Sun, M.A., Wang, B.E., Annoni, G., Degli Esposti, S., Biempica, L., and Zern, M.A., 1990, Two rat models of hepatic fibrosis. A morphologic and molecular comparison. *Lab Invest* 63: 467-475.

Ten Dijke, P., Ichijo, H., Franzén, P., Schulz, P., Saras, J., Toyoshima, H., Heldin C.-H., and Miyazono, K., 1993, Activin receptor-like kinases: a novel subclass of cell-surface receptors with predicted serine/threonine kinase activity. *Oncogene* 8: 2879-2887.

Thimme, R., Bukh, J., Spangenberg, H.C., Wieland, S., Pemberton, J., Steiger, C., Govindarajan, S., Purcell, R.H., and Chisari, F.V., 2002, Viral and immunological determinants of hepatitis C virus clearance, persistence, and disease. *Proc Natl Acad Sci U S A* 99: 15661-15668.

Tsuchida, K., Nakatani, M., Yamakawa, N., Hashimoto, O., Hasegawa, Y., and Sugino, H., 2004, Activin isoforms signal through type I receptor serine/threonine kinase ALK7. *Mol Cell Endo* 220: 59-65.

Ulsenheimer, A., Gerlach, J.T., Gruener, N.H., Jung, M.C., Schirren, C.A., Schraut, W., Zachoval, R., Pape, G.R., and Diepolder, H.M., 2003, Detection of functionally altered hepatitis C virus-specific CD4 T cells in acute and chronic hepatitis C. *Hepatology* 37: 1189-1198.

Underwood, D.C., Osborn, R.R., Grygielko, E.T., and Laping, N.J., 2003, SB-525334, a Novel TGF-β type 1 receptor kinase inhibitor, prevents and reverses bleomycin-induced pulmonary fibrosis in the Rat. *Abstract for the American Thoracic Society. Seattle WA.*

Von Boehmer, H., 2005, Mechanisms of suppression by suppressor T cells. *Nat Immunol* 6: 338-344.

Wang, Q., Wang, Y., Hyde, D.M., Gotwals, P.J., Koteliansky, V.E., Ryan, S.T., and Giri, S.N., 1999, Reduction of bleomycin induced lung fibrosis by transforming growth factor β soluble receptor in hamsters. *Thorax* 54: 805-812.

Westerhausen, D.R. Jr., Hopkins, W.E., and Billadello, J.J., 1991, Multiple transforming growth factor-β-inducible elements regulate expression of the plasminogen activator inhibitor type-1 gene in Hep G2 cells. *J Biol Chem* 266: 1092-1100.

Wieser, R., Wrana, J.L., and Massagué, J., 1995, GS domain mutations that constitutively activate TβR-I, the downstream signaling component in the TGF-β receptor complex. *EMBO J* 14: 2199-2208

Wrana, J.L., Attisano, L., Carcamo, J., Zentella, A., Doody, J., Laiho, M., Wang, X.-F., and Massagué, J., 1992, TGFβ signals through a heteromeric protein kinase receptor complex. *Cell* 71: 1003-1014.

Wu, J., and Norton, P., 1996, Animal models of liver fibrosis. *Scand J Gastroenterol* 31: 1137-1143.

Yakymovych, I., Engström, U., Grimsby, S., Heldin, C.-H., and Souchelnytskyi, S., 2002, Inhibition of transforming growth factor-β signaling by low molecular weight compounds interfering with ATP- or substrate-binding sites of the TGFβ type I receptor kinase. *Biochem* 41: 11000-11007.

Zhang, Y., Feng, X., We, R., and Derynck, R., 1996, Receptor-associated Mad homologues synergize as effectors of the TGF-β response. *Nature* 383: 168-172.

Ziyadeh, F., Hoffman, B., Han, D.C., Iglesias-de la Cruz, M., Hong, S.W., Isono, M., Chen, S., McGowan, T., and Sharma, K., 2000, Long term prevention of renal insufficiency, excess matrix gene expression, and glomerular mesangial matrix expansion by treatment with monoclonal antitransforming growth factor-β antibody in db/db diabetic mice. *Proc Natl Acad Sci U S A* 97: 8015-8020.

NAME INDEX

Proteins and Cell Regulation

1. R.A. Kahn (ed.): *ARF Family GTPases*. 2004 ISBN 1-4020-1719-7
2. J. Frampton (ed.): *Myb Transcription Factors: Their Role in Growth, Differentiation and Disease*. 2004. ISBN 1-4020-2779-6
3. E. Manser (ed.): *RHO Family GTPases*. 2005 ISBN 1-4020-3461-X
4. C. Der (ed.): *RAS Family GTPases*. 2006 ISBN 1-4020-4328-7
5. P. ten Dijke and C.-H. Heldin (eds.): *Smad Signal Transduction: Smads in Proliferation, Differentiation and Disease*. 2006 ISBN 1-4020-4542-5